WIND and EARTHQUAKE
RESISTANT BUILDINGS
STRUCTURAL ANALYSIS AND DESIGN

WIND and EARTHQUAKE RESISTANT BUILDINGS

STRUCTURAL ANALYSIS AND DESIGN

BUNGALE S. TARANATH Ph.D., S.E.
DESIMONE CONSULTING ENGINEERS, PLLC.
LAS VEGAS, NEVADA

CRC Press
Taylor & Francis Group
Boca Raton London New York

CRC Press is an imprint of the
Taylor & Francis Group, an **informa** business

CRC Press
Taylor & Francis Group
6000 Broken Sound Parkway NW, Suite 300
Boca Raton, FL 33487-2742

First issued in paperback 2019

© 2005 by Taylor & Francis Group, LLC
CRC Press is an imprint of Taylor & Francis Group, an Informa business

No claim to original U.S. Government works

ISBN-13: 978-0-8247-5934-6 (hbk)
ISBN-13: 978-0-367-39349-6 (pbk)

Library of Congress Cataloging-in-Publication Data
A catalog record for this book is available from the Library of Congress.

Visit the Taylor & Francis Web site at
http://www.taylorandfrancis.com

and the CRC Press Web site at
http://www.crcpress.com

Cosmopolitan Resort & Casino, Las Vegas, Nevada

This project consists of two towers containing approximately seven million square feet of gaming, hospitality, theatre, retail and residential space. The west tower is 662 feet tall above the mat foundation and contains 3,000 guest rooms. The second tower stands 654 feet tall above the mat foundation and provides time-share condominiums.

The lateral system for both towers incorporates cast-in-place reinforced concrete and composite steel plate shear walls. The floor framing for the west tower consists of a cast-in-place, post-tensioned concrete flat plate while a mild steel reinforced flat plate is used for the east tower.

(Client: Continuum Company; Architect: Arquitectonica – Friedmutter Group; Structural Engineer: DeSimone Consulting Engineers).

This book is dedicated to my wife
SAROJA
without whose patience and devotion, this book would not be.

Acknowledgments

I wish to express my sincere appreciation and thanks to the entire staff of John A. Martin and Associates (JAMA), Los Angeles, CA for their help in this endeavor. Special thanks are extended to John A. Martin, Sr. (Jack) and John A. Martin, Jr. (Trailer) for their support and encouragement during the preparation of this book.

Numerous JAMA engineers reviewed various portions of the manuscript and provided valuable comments. In particular, I am indebted to

Dr. Roger Di Julio, Chapters 2 and 6
Ryan Wilkerson, Chapters 1 and 2
Kai Chen Tu, Chapter 1
Kan B. Patel, Chapter 5
Louis Choi and Vernon Gong, Chapter 3
Brett W. Beekman, Ron Lee, and Filbert Apanay, Chapter 4
Farro Tofighi, Chapters 3 and 5
Chuck G. Whitaker, Chapter 8

Additionally, the text had the privilege of review from the following individuals. My sincere thanks to

> Dr. Hussain Bhatia, Senior Structural Engineer, OSHPD, Sacramento, CA, Chapters 2 and 6
> M. V. Ravindra, President, LeMessurier Consultants, Cambridge, MA, and Rao V. Nunna, Structural Engineer, S. B. Barnes Associates, Chapter 7
> Kenneth B. Wiesner, Principal (retired), LeMessurier Consultants, Cambridge, MA, Chapter 8

Appreciation is acknowledged to the following JAMA individuals who were helpful to the author at one or more times during preparation of the manuscript:

Margaret Martin for preparing artwork for the book cover
Marvin F. Mittelstaedt, Tony Galina, Richard Lubas, Murjani Oseguera, April Oseguera, and Nicholas Jesus Oseguera for their help in preparation of the artwork
Andrew Besirof, Evita Santiago- Oseguera, Ron Lee, Hung C. Lee, Chaoying Luo, and Walter Steimle, all of JAMA; Greg L. Clapp of Martin and Peltyon; and Gary Chock of Martin and Chock; and Charles D. Keyes of Martin and Martin, for providing photographs
Ron M. Tong, Robert Barker, Ahmad H. Azad, Dr. Farzad Naeim, Kal Benuska, Mike Baltay, Mark Day, Dan Pattapongse, and Eric D. Brown for their general help
Ivy Policar, Rima Roerish, Betty D. Cooper, and Rosie Nyenke for typing parts of the manuscript
Raul Oseguera, Andrew Gannon, Ferdinand Encarnacion, and Ignacio Morales for duplicating the manuscript

Sincere thanks are extended to

B. J. Clark and Brian Black, formerly of Marcel Dekker, for their guidance in preparation of the manuscript

Edwin Shlemon, Associate Principal, ARUP, Los Angeles, CA, for reviewing the book proposal and making valuable suggestions

Mark Johnson, International Code Council, for his help and encouragement

Jan Fisher, Project Manager, Publication Services, Inc., and editor Jennifer Putman for their cooperation, help, and patience in transforming the manuscript into this book

Srinivas Bhat, and S. Venkatesh of Kruthi Computer Services, Bangalore, India, for their artwork suggestions.

Special thanks to my family:

My daughter, Dr. Anupama Taranath; son-in-law, Dr. Rajesh Rao; and son, Abhiman Taranath, provided a great deal of help and support. My sincere thanks to them.

Most deserving of special gratitude is my wife, Saroja. My source of inspiration, she helped in all aspects of this venture—from manuscript's inception to final proofreading. Her companionship made the arduous task of writing this book a less formidable activity. My profound admiration and appreciation are extended to her for unconditional love, encouragement, support, and devotion. Without her patience and absolute commitment, this modest contribution to structural engineering could not have been made.

Preface

The primary objective of this book is to disseminate information on the latest concepts, techniques, and design data to structural engineers engaged in the design of wind- and seismic-resistant buildings. Integral to the book are recent advances in seismic design, particularly those related to buildings in zones of low and moderate seismicity. These stipulations, reflected in the latest provisions of American Society of Civil Engineers (ASCE) 7-02, International Building Code (IBC)-03, and National Fire Protection Association (NFPA) 5000, are likely to be adopted as a design standard by local code agencies. There now exists the unprecedented possibility of a single standard becoming a basis for earthquake-resistant design virtually in the entire United States, as well as in other nations that base their codes on U.S. practices. By incorporating these and the latest provisions of American Concrete Institute (ACI) 318-02, American Institute of Steel Construction (AISC) 341-02, and Federal Emergency Management Agency (FEMA) 356 and 350 series, this book equips designers with up-to-date information to execute safe designs, in accordance with the latest regulations.

Chapter 1 presents methods of determining design wind loads using the provisions of ASCE 7-02, National Building Code of Canada (NBCC) 1995, and 1997 Uniform Building Code (UBC). Wind-tunnel procedures are discussed, including analytical methods for determining along-wind and across-wind response.

Chapter 2 discusses the seismic design of buildings, emphasizing their behavior under large inelastic cyclic deformations. Design provisions of ASCE 7-02 (IBC-03, NFPA 5000) and UBC-97 that call for detailing requirements to assure seismic performance beyond the elastic range are discussed using static, dynamic, and time-history procedures. The foregone design approach—in which the magnitude of seismic force and level of detailing were strictly a function of the structure's location—is compared with the most recent provisions, in which these are not only a function of the structure's location, but also of its use and occupancy, and the type of soil it rests upon. This comparison will be particularly useful for engineers practicing in many seismically low- and moderate-risk areas of the United States, who previously did not have to deal with seismic design and detailing, but are now obligated to do so. Also explored are the seismic design of structural elements, nonstructural components, and equipment. The chapter concludes with a review of structural dynamic theory.

The design of steel buildings for lateral loads is the subject of Chapter 3. Traditional as well as modern bracing systems are discussed, including outrigger and belt truss systems that have become the workhorse of lateral bracing systems for super-tall buildings. The lateral design of concentric and eccentric braced frames, moment frames with reduced beam section, and welded flange plate connections are discussed, using provisions of ASCE 341-02 and FEMA-350 as source documents.

Chapter 4 addresses concrete structural systems such as flat slab frames, coupled shear walls, frame tubes, and exterior diagonal and bundled tubes. Basic concepts of structural behavior that emphasize the importance of joint design are discussed. Using design provisions of ACI 318-02, the chapter also details building systems such as ordinary, intermediate, and special reinforced concrete moment frames, and structural walls.

The design of buildings using a blend of structural steel and reinforced concrete, often referred to as composite construction, is the subject of Chapter 5. The design of composite beams, columns, and shear walls is discussed, along with building systems such as composite shear walls and megaframes.

Chapter 6 is devoted to the structural rehabilitation of seismically vulnerable buildings. Design differences between a code-sponsored approach and the concept of ductility trade-off for strength are discussed, including seismic deficiencies and common upgrade methods.

Chapter 7 is dedicated to the gravity design of vertical and horizontal elements of steel, concrete, and composite buildings. In addition to common framing types, novel systems such as haunch and stub girder systems are also discussed. Considerable coverage is given to the design of prestressed concrete members based on the concept of load balancing.

The final chapter is devoted to a wide range of topics. Chapter 8 begins with a discussion of the evolution of different structural forms particularly applicable to the design of tall buildings. Case studies of buildings with structural systems that range from run-of-the-mill bracing techniques to unique composite systems—including megaframes and external superbraced frames—are examined. Next, reduction of building occupants' motion perceptions using damping devices is considered, including tuned mass dampers, slashing water dampers, tuned liquid column dampers, and simple and nested pendulum dampers. Panel zone effects, differential shortening of columns, floor-leveling problems, and floor vibrations are studied, followed by a description of seismic base isolation and energy dissipation techniques. The chapter concludes with an explanation of buckling-restrained bracing systems that permit plastic yielding of compression braces.

The book speaks to a multifold audience. It is directed toward consulting engineers and engineers employed by federal, state, and local governments. Within the academy, the book will be helpful to educators and students alike, particularly as a teaching tool in courses for students who have completed an introductory course in structural engineering and seek a deeper understanding of structural design principles and practice. To assist readers in visualizing the response of structural systems, numerous illustrations and practical design problems are provided throughout the text.

Wind- and Earthquake-Resistant Buildings integrates the design aspects of steel, concrete, and composite buildings within a single text. It is my hope that it will serve as a comprehensive design reference for practicing engineers and educators.

October 2004

Bungale S. Taranath Ph.D., S.E.
John A. Martin & Associates
Structural Engineers
1212 S. Flower Street
Los Ageles, California 90015
www.johnmartin.com

Contents

New Bangkok International Airport
Bangkok, Thailand

Martin & Martin, Inc.
Structural Engineers
Denver, CO

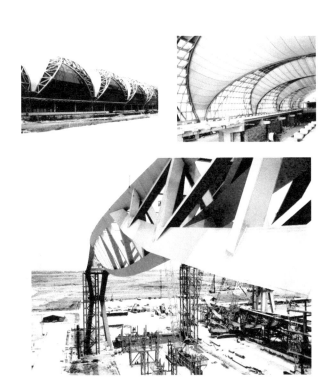

1
Wind Loads

1.1. DESIGN CONSIDERATIONS

Windstorms pose a variety of problems in buildings—particularly in tall buildings—causing concerns for building owners, insurers, and engineers alike. Hurricane winds are the largest single cause of economic and insured losses due to natural disasters, well ahead of earthquakes and floods. For example, in the United States between 1986 and 1993, hurricanes and tornadoes caused about $41 billion in insured catastrophic losses, compared with $6.18 billion for all other natural hazards combined, hurricanes being the largest contributor to the losses. In Europe in 1900 alone, four winter storms caused $10 billion in insured losses, and an estimated $15 billion in economic losses. According to one 1999 insurance industry estimate, the natural catastrophe resulting in the largest amount of insured losses up to that date was hurricane Andrew in 1992 ($16.5 billion). The runner-up, the 1994 Northridge earthquake, resulted in $12.5 billion in reported losses.

In designing for wind, a building cannot be considered independent of its surroundings. The influence of nearby buildings and land configuration on the sway response of the building can be substantial. The sway at the top of a tall building caused by wind may not be seen by a passerby, but may be of concern to those occupying its top floors. There is scant evidence that winds, except those due to a tornado or hurricane, have caused major structural damage to new buildings. However, a modern skyscraper, with lightweight curtain walls, dry partitions, and high-strength materials, is more prone to wind motion problems than the early skyscrapers, which had the weight advantage of masonry partitions, heavy stone facades, and massive structural members.

To be sure, all buildings sway during windstorms, but the motion in earlier tall buildings with heavy full-height partitions has usually been imperceptible and certainly has not been a cause for concern. Structural innovations and lightweight construction technology have reduced the stiffness, mass, and damping characteristics of modern buildings. In buildings experiencing wind motion problems, objects may vibrate, doors and chandeliers may swing, pictures may lean, and books may fall off shelves. If the building has a twisting action, its occupants may get an illusory sense that the world outside is moving, creating symptoms of vertigo and disorientation. In more violent storms, windows may break, creating safety problems for pedestrians below. Sometimes, strange and frightening noises are heard by the occupants as the wind shakes elevators, strains floors and walls, and whistles around the sides.

Following are some of the criteria that are important in designing for wind:

1. Strength and stability.
2. Fatigue in structural members and connections caused by fluctuating wind loads.
3. Excessive lateral deflection that may cause cracking of internal partitions and external cladding, misalignment of mechanical systems, and possible permanent deformations of nonstructural elements.

1

4. Frequency and amplitude of sway that can cause discomfort to occupants of tall, flexible buildings.
5. Possible buffeting that may increase the magnitude of wind velocities on neighboring buildings.
6. Wind-induced discomfort in pedestrian areas caused by intense surface winds.
7. Annoying acoustical disturbances.
8. Resonance of building oscillations with vibrations of elevator hoist ropes.

1.2. NATURE OF WIND

Wind is the term used for air in motion and is usually applied to the natural horizontal motion of the atmosphere. Motion in a vertical or nearly vertical direction is called a *current*. Movement of air near the surface of the earth is three-dimensional, with horizontal motion much greater than the vertical motion. Vertical air motion is of importance in meteorology but is of less importance near the ground surface. On the other hand, the horizontal motion of air, particularly the gradual retardation of wind speed and the high turbulence that occurs near the ground surface, are of importance in building engineering. In urban areas, this zone of turbulence extends to a height of approximately one-quarter of a mile aboveground, and is called the surface boundary layer. Above this layer, the horizontal airflow is no longer influenced by the ground effect. The wind speed at this height is called the *gradient wind speed*, and it is precisely in this boundary layer where most human activity is conducted. Therefore, how wind effects are felt within this zone is of great concern.

Although one cannot see the wind, it is a common observation that its flow is quite complex and turbulent in nature. Imagine taking a walk outside on a reasonably windy day. You no doubt experience the constant flow of wind, but intermittently you will experience sudden gusts of rushing air. This sudden variation in wind speed, called gustiness or turbulence, plays an important part in determining building oscillations.

1.2.1. Types of wind

Winds that are of interest in the design of buildings can be classified into three major types: prevailing winds, seasonal winds, and local winds.

1. *Prevailing winds*. Surface air moving toward the low-pressure equatorial belt is called prevailing winds or trade winds. In the northern hemisphere, the northerly wind blowing toward the equator is deflected by the rotation of the earth to become northeasterly and is known as the northeast trade wind. The corresponding wind in the southern hemisphere is called the southeast trade wind.
2. *Seasonal winds*. The air over the land is warmer in summer and colder in winter than the air adjacent to oceans during the same seasons. During summer, the continents become seats of low pressure, with wind blowing in from the colder oceans. In winter, the continents experience high pressure with winds directed toward the warmer oceans. These movements of air caused by variations in pressure difference are called seasonal winds. The monsoons of the China Sea and the Indian Ocean are an examples.
3. *Local winds*. Local winds are those associated with the regional phenomena and include whirlwinds and thunderstorms. These are caused by daily changes in temperature and pressure, generating local effects in winds. The daily variations in temperature and pressure may occur over irregular terrain, causing valley and mountain breezes.

All three types of wind are of equal importance in design. However, for the purpose of evaluating wind loads, the characteristics of the prevailing and seasonal winds are analytically studied together, whereas those of local winds are studied separately. This grouping is to distinguish between the widely differing scale of fluctuations of the winds; prevailing and seasonal wind speeds fluctuate over a period of several months, whereas the local winds vary almost every minute, The variations in the speed of prevailing and seasonal winds are referred to as *fluctuations* in mean velocity. The variations in the local winds, are referred to as *gusts*.

The flow of wind, unlike that of other fluids, is not steady and fluctuates in a random fashion. Because of this, wind loads imposed on buildings are studied statistically.

1.3. CHARACTERISTICS OF WIND

The flow of wind is complex because many flow situations arise from the interaction of wind with structures. However, in wind engineering, simplifications are made to arrive at design wind loads by distinguishing the following characteristics:

- Variation of wind velocity with height.
- Wind turbulence.
- Statistical probability.
- Vortex shedding phenomenon.
- Dynamic nature of wind–structure interaction.

1.3.1. Variation of Wind Velocity with Height

The viscosity of air reduces its velocity adjacent to the earth's surface to almost zero, as shown in Fig. 1.1. A retarding effect occurs in the wind layers near the ground, and these inner layers in turn successively slow the outer layers. The slowing down is reduced at each layer as the height increases, and eventually becomes negligibly small. The height at which velocity ceases to increase is called the gradient height, and the corresponding velocity, the gradient velocity. This characteristic of variation of wind velocity with height is a well-understood phenomenon, as evidenced by higher design pressures specified at higher elevations in most building codes.

At heights of approximately 1200 ft (366 m) aboveground, the wind speed is virtually unaffected by surface friction, and its movement is solely dependent on prevailing seasonal and local wind effects. The height through which the wind speed is affected by topography is called the *atmospheric boundary layer.* The wind speed profile within this layer is given by

$$V_z = V_g (Z/Z_g)^{1/\alpha} \tag{1.1}$$

where
V_z = mean wind speed at height Z aboveground
V_g = gradient wind speed assumed constant above the boundary layer
Z = height aboveground
Z_g = nominal height of boundary layer, which depends on the exposure (Values for Z_g are given in Fig. 1.1.)
α = power law coefficient

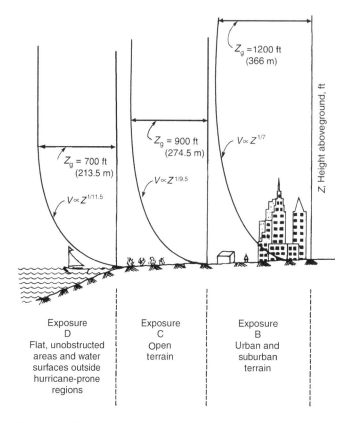

Figure 1.1.   Influence of exposure terrain on variation of wind velocity with height.

With known values of mean wind speed at gradient height and exponent α, wind speeds at height Z are calculated by using Eq. (1.1). The exponent $1/\alpha$ and the depth of boundary layer Z_g vary with terrain roughness and the averaging time used in calculating wind speed. α ranges from a low of 0.087 for open country of 0.20 for built-up urban areas, signifying that wind speed reaches its maximum value over a greater height in an urban terrain than in the open country.

1.3.2.   Wind Turbulence

Motion of wind is turbulent. A concise mathematical definition of turbulence is difficult to give, except to state that it occurs in wind flow because air has a very low viscosity—about one-sixteenth that of water. Any movement of air at speeds greater than 2 to 3 mph (0.9 to 1.3 m/s) is turbulent, causing particles of air to move randomly in all directions. This is in contrast to the laminar flow of particles of heavy fluids, which move predominantly parallel to the direction of flow.

For structural engineering purposes, velocity of wind can be considered as having two components: a mean velocity component that increases with height, and a turbulent velocity that remains the same over height (Fig. 1.1a). Similarly, the wind pressures, which are proportional to the square of the velocities, also fluctuate as shown in Fig. 1.2. The total pressure P_t at any instant t is given by the relation

$$P_t = \overline{P} + P' \tag{1.2}$$

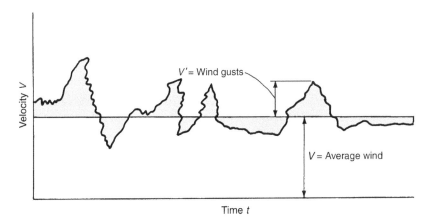

Figure 1.1a. Variation of wind velocity with time; at any instant t, velocity $V_t = V' + V$.

where

P_t = pressure at instant t
$\overline{P}$ = average or mean pressure
P' = instantaneous pressure fluctuation

1.3.3. Probabilistic Approach

In many engineering sciences the intensity of certain events is considered to be a function of the duration recurrence interval (return period). For example, in hydrology the intensity of rainfall expected in a region is considered in terms of a return period because the rainfall expected once in 10 years is less than the one expected once every 50 years. Similarly, in wind engineering the speed of wind is considered to vary with return periods. For example, the fastest-mile wind 33 ft (10 m) above ground in Dallas, TX, corresponding

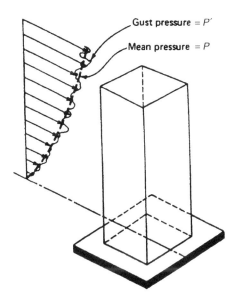

Figure 1.2. Schematic representation of mean and gust pressure. At any instant t, the pressure $P_t = P' + P$.

to a 50-year return period, is 67 mph (30 m/s), compared to the value of 71 mph (31.7 m/s) for a 100-year recurrence interval.

A 50-year return-period wind of 67 mph (30 m/s) means that on the average, Dallas will experience a wind faster than 67 mph within a period of 50 years. A return period of 50 years corresponds to a probability of occurrence of 1/50 = 0.02 = 2%. Thus the chance that a wind exceeding 67 mph (30 m/s) will occur in Dallas within a given year is 2%. Suppose a building is designed for a 100-year lifetime using a design wind speed of 67 mph. What is the probability that this wind will exceed the design speed within the lifetime of the structure? The probability that this wind speed will not be exceeded in any year is 49/50. The probability that this speed will not be exceeded 100 years in a row is $(49/50)^{100}$. Therefore, the probability that this wind speed will be exceeded at least once in 100 years is

$$1 - \left(\frac{49}{50}\right)^{100} = 0.87 = 87\%$$

This signifies that although a wind with low annual probability of occurrence (such as a 50-year wind) is used to design structures, there still exists a high probability of the wind being exceeded within the lifetime of the structure. However, in structural engineering practice it is believed that the actual probability of overstressing a structure is much less because of the factors of safety and the generally conservative values used in design.

It is important to understand the notion of probability of occurrence of design wind speeds during the service life of buildings. The general expression for probability P that a design wind speed will be exceeded at least once during the exposed period of n years is given by

$$P = 1 - (1 - P_a)^n \tag{1.3}$$

where

P_a = annual probability of being exceeded (reciprocal of the mean recurrence interval)
n = exposure period in years

Consider a building in Dallas designed for a 50-year service life instead of 100 years. The probability of exceeding the design wind speed at least once during the 50-year lifetime of the building is

$$P = 1 - (1 - 0.02)^{50} = 1 - 0.36 = 0.64 = 64\%$$

The probability that wind speeds of a given magnitude will be exceeded increases with a longer exposure period of the building and the mean recurrence interval used in the design. Values of P for a given mean recurrence interval and a given exposure period are shown in Table 1.1.

Wind velocities (measured with anemometers usually installed at airports across the country) are necessarily averages of the fluctuating velocities measured during a finite interval of time. The value usually reported in the United States, until the publication of the American Society of Civil Engineers' ASCE 7-95 standard, was the average of the velocities recorded during the time it takes a horizontal column of air 1 mile long to pass a fixed point. For example, if a 1-mile column of air is moving at an average velocity of 60 mph, it passes an anemometer in 60 seconds, the reported velocity being the average of the velocities recorded these 60 seconds. The fastest mile is the highest velocity in one day. The annual extreme mile is the largest of the daily maximums. Furthermore, since the annual extreme mile varies from year to year, wind pressures used in design are based on

TABLE 1.1 Probability of Exceeding Design Wind Speed During Design Life of Building

Annual probability P_a	Mean recurrence interval $(1/P_a)$ years	Exposure period (design life), n (years)					
		1	5	10	25	50	100
0.1	10	0.1	0.41	0.15	0.93	0.994	0.999
0.04	25	0.04	0.18	0.34	0.64	0.87	0.98
0.034	30	0.034	0.15	0.29	0.58	0.82	0.97
0.02	50	0.02	0.10	0.18	0.40	0.64	0.87
0.013	75	0.013	0.06	0.12	0.28	0.49	0.73
0.01	100	0.01	0.05	0.10	0.22	0.40	0.64
0.0067	150	0.0067	0.03	0.06	0.15	0.28	0.49
0.005	200	0.005	0.02	0.05	0.10	0.22	0.39

a wind velocity having a specific mean recurrence interval. Mean recurrence intervals of 20 and 50 years are generally used in building design, the former interval for determining the comfort of occupants in tall buildings subject to wind storms, and the latter for designing lateral resisting elements.

1.3.4.   Vortex Shedding

In general, wind buffeting against a bluff body gets diverted in three mutually perpendicular directions, giving rise to forces and moments about the three directions. Although all six components, as shown in Fig.1.3, are significant in aeronautical engineering, in civil and structural work, the force and moment corresponding to the vertical axis (lift and yawing moment) are of little significance. Therefore, aside from the uplift forces on large roof areas, the flow of wind is simplified and considered two-dimensional, as shown in Fig.1.4, consisting of *along wind* and *transverse wind*.

Along wind—or simply wind—is the term used to refer to drag forces, and transverse wind is the term used to describe crosswind. The crosswind response causing motion in a plane perpendicular to the direction of wind typically dominates over the along-wind response for tall buildings. Consider a prismatic building subjected to a smooth wind flow.

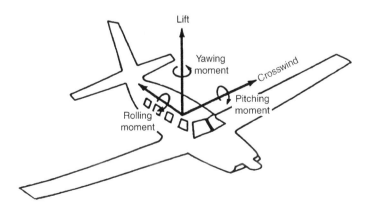

Figure 1.3.   Six components of wind.

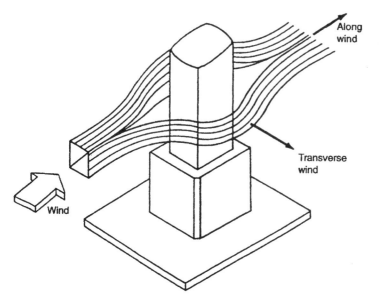

Figure 1.4. Simplified two-dimensional flow of wind.

The originally parallel upwind streamlines are displaced on either side of the building, Fig.1.5. This results in spiral vortices being shed periodically from the sides into the downstream flow of wind, called the *wake*. At relatively low wind speeds of, say, 50 to 60 mph (22.3 to 26.8 m/s), the vortices are shed symmetrically in pairs, one from each side. When the vortices are shed, i.e., break away from the surface of the building, an impulse is applied in the transverse direction.

At low wind speeds, since the shedding occurs at the same instant on either side of the building, there is no tendency for the building to vibrate in the transverse direction. It is therefore subject to along-wind oscillations parallel to the wind direction. At higher speeds, the vortices are shed alternately, first from one and then from the other side. When this occurs, there is an impulse in the along-wind direction as before, but in addition, there is an impulse in the transverse direction. The transverse impulses are, however, applied alternately to the left and then to the right. The frequency of transverse impulse is precisely half that of the along-wind impulse. This type of shedding, which gives rise to structural vibrations in the flow direction as well as in the transverse direction, is called *vortex shedding* or the *Karman vortex street*, a phenomenon well known in the field of fluid mechanics.

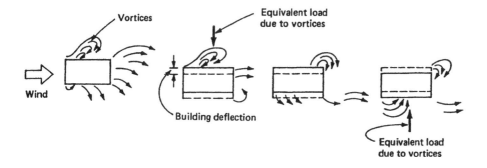

Figure 1.5. Vortex-shedding phenomenon.

There is a simple formula to calculate the frequency of the transverse pulsating forces caused by vortex shedding:

$$f = \frac{V \times S}{D} \tag{1.4}$$

where

f = frequency of vortex shedding in hertz
V = mean wind speed at the top of the building
S = a dimensionless parameter called the Strouhal number for the shape
D = diameter of the building

In Eq. (1.4), the parameters V and D are expressed in consistent units such as ft/s and ft, respectively.

The Strouhal number is not a constant but varies irregularly with wind velocity. At low air velocities, S is low and increases with the velocity up to a limit of 0.21 for a smooth cylinder. This limit is reached for a velocity of about 50 mph (22.4 m/s) and remains almost a constant at 0.20 for wind velocities between 50 and 115 mph (22.4 and 51 m/s).

Consider for illustration purposes, a circular prismatic-shaped high-rise building having a diameter equal to 110 ft (33.5 m) and a height-to-width ratio of 6 with a natural frequency of vibration equal to 0.16 Hz. Assuming a wind velocity of 60 mph (27 m/s), the vortex-shedding frequency is given by

$$f = \frac{V \times 0.2}{110} = 0.16 \text{ Hz}$$

where V is in ft/s.

If the wind velocity increases from 0 to 60 mph (27.0 m/s), the frequency of vortex excitation will rise from 0 to a maximum of 0.16 Hz. Since this frequency happens to be very close to the natural frequency of the building, and assuming very little damping, the structure would vibrate as if its stiffness were zero at a wind speed somewhere around 60 mph (27 m/s). Note the similarity of this phenomenon to the ringing of church bells or the shaking of a tall lamppost whereby a small impulse added to the moving mass at each end of the cycle greatly increases the kinetic energy of the system. Similarly, during vortex shedding an increase in deflection occurs at the end of each swing. If the damping characteristics are small, the vortex shedding can cause building displacements far beyond those predicted on the basis of static analysis.

When the wind speed is such that the shedding frequency becomes approximately the same as the natural frequency of the building, a resonance condition is created. After the structure has begun to resonate, further increases in wind speed by a few percent will not change the shedding frequency, because the shedding is now controlled by the natural frequency of the structure. The vortex-shedding frequency has, so to speak, locked in with the natural frequency. When the wind speed increases significantly above that causing the lock-in phenomenon, the frequency of shedding is again controlled by the speed of the wind. The structure vibrates with the resonant frequency only in the lock-in range. For wind speeds either below or above this range, the vortex shedding will not be critical.

Vortex shedding occurs for many building shapes. The value of S for different shapes is determined in wind tunnel tests by measuring the frequency of shedding for a range of wind velocities. One does not have to know the value of S very precisely because the lock-in phenomenon occurs within a range of about 10% of the exact frequency of the structure.

1.3.5. Dynamic Nature of Wind

Unlike the mean flow of wind, which can be considered as static, wind loads associated with gustiness or turbulence change rapidly and even abruptly, creating effects much larger than if the same loads were applied gradually. Wind loads, therefore, need to be studied as if they were dynamic in nature. The intensity of a wind load depends on how fast it varies and also on the response of the structure. Therefore, whether the pressures on a building created by a wind gust, which may first increase and then decrease, are considered as dynamic or static depends to a large extent on the dynamic response of the structure to which it is applied.

Consider the lateral movement of an 800-ft tall building designed for a drift index of $H/400$, subjected to a wind gust. Under wind loads, the building bends slightly as its top moves. It first moves in the direction of wind, with a magnitude of, say, 2 ft (0.61 m), and then starts oscillating back and forth. After moving in the direction of wind, the top goes through its neutral position, then moves approximately 2 ft (0.61 m) in the opposite direction, and continues oscillating back and forth until it eventually stops. The time it takes a building to cycle through a complete oscillation is known as a *period*. The period of oscillation for a tall steel building in the height range of 700 to 1400 ft (214 to 427 m) normally is in the range of 5 to 10 seconds, whereas for a 10-story concrete or masonry building it may be in the range of 0.5 to 1 seconds. The action of a wind gust depends not only on how long it takes the gust to reach its maximum intensity and decrease again, but on the period of the building itself. If the wind gust reaches its maximum value and vanishes in a time much shorter than the period of the building, its effects are dynamic. On the other hand, the gusts can be considered as static loads if the wind load increases and vanishes in a time much longer than the period for the building. For example, a wind gust that develops to its strongest intensity and decreases to zero in 2 seconds is a dynamic load for a tall building with a period of, say, 5 to 10 seconds, but the same 2-second gust is a static load for a low-rise building with a period of less than 2 seconds.

1.3.6. Cladding Pressures

The design of cladding for lateral loads is of major concern to architects and engineers. Although the failure of exterior cladding resulting in broken glass may be of less consequence than the collapse of a structure, the expense of replacement and hazards posed to pedestrians require careful consideration. Cladding breakage in a windstorm is an erratic occurrence, as witnessed in hurricane Alicia, which hit Galveston and downtown Houston on August 18, 1983, causing breakage of glass in several tall buildings. Wind forces play a major role in glass breakage, which is also influenced by other factors, such as solar radiation, mullion and sealant details, tempering of the glass, double- or single-glazing of glass, and fatigue. It is known with certainty that glass failure starts at nicks and scratches that may be made during manufacture, and by handling operations.

There appears to be no analytical approach available for a rational design of curtain walls of all shapes and sizes. Although most codes have tried to identify regions of high wind loads around building corners, the modern trend in architecture of using nonprismatic and curvilinear shapes combined with the unique topography of each site, has made experimental determination of wind loads even more necessary.

Thus it has become routine to obtain design information concerning the distribution of wind pressures over a building's surface by conducting wind tunnel studies. In the past two decades, curtain wall has developed into an ornamental item and has emerged as a significant architectural element. Sizes of window panes have increased considerably,

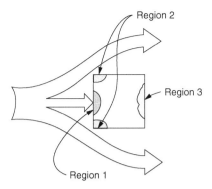

Figure 1.6. Distribution of pressures and suctions.

requiring that the glass panes be designed for various combinations of forces due to wind, shadow effects, and temperature movement. Glass in curtain walls must not only resist large forces, particularly in tall buildings, but must also be designed to accommodate the various distortions of the total building structure. Breaking of large panes of glass can cause serious damage to neighboring properties and can injure pedestrians.

1.3.6.1. *Distribution of Pressures and Suctions*

When air flows around edges of a structure, the resulting pressures at the corners are much in excess of the pressures on the center of elevation. This has been evidenced by damage caused to corner windows, eave and ridge tiles, etc., in windstorms. Wind tunnel studies conducted on scale models of buildings indicate that three distinct pressure areas develop around a building. These are shown schematically in Fig.1.6.

1. Positive-pressure zone on the upstream face (Region 1).
2. Negative pressure zones at the upstream corners (Regions 2).
3. Negative pressure zone on the downstream face (Region 3).

The highest negative pressures are created in the upstream corners designated as Regions 2 in Fig. 1.6. Wind pressures on a building's surface are not constant, but fluctuate continuously. The positive pressure on the upstream or the windward face fluctuates more than the negative pressure on the downstream or the leeward face. The negative-pressure region remains relatively steady as compared to the positive-pressure zone. The fluctuation of pressure is random and varies from point to point on the building surface. Therefore, the design of the cladding is strongly influenced by local pressures. As mentioned earlier, the design pressure can be thought of as a combination of the mean and the fluctuating velocity. As in the design of buildings, whether or not the pressure component arising from the fluctuating velocity of wind is treated as a dynamic or as a pseudostatic load is a function of the period of the cladding. The period of cladding on a building is usually on the order of 0.2 to 0.02 sec, which is much shorter than the period it takes for wind to fluctuate from a gust velocity to a mean velocity. Therefore, it is sufficiently accurate to consider both the static and the gust components of winds as equivalent static loads in the design of cladding.

The strength of glass, and indeed of any other cladding material, is not known in the same manner that the strengths of steel and concrete are known. For example, it is not possible to buy glass based on yield strength criteria as with steel. Therefore, the selection, testing, and acceptance criteria for glass are based on statistical probabilities rather than on absolute strength. The glass industry has addressed this problem, and commonly uses 8 failures per 1000 lights (panes) of glass as an acceptable probability of failure.

1.3.6.2.   *Local Cladding Loads and Overall Design Loads*

The overall wind load for lateral analysis consists of combined positive and negative pressures around the building. The local wind loads that act on specific areas of the building are required for the design of exterior cladding elements and their connections to the building. The two types of loads differ significantly, and it is important that these differences be understood. These are

1. Local winds are more influenced by the configuration of the building than the overall loading.
2. The local load is the maximum load that may occur at any location at any time on any wall surface, whereas the overall load is the summation of positive and negative pressures occurring simultaneously over the entire building surface.
3. The intensity and character of local loading for any given wind direction and velocity differ substantially on various parts of the building surface, whereas the overall load is considered to have a specific intensity and direction.
4. The local loading is sensitive to the momentary nature of wind, but in determining the critical overall loading, only gusts of about 2 sec or more are significant.
5. Generally, maximum local negative pressures, also referred to as suctions, are of greater intensity than the overall load.
6. Internal pressures caused by leakage of air through cladding systems have a significant effect on local cladding loads but are of no consequence in determining the overall load.

The relative importance of designing for these two types of wind loading is quite obvious. Although proper assessment of overall wind load is important, very few, if any, buildings have been toppled by winds. There are no classic examples of building failures comparable to the Tacoma bridge disaster. On the other hand, local failures of roofs, windows, and wall cladding are not uncommon.

The analytical determination of wind pressure or suction at a specific surface of a building under varying wind direction and velocity is a complex problem. Contributing to the complexity are the vagaries of wind action as influenced both by adjacent surroundings and the configuration of the wall surface itself. Much research is needed on the microeffects of common architectural features such as projecting mullions, column covers, and deep window reveals, etc. In the meantime, model testing of building wind tunnels is perhaps the only answer.

Probably the most important fact established by tests is that the negative or outward-acting wind loads on wall surfaces are greater and more critical than had formerly been assumed. These loads may be as much as twice the magnitude of positive loading. In most instances of local cladding failure, glass panels have been blown off of the building, not into it, and the majority of such failures have occurred in areas near building corners. Therefore it is important to give careful attention to the design of both anchorage and glazing details to resist outward-acting forces, especially near the corners.

Another feature that has come to light from model testing is that wind loads, both positive and negative, do not vary in proportion to height aboveground. Typically, the positive-pressure contours follow a concentric pattern as illustrated in Fig.1.7, with the highest pressure near the lower center of the facade, and pressures at the very top somewhat

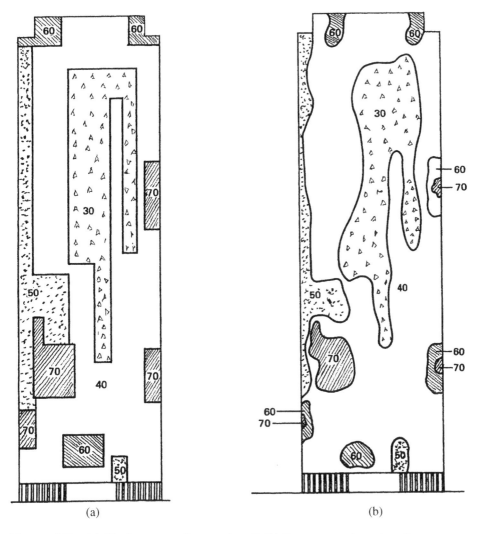

Figure 1.7. (a) Block pressure diagram, in psf; (b) Pressure countours in psf.

less than those a few stories below the roof. Figure 1.7a shows a pressure diagram for the design of cladding derived from pressure contours measured in wind tunnel tests shown in Fig.1.7b. The block pressure diagram shown in Fig.1.7a gives zones of design pressures based on the building grid system, to assist in the cladding design.

1.4. CODE PROVISIONS FOR WIND LOADS

In recent years, wind loads specified in codes and standards have been refined significantly. This is because our knowledge of how wind affects buildings and structures has expanded due to new technology and advanced research that have ensued in greater accuracy in predicting wind loads. We now have an opportunity to design buildings that will satisfy anticipated loads without excessive conservatism. The resulting complexity in the determination of wind loads may be appreciated by comparing the 1973 Standard Building Code (SBC), which contained only a page and one-half of wind load requirements, to the 2002 edition of the ASCE 7, which contains 97 pages of text, commentary, figures, and tables to predict wind loads for a particular structure. As compared to a single method given in the 1973 SBC, ASCE 7 contains three methods for determining winds: the simplified procedure, the analytical procedure, and the wind-tunnel procedure. The controlling equations for determining wind loads require calculating velocity pressure as before, but are now modified to account for several variables such as gusts, internal pressure, and aerodynamic properties of the element under consideration, as well as topographic effects. Using the low-rise buildings' analytical procedure in ASCE-7 and applying it to the simplest building requires the use of up to 11 variables. An important criterion that influences the calculation of wind loads is the enclosure classification of the building. Three classifications are used: 1) enclosed; 2) partially enclosed; or 3) open. A building classified as partially enclosed assumes that a large opening is on one side of a building and no (or minimal) openings are on the other walls. As openings on one wall reach a certain size with respect to openings on the other walls, the building is classified as partially enclosed. Depending upon the wind's direction, this type of situation allows two conditions to develop: internal pressure or internal suction. Internal pressure occurs when air enters a building opening on the windward wall and becomes trapped, exerting an additional force on the interior elements of the building. Typically the internal pressures act in the same directions as the external pressures on all walls except the windward wall. Internal suction is a condition that exists when there is an opening on the leeward wall allowing air to be pulled out of the building. This results in the internal forces acting in the same direction as the external forces on the windward wall. The additional forces produced by this type of pressurization are characterized by requiring an internal pressure coefficient that is more than three times greater than that required for an enclosed building.

Another criterion that significantly affects the magnitude of the wind pressures is the site's exposure category, which provides a way to define the relative roughness of the boundary layers at the site.

The ASCE 7-02 and IBC-03 define three exposure categories: B, C, and D. Exposure B is the roughest and D is the smoothest. Consequently, when all other conditions are equal, calculated wind loads are reduced as the exposure category moves from D to B. Exposure B is the most common category, consisting primarily of terrain associated with a suburban or urban site. Accordingly, B is the default exposure category in both ASCE 7 and IBC. Exposure C consists primarily of open terrain with scattered obstructions but also includes shoreline in hurricane-prone regions. Exposure D applies to shore lines (excluding those in hurricane-prone regions) with wind flowing over open water for a distance of at least one mile.

Buildings must also be classified based on their importance. The wind importance factor I_w specified in the codes is used to adjust the return period for a structure based on its relative level of importance. For example, the importance factor for structures housing critical national defense functions is 1.15, while the importance factor for an agricultural building not as critical as a defense facility, is 0.87.

The applicable wind speeds for the United States and some tropical islands specified in the wind speed maps are three-second gusts at 33 feet above ground for Exposure Category C. In the model codes that preceded the IBC (the National Building Code. Standard Building Code, and Uniform Building Code) and versions of ASCE 7 prior to 1995, wind speeds were shown as "fastest-mile winds," which is defined as the average speed of a one-mile column of air passing a reference point.

While the designated 3-sec gust wind speed for a particular site is higher than values on the fastest-mile map, the averaging times are also different. The averaging time for a fastest-mile wind speed is different for each wind speed, while the averaging time for the 3-sec gust speeds varies from 3 to 8 sec, depending upon the sensitivity of the instruments.

Wind load provisions given in three nationally and internationally recognized standards are discussed in this section. These are the

1. Uniform Building Code (UBC) 1997.
2. ASCE Minimum Design Loads for Buildings and Other Structures (ASCE 7-02).
3. National Building Code of Canada (NBCC) 1995.

1.4.1. Uniform Building Code, 1997: Wind Load Provisions

Wind load provisions of UBC 1997 are based on the ASCE 7-88 standard with certain simplifying assumptions to make calculations easier. The design wind speed is based on the fastest-mile wind speed as compared to the 3-sec gust speeds of the later codes. The prevailing wind direction at the site is not considered in calculating wind forces on the structures: The direction that has the most critical exposure controls the design. Consideration of shielding by adjacent buildings is not permitted because studies have shown that in certain configurations, the nearby buildings can actually increase the wind speed through funneling effects or increased turbulence. Additionally, it is possible that adjacent existing buildings may be removed during the life of the building being designed.

To shorten the calculation procedure, certain simplifying assumptions are made. These assumptions do not allow determination of wind loads for flexible buildings that may be sensitive to dynamic effects and wind-excited oscillations such as vortex shedding. Such buildings typically are those with a height-to-width ratio greater than 5, and over 400 ft (121.9 m) in height. The general section of the UBC directs the user to an approved standard for the design of these types of structures. The ASCE 7-02, adopted by IBC 2003 (discussed later in this chapter), is one such standard for determining the dynamic gust response factor required for the design of these types of buildings.

UBC provisions are not applicable to buildings taller than 400 ft (122 m) for normal force method, Method 1, and 200 ft (61 m) for projected area method, Method 2. Any building, including those not covered by the UBC, may be designed using wind-tunnel test results.

1.4.1.1. Wind Speed Map

The minimum basic wind speed at any site in the United States is shown in Fig. 1.8. The wind speed represents the fastest-mile wind speed in an exposure C terrain at 33 ft (10 m) above grade, for a 50-year mean recurrence interval. The probability of experiencing a wind speed faster than the value indicted in the map, in any given year is 1 in 50, or 2%.

1.4.1.2. Special Wind Regions

Although basic wind speeds are constant over hundreds of miles, some areas have local weather or topographic characteristics that affect design wind speeds. These special wind regions are defined in the UBC map. Because some jurisdictions prescribe basic wind speeds higher than the map, it is prudent to contact local building officials before commencing with the wind design.

1.4.1.3. Hurricanes and Tornadoes

The wind speeds shown in the UBC map come from data collected by meterological stations throughout the continental United States, Alaska, Hawaii, Puerto, Rico, and Virgin Islands. However, coastal regions did not have enough statistical measurements to predict hurricane wind speeds. Therefore data generated by computer simulations have been used to formulate basic hurricane wind speeds.

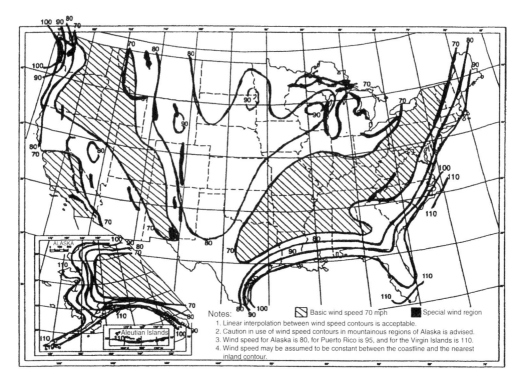

Figure 1.8. Minimum basic wind speeds in miles per hour ($\times$ 1.61 for km/h). (From UBC 1997.)

Tornado level winds are not included in the map because the mean recurrence intervals of tornadoes are in the range of 400–500 years, as compared to the 50 years interval typically used in wind design.

1.4.1.4. Exposure Effects

Every building site has its own unique characteristics in terms of surface roughness and length of upwind terrain associated with the roughness. Simplified code methods cannot account for the uniqueness of the site. Therefore the code approach is to assign broad exposure categories for design purposes.

Similar to the ASCE method, the UBC distinguishes between three exposure categories; B, C, and D. Exposure B is the least severe, representing urban, suburban, wooded, and other terrain with numerous closely spaced surface irregularities; Exposure C is for flat and generally open terrain with scattered obstructions; and the most severe, Exposure *D*, is four unobstructed coastal areas directly exposed to large bodies of water. Discussion of the exposure categories follows.

It should be noted that Exposure A (centers of large cities where over half the buildings have a height in excess of 70 feet), included in some standards, is not recognized in the UBC. The UBC considers this type of terrain as Exposure B, allowing no further decrease in wind pressure.

Exposure B has terrain with buildings, forest, or surface irregularities, covering at least 20% of the ground level area extending 1 mile (1.61 km) or more from the site.

Exposure C has terrain that is flat and generally open, extending one-half mile (0.81 km) or more from the site in any full quadrant.

Exposure D represents the most severe exposure in areas of basic wind speeds of 80 mph (129 km/h) or greater, and has terrain that is flat and unobstructed facing large bodies of water over one mile (1.61 km) or more in width relative to any quadrant of the building site. Exposure D extends inland from shoreline one-fourth mile (0.4 km) or 10 times the building height, whichever is greater.

1.4.1.5. *Site Exposure*

Even though a building site may have different exposure categories in different directions, the most severe exposure is used for all wind-load calculations regardless of building orientation or direction of wind.

Exposure D is perhaps the easiest to determine because it is explicitly for unobstructed coastal areas directly exposed to large bodies of water. It is not as easy to determine whether a site falls into Exposure B or C because the description of these categories is somewhat ambiguous. Morevoer, the terrain surrounding a site is usually not uniform and can be composed of zones that would be classified as Exposure B while others would be classified as Exposure C. When such a mix is encountered, the more severe exposure governs. The UBC classifies a site as Exposure C when open terrain exists for one full 90° quadrant extending outward from the building for at least one-half mile. If the quadrant is less than 90° or less than one-half mile, then the site is classified as Exposure B. It is essential to select the appropriate category because force levels could differ by as much as 65% between Exposure B and C. It is advisable to contact the local building official before embarking on a building design with a questionable site exposure category. If the site has a view of a cliff or hill, it may be prudent to assign Exposure C to D to account for higher wind velocity effects.

1.4.1.6. *Design Wind Pressures*

The design wind pressure p is given as a product of the combined height, exposure, and gust factor coefficient C_e; the pressure coefficient C_q; the wind stagnation pressure q_s; and building Importance Factor I_w.

$$p = C_e C_q q_s I_w \tag{1.5}$$

The pressure q_s manifesting on the surface of a building due to a mass of air with density ρ, moving at a velocity v is given by Bernoulli's equation:

$$q_s = \tfrac{1}{2} \rho V^2 \tag{1.6}$$

The density of air ρ is 0.0765 pcf, for conditions of standard atmosphere, temperature (59 °F), and barometric pressure (29.92 in. of mercury).

Since velocity given in the wind map is in mph, Eq. (1.6) reduces to

$$q_s = \frac{1}{2} \left[\frac{0.0765 \text{ pcf}}{32.2 \text{ ft/s}^2} \right] \left[\frac{5280 \text{ft}}{\text{mile}} \times \frac{1 \text{ hr}}{3600 \text{s}} \right] V^2 \tag{1.7}$$

$$q_s = 0.00256 V^2$$

For instance, if the wind speed is 80 mph, $q_s = 0.00256 \times 80^2 = 16.38$ psf, which the UBC rounds off to 16.4 psf (Table 2.10). Note UBC does not consider the effect of reduced air density at sites located at higher altitudes.

1.4.1.7. *The C_e Factor*

The effects of height, exposure, and gust factor are all lumped into one factor C_e in the interest of keeping the UBC method simple. Values of C_e shown in Table 1.2 (UBC, Table 16-G) are essentially equal to the product of two parameters—K_z, the velocity pressure exposure

TABLE 1.2   Combined Height, Exposure, and Gust Factor Coefficient $(C_e)^a$

Height above average level of adjoining ground (feet) × 304.8 for mm	Exposure D	Exposure C	Exposure B
0–15	1.39	1.06	0.62
20	1.45	1.13	0.67
25	1.50	1.19	0.72
30	1.54	1.23	0.76
40	1.62	1.31	0.84
60	1.73	1.43	0.95
80	1.81	1.53	1.04
100	1.88	1.61	1.13
120	1.93	1.67	1.20
160	2.02	1.79	1.31
200	2.10	1.87	1.42
300	2.23	2.05	1.63
400	2.34	2.19	1.80

[a] Values for intermediate heights above 15 feet (4572 mm) may be interpolated.
(From UBC 1997, Table 16-G.)

coefficient, and G_h, the gust response factor. Both these parameters are defined separately in ASCE 7-02, and hence are more appropriate for non-ordinary buildings.

The height and exposure factors account for the terrain effects on gradient heights and typically cause lower wind speeds in built-up terrain than in an open terrain. The gust factor accounts for air turbulence and dynamic building behavior.

For low-rise buildings with natural period of less than 1 sec, the wind response is essentially static with the lateral deflection proportional to the wind force. For tall buildings, on the other hand, the response is dynamic resulting in deflections greater than those estimated by simple procedures. Therefore for slender buildings a procedure such as the one given in the ASCE 7-02, which takes into account the dynamic characteristic of the building, would likely be more appropriate.

1.4.1.8.   *Pressure Coefficient C_q*

The C_q given in Table 1.3 is a function of building shape and location, and whether the wind load induces inward or outward pressures.

It is given in two parts. The first part, Primary Frames Systems, is for the design of the entire building. The second part, Elements and Components of Structure, is for the design of cladding.

Wind gusting around a building does not cause peak pressures and sections simultaneously over the entire surface of the building. Therefore, wind loads for design of primary frames and systems are calculated using average wind pressures and suctions. On the other hand the design of building components such as curtain walls and cladding is controlled by the instantaneous peak pressures and suction acting over relatively small localized areas. This is the reason why the pressures and suctions for building components are larger than those for the entire building.

Wind pressures and suctions for primary systems are mainly a function of the building height. Although these are influenced by the building's shape, the roughness of its exterior, and its plan aspect ratio, these are ignored. For example, even though wind load on a circular building is theoretically about 80% of that for a rectangular building, no reduction of forces is permitted in the UBC.

TABLE 1.3 Pressure Coefficients C_g for Primary Frames and Systems

Description	C_g
Method 1 (Normal force method) Maximum height 400 ft	
Walls	
Windward wall	0.8 inward
Leeward wall	0.5 outward
Roof	
Wind perpendicular to ridge	
Leeward roof or flat roof	0.7 outward
Windward roof	
Slope less than 2:12 (16.7%)	0.7 outward
Slope 2:12 (16.7%) to less than 9:12 (75%)	0.9 outward or
	0.3 inward
Slope 9:12 (75%) to 12:12 (100%)	0.4 inward
Slope > 12:12 (100%)	0.7 inward
Wind parallel to ridge and flat roofs	0.7 outward
Method 2 (Projected area method) Maximum height 200 ft	
On vertical projected area	
Structures 40 feet (12.19 m) or less in height	1.3 horizontal any direction
Structures over 40 feet (12.19 m) in height	1.4 horizontal any direction
On horizontal projected area	0.7 upward

(From UBC 1997.)

Two methods, are given in the UBC for determining wind loads for primary frames (Table 1.3). Method 1, the normal force method, is applicable to all structures, and is the only method permitted for the design of gable-roofed buildings. It assumes wind loads act perpendicular to the surfaces of the roof, and the walls. Method 2, the projected area method, is easier to use than Method 1. The wind pressures and suctions are integrated into a single value and are assumed to act on the entire projected area of the building, instead of on individual surfaces of roof and walls.

Another important difference between the two methods is that method 1 uses a constant value of C_e based on mean roof height to calculate wind suctions on leeward walls. Method 2 uses a C_e value that varies with height. Hence, method 2 underestimates the wind loads on taller structures. For this reason, use of method 2 is limited to structures less than 200 ft (61 m), in order to minimize the underestimated leeward forces.

1.4.1.9. *Importance Factor I_w*

Importance factor I_w is applied to increase the wind loads for certain occupancy categories. The 1997 UBC gives five separate occupancy categories: essential facilities, hazardous facilities, special occupancy structures, standard occupancy structures, and miscellaneous structures. Essential or hazardous facilities are assigned an importance factor $I_w = 1.15$, which has the effect of increasing the mean reference interval from a 50-year to a 10-year return period. Special structures, standard occupancy structures, and miscellaneous structures are assigned an importance factor I_w of 1.00. Office and residential buildings are typically assigned a standard occupancy factor of 1.00.

1.4.1.10. *Design Examples, UBC 1997*

Eleven-Story Building: UBC 1997.
Given.
- Eleven-story communication building deemed necessary for post-disaster emergency communications, $I_w = 1.15$

- Building height 120 ft (36.6 m) consisting of 2 bottom floors at 15 ft (4.6 m) and 9 typical floors at 10 ft (3.05 m)
- Exposure category = C
- Basic wind speed V = 100 mph
- Building width = 60 ft

Required.   Design wind pressures on primary wind-resisting system.
Solution.   The design pressure is given by the chain equation

$$p = C_e C_q q_s I_w$$

The values of C_e—the combined height, exposure, and gust factor coefficient tabulated in Table 1.4—are taken directly from Table 1.2. Note that for suction on the leeward face, C_e is at the roof hight, and is constant for the full height of the building. The wind pressure q_s corresponding to basic wind speed of 100 mph is given by

$$q_s = 0.00256V^2$$
$$q_s = 0.00256 \times 100^2 = 25.6 \; psf$$

The values of pressure coefficient C_q (Table 1.3), obtained using the normal force method (Method 1), are 0.8 for the inward pressure on the windward face, and 0.5 for the suction on the leeward face. Because the building is less than 200 ft (61 m), the combined value of 0.8 + 0.5 = 1.3 may be used throughout the height to calculate the wind load on the primary wind-resisting system. Observe that Method 2 (projected area method) yields the same value of C_q = 1.3.

Design pressures and floor-by-floor wind loads are shown in Table 1.4. Notice that the wind pressure and suction on the lower half of the first story (between the ground and 7.5 ft aboveground) is commonly considered to be transmitted directly into the ground. The wind load at each level is obtained by multiplying the tributary area for the level by the average of design pressures above and below that level. For example,

$$\text{wind force at level } 10 = \frac{60 \times 10(62.6 + 61.6)}{1000 \times 2} = 37.52 \text{ kips}$$

Thirty-Story Building: UBC 1997.
Given.

Basic wind speed	90 mph
Plan dimensions of building	98.5 × 164 ft
Height of building	394 ft
Importance Factor I_w	1.0
Exposure Category D	Flat unobstructed terrain facing a large body of water

Required.   Design wind pressures for lateral load analysis of the building.
Solution.   The design wind pressure is given by

$$P = C_e C_c Q_s I_w$$

The values of C_e given in Table 1.2. (UBC Table 1.2) are shown for the example problem in Table 1.5. Observe that the coefficient C_e for the leeward wall is the value at the roof level, and remains constant for the entire building height. The pressure of corresponding to V = 90 mph is given by

$$q_s = 0.00256 \times 90^2$$
$$= 20.8 \text{ psf}$$

TABLE 1.4 Design Example: Design Loads for Primary Wind-Resisting System; UBC 1997 Procedure

Level (1)	Height above ground, ft (2)	C_e (3) Windward	C_e (3) Leeward	C_q (4) Windward	C_q (4) Leeward	Windward pressure, psf $p = C_e C_q q_s I_w^{b,c}$ (5)	Leeward suction, psf $p = C_e C_g q_s I_w$ (6)	Design pressure, psf (5) + (6)	Floor-by-floor load, kips
Roof	120	1.67	1.67	+0.8	−0.5	39.40	24.6	64.0	19.2
11	110	1.64	1.67	+0.8	−0.5	38.70	24.6	63.30	38.18
10	100	1.61	1.67	+0.8	−0.5	38.00	24.6	62.6	37.77
9	90	1.57	1.67	+0.8	−0.5	37.00	24.6	61.6	37.26
8	80	1.53	1.67	+0.8	−0.5	36.00	24.6	60.6	36.66
7	70	1.48	1.67	+0.8	−0.5	35.00	24.6	59.6	36.06
6	60	1.43	1.67	+0.8	−0.5	33.80	24.6	58.40	35.40
5	50	1.37	1.67	+0.8	−0.5	32.40	24.6	57.0	34.02
4	40	1.31	1.67	+0.8	−0.5	31.0	24.6	55.6	33.78
3	30	1.23	1.67	+0.8	−0.5	29.0	24.6	53.60	32.76
2	15	1.06	1.67	+0.8	−0.5	25.0	24.6	49.62	38.40

[a]Building width perpendicular to wind = 60 ft.
[b]q_s = 25.60 psf.
[c]I_w = 1.15.

TABLE 1.5 Design Example: 30-Story Building, Design Wind Loads; UBC 1997 Procedure

Level (1)	Height above ground, ft (2)	C_e Windward (3)	C_e Leeward (4)	C_q Windward (5)	C_g Leeward (6)	Windward pressure $p = c_e c_q q_s I_w$, psf (7)	Leeward suction $p = c_e c_q q_s I_w$, psf (8)	Design pressure, psf (9) = (7) + (8)	Floor-by-floor lateral loads, kips (10)
ROOF	394	2.34	2.34	+0.8	−0.5	38.94	24.34	63.28	40.51
30	381	2.32	2.34	+0.8	−0.5	38.60	24.34	62.94	80.81
29	368	2.31	2.34	+0.8	−0.5	38.44	24.34	62.78	80.49
28	355	2.29	2.34	+0.8	−0.5	38.10	24.34	62.44	80.18
27	342	2.28	2.34	+0.8	−0.5	37.94	24.34	62.28	79.85
26	329	2.26	2.34	+0.8	−0.5	37.61	24.34	61.95	79.54
25	316	2.25	2.34	+0.8	−0.5	37.44	24.34	61.78	79.21
24	303	2.23	2.34	+0.8	−0.5	37.11	24.34	61.45	78.90
23	290	2.22	2.34	+0.8	−0.5	36.94	24.34	61.28	78.57
22	277	2.21	2.34	+0.8	−0.5	36.77	24.34	61.11	78.35
21	264	2.19	2.34	+0.8	−0.5	36.44	24.34	60.78	78.04
20	251	2.17	2.34	+0.8	−0.5	36.11	24.34	60.45	77.63
19	238	2.15	2.34	+0.8	−0.5	35.78	24.34	60.12	77.19
18	225	2.13	2.34	+0.8	−0.5	35.44	24.34	59.78	76.76
17	212	2.12	2.34	+0.8	−0.5	35.28	24.34	59.62	76.45
16	199	2.10	2.34	+0.8	−0.5	34.95	24.34	59.29	76.13
15	186	2.07	2.34	+0.8	−0.5	34.44	24.34	58.78	75.60
14	173	2.05	2.34	+0.8	−0.5	34.11	24.34	58.45	75.05
13	160	2.02	2.34	+0.8	−0.5	33.61	24.34	57.95	74.52
12	147	2.00	2.34	+0.8	−0.5	33.28	24.34	57.62	74.00
11	134	1.96	2.34	+0.8	−0.5	32.61	24.34	56.95	73.35
10	121	1.93	2.34	+0.8	−0.5	32.11	24.34	56.45	72.60
9	108	1.91	2.34	+0.8	−0.5	31.79	24.34	56.13	72.08
8	95	1.88	2.34	+0.8	−0.5	31.28	24.34	55.62	71.54
7	82	1.81	2.34	+0.8	−0.5	30.12	24.34	54.46	70.48
6	69	1.78	2.34	+0.8	−0.5	29.62	24.34	53.96	69.42
5	56	1.72	2.34	+0.8	−0.5	28.62	24.34	52.96	68.45
4	43	1.63	2.34	+0.8	−0.5	27.12	24.34	51.46	66.86
3	30	1.54	2.34	+0.8	−0.5	25.63	24.34	50.00	64.95
2	17	1.41	2.34	+0.8	−0.5	23.46	24.34	47.80	72.25

Notes: $V = 90$ mph, exposure category D.
Importance factor $I_w = 1.0$.
Wind pressure q_s at 33 ft for 90 mph basic wind speed = 20.8 psf.

Because the building height is more than 200 ft according to UBC 1997, use of method 2 is not permitted. Therefore method 1, with different values of C_q for the windward and leeward walls, is used.

$C_q = 0.8$ inward pressure for windward wall (Table 1.3)

$C_q = 0.5$ outward suction for leeward wall (Table 1.3)

Windward pressures are calculated using the tabulated values of C_e for various heights. Leeward suction is calculated only at the roof level. Therefore the suction on the leeward wall remains constant for the entire building height (Table 1.5, column 6). The combined design pressures and floor-by-floor wind loads for lateral design are tabulated in Fig. 1.8a.

It should be noted that the height of 394 ft chosen for the example problem is just under the 400-ft limit, the maximum permitted by the simple procedure of the UBC. If

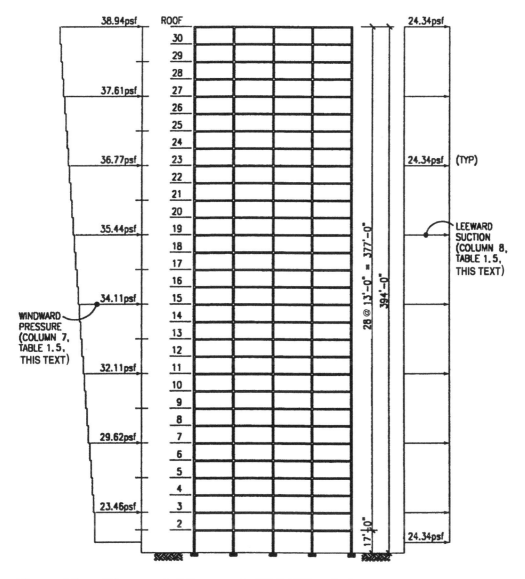

Figure 1.8a.   Thirty-story building example, UBC 1997 method.

the building were taller than 400 ft, we would be required by the UBC to use other notionally accepted standards for determining the wind loads. ASCE 7-02 is one such standard discussed later in this chapter

1.4.2. ASCE 7-02: Wind Load Provisions

The full title of this ASCE standard is American Society of Civil Engineers *Minimum Design Loads for Buildings and Other Structures.* In one of its 10 sections, ASCE 7-02 provides three procedures for calculating wind loads for buildings and other structures, including the main wind-force-resisting systems and all components thereof. The designer can use Method 1, the simplified procedure, to select wind pressures directly without calculation when the building is less than 60 ft in height and meets all requirements given in Section 6.4 of the standard. Method 2 can be used for buildings and structures of any height that are regular in shape, provided the buildings are not sensitive to across-wind loading, vortex shedding, or instability due to galloping or flutter; or do not have a site for which channeling effects warrant special consideration. Method 3 is a wind-tunnel test procedure that can be used in lieu of methods 1 and 2 for any building or structure. Method 3 is recommended for buildings that possess any of the following characteristics:

- Have nonuniform shapes.
- Are flexible with natural frequencies less than 1 Hz.
- Are subject to significant buffeting by the wake of upwind buildings or other structures.
- Are subject to accelerated flow of wind by channeling or local topographic features.

Basic wind speeds for any location in the continental United States and Alaska are shown on a map having isotachs representing a 3-sec gust speed at 33 ft (10 m) above the ground (see Fig. 1.9). For certain locations, such as Hawaii and Puerto Rico, basic wind speeds are given in a table as 105 and 145 mph (47 and 65 m/s), respectively. The map is standardized to represent a 50-year recurrence interval for exposure C topography (flat, open, country and grasslands with open terrain and scattered obstructions generally less than 30 ft (9 m) in height). The minimum basic wind speed provided in the standard is 85 mph (38 m/s). Increasing the minimum wind speed for special topographies such as mountain terrain, gorges, and ocean fronts is recommended.

The wind speed map for the United States and adjoining landmasses is based on data collected over a long period of time at weather stations located throughout the country. The maximum wind velocity expected at any location can be found simply by referring to the map.

The abandonment of the fastest-mile speed in favor of a 3-sec-gust speed first took place in the ASCE 7-1995 edition. The reasons are: 1) modern weather stations no longer measure wind speeds using the fastest-mile method; 2) a 3-sec-gust speed is closer to the sensational wind speeds often quoted by news media; and 3) it matches closely the wind speeds experienced by small buildings and by components of all buildings.

Method 1, the simplified procedure, and Method 3, the wind tunnel procedure, are not discussed here. The emphasis is on Method 2.

Method 2, the analytical procedure covered in this section, applies to a majority of buildings. It accounts for the following factors that influence the design wind forces:

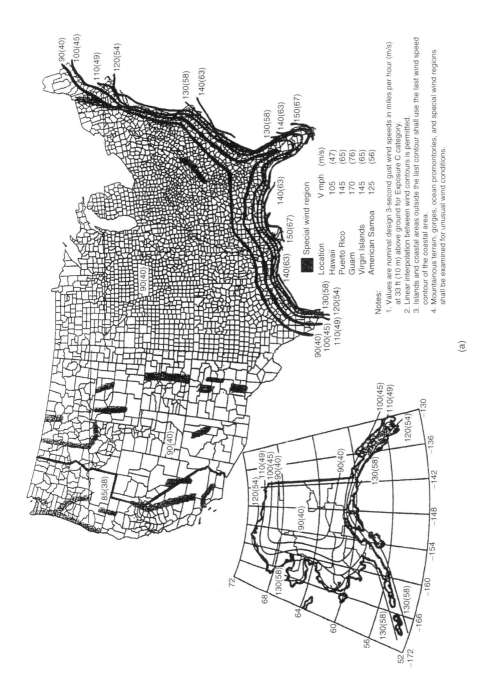

Figure 1.9. Basic wind speed map, 3-sec gust wind speed: (a) map of the United States; (b) western Gulf of Mexico hurricane coastline (enlarged); (c) eastern Gulf of Mexico and southeastern U.S. hurricane coastline (enlarged); (d) mid- and north-Atlantic hurricane coastline (enlarged). (Adapted from ASCE 7-02.)

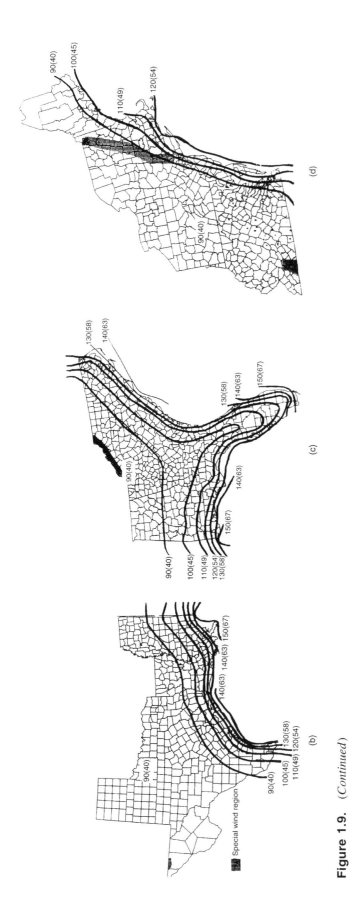

Figure 1.9. (*Continued*)

1. The basic wind speed.
2. The mean recurrence interval of the wind speed considered appropriate for the design.
3. The characteristics of the terrain surrounding the building.
4. The height at which the wind load is being determined.
5. Directional properties of the wind climate.
6. The size, geometry, and aerodynamics of the building.
7. The positions of the area acted on by the wind flow.
8. The magnitude of the area of interest.
9. The porosity of the building envelope.
10. The structural properties that may make the building susceptible to dynamic effects.
11. The speed-up effect of certain topographic features such as hills and escarpments.

1.4.2.1. *Wind Loads on Main Wind-Force-Resisting System: Overview of Analytical Procedure*

The analytical procedure has two steps. The first step considers the properties of the wind flow and the second accounts for the properties of the structure and its dynamic response to the longitudinal (along-wind) wind turbulence. The effects of across-wind response are not explicitly considered in the ASCE 7-02, Methods 1 and 2.

The velocity pressure at elevation z is given by the equation

$$q_z = 0.00256 K_z K_{zt} K_d V^2 I \qquad (q_z \text{ in psf, } V \text{ in mph}) \tag{1.8}$$

The basic wind speed V corresponds to a 50-year mean recurrence interval. It represents the speed from any direction at an elevation 33 ft (10 m) aboveground in flat open country (exposure C).

The velocity pressure exposure coefficient K_z depends on the velocity, terrain roughness (i.e., exposure category), and the height aboveground.

Three exposure categories—B, C, and D—are defined. Exposure A, at one time intended for heavily built-up city centers, was deleted in the 2002 edition of ASCE 7. The exposure for each wind direction is now defined as the worst case of the two 45° sectors on either side of the wind direction being considered.

In summary, exposure B corresponds to surface roughness B typical of urban and suburban areas, exposure C to surface roughness C in flat open country, and exposure D to surface roughness D representative of flat unobstructed area and water surfaces outside hurricane-prone regions. Exposure C applies to all cases where exposures B and D do not apply. Interpolation between exposure categories is now permitted for the first time in the ASCE 7-02. Formal definitions of exposure categories are given later in Section 1.4.2.9.

The importance factor I is a factor that accounts for the degree of hazard to human life and damage to property. For category II buildings (See Table 1.7), or other structures representative of typical occupancy, $I = 1.0$. For category I buildings or other structures representing low hazard in the event of failure (e.g., agriculture facilities), $I = 0.87$ or 0.77, depending upon whether the building site is located in hurricane-prone regions. For buildings and other structures in category III posing a substantial hazard to human life in the event of failure (e.g., buildings where more than 300 people congregate in one area, and essential facilities such as fire stations), $I = 1.15$. For category IV buildings or other structures deemed as essential facilities, $I = 1.15$, the same as for category III.

The topographic factor K_{zt} is given by

$$K_{zt} = (1 + K_1 K_2 K_3)^2 \tag{1.9}$$

It reflects the speed-up effect over hills and escarpments. The multipliers K_1, K_2, and K_3 are given in Fig. 6.4 of the Standard (Figs. 1.11 and 1.11a of this text).

Wind directionality is explicitly accounted for by introducing a new factor K_d. It is no longer a component of the wind load factor. K_d varies depending upon the type of structure. Prior to introduction of exposure factor K_d, the load factor for wind was 1.3. Now it is 1.6, obtained by dividing 1.3 by the K_d factor equal to 0.85 for most buildings. Thus the new load factor = 1.3/0.85 = 1.53 rounded to 1.6.

Internal pressures and suctions on side walls and the roof of buildings do not affect the value of wind load for the main wind-force-resisting system (MWFRS). Therefore, pressures and suctions, both denoted by P_z, are calculated for the MWFRS using the following equations:

$$P_z = q_z G_f C_p \qquad \text{(for positive pressures)} \tag{1.10}$$

$$P_z = q_h C_f C_p \qquad \text{(for negative pressures)} \tag{1.11}$$

instead of the more general equation:

$$P = q(GC_p) - q_i (GC_{pi}) \qquad \text{[ASCE 7-02 Eq. (6.23)]} \tag{1.11a}$$

The overall wind load is the summation of positive pressures on the windward wall, and negative pressure or suction, on the leeward wall. In the above equations G_f is a gust factor equal to 0.85 for rigid buildings, and C_p is an external coefficient, typically equal to 0.8 and 0.5 for the windward and leeward walls, respectively.

Thus, for a typical rigid buildings, the total design wind pressure at height Z above ground level is given by

$$P_z = 0.85(0.8q_z + 0.5q_h) \tag{1.12}$$

1.4.2.2. Analytical Procedure: Step-by-Step Process

Design wind pressure or suction on a building surface is given by the equation:

$$P_z = q_z \times G_f \times C_p \tag{1.13}$$

where
  P_z = design wind pressure or suction, in psf, at height z, above ground level
  q_z = velocity pressure, in psf, determined at height z above ground
  G_f = gust effect factor, dimensionless
  C_p = external pressure coefficient, which varies with building height acting as pressure (positive load) on windward face, and as suction (negative load) on nonwindward faces and roof. The values of C_p, unchanged from the previous edition of the Standard, are shown in Figs. 1.10 and 1.10a for various ratios of building width to depth.

The velocity pressure and suction q_z and q_h are given by

$$q_z = 0.00256 K_z K_{zt} K_d V^2 I \tag{1.14}$$

$$q_h = 0.00256 K_h K_{zt} K_d V^2 I \tag{1.15}$$

where
  K_h and K_z = combined velocity pressure exposure coefficients (dimensionless), which take into account changes in wind speed aboveground and the nature of the terrain (exposure category B, C, or D). (See Fig. 1.11b and Table 1.6.)

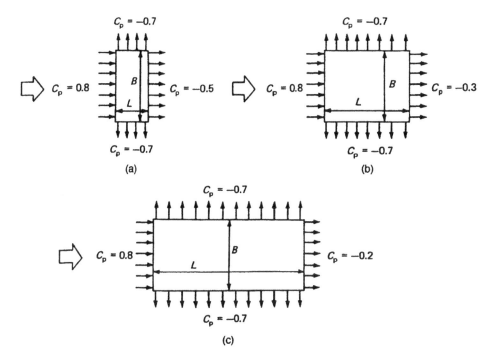

Figure 1.10. Horizontal variation of external wind pressure coefficient C_p with respect to plan aspect ratio L/B: (a) $0 \leq L/B \leq 1$; (b) $L/B = 2$; (c) $L/B > 4$. (Adapted from ASCE 7-02)

K_{zt} = topographic factor, introduced in ASCE 7-95 for the first time

I = importance factor, a dimensionless parameter that accounts for the degree of hazard to human life and damage to property (Tables 1.7 and 1.7a)

V = basic wind speed, Fig. 1.10 in miles per hour that corresponds to a 3-sec gust speed at 33-ft (10 m) aboveground, exposure category C, for a 50-year mean recurrence interval

K_d = wind directionality factor that varies from 0.85 to 0.95 depending on the structure type (Table 1.8, ASCE 7-02 Table 6.4)

The wind directionality factor identified as K_d in ASCE 7-02 accounts for two effects:

- The reduced probability of maximum winds flowing from any given direction
- The reduced probability of the maximum pressure coefficient occurring for any given direction

This factor, which was hidden in the load factors of the previous editions of the Standard, is now explicity included in the equation for velocity pressure:

$$q_z = 0.00256 K_z K_{zt} K_d V^2 I$$

The value of K_d is equal to 0.85 for most types of structures, including buildings. Therefore, q_z calculated from the previous equation is equal to 85% of the value designers were used to, prior to publication of ASCE 7-02. However, the load factors specified in ASCE 7-02 have been

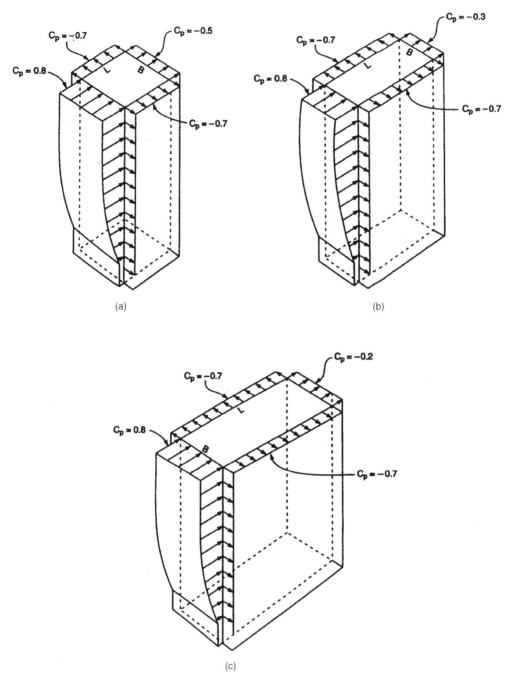

Figure 1.10a. Vertical variation of external wind pressure coefficient C_p with respect to plan aspect ratio L/B. (a) $0 \leq L/B \leq 1$; (b) $L/B = 2$; (c) $L/B > 4$.

adjusted upward, so the wind loads are about the same as before. Thus, for LRFD or strength design, the new load factor is 1.6, which previously was 1.3. The factor 1.6, when multiplied by the directionality factor K_d = 0.85, gives an effective load factor equal to $1.6 \times 0.85 = 1.36$ approximately equal to the previous factor of 1.3.

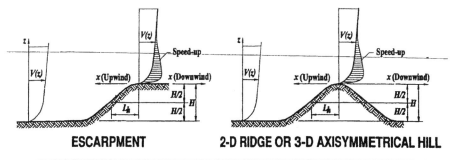

ESCARPMENT   2-D RIDGE OR 3-D AXISYMMETRICAL HILL

Topographic Multipliers for Exposure C									
H/L_h	K_1 Multiplier			x/L_h	K_2 Multiplier		z/L_h	K_3 Multiplier		
	2-D Ridge	2-D Escarp.	3-D Axisym. Hill		2-D Escarp.	All Other Cases		2-D Ridge	2-D Escarp.	3-D Axisym. Hill
0.20	0.29	0.17	0.21	0.00	1.00	1.00	0.00	1.00	1.00	1.00
0.25	0.36	0.21	0.26	0.50	0.88	0.67	0.10	0.74	0.78	0.67
0.30	0.43	0.26	0.32	1.00	0.75	0.33	0.20	0.55	0.61	0.45
0.35	0.51	0.30	0.37	1.50	0.63	0.00	0.30	0.41	0.47	0.30
0.40	0.58	0.34	0.42	2.00	0.50	0.00	0.40	0.30	0.37	0.20
0.45	0.65	0.38	0.47	2.50	0.38	0.00	0.50	0.22	0.29	0.14
0.50	0.72	0.43	0.53	3.00	0.25	0.00	0.60	0.17	0.22	0.09
				3.50	0.13	0.00	0.70	0.12	0.17	0.06
				4.00	0.00	0.00	0.80	0.09	0.14	0.04
							0.90	0.07	0.11	0.03
							1.00	0.05	0.08	0.02
							1.50	0.01	0.02	0.00
							2.00	0.00	0.00	0.00

Figure 1.11.   Topographic factor K_{zt}.

Notes:

1. For values of H/L_h, x/L_h, and z/L_h other than those shown, linear interpolation is permitted.
2. For $H/L_h > 0.5$, assume $H/L_h = 0.5$ for evaluating K_1 and substitute 2H for L_h for evaluating K_2 and K_3.
3. Multipliers are based on the assumption that wind approaches the hill or escarpment along the direction of maximum slope.
4. Notation:

   H: Height of hill or escarpment relative to the upwind terrain, in feet (meters).

   L_h: Distance upwind of crest to where the difference in ground elevation is half the height of hill or escarpment, in feet (meters).

   K_1:   Factor to account for shape of topographic feature and maximum speed-up effect.

   K_2:   Factor to account for reduction in speed-up with distance upwind or downwind of crest.

   K_3:   Factor to account for reduction in speed-up with height above local terrain.

   x: Distance (upwind or downwind) from the crest to the building site, in feet (meters).

   z: Height above local ground level, in feet (meters).

   μ: Horizontal attenuation factor.

   γ: Height attenuation factor.

(From ASCE 7-02, Fig. 6.4.)

For allowable stress design, the ASCE 7-02 load factor is still equal to 1.0. However, since one-third increase in allowable stress is not permitted, the overall effect is the same as before.

The basic wind speed is converted to the design speed at any height z for a given exposure category by using the velocity exposure coefficient K_z, evaluated at height z.

Parameters for Speed-Up Over Hills and Escarpments						
Hill Shape	$K_1/(H/L_h)$ Exposure			γ	μ Upwind of Crest	μ Downwind of Crest
	B	C	D			
2-Dimensional ridges (or valleys with negative H in $K_1/(H/L_h)$)	1.30	1.45	1.55	3	1.5	1.5
2-Dimensional escarpments	0.75	0.85	0.95	2.5	1.5	4
3-Dimensional axisym. hill	0.95	1.05	1.15	4	1.5	1.5

Figure 1.11a. Topographic factor K_{zt} based on equations

$K_{zt} = (1 + K_1 K_2 K_3)^2$
K_1 determined from table

$$K_2 = \left(1 - \frac{|x|}{\mu L_h}\right)$$

$K_3 = e^{-\gamma z/L_h}$

(From Fig. 6.4 in ASCE 7-02.)

K_z is given by

$$K_z = 2.01\left(\frac{z}{z_g}\right)^{2/\alpha} \qquad \text{(for 15 ft} < z < z_g) \qquad\qquad (1.16)$$

$$K_z = 2.01\left(\frac{15}{z_g}\right)^{2/\alpha} \qquad \text{(for } z < 15 \text{ ft)} \qquad\qquad (1.17)$$

where z_g is the gradient height above which the frictional effect of terrain becomes negligible. It varies with the characteristics of the ground surface irregularities at the building site that arise as a result of natural topographic variations as well as human-made features. In the ASCE 7-02 standard, the K_z expressions are unchanged from ASCE 7-98. However, interpolation of the K_z values between standard exposures is permitted for the first time in the 2002 edition of ASCE 7.

The power coefficient α (Table 1.9) is the exponent for velocity increase in height, and has values of 7.0, 9.5, and 11.5, respectively, for exposure B, C, and D. The values of K_z for various exposures up to a height of 500 ft (152.6 m) are given in ASCE 7-02. An extended version up to a height of 1500 ft (457 m) is given in Table 1.6 and in Fig. 1.11b. The values of the gradient height z_z, given in ASCE 7.02, are of course identical to those given in the previous ASCE-7 editions. This should be obvious because the gradient height z_z for a given exposure does not vary with the reference wind speed used in design. As with the previous ASCE-7 editions, the values of K_z are assumed to be constant for heights less than 15 ft (4.6 m), and for heights greater than the gradient height z. The variation of velocity pressure q_z for exposure categories B, C, and D is given in Fig. 1.12.

1.4.2.3. Wind Speed-Up Over Hills and Escarpments: K_{zt} Factor

The topographic factor K_{zt} accounts for the effect of isolated hills or escarpments located in exposures B, C, and D. Buildings sited on the upper half of an isolated hill

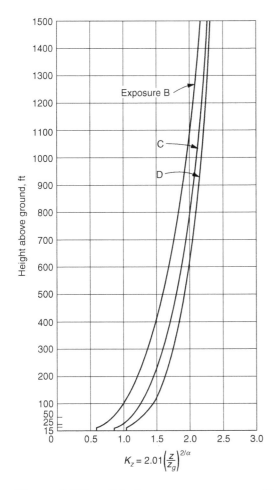

$$K_z = 2.01\left(\frac{z}{z_g}\right)^{2/\alpha}$$

Figure 1.11b. Velocity pressure exposure coefficients K_h and K_z. The graphical representation of K_z values is given in Table 1.6. (From ASCE 7-02.)

or escarpment may experience significantly higher wind speeds than buildings situated on level ground. To account for these higher wind speeds, the velocity pressure exposure coefficients are multiplied by a topographic factor K_{zt}, determined from the three multipliers K_1, K_2, and K_3 (Fig. 1.11), K_1 is related to the shape of the topographic feature and the maximum speed-up with distance upwind or downward of the crest, K_2 accounts for the reduction in speed-up with distance upwind or downwind of the crest, and K_3 accounts for the reduction in speed-up with height above the local ground surface.

1.4.2.4. *Design Wind Load Cases*

This requirement, first introduced in ASCE 7-95 under the heading "Full and Partial Loading," was for including the torsional response of buildings. Now the design requirements have become more stringent under a new heading, "Design Wind Load Cases."

Recent wind tunnel research has shown that torsional load requirements previously given in ASCE 7-98 often grossly underestimated the actual torsion on a building under wind, even those that are symmetric in geometric form and stiffness. This torsion is a result of nonuniform pressures on the different faces of the building as wind flows around

TABLE 1.6　Velocity Pressure Exposure Coefficients, $K_z^{a,b}$

Height above ground level, z		Exposure category		
ft	(m)	B	C	D
0–15	(0–4.6)	0.57	0.85	1.03
20	(6.1)	0.62	0.90	1.08
25	(7.6)	0.66	0.94	1.12
30	(9.1)	0.70	0.98	1.16
40	(12.2)	0.76	1.04	1.22
50	(15.2)	0.81	1.09	1.27
60	(18)	0.85	1.13	1.31
70	(21.3)	0.89	1.17	1.34
80	(24.4)	0.93	1.21	1.38
90	(27.4)	0.96	1.24	1.40
100	(30.5)	0.99	1.26	1.43
120	(36.6)	1.04	1.31	1.48
140	(42.7)	1.09	1.36	1.52
160	(48.8)	1.13	1.39	1.55
180	(54.9)	1.17	1.43	1.58
200	(61.0)	1.20	1.46	1.61
250	(76.2)	1.28	1.53	1.68
300	(91.4)	1.35	1.59	1.73
350	(106.7)	1.41	1.64	1.78
400	(121.9)	1.47	1.69	1.82
450	(137.2)	1.52	1.73	1.86
500	(152.4)	1.56	1.77	1.89
550	(167.6)	1.61	1.81	1.93
600	(182.9)	1.65	1.85	1.96
650	(191.1)	1.69	1.88	1.98
700	(213.3)	1.72	1.91	2.01
750	(228.6)	1.76	1.93	2.03
800	(243.8)	1.79	1.96	2.06
850	(259.1)	1.82	1.99	2.08
900	(274.3)	1.85	2.01	2.10
950	(289.5)	1.88	2.03	2.12
1000	(304.8)	1.91	2.06	2.14
1050	(320)	1.93	2.08	2.16
1100	(335.3)	1.96	2.10	2.17
1150	(350.5)	1.99	2.12	2.19
1200	(365.7)	2.01	2.14	2.21
1250	(381)	2.03	2.15	2.22
1300	(396.2)	2.06	2.17	2.24
1350	(411.5)	2.08	2.19	2.26
1400	(426.7)	2.10	2.21	2.27
1450	(441.9)	2.12	2.22	2.28
1500	(457.2)	2.14	2.24	2.29

[a] The velocity pressure exposure coefficient K_z may be determined from the following formula:

For 15 ft $\leq z \leq z_g$, $K_z = 2.01\ (z/z_g)^{2/\alpha}$.

For $z < 15$ ft, $K_z = 2.01\ (15/z_g)^{2/\alpha}$.

[b] All main wind force resisting systems in buildings and in other structures except those in low-rise buildings.

(Adapted from Table 6.3 of ASCE 7-02.)

TABLE 1.7 Classification of Buildings for Flood, Wind, Snow, Earthquake, and Ice Loads

Nature of occupancy	Category
Buildings that represent a low hazard to human life in the event of failure including, but not limited to: Agricultural facilities Certain temporary facilities Minor storage facilities	I
All buildings except those listed in Categories I, III, and IV	II
Buildings that represent a substantial hazard to human life in the event of failure including, but not limited to Buildings where more than 300 people congregate in one area. Buildings with day care facilities with capacity greater than 150. Buildings with elementary school or secondary school facilities with capacity greater than 250. Buildings with a capacity greater than 500 for colleges or adult education facilities. Health care facilities with a capacity of 50 or more resident patients but not having surgery or emergency treatment facilities. Jails and detention facilities. Power-generating stations and other public utility facilities not included in Category IV.	III
Buildings not included in Category IV (including, but not limited to, facilities that manufacture, process, handle, store, use, or dispose of such substances as hazardous fuels, chemicals, and waste, or explosives) containing sufficient quantities of hazardous materials to be dangerous to the public if released.	
Buildings containing hazardous materials shall be eligible for classification as Category II structures if it can be demonstrated to the satisfaction of the authority having jurisdiction by a hazard assessment as described in Section 1.5.2 that a release of the hazardous material does not pose a threat to the public.	
Buildings designated as essential facilities including, but not limited to Hospitals and other health care facilities having surgery or emergency treatment facilities. Fire, rescue, ambulance, and police stations and emergency vehicle garages. Designated earthquake, hurricane, or other emergency shelters. Designated emergency preparedness, communication, and operation centers and other facilities required for emergency response. Power-generating stations and other public utility facilities required in an emergency. Ancillary structures (including, but not limited to, communication towers, fuel storage tanks, cooling towers, electrical substation structures, fire water storage tanks or other structures housing or supporting water, or other fire-suppression material or equipment) required for operation of Category IV structures during an emergency. Aviation control towers, air traffic control centers, and emergency aircraft hangars. Water storage facilities and pump structures required to maintain water pressure for fire suppression. Buildings and other structures having critical national defense functions.	IV

(Continued)

TABLE 1.7   (Continued)

Nature of occupancy	Category
Buildings (including, but not limited to, facilities that manufacture, process, handle, store, use, or dispose of such substances as hazardous fuels, chemicals, and waste, or explosives) containing extremely hazardous materials where the quantity of the material exceeds a threshold quantity established by the authority having jurisdiction.	IV
Buildings containing extremely hazardous materials shall be eligible for classification as Category II structures if it can be demonstrated to the satisfaction of the authority having jurisdiction that a release of the extremely hazardous material does not pose a threat to the public. This reduced classification shall not be permitted if the buildings also function as essential facilities.	

(From ASCE 7-02 Table 1.1.)

TABLE 1.7a   Importance Factor, I (Wind Loads)

Category[a]	Non-hurricane-prone regions and hurricane-prone regions with $V = 85$–100 mph and Alaska	Hurricane-prone regions with $V > 100$ mph
I	0.87	0.77
II	1.00	1.00
III	1.15	1.15
IV	1.15	1.15

[a] The building and structure classification categories are listed in Table 1.7, ASCE 7-02, Table 1.1.
(From ASCE 7-02 Table 6.1.)

TABLE 1.8   Wind Directionality Factor K_d

Structure type	Directionality factor K_d^{a}
Buildings	
Main wind-force-resisting system	0.85
Components and cladding	0.85
Arched roofs	0.85
Chimneys, tanks, and similar structures	
Square	0.90
Hexagonal	0.95
Round	0.95
Solid signs	0.85
Open signs and lattice framework	0.85
Trussed towers	
Triangular, square, rectangular	0.85
All other cross sections	0.95

[a] Directionality factor K_d shall only be applied when used in conjunction with load combinations specified in ASCE 7-02 Sections 2.3 and 2.4.
(From ASCE 7-02 Table 6.4.)

TABLE 1.9 Terrain Exposure Constants

Exposure	α	z_g (ft)	$\hat{a}$	$\hat{b}$	$\bar{\alpha}$	$\bar{b}$	c	ℓ(ft)	$\bar{\varepsilon}$	z_{min} (ft)[a]
B	7.0	1200	1/7	0.84	1/4.0	0.45	0.30	320	1/3.0	30
C	9.5	900	1/9.5	1.00	1/6.5	0.65	0.20	500	1/5.0	15
D	11.5	700	1/11.5	1.07	1/9.0	0.80	0.15	650	1/8.0	7

[a] z_{min} = minimum height used to ensure that the equivalent height $\bar{z}$ is the greater of $0.6h$ or z_{min}. For buildings with $h \le z_{min}$, $\bar{z}$ shall be taken as z_{min}.
(From ASCE 7-02, Table 6.2.)

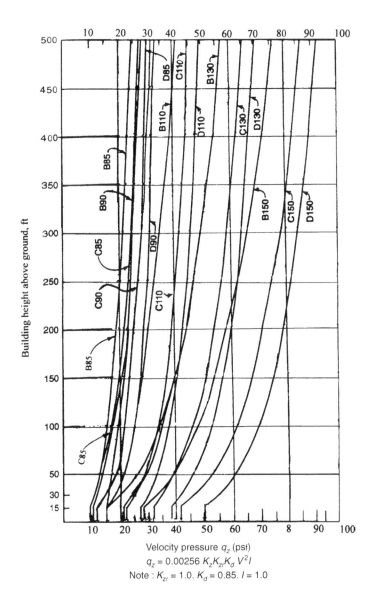

Velocity pressure q_z (psf)

$$q_z = 0.00256 \, K_z K_{zt} K_d \, V^2 I$$

Note : $K_{zt} = 1.0$. $K_d = 0.85$. $I = 1.0$

Figure 1.12. Building height h, velocity pressure q_z. (Adapted from ASCE 7-02.)

the building. These irregular pressures are the results of interference effects of nearby buildings and terrain, and dynamic effects on more flexible buildings. The ASCE 7-02 requirements represent of square and rectangular buildings with aspect ratios up to about 2.5. They may not cover all cases, even for symmetric and common building shapes where larger torsions have been observed. Therefore, the designer may wish to apply this level of eccentricity at full, rather than reduced, wind loading for certain more critical buildings, even though it is not required by the Standard.

In buildings with unusual structural systems, such as the one used for the City Corp. Tower in New York, more severe loading can occur when the resultant wind load acts diagonally to the building. To account for this effect and the fact that many buildings exhibit maximum response in the across-wind direction, a structure should be capable of resisting 75% of the design wind load applied simultaneously along each principal axis, as required by case 3 in Fig. 6.9 of ASCE 7-02.

For flexible buildings, dynamic effects can increase torsional loading. Additional torsional loading can occur because of eccentricity between the elastic shear center and the center of mass at each level of the structure. The new Eq. (1.18) given below accounts for this effect.

$$e = \frac{e_Q + 1.7I_{\bar{z}}\sqrt{(g_Q Q e_Q)^2 + (g_R R e_R)^2}}{1.7I_{\bar{z}}\sqrt{(g_Q Q)^2 + (g_R R)^2}} \tag{1.18}$$

where

$\quad e_Q$ = eccentricity e as determined for rigid structures in Fig. 6.9 of ASCE 7-02
$\quad e_R$ = distance between the elastic shear center and center of mass of each floor
$\qquad I_{\bar{z}}$, g_Q, Q, g_R shall be as defined in 6.5.8 of ASCE 7-02

The sign of the eccentricity e shall be plus or minus, whichever causes the more severe load effect.

The eccentricity e for flexible structures shall be considered for each principal axis (e_X, e_Y).

The eccentricity is used for calculating torsional moment M_T per unit height acting about a vertical axis of the building. The designer is referred to ASCE 7-02, Fig. 6.9 for additional information.

1.4.2.5. *Gust Effect Factor*

The gust effect factor accounts for additional dynamic amplification of loading in the along-wind direction due to wind turbulence and structure interaction. It does not include allowances for across-wind loading effects, vortex shedding, instability due to galloping or flutter, or dynamic torsional effects. Buildings susceptible to these effects should be designed using wind tunnel results.

Three methods are permitted for calculating G. The first two are for rigid structures and the third is for flexible or dynamically sensitive structures.

Gust Effect Factor G for Rigid Structure: Simplified Method.   For rigid structures (defined as those having a natural frequency of vibration greater than 1 Hz), the engineer can use a single value of $G = 0.85$, irrespective of exposure category.

Gust Effect Factor G for Rigid Structure: Improved Method.   As an option to using $G = 0.85$ the designer may calculate a more accurate value by including specific features of the wind environment at the building site. The procedure is as follows:

The gust effect factor G is given by

$$G = 0.925\left(\frac{(1 + 1.7g_Q I_{\bar{z}} Q)}{1 + 1.7g_v I_{\bar{z}}}\right) \tag{1.19}$$

$$I_{\bar{z}} = c(33/\bar{z})^{1/6} \tag{1.20}$$

where $I_{\bar{z}}$ = the intensity of turbulence at height $\bar{z}$ and where $\bar{z}$ = the equivalent height of the structure defined as $0.6\,h$ but not less than z_{min} for all building heights h. z_{min} and c are listed for each exposure in Table 1.9; g_Q and g_v shall be taken as 3.4. The background response Q is given by

$$Q = \sqrt{\frac{1}{1 + 0.63\left(\dfrac{B + h}{L_{\bar{z}}}\right)^{0.63}}} \tag{1.21}$$

where B, h are defined in Section 6.3; and $L_{\bar{z}}$ = the integral length scale of turbulence at the equivalent height given by

$$L_{\bar{z}} = l(\bar{z}/33)^{\bar{\varepsilon}} \tag{1.22}$$

in which l and $\bar{\varepsilon}$ are constants listed in Table 6.2 of ASCE 7-02 (Table 1.9 of this chapter).

Gust Effect Factor G_f for Flexible or Dynamically Sensitive Structures. Flexible buildings are those that have a frequency less than 1 Hz (i.e., buildings with a fundamental period greater than 1 sec.). Included are buildings with heights in excess of four times their least plan dimension.

The formula for calculating G_f is as follows:

$$G_f = 0.925\left(\frac{1 + 1.7I_{\bar{z}}\sqrt{g_Q^2 Q^2 + g_R^2 R^2}}{1 + 1.7g_v I_{\bar{z}}}\right) \tag{1.23}$$

where g_Q and g_v shall be taken as 3.4 and g_R is given by

$$g_R = \sqrt{2\ln(3,600n_1)} + \frac{0.577}{\sqrt{2\ln(3,600n_1)}} \tag{1.24}$$

and where R, the resonant response factor, is given by

$$R = \sqrt{\frac{1}{\beta} R_n R_h R_B (0.53 + 0.47 R_L)} \tag{1.25}$$

$$R_n = \frac{7.47 N_1}{(1 + 10.3 N_1)^{5/3}} \tag{1.26}$$

$$N_1 = \frac{n_1 L_{\bar{z}}}{\bar{V}_{\bar{z}}} \tag{1.27}$$

$$R_\ell = \frac{1}{\eta} - \frac{1}{2\eta^2}(1 - e^{-2\eta}) \quad \text{for } \eta > 0 \tag{1.28}$$

$$R_\ell = 1 \text{ for } \eta = 0 \tag{1.29}$$

where the subscript ℓ in Eq. 1.28 shall be taken as h, B, and L respectively, and where

n_1 = building natural frequency
$R_\ell = R_h$ setting $\eta = 4.6n_1 h/\bar{V}_{\bar{z}}$

$R_\ell = R_B$ setting $\eta = 4.6n_1 B/\bar{V}_{\bar{z}}$

$R_\ell = R_L$ setting $\eta = 15.4n_1 L/\bar{V}_{\bar{z}}$

β = damping ratio, percent of critical h, B, L are defined in Section 6.3

$\bar{V}_{\bar{z}}$ = mean hourly wind speed (ft/s) at height $\bar{z}$ determined from Eq. 1.30,

$$\bar{V}_{\bar{z}} = \bar{b}\left(\frac{\bar{z}}{33}\right)^{\bar{\alpha}} V\left(\frac{88}{60}\right) \qquad (1.30)$$

where $\bar{b}$ and $\bar{\alpha}$ are constants listed in Table 1.9 and V is the basic wind speed in mph

1.4.2.6. Along-Wind Response

A typical modern building that is light and flexible is more prone to dynamic motions than its earlier counterpart with heavy masonry cladding and partition walls. Dynamic motions are those cause by time-dependent forces such as seismic accelerations and short-period wind loads, or gusts. Building dynamic excitations during earthquakes, insofar as perception of motion by the occupants is concerned, is irrelevant because occupants are thankful to have survived the trauma and are less prone to complain about motion perception. However, the sentiment when estimating peak dynamic response of buildings to fluctuating wind forces is quite different, because windstorms occur more frequently and are not as traumatic as earthquakes. Consequently, it is necessary to determine whether the building is prone to wind-induced problems related to the comfort of the occupants.

When considering the response of a tall building to wind gusts, both along-wind and across-wind responses must be considered. These arise from different effects of wind, the former being primarily due to buffeting effects caused by turbulence; the latter being primarily due to alternate-side vortex shedding. The cross-wind response may be of particular importance with regard to the comfort of the occupants because it is likely to exceed along-wind accelerations if the building is slender about both axes, such that the geometric ratio $\sqrt{WD/H}$ is less than one-third, where W and D are the across- and along-wind plan dimensions, and H is the building height.

The most important criterion for verifying the comfort of the building's occupants is the peak acceleration they experience. It is thus important to be able to estimate the probable maximum accelerations in both the along-wind and across-wind directions. ASCE 7-02 gives a method for predicting along-wind responses, including peak acceleration, but does not provide a procedure for estimating across-wind response. However, the National Building Code of Canada (NBCC), addressed presently in Section 1.4.3, provides such a procedure.

Maximum Along-Wind Displacement. The maximum along-wind displacement $X_{max}(z)$ as a function of height above the ground surface is given by

$$X_{max}(z) = \frac{\phi(z)\rho BhC_{fx}\hat{V}_{\bar{z}}^2}{2m_1(2\pi n_1)^2} KG \qquad (1.31)$$

where

$\phi(z)$ = the fundamental model shape = $(z/h)^\xi$

ξ = the mode exponent

ρ = air density

C_{fx} = mean alongwind force coefficient

$$m_1 = \text{modal mass} = \int_o^h \mu(z)\phi^2(z)dz \qquad (1.32)$$

where $\mu(z)$ = mass per unit height.

$$K = (1.65)^{\hat{\alpha}}/(\hat{\alpha} + \xi + 1) \tag{1.33}$$

$\bar{V}_{\bar{z}}$ is the 3-sec gust speed at height $\bar{z}$. This can be evaluated as $\hat{V}_{\bar{z}} = \hat{b}(z/33)^{\hat{\alpha}}V$, where V is the 3-sec gust speed in exposure C at the reference height (obtained from Fig. 1.10), $\hat{b}$ and $\hat{\alpha}$ are given in Table 1.9.

RMS Along-Wind Acceleration. The rms along-wind acceleration $\sigma_{\ddot{x}}(z)$ as a function of height above the ground surface is given by

$$\sigma_{\ddot{x}}(z) = \frac{0.85\phi(z)\rho BhC_{fx}\bar{V}_{\bar{z}}^2}{m_1} I_{\bar{z}}KR \tag{1.34}$$

where $\bar{V}_{\bar{z}}$ is the mean hourly wind speed at height $\bar{z}$, ft/s

$$\bar{V}_{\bar{z}} = \bar{b}\left(\frac{\bar{z}}{33}\right)^{\bar{\alpha}} V \tag{1.35}$$

where $\bar{b}$ and $\bar{\alpha}$ are defined in Table 1.7a.

Maximum Along-Wind Acceleration. The maximum along-wind acceleration as a function of height above the ground surface is given by

$$\ddot{X}_{\max}(z) = g_{\ddot{x}}\sigma_{\ddot{x}}(z) \tag{1.36}$$

$$g_{\ddot{x}} = \sqrt{2\ln(n_1 T)} + \frac{0.5772}{\sqrt{2\ln(n_1 T)}} \tag{1.37}$$

where T = the length of time over which the acceleration is computed, usually taken to be 3600 s to represent 1 h.

1.4.2.7. *Worksheet for Calculation of Gust Effect Factor G_f and Along-Wind Displacement and Acceleration*

The formulas given in the ASCE Standard are in a concise format. They may be harder to use without rewriting many of the formulas in an expanded manner. Therefore, to make the calculation of G_f somewhat less forbidding, the ASCE formulas have been expanded and given in a worksheet format in the following section. The worksheets also include formulas for calculating along-wind response, given in the commentary section of the Standard. Also included are comments that may be helpful in going through various calculations.

Worksheet for Calculating Gust Effect Factor, Along-Wind Displacement, and Accelerations

ASCE 7-02 formulas	Commentary
V = wind speed in ft/s $= V_{mph} \times 1.467$	V from wind map, converted from mph to ft/s
$\bar{Z} = 0.6\,h,$ but not less than z_{min}	Z_{min} from Table 6.4 h = building height, ft
$I_{\bar{z}} = C\left(\dfrac{33}{\bar{z}}\right)^{\frac{1}{6}}$	C from Table 6.4
$L_{\bar{z}} = l\left(\dfrac{\bar{z}}{33}\right)^{\varepsilon}$	l and ε from Table 6.4

ASCE 7-02 formulas	Commentary
$Q = \sqrt{\dfrac{1}{1 + 0.63\left(\dfrac{B+h}{L_{\bar{z}}}\right)^{0.63}}}$	B = building width perpendicular to wind h = building height Q = background response, a term used in random vibration theory.
$\bar{V}_{\bar{z}} = \bar{b}\left(\dfrac{\bar{z}}{33}\right)^{\bar{\alpha}} V$	$\bar{b}$ and $\bar{\alpha}$ from Table 6.4
$\hat{V}_{\bar{z}} = \hat{b}\left(\dfrac{\bar{z}}{33}\right)^{\hat{\alpha}}$	$\hat{b}$ and $\hat{\alpha}$ from Table 6.4
$N_1 = \dfrac{n_1 L_{\bar{z}}}{\bar{V}_{\bar{z}}}$	n_1 = natural frequency of the building
$R_n = \dfrac{7.47 N_1}{(1 + 10.3\, N_1)^{5/3}}$	R_n is a parameter required for calculating R^2
$\eta_b = \dfrac{4.6 n_1 \mathrm{B}}{\bar{V}_{\bar{z}}}$ $R_B = \dfrac{1}{\eta} - \dfrac{1}{2\eta_b^2}(1 - e^{-2\eta_b})$ for $\eta_b > 0$ $R_B = 1$, for $\eta_b = 0$	R_b is a parameter required for calculating R^2
$\eta_h = \dfrac{4.6 n_1 h}{\bar{V}_{\bar{z}}}$ $R_h = \dfrac{1}{\eta_h} - \dfrac{1}{2\eta_h^2}(1 - e^{-2\eta_h})$ for $\eta_h > 0$ $R_h = 1$, for $\eta_h = 0$	R_h is a parameter required for calculating R^2
$\eta_L = \dfrac{15.4 n_1 L}{\bar{V}_{\bar{z}}}$ $R_L = \dfrac{1}{\eta_L} - \dfrac{1}{2\eta_L^2}(1 - e^{-2\eta_L})$ for $\eta_L > 0$ $R_L = 1$ for $\eta_L = 0$	R_L is a parameter required for calculating R^2
$R^2 = \dfrac{1}{\beta} R_n R_h R_B (0.53 + 0.47\, R_L)$ $G = 0.925\left(\dfrac{1 + 1.7\, I_{\bar{z}}\sqrt{g_Q^2 Q^2 + g_R^2 R^2}}{1 + 1.7 g_v\, I_{\bar{z}}}\right)$	β = damping ratio, percent of critical g_Q = peak factor for background response g_R = peak factor for resonance response G = gust factor g_v = peak factor for wind response $g_Q = g_v$ always taken $= 3.4$
$k = \dfrac{1.65^{\bar{\alpha}}}{(\bar{\alpha} + \xi + 1)}$	ξ = mode exponent
$g_{\bar{x}} = 2 \ln{(n_1 T)} + \dfrac{0.5772}{\sqrt{2 \ln{(n_1 T)}}}$	ln means logarithm to base $e = 2.71$.
m_1 = modal mass $= \displaystyle\int_0^h \mu_z Q_z^2 dz$ $= \mu_z \times \dfrac{h}{3}$ as shown below.	μ_z = mass per unit height, slugs/ft h = building height

For a linear first-mode shape, $Q = x/h$, where x is the displacement of the building at top. If we assume that μ_z is constant for the full height of the building (meaning that the building is uniform with a constant density = μ slugs/ft³), the modal mass is given by

$$m = \int_{o}^{h} \mu_z Q^2 dz$$

$$= \mu_z \int_{o}^{h} \frac{x^2}{h^2} dz$$

$$= \mu_z \left[\frac{x^3}{3h^2} \right]_{o}^{h}$$

$$= \mu_z \times \frac{h}{3}$$

Maximum Along-Wind Acceleration

$$X_{max(z)} = \frac{\phi_{(2)}\rho BhC_{fx}\hat{V}_{\bar{z}}^2}{2m_1(2\pi n_1)^2} KG$$

$$g_{\ddot{x}} = \sqrt{2 \ln (n_1 T)} + \frac{0.5772}{\sqrt{2 \ln (n_1 T)}}$$

$$\sigma_{\ddot{x}(z)} = \frac{0.85 \, \phi_z \rho BhC_{fx}\bar{V}_{\bar{z}}^2}{m_1} I_{\bar{z}} KR$$

$$\ddot{X}_{max(z)} = g_{\ddot{x}}\sigma_{\ddot{x}(z)}$$

$$\ddot{X}_{max(h)} = g_{\ddot{x}}\sigma_{\ddot{x}(h)}$$

At $z = h$, $X_{max(h)}$ gives the maximum lateral load deflection at top

ρ = air density = 0.0024 slugs/ft³

C_{fx} = mean along-wind force coefficient, typically equal to 1.3.

T = time in seconds over which acceleration is computed, usually taken to be 1 hour = 3600 seconds.

$\sigma_{\ddot{x}(z)}$ = the root-mean-square along-wind acceleration above the ground surface

$\ddot{X}_{max(z)}$ = the maximum along-wind acceleration as a function of height above the ground surface.

$\ddot{X}_{max(h)}$ = the maximum acceleration at the building top—the item of interest. If greater than 20 milli-g, further investigation is recommended.

1.4.2.8. *Design Examples: ASCE 7-02*

Several examples are given in this section. The first demonstrates calculations for gust effect factor G for a rigid structure. The second is for a flexible structure. The third shows calculations for design wind pressures for a 450-ft tall building using the graphs given in Fig. 1.12. Also given in this example are calculations for gust effect factor and along-wind response, using the worksheets given in Section 1.4.1.7.

In the fourth and fifth examples, design wind pressure calculations for a 10- and 30-story building are given. The final example gives a comparison of gust effect factor and along-wind displacements and acceleration for four randomly chosen buildings.

Calculations for Gust Effect Factor G: Rigid Structure.
Given. A 10-story concrete building with the following characteristics:
Height $h = 112$ ft
Width perpendicular to wind, $B = 90$ ft

Exposure category = C
Basic wind speed $V = 100$ mph
Topographic factor $K_{zt} = 1.0$
Building depth parallel to wind, $L = 95$ ft
Building natural frequency $n_1 = 1.1$ Hz

Required.   Gust effect factor G, using the improved method.
Solution.

ASCE 7-02 formulas	Commentary
$\bar{z} = 0.6 \times h = 0.6 \times 112 = 67.2$ ft	h = building height = 112 ft, given
$I_{\bar{z}} = C\left(\dfrac{33}{\bar{z}}\right)^{1/6} = 0.2\left(\dfrac{33}{67.2}\right)^{1/6} = 0.1776$	c from Table 6.4 l and ε from Table 6.4
$L_{\bar{z}} = l\left(\dfrac{\bar{z}}{33}\right)^{\varepsilon}$	
$\quad = 500\left(\dfrac{67.2}{33}\right)^{1/5}$	
$\quad = 576$	
$Q = \sqrt{\dfrac{1}{1+0.63\left(\dfrac{B+h}{L_{\bar{z}}}\right)^{0.63}}}$	Q = background response B = building width $\quad$ perpendicular to wind $\quad = 112$ ft, given
$\quad = \sqrt{\dfrac{1}{1+0.63\left(\dfrac{90+112}{576}\right)^{0.63}}}$	
$\quad = 0.87$	
$G = 0.925\left(\dfrac{(1+1.7g_Q I_{\bar{z}} Q)}{1+1.7g_v I_{\bar{z}}}\right)$	G = gust effect factor
$\quad = 0.925\left(\dfrac{1+1.7\times 3.4 \times 0.1776 \times 0.86}{1+1.7\times 3.4 \times 0.1776}\right)$	
$\quad = 0.86$	

Observe this is not much different from $G = 0.85$ permitted for rigid structures.

Calculations for Gust Effect Factor G_f: Flexible Structure
Given.
Building height $h = 600$ ft
Building width perpendicular to wind, $B = 100$ ft
Building depth parallel to wind, $L = 100$ ft
Building natural frequency $n_1 = 0.2$ Hz
Damping ratio = 0.015
Exposure category = C
Basic windspeed $V = 140$ mph

Required.   Gust effect factor G_f

ASCE 7-02 formulas	Commentary

V = wind speed in ft/s

$\quad = V_{mph} \times 1.467$

$\quad = 140 \times 1.467 = 205$ ft/s

$\bar{z} = 0.6\,h,$ but not less than z_{min}

$\quad = 0.6 \times 600 = 360$ ft $> \bar{z}_{min} = 15$ ft   OK

$$I_{\bar{z}} = C\left(\frac{33}{\bar{z}}\right)^{\frac{1}{6}} = 0.20\left(\frac{33}{360}\right)^{\frac{1}{6}} = 0.134$$

$$L_{\bar{z}} = l\left(\frac{\bar{z}}{33}\right)^{\varepsilon} = 500\left(\frac{360}{33}\right)^{\frac{1}{5}} = 806 \text{ ft}$$

$$Q = \sqrt{\frac{1}{1 + 0.63\left(\frac{B+h}{L_{\bar{z}}}\right)^{0.63}}}$$

$$= \sqrt{\frac{1}{1 + 0.63\left(\frac{100+600}{761}\right)^{0.63}}}$$

$$= 0.796, \quad Q^2 = 0.634$$

$$\bar{V}_{\bar{z}} = \bar{b}\left(\frac{\bar{z}}{33}\right)^{\bar{\alpha}} V = 0.65\left(\frac{360}{33}\right)^{\frac{1}{6.5}} \times 205 = 192 \text{ ft/s}$$

$$\hat{V}_{\bar{z}} = \hat{b}\left(\frac{\bar{z}}{33}\right)^{\hat{\alpha}} V = 1.0 \times \left(\frac{360}{33}\right)^{\frac{1}{9.5}} \times 205 = 264 \text{ ft/s}$$

$$N_1 = \frac{n_1 L_{\bar{z}}}{\bar{V}_{\bar{z}}} = \frac{0.2 \times 806}{192} = 0.84$$

$$R_n = \frac{7.47\,N_1}{(1 + 10.3\,N_1)^{\frac{5}{3}}}$$

$$= \frac{7.47 \times 0.84}{(1 + 10.3 \times 0.84)^{\frac{5}{3}}} = 0.143$$

$$\eta_b = \frac{4.6 n_1 B}{\bar{V}_{\bar{z}}} = \frac{4.6 \times 0.2 \times 100}{192} = 0.479$$

$$R_B = \frac{1}{\eta_b} - \frac{1}{2\eta_b^2}(1 - e^{-2\eta_b})$$

$$= \frac{1}{0.479} - \frac{1}{2 \times 0.479^2}(1 - 2.71^{-2 \times 0.479})$$

$$= 0.739$$

$$\eta_h = \frac{4.6 n_1 h}{\bar{v}_{\bar{z}}} = \frac{4.6 \times 0.2 \times 600}{192} = 2.875$$

Commentary column:

V = 140 mph, given

h = building height

$\quad = 600$ ft, given

$\bar{z}_{mir}$ = 15 ft, from Table 6.4

c = 0.20, from Table 6.4

l = 500 ft, $\varepsilon = \frac{1}{5}$, from Table 6.4

Q = background response

B = building with perpendicular to wind = 100 ft, given

$\bar{b}$ = 0.65, $\bar{\alpha} = \frac{1}{6.5}$, from Table 6.4

$\hat{b}$ = 1.0, $\hat{\alpha}$ = 1.0, from Table 6.4

n_1 = natural frequency = 0.2 Hz, given

e = 2.71

$$R_h = \frac{1}{\eta_h} - \frac{1}{2\eta_h^2}(1 - e^{-2\eta_h})$$

$$= \frac{1}{2.875} - \frac{1}{2 \times 2.875^2}(1 - 2.71^{-2 \times 2.875})$$

$$= 0.288$$

$$\eta_L = \frac{15.4n_1 L}{\bar{v}_{\bar{z}}} = \frac{15.4 \times 0.2 \times 100}{192} = 1.604 \qquad \begin{array}{l} L = \text{building breadth parallel} \\ \text{to wind} = 100 \text{ ft, given} \end{array}$$

$$R_L = \frac{1}{\eta_L} - \frac{1}{2\eta_L^2}(1 - 2.71^{-2\eta_L})$$

$$= \frac{1}{1.604} - \frac{1}{2 \times 1.604^2}(1 - 2.71^{-2 \times 1.604})$$

$$= 0.618$$

$$R^2 = \frac{1}{\beta}R_n R_h R_B(0.53 + 0.47 R_L) \qquad \begin{array}{l} \beta = \text{damping ratio} = 0.015, \\ \text{given} \end{array}$$

$$= \frac{1}{0.015} \times 0.143 \times 0.288 \times 0.739(0.53 + 0.47 \times 0.618)$$

$$= 1.665 \qquad R = \sqrt{1.665} = 1.29$$

$$g_R = \sqrt{2\ln(3600 \times n_1)} + \frac{0.577}{\sqrt{2\ln(3600 \times n_1)}} \qquad \begin{array}{l} \text{ln means logarithm to base} \\ e = 2.71. \end{array}$$

$$= \sqrt{2\ln(3600 \times 0.2)} + \frac{0.577}{\sqrt{2\ln(3600 \times 0.2)}}$$

$$= 3.789$$

$$G = 0.925\left(\frac{1 + 1.7 I_{\bar{z}}\sqrt{g_Q^2 Q^2 + g_R R^2}}{1 + 1.7 g_V I_{\bar{z}}}\right) \qquad \begin{array}{l} G \text{ is the gust factor.} \\ g_Q = g_V = 3.4 \text{ (defined in the} \\ \text{equation for } G). \end{array}$$

$$= 0.925\left(\frac{1 + 1.7 \times 0.134\sqrt{3.4^2 \times 0.634 + 3.789^{-2} \times 1.665}}{1 + 1.7 \times 3.4 \times 0.134}\right)$$

$$= 1.185$$

Calculations for Design Wind Pressures: Graphical Procedure.

Given. A concrete building located in a hurricane prone region with the following characteristics:

- Building height = 450 ft (137.15 m)
- Building plan dimensions = 185 × 125 ft (56.396 × 38.10 m)
- Exposure category = C
- Basic wind speed = 110 mph (49 m/s)
- The building is sited on the upper half of a 2-D ridge and has the following topographic parameters:

$$L_h = 200 \text{ ft}, \quad H = 200 \text{ ft}, \quad x = 50 \text{ ft}$$

(See Fig. 1.11 for definitions.)

- It is anticipated that design will be performed by using basic load combinations specified in Sections 2.3.2 and 2.4.1 of ASCE 7-02. Observe that load factors associated with wind load combinations do not account for the directionality factor K_d. Therefore, the values of q_z that account for $K_d = 0.85$, as shown in Fig. 1.2, may be used directly in the ASCE 7-02 load combinations.
- The building is for typical office occupancy. However, it does have designated areas where more than 300 people congregate in one area.
- Damping ratio = 0.02 (2% of critical)

Required.   Using the graph given in Fig. 1.12, determine wind pressure for the main wind-force-resisting system (MWFRS) of the building. Use case 1 given in ASCE 7-02, Section 6.5.12.3. It should be noted that ASCE 7-02 specifies four distinct wind load cases. These include: 1) torsional effects caused by nonuniform pressure; 2) wind loads acting diagonally to the building; 3) torsion due to eccentricity between the elastic shear center and the center of mass at each level of the structure; and 4) wind distribution to capture possible across-wind response. In this example, we will calculate the wind loads in the x-direction for case 1, which consists of full design pressure acting on the projected area perpendicular to each principal axis of the building, considered separately along each principle axis. The designer is directed to Fig. 6.9 of ASCE 7-02 for a full description of the load cases.

Solution.

- The building is for office occupancy with certain areas designated for the congregation of more than 300 people. From Tables 1.1 and 6.1 of ASCE 7-02 (Tables 1.7 and 1.7a of this text), the classification of the building for wind load is category III, and importance factor for wind $I_w = 1.15$.
- Exposure category is C and basic wind speed $V = 110$ mph, as given in the statement of the problem. We select the curve designated as C110 in Fig. 1.12 to read the positive and negative pressures up the building height.
- The building's height-to-least-horizontal dimension is 450/125 = 3.6, less than 4.

Therefore, the building may be considered rigid from the first definition given in ASCE 7-02, Section C6.2. The second definition refers to the fundamental period T of the building. Using the formula

$T = C_t h_n^{3/4}$, determine T

where

T = fundamental period of the building, in secs
h_n = height of the building, in feet
C_T = coefficient equal to 0.030 for concrete moment frame buildings
$T = 0.030 \times 450^{3/4} = 2.93$ sec (say, 3 sec)

The natural frequency, n, which is the reciprocal of the period, is equal to $1/T = 1/3 = 0.33$ Hz. This is less than 1 Hz, the limiting frequency that delineates a rigid structure from a flexible structure. Therefore gust effect factor G_f must be determined using the procedure given in ASCE 7-02, Section 6.5.8.2. However, to emphasize the graphical procedure, for now we will assume $G_f = 0.910$, a value that will be determined shortly. Observe that if the building is considered rigid $G_f = G$, would have been 0.85.

- Because the building is located on a 2-D ridge, it may experience higher winds than buildings situated on level ground. Therefore, consider topographic effects in the determination of design wind pressures.

For the given values of L_h, H, and x, the multipliers K_1, K_2, and K_3 are obtained from Fig. 1.11. Observe that for $H/L_h > 0.5$, Note 2 for Fig. 1.11 alerts us to assume $H/L_h = 0.5$ for evaluating K_1 and to substitute $2\,H$ for L_h for evaluating K_2 and K_3. Therefore, for $H/L_h = 200/200 = 1.0$, which is greater than 0.5, from Fig. 1.11, for exposure C, for a 2-D ridge, $K_1 = 0.725$.

Substituting $2\,H$ for H, $x/H = x/2H = 50/400 = 0.125$, and from Fig. 1.11, $K_2 = 0.92$. Instead of the values tabulated in Fig. 1.11, we may also use the formulas in Fig. 1.11a to calculate K_2 and K_3. Thus,

$$K_2 = \left(1 - \frac{|x|}{\mu L_h}\right)$$

$$= \left(1 - \frac{50}{1.5 \times 400}\right) = 0.92 \qquad \text{(Note that for } H/L_h > 0.5,}{}$$
$$L_H = 2\,H, \text{ for calculating } K_2)$$

The parameter K_3 varies as the ratio x/L_h. It may be obtained by using either the tabulated values in Fig. 1.11 or the formula given in Fig. 1.11a.

$$K_3 = e^{-\gamma \frac{z}{L_h}}$$

Again, substituting $2\,H$ for L_h, and $\gamma = 3$,

$$K_3 = e^{-3 \frac{z}{400}}$$

We use the preceding formula to calculate K_3 for the selected z/L_h values shown. Note that $\gamma = 3$ for 2-D ridges, which is the topography for our building.

z(ft)	450	350	250	150	100	50	30	15
z/L_h	2.25	1.75	1.25	0.75	0.50	0.25	0.15	0.08
K_3	0.001	0.005	0.023	0.106	0.224	0.473	0.638	0.78

Wind Parallel to X-Axis. From the building's plan dimensions, $L/B = 125/180 = 0.694 < 1.0$. Therefore, from Fig. 1.9, C_p for the windward face $= 0.8$, and C_p for the leeward face $= -0.5$. From Fig. 1.12, select the curve identified as C110. C stands for exposure C, and 110 stands for $V = 110$ mph. Use the graph to read the values of q_z at various heights. For example, at $h = 150$ ft, $q_z = 36.3$ psf.

However, since the q_z and q_h values in Fig. 1.12 are normalized for $K_{zt} = 1.0$, $K_d = 0.85$, and $I_w = 1.0$, we multiply these values by the K_{zt} and I_w values of the example problem before recording the corresponding values in columns (7) and (8) of Table 1.10. For example, $q_z = 36.3$ psf at $z = 150$ ft, obtained from the graph is multiplied by $K_{zt} = 1.145$ and $I_w = 1.15$ to get a value of $q_z = 47.79$ psf, shown in column (7).

Observe that K_{zt} varies up the height. Similarly, values of q_z for different heights are recorded in column (7) of Table 1.10 after multiplication by K_{zt} and I_w. The suction q_h in column (8) is the value from the graph at $z = h = 450$ ft multiplied by $K_{zt} = 1.002$ and $I_w = 1.15$. Observe that the suction q_h referenced at roof height remains constant for the entire height of leeward wall. Column (9) gives the total design wind pressure throughout the building height. It is the summation of $0.8q_z$, the positive pressure on the windward wall, plus $0.5q_h$, the suction on the leeward wall, multiplied by the gust effect factor $G_f = 0.91$.

TABLE 1.10 Design Wind Pressures, Graphical Procedure; ASCE 7-02 Procedure

Height (ft)	K_z (1)	K_h (2)	K_1 (3)	K_2 (4)	K_3 (5)	$K_{zt} = (1 + K_1 K_2 K_3)^2$ (6)	q_z (psf) (From Fig 1.12) $\times I_w \times K_{zt}$ (7)	q_h (psf) (8)	Design pressure $P = (0.8q_z + 0.5q_h) \times G_f$ (9)	Design pressure (psf) without topographic factors, is $K_{zt} = 1.0$ (10)
450	1.74	1.74	0.725	0.92	0.00	1.002	52.68	52.68	63.4	62.2
400	1.69	1.74	0.725	0.92	0.00	1.003	54.48	52.68	62.5	61.3
350	1.65	1.74	0.725	0.92	0.01	1.007	50.24	52.68	61.6	60.2
300	1.59	1.74	0.725	0.92	0.01	1.014	49.01	52.68	60.7	59.1
250	1.53	1.74	0.725	0.92	0.02	1.031	47.94	52.68	59.9	57.7
200	1.46	1.74	0.725	0.92	0.05	1.067	47.33	52.68	59.5	56.2
150	1.38	1.74	0.725	0.92	0.11	1.145	47.79	52.68	59.8	54.3
100	1.26	1.74	0.725	0.92	0.22	1.318	50.53	52.68	61.8	51.8
50	1.09	1.74	0.725	0.92	0.47	1.726	57.18	52.68	66.6	48.0
30	0.98	1.74	0.725	0.92	0.64	2.027	60.29	52.68	68.9	45.6
15	0.85	1.74	0.725	0.92	0.80	2.343	60.22	52.68	68.9	42.6

Notes: $G_f = 0.91$, $I_w = 1.15$, $K_d = 0.85$.

For comparative purposes, column (10) of Table 1.10 gives the design pressures P for the building assuming that it is located on a flat terrain, i.e., $K_{zt} = 1.0$.

Calculations for Gust Effect Factor G_f, and Along-Wind Displacements and Accelerations. As mentioned previously, to place emphasis on the graphical solution, we assumed $G_f = 0.910$ in our calculations for design wind pressures. We will now calculate this value using the worksheet format given in Section 1.4.2.7. We will also calculate the maximum values for along-wind displacement and acceleration using the worksheet.

Given. This is a continuation of the illustrative problem stated in the section titled "Calculations for Design Wind Pressures: Graphical Procedure." Therefore we use the same building characteristics and wind environment data given therein.

Required. Gust effect factor G_f and maximum values for along-wind displacement and accelerations using the worksheet given in Section 1.4.2.7.

ASCE 7-02 Formulas	Commentary

V = wind speed in ft/s

$\quad = V_{mph} \times 1.467$

$\quad = 110 \times 1.467 = 161$ ft/s

$V = 110$ mph, given

$\bar{z} = 0.6\ h$, but not less than z_{min}

$\quad = 0.6 \times 450 = 270$ ft $> \bar{z}_{min} = 15$ ft $\quad$ OK

h = building height =
    450 ft, given

$\bar{z}_{min} = 15$ ft, from Table 6.4

$I_{\bar{z}} = C\left(\dfrac{33}{\bar{z}}\right)^{\frac{1}{6}} = 0.20\left(\dfrac{33}{270}\right)^{\frac{1}{6}} = 0.141$

$c = 0.20$, from Table 6.4

$L_{\bar{z}} = l\left(\dfrac{\bar{z}}{33}\right)^{\varepsilon} = 500\left(\dfrac{270}{33}\right)^{\frac{1}{5}} = 761.3$ ft

$l = 500$ ft, $\varepsilon = \frac{1}{5}$, from Table 6.4

$Q = \sqrt{\dfrac{1}{1 + 0.63\left(\frac{B+h}{L_{\bar{z}}}\right)^{0.63}}}$

$\quad = \sqrt{\dfrac{1}{1 + 0.63\left(\frac{185+450}{761}\right)^{0.63}}}$

$\quad = 0.80, \quad Q^2 = 0.64$

Q = the background response

B = building width
    perpendicular to wind
    = 185 ft, given

$\bar{V}_{\bar{z}} = \bar{b}\left(\dfrac{\bar{z}}{33}\right)^{\bar{\alpha}} V = 0.65\left(\dfrac{270}{33}\right)^{\frac{1}{6.5}} \times 161 = 144.9$ ft/s

$\bar{b} = 0.65, \bar{\alpha} = \frac{1}{6.5}$, from Table 6.4

$\hat{V}_{\bar{z}} = \hat{b}\left(\dfrac{\bar{z}}{33}\right)^{\hat{\alpha}} V = 1.0 \times \left(\dfrac{270}{33}\right)^{\frac{1}{9.5}} \times 161 = 201$ ft/s

$\hat{b} = 1.0, \hat{\alpha} = 1.0$, from Table 6.4

$N_1 = \dfrac{n_1 L_{\bar{z}}}{\bar{V}_{\bar{z}}} = \dfrac{0.33 \times 761}{144.9} = 1.751$

n_1 = natural frequency
    = 0.33 Hz, given

$R_n = \dfrac{7.47\ N_1}{(1 + 10.3\ N_1)^{\frac{5}{3}}}$

$\quad = \dfrac{7.47 \times 1.751}{(1 + 10.3 \times 1.751)^{\frac{5}{3}}} = 0.096$

$$\eta_b = \frac{4.6 n_1 B}{\bar{V}_z} = \frac{4.6 \times 0.33 \times 185}{144.9} = 1.958$$

$e = 2.71$

$$R_B = \frac{1}{\eta_b} - \frac{1}{2\eta_b^2}(1 - e^{-2\eta_b})$$

$$= \frac{1}{1.958} - \frac{1}{2 \times 1.958^2}(1 - 2.71^{-2 \times 1.958})$$

$$= 0.383$$

$$\eta_h = \frac{4.6 n_1 h}{\bar{V}_z} = \frac{4.6 \times 0.33 \times 450}{144.9} = 4.762$$

$$R_h = \frac{1}{\eta_h} - \frac{1}{2\eta_h^2}(1 - e^{-2\eta_h})$$

$$= \frac{1}{4.762} - \frac{1}{2 \times 4.762^2}(1 - 2.71^{-2 \times 4.762})$$

$$= 0.188$$

$$\eta_L = \frac{15.4 n_1 L}{\bar{V}_z} = \frac{15.4 \times 0.33 \times 125}{144.9} = 4.429$$

L = building breadth parallel to wind = 125 ft, given

$$R_L = \frac{1}{\eta_L} - \frac{1}{2\eta_L^2}(1 - 2.71^{-2\eta_L})$$

$$= \frac{1}{4.429} - \frac{1}{2 \times 4.429^2}(1 - 2.71^{-2 \times 4.429})$$

$$= 0.200$$

$$R^2 = \frac{1}{\beta} R_n R_h R_B (0.53 + 0.47 R_L)$$

β = damping ratio = 0.02, given

$$= \frac{1}{0.02} \times 0.096 \times 0.188 \times 0.388(0.53 + 0.47 \times 0.200)$$

$$= 0.216 \qquad R = \sqrt{0.216} = 0.465$$

$$g_R = \sqrt{2 \ln(3600 \times n_1)} + \frac{0.577}{\sqrt{2 \ln(3600 \times n_1)}}$$

ln means logarithm to base $e = 2.71$

$$= \sqrt{2 \ln(3600 \times 0.33)} + \frac{0.577}{\sqrt{2 \ln(3600 \times 0.33)}}$$

$$= 3.919$$

$$G = 0.925 \left(\frac{1 + 1.7 I_{\bar{z}} \sqrt{g_Q^2 Q^2 + g_R R^2}}{1 + 1.7 g_V I_{\bar{z}}} \right)$$

G is the gust factor
$g_Q = g_V = 3.4$
(defined in the equation for G)

$$= 0.925 \left(\frac{1 + 1.7 \times 0.141\sqrt{3.4^2 \times 0.64 + 3.919^2 \times 0.216}}{1 + 1.7 \times 3.4 \times 0.141} \right)$$

$$= 0.910$$

For comparative purposes, it may be of interest to calculate the gust factor G_f for this building using the improved method for rigid structures given in Section 1.4.2.5.

$$G = 0.925 \left(\frac{1 + 1.7 g_q I_{\bar{z}} Q}{1 + 1.7 g_v I_z} \right)$$

$$I_{\bar{z}} = C \left(\frac{33}{z} \right)^{\frac{1}{6}} = 0.141$$

$$L_{\bar{z}} = 761 \quad Q = 0.80 \tag{1.38}$$

$$G = 0.925 \left(\frac{1 + 1.7 \times 3.4 \times 0.141 \times 0.80}{1 + 1.7 \times 3.4 \times 0.141} \right)$$

$$= 0.912$$

Observe that this value of 0.912 is not much different from 0.910 calculated using the more complex procedure.

$$K = \frac{1.65 \hat{\alpha}}{\hat{\alpha} + \xi + 1} = \frac{1.65^{\frac{1}{9.5}}}{\frac{1}{9.5} + 1 + 1} = 0.50 \qquad \xi = \text{the first mode exponent taken} = 1.0$$

$$g_{\ddot{x}} = g_R = 3.91$$

Maximum Along-Wind Displacement.

$$\begin{aligned}
X_{\max(z)} &= \frac{\phi_{(z)} \rho B h C_{fx} \hat{V}^2}{2 m_1 (2 \pi n_1)^2} \\
&= \frac{1 \times 0.0024 \times 185 \times 450 \times 1.3 \times 201^2}{2 \times 1,618,200 \, (2 \times \pi \times 0.3)^2} \qquad \rho = \text{air density} = 0.0024 \text{ slugs} \\
&\quad \times 0.5 \times 0.925 \\
&= 0.4221 \text{ ft}
\end{aligned} \tag{1.39}$$

Note: $X_{\max(z)}$ is also commonly reffered to as lateral drift, Δ. Tall buildings are usually designed for a drift index $\frac{\Delta}{h} \cong \frac{1}{500}$. In our case, $\frac{\Delta}{h} = \frac{0.4221}{450} = \frac{1}{1056}$, indicating that the example building is quite stiff.

$$\begin{aligned}
\sigma_{\ddot{x}_{(z)}} &= \frac{0.85 \phi_z \rho B h C_{fx} \bar{V}^2}{m_1} \times I_{\bar{z}} K R \\
&= \frac{0.85 \times 1 \times 0.0024 \times 185 \times 450 \times 1.3 \times 144.6^2}{1,618,200} \times 0.141 \times 0.5 \times 0.5196 \\
&= 0.104 \text{ ft/sec}^2
\end{aligned} \tag{1.40}$$

Maximum Along-Wind Acceleration.

$$\begin{aligned}
\ddot{X}_{\max(z)} &= g_{\ddot{x}} \sigma_{\ddot{x}_{(z)}} \\
&= 3.89 \times 0.1040 \\
&= 0.40 \text{ ft/sec}^2 \\
&= 0.40 \times 31.11 = 12.58 \text{ milli-g}
\end{aligned} \tag{1.41}$$

This is well below the normally accepted limit of 20 milli-g, warranting no further investigation.

Calculations for Wind Pressure (ASCE 7-02): 10-Story Building.

Given.   An office building with the following structural characteristics:

Plan dimensions	60 ft × 120 ft (18.28 m × 36.57 m)
Building height	10 floors at 14 ft floor-to-floor = 10 × 14 = 140 ft (42.67 m)
Fundamental frequency	1.1 Hz
Building classification	Category II
Basic wind speed	90 mph, 3-sec gust speed for Las Vegas, from wind speed map, Fig. 1.9
Exposure category	Urban terrain
Topographic factor K_{zt}	1.0
Load combinations	ASCE 7-02 ultimate strength design. Therefore, use directionality factor $K_d = 0.85$.

Required.   Wind pressures for the design of primary lateral system.

Solution.

Step 1.   Building Classification. A typical office building is not generally considered an essential facility in the aftermath of windstorm, nor is its primary function for occupancy by more than 300 persons in one area. Therefore, the example building is judged to be type-II category. However, before a building is classified into a category, it is good practice to ascertain with building owners and plan-check officials that the category is consistent with their policies.

> Results of Step 1: Category II, $I_w = 1.0$ (ASCE 7-02 Tables 1.1 and 6.1)
>   (Tables 1.7 and 1.7a of this text)

Step 2.   Basic Wind Speed V. The ASCE wind speed map shown in Fig. 1.10 indicates that Las Vegas is in a wind contour of 90 mph. As indicated previously, it is good practice to confirm the design wind speed with local plan-check officials.

> Results of Step 2: $V = 90$ mph

Step 3.   Determination of Gust Response Factor G. The building height-to-width ratio of $140/60 = 2.3$ is less than 4, and its natural frequency of 1.1 Hz is more than 1.0 Hz. Therefore, the building may be considered a nonflexible building and a value of 0.85 may be used for the gust response factor G.

> Results of Step 3: $G = 0.85$

Step 4.   Directionality Factor K_d. Since ASCE 7-02 load combinations are anticipated, we use $K_d = 0.85$.

Step 5.   External Pressure Coefficient C_p. From the given plan dimensions, the building width-to-depth ratio of $L/B = 60/120 = 0.5$ for wind parallel to the 60-ft face. For wind parallel to the 120-ft face, the ratio $= 120/60 = 2.0$.

From Figs. 1.10 and 1.10a the following values of C_p are obtained:

> $C_p = 0.8$ for the windward wall
> $C_p = 0.5$ for the leeward wall, wind parallel to 60-ft face
> $C_p = 0.3$ for leeward wall, wind parallel to 120-ft face

Roof and internal pressures and suctions are not relevant in determining wind loads for primary lateral system: Internal pressures and suctions acting on the windward and leeward walls cancel out without adding or subtracting to the overall wind loads. Roof suction, which results in uplift forces, is generally neglected in the design of a primary lateral system.

Results of Step 4 are shown as C_p values in Table 1.11.

TABLE 1.11 Example Problem, 10-Story Building: Design Pressures for Main Wind-Force-Resisting Frame; ASCE 7-02 Procedure

Height, ft	Value of K_z and q_z Windward K_z	q_z, psf	Leeward K_h	q_h	Design pressures for main wind-force-resisting system X-wind, psf $p_z = q_z G \times 0.8$ $+ q_h G \times 0.5$	Y-wind, psf $p_z = q_z G \times 0.8$ $+ q_h G \times 0.3$
140	1.09	19.2	1.09	19.2	21.2	18.0
100	0.99	17.4	1.09	19.2	20.0	16.7
80	0.93	16.0	1.09	19.2	19.0	15.8
60	0.85	15.0	1.09	19.2	18.4	15.1
40	0.76	13.0	1.09	19.2	17.0	13.7
30	0.70	12.3	1.09	19.2	16.5	13.3
20	0.62	11.4	1.09	19.2	15.9	12.6
0–15	0.57	10.2	1.09	19.2	15.1	11.8

X-Wind: $c_p = +0.8$ Windward; $c_p = -0.5$ Leeward. Gust factor $G = 0.85$. Exposure category = B. Directionality factor $K_d = 0.85$. Wind importance factor $I_w = 1.0$.
Y-Wind: $c_p = +0.8$ Windward; $c_p = -0.3$ Leeward. $V = 90$ mph (3-sec. gust). Topographic factor $K_{zt} = 1$.
$q_z = 0.00256 \, K_z \, K_{zt} \, K_d \, V^2 \, I_w$

Step 6. Building Exposure. Since the building is located in an urban terrain, the exposure category is judged to be B.

Results of Step 5: Exposure B

Step 7. Combined Velocity Pressure, Exposure Coefficient K_z. The gradient height Z_g and the power coefficient α for exposure B are 1200 ft and 7.5, respectively (Table 1.7a). Below 15 ft, the value of K_z is taken as a constant determined at height 15 ft.

$$K_z = 2.01 \left(\frac{z}{1200} \right)^{2/7.5} \quad \text{for} \quad 15 \text{ ft} \leq z \leq z_g \tag{1.42}$$

$$K_z = 2.01 \left(\frac{15}{1200} \right)^{2/7.5} \quad \text{for} \quad z \leq 15 \text{ ft} \tag{1.43}$$

Instead of calculating values of Kz from the preceding equations, they can be obtained directly from Table 1.8 or Fig. 1.11. The results of Step 7 are shown in Table 1.11.

Step 8. Velocity Pressure q_z. Values for q_z are obtained from Eqs. (1.14) and (1.15). The results of Step 8 are shown in column 3 and 5 of Table 1.11.

Step 9. Design Pressure p. With the known values of q, G, and C_p, the design pressure is obtained by the chain equation

$$P = q \times G \times C_p$$

Wind pressures and suctions on windward and leeward walls are shown in Table 1.11.
Summation of the two is used is used in determining the pressure for the design of main wind-force-resisting system.

Step 10. Floor-by-Floor Wind Load. This is obtained by multiplying the exposed area tributary to the level by the corresponding value of design pressure at floor height.

Calculations for Wind Pressures (ASCE 7-02): 30-Story Building.
Given.

- Location of building: Houston, Texas, use $V = 110$ mph
- Terrain: Flat, open country with scattered obstructions having the size of single-family dwellings, height generally less than 30 ft
- Plan dimensions: 98.5 ft × 164 ft
- Building height: 394 ft
- Building lateral-load-resisting system: moment frame with shear walls. Fundamental period $T = 3.12$ secs. Therefore, frequency $f = 1/3.12 = 0.32$ Hz.
- The building is regular, as defined in ASCE 7-02 Section 6.2. It does not have unusual geometric irregularities.
- It does not have response characteristics that would subject the building to across-wind loading, vortex shedding, or instability due to galloping or flutter. The building does not have a site location for which channeling effects or buffeting in the wake of upwind obstructions warrant special consideration.
- Damping factor: 1.5% of critical
- Topographic factor $K_{zt} = 1.0$
- The building is for typical office occupancy

Required. Wind pressures for the design of MWFRS using ASCE 7-02. Include ample explanation to emphasize the essential requirements of ASCE provisions. Use $K_{zt} = 1.0$ for the basic problem and compare results by using the following topographic factors for a 2-D ridge. (Refer to Fig. 1.11 for definitions.)

$L_h = 100$ ft, $H = 100$ ft, $x = 50$ ft

Solution. The design wind loads for lateral analysis of the building will be based on method 2, the analytical procedure of ASCE 7-02 Section 6.5. This method is applicable to the example building, as it satisfies the two conditions set forth in Section 6.5.1. of ASCE 7.
The design wind load is calculated after determining the following quantities:

- The basic wind speed V (6.5.4)
- A wind directionality factor K_d (6.5.4.4)
- An importance factor I_w (6.5.5)
- An exposure category and velocity pressure coefficient K_z or K_h, as applicable (6.5.6)
- A topographic factor K_{zt} (6.5.7)
- A gust for G or G_f as applicable (6.5.8)
- An enclosure classification (6.5.9)
- Internal pressure coefficient GC_{pi} (6.5.11.1)
- External pressure coefficients C_p or GC_{pf}, or force coefficients C_f as applicable, (6.5.11.2 or 6.5.11.3)
- Velocity pressure q_z or q_h as applicable (6.5.10)
- Design wind pressure P (6.5.12)

The numbers in parentheses indicate section numbers of the ASCE 7-02 Standard.
Enclosure Classification Determination of enclosure classification and internal pressure coefficient is not required for the calculation of overall wind loads for typical diaphragmed buildings, because internal pressures acting in opposite directions on the windward and leeward walls cancel out.

Basic Wind Speed   From Figure 1.10d, the wind velocity $V = 110$ mph for Houston, Texas. The wind must be assumed to come from any direction.

Wind Directionality Factor   From ASCE 7-02 Table 6-6 (Table 1.8 of this text), $K_d = 0.85$.

Importance Factor   The example office building can be placed under category II.

Importance factor $I_w = 1.00$ (Table 1.7a)

Exposure Category   The terrain for the example building consists of flat, open country with scattered obstructions having heights generally less than 30 ft (9.1 m).

Therefore, exposure category = C (ASCE 7-02, Section 6.5.9)

Velocity Pressure Exposure Coefficients K_z and K_h   Case 2 of ASCE 7-02, Table 6-3, is applicable for determining K_z and K_h. These may be calculated using the general equations

$$K_z = 2.01 \ (z/_{zg})^{2/\alpha} \qquad \text{for} \quad 15 \text{ ft} \leq z \leq z_g \qquad\qquad\qquad (1.44)$$

and

$$K_z = 2.01 \ (15/Z_g)^{2/\alpha} \qquad \text{for} \quad Z \leq 15 \text{ ft} \qquad\qquad\qquad (1.44a)$$

where

     Z = height above ground level, ft
     Z_g = gradient height
     α = 3-sec gust speed power law exponent

The values for Z_g and α are given in Table 6.2 of ASCE 7-02 (Table 1.9 of this text). From this table, for exposure category C:

     α = 3-sec. Gust speed pow\er law exponent = 9.5

     Z_g = nominal height of the boundary layer = 900 ft

The magnitudes of K_z at different heights and K_h at the roof are shown in Table 1.6. These values are the same as those given in Table 6.3 of AISC 7-02, except only the values required for determining wind loads on MWFRS are given, and the height z aboveground has been extended to 1500 ft.

Topographic Effects   This effect is given by a factor K_{zt}

$$K_{zt} = (1 + K_1K_2K_3)^2 \qquad\qquad\qquad \text{[ASCE 7-02 Eq. (6.13)]} \quad (1.44b)$$

where K_1, K_2, and K_3 are given in Fig. 1.11 for various topographic factors. The example building is situated on level ground. Therefore, the K_{zt} factor may be taken equal to 1.

Calculation of Flexibility of Structure   A structure is considered flexible per commentary Section 6.2 of ASCE if it has a fundamental natural frequency of less than 1 Hz (i.e., a fundamental period $T > 1$ sec.). To find the period of our building, we will use an approximate formula normally used in seismic design. This formula gives the period T, in terms of number of stories N present in the building by the relation: $T = 0.1N$. For the subject building, then, $T = 0.1 \times 30 = 3.0$ sec with a corresponding fundamental frequency of 0.333 Hz. This is considerably less than 1 Hz. Therefore, the building is considered flexible for purpose of determining the gust effect factor. It should be noted that a dynamic analysis of buildings typically gives building periods longer than those from the approximate formula. For a moment frame building, for example, the approximate period is usually in the range of $T = 0.15N$. However, for our building, which has a combination of shear walls and moment frames, we use the given fundamental period of $T = 3.12$ sec. with a corresponding frequency $f = 0.32$ Hz.

TABLE 1.12 Internal Pressure Coefficient GC_{pi} for Main Wind-Force-Resisting System/Components and Cladding (Walls and Roofs)

Enclosure classification	GC_{pi} [b,c]
Open buildings	0.00
Partially enclosed buildings	+0.55[a]
	−0.55
Enclosed buildings	+0.18[a]
	−0.18

[a] Plus and minus signs signify pressures acting toward and away from the internal surfaces, respectively.
[b] Values of GC_{pi} shall be used with q_z or q_h as specified in 1.4.2.2.
[c] Two cases shall be considered to determine the critical load requirements for the appropriate condition:
    1. A positive value of GC_{pi} applied to all internal surfaces.
    2. A negative value of GC_{pi} applied to all internal surfaces.
(From ASCE 7-02 Fig. 6.5.)

Enclosure Classifications and Internal Pressure Coefficients For purposes of determining internal pressure coefficients, all buildings are classified as enclosed, partially enclosed, or open (see Table 1.12, ASCE 7-02 Fig. 6.5). However, internal pressures do not come into play in the determination of overall wind loads for lateral load analysis of buildings. The overall wind loads can therefore be determined using only the external pressures and suctions on the windward and leeward faces.

Velocity pressure Velocity pressure q_z, evaluated at height z, is calculated by the following equation:

$$q_z = 0.00256\, K_z\, K_{zt} K_d V^2\, I \tag{1.45}$$

Design Wind Pressure P for Enclosed Flexible Buildings is determined from the following general equation:

$$P = qG_f C_p - q_i\,(GC_{pi}) \tag{1.46}$$

where

    $q = q_z$ for windward walls evaluated at height z
    $q = q_h$ for leeward walls, side walls, and roofs, evaluated at height h
    $q_i = q_h$ for internal pressure evaluation in enclosed buildings
GC_{pi} = internal pressure coefficient (Table 1.12)

Since internal pressures do not effect the overall wind loads, q_i may be eliminated from the preceding equation, giving the wind pressure and suction for MWFRS, as follows.

$P_{\text{pressure}} = q_z C_f C_p$ (pressure on windward wall calculated at height z)

$P_{\text{suction}} = q_h G_f C_p$ (suction on leeward wall, side walls, and roof calculated at height h)

With this explanation, we now proceed to calculate the design pressures for the example building. Instead of hand calculations, a spreadsheet[a] has been used to obtain the results shown in Tables 1.13 and 1.14.

---

[a] The author wishes to acknowledge gratitude to his colleague Mr. Ryan Wilkerson, S.E., who developed the spreadsheets and reviewed this chapter and made valuable suggestions.

TABLE 1.13 Main Wind-Resisting-System (No Topographic Effects)

Exposure category	C	Bldg. frequency	0.32	(Flexible)
Building height	394	ft		
Wind speed	110	mph		
Width (B)	164	ft		
Length (L)	98.5	ft		
Importance	1			
K_d	0.85			
Bldg. period	3.1	(seconds)		
β	0.015	(damping ratio)		

Terrain Exposure Constants

α	z_g	$\hat{a}$	$\hat{b}$	$\bar{\alpha}$
9.5	900	0.1053	1	0.1538

$\bar{b}$	c	ℓ	$\bar{\varepsilon}$	z_{min}
0.65	0.2	500	0.2	15

$L_{\bar{z}}$	$\bar{z}$	$l_{\bar{z}}$	$V_{\bar{z}}$	g_r	N_1	R_n			
741.3	236.4	0.144	142.0	3.911	1.685	0.099	R	G_t	G
Q	η_h	R_h	η_a	R_a	η_L	R_L	0.616	0.96	0.84
0.809	4.118	0.213	1.714	0.419	3.447	0.248			

Height	K_3	K_{zt}	K_z	q_z
0	1	1	0.85	22.35
17	0.77	1	0.87	22.95
30	0.64	1	0.98	25.86
43	0.52	1	1.06	27.90
56	0.43	1	1.12	29.49
69	0.36	1	1.17	30.82
82	0.29	1	1.21	31.96
95	0.24	1	1.25	32.97
108	0.20	1	1.29	33.87
121	0.16	1	1.32	34.69
134	0.13	1	1.35	35.44
147	0.11	1	1.37	36.14
160	0.09	1	1.40	36.79
173	0.07	1	1.42	37.40
186	0.06	1	1.44	37.97
199	0.05	1	1.46	38.52
212	0.04	1	1.48	39.03
225	0.03	1	1.50	39.53
238	0.03	1	1.52	40.00
251	0.02	1	1.54	40.45
264	0.02	1	1.55	40.88
277	0.02	1	1.57	41.30
290	0.01	1	1.58	41.70

(Continued)

TABLE 1.13 (Continued)

Height	K_3	K_{zt}	K_z	q_z
303	0.01	1	1.60	42.08
316	0.01	1	1.61	42.46
329	0.01	1	1.63	42.82
342	0.01	1	1.64	43.17
355	0.00	1	1.65	43.51
368	0.00	1	1.67	43.84
381	0.00	1	1.68	44.16
394	0.00	1	1.69	44.47

	Windward wind				Leeward wind				Total wind		
Height	C_p	Wind, psf	Load, plf	OTM, kip-ft	C_p	Wind, psf	Load, plf	OTM, kip-ft	Wind, psf	Load, plf	OTM, kip-ft
0	0.8	17.1	146.4	0.0	0.5	21.3	180.9	0.0	38.4	327.2	0.0
17	0.8	17.6	266.1	4.5	0.5	21.3	319.1	5.4	38.8	585.3	9.9
30	0.8	19.8	256.3	7.7	0.5	21.3	276.6	8.3	41.1	532.8	16.0
43	0.8	21.4	277.1	11.9	0.5	21.3	276.6	11.9	42.6	553.7	23.8
56	0.8	22.6	293.1	16.4	0.5	21.3	276.6	15.5	43.9	569.7	31.9
69	0.8	23.6	306.4	21.1	0.5	21.3	276.6	19.1	44.9	583.0	40.2
82	0.8	24.5	317.9	26.1	0.5	21.3	276.6	22.7	45.7	594.4	48.7
95	0.8	25.2	327.9	31.2	0.5	21.3	276.6	26.3	46.5	604.5	57.4
108	0.8	25.9	336.9	36.4	0.5	21.3	276.6	29.9	47.2	613.5	66.3
121	0.8	26.6	345.1	41.8	0.5	21.3	276.6	33.5	47.8	621.7	75.2
134	0.8	27.1	352.6	47.2	0.5	21.3	276.6	37.1	48.4	629.2	84.3
147	0.8	27.7	359.5	52.9	0.5	21.3	276.6	40.7	48.9	636.1	93.5
160	0.8	28.2	366.0	58.6	0.5	21.3	276.6	44.3	49.4	642.6	102.8
173	0.8	28.6	372.1	64.4	0.5	21.3	276.6	47.9	49.9	648.7	112.2
186	0.8	29.1	377.8	70.3	0.5	21.3	276.6	51.4	50.3	654.4	121.7
199	0.8	29.5	383.2	76.3	0.5	21.3	276.6	55.0	50.8	659.8	131.3
212	0.8	29.9	388.4	82.3	0.5	21.3	276.6	58.6	51.2	665.0	141.0
225	0.8	30.3	393.3	88.5	0.5	21.3	276.6	62.2	51.5	669.9	150.7
238	0.8	30.6	398.0	94.7	0.5	21.3	276.6	65.8	51.9	674.6	160.5
251	0.8	31.0	402.5	101.0	0.5	21.3	276.6	69.4	52.2	679.0	170.4
264	0.8	31.3	406.8	107.4	0.5	21.3	276.6	73.0	52.6	583.4	180.4
277	0.8	31.6	410.9	113.8	0.5	21.3	276.6	76.6	52.9	687.5	190.4
290	0.8	31.9	414.9	120.3	0.5	21.3	276.6	80.2	53.2	691.5	200.5
303	0.8	32.2	418.7	126.9	0.5	21.3	276.6	83.8	53.5	695.3	210.7
316	0.8	32.5	422.5	133.5	0.5	21.3	276.6	87.4	53.8	699.1	220.9
329	0.8	32.8	426.1	140.2	0.5	21.3	276.6	91.0	54.1	702.7	231.2
342	0.8	33.0	429.5	146.9	0.5	21.3	276.6	94.6	54.3	706.1	241.5
355	0.8	33.3	432.9	153.7	0.5	21.3	276.6	98.2	54.6	709.5	251.9
368	0.8	33.6	436.2	160.5	0.5	21.3	276.6	101.8	54.8	712.8	262.3
381	0.8	33.8	439.4	167.4	0.5	21.3	276.6	105.4	55.1	716.0	272.8
394	0.5	34.0	220.9	87.0	0.5	21.3	138.3	54.5	55.3	359.2	141.5
		Totals:	11125.4	2390.8		Totals:	8383.0	1651.4	Totals:	19508.4	4042.3

Comparison of Gust Effect Factors and Along-Wind Responses. ASCE 7-02 permits a single gust effect factor of 0.85 for rigid buildings and as an option, specific features of the building size and wind environment may be incorporated to more accurately calculate a gust effect factor. The gust effect factor accounts for the loading effects due to wind turbulence structure interaction, and along-wing loading effects due to dynamic amplification for flexible buildings.

TABLE 1.14 Main Wind-Resisting-System (Topographic Factors: $L_h = 100$ ft, $H = 100$ ft, $x = 50$ ft)

Exposure category	C		Bldg. frequency	0.32	(Flexible)
Building height	395	ft	K_h	1.69	
Wind speed	110	mph	K_1	0.725	
Width (B)	164	ft	K_2	0.83	
Length (L)	98.5	ft	K_{3h}	0.00	
Importance	1		K_{ht}	1.003231	
K_d	0.85		q_i	44.64	psf
Bldg. period	3.1	(seconds)			
β	0.015	(Damping ratio)			
GC_{pi}	0	(Internal pressure)			

Terrain Exposure Constants

α	z_g	$\hat{a}$	$\hat{b}$	$\bar{\alpha}$
9.5	900	0.1053	1	0.1538

$\bar{b}$	c	ℓ	$\bar{\varepsilon}$	z_{min}
0.65	0.2	500	0.2	15

Topographic factors 2-D ridge/valley ▼

— Bldg location relative to crest —
○ Upwind   ⦿ Downwind

L_h	100	ft	γ	3	
H	100	ft (+/− ridge/valley)	μ	1.5	
x	50	ft	$K_1/(H/L_h)$	0.725	

$L_{\bar{z}}$	$\bar{z}$	$I_{\bar{z}}$	$V_{\bar{z}}$	g_r	N_1	R_n			
741.7	237	0.144	142.0	3.911	1.685	0.099	R	G_f	G
Q	η_h	R_h	η_B	R_B	η_L	R_L	0.616	0.957	0.84
0.809	4.127	0.213	1.714	0.419	3.446	0.248			

Height	K_3	K_{zt}	K_z	q_z
0	1	2.573351	0.85	57.52
17	0.60	1.857222	0.87	42.62
30	0.41	1.551609	0.98	40.13
43	0.28	1.360278	1.06	37.95
56	0.19	1.237881	1.12	36.51
69	0.13	1.158287	1.17	35.70
82	0.09	1.105898	1.21	35.34
95	0.06	1.071117	1.25	35.31
108	0.04	1.047883	1.29	35.49
121	0.03	1.032297	1.32	35.81
134	0.02	1.021811	1.35	36.21
147	0.01	1.014741	1.37	36.67
160	0.01	1.009969	1.40	37.16
173	0.01	1.006744	1.42	37.65
186	0.00	1.004564	1.44	38.15
199	0.00	1.003089	1.46	38.64
212	0.00	1.002091	1.48	39.12
225	0.00	1.001415	1.50	39.58
238	0.00	1.000958	1.52	40.04
251	0.00	1.000649	1.54	40.47

(Continued)

TABLE 1.14 (Continued)

Height	K_3	K_{zt}	K_z	q_z
264	0.00	1.000439	1.55	40.90
277	0.00	1.000297	1.57	41.31
290	0.00	1.000201	1.58	41.70
303	0.00	1.000136	1.60	42.09
316	0.00	1.000092	1.61	42.46
329	0.00	1.000062	1.63	42.82
342	0.00	1.000042	1.64	43.17
355	0.00	1.000029	1.65	43.51
368	0.00	1.000019	1.67	43.84
381	0.00	1.000013	1.68	44.16
395	0.00	1.000009	1.69	44.48

		Windward wind				Leeward wind			Total wind		
Height	C_p	Wind, psf	Load, plf	OTM, kip-ft	C_p	Wind, psf	Load, plf	OTM, kip-ft	Wind, psf	Load, plf	OTM, kip-ft
0	0.8	44.0	349.9	0.0	0.5	21.4	181.5	0.0	65.4	531.4	0.0
17	0.8	32.6	510.4	6.7	0.5	21.4	320.3	5.4	54.0	830.6	14.1
30	0.8	30.7	399.6	12.0	0.5	21.4	277.6	8.3	52.1	677.2	20.3
43	0.8	29.0	378.5	16.3	0.5	21.4	277.6	11.9	50.4	656.0	28.2
56	0.8	27.9	364.0	20.4	0.5	21.4	277.6	15.5	49.3	641.6	35.9
69	0.8	27.3	355.7	24.5	0.5	21.4	277.6	19.2	48.7	633.3	43.7
82	0.8	27.0	352.0	28.9	0.5	21.4	277.6	22.8	48.4	629.6	51.6
95	0.8	27.0	351.5	33.4	0.5	21.4	277.6	26.4	48.4	629.1	59.8
108	0.8	27.2	353.2	38.1	0.5	21.4	277.6	30.0	48.5	630.8	68.1
121	0.8	27.4	356.3	43.1	0.5	21.4	277.6	33.6	48.8	633.9	76.7
134	0.8	27.7	360.3	48.3	0.5	21.4	277.6	37.2	49.1	637.9	85.5
147	0.8	28.1	364.9	53.6	0.5	21.4	277.6	40.8	49.4	642.4	94.4
160	0.8	28.4	369.7	59.1	0.5	21.4	277.6	44.4	49.8	647.2	103.6
173	0.8	28.8	374.6	64.8	0.5	21.4	277.6	48.0	50.2	652.2	112.8
186	0.8	29.2	379.5	70.6	0.5	21.4	277.6	51.6	50.5	657.1	122.2
199	0.8	29.6	384.4	76.5	0.5	21.4	277.6	55.2	50.9	661.9	131.7
212	0.8	29.9	389.1	82.5	0.5	21.4	277.6	58.8	51.3	666.7	141.3
225	0.8	30.3	393.6	88.6	0.5	21.4	277.6	62.5	51.6	671.4	151.1
238	0.8	30.6	398.3	94.8	0.5	21.4	277.6	66.1	52.0	675.9	160.9
251	0.8	31.0	402.6	101.1	0.5	21.4	277.6	69.7	52.3	680.2	170.7
264	0.8	31.3	406.9	107.4	0.5	21.4	277.6	73.3	52.7	684.4	180.7
277	0.8	31.6	410.9	113.8	0.5	21.4	277.6	76.9	53.0	688.5	190.7
290	0.8	31.9	414.9	120.3	0.5	21.4	277.6	80.5	53.3	692.5	200.8
303	0.8	32.2	418.7	126.9	0.5	21.4	277.6	84.1	53.6	696.3	211.0
316	0.8	32.5	422.4	133.5	0.5	21.4	277.6	87.7	53.8	700.0	221.2
329	0.8	32.6	426.0	140.2	0.5	21.4	277.6	91.3	54.1	703.6	231.5
342	0.8	33.0	429.5	146.9	0.5	21.4	277.6	94.9	54.4	707.1	241.8
355	0.8	33.3	432.9	153.7	0.5	21.4	277.6	98.5	54.6	710.4	252.2
366	0.8	33.6	436.1	160.5	0.5	21.4	277.6	102.1	54.9	713.7	262.7
381	0.8	33.8	439.3	167.4	0.5	21.4	277.6	105.8	55.1	716.9	273.1
395	0.8	34.0	220.8	87.0	0.5	21.4	138.8	54.7	55.4	359.6	141.7
395	0.8	0.0	—	—	0.5	0.0	—	—	0.0	—	—
395	0.8	0.0	—	—	0.5	0.0	—	—	0.0	—	—
395	0.8	0.0	—	—	0.5	0.0	—	—	0.0	—	—
395	0.8	0.0	—	—	0.5	0.0	—	—	0.0	—	—
		Totals:	12046.8	2422.8		Totals:	8412.8	1657.3	Totals:	20459.6	4080.1

TABLE 1.15 Buildings' Characteristics and Wind Environment

Problem #	Exposure category	Basic wind speed at exposure C V, mph	Height h, ft	Base B, ft	Depth L, ft	Frequency Hz (period, sec)	Damping ratio β	Building density, slugs/cu ft[a]
1	A	90	600	100	100	0.2 Hz (5 sec)	0.01	0.3727
2	C	90	600	100	100	0.2 Hz (5 sec)	0.01	0.3727
3	B	120	394	98.5	164	0.222 Hz (4.5 sec)	0.01	0.287
4	C	130	788	164	164	0.125 Hz (8 sec)	0.015	0.3346

[a] 1 slug = 32.17 lbs.

The two along-wind responses—the along-wind displacements and along-wind accelerations of typical buildings—are due entirely to the action of the turbulence of the longitudinal component of the wind velocity, superimposed on their corresponding mean values. As discussed presently, the most important criterion for the comfort of the building's occupants is the peak or maximum accelerations they are likely to experience in a windstorm. Human perception of building motion is influenced by many cues, such as the movement of suspended objects; noise due to ruffling between building components; and, if the building twists, apparent movement of objects at a distance viewed by the occupants. Although at present there are no comprehensive comfort criteria, a generally accepted benchmark value in North American practice is to limit the acceleration at the upper floor of a building to 20 milli-g. This limit applies to both human comfort and motion perception.

A windstorm postulated to occur at a frequency of once every 10 years is used as the design event. The threshold of accelerations for residential occupancies is somewhat more stringent—about 15 milli-g for a 10-year windstorm. The rationale is that occupants are likely to remain longer in a given location of a residence, than in a typical office setting.

With this background, it is of interest to evaluate the gust effect factors and along-wind responses for some example buildings. Four buildings are considered here. Example 1 is a building located in wind terrain exposure category A, a category that is no longer recognized in ASCE 7-02, but included in its commentary as a numerical example. We use the results of this example solely to compare the wind response characteristics with the three other buildings.

Table 1.15 gives in summary form the buildings' characteristics and their wind environment. Given in Table 1.16 are the values of various parameters such as z_{min}, $\bar{\varepsilon}$, etc., obtained from ASCE 7-02 Table 6.2 (Table 1.9 of this text). These values serve as starting points for the determination of gust effect factor, maximum lateral displacements, and accelerations.

Instead of presenting all examples in excruciating detail, only the final values of the derived parameters (as many as 24 for each example) are given in Table 1.17. However, for Building No. 3, the worked example follows the step-by-step procedure using the worksheet introduced in Section 1.4.2.7.

Discussions of Results. Because the example buildings are chosen randomly, it is impractical to make a comprehensive qualitative comparison. However, it may be appropriate to record the following observations regarding their wind-induced response characteristics.

TABLE 1.16 Design Parameters for Example Buildings

Problem #	1	2	3	4
z_{min}	60 ft	15 ft	30 ft	15 ft
$\bar{\varepsilon}$	0.5	0.2	0.333	0.2
c	0.45	0.20	0.30	0.20
$\bar{b}$	0.3	0.65	0.45	0.65
$\bar{\alpha}$	0.33	0.1538	0.25	0.1538
b	0.64	1.0	0.84	1.0
α	0.2	0.1053	0.143	0.1053
ℓ	180	500	320	500
C_{fx}	1.3	1.3	1.3	1.3
ξ	1	1.0	1.0	1.0

(Values obtained from ASCE 7-02, Table 6.2, or Table 1.9 of this text.)

- Building No. 1 has a rather large height-to-width ratio of 600/100 = 6. Yet, because it is located in exposure category A, the most favorable wind terrain (per ASCE 7-98), and is subjected to a relatively low wind velocity of 90 mph, its lateral response to wind does not appear to be overly sensitive. The calculated acceleration at the top floor is 26 milli-g, as compared to the threshold value of 20 milli-g (2% of g).

TABLE 1.17 Comparison of Dynamic Response to Wind Loads

Calculated values	Problem 1	Problem 2	Problem 3	Problem 4
V	132 ft/s	132 ft/sec	176 ft/sec	191 ft/sec
$\bar{z}$	360	360 ft	236 ft	473 ft
$I_{\bar{z}}$	0.302	0.1343	0.216	0.128
$L_{\bar{z}}$	594.52 ft	806 ft	616 ft	852 ft
Q^2	0.589	0.634	0.64	0.596
$\bar{V}_{\bar{z}}$	87.83 ft/s	124 ft/s	130 ft/s	187 ft/s
$\hat{V}_{\bar{z}}$	136.24 ft/s	170 ft/s	195 ft/s	253 ft/s
N_1	1.354	1.30	1.051	0.5695
R_n	0.111	0.114	0.128	0.171
η	1.047	0.742	0.773	0.504
R_B	0.555	0.646	0.6360	0.74
η	6.285	4.451	3.095	2.423
R_h	0.146	0.1994	0.271	0.328
η	3.507	2.484	4.31	1.688
R_L	0.245	0.322	0.205	0.423
R^2	0.580	1.00	1.381	2.01
G	1.055	1.074	1.20	1.204
K	0.502	0.50	0.501	0.50
m_1	745,400 slugs	745,400 slugs	608,887 slugs	236,4000 slugs
g_R	3.787	3.787	3.813	3.66
X_{max}	0.78 ft	1.23 ft	1.16 ft	5.3 ft
$g_{\ddot{x}}$	3.786	3.786	3.814	3.66
$\sigma_{\ddot{x}}$	0.19	0.22	0.363	0.463
$\ddot{X}_{max}$	0.72 ft/sec^2 (22.39 milli-g)	0.834 ft/sec^2 (26 milli-g)	1.385 ft/sec^2 (43 milli-g)	1.68 ft/sec^2 (52.26 milli-g)

- Building No. 2 has the same physical characteristics as Building No. 1, but now is sited in exposure category C, the second most severe exposure category, consisting of open terrain with scattered obstructions. Because the basic wind is the same as for the first, (90 mph at exposure C), its peak acceleration is only slightly higher than for Building 1. Pushing exposure category from A to C does not appear to unduly alter the wind sensitivity of the building.

- Building No. 3, in contrast to the other three, is not that tall. It is only 394 ft, in height, equivalent to a 30-story office building at a floor-to-floor height of 12 ft-6 in. At a fundamental frequency of 4.5 sec, its lateral stiffness is quite in line with buildings designed in high seismic zones. But because it is subjected to hurricane winds of 120 mph, its peak acceleration is a head-turning 43 milli-g—more than twice the threshold value of 20 milli-g.

- Building No. 4 is the tallest of the four, equivalent to a 60-plus story building. Its height-to-width ratio is not very large (only 4.8) but it appears to be quite flexible at a fundamental frequency of 8 sec and a calculated peak acceleration of 52.26 milli-g. It is doubtful that even with the addition of a supplemental damping system such as a tuned mass damper (TMD) or a simple pendulum damper (discussed in Chapter 8), the building oscillations can be tamed. Consulations with an engineering expert who specializes in performing wind-tunnel tests and in designing damping systems for dynamically sensitive structures would be recommended before finalizing the structural system.

Although in North American practice, the determination of wind-motion characteristics of buildings is primarily the domain of wind engineering consultants, the author strongly recommends that an analytical study be undertaken, as given in this section. The result will help in communication with owners of buildings, architects, and wind engineering experts in identifying problems associated with motion perception and human comfort.

Building No. 3: Calculations for Gust Effect Factor, Maximum Along-Wind Deflection, and Acceleration.

Given.

Building height h	= 394 ft
Building depth L	= 164 ft
Building width B	= 98.5 ft
Building natural frequency	= 0.222 Hz (period $T = 4.5$ s)
Damping ratio	= 0.01
Building density	= 0.287 slugs/ft^3
Exposure category	= B
Basic wind speed	= 120 mph
Air density	= 0.0024 slugs/cu ft
Mode exponent φ	= 1.0
Coefficient C_{fx}	= 1.3

Required.

Gust factor G

Maximum lateral load deflection at top, $X_{\max\ (h)}$

Maximum along-wind acceleration top, $\ddot{X}_{\max(h)}$

Solution.

ASCE 7-02 Formulas	Commentary
V = wind speed in ft/s $= V_{mph} \times 1.467$ $= 120 \times 1.467$ $= 176$ ft/s	V = basic wind speed = 120 mph, given
$\bar{z} = 0.6\,h$ $= 0.6 \times 394$ $= 236$ ft $> \bar{z}_{min} = 15$ ft	h = building height = 394 ft $\bar{z}_{min} = 15$ ft from Table 6.4 $c = 30$ from Table 6.4
$I_{\bar{z}} = C\left(\dfrac{33}{\bar{z}}\right)^{\frac{1}{6}}$ $= 0.30\left(\dfrac{33}{236}\right)^{\frac{1}{6}}$ $= 0.216$	
$L_{\bar{z}} = l\left(\dfrac{\bar{z}}{33}\right)^{\varepsilon}$ $= 320\left(\dfrac{236}{33}\right)^{0.333}$ $= 616$	$l = 320$ ft, $\varepsilon = \frac{1}{5}$ both from Table 6.4
$Q = \sqrt{\dfrac{1}{1 + 0.63\left(\dfrac{B+h}{L_{\bar{z}}}\right)^{0.63}}}$ $= \sqrt{\dfrac{1}{1 + 0.63\left(\dfrac{98.5+394}{616}\right)^{0.63}}}$ $= 0.804$	Q = is called background response B = building width perpendicular to wind = 98.5 ft, given
$\bar{V}_{\bar{z}} = \bar{b}\left(\dfrac{\bar{z}}{33}\right)^{\bar{\alpha}} V$ $= 0.45\left(\dfrac{236}{33}\right)^{0.25} \times 176$ $= 130$ ft/s	$\bar{b} = 0.65, \bar{\alpha} = \frac{1}{6.5}$, both from Table 6.4
$\hat{V}_{\bar{z}} = \hat{b}\left(\dfrac{\bar{z}}{33}\right)^{\hat{\alpha}} V$ $= 0.84\left(\dfrac{236}{33}\right)^{0.142} \times 176$ $= 195$ ft/s	$\hat{b} = 1.0, \hat{\alpha} = 1.0$, both from Table 6.4
$N_1 = \dfrac{n_1 L_{\bar{z}}}{\bar{V}_{\bar{z}}}$ $= \dfrac{0.222 \times 616}{130}$ $= 1.051$	n_1 = natural frequency of the building = 0.222 Hz, given

$$R_n = \frac{7.47 N_1}{(1 + 10.3 N_1)^{5/3}}$$

$$= \frac{7.47 \times 1.051}{(1 + 10.3 \times 1.051)^{5/3}}$$

$$= 0.128$$

$$\eta_b = \frac{4.6 n_1 B}{\bar{v}_z}$$

$$= \frac{4.6 \times 0.222 \times 98.5}{130}$$

$$= 0.773$$

$$R_B = \frac{1}{\eta_b} - \frac{1}{2\eta_b^2}(1 - e^{-2\eta_b})$$ $e = 2.71$

$$= \frac{1}{0.773} - \frac{1}{2 \times 0.773^2}(1 - 2.71^{-2 \times 0.773})$$

$$= 0.6370$$

$$\eta_h = \frac{4.6 n_1 h}{\bar{v}_z}$$

$$= \frac{4.6 \times 0.222 \times 394}{130}$$

$$= 3.095$$

$$R_h = \frac{1}{\eta_h} - \frac{1}{2\eta_h^2}(1 - e^{-2\eta_h})$$

$$= \frac{1}{3.095} - \frac{1}{2 \times 3.095^2}(1 - 2.71^{-2 \times 3.095})$$

$$= 0.271$$

$$\eta_L = \frac{15.4 n_1 L}{\bar{v}_z}$$ L = building breadth parallel

to wind = 164 ft given

$$= \frac{15.4 \times 0.222 \times 164}{130}$$

$$= 4.31$$

$$R_L = \frac{1}{\eta_L} - \frac{1}{2\eta_L^2}(1 - e^{-2\eta})$$

$$= \frac{1}{4.31} - \frac{1}{2 \times 4.31^2}(1 - 2.71^{-2 \times 4.31})$$

$$= 0.205$$

$$R^2 = \frac{1}{\beta} R_n R_h R_B (0.53 + 0.47 R_L)$$ β = damping ratio = 0.015,

given

$$= \frac{1}{0.01} \times 0.128 \times 0.271 \times 0.6360(0.53 + 0.47 \times 0.205)$$

$$= 1.381$$

$$R = \sqrt{1.381} = 1.175$$

$$g_R = \sqrt{2 \ln(3600 n_1)} + \frac{0.577}{\sqrt{2 \ln(3600 n_1)}}$$ ln means logarithm to

base $e = 2.71$.

$$= \sqrt{2 \ln(3600 \times 0.222)} + \frac{0.577}{\sqrt{2 \ln(3600 \times 0.222)}}$$

$$= 3.813$$

$$G_f = 0.925 \left(\frac{1 + 1.7 \, I_{\bar{z}} \sqrt{g_Q^2 Q^2 + g_R^2 R^2}}{1 + 1.7 \, g_v I_{\bar{z}}} \right)$$

G_f is gust effect factor

$$= 0.925 \left(\frac{1 + 1.7 \times 0.216 \sqrt{3.4^2 \times 0.64 + 3.813^2 \times 1.381}}{1 + 1.7 \times 3.4 \times 0.216} \right)$$

$$= 1.203$$

This is the gust factor required for calculating design wind pressures for the main wind-force-resisting system of the building.

$$m_1 = \mu_z \times \frac{h}{3}$$

$$= 4636 \times \frac{394}{3}$$

$$= 608,887 \text{ slugs}$$

$$K = \frac{1.65^{\hat{\alpha}}}{(\hat{\alpha} + \xi + 1)}$$

$\xi = $ first mode exponent $= 1.0$

$$= \frac{1.65^{0.142}}{(0.142 + 1 + 1)}$$

$$= 0.501$$

Maximum Along-Wind Displacement.

$$X_{\max(z)} = \frac{\phi_{(z)} \rho B h C_{fx} \hat{V}_{\bar{z}}^2}{2 m_1 (2 \pi n_1)^2} KG$$

$$= \frac{1 \times 0.0024 \times 98.5 \times 394 \times 1.3 \times 195^2}{2 \times 608,887 (2 \pi \times 0.222)^2} \times 0.501 \times 1.203 \qquad (1.47)$$

$= 1.16$ ft. This is the maximum deflection at the building top due to wind pressures.

$$g_{\ddot{x}} = \sqrt{2 \ln(n_1 T)} + \frac{0.5772}{\sqrt{2 \ln(n_1 T)}}$$

$$= \sqrt{2 \ln (0.222 \times 3600)} + \frac{0.5772}{\sqrt{2 \ln (0.222 \times 3600)}}$$

$$= 3.6567 + 0.1579$$

$$= 3.813 \text{ (Note: } g_{\ddot{x}} \text{ is the same as } g_R)$$

$$\sigma_{\ddot{x}(z)} = \frac{0.85 \phi_z \rho B h C_{fx} \bar{V}_{\bar{z}}^2}{m_1} I_{\bar{z}} KR$$

$$= \frac{0.85 \times 0.0024 \times 98.5 \times 394 \times 1.3 \times 130^2}{608,887} \times 0.216 \times 0.501 \times 1.175$$

$$= 0.363 \qquad (1.48)$$

Maximum Along-Wind Acceleration.

$$
\begin{aligned}
X_{\max(z)} &= g_{\ddot{x}}\sigma_{\ddot{x}(z)} \\
&= 3.814 \times 0.363 \\
&= 1.384 \text{ ft/sec}^2 \\
&= 1.384 \times 31.11 \text{ milli-g} \\
&= 43 \text{ milli-g} \quad \text{This is the maximum along-wind acceleration} \qquad (1.49)
\end{aligned}
$$

This is more than twice the threshold value of 20 milli-g. Therefore, it is important to verify the results by conducting wind tunnel tests. If the test results confirm the likelihood of the building experiencing high accelerations, adding dampers to the building to reduce building oscillations is an option.

This is discussed in Chapter 8.

1.4.2.9. *Formal Definitions of Exposure Categories*

Exposure B: Exposure B applies to urban and suburban areas or other terrain with numerous closely spaced obstructions the size of single-family dwellings or larger. To appropriately assign this exposure category, this topography must prevail in the upwind direction for a distance of at least 2630 ft (800 m) or 10 times the height of the buildings, whichever is greater.

Exposure C: Exposure C applies to terrain that consists of scattered obstructions of height generally less than 30 ft (9.1 m). This category includes flat open country, grasslands, and all water surfaces in hurricane-prone regions. It also applies to all areas where exposures B and D do not apply.

Exposure D: Exposure D consists of unobstructed areas and water surfaces outside hurricane-prone regions. This category includes smooth mud flats, salt flats, and unbroken ice extending in the upwind direction for a distance of at least 5000 ft (1524 m) or 10 times the buildings' height, whichever is greater. Exposure D extends inland from the shoreline for a distance of 660 ft (200 m) or 10 times the height of the building, whichever is greater.

The proper assessment of exposure is a matter of good engineering judgment, particularly because the exposure may change in one, wind direction or more as a result of future development and/or demolition. Figures 1.13a–e are aerial photographs representative of some of the exposure types.

1.4.3. National Building Code of Canada (NBCC 1995): Wind Load Provisions

The reader may be wondering why, after an arguably extensive coverage of the ASCE 7-02 wind load provisions, the author would burden the text with yet another building code provision. The reason is simple: Although extensive in its treatment of wind, the ASCE 7-02 does not provide an analytical procedure for estimating across-wind response of tall, flexible buildings. To the best of the author's knowledge, NBCC is the only code in North America that presents an analytical method for computing across-wind response. It is perhaps the most comprehensive standard for wind because it takes into consideration characteristics such as building dimensions, shape, stiffness, damping ratios, site topography, climatology, boundary layer meteorology, bluff body aerodynamics, and probability theory.

Figure 1.13a. Exposure B; suburban residential area predominated by single-family dwellings.

Figure 1.13b. Exposure B; urban area with numerous closely spaced buildings the size of single-family homes or larger.

Figure 1.13c. Exposure B; urban area with numerous closely spaced buildings the size of single-family homes or larger.

Figure 1.13d. Structure in the foreground is located in exposure B. Structures in the rear, adjacent to the clearing, are located in exposure C when wind flows from the left over the clearing.

Figure 1.13e. Structures on a shoreline with wind flowing over open water for a distance of at least one mile are located in exposure D.

Three different approaches for determining wind loads on buildings are given: 1) simple procedure; 2) experimental procedure; and 3) detailed procedure.

1.4.3.1. Simple Procedure

The simple procedure is applicable for determining structural wind loads for a majority of low- and medium-rise buildings and also for cladding design of low-, medium-, and high-rise buildings. The method is similar to other code approaches in which the dynamics action of wind is dealt with by equivalent static loads defined independently of the dynamic properties of wind.

The recurrence intervals used for evaluating wind loads are

1. 1 in 10 years for the design of cladding and structural members designed for deflection and vibration limits.
2. 1 in 30 years for the design of structural members of all, except post-disaster buildings, for strength.
3. 1 in 100 years for the design of structural members of post-disaster buildings for strength.

The external pressure or suction on the building surface is given by the equation

$$P = qC_eC_gC_p \tag{1.50}$$

where

P = design static pressure or suction, acting normal to the surface: kilo pascals

q = reference velocity pressure; kilo pascals

C_e = exposure factor that reflects the changes in wind speed with height and variations in the surrounding terrain: dimensionless

C_g = gust factor, with a value of 2.0 for the primary structural system, and 2.50 for cladding: dimensionless

C_p = external pressure coefficient averaged over the area of the surface considered: dimensionless

Reference Pressure *q*.   The reference velocity pressure q, in kilo pascals, is determined from referenced wind speed V by the equation:

$$q = C\overline{V}^2 \tag{1.51}$$

The factor C depends on the atmospheric pressure and air temperature. If the wind speed $\overline{V}$ is in meters per second, the design pressure, in kilo pascals, is obtained by using a value of $C = 650 \times 10^{-6}$. The reference wind pressure q, is given for three different levels of probability being exceeded per year (1/10, 1/30, and 1/100), that is, for return periods for 10, 30, and 100 years, respectively. A 10-year recurrence pressure is used for the design of cladding an for the serviceability check of structural members for deflection and vibration. A 30-year wind pressure is used for the strength design of structural members of all buildings except those classified as post-disaster buildings. A 100-year wind is used for the design of post-disaster buildings such as hospitals, fire stations, etc. The 10-, 30-, and 100-year mean hourly wind pressures in Montreal, Quebec are 0.31 kPa (6.5 psf), 0.37 kPa (7.72 psf), and 0.44 kPa (9.2 psf), respectively, with corresponding wind speeds of 22 m/s (49.2 mph), 24 m/s (54 mph), and 26 m/s (58 mph).

Exposure Factor *C_e*.   The exposure factor C_e is based on the 1/5 power law corresponding to wind gust pressures in open terrain. An averaging period of 3 to 5 seconds is used in determining the gust factor. It represents a 'parcel' of wind assumed to be effective over the entire building. For tall buildings, the reference height for pressures on the windward face corresponds to the actual height aboveground, and for suctions on the leeward face, the reference height is half the height of the structure.

The exposure factor C_e reflects the changes in wind speed and height, and the effects of variations in the surrounding terrain and topography. Hills and escarpments that can significantly amplify wind speeds are reflected in the exposure factor.

The exposure factor C_e may be obtained from any of the following three methods:

1. The value shown in Table 1.

2. The value of the function $(h/10)^{\frac{1}{5}}$ but not less than 0.9, where h is the reference height above grade, in meters.

3. If a dynamic approach is used, an appropriate value depending on both the height and shielding.

Gust Effect Factor (Dynamic Response Factor) *C_g*.   This factor accounts for the increase in the mean wind loads due to the following factors:

- Random wind gusts acting for short durations over entire or part of structure.
- Fluctuating pressures induced in the wake of a structure, including vortex shedding forces.
- Fluctuating forces induced by the motion of a structure.

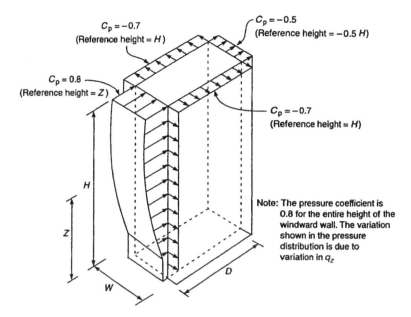

Figure 1.14.   External wind pressure coefficient C_p; flat-roofed buildings $H > W$. (Adapted from NBCC 1995.)

All buildings are affected to some degree by their dynamic response. The total response may be considered as a summation of the mean component without any structural dynamic magnification, and a resonant component due to building vibrations close to its natural frequency. For the majority of buildings less than 120 m (394 ft) tall, and with height-to-width ratio less than 4, the resonant component is small. The only added loading is due to gusts that can be dealt with in a simple static manner.

For buildings and components that are not particularly tall, long, slender, light-weight, flexible, or lightly damped, a simplified set of dynamic gust factors is given as follows:

C_g = 2.5 for building components and cladding

C_g = 2.0 for the primary structural system including anchorages to foundation

Pressure coefficient C_p. C_p is a nondimensional ratio of wind-induced pressure on a building to the velocity pressure of the wind speed at the reference height (see Fig. 1.14). It depends on the shape of the building, wind direction, and profile of the wind velocity, and can be determined most reliably from wind-tunnel tests. However, for the simple procedure, based on some limited measurements on full-scale buildings supplemented by wind-tunnel tests, NBC gives the following values of C_p for simple building shapes:

Windward wall:	C_p = +0.8 (positive pressure)
	Reference height = Z aboveground
Side wall and roof:	C_p = −1.0 (negative pressure, suction)
	Reference height = H aboveground
Leeward wall:	C_p = −0.5 (negative pressure, suction)
	Reference pressure = $0.5H$ aboveground

1.4.3.2. *Experimental Procedure*

The second approach is to use the results of wind-tunnel or other experimental procedures for buildings likely to be susceptible to wind-induced vibrations. Included in this category are tall, slender structures for which wind loading plays a major role in the structural design. A wind-tunnel test is also recommended for determining exterior pressure coefficients for cladding design of buildings whose geometry deviates markedly from more common shapes for which information is already available.

1.4.3.3. *Detailed Procedure*

In this method, a series of calculations is performed to determine more accurate values for the gust factor C_g, the exposure factor C_e, and the pressure coefficient C_p. The end product of the calculations yields a static design pressure, which is expected to produce the same peak effect as the actual turbulent wind, with due consideration for building properties such as height, width, natural frequency of vibration, and damping. This approach is primarily for determining the overall wind loading and response of tall slender structures, and is not intended for determining exterior pressure coefficients for cladding design.

The code gives procedures for calculating the dynamic effects of vortex shedding for slender cylindrical towers and for tapered structures. Since the available data are limited for slender structures with cross sections other than circular, wind-tunnel tests are recommended for estimating the likely response. To limit the cracking of masonry and interior finishes, the total drift per story under specified wind and gravity loads is limited to 1/500 of the story height, unless a detailed analysis is made and precautions taken to permit larger movements.

The code recognizes that maximum accelerations of a building leading to possible human perception of motion or discomfort may occur in a direction perpendicular to the wind. A tentative acceleration limit of 1 to 3% of gravity for a 10-year return wind is recommended to limit the possibility of perception of motion.

Exposure Factor C_e. The exposure factor C_e is based on the mean wind speed profile, which depends on the roughness of terrain over which the wind has traveled before reaching the building. Three wind profile categories are used in building design.

Exposure A. This is the exposure on which the reference wind speeds are based. The exposure is defined as open, level terrain with only scattered buildings, trees or other obstructions, and open water or shorelines. C_e is given by

$$C_e = \left(\frac{z}{10}\right)^{0.28}, \qquad C_e \geq 1.0 \tag{1.52}$$

Exposure B. Suburban and urban areas, wooded terrain, or centers of large towns with terrain roughness extending in the upwind direction for at least 1.5 km. C_e is given by

$$C_e = 0.5\left(\frac{z}{12.7}\right)^{0.50}, \qquad C_e \geq 0.5 \tag{1.53}$$

Exposure C. Centers of large cities with heavy concentrations of buildings extending in the upwind direction for at least 1.5 km, with at least 50% of the buildings exceeding four stories in height. C_e is given by

$$C_e = 0.4\left(\frac{z}{30}\right)^{0.72}, \qquad C_e \geq 0.4 \tag{1.54}$$

Exposure factor C_e can be calculated from Eq. (1.54) or obtained directly from the graph in Fig. 1.15.

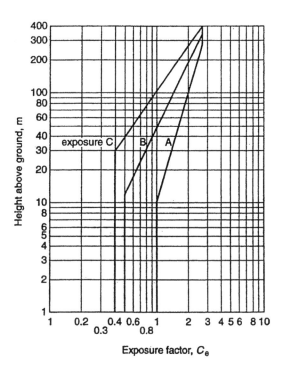

Figure 1.15. Exposure factor C_e as a function of terrain roughness and height aboveground. (From NBCC 1995.)

Gust Effect Factor C_g (Detailed Procedure). A general expression for the maximum or peak load effect, denoted W_p, is given by

$$W_p = \mu + g_p\sigma \tag{1.55}$$

where

μ = the mean loading effect
σ = the root-mean square loading effect
g_p = a peak factor for the loading effect

The dynamic gust response factor is defined as the ratio of peak loading to mean loading,

$$C_g = W_p/\mu$$

$$= 1 + g_p\left(\frac{\sigma}{\mu}\right) \tag{1.56}$$

The parameter σ/μ is given by the expression

$$\frac{\sigma}{\mu} = \sqrt{\frac{K}{C_{eH}}\left(B + \frac{sF}{\beta}\right)} \tag{1.57}$$

where

K = a factor related to the surface roughness coefficient of the terrain
$K = 0.08$ for exposure A
$K = 0.10$ for exposure B
$K = 0.14$ for exposure C

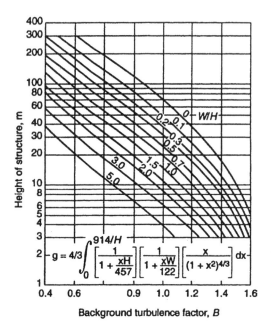

Figure 1.16. Background turbulence factor as a function of width and height of structure. (From NBCC 1995.)

C_{eH} = exposure factor at the top of the building, H, evaluated using Fig. 1.15
  B = background turbulence factor obtained from Fig. 1.16 as a function of building width-to-height ratio W/H
  H = height of the building
  W = width of windward face of the building
  s = size reduction factor obtained from Fig. 1.17 as a function of W/H and reduced frequency $n_0 H/V_H$
  n_0 = natural frequency of vibration, Hz
  V_H = mean wind speed (m/s) at the top of structure, H

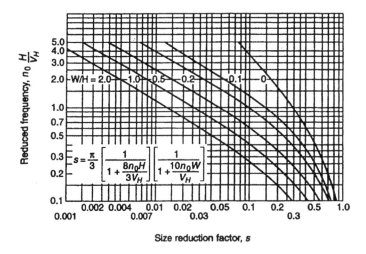

Figure 1.17. Size reduction factor as a function of width, height, and reduced frequency of structure. (From NBCC 1995.)

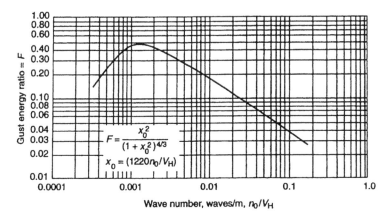

Figure 1.18. Gust energy ratio as a function of wave number. (From NBCC 1995).

F = gust energy ratio at the natural frequency of the structure obtained from Fig. 1.18 as a function of wave number n_0/V_H

β = critical damping ratio, with commonly used values of 0.01 for steel, 0.015 for composite, and 0.02 for cast-in place concrete buildings

Design Example: Calculations for Gust Effect Factor C_g.
Given.
Height H = 240 m (787.5 ft)
Width W (across-wind) = 50 m (164 ft)
Depth D (along-wind) = 50 m (164 ft)
Fundamental frequency n_0 = 0.125 Hz (period = 8 sec)
Critical damping ratio β = 0.010
Average density of the building = 195 kg/m³ (12.2 pcf)
Terrain for site = exposure B
Reference wind speed at 10 m, open terrain (exposure A) = 26.4 m/s (60 mph)

Required.   Gust factor C_g
Solution.   From Fig. 1.15, for H = 240 m and exposure category B, exposure factor C_{CH} = 2.17
Mean wind speed V_H at top

$$V_H = \overline{V}\sqrt{C_{CH}}$$
$$= 26.4\sqrt{2.17}$$
$$= 38.88 \text{ m/s}$$

Aspect ratio $\dfrac{W}{H} = \dfrac{50}{240} = 0.208$

Wave number $F = \dfrac{n_0}{V_H}$

$$= \frac{0.125}{38.8} = 0.00322$$

$$\frac{n_0 H}{V_H} = \frac{0.125 \times 240}{38.80} = 0.772$$

Calculate $\dfrac{\sigma}{\mu}$ using the following parameters:

1. $K = 0.10$ for exposure B

2. $B = 0.50$ from Fig. 1.16, for $\dfrac{W}{H} = 0.208$

3. $s = 0.14$ from Fig 1.17, for $\dfrac{n_0 H}{V_H} = 0.772$

   and $\dfrac{W}{H} = 0.208$

4. $F = 0.36$ from Fig. 1.18, for $\dfrac{n_0}{V_H} = 0.0032$

5. $\beta = 0.010$, given value of damping

6. $\dfrac{\sigma}{\mu} = \sqrt{\dfrac{K}{C_{eH}}\left(B + \dfrac{sF}{\beta}\right)}$

   $= \sqrt{\dfrac{0.10}{2.17}\left(0.50 + \dfrac{0.14 \times 0.36}{0.010}\right)}$

   $= 0.505$

Calculate v from the equation

$$v = n_0 \sqrt{\dfrac{sF}{sF + \beta B}}$$

$$= 0.125 \sqrt{\dfrac{0.14 \times 0.36}{0.14 \times 0.36 + 0.01 \times 0.5}}$$

$$= 0.119 \text{ cycles/sec}$$

Using Fig 1.19, read the peak factor g_p corresponding to $v = 0.119$

$g_p = 3.6$

Calculate the required gust response factor C_g from the formula

$$C_g = 1 + g_p\left(\dfrac{\sigma}{\mu}\right)$$

$$= 1 + 3.6 \times 0.505$$

$$= 2.82$$

With the known gust effect factor C_g peak dynamic forces are determined by multiplying mean wind pressures by C_g.

1.4.3.4. *Wind-Induced Building Motion*

Although the maximum lateral deflection is generally in a direction parallel to wind (along-wind direction), the maximum acceleration leading to possible human perception of motion or even discomfort may occur in a direction perpendicular to the wind (across-wind direction). Across-wind accelerations are likely to exceed along-wind accelerations if

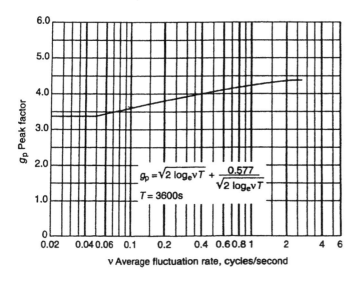

Figure 1.19. Peak factor g_p as a function of average fluctuation rate. (From NBCC 1995.)

the building is slender about both axes, with the aspect ratio $\sqrt{WD}/H$ less than one-third, where W and D are the across-wind and along-wind plan dimensions and H is the height of the building.

Based on wind tunnel studies, NBC gives two expressions for determining the across- and along-wind accelerations.

The across-wind acceleration a_w is given by

$$a_w = n_W^2 g_p \sqrt{WD}\left(\frac{a_r}{\rho_B g \sqrt{\beta_W}}\right) \tag{1.58}$$

The along-wind acceleration a_D is given by

$$a_D = 4\pi^2 n_D^2 g_p \sqrt{\frac{KsF}{C_e \beta_D}}\frac{\Delta}{C_g} \tag{1.59}$$

Observe that Δ, the maximum wind-induced lateral displacement in the along-wind direction is typically obtained from a computer analysis. Substitution of this value in Eq. (1.20) yields the best estimation of a_D. However, as a rough guess for: preliminary evaluations, Δ can be assumed equal to $H/450$, the drift index normally used in wind-design of tall buildings.

Using a linear modal representation for the building motion, the maximum deflection, Δ can be related to the fundamental frequency of the building. The resulting expression is shown in Eq. (1.60) for the ratio a_D/g.

$$\frac{a_D}{g} = g_P \sqrt{\frac{KsF}{C_e \beta_D}}\left(\frac{3.9}{2+\alpha}\right)\left(\frac{C_e q}{D_g \rho_B}\right) \tag{1.60}$$

where

$\quad a_D$ = acceleration in the along-wind direction

$\quad g$ = acceleration due to gravity = 9.81 m/sec^2

g_P = a statistical peak factor for the loading effect

K = a factor related to surface roughness coefficient of terrain

    = 0.08 for exposure A

    = 0.10 for exposure B

    = 0.14 for exposure C

s = size reduction factor, from Fig. 1.17

F = gust energy ratio, from Fig 1.18

C_e = exposure factor

β_D = critical damping ratio, in the along-wind direction

α = power coefficient related to C_e

    = 0.28 for exposure A

    = 0.50 for exposure B

    = 0.72 for exposure C

q = reference wind pressure, kpa

    = $650 \times 10^{-6} \times \overline{V}^2$, ($\overline{V}$ in meters per second)

D = building depth parallel to wind, meters

ρ_B = mass density of building, kg/m³

Design Example.   A representative calculation for a_w and a_D using Eq. (1.58) and (1.59) will be made for the sample problem worked earlier to illustrate the calculation of a gust factor.

Given.

Building frequency $n_w = n_D = 0.125$ Hz

Damping coefficient $\beta_w = \beta_D = 0.01$

Building density $\rho_B = 195$ kg/m³ (12.2 pcf)

All other data as given for the previous illustrative problem.

Required.   Building accelerations in both across-wind along-wind directions.

Solution.

Step 1.   Calculate a_r

$$a_r = 78.5 \times 10^{-3} \left[\frac{V_H}{n_w \sqrt{WD}} \right]^{3.3}$$

$$= 78.5 \times 10^{-3} \left[\frac{38.88}{0.125 \times 50} \right]^{3.3}$$

$$= 32.7 \ \text{m/s}^2$$

Step 2.   Calculate a_W (across-wind response)

In our case, $n_O = n_W = n_D = 0.125$ and $\beta_W = \beta_D = 0.10$

$$a_W = n_W^2 g_P \sqrt{WD} \left(\frac{a_r}{\rho_{Bg} \sqrt{\beta_W}} \right)$$

$$= 0.125^2 \times 3.6 \times \sqrt{50 \times 50} \left(\frac{32.7}{195 \times 9.81 \sqrt{0.01}} \right)$$

$$= 0.482 \ \text{m/s}^2$$

$$\frac{a_W}{g} = \frac{0.482}{9.81} \times 100 = 4.91\% \ \text{of gravity}$$

The calculated value of across-wind acceleration a_W exceeds the acceptable limit of 3% of gravity for office buildings, warranting a detailed boundary layer wind-tunnel study.

Step 3. Calculate q (reference wind pressure)

$$q = C\bar{V}^2$$
$$= 650 \times 10^{-6} \times 26.4^2$$
$$= 0.453$$

Step 4. Calculate along-wind response a_D

$$a_D = g_P \sqrt{\frac{KsF}{C_c\beta_D}\left(\frac{3.9}{2+\alpha}\right)\left(\frac{C_eq}{D\rho_B}\right)}$$

$$= 3.6\sqrt{\frac{0.10 \times 0.14 \times 0.36}{2.17 \times 0.010}\left(\frac{3.9}{2+0.5}\right)\left(\frac{2.17 \times 0.453}{50 \times 195}\right)}$$

For the example problem we have:

$g_P = 3.6$ $\beta_D = 0.010$ $q = C\bar{V}^2$
$K = 0.10$ $\alpha = 0.5$ $= 650 \times 10^{-6} \times 26.4^2$
$s = 0.14$ $D = 50$ m $= 0.453$ k Pa
$F = 0.36$ $g = 9.81$ m/sec^2
$C_e = C_H = 2.17$ $\rho_B = 195$ kg/m^3

Substituting the preceding values in Eq. (1.60)

$$\frac{a_D}{g} = 3.6\sqrt{\frac{0.10 \times 0.14 \times 0.36}{2.17 \times 0.010}\left(\frac{3.9}{2+0.5}\right)\left(\frac{2.17 \times 0.453}{50 \times 0.00981 \times 195}\right)}$$

$$= 0.027 = 2.7\% \text{ of } g$$

The calculated value is below the 3% limit. Its along-wind response is unlikely to disturb the comfort and equanimity of building's occupants.

Comparison of Along-wind and Across-wind Accelerations. To get a sense for along-wind and across-wind, the results for two buildings are given in summary form in Fig. 1.20. One is representative of a 30-story rectangular building, shown in Fig.1.20a and 1.20b, and is examined for wind along both its principle axes. The other, shown in Fig.1.20c, is square with a height corresponding approximately to a 60-story-plus building. Results for both are given for suburban exposure B.

Response characteristics were also evaluated for the other two types of exposure categories. From the calculations performed but not shown here, it appears that the type of exposure has a significant effect on both along-wind and across-wind response. Accelerations were about 20 to 50% greater for an open-terrain exposure A. The reductions for an urban setting, exposure C, were of the same order of magnitude.

Observe that in Fig.1.20, the maximum acceleration of the building occurs in a direction perpendicular to the wind (across-wind direction) because the building is considerably more slender in the across-wind than in the along-wind direction. Across-wind accelerations control the design if the building is slender about both axes, that is, if $\sqrt{WD/H}$ is less than one-third, where W and D are the across-wind and along-wind plan dimensions and H is the building height.

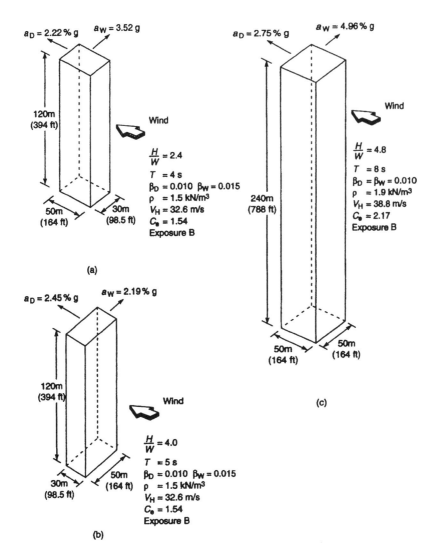

Figure 1.20. Wind-induced peak accelerations; 1995 NBCC procedure: (a) 30-story building, wind on narrow face; (b) 30-story building, wind on broad face; (c) 60-story building.

Since Eqs (1.58) and (1.59) for along-wind and across-wind accelerations are sensitive to the natural frequency of the building, use of approximate formulas for period calculations are not appropriate. Therefore, results of more rigorous methods such as computer dynamic analyses are recommended for use in these formulas.

In addition to acceleration, many other factors such as visual cues, body position and orientation, and state of mind of occupants during windstorms influence human perception of motion. However, research has shown that when the amplitude of acceleration is in the range of 0.5 to 1.5% of acceleration due to gravity, movement of buildings becomes perceptible to most building occupants. Based on this and other information, a tentative acceleration limit of 1 to 3% of gravity is recommended. The lower value is considered appropriate for apartment buildings, the higher values for office buildings.

1.5. WIND-TUNNEL ENGINEERING

Wind-tunnel testing of buildings has been an offshoot of aeronautical engineering, in which the flow of wind is duplicated at high altitudes. The tunnels for testing airplanes are designed to minimize the effects of turbulence, and as such, they do not duplicate atmospheric boundary layer or wind turbulence. This is because majority of airplane flights, except for brief periods of landing and takeoff, occur at a height well above the boundary layer. Building activity, on the other hand, occurs precisely within this atmospheric boundary layer, characterized by a gradual retardation of wind speed and high turbulence near the surface of the earth. Therefore, for testing of buildings, aeronautical wind tunnels have been modified and entirely new facilities have been built to reproduce turbulence and natural flow of wind within the boundary layer.

Wind-tunnel tests (or similar tests employing fluids other than air) are considered to be properly conducted only if the following conditions are satisfied:

1. The natural atmospheric boundary layer has been modeled to account for the variation of wind speed with height.
2. The length scale of the longitudinal component of atmospheric turbulence is modeled to approximately the same scale as that used to model the building.
3. The modeled building and surrounding structures and topography are geometrically similar to their full-scale counterparts.
4. The projected area of the modeled building and surroundings is less than 8% of the test section cross-sectional area unless correction is made for blockage.
5. The longitudinal pressure gradient in the wind tunnel test section is accounted for.
6. Reynolds number effects on pressures and forces are minimized.
7. Response characteristics of the wind-tunnel instrumentation are consistent with the required measurements.

Boundary-layer wind tunnels capable of developing flows that meet the conditions stipulated above typically have test-section dimensions in the following ranges: width, 6 to 12 ft (2 to 4 m); height, 6 to 10 ft (2 to 3 m); and length, 50 to 100 ft (15 to 30 m). Maximum wind speeds are ordinarily in the range of 25 to 100 mph (10 to 45 m/s).

Three basic types of wind-tunnel test models are commonly used:

1. Rigid pressure model (PM)
2. Rigid high-frequency base balance model (H-FBBM)
3. Aeroelastic model (AM)

One or more of the models may be employed to obtain design loads for a particular building or structure. The pressure model provides local peak pressures for design of elements such as cladding and mean pressures for the determination of overall mean loads. The high-frequency model measures overall fluctuating loads for the determination of dynamic responses. The aeroelastic model is employed for direct measurement of overall loads, deflections, and accelerations, when the lateral motions of a building are considered to have a large influence on wind loading.

Various techniques are used in aeronautical tunnels to generate turbulence and atmospheric boundary layer by using devices such as screens, spires, and grids. In special wind tunnels with long test sections, turbulent boundary layer is generated by installing appropriate roughness elements in the upstream flow. Another approach is to use a counterjet technique. In every case there is some question whether the natural wind turbulence characteristic is appropriately modeled and proper gust simulation is included. The degree of scaling required to appropriately account for these may yield a very extreme scale for the building, on the order of 1:500 or even more for urban environment studies.

1.5.1. Rigid Model

Although the primary purpose of the rigid-model test is for obtaining cladding design pressures, the data acquired from the wind-tunnel tests may be extrapolated to get the floor-by-floor shear forces for the design of the overall main wind-force-resisting frame.

Most commonly, pressure study models are made from methyl methacrylate sheets commonly known as Plexiglas, Lucite, and Perspex. This material has several advantages over wooden or aluminum alloy models because it can be easily and accurately machined and drilled and is transparent, facilitating observation of the instrumentation inside the model. It can also be formed into curved shapes by heating the material to about 200°C. Model panels can either be cemented together or joined, using flush-mount screws.

A scale model of the prototype in a 1:300 and 1:500 range is constructed at the testing facility by using architectural drawings provided by the project architect. In a rigid model, building features that have significance in regard to the wind flow, such as building profile, protruding mullions, and overhangs are simulated to the correct length scale. Wind measurements obtained are only for the mean and fluctuating pressures acting on the building.

The model typically instrumented with a large number of pressure taps (sometimes as many as 500 to 700), is tested surrounded by a detailed modeling of nearby surroundings within a radius of 1500 ft (457 m), as shown in Fig.1.21. Flexible, transparent vinyl or polyethylene tubing of about 1/16 in (1.5 mm) internal diameter is used as pressure tappings around the exterior of the model. Pressure tap locations are generally more concentrated in regions of high pressure gradients such as around corners.

The wind-tunnel test is run for a duration of about 60 sec which corresponds to approximately 1 hr in real time. Sufficient numbers of readings are obtained from each port to obtain a stationary value such that fluctuations become independent of time. From the values thus obtained, the mean pressure and the root-mean-square value of the pressure fluctuations are evaluated.

The boundary-layer wind tunnel, by virtue of having a long working section with roughened floor and turbulence generators at the upwind end, is expected to correctly simulate the mean wind speed profile and turbulence of natural wind. The model is mounted on a turntable to allow measurement in any wind direction. Near-field characteristics around the building are duplicated, typically using polystyrene foam models.

1.5.1.1. Cladding Pressures

Measurements are taken for representative wind directions, generally spaced about 10 to 20° apart. From the data acquired, full-scale peak exterior pressures and suctions at each tap location are derived by combining the wind-tunnel data with a statistical model of windstorms expected at the building site. The results are typically given for 25-, 50-, and 100-year return periods.

In evaluating peak wind loads on the exterior of the prototype, the effects of internal pressures arising from air leakage, mechanical equipment, and stack effect should be included. The possibility of window breakage caused by roof gravel scoured from roofs of adjacent buildings and other flying debris during a windstorm should also be included. As a rough guide, the resulting internal pressure can be considered to be in the range of

(a)

(b)

Figure 1.21. (a) Rigid models of high-rise buildings in a wind tunnel; (b) close-up view of a pressure model. (Photographs courtesy of Dr. Peter Irwin, Rowan, Williams, Davis & Irwin, Inc.)

± 5 psf (25 kg/m^2) at the base, to as much as ± 20 psf (100 kg/m^2) a the roof of a 50-story building.

In the design of glass, a 1-minute loading is commonly used. The duration of measured peak pressure in a wind tunnel is different from the 1-min interval; usually it corresponds to 5 to 10 seconds or less in terms of real time. Therefore, it is necessary to reduce the peak loads measured in wind-tunnel tests. Empirical reduction factors of 0.80, 0.94, and 0.97 have been given in glass manufacturers' recommendations for three different types of glass—annealed float glass, heat-strengthened glass, and tempered glass.

1.5.1.2.   Overall Building Loads

The results of rigid-model tests are used to predict the design wind loads for glass and cladding. For buildings that are not dynamically sensitive to wind, the results can nevertheless be extrapolated to obtain lateral loads for the design of the main wind-force-resisting system of the building. The procedure entails introducing a gust factor for converting the mean wind load to gust loads. An appropriate gust factor estimation should take into account:

- Averaging period of the mean wind load
- Terrain roughness in relation to the building height
- Peak gust factor, which depends on the natural frequency of the building
- Effect of turbulence
- Critical damping ratio of the building.

In spite of the fact that rigid-model wind study does not take into account all of the preceding factors, it is still considered to provide adequate design data for buildings with height-to-width ratios of less than 5.

1.5.2.   Aeroelastic Study

Aeroelastic model study attempts to take the guesswork out of the gust factor computation by measuring directly the magnitude of dynamic loads. These are measured using a variety of models ranging from very simple rigid models mounted on flexible supports to models exhibiting the multimode vibration characteristics of tall buildings. The more common types of models used in aeroelastic studies can be broadly classified into two categories: 1) stick models; and 2) multi degree-of-freedom models.

In addition to the similarity of the exterior geometry, the aeroelastic studies require similarity of the inertia, stiffness, and damping characteristics of the building. Although a building in reality responds dynamically to wind loads in a multimode configuration. Enough evidence exists to show that the dynamic response occurs primarily in the lower modes of vibration. As a result, it is possible to study the dynamic behavior of buildings by using simple dynamic models.

Aeroelastic study basically examines the wind-induced sway response, in addition to providing information on the overall wind-induced mean and dynamic loads. These tests are important for slender, flexible, and dynamically sensitive structures where aeroelastic or body-motion-induced effects are of significance. When a tall building sways and twists under wind action, the resulting acceleration generates inertial loads, causing fluctuating stresses. At any given instant, the amplitude of twisting and swaying motion is not just a function of the magnitude of wind load at the instant but also depends on the integrated effect of the wind over the several previous minutes. Therefore, for slender buildings it is important to consider the dynamic response when predicting design wind loads. In addition to providing an accurate assessments of load for structural design, an aeroelastic model test provides one of the most reliable approaches to predicting building response to wind which can be used by the designer to ensure that the predicted motion will not cause discomfort to the building occupants.

Typically, aeroelastic measurements are carried out at several wind speeds covering a range selected to provide information on both relatively common events, such as 10-year wind loads, which may influence the serviceability and occupant comfort, and relatively rare events, such as 100-year winds, which govern the strength design. The modeling of dynamic properties requires the simulation of inertial, stiffness, and damping characteristics. It is necessary, however, to simulate these properties for only those modes of vibration which are susceptible to wind excitation.

It is often difficult to determine quantitatively when an aeroelastic study is required on a building project. The following factors may be used as a guide in making a decision:

1. The building height-to-width ratio is greater than about 5; i.e., the building is slender.
2. Approximate calculations show that there is a likelihood of vortex shedding phenomenon.
3. The structure is light in density on the order of 8 to 10 lb/ft^3 (1.25 to 1.57 kN/m^3).
4. The structure has very little inherent damping, such as a building with welded steel construction.
5. The structural stiffness is concentrated in the interior of the building, making it torsionally flexible. A building with a braced central core is one such example.
6. The calculated period of oscillation of the building is long, in excess of 4 or 5 sec.
7. Existence of nearby buildings that could create unusual channeling of wind, resulting in torsional loads and strong buffeting action.
8. The building is sited such that predominant winds blow from a direction most sensitive to the building oscillations.
9. The building is a high-rise apartment, condominium, or hotel whose occupants are more likely than occupants of office buildings to experience discomfort from building oscillations. This is because residents are likely to remain longer in a given location than they would in a typical office setting.

1.5.2.1. Rigid Aeroelastic Model

The main objective of conducting aeroelastic model study is to determine a more accurate design wind load and to predict building oscillations to get an idea of the degree of occupant sensitivity to building motions.

Rigid-model study is based on the premise that the fundamental displacement mode of a tall building can be approximated by a straight line. In terms of aerodynamic modeling, it is not necessary to achieve the correct density distribution along the building height as long as the mass moment of inertia about a chosen pivot point is the same as that of the correct density distribution. It should be noted that the pivot point is chosen to obtain a mode shape that provides the best agreement with the calculated fundamental mode shapes of the prototype. For example, modal calculations for a tall building with a relatively stiff podium may show that the pivot point is located at the intersection of podium and the tower and not at the ground level. Therefore the pivot point for the model should be at a location corresponding to this intersection point rather than at the base of the building.

Figure 1.22 shows a rigid aeroelastic model mounted on gimbals. The springs located near the gimbals are chosen to achieve the correct frequencies of vibration in the two fundamental sway modes. An electromagnet or oil dashpot provides the model with a damping corresponding to that of the full-scale building.

An alternative method is to mount the model on a flexible steel bar attached to a vibration-free table. The width and thickness of the bar are chosen to simulate the building stiffness in two horizontal directions. Damping is simulated by using dashpots. Figure 1.23 shows a schematic elevation of a rigid aeroelastic model mounted to a flexible steel bar. Shown in Fig. 1.24 is a photograph of an aeroelastic model of a 62-story building. In either

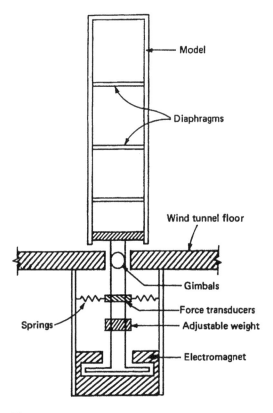

Figure 1.22. Rigid aeroelastic model with gimbal.

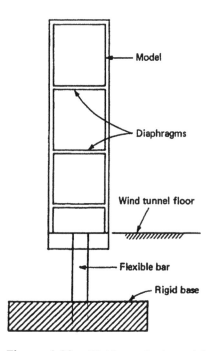

Figure 1.23. Rigid aeroelastic model mounted to flexible steel bar.

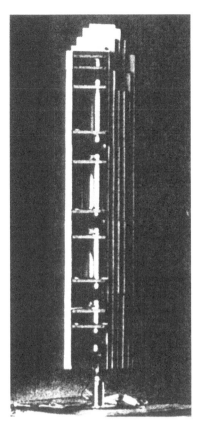

Figure 1.24. Cutaway view of an aeroelastic model of a tall building. (Photo courtesy of Dr. Peter Irwin, Rowan, Williams, Davis & Irwin, Inc.)

type, torsional modes are not simulated because the model effectively rotates as a rigid body about the vertical axis.

1.5.2.2. *Rigid Model Simulating Torsion*

Torsion is a consequence of an unsymmetrical distribution of building stiffness about its shear center, or it may occur because of eccentric disposition of lateral loads with respect to the center of stiffness of the building. Centrally supported concrete-core buildings often use open section shear walls that may have their shear centers located a considerable distance from the geometrical center of the core. Unless additional lateral resisting elements such as moment frames, braces, or shear walls are used on the building perimeter, torsional characteristics of the building may play an important role in its design. When such characteristics are present it is necessary to simulate not only the bending characteristics of the building but also its torsional behavior. This is achieved by introducing torsional springs in the aeroelastic model at appropriate locations along the height. To allow one section of the model to rotate relative to the next, the model shell is cut around the periphery. Figure 1.25 shows a schematic representation of a model with two cuts. The resulting model with three vertical segments behaves as a three-degrees-of-freedom system in torsion, and can therefore capture the dynamic behavior of the three lowest torsional modes.

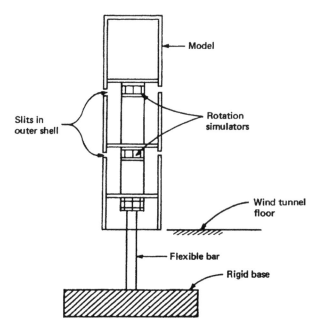

Figure 1.25.   Rigid aeroelastic model with provisions for simulating torsion.

1.5.2.3.   Flexible Model

If a building geometry is uniform for the entire height, it is reasonable to assume that the sway modes of vibration vary linearly along the height. However, for buildings of complex shapes with stepbacks and similar major variations in stiffness, this assumption may not yield acceptable results because fundamental mode shapes may not be linear, and higher modes could contribute significantly to the dynamic behavior.

In such cases it is essential to simulate the multimode behavior of the building. This is achieved using a model with several lumped masses interconnected with elastic columns. A schematic representation of such a model is shown in Fig. 1.26, in which the building is divided into three zones, with the mass of each zone located at the center. The masses are concentrated in diaphragms representing the floor system and are interconnected by flexible

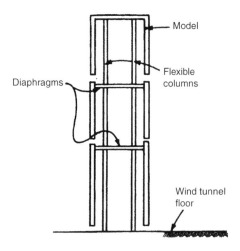

Figure 1.26.   Flexible aeroelastic model (schematic cross section).

columns. A lightweight shell simulating the building shape encloses the assembly of the floor system, masses, and columns. The shell is cut out at these zones to allow for relative movements between the masses. Similarity among elastic properties of the prototype and the model is achieved to varying degrees depending upon the predominant characteristics of the building. For example, a building in which girder rotations and column axial deformations are negligible can be duplicated by using rigid diaphragms and flexible columns. Diaphragm flexibilities can, however, be simulated at considerable fabrication effort and cost.

1.5.2.4. Prediction of Building Acceleration and Human Comfort

One of the basic reasons for undertaking aeroelastic study is to evaluate the effect of building motions on the comfort of its occupants. It is generally known that quantitative prediction of human discomfort is difficult if not impossible to define in absolute terms because perception of motion and associated discomfort are subjective by their very nature. However, in practice certain thresholds of comfort have been established by relating acceleration due to building motion at top floors to the frequency of windstorms. One such criterion is to limit accelerations of top floors to 20-milli-g (2% of acceleration due to gravity).

In wind-tunnel tests, accelerations are measured directly by accelerometers. Two are typically used to measure components in the x and y directions, while a third records the torsional component. Peak acceleration is evaluated from the expression

$$a = G_p \sqrt{a_x^2 + a_y^2 + a_z^2} \tag{1.61}$$

where

a = peak acceleration
G_p = a peak factor for acceleration, usually in the range of 3.0 to 3.5
a_x and a_y = accelerations due to the sway components in the x and y directions
a_z = acceleration due to torsional component

The peak accelerations measured for a series of wind directions and speeds are combined with the meteorological data to predict frequency of occurrence of human discomfort, for various levels of accelerations. A commonly accepted criterion is that for human comfort, the maximum acceleration in upper floors should not exceed 2.0% of gravitational acceleration for a 10-year return period storm.

1.5.3. High-Frequency Base Force Balance Model

The effect of wind load on a flexible building can be considered as an integrated action resulting from three distinct sources. First is the mean wind load, that bends and twists a building, which returns to its normal undeflected position upon load removal. Second is the fluctuating load from the unsteady nature of the wind that results in oscillation of the building about a steady deflected shape. The third contribution comes from the inertia forces similar to the lateral forces induced in a building during earthquakes. However, for design purposes, the inertial effects can be considered as an additional equivalent wind load.

A rigid model is convenient for measuring local wind pressures consisting of positive and negative pressures distributed uniquely around a building. These local pressures are integrated to derive net lateral forces in two perpendicular directions and a torsional moment about a vertical axis, at each level. The cumulative shear, and the overturning and torsional moments at each floor are obtained from simple statics, as are the base shear and overturning moments. These values derived from the mean measurements would have been sufficient for the design of buildings bracing system, except for the drawback that they ignore the influence of gust factor. Therefore, when using rigid-model pressure studies, it is necessary to assume a conservative gust factor to increase the mean values. An alternative and better approach is to take the guesswork out of gust factor by experimentally determining it.

An aeroelastic model provides one such procedure. It furnishes comprehensive information on dynamic loads and motions because the essential structural features such as flexibility, mass, and damping of the prototype are simulated in the model. However, an aeroelastic model is quite complex to design and build, and takes 10 to 12 weeks to complete the required tests.

A high-frequency base force balance model provides an alternative, more economical, and time-efficient method of furnishing the same design information as provided by aeroelastic models.

Two basic types of force balance models are in vogue. In the first type, the outer shell of the model is connected to a flexible metal cantilever bar. Accelerometers and strain gauges are fitted into the model, and the aerodynamic forces are derived from the acceleration and strain measurements. In the second type, a simple foam model of the building is mounted on a five-component, high-sensitivity force balance that measures bending moments and shear forces in two orthogonal directions and torsion about a vertical axis. In both of these force balance models, the resulting overall fluctuating loads are determined, and by making certain simplifying assumptions, the information of interest to the structural engineer—floor-by-floor lateral loads—and the expected acceleration at top floors is deduced. A brief description of the concept behind each of the two is given in the following.

1.5.3.1.   *Flexible Support Model*

It consists of a lightweight rigid model mounted on a high-frequency-response force balance. Design lateral loads and expected building motions are computed from the results acquired from the tests. The method is suitable when building motion does not, itself, affect the aerodynamic forces, and when torsional effects are not of prime concern. In practice, this method is applicable to many tall buildings.

The high-frequency force balance model is typically constructed to a scale on the order of 1:500. Shown in Figs. 1.27 and 1.28 is a model mounted on a rectangular steel bar. The model itself is constructed of a lightweight material such as balsa wood and is mounted on top of a torsion spring through a relatively rigid plate. Strain gauges attached to the bar measure the instantaneous overturning and torsional moments at the base.

From the measured bending and twisting moments and known frequency and mass distribution of the prototype, wind forces at each floor and the expected peak acceleration are derived.

1.5.3.2.   *Five-Component High-Frequency Base: Force Balance Model*

In this model, a prototype building is represented as a rigid model. Made of lightweight material such as polystyrene foam, the model is attached to a measuring device consisting of a set of five highly sensitive load cells attached to a three-legged miniature frame and an interconnecting rigid beam. A typical configuration is shown in Fig. 1.29, in which the load cells are schematically represented as extension springs. Horizontal forces acting in the x direction produce extension of the vertical spring at 1, that can be related to the base overturning moment M_y, with the known extension of the spring and the pivotal distance P_x. Similarly, the base-overturning moment M_y can be calculated from a knowledge of extension of the spring at 2 and the pivotal distance P_y. The horizontal spring at 3 measures the shear force in the x direction, while those at 4 and 5 measure the shear force in the y

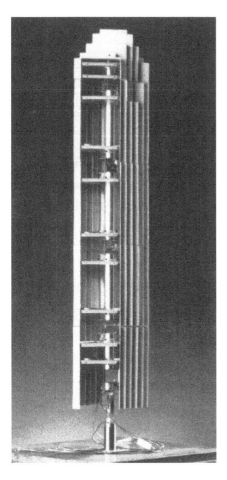

Figure 1.27. High-frequency force balance model of a medium-rise building. (Photo courtesy of Dr. Peter Irwin, Rowan, Williams, Davis & Irwin, Inc.)

direction. The difference in the measurements of springs at 4 and 5 serves to compute the torsional moment at the base about the z-axis. However, the results obtained for torsion are an approximation of the true response because the model does not account for the relative twist present in the prototype.

1.5.4. Pedestrian Wind Studies

A sheet of air moving over the earth's surface is reluctant to rise when it meets an obstacle such as a tall building. If the topography is suitable, it prefers to flow around the building rather than over it. Some examples are shown in Fig.1.30. There are good physical reasons for this tendency, the predominant one being that wind, if it has to pass an obstacle, will find the path of least resistance, i.e., a path that requires minimum expenditure of energy. As a rule, it requires less energy for wind to flow around an obstacle at the same level than for it to rise. Also, if wind has to go up or down, additional energy is required to compress the column of air above or below it. Generally, wind will try to seek a gap at the same level. However, during high winds when the air stream is blocked by the broadside of a tall, flat building, its tendency is to drift in a vertical direction rather than to go around the building at the same level; the circuitous path around the building would require expenditure of more energy. Thus, wind is driven in two directions. Some of it will be

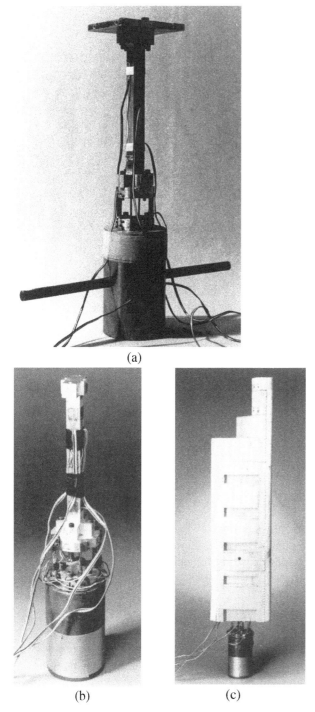

(a)

(b)   (c)

Figure 1.28.   High-frequency force balance model: (a, b) Close-up views of base; (c) high-rise model atop the base. (Photo courtesy of Dr. Peter Irwin, Rowan, Williams, Davis & Irwin, Inc.)

deflected upward, but most of it will spiral to the ground, creating a so-called standing vortex or mini tornado at sidewalk level.

Buildings and their smooth walls are not the only victims of wind buffeting. Pedestrians who walk past tall, smooth-skinned skyscrapers may be subjected to what is called the Mary Poppins syndrome, referring to the tendency of the wind to lift the pedestrian

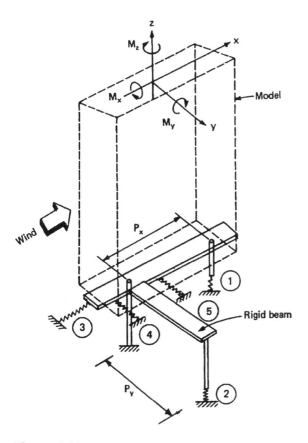

Figure 1.29. Schematic representation of five-component force balance model.

literally off his or her feet. Another effect, known as the Marilyn Monroe effect, refers to the billowing action of women's skirts in the turbulence of wind around and in the vicinity of a building. The point is that during windy days, even a simple activity such as crossing a plaza or taking an afternoon stroll becomes an extremely unpleasant experience to pedestrians, especially during winter months around buildings in cold climates. Walking may become irregular, and the only way to keep walking in the direction of the wind is to bend the upper body windward (see Fig.1.31).

Although one can get some idea of wind flow patterns from the preceding examples, analytically it is impossible to estimate pedestrian-level wind conditions in the outdoor areas of building complexes. This is because there are innumerable variations in building location, orientation, shape, and topography, making it impossible to formulate an analytical solution. Based on actual field experience and results of wind-tunnel studies, it is, however, possible to qualitatively recognize situations that adversely affect pedestrian comfort within a building complex.

Model studies can provide reliable estimates of pedestrian-level wind conditions based on considerations of both safety and comfort. From pedestrian-level wind speed measurements taken at specific locations of the model, acceptance criteria can be established in terms of how often wind speed occurrence is permitted to occur for various levels of activity. The criterion is given for both summer and winter seasons, with the acceptance criteria being more severe during the winter months. For example, the occurrence once a week of a mean speed of 15 mph (6.7 m/s) is considered acceptable for walking during the summer, whereas only 10 mph (4.47 m/s) is considered acceptable during winter months.

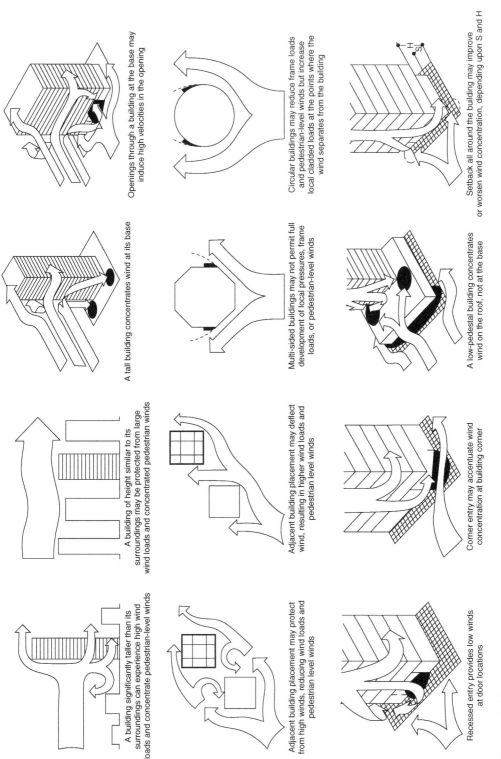

Figure 1.30. Near wind climate.

Figure 1.31.   Pedestrian discomfort. A simple task such as crossing a plaza may become extremely unpleasant during winter months in cold climates. (Photo courtesy of Dr. Peter Irwin, Rowan, Williams, Davis & Irwin Inc.)

1.5.5.   Motion Perception: Human Response to Building Motions

Every building or other structure must satisfy a strength criterion, in which each member is sized to carry its design load without buckling, yielding, or fracture. It should also satisfy the intended function (serviceability) without excessive deflection and vibration. While strength requirements are traditionally specified, serviceability limit states are generally not included in building codes. The reasons for not codifying the serviceability requirements are several: Failure to meet serviceability limits is generally noncatastrophic, is a matter of judgment as to the requirements' application, and entails the perceptions and expectations of the user or owner, and because the benefits themselves are often subjective and difficult to quantify. However, the fact that serviceability limits are not codified should not diminish their importance. A building that is designed for code loads may nonetheless be too flexible for its occupants, due to lack of deflection criteria. Excessive building drifts can cause safety-related frame stability problems because of large $P\Delta$ effects. It can also cause portions of building cladding to fall, potentially injuring pedestrians below.

Perception of building motion under the action of wind is a serviceability issue. In locations where buildings are close together, the relative motion of an adjacent building may make occupants of the other buildings more sensitive to an otherwise imperceptible motion. Human response to building motions is a complex phenomenon encompassing many physiological and psychological factors. Some people are more sensitive than others to building motions. Although building motion can be described by various physical quantities, including maximum values of velocity, acceleration, and rate of change of acceleration—sometimes called jerk—it is generally agreed that acceleration, especially when associated with torsional rotations, is the best standard for evaluation of motion perception in tall buildings. A commonly used criterion is to limit the acceleration of a building's upper floors to no more than 2.0% of gravity (20 mg) for a 10-year return period. The building motions associated with this acceleration are believed to not seriously affect the comfort and equanimity of the building's occupants.

The Bellagio Hotel & Casino
Las Vegas, Nevada

Martin & Peltyn, Inc.
Structural Engineers
Las Vegas, NV

Photo Credit:
Martin & Peltyn / Greg L. Clapp, PE

2
Seismic Design

Structural design of buildings for seismic loading is primarily concerned with structural safety during major earthquakes, but serviceability and the potential for economic loss are also of concern. Seismic loading requires an understanding of the structural behavior under large inelastic, cyclic deformations. Behavior under this loading is fundamentally different from wind or gravity loading, requiring much more detailed analysis, and application of a number of stringent detailing requirements to assure acceptable seismic performance beyond the elastic range. Some structural damage can be expected when the building experiences design ground motions because almost all building codes allow inelastic energy dissipation in structural systems.

The seismic analysis and design of buildings has traditionally focused on reducing the risk of loss of life in the largest expected earthquake. Building codes have based their provisions on the historic performance of buildings and their deficiencies and have developed provisions around life safety concerns, i.e., to prevent collapse under the most intense earthquake expected at a site during the life of a structure. These provisions are based on the concept that the successful performance of buildings in areas of high seismicity depends on a combination of strength, ductility manifested in the details of construction, and the presence of a fully interconnected, balanced, and complete lateral-force-resisting system. In regions of low seismicity, the need for ductility reduces substantially. In fact, in some instances, strength may even substitute for a lack of ductility. Very brittle lateral-force-resisting systems can be excellent performers as long as they are never pushed beyond their elastic strength.

Most seismic codes specify criteria for the design and construction of new structures subjected to earthquake ground motions with three goals: 1) minimize the hazard to life for all structures; 2) increase the expected performance of structures having a substantial public hazard due to occupancy or use; and 3) improve the capability of essential facilities to function after an earthquake.

Some structural damage can be expected as a result of design ground motion because the codes allow inelastic energy dissipation in the structural system. For ground motions in excess of the design levels, the intent of the codes is for structures to have a low likelihood of collapse.

In most structures that are subjected to moderate-to-strong earthquakes, economical earthquake resistance is achieved by allowing yielding to take place in some structural members. It is generally impractical as well as uneconomical to design a structure to respond in the elastic range to maximum expected earthquake-induced inertia forces. Therefore, in seismic design, yielding is permitted in predetermined structural members or locations, with the provision that the vertical load-carrying capacity of the structure is maintained even after strong earthquakes. However, for certain types of structures such as nuclear facilities, yielding cannot be tolerated and as such, the design needs to be elastic.

Structures that contain facilities critical to postearthquake operations—such as hospitals, fire stations, power plants, and communication centers—must not only survive without collapse, but must also remain operational after an earthquake. Therefore, in addition to life safety, damage control is an important design consideration for structures deemed vital to postearthquake functions.

In general, most earthquake code provisions implicity require that structures be able to resist

1. Minor earthquakes without any damage.
2. Moderate earthquakes with negligible structural damage and some nonstructural damage.
3. Major earthquakes with some structural and nonstructural damage but without collapse. The structure is expected to undergo fairly large deformations by yielding in some structural members.

It is important to distinguish between forces due to wind and those induced by earthquakes. Earthquake forces result directly from the distortions induced by the motion of the ground on which the structure rests. The magnitude and distribution of forces and displacements resulting from ground motion is influenced by the properties of the structure and its foundation, as well as the character of the ground motion.

An idea of the behavior of a building during an earthquake may be grasped by considering the simplified response shape shown in Fig. 2.1. As the ground on which the building rests is displaced, the base of the building moves with it. However, the building above the base is reluctant to move with it because the inertia of the building mass resists motion and causes the building to distort. This distortion wave travels along the height of the structure, and with continued shaking of the base, causes the building to undergo a complex series of oscillations.

Although both wind and seismic forces are essentially dynamic, there is a fundamental difference in the manner in which they are induced in a structure. Wind loads, applied as external loads, are characteristically proportional to the exposed surface of a structure, while the earthquake forces are principally internal forces resulting from the distortion produced by the inertial resistance of the structure to earthquake motions.

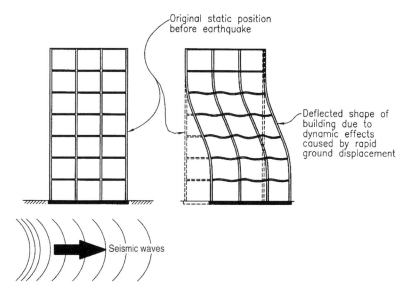

Figure 2.1.   Behavior of a building during earthquakes.

The magnitude of earthquake forces is a function of the mass of the structure rather than its exposed surface. Whereas in wind design, one would feel greater assurance about the safety of a structure made up of heavy sections, in seismic design, this does not necessarily produce a safer design.

2.1. BUILDING BEHAVIOR

The behavior of a building during an earthquake is a vibration problem. The seismic motions of the ground do not damage a building by impact, as does a wrecker's ball, or by externally applied pressure such as wind, but by internally generated inertial forces caused by vibration of the building mass. An increase in mass has two undesirable effects on the earthquake design. First, it results in an increase in the force, and second, it can cause buckling or crushing of columns and walls when the mass pushes down on a member bent or moved out of plumb by the lateral forces. This effect is known as the $p\Delta$ effect and the greater the vertical forces, the greater the movement due to $p\Delta$. It is almost always the vertical load that causes buildings to collapse; in earthquakes, buildings very rarely fall over—they fall down. The distribution of dynamic deformations caused by the ground motions and duration of motion are of concern in seismic design. Although duration of strong motion is an important design issue, it is not presently (2004) explicitly accounted for in design.

In general, tall buildings respond to seismic motion differently than low-rise buildings. The magnitude of inertia forces induced in an earthquake depends on the building mass, ground acceleration, the nature of the foundation, and the dynamic characteristics of the structure (Fig. 2.2). If a building and its foundation were infinitely rigid, it would have the same acceleration as the ground: the inertia force F for a given ground acceleration a may be calculated by Newton's law $F = Ma$, where M is the building mass. For a structure that deforms only slightly, thereby absorbing some energy, the force F tends to be less than the product of mass and ground acceleration. Tall buildings are invariably more flexible than low-rise buildings, and in general, experience much lower accelerations than

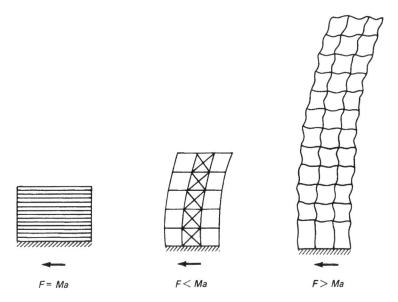

$F = Ma$ $F < Ma$ $F > Ma$

Figure 2.2. Schematic representation of seismic forces.

low-rise buildings. But a flexible building subjected to ground motions for a prolonged period may experience much larger forces if its natural period is near that of the ground waves. Thus, the magnitude of lateral force is not a function of the acceleration of the ground alone, but is influenced to a great extent by the type of response of the structure itself and its foundation as well. This interrelationship of building behavior and seismic ground motion also depends on the building period as formulated in the so-called response spectrum, explained later in this chapter.

Consider, for example, the behavior of a 30-story building during an earthquake. Although the motion of the ground is erratic and three-dimensional, the horizontal components in two mutually perpendicular directions are of importance. The fundamental period T_1 of a tall building is a function of its stiffness, mass, and damping characteristics, and can vary over a broad range anywhere from 0.05 to 0.30 times the number of stories, depending upon the materials used in the construction and the structural system employed. As a preliminary approximation for steel-framed buildings, the period T_1 is approximately equal to $0.15N$, where N is the number of stories. A typical 30-story building would have a fundamental period of 4.5 sec, with the periods of the next two higher modes, T_2 and T_3, approximately equal to one-third and one-fifth of T_1.

The second and third modes of vibration for the 30-story building are thus approximately equal to 1.5 and 0.9 sec. During the first few seconds of earthquake, the acceleration of the ground reaches a peak and is associated with relatively short-period components of the range 0 to 0.5 sec, which have little influence on the fundamental response of the building. On the other hand, the long-period components that occur at the tail end of earthquakes, with periods closer to the fundamental period of the building, have a profound influence on its behavior.

The intensity of ground motion reduces with the distance from the epicenter of the earthquake. The reduction, called attenuation, occurs at a faster rate for higher-frequency (short-period) components than for lower-frequency (long-period) components. The cause of the change in attenuation rate is not understood, but its existence is certain. This is a significant factor in the design of tall buildings, because a tall building, although situated farther from a causative fault than a low-rise building, may experience greater seismic loads because long-period components are not attenuated as fast as the short-period components. Therefore, the area influenced by ground shaking potentially damaging to, say, a 50-story building is much greater than for a 1-story building.

2.1.1. Influence of Soil

The seismic motion that reaches a structure on the surface of the earth is influenced by local soil conditions. The subsurface soil layers underlying the building foundation may amplify the response of the building to earthquake motions originating in the bedrock. It is possible that a number of underlying soil layers can have a period similar to the period of vibration of the structure. Low- to mid-rise buildings typically have periods in the 0.10 to 1.0 sec range, whereas taller, more flexible buildings have periods between 1 and 5 sec or greater. Harder soils and bedrock will efficiently transmit short-period vibrations (caused by near field earthquakes) while filtering out longer-period vibrations (caused by distant earthquakes), whereas softer soils will transmit longer-period vibrations.

As a building vibrates due to ground motion, its acceleration will be amplified if the fundamental period of the building coincides with the period of vibrations being transmitted through the soil. This amplified response is called resonance. Natural periods of soil are in the range of 0.5 to 1.0 sec. Therefore, it is entirely possible for the building and ground to have the same fundamental period. This was the case for many 5- to 10-story buildings in the September, 1985 earthquake in Mexico City. An obvious design strategy is to ensure that buildings have a natural period different from that of the expected ground vibration to prevent amplification.

2.1.2. Damping

Buildings do not resonate with the purity of a tuning fork because they are damped; the extent of damping depends upon the construction materials, type of connections, and the influence of nonstructural elements on the stiffness characteristics of the building. Damping is measured as a percentage of critical damping.

In a dynamic system, critical damping is defined as the minimum amount of damping necessary to prevent oscillation altogether. To visualize critical damping, imagine a tensioned string immersed in water. When the string is plucked, it oscillates about its rest position several times before stopping. If we replace water with a liquid of higher viscosity, the string will oscillate, but certainly not as many times as it did in water. By progressively increasing the viscosity of the liquid, it is easy to visualize that a state can be reached where the string, once plucked, will return to its neutral position without ever crossing it. The minimum viscosity of the liquid that prevents the vibration of the string altogether can be considered equivalent to the critical damping.

The damping of structures is influenced by a number of external and internal sources. Chief among them are

1. External viscous damping caused by air surrounding the building. Since the viscosity of air is low, this effect is negligible in comparison to other types of damping.
2. Internal viscous damping associated with the material viscosity. This is proportional to velocity and increases in proportion to the natural frequency of the structure.
3. Friction damping, also called Coulomb damping, occurring at connections and support points of the structure. It is a constant, irrespective of the velocity or amount of displacement.
4. Hysteretic damping which contributes to a major portion of the energy absorbed in ductile structures.

It is a common practice to lump different sources of damping into a single viscous type of damping. For nonbase-isolated buildings, analyzed for code-prescribed loads, the damping ratios used in practice vary anywhere from 1 to 10% of critical. The low-end values are for wind, while those for the upper end are for seismic design.

The damping ratio used in the analysis of seismic base-isolated buildings is rather large compared to values used for nonisolated buildings, and varies from about 0.20 to 0.35 (20 to 35% of critical damping).

Base isolation, discussed in Chapter 8, consists of mounting a building on an isolation system to prevent horizontal seismic ground motions from entering the building. This strategy results in significant reductions in interstory drifts and floor accelerations, thereby protecting the building and its contents from earthquake damage.

A level of ground acceleration on the order of $0.1g$, where g is the acceleration due to gravity, is often sufficient to produce some damage to weak construction. An acceleration of $1.0g$, or 100% of gravity, is analytically equivalent, in the static sense, to a building that cantilevers horizontally from a vertical surface (Fig. 2.3).

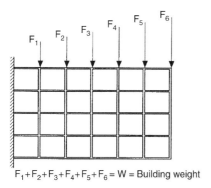

$F_1 + F_2 + F_3 + F_4 + F_5 + F_6 = W$ = Building weight

Figure 2.3.   Concept of 100% *g*. A building subjected to an acceleration of 100% *g* conceptually behaves as if it cantilevers horizontally from a vertical surface.

2.1.3.   Building Motions and Deflections

Earthquake-induced motions, even when they are more violent than those induced by wind, evoke a totally different human response—first, because earthquakes occur much less frequently than windstorms, and second, because the duration of motion caused by an earthquake is generally short. People who experience earthquakes are grateful that they have survived the trauma and are less inclined to be critical of the building motion. Earthquake-induced motions are, therefore, a safety rather than a human discomfort phenomenon.

Lateral deflections that occur during earthquakes should be limited to prevent distress in structural members and architectural components. Nonload-bearing in-fills, external wall panels, and window glazing should be designed with sufficient clearance or with flexible supports to accommodate the anticipated movements.

2.1.4.   Building Drift

Drift is generally defined as the lateral displacement of one floor relative to the floor below. Drift control is necessary to limit damage to interior partitions, elevator and stair enclosures, glass, and cladding systems. Stress or strength limitations in ductile materials do not always provide adequate drift control, especially for tall buildings with relatively flexible moment-resisting frames or narrow shear walls.

Total building drift is the absolute displacement of any point relative to the base. Adjoining buildings or adjoining sections of the same building may not have identical modes of response, and therefore may have a tendency to pound against one another. Building separations or joints must be provided to permit adjoining buildings to respond independently to earthquake ground motion.

2.2.   SEISMIC DESIGN CONCEPT

An effective seismic design generally includes

1.   Selecting an overall structural concept including layout of a lateral-force-resisting system that is appropriate to the anticipated level of ground shaking. This includes providing a redundant and continuous load path to ensure that a building responds as a unit when subjected to ground motion.

2. Determining code-prescribed forces and deformations generated by the ground motion, and distributing the forces vertically to the lateral-force-resisting system. The structural system, configuration, and site characteristics are all considered when determining these forces.
3. Analysis of the building for the combined effects of gravity and seismic loads to verify that adequate vertical and lateral strength and stiffness are achieved to satisfy the structural performance and acceptable deformation levels prescribed in the governing building code.
4. Providing details to assure that the structure has sufficient inelastic deformability to undergo fairly large deformations when subjected to a major earthquake. Appropriately detailed members possess the necessary characteristics to dissipate energy by inelastic deformations.

2.2.1. Structural Response

If the base of a structure is suddenly moved, as in a seismic event, the upper part of the structure will not respond instantaneously, but will lag because of the inertial resistance and flexibility of the structure. The resulting stresses and distortions in the building are the same as if the base of the structure were to remain stationary while time-varying horizontal forces are applied to the upper part of the building. These forces, called inertia forces, are equal to the product of the mass of the structure times acceleration, i.e., $F = ma$ (the mass m is equal to weight divided by the acceleration of gravity, i.e., $m = w/g$). Because earthquake ground motion is three-dimensional (one vertical and two horizontal), the structure, in general, deforms in a three-dimensional manner. Generally, the inertia forces generated by the horizontal components of ground motion require greater consideration for seismic design since adequate resistance to vertical seismic loads is usually provided by the member capacities required for gravity load design. In the equivalent static procedure, the inertia forces are represented by equivalent static forces.

2.2.2. Load Path

Buildings are generally composed of vertical and horizontal structural elements. The vertical elements commonly used to transfer lateral forces to the ground are: 1) shear walls; 2) braced frames; and 3) moment-resisting frames. The horizontal elements that distribute lateral forces to the vertical elements are: 1) diaphragms, such as floor and roof slabs; and 2) horizontal bracing that transfers large shears from discontinuous walls or braces. The seismic forces that are proportional to the mass of the building elements are considered to act at their centers of mass. All of the inertia forces originating from the masses on and off the structure must be transmitted to the lateral force-resisting elements, and then to the base of the structure and into the ground.

A complete load path is a basic requirement for all buildings. There must be a complete lateral-force-resisting system that forms a continuous load path between the foundation, all diaphragm levels, and all portions of the building for proper seismic performance. The general load path is as follows. Seismic forces originating throughout the building, mostly in the heavier mass elements such as diaphragms, are delivered through connections to horizontal diaphragms; the diaphragms distribute these forces to vertical force-resisting elements such as shear walls and frames; the vertical elements transfer the forces into the foundation; and the foundation transfers the forces into the supporting soil.

If there is a discontinuity in the load path, the building is unable to resist seismic forces regardless of the strength of the elements. Interconnecting the elements needed to complete the load path is necessary to achieve good seismic performance. Examples of gaps in the load path would include a shear wall that does not extend to the foundation, a missing shear transfer connection between a diaphragm and vertical elements, a discontinuous chord at a diaphragm's notch, or a reentrant corner, or a missing collector.

A good way to remember this important design strategy is to ask yourself the question, "How does the inertia load get from here (meaning the point at which it is generated) to there (meaning the shear base of the structure, typically the foundations)?"

2.2.3. Demands of Earthquake Motions

Seismic loads result directly from the distortions induced in the structure by the motion of the ground on which it rests. Base motion is characterized by displacements, velocities, and accelerations that are erratic in direction, magnitude, duration, and sequence. Earthquake loads are inertia forces related to the mass, stiffness, and energy-absorbing (e.g., damping and ductility) characteristics of the structure. During its life, a building located in a seismically active zone is generally expected to go through many small, some moderate, one or more large, and possibly one very severe earthquake. In general, it is uneconomical or impractical to design buildings to resist the forces resulting from large or severe earthquakes within the elastic range of stress. In severe earthquakes, most buildings are designed to experience yielding in at least some of their members. The energy-absorption capacity of yielding will limit the damage to properly designed and detailed buildings. These can survive earthquake forces substantially greater than the design forces associated with an allowable stress in the elastic range.

2.2.4. Response of Elements Attached to Buildings

Elements attached to the floors of buildings (e.g., mechanical equipment, ornamentation, piping, nonstructural partitions) respond to floor motion in much the same manner as the building responds to ground motion. However, the floor motion may vary substantially from the ground motion. The high-frequency components of the ground motion tend to be filtered out at the higher levels in the building, whereas the components of ground motion that correspond to the natural periods of vibrations of the building tend to be magnified. If the elements are rigid and are rigidly attached to the structure, the forces on the elements will be in the same proportion to the mass as the forces on the structure. But elements that are flexible and have periods of vibration close to any of the predominant modes of the building vibration will experience forces in proportion substantially greater than the forces on the structure.

2.2.5. Adjacent Buildings

Buildings are often built right up to property lines in order to make maximum use of space. Historically, buildings have been built as if the adjacent structures do not exist. As a result, the buildings may pound during an earthquake. Building pounding can alter the dynamic response of both buildings, and impart additional inertial loads to them.

Buildings that are the same height and have matching floors are likely to exhibit similar dynamic behavior. If the buildings pound, floors will impact other floors, so damage usually will be limited to nonstructural components. When floors of adjacent buildings are at different elevations, the floors of one building will impact the columns of the adjacent building, causing structural damage. When buildings are of different heights, the shorter building may act as a buttress for the taller neighbor. The shorter building receives an unexpected load while the taller building suffers from a major discontinuity that alters its dynamic response. Since neither is designed to weather such conditions, there is potential for extensive damage and possible collapse.

One of the basic goals in seismic design is to distribute yielding throughout the structure. Distributed yielding dissipates more energy and helps prevent the premature failure of any one element or group of elements. For example, in moment frames, it is desirable to have strong columns relative to the beams to help distribute the formation of plastic hinges throughout the building and prevent a story collapse mechanism.

2.2.6. Irregular Buildings

The seismic design of regular buildings is based on two concepts. First, the linearly varying lateral force distribution is a reasonable and conservative representation of the actual response distribution due to earthquake ground motions. Second, the cyclic inelastic deformation demands are reasonably uniform in all of the seismic force-resisting elements. However, when a structure has irregularities, these concepts may not be valid, requiring corrective factors and procedures to meet the design objectives.

The impact of irregular parameters in estimating seismic force levels, first introduced into the Uniform Building Code (UBC) in 1973, long remained a matter of engineering judgment. Beginning in 1988, however, some configuration parameters have been quantified to establish the condition of irregularity, and specific analytical treatments have been mandated to address these flaws.

Typical building configuration deficiencies include an irregular geometry, a weakness in a story, a concentration of mass, or a discontinuity in the lateral-force-resisting system. Vertical irregularities are defined in terms of strength, stiffness, geometry, and mass. Although these are evaluated separately, they are related and may occur simultaneously. For example, a building that has a tall first story can be irregular because of a soft story, a weak story, or both, depending on the stiffness and strength of this story relative to those above.

Those who have studied the performance of buildings in earthquakes generally agree that the building's form has a major influence on performance. This is because the shape and proportions of the building have a major effect on the distribution of earthquake forces as they work their way through the building. Geometric configuration, type of structural members, details of connections, and materials of construction all have a profound effect on the structural-dynamic response of a building. When a building has irregular features, such as asymmetry in plan or vertical discontinuity, the assumptions used in developing seismic criteria for buildings with regular features may not apply. Therefore, it is best to avoid creating buildings with irregular features. For example, omitting exterior walls in the first story of a building to permit an open ground floor leaves the columns at the ground level as the only elements available to resist lateral forces, thus causing an abrupt change in rigidity at that level. This condition is undesirable. It is advisable to carry all shear walls down to the foundation. When irregular features are unavoidable, special design considerations are required to account for the unusual dynamic characteristics and the load transfer and stress concentrations that occur at abrupt changes in structural resistance. Examples of plan and elevation irregularities are illustrated in Figs. 2.4 and 2.5.

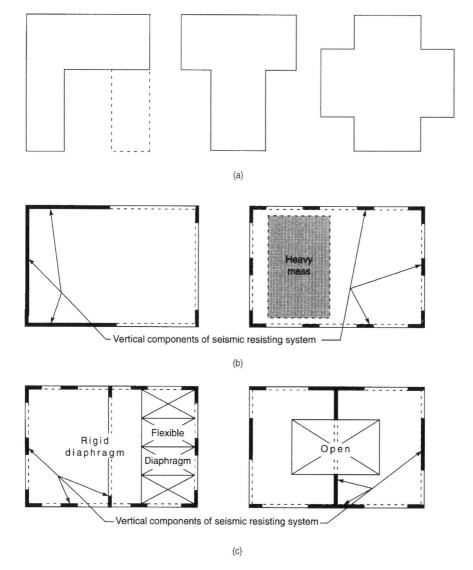

Figure 2.4.   Plan irregularities: (a) geometric irregularities; (b) irregularity due to mass-resistance eccentricity; (c) irregularity due to discontinuity in diaphragm stiffness.

2.2.7. Lateral-Force-Resisting Systems

There are several systems that can be used effectively for providing resistance to seismic lateral forces. Some of the more common systems are shown in Fig. 2.6. All of the systems rely on a complete, three-dimensional space frame; a coordinated system of moment frames, shear walls, or braced frames with horizontal diaphragms; or a combination of the systems.

1. In buildings where a space frame resists the earthquake forces, the columns and beams act in bending. During a large earthquake, story-to-story deflection (story drift) may be accommodated within the structural system without causing failure of columns or beams. However, the drift may be sufficient to

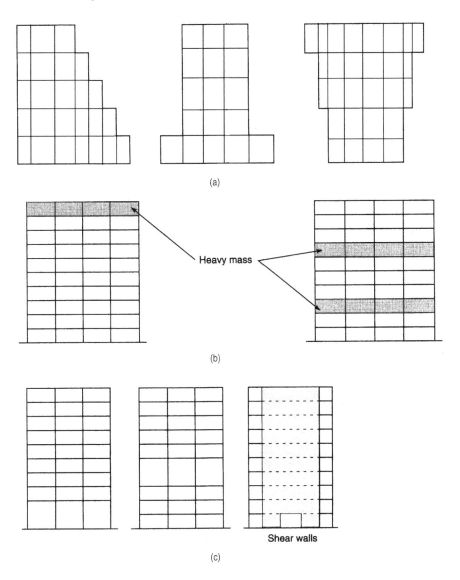

Figure 2.5. Elevation irregularities: (a) abrupt change in geometry; (b) large difference in floor masses; (c) large difference in story stiffnesses.

damage elements that are rigidly tied to the structural system such as brittle partitions, stairways, plumbing, exterior walls, and other elements that extend between floors. Therefore, buildings can have substantial interior and exterior nonstructural damage and still be structurally safe. Although there are excellent theoretical and economic reasons for resisting seismic forces by frame action, for particular buildings, this system may be a poor economic risk unless special damage-control measures are taken.

2. A shear wall (or braced frame) building is normally more rigid than a framed structure. With low design stress limits in shear walls, deflection due to shear forces is relatively small. Shear wall construction is an economical method of bracing buildings to limit damage, and this type of construction is normally economically feasible up to about 15 stories. Notable exceptions to the

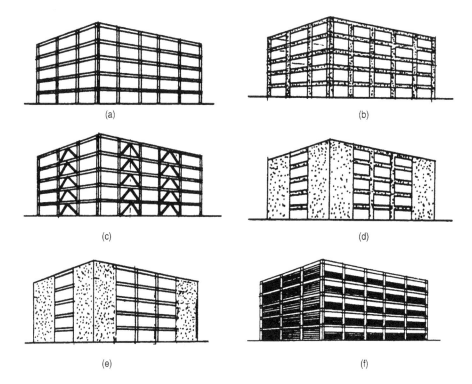

Figure 2.6. Lateral-force-resisting systems: (a) steel moment-resisting frame; (b) reinforced concrete moment-resisting frame; (c) braced steel frame; (d) reinforced concrete shear walls; (e) steel frame building with cast-in-place concrete shear walls; (f) steel frame building with in-filled walls of nonreinforced masonry.

   excellent performance of shear walls occur when the height-to-width ratio becomes great enough to make overturning a problem and when there are excessive openings in the shear walls. Also, if the soil beneath its footings is relatively soft, the entire shear wall may rotate, causing localized damage around the wall.

3. The structural systems just mentioned may be used singly or in combination with each other. When frames and shear walls interact, the system is called a dual system if the frame alone can resist 25% of the lateral load. Otherwise, it is referred to as a combined system. The type of structural system and the details related to the ductility and energy-absorbing capacity of its components will establish the minimum R-value, a seismic coefficient defined later, used for calculating the total base shear.

The design engineer must be aware that a building does not merely consist of a summation of parts such as walls, columns, trusses, and similar components, but is a completely integrated system or unit that has its own properties with respect to lateral force response. The designer must follow the flow of forces through the structure into the ground and make sure that every connection along the path of stress is adequate to maintain the integrity of the system. It is necessary to visualize the response of the complete structure and to keep in mind that the real forces involved are not static but dynamic, are usually erratic and repetitive, and can cause deformations well beyond those determined from the elastic design.

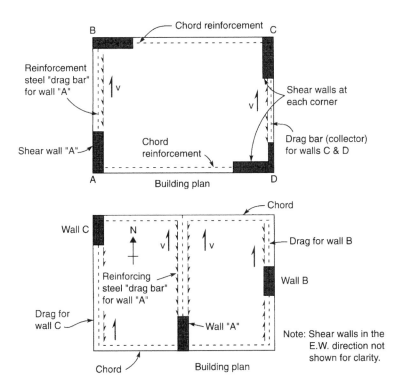

Figure 2.7. Diaphragm drag and chord reinforcement for north–south seismic loads.

2.2.8. Diaphragms

Earthquake loads at any level of a building will be distributed to the vertical structural elements through the floor and roof diaphragms. The roof/floor deck or slab responds to loads like a deep beam. The deck or slab is the web of the beam carrying the shear, and the perimeter spandrel or wall is the flange of the beam resisting bending.

Three factors are important in diaphragm design:

1. The diaphragm must be adequate to resist both the bending and shear stresses and be tied together to act as one unit.
2. The collectors and drag members (see Fig. 2.7) must be adequate to transfer loads from the diaphragm into the lateral-load-resisting vertical elements.
3. Openings or reentrant corners in the diaphragm must be properly placed and adequately reinforced.

Inappropriate location or large-size openings (stair or elevator cores, atriums, skylights) create problems similar to those related to cutting a hole in the web of a beam. This reduces the ability of the diaphragm to transfer the forces and may cause failure (Fig. 2.8).

2.2.9. Ductility

Ductility is the capacity of building materials, systems, or structures to absorb energy by deforming into the inelastic range. The capability of a structure to absorb energy, with acceptable deformations and without failure, is a very desirable characteristic in any

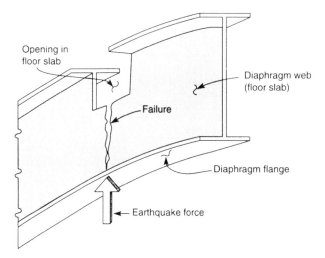

Figure 2.8.   Diaphragm web failure due to large opening.

earthquake-resistant design. Concrete, a brittle material, must be properly reinforced with steel to provide the ductility necessary to resist seismic forces. In concrete columns, for example, the combined effects of flexure (due to frame action) and compression (due to the action of the overturning moment of the structure as a whole) produce a common mode of failure; buckling of the vertical steel and spalling of the concrete cover near the floor levels. Columns must, therefore, be detailed with proper spiral reinforcing or hoops to have greater reserve strength and ductility.

Ductility is measured by the hysteretic behavior of critical components such as a column-beam assembly of a moment frame. It is obtained by cyclic testing of moment-rotation (or force-deflection) behavior of the assembly. The slope of the curves shown in Figs. 2.9a and b represents the stiffness of the structure, and the enclosed areas the dissipated energy. The areas may be full and fat, or lean and pinched. Structural assemblies with curves enclosing a large area representing large dissipated energy are regarded as superior systems for resisting seismic loading.

2.2.10. Damage Control Features

The design of a structure in accordance with seismic provisions will not fully ensure against earthquake damage because the horizontal deformations that can be expected during a major earthquake are several times larger than those calculated under design loads. A list of features that can minimize earthquake damage follows:

1. Provide details that allow structural movement without damage to nonstructural elements. Damage to such items as piping, glass, plaster, veneer, and partitions may constitue a major financial loss. To minimize this type of damage, special care in detailing, either to isolate these elements or to accommodate the movement, is required.
2. Breakage of glass windows can be minimized by providing adequate clearance at edges to allow for frame distortions.
3. Damage to rigid nonstructural partitions can be largely eliminated by providing a detail at the top and sides, which will permit relative movement between the partitions and the adjacent structural elements.

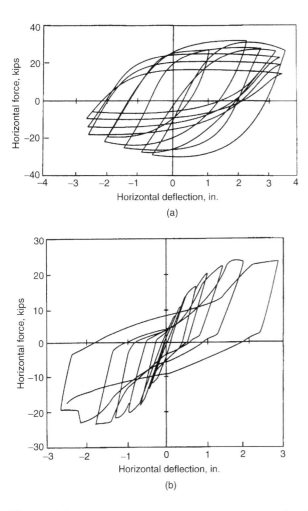

Figure 2.9. Hysteretic behavior: (a) curve representing large energy dissipation, (b) curve representing limited energy dissipation.

4. In piping installations, the expansion loops and flexible joints used to accommodate temperature movement are often adaptable to handling the relative seismic deflections between adjacent equipment items attached to floors.
5. Fasten freestanding shelving to walls to prevent toppling.
6. Concrete stairways often suffer seismic damage due to their inhibition of drift between connected floors. This can be avoided by providing a slip joint at the lower end of each stairway to eliminate the bracing effect of the stairway or by tying stairways to stairway shear walls.

2.2.11. Continuous Load Path

A continuous load path, or preferably more than one path, with adequate strength and stiffness should be provided from the origin of the load to the final lateral-load-resisting elements. The general path for load transfer is in reverse to the direction in which seismic loads are delivered to the structural elements. Thus, the path for load transfer is as follows: Inertia forces generated in an element, such as a segment of exterior curtain wall, are delivered through structural connections to a horizontal diphragm (i.e., floor slab or roof);

the diaphragms distribute these forces to vertical components such as moment frames, braces, and shear walls; and finally, the vertical elements transfer the forces into the foundations. While providing a continuous load path is an obvious requirement, examples of common flaws in load paths are: a missing collector, or a discontinuous chord because of an opening in the floor diaphragm, or a connection that is inadequate to deliver diaphragm shear to a frame or shear wall.

2.2.12. Redundancy

Redundancy is a fundamental characteristic for good performance in earthquakes. It tends to mitigate high demands imposed on the performance of members. It is a good practice to provide a building with a redundant system such that failure of a single connection or component does not adversely affect the lateral stability of the structure. Otherwise, all components must remain operative for the structure to retain its lateral stability.

2.2.13. Configuration

A building with an irregular configuration may be designed to meet all code requirements, but it will not perform as well as a building with a regular configuration. If the building has an odd shape that is not properly considered in the design, good details and construction are of a secondary value.

Two types of structural irregularities are typically defined in most seismic standards: vertical irregularities and plan irregularities. These irregularities result in building responses significantly different from those assumed in the equivalent static force procedure. Although most codes give certain recommendations for assessing the degree of irregularity and corresponding penalties and restrictions, it is important to understand that these recommendations are not an endorsement of their design; rather, the intent is to make the designer aware of the potential detrimental effects of irregularities. Consider, for example, a reentrant corner, resulting from an irregularity characteristic of a building's plan shape. If the configuration of a building has an inside corner, then it is considered to have a reentrant corner. It is the characteristic of buildings with an L, H, T, X, or variations of these shapes (see Fig. 2.10).

Two problems related to seismic performance are created by these shapes: 1) differential vibrations between different wings of the building may result in a local stress concentration at the reentrant corner; and 2) torsion may result because the center of rigidity and center of mass for this configuration do not coincide.

There are two alternative solutions to this problem: Tie the building together at lines of stress concentration and locate seismic-resisting elements at the extremity of the wings to reduce torsion, or separate the building into simple shapes. The width of the separation joint must allow for the estimated inelastic deflections of adjacent wings. The purpose of the separation is to allow adjoining portions of buildings to respond to earthquake ground motions independently without pounding on each other. If it is decided to dispense with the separation joints, collectors at the intersection must be added to transfer forces across the intersection areas. Since the free ends of the wings tend to distort most, it is beneficial to place seismic-resisting members at these locations.

2.2.14. Dynamic Analysis

Symmetrical buildings with uniform mass and stiffness distribution behave in a fairly predictable manner, whereas buildings that are asymmetrical or with areas of discontinuity

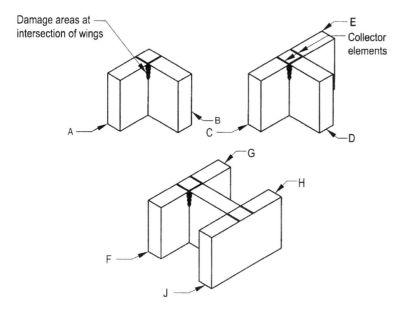

A, B, C, D, E, F, H, and J: end walls

Figure 2.10.   Reentrant corners in L-, T-, and H-shaped buildings. (As a solution, add collector elements and/or stiffen end walls.)

or irregularity do not. For such buildings, dynamic analysis is used to determine significant response characteristics such as: 1) the effects of the structure's dynamic characteristics on the vertical distribution of lateral forces; 2) the increase in dynamic loads due to torsional motions; and 3) the influence of higher modes, resulting in an increase in story shears and deformations.

Static methods specified in building codes are based on single-mode response with simple corrections for including higher mode effects. While appropriate for simple regular structures, the simplified procedures do not take into account the full range of seismic behavior of complex structures. Therefore, dynamic analysis is the preferred method for the design of buildings with unusual or irregular geometry.

Two methods of dynamic analysis are permitted: 1) elastic response spectrum analysis; and 2) elastic or inelastic time-history analysis. The response spectrum analysis is the preferred method because it is easier to use. The time-history procedure is used if it is important to represent inelastic response characteristics or to incorporate time-dependent effects when computing the structure's dynamic response.

Structures that are built into the ground and extended vertically some distance above-ground respond as either simple or complex oscillators when subjected to seismic ground motions. Simple oscillators are represented by single-degree-of-freedom systems (SDOF), and complex oscillators are represented by multidegree-of-freedom (MDOF) systems.

A simple oscillator is represented by a single lump of mass on the upper end of a vertically cantilevered pole or by a mass supported by two columns, as shown in Fig. 2.11.

The idealized system represents two kinds of structures: 1) a single-column structure with a relatively large mass at its top; and 2) a single-story frame with flexible columns and a rigid beam. The mass M is the weight W of the system divided by the acceleration of gravity g, i.e., $M = W/g$.

The stiffness K of the system is the force F divided by the corresponding displacement Δ. If the mass is deflected and then suddenly released, it will vibrate at a certain

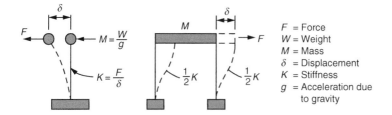

Figure 2.11.   Idealized single-degree-of-freedom system.

frequency, called its natural or fundamental frequency of vibration. The reciprocal of frequency is the period of vibration. It represents the time for the mass to move through one complete cycle. The period T is given by the relation

$$T = 2\pi \sqrt{\frac{M}{K}} \tag{2.1}$$

In an ideal system having no damping, the system would vibrate forever (Fig. 2.12). In a real system, where there is always some damping, the amplitude of motion will gradually decrease for each cycle until the structure comes to a complete stop (Fig. 2.13). The system responds in a similar manner if, instead of displacing the mass at the top, a sudden impulse is applied to the base.

Buildings may be analyzed as multidegree-of-freedom (MDOF) systems by lumping story-masses at intervals along the length of a vertically cantilevered pole. During vibration, each mass will deflect in one direction or another. For higher modes of vibration, some masses may move in opposite directions. Or all masses may simultaneously deflect in the same direction as in the fundamental mode. An idealized MDOF system has a number of modes equal to the number of masses. Each mode has its own natural period of vibration with a unique mode shaped by a line connecting the deflected masses. When ground motion is applied to the base of the multimass system, the deflected shape of the system is a combination of all mode shaped, but modes having periods near predominant periods of the base motion will be excited more than the other modes. Each mode of a multimass system can be represented by an equivalent single-mass system having generalized values M and K for mass and stiffness. The generalized values represent the equivalent combined effects of story masses m_1, m_2, . . . and stiffness k_1, k_2, ... This concept, shown in Fig. 2.14, provides a computational basis for using response spectra based on single-mass systems for analyzing multistoried buildings. Given the period, mode shape, and mass distribution of a multistoried building, we can use the response spectra of a single-degree-of-freedom (SDOF) system for computing the deflected shape, story

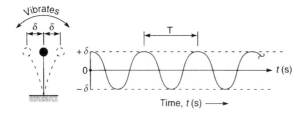

Figure 2.12.   Undamped free vibrations of a single-degree-of-freedom system.

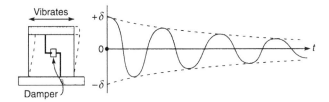

Figure 2.13. Damped free vibration of a single-degree-of-freedom system.

accelerations, forces, and overturning moments. Each predominant mode is analyzed separately and the results are combined statistically to compute the multimode response.

Buildings with symmetrical shape, stiffness, and mass distribution and with vertical continuity and uniformity behave in a fairly predictable manner whereas when buildings are eccentric or have areas of discontinuity or irregularity, the behavioral characteristics are very complex. The predominant response of the building may be skewed from the apparent principal axes of the building. The torsional response as well as the coupling or interaction of the two translational directions of response must be considered. This is similar to the Mohr's circle representation of principal stresses.

Thus, three-dimensional methods of analysis are required as each mode shape is defined in three dimensions by the longitudinal and transverse displacement and the rotation about a vertical axis. Thus, building irregularities complicate not only the method of dynamic analysis, but also the methods used to combine modes.

For a building that is regular and essentially symmetrical, a two-dimensional model is generally sufficient. Note that when the floor plan aspect ratio (length-to-width) of the building is large, torsion response may be predominant, thus requiring a 3-D analysis in an otherwise symmetrical and regular building.

For moderate- to-high-rise buildings, the effects of higher modes may be significant. For a fairly uniform building, the dynamic characteristics can be approximated using the general modal relationship shown in Table 2.1. The fundamental period of vibration may be estimated by using code formulas, and the periods for the second through fifth modes

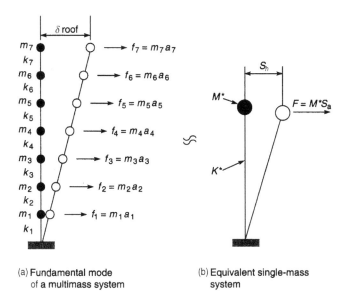

(a) Fundamental mode
of a multimass system

(b) Equivalent single-mass
system

Figure 2.14. Representation of a multimass system by a single-mass system.

TABLE 2.1   General Modal Relationships

Mode	1	2	3	4	5
Ratio of period to 1st mode period	1.000	0.327	0.186	0.121	0.083
Participation factor at roof	1.31	−0.47	0.24	−0.11	0.05
Base shear participation factor	0.828	0.120	0.038	0.010	0.000

may be estimated using the relationship shown in Table 2.1. The table can also be used to estimate modal participation factors at building roof levels and base shear participation factors corresponding to the first five modes.

For most buildings, inelastic response can be expected to occur during a major earthquake, implying that an inelastic analysis is more proper for design. However, in spite of the availability of nonlinear inelastic programs, they are not used in typical design practice because: 1) their proper use requires knowledge of their inner workings and theories; 2) results produced are difficult to interpret and apply to traditional design criteria; and 3) the necessary computations are expensive. Therefore, analyses in practice typically use linear elastic procedures based on the response spectrum method.

2.2.14.1.   *Response Spectrum Method*

The word spectrum in seismic engineering conveys the idea that the response of buildings having a broad range of periods is summarized in a single graph. For a given earthquake motion and a percentage of critical damping, a typical response spectrum gives a plot of earthquake-related responses such as acceleration, velocity, and deflection for a complete range, or spectrum, of building periods.

Thus, a response spectrum (Figs. 2.15 and 2.16) may be visualized as a graphical representation of the dynamic response of a series of progressively longer cantilever

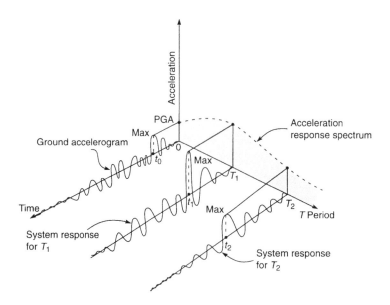

Figure 2.15.   Graphical description of response spectrum.

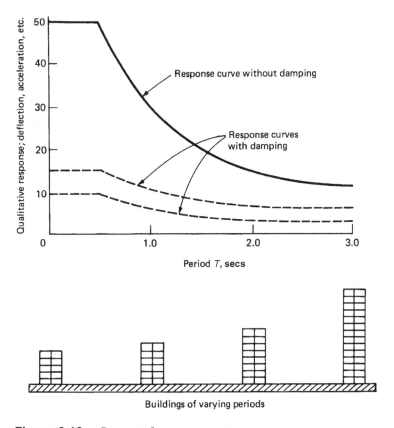

Figure 2.16. Concept of response spectrum.

pendulums with increasing natural periods subjected to a common lateral seismic motion of the base. Imagine that the fixed base of the cantilevers is moved rapidly back and forth in the horizontal direction, its motion corresponding to that occurring in a given earthquake. A plot of maximum dynamic response, such as accelerations versus the periods of the pendulums, gives us an acceleration response spectrum as shown in Fig. 2.15 for the given earthquake motion. In this figure, the absolute value of the peak acceleration response occurring during the excitation for each pendulum is represented by a point on the acceleration spectrum curve. As an example, an acceleration response spectra for the 1940 El Centro earthquake is illustrated in Fig. 2.17. Using ground acceleration as an input, a family of response spectrum curves can be generated for various levels of damping, where higher values of damping result in lower spectral response.

To establish the concept of how a response spectrum is used to evaluate seismic lateral forces, consider two single-degree-of-freedom structures: 1) an elevated water tank supported on columns; and 2) a revolving restaurant supported at the top of a tall concrete core (see Fig. 2.17a). To simplify, we will neglect the mass of the columns supporting the tank, and consider only the mass m_1 of the tank in the dynamic analysis. Similarly, the mass m_2 assigned to the restaurant is the only mass considered in the second structure. Given the simplified models, let us examine how we can calculate the lateral loads for both these structures resulting from an earthquake, for example, one that has the same ground motion characteristics as the 1940 El Centro earthquake shown in Fig. 2.18. To evaluate the seismic lateral loads, we shall use the recorded ground acceleration for the first 30 seconds. Observe that the maximum acceleration recorded is $0.33g$. This occurred about 2 seconds after the start of the record.

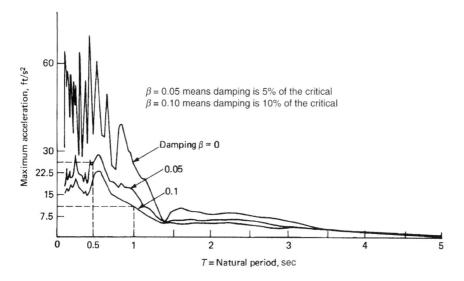

Figure 2.17.   Acceleration spectrum: El Centro earthquake.

As a first step, the base of the two structures is analytically subjected to the same acceleration as the El Centro recorded acceleration. The purpose is to calculate the maximum dynamic response experienced by the two masses during the first 30 seconds of the earthquake. The maximum response such as displacement, velocity, and acceleration response of an SDOF system such as the two examples considered here may be obtained by considering the earthquake effects as a series of impulsive loads, and then integrating the effect of individual impulses over the duration of the earthquake. This procedure, the Duhamel Integration Method, requires considerable analytical effort. However, in seismic design, it is generally not necessary to carry out the integration because the maximum response for many previously recorded and synthetic earthquakes are already established. The spectral acceleration response for the north–south component of the El Centro earthquake, shown in Fig. 2.17, is one such example.

To determine the seismic lateral loads, assume the tank and restaurant structures weigh 720 kips (3202 kN) and 2400 kips (10,675 kN), with corresponding periods of vibration of 0.5 sec and 1 sec, respectively. Since the response of a structure is strongly influenced by damping, it is necessary to estimate the damping factors for the two structures. Let us assume that the percentage of critical damping β for the tank and restaurant are 5 and 10% of the critical damping, respectively. From Fig. 2.17, the acceleration for the tank structure is 26.25 ft/s^2, giving a horizontal force in kips equal to the mass of the tank, times the acceleration. Thus,

$$F = \frac{720}{32.2} \times 26.25 = 587 \text{ kips}$$

The acceleration for the second structure from Fig. 2.17 is 11.25 ft/s^2, and the horizontal force in kips would be equal to the mass at the top times the acceleration.

$$F = \frac{2400}{32.2} \times 11.25 = 838.51 \text{ kips}$$

The two structures can then be designed by applying the seismic loads at the top and determining the associated forces, moments, and deflection. The lateral load, obtained

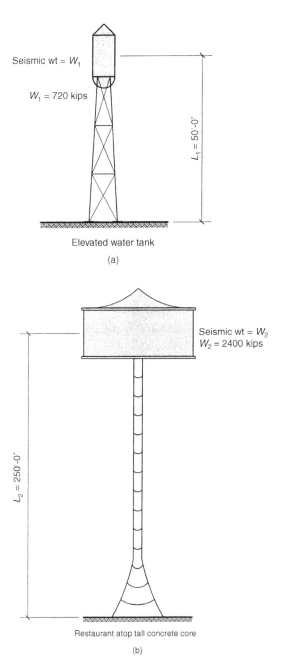

Seismic wt = W_1

W_1 = 720 kips

L_1 = 50'-0"

Elevated water tank

(a)

L_2 = 250'-0"

Seismic wt = W_2
W_2 = 2400 kips

Restaurant atop tall concrete core

(b)

Figure 2.17a. Examples of single-degree-of-freedom systems: (a) elevated water tank (b) restaurant atop tall concrete core. Note from Fig. 2.17, the acceleration = 26.25 ft/s² for $T = 0.5$ s and $\beta = 0.05$ (water tank) and the acceleration = 11.25 ft/s² for $T = 1.00$ and $\beta = 0.10$ (restaurant).

by multiplying the response spectrum acceleration by the effective mass of the system, is referred to as base shear, and its evaluation forms one of the major tasks in earthquake analysis.

In the examples, SDOF structures were chosen to illustrate the concept of spectrum analysis. A multistory building, however, cannot be modeled as an SDOF system because it will have as many modes of vibration as its degrees-of-freedom which are infinite for a real system. However, for practical purposes, the distributed mass of a building may be

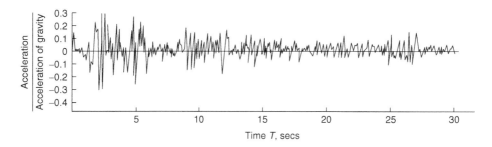

Figure 2.18.   Recorded ground acceleration: El Centro earthquake.

lumped at discrete levels to reduce the degrees-of-freedom to a manageable number. In multistory buildings, the masses are typically lumped at each floor level.

Thus, in the 2-D analysis of a building, the number of modes of vibration corresponds to the number of levels, with each mode having its own characteristic frequency. The actual motion of a building is a linear combination of its natural modes of vibration. During vibration, the masses vibrate in phase with the displacements as measured from their initial positions, always having the same relationship to each other. Therefore, all masses participating in a given mode pass the equilibrium position at the same time and reach their extreme positions at the same instant.

Using certain simplifying assumptions, it can be shown that each mode of vibration behaves as an independent SDOF system with a characteristic frequency. This method, called the modal superposition method, consists of evaluating the total response of a building by statistically combining the response of a finite number of modes of vibration.

A building, in general, vibrates with as many mode shapes and corresponding periods as its degrees-of-freedom. Each mode contributes to the base shear, and for elastic analysis, this contribution can be determined by multiplying a percentage of the total mass, called effective mass, by an acceleration corresponding to that modal period. The acceleration is typically read from the response spectrum modified for a damping associated with the structural system and the assumed return period of the design earthquake. Therefore, the procedure for determining the contribution of the base shear for each mode of an MDOF structure is the same as that for determining the base shear for an SDOF structure, except that an effective mass is used instead of the total mass. The effective mass is a function of the lumped mass and deflection at each floor with the largest value for the fundamental mode, becoming progressively less for higher modes. The mode shape must therefore be known in order to compute the effective mass.

Because the actual deflected shape of a building consists of a linear combination of its modal shapes, higher modes of vibration also contribute, although to a lesser degree, to the structural response. These can be taken into account through use of the concept of a participation factor. Further mathematical explanation of this concept is deferred to a later section, but suffice it to note that the base shear for each mode is determined as the summation of products of effective mass and spectral acceleration at each level. The force at each level for each mode is then obtained by distributing the base shear in proportion to the product of the floor weight and displacement. The design values are then computed using modal combination methods, such as CQC or SRSS.

Types of Response Spectrum.   Three types of response spectra are used in practice.

1. Response spectra from actual earthquake records.
2. Smoothed design response spectra.
3. Site-specific response spectra.

Response Spectra from Actual Earthquake Records.   To develop these response spectra, a series of damped SDOF mass-spring systems is subjected to an actual earthquake ground excitation, and by numerical integration of the maximum values for a range of periods of vibration is determined. However, the resulting spectral curves are quite jagged, being characterized by sharp peaks and troughs. Because the magnitude of these troughs and peaks varies significantly for different earthquakes, several possible earthquake spectra are used in the evaluation of the structural response.

Smooth Response Spectrum.   As an alternative to the use of several earthquake spectra, a smooth spectrum representing an upper-bound response to several ground motions may be generated. The sharp peaks in earthquake records may indicate the resonant behavior of the system when the natural period of the system approaches a period of the forcing function, especially for systems with little or no damping. However, even a moderate amount of damping has a tendency to smooth out the peaks and reduce the spectral response.

Because buildings have some degree of damping, the peaks in the response spectra are of limited significance and therefore are smoothed out, as shown in Fig. 2.19. The other two response spectra for the velocity and displacement, shown in Figs. 2.20 and 2.21, are obtained from the acceleration spectrum, since they are related to one another. The three spectra can be represented in one graph, as shown in Fig. 2.22, in which the horizontal axis denotes the natural period and the ordinate the spectrum velocity, both on a logarithmic scale. The acceleration and displacement are represented on diagonal axes inclined at 45° to the horizontal. The plot, which presents all three spectral parameters, is called a tripartite response spectrum.

Consider, for example, that we wish to calculate the first-mode displacement of a building having a fundamental period $T = 2$ sec, subjected to a given base shear evaluated from the response spectrum given in Fig 2.22. One method is, of course, to perform a stiffness analysis of the building by defining the geometry, material and stiffness properties, and then subjecting it to the lateral loads evaluated from the response spectrum. However, an easier method, without having to go through an analysis, is to simply read off the lateral deflection from the tripartite diagram (Fig. 2.22), as will be shown presently.

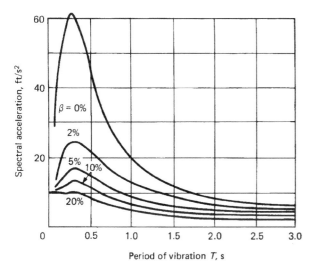

Figure 2.19.   Smoothed acceleration spectra for the El Centro earthquake.

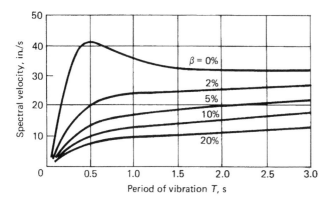

Figure 2.20.   Smoothed velocity spectra for the El Centro earthquake.

The displacement curve is plotted on the tripartite diagram using the relation

$$s_d = \frac{T^2}{4\Pi^2} S_a \qquad\qquad\qquad (2.2)$$

where
   S_d = spectral response displacement
   S_a = spectral response acceleration
   T = structural period

Revisiting the response spectrum plot, Fig. 2.22, we can make the following observations:

- For very stiff structures, the spectral acceleration approaches the maximum ground acceleration. Structures in this period range would behave like rigid bodies attached to the ground.
- For moderately short periods on the order of 0.1 to 0.3 sec, the spectral accelerations are about 2 to $2\frac{1}{2}$ times as large as the maximum ground acceleration.

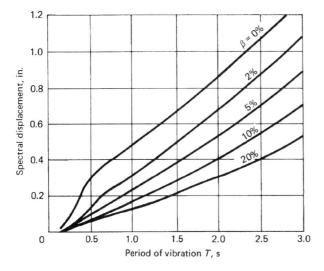

Figure 2.21.   Smoothed displacement spectra for El Centro earthquake.

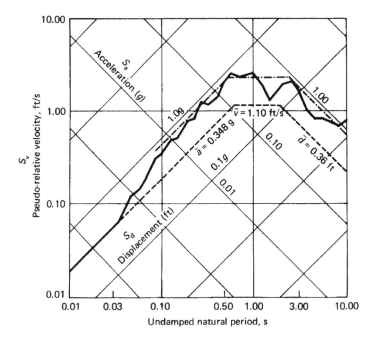

Figure 2.22. Tripartite response spectra for El Centro earthquake (5% damping, north–south component).

- For long-period buildings, the maximum spectral displacements approach the maximum ground displacements.
- For intermediate values of period, the maximum spectral velocity is several times the input velocity.

Thus, in the short-period range, the variation of the spectrum curve tends to show correlation with the line of maximum ground acceleration. In the medium-period range, the correlation is with maximum ground velocity while in the higher-period range, the correlation is with the displacement.

Because of the aforementioned characteristics, it is possible to represent an idealized upper-bound response spectrum by a set of three straight lines, as shown in Fig. 2.22. Also shown in the same figure are the values of ground acceleration ($\bar{a} = 0.348g$), maximum velocity ($\bar{V} = 1.10$ ft), and displacement ($\bar{d} = 0.36$ ft/sec) experienced during the El Centro earthquake.

Site-Specific Response Spectrum. For especially important structures or where local soil conditions are not amenable to simple classification, the use of smooth spectra curves is inadequate. In such cases, site-specific studies are performed to determine more precisely the expected intensity and character of seismic motion. The development of site-specific ground motions is generally the responsibility of geotechnical consultants. However, it is important for the structural engineer to be aware of the procedure used in the generation of site-specific response spectrum. This is considered next.

Procedure for Developing Site-Specific Response Spectra. The seismicity of the region surrounding the site is determined from a search of an earthquake database. A list of active, potentially active, and inactive faults is compiled from the database along with their nearest distance from the site.

The predicted response of the deposits underlying the site and the influence of local soil and geologic conditions during earthquakes are determined based on statistical results of studies of site-dependent spectra developed from actual time-histories recorded by strong motion instruments.

Several postulated design earthquakes are selected for study based on the characteristics of the faults. The peak ground motions generated at the site by the selected earthquakes are estimated from empirical relationships.

The dynamic characteristics of the deposits underlying the site are estimated from the results of a nearby downhole seismic survey, from the logs of borings, static test data, and dynamic test data.

The causative faults are selected from a list as the most significant ones along which earthquakes are expected to generate motions affecting the site.

Several earthquakes with different probabilities of occurrence that may be generated along the causative faults are selected. The maximum capable earthquake (MCE), for example, constitutes the largest earthquake reasonably likely to occur. Since the probability of such an earthquake occurring during the lifetime of the subject development is low, the ground motions associated with the MCE events are estimated to have 10% probability of being exceeded in 250 years.

The slip rates of the faults are estimated from published data. Using the slip rates, the accumulated slip over an approximate 475-year period (corresponding to 10% probability of being exceeded in 50 years) and over an approximate 72-year period (corresponding to 50% probability of being exceeded in 50 years) are determined.

Using a statistical analysis approach, the peak ground motion values (acceleration, velocity, and displacement) anticipated at the site are estimated. By applying structural amplification factors to these values, the spectral bounds for acceleration, velocity, and displacement are obtained for each desired value of structural damping, most often 2, 5, and 10% of critical damping. The ground motion values for a given site thus vary with the magnitude of the earthquake and distance of the site from the source of energy release.

These values provide a basis by which site-dependent response spectra are computed. For each of the six site classes, spectral bounds are obtained by multiplying the ground motion values by damping-dependent amplification factors.

Use of Tripartite Response Spectra.   Site-specific spectra are shown in Fig. 2.23. Tripartite response spectra for four seismic events characterized as earthquakes A, B, C, and D for a downtown Los Angeles site are shown in Fig. 2.24a–d. Response spectrum A is for a maximum capable earthquake of magnitude 8.25 occurring at San Andreas fault at a distance of 34 miles while B is for a magnitude 6.8 earthquake occurring at Santa Monica–Hollywood fault at a distance of 3.7 miles from the site. Response spectra C and D are for earthquakes with a 10 and 50% probability of being exceeded in 50 years.

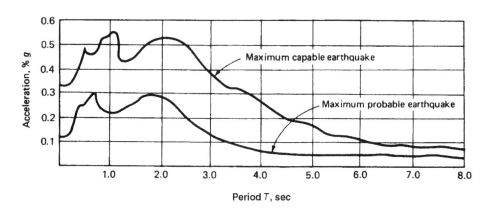

Figure 2.23.   Unique site-specific design spectra.

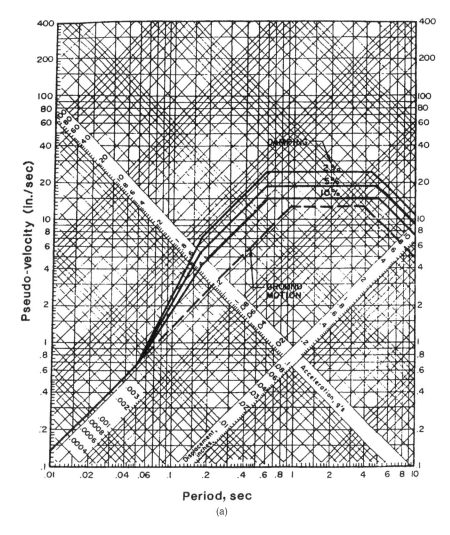

Period, sec

(a)

Figure 2.24. Tripartite site-specific response spectra: (a) earthquake A; (b) earthquake B; (c) earthquake C; (d) earthquake D.

The response spectrum tells us that the forces experienced by buildings during an earthquake are not just a function of the quake, but are also their dynamic response characteristics to the quake. The response primarily depends on the period of the building being studied. A great deal of single-mode information can be read directly from the response spectrum. Referring to Fig. 2.25, the horizontal axis of the response spectrum expresses the period of the building affected by the quake. The vertical axis shows the velocity attained by this building during the quake. The diagonal axis running up toward the left-hand corner reads the maximum accelerations to which the building is subjected. The axis at right angles to this will read the displacement of the building in relation to the support. Superimposed on these tripartite scales are the response curves for an assumed 5% damping of critical. Now let us see how various buildings react during an earthquake described by these curves.

If the building to be studied had a natural period of 1 second, we would start at the bottom of the chart and reference vertically until we intersect the response curve. From this intersection, point A, we travel to the extreme right and read a velocity of 16-in. per second. Following a displacement line diagonally down to the right, we find a displacement

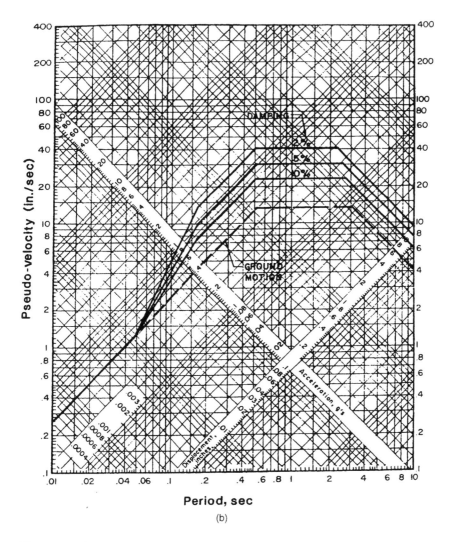

Figure 2.24. (*Continued*)

of 2.5-in. Following an acceleration line down to the left, we see that it will experience an acceleration of 0.25g. If we then move to the 2-second period, point B, in the same sequence, we find that we will have the same maximum velocity of 16-in. per second, a displacement of 4-in., and a maximum acceleration of 0.10g. If we then move to 4 seconds, point C, we see a velocity of 16-in. per second, a displacement of 10-in., and an acceleration of 0.06g. If we run all out to 10 seconds, point D, we find a velocity of 7 in. per second, a displacement of 10-in. the same as for point C, and an acceleration of 0.01g. Notice that the values vary widely, as stated earlier, depending on the period of the building exposed to this particular quake.

2.2.14.2. *Time-History Analysis*

The mode superposition, or spectrum method, outlined in the previous section is a useful technique for the elastic analysis of structures. It is not directly transferable to inelastic analysis because the principle of superposition is no longer applicable. Also, the analysis is subject to uncertainties inherent in the modal superimposition method. The actual process of combining the different modal contributions is, after all, a probabilistic technique and

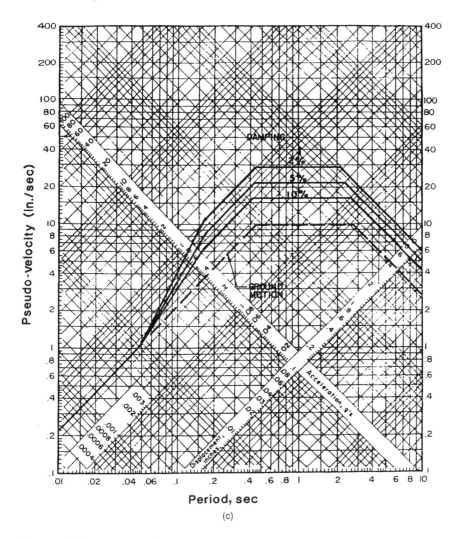

Figure 2.24. (*Continued*)

in certain cases, may not be entirely representative of the actual behavior of the structure. Time-history analysis overcomes these two uncertainties, but it requires a large computational effort. It is not normally employed as an analysis tool in practical design of buildings. The method consists of a step-by-step direct integration in which the time domain is discretized into a number of small increments δt; for each time interval, the equations of motion are solved with the displacements and velocities of the previous step serving as initial functions. The method is applicable to both elastic and inelastic analyses. In elastic analysis, the stiffness characteristics of the structure are assumed to be constant for the duration of the earthquake. In inelastic analysis, however, the stiffness is assumed to be constant through the incremental time δt only. Modifications to structural stiffness caused by cracking, formation of plastic hinges, etc., are incorporated between the incremental solutions. A brief outline of the analysis procedure applicable to both elastic and inelastic analysis is given in the following discussion.

Analysis Procedure. The method consists of applying a specific earthquake motion directly to the base of a computer model of a structure. Instantaneous stresses throughout the structure are calculated at small intervals of time for the duration of the

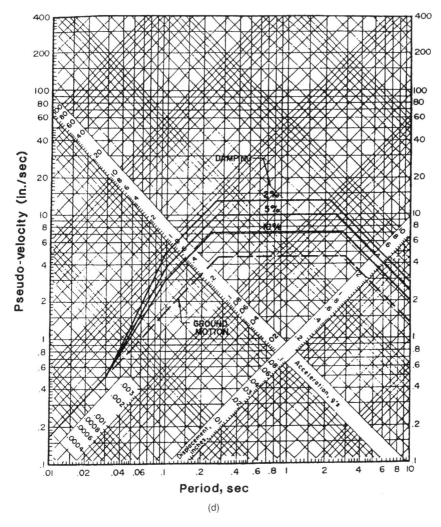

Figure 2.24. (*Continued*)

earthquake or a significant portion of it. The maximum stresses that occur within the entire analysis period are found by scanning the computer results.

The procedure includes the following steps:

1. An earthquake record representing the design earthquake is selected.
2. The record is digitized as a series of small time intervals of about 1/50 to 1/100 of a second.
3. A mathematical model of the building is set up, usually consisting of a lumped mass at each floor.
4. The digitized record is applied to the model as accelerations at the base of the structure.
5. The computer integrates the equations of motions and gives a complete record of the acceleration, velocity, and displacement of each lumped mass at each interval.

The accelerations and relative displacements of the lumped masses are translated into member stresses. The maximum values are found by scanning the output record.

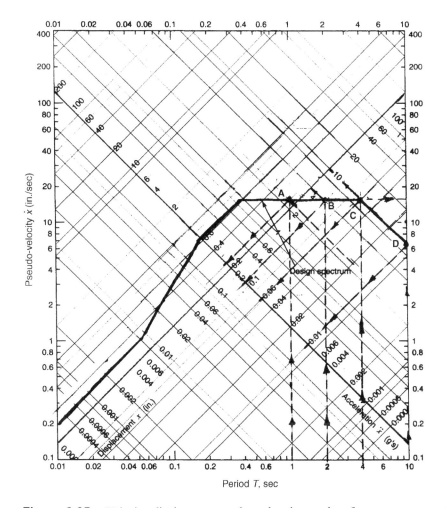

Figure 2.25. Velocity, displacement, and acceleration readout from response spectra.

This procedure automatically includes various modes of vibration by combining their effect as they occur, thus eliminating the uncertainties associated with modal combination methods.

The time-history technique represents one of the most sophisticated methods of analysis used in building design. However, it has the following sources of uncertainty:

1. The design earthquake must still be assumed.
2. If the analysis used unchanging values for stiffness and damping (i.e., linear analysis), it will not reflect the cumulative effects of stiffness variation and progressive damage.
3. There are uncertainties related to the erratic nature of earthquakes. By pure coincidence, the maximum response of the calculated time-history could fall at either a peak or a valley of the digitized spectrum.
4. Small inaccuracies in estimating properties of the structure will have considerable effect on the maximum response.
5. Errors latent in the magnitude of the time step chosen are difficult to assess unless the solution is repeated with several time steps.

2.3. UNIFORM BUILDING CODE, 1997 EDITION: SEISMIC PROVISIONS

The 1997 UBC incorporates a number of important lessons learned from the 1994 Northridge and the 1995 Kobe earthquakes. Strong motion measurements in these events showed that ground motions are significantly greater near the earthquake source. These earthquakes had near-source ground motions that greatly exceeded the effective peak accelerations (EPA) given for zone 4 in the 1994 UBC. It was also observed that amplification of long-period ground motions was greater for less ideal site soil conditions. As a result, the 1997 UBC incorporated two near-source factors N_A and N_V in seismic zone 4 to recognize the amplified ground motions that occur at distances close to the fault. The near-source factors apply one in zone 4 because it is believed that the near-source effects are only significant for large earthquakes.

The following key ideas are contained in the 1997 UBC:

1. Earthquake loads are specified for use with strength or load factor resistance design (LFRD), although allowable stress design (ASD) is also permitted. This is a departure from previous editions where the seismic loads were specified at the working stress level.

2. The structural system coefficient, R, which is a measure of the ductility and overstrength of the structural system, has been adjusted to provide a strength level base shear. It is essentially equal to R_W, the seismic coefficient specified in previous edition, divided by 1.4.

3. Two near-source factors N_a and N_r, new for the 1997 UBC, have been incorporated in seismic zone 4 to amplify ground motions that occur at close distances to the fault.

4. A redundancy-reliability factor, ρ, also new in the 1997 UBC, has been incorporated to promote redundant lateral-force-resisting systems. Nonredundant systems are penalized through higher lateral load requirements, while super-redundant systems are not rewarded with less stringent seismic design requirements.

5. A set of soil profile categories, S_A through S_F, have been incorporated. These are used in combination with seismic zone factor Z, and near-source factors N_a and N_r, to provide the site-dependent ground motion coefficients C_a and C_r.

    The concept of soil factor was first acknowledged by recognizing the importance of local site effects in the 1976 edition of the UBC. At that time, the soil-structure resonance factor, S, was part of the base shear equation. It varied from a minimum of 1.0 to a maximum of 1.5, depending on the ratio T/T_S, where T is the fundamental building period and T_S the characteristic site period.

    In the 1985 edition, instead of the T/T_S ratio, the code defined three soil types, S_1, S_2, and S_3, to designate rock, deep soil, and soft soil with soil factors of 1.0, 1.2, and 1.5, respectively. In response to the 1985 Mexico City earthquake, a fourth soil profile type, S_4, for very deep soft soils was added to the 1988 UBC, with the S_4 factor equal to 2.0.

    The 1997 UBC adapted a new set of six soil types, S_A through S_F. Five of these, S_A through S_E, are considered stable profiles representing hard rock (S_A), rock (S_B), very dense and soft rock (S_C), stiff soil (S_D), and soft soil (S_E). Soil categories are based on the average shear wave velocity in the upper 100 feet or below count of a standard penetration test. Type S_F is a soft soil profile requiring a site-specific evaluation. The default profile is S_D, probably the most common soil profile in most of California.

Unlike the previous editions, the 1997 UBC does not use soil profiles directly in the base shear equations. Instead, S_A, S_B, S_C, S_D, S_E, or S_F are used in combination with the seismic zone factor Z, and the near-source factors N_a and N_V, to determine the site-dependent coefficients C_A and C_V. C_A and C_V define ground motion response within the acceleration and velocity-controlled range of the response spectrum.

6. Substantial revisions have been made to the lateral force requirements for nonbuilding structures, equipment supported by structures, and nonstructural components. These are discussed in section 2.5.

7. A simple procedure is permitted for the calculation of base shears for one- and two-story dwellings, one- to three-story light frame construction, and other one- and two-story buildings.

2.3.1. Building Irregularities

The impact of irregularities in estimating seismic force levels, first introduced into the Uniform Building Code in 1973, long remained a matter of engineering judgment. Beginning in 1988, however, some configuration parameters have been quantified to establish the condition of irregularity, and specific analytical treatments have been mandated to address these conditions.

Typical building configuration deficiencies include an irregular geometry, a weakness in a story, a concentration of mass, or a discontinuity in the lateral-force-resisting system. Although these are evaluated separately, they are related and may occur simultaneously. For example, a tall first story can be a soft story, a weak story, or both, depending on its stiffness and strength relative to those above.

The 1997 UBC quantifies the idea of irregularity by defining geometrically or by use of dimensional ratios the points at which the specific irregularity becomes an issue requiring remedial measures (see Figs. 2.26 through 2.35). It should be noted that not all irregularities

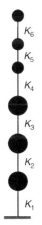

Figure 2.26.   Stiffness irregularity; soft story. Soft story exists if at any given story;

1. The story stiffness is $<70\%$ of that of the story above

$$K_2 < 0.7\, K_3, \text{ or}$$

2. The story stiffness $<80\%$ of the average stiffness of the three stories above

$$K_2 < 0.8\, \frac{(K_3 + K_4 + K_5)}{3}$$

(Use story drifts for stiffness comparison.)

Figure 2.27. Mass irregularity. Mass irregularity exists if mass of any story is more than 150% of the mass of the story below or above. $M_2 > 1.5M_1$ or $1.5M_3$.

require remedial measures. Some, such as stiffness, mass, and geometric irregularities, may be accounted for by performing dynamic analysis. See Tables 2.2 and 2.3.

The irregularities are divided into two broad categories: 1) vertical; and 2) plan irregularities. Vertical irregularities include soft or weak stories, large changes in mass from floor to floor, and large discontinuities in the dimensions or in-plane locations of lateral-load-resisting elements. Buildings with plan irregularities include those that undergo substantial torsion when subjected to seismic loads or have reentrant corners, discontinuities in floor diaphragms, discontinuity in the lateral force path, or lateral-load-resisting elements that are not parallel to each other or to the axes of the building. The definitions of these irregularities are found in Tables 2.2 and 2.3.

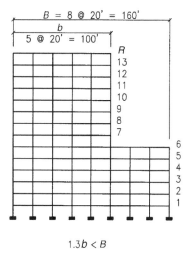

Figure 2.28. Vertical geometric irregularity. Vertical geometric irregularity exists if the horizontal dimension of the lateral-force-resisting system is more than 130% of that in the adjacent story: $1.3b < B$. For example, the width of frame at level 6 = 160 ft, the width of frame at level 7 = 100 ft. 160 ft > 1.3 × 100. Therefore, vertical geometric irregularity exists.

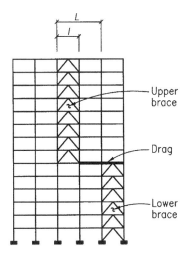

Figure 2.29. In-plane discontinuity; type-4 irregularity. A type 4-irregularity exists when there is an in-plane offset of the lateral system greater than the width of the system.

$$L > l$$

Assume $L = 60$ ft and $l = 30$ ft. Then the left side of upper brace is offset 60 ft from the left side of lower brace, greater than the 30 ft, the width of the offset brace. Therefore, a type-4 irregularity exists.

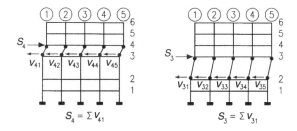

Figure 2.30. Discontinuity in capacity (weak story); story strength at levels 4 and 3. A weak story condition exists when the story strength is less than 80% of that of the story above:

$$S_3 < 0.8 S_4$$

For a moment frame, compare the strength of column shears in the two stories. For shear walls, compare the smaller of nominal shear strength V_n or V_m, the shear strength corresponding to the nominal flexural strength.

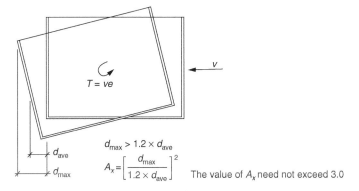

$$d_{max} > 1.2 \times d_{ave}$$

$$A_x = \left[\frac{d_{max}}{1.2 \times d_{ave}} \right]^2$$

The value of A_x need not exceed 3.0

Figure 2.31. Torsional irregularity. Torsional irregularity exists when d_{max}, the maximum story drift at a corner including accidental torsion, is more than 1.2 times the average of the drifts at the corresponding corners.

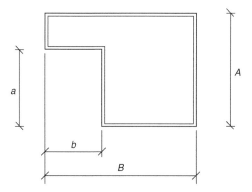

Figure 2.32.　Irregularity due to reentrant corners. This irregularity exists when plan configuration of the building and its lateral system have projections greater than 15% of the plan dimension in the corresponding direction. $b > 0.15B$ or $a > 0.15A$.

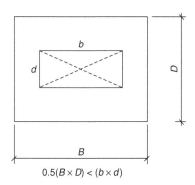

$$0.5(B \times D) < (b \times d)$$

Figure 2.33.　Irregularity due to diaphragm discontinuity. This irregularity exists if: 1) the area $b \times d$ of the opening is more than 50% of the area $B \times D$ of the diaphragm; or 2) the change in diaphragm stiffness is more than 50% from one story to the next.

2.3.2.　Design Base Shear, *V*

The strength level design base shear is given by the formula

$$V = \frac{C_v I}{RT} W \qquad\qquad (2.3) \text{ [UBC (Eq. 30.4)]}$$

where

T = fundamental period of the structure in the direction under consideration

I = seismic importance factor

C_v = a numerical coefficient dependent on the soil conditions at the site and the seismicity of the region, as set forth in Table 2.5 (UBC Table 16-R)

W = seismic dead load

R = a factor that accounts for the ductility and overstrength of the structural system, as set forth in Table 2.4 (UBC Table 16-N)

Z = seismic zone factor, as set forth in Table 2.3a (UBC Table 16-I). Note that Z does not directly appear in the base shear formula. It does, however, affect the seismic coefficients C_a and C_v.

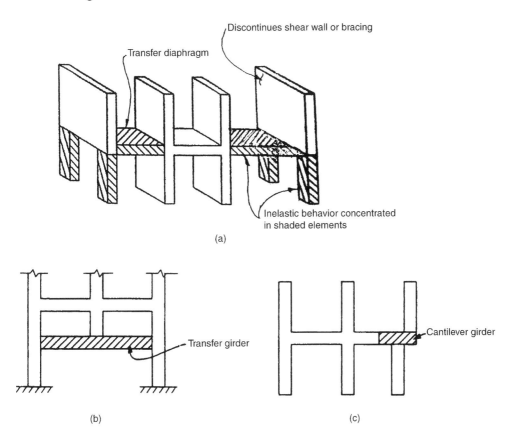

Figure 2.34. Irregularity due to out-of-plane offsets of vertical elements. This irregularity exists when there is discontinuity in the lateral-load-resisting path. In this figure, the columns and other elements such as the transfer girder and the cantilever girder support a discontinuous wall or a frame-column. They must be designed for special seismic load combination given in the 1997 UBC and ASCE 7-02.

$$(1.2 + 0.2S_{DS})D + f_1L + \Omega_oE$$
$$(0.9 - 0.2S_{DS})D + \Omega_oE$$

Since collector elements of diaphragms and columns supporting stiff elements are particularly vulnerable to earthquake damage, these elements must be designed for the estimated maximum axial forces that can realistically develop in these elements. Design of the diaphragm itself is not addressed in seismic codes. Typically it is designed for the above-load combinations or as a special reinforced concrete shear wall.

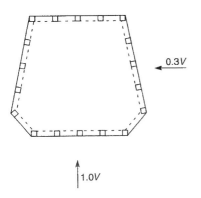

Figure 2.35. Irregularity due to nonparallel system. A nonparallel system irregularity exists when a lateral system is not parallel to or symmetric about the building's orthogonal axes.

TABLE 2.2   Vertical Irregularities

Irregularity type and definition	Reference fig.	Requirement
1. Stiffness irregularity — Soft story A soft story is one in which the lateral stiffness is less than 70% of that in the story above or less than 80% of the average stiffness of the three stories above.	Fig. 2.26	No penalty. Use dynamic analysis to determine lateral-force distribution.
2. Weight (mass) irregularity Mass irregularity shall be considered to exist where the effective mass of any story is more than 150% of the effective mass of an adjacent story. A roof that is lighter than the floor below need not be considered.	Fig. 2.27	No penalty. Use dynamic analysis to determine lateral-force distribution.
3. Vertical geometric irregularity Vertical geometric irregularity shall be considered to exist where the horizontal dimension of the lateral-force-resisting system in any story is more than 130% of that in an adjacent story. One-story penthouses need not be considered.	Fig. 2.28	No penalty. Use dynamic analysis to determine lateral-force distribution.
4. In-plane discontinuity in vertical lateral-force-resisting element An in-plane offset of the lateral-load-resisting elements is greater than the length of those elements.	Fig. 2.29	Use special seismic load combinations for members below discontinuity.
5. Discontinuity in capacity—weak story A weak story is one in which the story strength is less than 80% of that in the story above. The story strength is the total strength of all seismic-resisting elements sharing the story shear for the direction under consideration.	Fig. 2.30	Increase seismic loads for members below discontinuity by a factor = Ω_o.

(From UBC 1997, Table 16-L.)

The base shear as specified by Eq. (2.3) is subject to three limits:

1. The design base shear need not exceed

$$V = \frac{2.5 C_a I}{R} W \qquad \text{(2.4)[UBC Eq. (30.5)]}$$

2. It cannot be less than

$$V = 0.11 C_a I W \qquad \text{(2.5)[UBC Eq. (30.6)]}$$

where C_a is a seismic coefficient dependent on soil conditions at the site and on regional seismicity.

3. In the zone of highest seismicity (zone 4), the design base shear must be equal to or greater than

$$V = \frac{0.8 Z N_v I}{R} W \qquad \text{(2.6)}$$

TABLE 2.3 Plan Irregularities

Irregularity type and definition	Reference fig.	Requirement
1. **Torsional irregularity—to be considered when diaphragms are not flexible** Torsional irregularity shall be considered to exist when the maximum story drift, computed including accidental torsion, at one end of the structure transverse to an axis is more than 1.2 times the average of the story drifts of the two ends of the structure.	Fig. 2.31	Increase torsional forces by an amplification factor A_x.
2. **Reentrant corners** Plan configurations of a structure and its lateral-force-resisting system contain reentrant corners, where both projections of the structure beyond a reentrant corner are greater than 15% of the plan dimension of the structure in the given direction.	Fig. 2.32	Provide structural elements in diaphragms to resist flapping actions.
3. **Diaphragm discontinuity** Diaphragms with abrupt discontinuities or variations in stiffness, including those having cutout or open areas greater than 50% of the gross enclosed area of the diaphragm, or changes in effective diaphragm stiffness of more than 50% from one story to the next.	Fig. 2.33	Provide structural elements to transfer forces into the diaphragm and structural system. Reinforce boundaries at openings.
4. **Out-of-plane offsets** Discontinuities in a lateral-force path, such as out-of-plane offsets of the vertical elements.	Fig. 2.34	Use special seismic load combinations. One-third increase in stress not permitted. Design braced frames per UBC 2213.8.
5. **Nonparallel systems** The vertical lateral-load-resisting elements are not parallel to or symmetric about the major orthogonal axes of the lateral-force-resisting system.	Fig. 2.35	Design for orthogonal effects.

(From UBC 1997, Table 16-M.)

where N_v is a near-source factor that depends on the proximity to and activity of known faults near the structure. Faults are identified by seismic source type, which reflects the slip rate and potential magnitude of earthquake generated by the fault.

The near-source factor, N_v, is also used in determining the seismic coefficient C_v for buildings located in seismic zone 4.

2.3.3. Seismic Zone Factor *Z*

Five seismic zones—numbered 1, 2A, 2B, 3, and 4—are defined. The zone for a particular site is determined from a seismic zone map (Fig. 2.36). The map accounts for the geographical variations in the expected levels of earthquake ground shaking, and gives the an estimated peak horizontal acceleration on rock having a 10% chance of being exceeded in a 50-year period. The numerical values of Z are

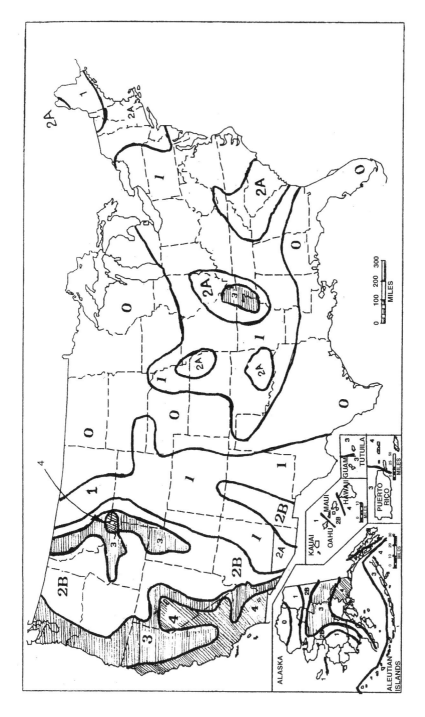

Figure 2.36. 1997 UBC Seismic Zone Map of the United States. For areas outside of the United States, see Appendix Chapter 16 of 1997 UBC. The map is based on a 10% probability of exceedence in 50 years.

Zone	1	2A	2B	3	4
Z	0.075	0.15	0.2	0.3	0.4

(From Table 16-I, UBC 1997.)

The value of the seismic zone coefficient Z can be considered the peak ground acceleration in percentage of gravity. For example, $Z = 0.4$ indicates a peak ground acceleration of $0.4g$ equal to 40% of gravity.

2.3.4. Seismic Importance Factor I_E

In seismic design, the importance factor I is used to increase the margin of safety against collapse. For example, $I = 1.50$ for essential facilities, $I = 1.25$ for hazardous facilities, and $I = 1.15$ for special occupancy structures. Essential structures are those that must remain operative immediately following an earthquake such as emergency treatment areas and fire stations. Hazardous facilities include those housing toxic or explosive substances. Examples of special occupancy structures are those not classified as special or hazardous, and required for continuous operation. Standard occupancy structures such as office buildings, hotels, and residences are designed for $I = 1.0$. The above values of I apply to facilities designed under the regulations of the 2001 California Building Code (2001 CBC). For buildings designed under the 1997 UBC, the values of I are as follows (Table 16-K, UBC 1997):

- Essential facilities $\quad I = 1.25$
- Hazardous facilities $\quad I = 1.25$
- Special occupancy structures $\quad I = 1.0$
- Standard occupancy structures $\quad I = 1.0$
- Miscellaneous structures $\quad I = 1.0$

2.3.5. Building Period T

The building period T may be determined by analysis or by using empirical formulas. It is denoted T_A if determined by empirical formulas, and T_B if determined by analysis. The following single empirical formula may be used for all framing systems:

$$T_A = C_t h_n^{3/4} \tag{2.7}$$

where
$\quad C_t = 0.035$ for steel moment frames
$\qquad = 0.030$ for concrete moment frames
$\qquad = 0.030$ for eccentric braced frames
$\qquad = 0.020$ for all other buildings
$\quad h_n = $ the height of the building in feet

If the period is determined more accurately using Rayleigh's formula (see Figs. 2.37 and 2.38) or a computer analysis, the value of T_B that can be used in calculating the base shear has certain limitations. In seismic zone 4, T_B cannot be more than 30% greater than that determined by Eq. (2.7) and in zones 1, 2A, 2B, and 3, it cannot be more than 40% greater. This provision is included to eliminate the possibility of using an excessively long period to justify an unreasonably low base shear. This limitation does not apply when checking drifts.

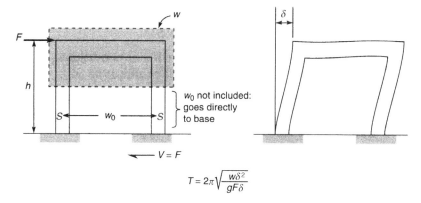

Figure 2.37.   Period calculation for a single-degree-of-freedom system. Rayleigh's formula.

2.3.6.   Structural System Coefficient *R*

The coefficient R shown in Table 2.4 is a measure of ductility and overstrength of a structural system, based primarily on performance of similar systems in past earthquakes.

A higher value of R has the effect of reducing the design base shear. For example, for a steel special moment-resisting frame, the factor has a value of 8.5, whereas for ordinary moment-resisting frame, the value is 4.5. This reflects the fact that a special moment-resisting frame performs better during an earthquake.

The values of R are the same as UBC 1994 values except they are divided by a load factor of 1.4 to account for the strength level design values. For example, the new value of R for a special moment-resisting frame is equal to the old $R_w = 12$, divided by 1.4. Thus, $R = 12/1.4 = 8.57$, which is rounded to 8.5 in 1997 UBC.

2.3.7.   Seismic Dead Load *W*

The dead load W, used for calculating the base shear, includes the total dead load of the structure, the actual weight of partitions with a minimum allowance of 10 psf of floor area, 25% of the floor live load in storage and warehouse occupancies, and the weight of

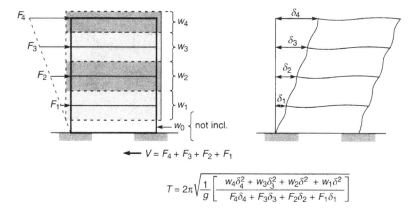

Figure 2.38.   Period calculation for multidegree-of-freedom system. Rayleighs formula.

TABLE 2.4 Structural Systems[a]

Basic structural system[b]	Lateral-force-resisting system description	R	Ω_o	Height limit for seismic zones 3 and 4
1. Bearing wall system	1. Shear walls			
	a. Concrete	4.5	2.8	160
	b. Masonry	4.5	2.8	160
	2. Braced frames where bracing carries gravity load			
	a. Steel	4.4	2.2	160
	b. Concrete[e]	2.8	2.2	N.P.
2. Building frame system	1. Steel eccentrically braced frame (EBF)	7.0	2.8	240
	2. Shear walls			
	a. Concrete	5.5	2.8	240
	b. Masonry	5.5	2.8	160
	3. Ordinary braced frames			
	a. Steel	5.6	2.2	160
	b. Concrete[c]	5.6	2.2	N.P.
	4. Special concentrically braced frames			
	a. Steel	6.4	2.2	240
3. Moment-resisting frame system	1. Special moment-resisting frame (SMRF)			
	a. Steel	8.5	2.8	N.L.
	b. Concrete[d]	8.5	2.8	N.L.
	2. Masonry moment-resisting wall frame (MMRWF)	6.5	2.8	160
	3. Concrete intermediate moment-resisting frame (IMRF)[e]	5.5	2.8	N.P.
	4. Ordinary moment-resisting frame (OMRF)			
	a. Steel[f]	4.5	2.8	160
	b. Concrete[h]	3.5	2.8	N.P.
	5. Special truss moment frames of steel (STMF)	6.5	2.8	240
4. Dual systems (frame resists at least 25% of seismic shear)	1. Shear walls			
	a. Concrete with SMRF	8.5	2.8	N.L.
	b. Concrete with steel OMRF	4.2	2.8	160
	c. Concrete with concrete IMRF[e]	6.5	2.8	160
	d. Masonry with SMRF	5.5	2.8	160
	e. Masonry with steel OMRF	4.2	2.8	160
	f. Masonry with concrete IMRF[c]	4.2	2.8	N.P.
	g. Masonry with masonry MMRWF	6.0	2.8	160
	2. Steel EBF			
	a. With steel SMRF	8.5	2.8	N.L.
	b. With steel OMRF	4.2	2.8	160
	3. Ordinary braced frames			
	a. Steel with steel SMRF	6.5	2.8	N.L.
	b. Steel with steel OMRF	4.2	2.8	160
	c. Concrete with concrete SMRF[c]	6.5	2.8	N.P.
	d. Concrete with concrete IMRF[e]	4.2	2.8	N.P.

TABLE 2.4 Structural Systems[a] (Continued)

Basic structural system[b]	Lateral-force-resisting system description	R	Ω_o	Height limit for seismic zones 3 and 4
	4. Special concentrically braced frames			
	a. Steel with steel SMRF	7.5	2.8	N.L.
	b. Steel with steel OMRF	4.2	2.8	160
5. Cantilevered column building systems	1. Cantilevered column elements[g]	2.2	2.0	357
6. Shear wall-frame interaction systems	1. Concrete[h]	5.5	2.8	160
7. Undefined systems	See Sections 1629.6.7 and 1629.9.2 (UBC 1997)	—	—	—

N.L.—no limit   N.P.—not permitted.

[a] See Section 1630.4 for combination of structural systems.

[b] Basic structural systems are defined in Section 1629.6, 1997 UBC.

[c] Prohibited in seismic zones 3 and 4.

[d] Includes precast concrete conforming to Section 1921.2.7, 1997 UBC.

[e] Prohibited in seismic zones 3 and 4, except as permitted in Section 1634.2, 1997 UBC.

[f] Ordinary moment-resisting frames in seismic zone 1 meeting the requirements of Section 2211.6 may use an R value of 8. See UBC 1997.

[g] Total height of the building including cantilevered columns.

[h] Prohibited in seismic zones 2A, 2B, 3, and 4. See Section 1633.2.7, 1997 UBC.

(From UBC 1997, Table 16-N.)

snow when the design snow load is greater than 30 psf. The snow load may be reduced by up to 75% if its duration is short.

The rationale for including a portion of the snow load in heavy snow areas is the fact that in these areas, a significant amount of ice can build up and remain on roofs.

The total seismic load W represents the total mass of the building and includes the weights of structural slabs, beams, columns, and walls; and nonstructural components such as floor topping, roofing, fireproofing material, fixed electrical and mechanical equipment, partitions, and ceilings. When partition locations are subject to change (as in office buildings), a uniform distributed dead load of at least 10 psf of floor area is used in calculating W. Typical miscellaneous items such as ducts, piping, and conduits can be accounted for using an additional 2 to 5 psf. In storage areas, 25% of the design live load is included in the seismic weight W. In areas of heavy snow, a load of 30 psf should be used where the snow load is greater than 30 psf. However, it may be reduced to as little as 7.5 psf when approved by building officials.

In addition to determining the overall weight W, it is necessary to evaluate tributary weight W_x at each floor for both vertical and horizontal distribution of loads (Fig. 2.39). Therefore, the calculations for W must be done in an orderly tabular form so that overall weights as well as tributary weights can be properly accounted for.

2.3.8. Seismic Coefficients C_v and C_a

The seismic coefficients C_v and C_a, given in Tables 2.5 and 2.6, are site-dependent ground motion coefficients that define the seismic response throughout the spectral range. They are measures of expected ground acceleration at a site.

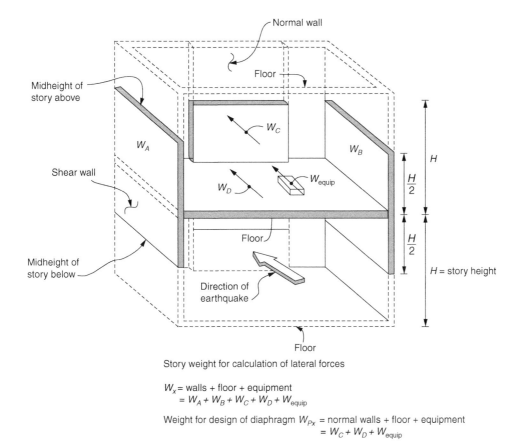

Story weight for calculation of lateral forces

W_x = walls + floor + equipment
$\quad$ = $W_A + W_B + W_C + W_D + W_{equip}$

Weight for design of diaphragm W_{px} = normal walls + floor + equipment
$\quad\quad\quad\quad\quad\quad$ = $W_C + W_D + W_{equip}$

Note: Floor weight W_D includes floor structure, suspended ceiling, mechanical equipment, and an allowance for partitions.

Figure 2.39. Tributary weights for seismic dead load calculation.

The coefficients, and hence the expected ground accelerations, are dependent on the seismic zone and soil profile type. They therefore reflect regional seismicity and soil conditions at the site.

Additionally, in seismic zone 4, they also depend on the seismic source type and near-source factors N_a and N_v.

TABLE 2.5 Seismic Coefficient C_v

Soil profile type	Seismic zone factor, Z				
	$Z = 0.075$	$Z = 0.15$	$Z = 0.2$	$Z = 0.3$	$Z = 0.4$
S_A	0.06	0.12	0.16	0.24	$0.32N_v$
S_B	0.08	0.15	0.20	0.30	$0.40N_v$
S_C	0.13	0.25	0.32	0.45	$0.56N_v$
S_D	0.18	0.32	0.40	0.54	$0.64N_v$
S_E	0.26	0.50	0.64	0.84	$0.96N_v$
S_F	Site-specific geotechnical investigation and dynamic site response analysis shall be performed for soil type S_F.				

(From UBC 1997, Table 16-R.)

TABLE 2.6   Seismic Coefficient C_a

Soil profile type	Seismic zone factor, Z				
	$Z = 0.075$	$Z = 0.15$	$Z = 0.2$	$Z = 0.3$	$Z = 0.4$
S_A	0.06	0.12	0.16	0.24	$0.32N_a$
S_B	0.08	0.15	0.20	0.30	$0.40N_a$
S_C	0.09	0.18	0.24	0.33	$0.40N_a$
S_D	0.12	0.22	0.28	0.36	$0.44N_a$
S_E	0.19	0.30	0.34	0.36	$0.36N_a$
S_F	Site-specific geotechnical investigation and dynamic site response analysis shall be performed for soil profile S_F.				

(From UBC 1997, Table 16-Q.)

For a given earthquake, a building on soft soil types such as S_C or S_D experiences a greater force than if the same building were located on rock, type S_A or S_B. This is addressed in the UBC through the C_a and C_v coefficients, which are calibrated to soil type S_B with a value of unity. Instead of a single coefficient, two coefficients, C_a and C_v, are used to distinguish the response characteristics of short-period and long-period buildings. Long-period buildings are more affected by soft soils than short-period buildings.

2.3.9.   Soil Profile Types

The soil profile types labeled S_A through S_F (Table 2.7) represent the effect of soil conditions on ground motion. Seismic ground motion can be amplified by site geology and soil characteristics. The value of Z, given in the seismic zone map, is for the rock, type S_B soil. Therefore, except for hard rock, type S_A soil, the value of Z increases for soil types S_C, S_D, S_E, and S_F. When soil properties are not known, type S_D must be used. S_E need not be assumed unless the building official determines that soil type S_E is present or it is established by geotechnical data.

TABLE 2.7   Soil Profile Types

Soil profile type	Soil profile name/generic description	Average soil properties for top 100 feet (30,480 mm) of soil profile		
		Shear wave velocity, fsseet/second (m/s)	Standard penetration test (blows/foot)	Undrained shear strength, psf (kPa)
S_A	Hard rock	>5,000 (1,500)	—	—
S_B	Rock	2,500 to 5,000 (760 to 1,500)		
S_C	Very dense soil and soft rock	1,200 to 2,500 (360 to 760)	>50	>2,000 (100)
S_D	Stiff soil profile	600 to 1,200 (180 to 360)	15 to 50	1,000 to 2,000 (50 to 100)
S_E	Soft soil profile	<600 (180)	<15	<1,000 (50)
S_F		Soil requiring site-specific evaluation.		

(From UBC 1997, Table 16-J.)

TABLE 2.8 Seismic Source Type

Seismic source type	Seismic source description	Seismic source definition	
		Maximum moment magnitude, M	Slip rate, SR (mm/year)
A	Faults that are capable of producing large-magnitude events and that have a high rate of seismic activity.	$M \geq 7.0$	$SR \geq 5$
B	All faults other than types A and C.	$M \geq 7.0$ $M < 7.0$ $M \geq 6.5$	$SR < 5$ $SR > 2$ $SR < 2$
C	Faults that are not capable of producing large-magnitude earthquakes and that have a relatively low rate of seismic activity.	$M < 6.5$	$SR \leq 2$

(From UBC 1997, Table 16-U.)

2.3.10. Seismic Source Type A, B, and C

The seismic source types labeled A, B, or C (Table 2.8) are used to identify earthquake potential and activity of faults in the immediate vicinity of the structure.

They are defined in terms of the slip rate of the fault and the maximum magnitude of earthquake that may be generated at the fault. The highest seismic risk is posed by seismic source type A, which is defined by a maximum moment magnitude of 7.0 or greater and a slip rate of 5 mm/year or greater.

Type A signifies active faults such as the San Andreas capable of producing large-magnitude events. Most faults in California are classified as type B, while those outside of California, not capable of producing large magnitude events, are classified as inactive, type C faults.

2.3.11. Near Source Factors N_a and N_v

The near-source factors N_a and N_v are given in Tables 2.9 and 2.10. In seismic zone 4, they are used in conjunction with the soil profile type to determine the seismic coefficients C_v and C_a (Tables 2.5 and 2.6). For example, for seismic source type B at a distance to the fault of less than 2 km, $N_a = 1.3$ (Table 2.9). This is then used with Table 2.6 to determine the seismic coefficient C_a.

Similarly, $N_v = 1.6$ (Table 2.10) for seismic source type B at a distance less than 2 km. This is used with Table 2.5 to determine C_v. The purpose of N_a and N_v is to increase the soil-modified ground motion parameters, C_a and C_v, when there are active faults capable of generating large-magnitude earthquakes within 15 kilometers (9 miles) of a seismic zone 4 site.

2.3.12. Distribution of Lateral Force F_x

The base shear V, as determined from Eqs. (2.3) through (2.6), is distributed over the height of the structure as a force at each level F_i, plus an extra force F_t at the top:

$$V = F_t + \sum_{i=1}^{n} F_i \qquad (2.8)$$

TABLE 2.9 Near-Source Factor $N_a{}^a$

Seismic source type	Closest distance to known seismic source[b,c]		
	≤2 km	5 km	≥10 km
A	1.5	1.2	1.0
B	1.3	1.0	1.0
C	1.0	1.0	1.0

[a] The near-source factor may be based on the linear interpolation of values for distances other than those shown in the table.

[b] The location and type of seismic sources to be used for design shall be established based on approved geotechnical data (e.g., most recent mapping of active faults by the U.S. Geological Survey or the California Division of Mines and Geology).

[c] The closest distance to seismic source shall be taken as the minimum distance between the site and the area described by the vertical projection of the source on the surface (i.e., surface projection of fault plane). The surface projection need not include portions of the source at depths of 10 km or greater. The largest value of the near-source factor considering all sources shall be used for design.

(From UBC 1997, Table 16-S.)

The extra force at the top is

$$F_t = 0.07TV \le 0.25V \qquad \text{if } T > 0.7 \text{ sec}$$

$$F_t = 0 \qquad \text{if } T \le 0.7 \text{ sec}$$

F_t accounts for the greater participation of the higher-mode responses of longer-period structures.

The remaining portion of the total base shear $(V - F_t)$ is distributed over the height, including the top, by the formula

$$F_x = \frac{(V - F_t)(w_x h_x)}{\sum_{i=1}^{n} w_i h_i} \tag{2.9}$$

where w is the weight at a particular level, and h is the height of that level above the shear base.

For equal story heights and weights, Eq. (2.9) distributes the force linearly, increasing toward the top. Any significant variation from this triangular distribution indicates an irregular structure.

TABLE 2.10 Near-Source Factor N_v

Seismic source type	Closest distance to known seismic source			
	≤2 km	5 km	10 km	≥15 km
A	2.0s	1.6	1.2	1.0
B	1.6	1.2	1.0	1.0
C	1.0	1.0	1.0	1.0

(From UBC 1997, Table 16-T.)

2.3.13. Story Shear V_x and Overturning Moment M_x

The story shear at level x is the sum of all the story forces at and above that level:

$$V_x = F_t + \sum_{i=x}^{n} F_i \tag{2.10}$$

The overturning moment at a particular level M_x is the sum of the moments of the story forces above, about that level. Hence

$$M_x = F_t(h_n - h_x) + \sum_{i=x}^{n} F_i(h_i - h_x) \tag{2.11}$$

2.3.14. Torsion

Accidental torsion that occurs due to uncertainties in the building's mass and stiffness distribution must be added to the calculated eccentricity. This is done by adding a torsional moment at each floor equal to the story force multiplied by 5% of the floor dimension, perpendicular to the direction of the force. This procedure is equivalent to moving the center of mass by 5% of the plan dimension, in a direction perpendicular to the force.

If the lateral deflection at either end of a building is more than 20% greater than the average deflection, the building is classified as torsionally irregular and the accidental eccentricity must be amplified using the formula

$$A_x = \left[\frac{\delta_{max}}{1.2\delta_{avg}} \right]^2 \leq 3.0 \tag{2.12}$$

where

δ_{avg} = the average displacement at level x
δ_{max} = the maximum displacement at level x
A_x = the torsional amplification factor at level x

Torsional shears may be subtracted from direct shears if the torsional shear is reduced by the effects of accidental torsion. However, torsional shears that are increased by the effects of accidental torsion must be added to direct shears.

2.3.15. Reliability/Redundancy Factor ρ

The seismic base shear, as determined from the preceding equations, must be multiplied by a reliability/redundancy factor, ρ, for the design of a lateral load-resisting system. It is given by

$$1 \leq \rho = 2 - \frac{20}{r_{max}\sqrt{A_B}} \leq 1.5 \tag{2.13}$$

where

A_B = the ground floor area of the structure in square feet
r_{max} = the maximum element-story shear ratios

The element-story shear ratio, r_i, at a particular level is the ratio of the shear in the most heavily loaded member to the total story shear. The maximum ratio, r_{max}, is defined as the largest value of r_i in the lower two-thirds of the building.

For special moment-resisting frames, if ρ exceeds 1.25, additional bays must be added. For the purposes of determining drift and in seismic zones 0, 1, and 2, $\rho = 1.0$.

The redundancy factor provides for multiple load paths for resisting earthquake forces. More redundancy means better reliability because there is increased opportunity for inelastic deformations. It takes into account the number of lateral-force-resisting elements, plan area of building, and distribution of forces to the lateral-force-resisting elements. For a shear wall building, ρ depends on floor area of the building, number of shear walls resisting the story shear, and the length of shear walls. For moment frames, it depends on the floor area of the building and the number of columns. For braced frames, it depends on the number of brace elements resisting the story shear. For dual systems, ρ is evaluated by calculating

- r_{max} for the portion of the story shear carried by moment frames.
- r_{max} for the portion of the story shear carried by shear walls or braced frames.
- ρ_{max} using the ρ_{max} value in steps 1 and 2, and multiplying it by 0.8.

The redundancy factor $\rho = 1.0$ for the following components:

- Diaphragms, except for transfer diaphragms between offset lateral force-resisting elements
- Parts and portions of structures.
- Nonbuilding structures.

2.3.16. Drift Limitations

The elastic deflections due to strength-level design seismic forces are called design-level response displacements, Δ_S. The subscript S in Δ_S stands for strength design. The seismic forces used to determine Δ_S may be calculated using a reliability/redundancy factor equal to 1.0, ignoring the previously mentioned limitations on the period used in the calculation of base shear. An elastic static or dynamic analysis may be used to determine Δ_S.

The maximum inelastic response is defined as

$$\Delta_M = 0.7R\Delta_S \tag{2.14}$$

where R is the structural system coefficient defined earlier. The subscript M in Δ_M signifies that we are calculating a maximum value for the deflection due to seismic response that includes inelastic behavior.

Deflection control is specified in terms of the story drift defined as the lateral displacement of one level relative to the level below. The story drift is determined from the maximum inelastic response, Δ_M, as defined in Eq. (2.14).

The calculated displacement must include the effects of both translation and torsion. Hence, drift must be checked in the plane of the lateral-load-resisting elements, generally at the building corners. Effects of $P\Delta$ must be included in the calculation of Δ_M unless it is shown by calculation that the effects are insignificant.

For structures with a period less than 0.7 seconds, the maximum story drift is limited to

$$\Delta_S \leq 0.025\, h\ (T \leq 0.7 \text{ seconds})$$

where h is the story height.

For structures with a period greater than 0.7 seconds, the story drift limit is

$$\Delta_S \le 0.020\, h \; (T \ge 0.7 \text{ seconds})$$

Observe that in the 1997 UBC, and for that matter in all subsequent codes, the inelastic response drift, Δ_M, rather than code force-level drift, Δ_s is used for verifying the performance of buildings. The drift limits are 2.5% and 2% of the story height for short- and long-period buildings, respectively. Thus, for a typical office building with a floor-to-floor height, h, equal to, say, 13 feet, the maximum allowable inelastic drift is equal to 2% of 13 feet = $1/50 \times 13 \times 12$ inches = 3.12".

Compare this to an allowable drift of $h/400 = 13 \times 12/400 = 0.39"$ for a similar building under wind loads. The large difference in the allowable drift values serves as a reminder that the seismic inelastic deflections of a building, should the postulated earthquake ever hit it, are indeed very large, about 10 times as large as the the drift limit in common usage for wind design of buildings.

Strength design load combinations, as given by the following equations, are used in the determination of Δ_S.

$$1.2D + 1.0E + 0.5L \tag{2.15}$$

$$0.9D + 1.0E \tag{2.16}$$

It should be noted that the calculations of Δ_S must include rotational components also.

For reinforced concrete buildings, it is mandatory to use cracked section properties, I_{cr} to compute displacements. Typical values are given below.

Walls $I_{cr} = 0.5\, E_c I_g$
Beams $I_{cr} = 0.5\, E_c I_g$
Columns $I_{cr} = 0.5\, E_c I_g$ to $0.7\, E_c I_g$

The designer is referred to Table 6.5, Federal Emergency Management Agency (FEMA) Publication 356, for additional stiffness values.

2.3.17. Deformation Compatibility

The 1994 Northridge earthquake taught a number of lessons in seismic design, one of which is the importance of satisfying the so-called deformation compatibility requirement. This requirement, extensively revised in 1997 UBC after observations from the Northridge earthequake, strives to achieve parity in seismic performance of framing elements and connections not required by design to be part of lateral-force-resisting systems, with those required by design. This is because we know now, from the Northridge earthquake, that even in a building with a properly designed and detailed lateral system, collapse can occur if all structural elements are not capable of deforming with the building during the event. Likewise, if certain nonstructural elements in the building are not capable of deforming with the building, the resulting falling hazards may threaten life safety or impede egress from the building.

Designing for deformation compatibility consists of

- Establishing deformation demands.
- Assessing individual elements and their connections for their capacity to deform.

Deformations include interstory drift, but any other deformation of the structure caused by seismic forces may also be of concern (see Figs. 2.40 and 2.41). While interstory

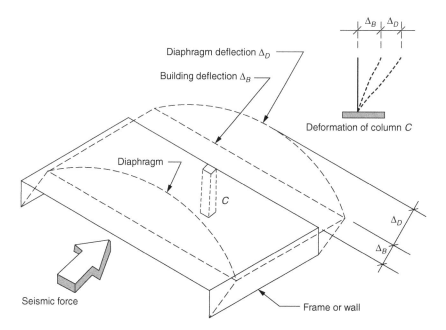

Figure 2.40. Column deformation for use in compatibility considerations. Deformation of column = building deflection Δ_B + diaphragm deflection Δ_D. (Adapted from SEAOC Blue Book, 1999 Edition.)

drift is the most common deformation used in practice, other types of deformation need to be considered, such as

1. Vertical racking of structural framing in eccentrically braced frames.
2. Shear distortions of concrete coupling beams.
3. Vertical racking of structural bays in dual systems.

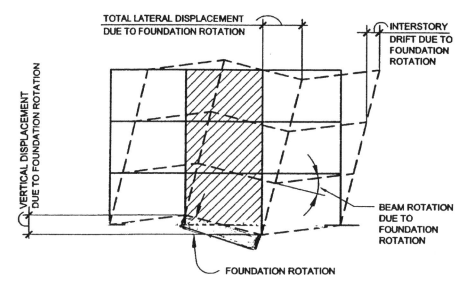

Figure 2.41. Deformation compatibility consideration of foundation flexibility. (Adapted from SEAOC Blue Book, 1999 Edition.)

The maximum expected seismic deformations are computed for a design ground motion representing a 10% probability of being exceeded in 50 years. For most structures, inelastic response of the structure will occur for this level of ground motion. This inelastic response should be recognized in the computation of the expected deformation. Current code provisions stipulate that all elements and their connections shall be investigated for deformation compatibility. The elements included are structural elements such as columns, beams, walls, slabs, trusses, and bracing that were considered in the design as not being part of the lateral systems. Nonstructural elements such as stairs, cladding, finishes, utilities, and equipment should also be investigated. The intent is to ensure that structural stability and/or life safety are not compromised because of failure of these elements.

It is recognized that many nonstructural elements are not designed directly by the engineer of record. In these cases, the deformation compatibility requirements must nonetheless be satisfied. Specific design requirements, including the maximum expected deformation values, must be included in the performance design specification. The engineer is obligated to notify the client of

1. The predicted deformations of the structure.
2. The code requirements for deformation compatibility.
3. The owner's obligation to comply with the governing building code.

The deformation demand is the greater of the maximum inelastic displacement Δ_M, considering $P\Delta$ effects or the deformation induced by a story drift 0.0025 times the story height. This value of Δ_M must be computed using a structure model that neglects the stiffening effects of those elements not part of the lateral-force-resisting system. This method of evaluating $\Delta_M = 0.7R\Delta_S$ assumes that the inelastic deformation can be estimated in terms of the deformation of the elastic structure model. When an engineering analysis is made to determine the maximum expected seismic deformations, proper modeling of the structure is needed. Because deformation incompatibility can have profound life safety implications, it is essential that the deformations not be underestimated.

A case in point is the deformation of diaphragms. Although it is common practice in an analysis to ignore deformations of the diaphragm, significant demands can result from diaphragm deformations. Including these effects may require supplemental hand calculations. The deformations of foundations should also be included in the deformation compatibility analysis. Although it is common practice to ignore sources of deformation such as rotation of the foundation, significant increase in deformation demands can result due to these effects.

Structural and nonstructural elements that are not part of the lateral system may initially contribute to structural stiffness, but because this stiffness may degrade when subject to cyclic loads, these stiffness properties must be neglected in the demand analysis for deformation compatibility.

For concrete and masonry elements, flexural and shear stiffness properties should, as a maximum, be computed as one-half of the gross section elastic stiffness. While it may be considered conservative to use gross section properties when computing the period of the structure for purposes of determining the minimum base shear, this practice is not conservative when analyzing the structure for deformation compatibility demands. Concrete and masonry elements generally crack before code force levels are reached. Further increase in load results in a reduction in effective stiffness. When computing the deformation of the structure, realistic effective stiffness properties must be used. It is generally accepted that one-half of the gross section properties adequately reflects the effective stiffness of a cracked structural member. Other, more accurate stiffness properties can be used if substantial by a rational analysis. These reduced stiffness properties must be used for all parts of the lateral-force-resisting system, including beam and column frame-type elements and for shear wall-type elements.

Once the maximum expected seismic deformations have been determined, the adequacy of all structural elements for the imposed deformation demands must be verified. As an alternate, conservative ductile detail for reinforced concrete and structural steel can usually be accepted as compliant measures.

Concrete columns pose a high risk if the design does not address deformation compatibility. The forces induced by interstory drift in the building can result in nonductile shear failures and/or compressive strain failures. Either mode of failure, accompanied by cyclic reversals of load, can destroy the column's ability to support vertical gravity load and can result in partial collapse of the structure. Such behavior can usually be avoided if: 1) the shear strength is in excess of the shear corresponding to the development of flexural strength of column, and 2) the column is confined over the potential plastic hinge region with hoops and crossties in lieu of the minimum lateral ties, to minimize compressive strain failures. Current seismic codes contain specific provisions for concrete members not part of the lateral system and require the aforementioned measures for heavily loaded columns.

Engineering judgment must be exercised when assessing nonstructural elements. It is generally accepted that for minor and moderate earthquakes, damage to nonstructural elements should be minimal; this normally requires that nonstructural elements be capable of accommodating code force-level deformations without damage. To accomplish this, particular attention to and specification of appropriate details is sufficient. Engineering analysis and computations (other than normal drift calculations) are not required.

However, some nonstructural elements of a building that provide a life safety function, or that if damaged become a life safety threat (including blocking emergency exits from the building), require special analysis, design, and detailing for deformation compatibility. Examples include: stair stringers rigidly connected at each end to the building, and cladding systems enveloping a building as a rigid skin.

When the structure deforms laterally, the stringer will act as a brace until the stringer and/or its connections fail. In the case of cladding, failure to accommodate deformations of the structure can result in loss of cladding supports, which, in case of heavy cladding systems such as stone, masonry, and concrete finishes, result in falling hazards with serious life safety threats.

Common elements considered in deformation compatibility checks are

- Concrete columns.
- Connections.
- Deep girders.
- Flat slabs.
- Trusses.
- Miscellaneous bracings.
- Stairs.
- Cladding.

As mentioned previously, the expected deformations are determined as the greater of the maximum inelastic response displacement Δ_M, considering $P\Delta$ effects, or the deformation induced by a story drift of 0.0025 times the story height. The maximum inelastic response displacement that occurs when the structure is subject to the design basis ground motion (10% probability of exceedence in 50 years) is given by

$$\Delta_M = 0.7\, R\Delta_S \tag{2.17}$$

The quantity Δ_S is the design-level response displacement, which is the elastic displacement that occurs when the structure is subjected to the design seismic forces.

The displacements Δ_S can be obtained from either a static or dynamic analysis. For concrete and masonry elements that are part of the lateral-force-resisting system, the assumed flexural and shear stiffness properties must not exceed one-half of the gross section properties unless a rational cracked section analysis is performed.

The 1997 UBC estimate of the maximum inelastic response displacement, Δ_M, is the equivalent of $3/8 \, R_w \, \Delta$ of the 1994 UBC. The deflection Δ of 1994 UBC used to be computed under service-level design earthquake forces, whereas Δ_S of 1997 UBC must be computed under strength-level design earthquake forces. Therefore, an approximate comparison between the 1994 and 1997 UBC estimates of Δ_M is given by

$$\frac{(\Delta_M)1997}{(\Delta_M)1994} = \frac{0.7 \, R\Delta_S}{3/8 \, R_W\Delta} = \frac{0.7R(1.4\Delta)}{0.375(1.4R)\Delta} = \frac{0.7}{0.375} = 1.87, \text{ approximately equal to 2.0}$$

Thus, gravity framing is now required to sustain design gravity loads under twice as much imposed lateral displacement as prescribed in 1994 UBC. Also, the lower-bound value of 0.25% of story drift was not included in 1994 UBC. Finally, 1994 UBC did not specifically require that the flexural and shear stiffness properties of concrete and masonry elements that are part of the lateral-force-resisting system be taken no more than one-half of the gross section properties. The stringent requirements come from the experience of the 1994 Northridge earthquake that caused the collapse or partial collapse of at least two parking structures that could be attributed primarily to the failure of interior columns designed to carry gravity loads only. Following the experience, the detailing requirements for frame members not proportioned to resist forces induced by earthquake motions have been extensively rewritten in 1997 UBC.

As noted earlier, the imposed displacement under which gravity frame members must sustain their design loads has gone up by a factor of 1.87 as compared with the 1994 UBC. In addition, it was not a violation of 1994 UBC provisions to compute lateral deflections of the lateral-force-resisting systems using gross section properties. However, under 1997 UBC, cracked-section properties must be used in such computations. This may account for another twofold increase in the imposed displacement under which full design loads must be sustained by the gravity frame see Ref. 98, 99, and 100 for further discussion.

2.3.18. Load Combinations

The emphasis in current design practice is on LRFD, also called strength design. The 1997 UBC is the first building code to make the fundamental change in the seismic loads from allowable stress to strength design level. The basic load factor combinations are thus intended for strength design. However, since many engineers prefer to work using the allowable stress basis, load combinations for allowable stress design are also given, in two formats, i.e., two sets of equations. One set does not permit one-third increase for wind or seismic, while the other does.

2.3.18.1. *Basic Load Combinations for Strength Design (LRFD)*

All Materials Except Concrete (No One-Third Increase for Wind or Seismic).

$$\begin{aligned}
U &= 1.4D \\
U &= 1.2D + 1.6L + 0.5(L_r \text{ or } S) \\
U &= 1.2D + 1.6(L_r \text{ or } S) + (f_1L \text{ or } 0.8W) \\
U &= 1.2D + 1.3W + f_1L + 0.5(L_r \text{ or } S) \\
U &= 1.2D + 1.0E + (f_1L + f_2S) \\
U &= 0.9D \pm (\rho E \text{ or } 1.3W)
\end{aligned} \qquad (2.18)$$

where

U = ultimate load resulting from load combinations

D = dead load

E = earthquake load resulting from the combination of the horizontal component, E_h, and the vertical component E_v;

$E = \rho E_h \pm E_V$

$E_V = 0.5 C_a I$

L = live load

L_r = roof live load

S = snow load

f_1 = 1.0 for floors in places of public assembly, for live loads in excess of 100 psf (4.9 kN/m^2), and for garage live load.

= 0.5 for other live loads.

f_2 = 0.7 for roof configurations (such as saw tooth) that do not shed snow off the structure.

= 0.2 for other roof configurations.

ρ = redundancy/reliability factor

W = load due to wind pressure

E_h = earthquake load due to the base shear, V.

2.3.18.2.   *Basic Load Combinations*

Concrete Structures
Ultimate Load Combinations (No One-Third Increase for Wind or Seismic)

$$U = 1.4D + 1.7L$$
$$U = 0.75(1.4D + 1.7L + 1.7W)$$
$$U = 0.9D + 1.3W \tag{2.19}$$
$$U = 1.2D + 1.0E + f_1 L + f_2 S$$
$$U = 0.9D + 1.0E$$

The designer is referred to ACI 318-02 for load combinations that include earth pressure, H, and structural effects, T, due to differential settlement, creep, shrinkage, temperature change, etc.

2.3.18.3.   *Basic Load Combinations Using ASD*

No One-Third Increase for Wind or Seismic

$$D$$
$$D + L + (L_r \text{ or } S)$$
$$D + \left(W \text{ or } \frac{E}{1.4} \right)$$
$$0.9D \pm \frac{E}{1.4} \tag{2.20}$$
$$D + 0.75\left[L + (L_r \text{ or } S) + \left(W \text{ or } \frac{E}{1.4} \right) \right]$$

It should be noted that the preceding load combinations given by Eqs. (2.19) reflect the amendments to UBC-97 recommended by the seismology committee of the Structural

Engineers Association of California (SEAOC). The recommendations are published in SEAOC's 1999 edition of *Recommended Lateral Force Requirements and Commentary*, commonly referred as the Blue Book in California. As always, it is prudent to verify with the presiding building official that SEAOC-recommended revisions have been approved by the authority having jurisdiction over building permits.

The load factor of 1.0 on the earthquake load E reflects the fact that the specified forces in UBC-97 are at strength design levels, without further amplification. They are typically about 1.4 times as high as the allowable stress design levels given in previous provisions. The earthquake load E is a function of both horizontal and vertical components of ground motion (E_h and E_v). This is represented in the earthquake load equation.

The component E_h is due to the horizontal forces corresponding to the prescribed base shear V. The component E_v is due to an additional increment of dead load D and may be taken as zero for allowable stress design. In seismic zone 4, this component is at least equal to $0.2D$, which, when added to $1.2D$, results in parity with the $1.4D$ factored load in the previous provisions for strength design of reinforced concrete and masonry elements. In allowable stress design, the vertical component is indirectly included in the load combinations, which would need to be multiplied by about 1.4, compared to directly with strength design requirement.

2.3.18.4.   Alternate Basic Load Combinations Using ASD

One-Third Increase for Wind or Seismic Allowed

$$D + L + (L_r \text{ or } S)$$
$$D + L + \left(W \text{ or } \frac{E}{1.4} \right)$$
$$D + L + W + \frac{S}{2}$$
$$D + L + S + \frac{W}{2} \qquad (2.21)$$
$$D + L + S + \frac{E}{1.4}$$
$$0.9D \pm \frac{E}{1.4}$$

A one-third increase in allowable stress values is typically permitted for all the load combinations that include wind or earthquake loads. However, this one-third increase should not be applied to the allowable shear values in UBC-97 Tables 23-11-H, 23-II-I-I, and 23-II 1–2 since these values have already been increased for short-time wind or earthquake load. Also the one-third increase should not be used concurrently with a permitted load duration factor for wood design. Neither should the general one-third increase be applied to allowable soil pressure that has already been increased for short-time wind or earthquake loads as permitted by UBC Section 1809.2.

It should be noted that two sets of load combinations for the LRFD and ASD designs are based on entirely different philosophies and specifically are not intended to be equivalent to each other. The LRFD set of load combinations is based on the premise that the design strength resulting from the allowable stress method should, in general, not be less than that resulting from the basic strength design method. The alternative basic set of load combinations is based on the premise that for seismic zone 4, away from near-source zones, the designs should be about the same as those resulting from the previous provisions.

2.3.18.5. *Special Seismic Load Combinations (for Both ASD and LRFD)*

$$1.2D + f_1L + 1.0E_m \qquad\qquad (2.22)$$

$$0.9D \pm 1.0E_m \qquad\qquad (2.23)$$

where

$E_m = \Omega_o E_h$, the estimated maximum earthquake force that can be developed in the structure

These combinations are intended to cover conditions where uniform ductility in the structural system is lacking due to vertical discontinuities (see Fig. 2.42 for some examples).

The parameter E_m represents the maximum earthquake force that can be developed in a structure. The overstrength factor Ω_o is equivalent to $3/8R_w$ factor that appears in previous UBC requirements to address nonductile issues as columns supporting discontinuous shear walls. It should be noted that UBC special load combinations do not consider the effect of vertical accelerations, whereas ASCE 7-02, dicussed presently does include this effect.

2.3.19. Design Example, 1997 UBC: Static Procedure

For convenience, before working through a design example, the 1997 UBC Provisions for determining static base shear are given here in a summary format.

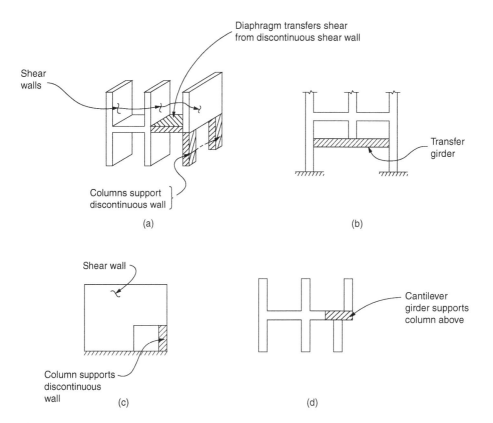

Figure 2.42. Examples of nonuniform ductility in structural systems due to vertical discontinuities. (Adapted from SEAOC Blue Book, 1999 Edition.)

The base shear is given by

$$V = \frac{C_v I}{RT}$$

However, V need not exceed

$$V = \frac{2.5\, C_a I}{R} W$$

Also, V cannot be less than

$V = 0.11\, C_a IW$, and additionally, in zone 4,

V cannot be less than

$$V = \frac{0.8\, ZN_v I}{R} W$$

C_a and C_v are given in terms of

1. Seismic zone factor, Z.
2. Soil profile type, S_A through S_E.
3. Near-source factors, N_a and N_v for zone 4.

The seismic zone factor, Z, has the following values:

$Z = 0.4$ for zone 4

$Z = 0.3$ for zone 3

$Z = 0.2$ for zone 2B

$Z = 0.15$ for zone 2A

$Z = 0.075$ for zone 1

$Z = 0.0$ for zone 0

In the preceding equations

V = total base shear in a given direction

W = total seismic weight

I = importance factor = 1.25 for essential and hazardous facilities

= 1.0 for special and standard occupancy

C_a = acceleration-based seismic coefficient

C_v = velocity-based seismic coefficient

R = numerical coefficient representative of the inherent overstrength and global ductility capacity of the lateral force-resisting systems (Table 2.4).

Since the 1997 UBC is formatted using strength-level earthquakes, the new $R = $ old $\dfrac{R_W}{1.4}$

T = elastic fundamental period of vibration, in seconds, of the structure in the direction under consideration. The period T is commonly noted as T_A when determined by approximate methods, and T_B when determined by more accurate methods such as dynamic analysis.

T_A may be determined by

$T_A = 0.035(h_n)^{3/4}$ for steel moment-resisting frames

$\quad = 0.030(h_n)^{3/4}$ for reinforced concrete moment frames and eccentrically braced frames

$\quad = 0.020(h_n)^{3/4}$ for all other buildings

T_B calculated using more accurate methods shall not exceed the following values:

T_B should not be more than 1.30 T_A for zone 4

T_B should not be more than 1.40 T_A for zones 1, 2A, 2B, and 3

N_a = near-source factor used in the determination of C_a in seismic zone 4, related to the proximity of the building to known faults with maximum moment magnitude and slip rates (Tables 2.6 and 2.8) (UBC Tables 16-S and 16-U)

N_v = near-source factor used in the determination of C_v in seismic zone 4, related to the proximity of the building to known faults with maximum moment magnitude and slip rates (Tables 2.10 and 2.8) (UBC Tables 16-T and 16-U)

S_A, S_B, S_C, S_D, S_E, and S_F = soil profile types (Table 2.7)   (UBC Table 16-J)

where

S_A = hard rock
S_B = soft rock, normally found in California
S_C = very dense soil and soft rock
S_D = stiff soil profile
S_E = soft soil profile
S_F = Soil profile requiring site-specific evaluation. Types of soils include soils vulnerable to potential failure under seismic loading, peats, organic clays, very high-plasticity clays, and very thick soft-to-medium stiff clays with depths in excess of 120 feet.

2.3.19.1. *Example Building*

Given. A 12-story steel building located in downtown Los Angeles, California. The lateral-load-resisting system consists of special moment-resisting space frames (SMRFs), interacting with eccentric braced frames (ECBs). The far ends of beams in the ECBs are moment-connected to columns. The building has two, 12-feet-high basement levels. The shear base is at ground level. (This example building is also analyzed using ASCE 7-02 Seismic provisions in Sect. 2.4.6. See Fig. 2.46.)

Building height h_n = Two @ 15 + 10 @ 13 = 160 ft

Plan dimensions = 90 feet × 210 feet

Fundamental period T_B from a computer analysis = 2 secs

Seismic zone factor $Z = 0.4$

Near-source factor $N_a = 1.0$

Near-source factor $N_v = 1.0$

Importance factor $I = 1.0$

Basic structural system = dual system; steel EBF with steel SMRF.

Soil type = S_D

Redundancy/reliability factor $\rho = 1.0$

Required. Using the Equivalent Lateral Procedure of UBC 1997, determine the following:

- Base shear V.
- Seismic forces, i.e., the F_x forces.
- Overturning moments.
- Diaphragm design forces.
- Maximum allowable inelastic response displacement, Δ_M. This is the total drift that occurs when the structure is subjected to design-basis ground motion, including estimated elastic and inelastic contributions to the deformation.
- Seismic force amplification factor, Ω_o required for the design of critical elements such as columns below structural discontinuity.

Solution. The purpose of the example is to illustrate the equivalent static procedure. Therefore, we make a daring assumption that none of the irregularities requiring dynamic analysis occurs in our building, an unlikely scenario in real-world building engineering.

The seismic weight W of the building is the summation of story weights w_x of each floor, with an allowance for exterior cladding, partitions, finishes, etc. This is not stated in the problem. Therefore, we calculate W by assuming that the building is of a construction typical in the Los Angeles area. These assumptions will have to be refined as more detail information becomes available during the final design phase.

Building Seismic Weight W. W is calculated as the summation of story weights tributary to each floor. It includes the weight of the floor system with an allowance for finishes, ceiling, mechanical and air conditioning duct work, weights of walls, columns, exterior cladding, and a code-required allowance of a minimum of 10 psf for partitions. As an approximation, for preliminary design, the weight W may be estimated as follows.

The unit weight of structural steel for a 12-story building including lateral-load-resisting frames, gravity columns, and beams from Fig. 8.29 of Chapter 8, is equal to 13 psf. Making an allowance for connections at 7%, additional steel weight = $7/100 \times 13 = 0.91$ psf, say, 1 psf.

Total steel weight = unit weight of steel framing + connections = 13 + 1 = 14 psf
Unit weight of $3^1/_4$ lightweight concrete topping on a 3"-deep metal deck, including the weight of deck = 50 psf
Allowance for finishes, partitions, and exterior cladding = 10 + 10 + 10 = 30 psf
Total seismic weight = 14 + 50 + 30 = 94 psf, use 100 psf
Building area: floors two through roof = $12 \times 90 \times 210 = 226,800$ ft^2
Seismic weight $W = 226,800 \times 100/1000 = 22,680$ kips

Assume, for purposes of preliminary design, all floors including the roof have the same seismic weight. Hence, seismic weight tributary to each floor and roof:

$$W_x = \frac{W}{12} = \frac{22,680}{12} = 1890 \text{ kips}$$

Seismic Data

The building is located in seismic zone 4   (Fig. 2.36, UBC Fig. 16.2)

Soil profile $= S_D$ (Given)

Lateral-load-resisting system is given as a dual system consisting of SMRFs with EBFs.

The structural system is permitted in zone 4   (Table 2.4, UBC Table 16-N)

Building height above shear base, $h_n = 15 + 15 + 10 @ 13 = 160$ ft

Building Period T_A from Method A

$$T_A = C_t(h_n)^{3/4} \qquad \text{[Eq. (2.24), UBC Eq. (30.8)]}$$

$$\begin{aligned} T_A &= 0.035(160)^{3/4} \\ &= 1.57 \text{ sec} \end{aligned}$$

$$T_B = 2 \text{ secs} \qquad \text{(Given)}$$

Period T for determining the base shear must not exceed

$$1.3\ T_A = 1.3 \times 1.57 = 2.04 \text{ secs} > T_B = 2.0 \text{ secs}$$

Therefore, T for design $= 2.0$ secs

Seismic Coefficients

$Z = 0.4$ (Sect. 2.3.3, UBC Table 16-I)

$I = 1$ (Sect. 2.3.4, UBC Table 16-K)

$R = 8.5$ (Table 2.4, UBC Table 16-N)

$C_a = 0.44N_a = 0.44 \times 1 = 0.44$ (Table 2.6, UBC Table 16-Q)

$C_v = 0.64N_v = 0.64 \times 1 = 0.64$ (Table 2.5, UBC Table 16-R)

$C_t = 0.035$ [Eq. (2.24), UBC Eq. (30.8)]

Design Base Shear V

$$\begin{aligned} V &= \frac{C_v I}{RT} W \\ &= \frac{0.64 \times 1}{8.5 \times 2} \times 22,680 \\ &= 854 \text{ kips} \end{aligned} \qquad \text{[UBC Eq. (30.4)]}$$

$$\begin{aligned} \text{Maximum}\ \ V &= \frac{2.5\, C_a I}{R} W \\ &= \frac{2.5 \times 0.44 \times 1}{8.5} \times 22,680 \\ &= 2935 \text{ kips} \end{aligned} \qquad \text{[UBC Eq. (30.5)]}$$

$$\begin{aligned} \text{Minimum } V &= 0.11\ C_a IW \\ &= 0.11 \times 0.44 \times 1 \times 22,680 \\ &= 1098 \text{ kips} \leftarrow \text{controls} \end{aligned} \qquad \text{[UBC Eq. (30.6)]}$$

Minimum V for buildings in zone 4

$$V = \frac{0.8ZN_V I}{R} W$$
$$= \frac{0.8 \times 0.4 \times 0.64}{8.5} \times 22{,}680$$
$$= 546 \text{ kips} \qquad\qquad\qquad\qquad [\text{UBC Eq. (30.7)}]$$

Controlling Design Base Shear V = 1098 kips
Base shear as a percentage of gravity is

$$\frac{1098}{22{,}680} \times 100 = 4.84\% \text{ of } g$$

Seismic Forces (F_x Forces)
Horizontal concentrated seismic force F_t at top in addition to F_x

$$F_t = 0.07\ TV$$
$$= 0.07 \times 2 \times 1098$$
$$= 154 \text{ kips}$$

Seismic forces at level x, i.e., the F_x forces, are given by

$$F_x = \frac{(V - F_t)W_x h_n}{\sum\limits_{i=1}^{n} W_i h_i}$$

For the example,

$$F_x \text{ at roof } = \frac{(1098 - 154) \times 302400}{2003400}$$
$$= 142 \text{ kips}$$

Adding $F_t = 154$ kips, total shear at the roof level,

$$F_{\text{roof}} = 154 + 142 = 296 \text{ kips}$$

The values of F_x for other floors are shown in Table 2.11.
Overturning Moments. Overturning moments at various story levels are given by the product of the story shear and the story height.

$$M_x = F_t(h_n - h_x) + \sum\limits_{i=x}^{n} F_i(h_i - h_x)$$

For the example, M_x at level 11 is equal to

$$M_n = 154(160 - 134) + 142(160 - 134) + 131(147 - 134)$$
$$= 4004 + 3692 + 1703$$
$$= 9399 \text{ kip-ft}$$

The values of M_x for other floors are shown in Table 2.11.

TABLE 2.11 Static Procedure: Design Example, UBC 1997

Level (1)	Height h above shear base, ft (2)	Weight w at height h kips (3)	Σ weight Σ w kips (4) = Σ (3)	wh (5) = (2) × (3)	Vertical distribution factor wh/Σwh (6)	Lateral seismic force F_x, kips (7)	Lateral seismic story shear kips $\Sigma F_x \, \Sigma$ (7) (8)	Overturning moment (7) × (2) (9)	Diaphragm design seismic coefficient (10)	Allowable maximum inelastic displacement Δ_M(in.) (11)
Roof	160.0	1,890	1,890	302,400	0.151	296	296	—	0.156	3.12
12	147.0	1,890	3,780	277,830	0.139	131	427	3,848	0.113	3.12
11	134.0	1,890	5,670	253,260	0.126	119	546	9,399	0.0963	3.12
10	121.0	1,890	7,560	228,690	0.114	108	654	16,497	0.086	3.12
9	108.0	1,890	9,450	204,120	0.102	96	750	24,999	0.079	3.12
8	95.0	1,890	11,340	179,550	0.090	85	835	34,749	0.074	3.12
7	82.0	1,890	13,230	154,980	0.077	73	908	45,604	0.0686	3.12
6	69.0	1,890	15,120	130,410	0.065	61	969	57,408	0.0641	3.12
5	56.0	1,890	17,010	105,840	0.053	50	1,019	70,005	0.060	3.12
4	43.0	1,890	18,900	81,270	0.041	38	1,058	83,252	0.056	3.12
3	30.0	1,890	20,790	56,700	0.028	27	1,084	97,006	0.052	3.6
2	15.0	1,890	22,680	28,350	0.014	13	1,098	113,266	0.048	3.6
1	0	—	—	—	—	—	—	129,736	—	—
Σ		23,680		2,003,400	1.00	1,098	—	129,736		

Note: Seismic force at roof = $F_t + F_x = 154 + 142 = 296$ kips

Δ_M = maximum inelastic displacement = $0.7R\Delta_S$

Δ_S = elastic displacement computed from analysis for code-prescribed seismic forces

R = response modification factor

Diaphragm Design Forces. Floor and roof diaphragm forces due to earthquake loads are computed by the equation

$$F_{px} = \frac{F_t + \sum\limits_{i=x}^{n} F_i}{\sum\limits_{i=x}^{n} W_i} W_{px}$$

For example, the diaphragm force at level 11 is given by

$$F_{pn} = \frac{154 + 142 + 131 + 119}{3 \times 1890} W_{px}$$
$$= 0.0963\, W_{px}$$

The multiplier 0.0963 may be considered as an effective acceleration for computing the diaphragm forces. Values for other floors are tabulated in Table 2.11.

Maximum Allowable Inelastic Drift Δ_M. The building period T is greater than 0.7 secs.

Therefore, the maximum allowable Δ_M, between ground and first level, is limited to

$$\Delta_M = 0.02h$$
$$= 0.02 \times 15 \times 12$$
$$= 3.6 \text{ in.} \qquad \text{(shown in column II of Table 2.11)}$$

The values of Δ_M for other floors are shown in Table 2.11.

Seismic Force Amplication Factor Ω_o. From Table 2.4 (UBC Table 16-N), for the example building with a dual system of lateral bracing consisting of EBFs and SMRFs, the value for Ω_o is equal to 2.8.

2.3.20. OSHPD and DSA Seismic Design Requirements

These bodies operate in California, not in 49 other states. The acronym OSHPD stands for Office of Statewide Health Planning and Development, while DSA stands for Division of State Architect, Office of Regulatory Services.

These two entities dictate the structural design of certain types of structures such as schools and hospitals. The requirements are given in the 2001 California Building Code (CBC 2001). The latest supplement to this document is dated January 30, 2004.

Building regulated by DSA include community colleges, public schools, and state-owned or -leased essential service buildings. Buildings regulated by OSHPD include general acute-care hospitals such as skilled nursing facilities, intermediate-care facilities, and correction treatment centers.

In addition to safeguarding buildings against major failures and loss of life, buildings designed in accordance with CBC requirements are expected to be functional during and immediately after a strong earthquake.

A brief description of some of the design considerations follows. The designer is referred to the source documents for a detailed description.

- Irregular features include, but are not limited to, those described in the 1997 UBC.
- Ordinary moment-resisting frames, (OMRFs) continue to be recognized by DSA, but not by OSHPD. Although their use in dual systems is not permitted for buildings such as hospitals, they are for DSA-controlled buildings.

- In dual systems, if the moment frame resists less than 25% of the base shears, the force in the moment frame are required to be proportioned by a factor equal to

  0.25 V/V_F

  where

  V = the total design base shear
  V_F = the portion of the base shear carried by the moment frame

- The analysis of regular or irregular, structures located on soil profile type S_F that have a period greater than 0.5 sec shall include the effects of the soils at the site.
- Structures with a discontinuity in capacity, vertical irregularity type 5, are not permitted.
- Structures with severe soft story, vertical irregularity, are not permitted.
- Structures with severe torsional irregularity, plan irregularity, are not permitted.
- Where buildings provide lateral support for walls retaining earth, and the exterior grades on opposite sides of the building differ by more than 6 feet, the load combination of the seismic increment of earth pressure due to earth-quake action on the higher side, as determined by a civil engineer qualified in soils engineering, plus the difference in earth pressures shall be added to the lateral forces discussed in this section.
- The period T_A calculated from method A is given by the formula

$$T_A = \frac{C_t(h_n)^{3/4}}{IN_v}$$

- T_A shall not be greater than T_B. If method B is not used to compute T, then the value of T shall be taken to be

$$T = \frac{T_A}{IN_v}$$

- Additional requirements for a two-stage analysis are: 1) where the design of elements of the upper portions is governed by special seismic load combinations, the special loads shall be considered in the design of the lower portion; and 2) The detailing requirements required by lateral system of the upper portion shall be used for structural components common to the structural system of the lower portion.
- The elastic design response spectrum constructed using C_a and C_v values may be used for regular structures only. A site-specific response spectrum shall be used for irregular structures and for all structures located on soil profile type S_F.
- Upper-bound earthquake ground motion is defined as motion having a 10% probability of being exceeded in a 100-year period or as the maximum level of motion that may ever be expected at the building site within the known geological framework. Structures shall be designed to sustain upper-bound earthquake motion, including $P\Delta$ effects, without forming a story collapse mechanism along any frameline. Every structure shall have sufficient ductility and strength to undergo the displacement caused by upper-bound earthquake motion without collapse. For irregular or unusual structures located in an area having large site-specific ground motion, criteria as determined by the project architect or structural engineer and approved by the enforcement agency will be required to demonstrate safety against collapse from upper-bound earthquake motion.
- The mathematical model of buildings with diaphragm discontinuities shall explicitly include the effect of diaphragm stiffness.

- For regular structures, the dynamic base shear shall not be less than 100% of the static base shear.
- For irregular structures with vertical irregularity types 1a, 2, or 5, dynamic base shear shall not be less than 125% of static base shear. (Refer to the CBC Addendum for exceptions.)
- Foundations shall be capable of transmitting the design base shear and the overturning forces from the structure into the supporting soil.
- The foundation and the connection of the superstructure elements to the foundation shall have the strength to resist—in addition to gravity loads—the lesser of the following seismic loads: 1) the strength of the superstructure elements; 2) the maximum forces that would occur in the fully yielded structural system; or 3) Ω_o times the forces in the superstructure elements resulting from the seismic forces as prescribed in this chapter. Exceptions: 1) where structures are designed using $R \leq 2.2$, as for inverted pendulum-type structures; 2) when it can be demonstrated that inelastic deformation of the foundation and superstructure-to-foundation connection will not result in a weak story or cause collapse of the structure.
- Where moment resistance is assumed at the base of the superstructure elements, the rotation and flexural deformation of the foundation as well as deformation of the superstructure-to-foundation connection shall be considered in the drift and deformation compatibility analyses.

2.3.2.1. *LARUCP Amendment to CBC Drift Limitations*

The Los Angeles Regional Uniform Code Program (LARUCP) has amended certain CBC sections related to drift calculations, discussed in the following section.

1630.10.2 Calculated. Calculated story drift using Δ_M shall not exceed 0.025 times the story height for structures having a fundamental period less than 0.5 sec. For structures having a fundamental period of 0.5 sec or greater, the calculated story drift shall not exceed $0.020/T^{1/3}$ times the story drift. (Note: The exceptions remain unchanged.)

1630.10.3 Limitations. The design lateral forces used to determine the calculated drift may disregard the limitations of Formula (30-6) and (30-7) and may be based on the period determined from Formula (30-10), neglecting the 30 to 40% limitations of Sec. 1630.2.2, Item 2. Figure 2.42a shows a graph for determining the allowable inelastic drift Δ_M based on the amended drift limitation.

Example.

Given. A 30-story building in Los Angeles with fundamental period $T = 3.5$ secs. The floor-to-floor height $h = 13'\text{-}0"$. The seismic coefficient $R = 8.5$.

Required. Using the LARUCP amendment, determine the maximum permissible inelastic displacement Δ_M and elastic displacement Δ_S.

Solution. From the graph shown in Fig. 2.42a, for $T = 3.5$ secs

$$\text{the maximum allowable } \Delta_M = \frac{h}{76}$$

$$= \frac{13}{76} = 0.171 \text{ ft} = 2.07 \text{ in.}$$

Since $\Delta_M = 0.7R\Delta_S$

$\Delta_S = \Delta_M/0.7R = 2.07/0.7 \times 8.5 = 0.348$ in.

Note: 1997 UBC Eq. (30.7) does not apply to drift calculations.

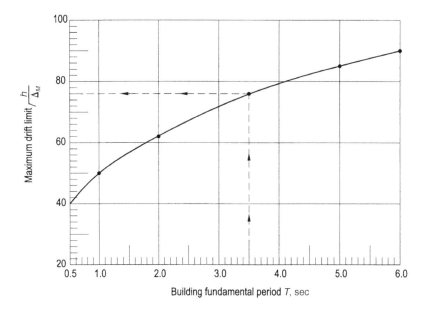

Figure 2.42a.   Permissible drift ratio h/Δ_M versus building fundamental period T. (LARUCP 2004.)

2.3.2.2.   The New Blue Book[a]

First published in 1959, the SEAOC Blue Book, *Recommended Lateral Force Requirements and Commentary,* has been the vanguard of earthquake engineering in California and the world. The new edition will first be published as an online serial, beginning in summer, 2004. Like its predecessors, it will offer background and commentary on the California Building Code (CBC), and be the vehicle for formalizing positions of SEAOC's Seismology Committee.

Blue Book commentary will be based on ASCE 7-02, expected to be the basis of the earthquake provisions in the next California Code. However, to ease the transition from existing codes, ample reference will be made to the 1977 Uniform Building Code and the 2001 CBC.

The new Blue Book will address how the new research findings by SAC Joint Venture and others should be incorporated into new codes. Publication as a serial will allow more frequent and timely revisions that address new research. However, the new format also means that some topics will not be revisited right away. Meanwhile, the 1999, 7th Edition of the Blue Book will continue to serve as a reference document.

According to its editor, David Bonowitz, the new Blue Book, like its predecessors, offers an "objective explanation of why we design the way we do." In doing so, it takes on a new role, "concerned less with writing code provisions and more with improving actual engineering practice." Its new role is "no longer to justify a code provision but to assess its impact on practice and to confirm whether it is supported by relevant analyses, testing, experience or judgment." The Blue Book's primary goal will be to guide earthquake-engineering practice by "illuminating the Code with history and context," thereby representing an "independent and authoritative statewide consensus."

The new Blue Book will comprise a series of technical articles, each covering a specific topic that includes the historical importance of the topic, the Code approach to

---

[a] This section is condensed from "The New Blue Book," David Bonowitz, S.E., Technical Editor, 2003 SEAOC Blue Book, published in *Proceedings, SEAOC 72nd Annual Convention.*

it, and proposed short-term changes and long-term studies related to the topic. Most articles will be similar to previous Blue Book commentary, offering technical explanations and Code interpretations that affect current practice. New topics will address the status of earthquake design codes and issues that will influence earthquake engineering of the future.

2.4. ASCE 7-02, IBC 2003, AND NFPA 5000: SEISMIC PROVISIONS

Before discussing the seismic provisions of the above referenced standards, it is instructive to briefly dwell on their evolution, particularly in the United States. The code development process for seismic provisions is less than 80 years old. In 1926, the Pacific Coast Building Officials published the first edition of the Uniform Building Code (UBC) with nonmandatory seismic provisions that appeared only in an appendix. They included only a few technical requirements consisting of design for a minimum base shear equal to approximately 10% of the building's weight on soft soil sites, and 3% of the building's weight on rock or firm soil sites.

Since then, building code provisions for seismic resistance have evolved on a largely empirical basis. Following the occurrence of damaging earthquakes, engineers investigated the damage, tried to understand why certain buildings and structures performed in an unsatisfactory manner, and developed recommendations on how to avoid similar vulnerabilities. Examples include limitations on the use of unreinforced masonry in regions anticipated to experience strong ground shaking, requirements to positively anchor concrete and masonry walls to floor and roof diaphragms, and limitations on the use of certain irregular building configurations.

The focus of seismic code development has traditionally been on California, the region where most U.S. earthquakes have occurred. Periodically, recommendations were published in the form of a best practice guide, the *Recommended Lateral Force Requirements and Commentary*, or more simply, the blue book, because it traditionally had a blue cover.

In 1971, the San Fernando earthquake demonstrated that the code provisions in place at the time were inadequate and that major revision was necessary. To accomplish this, the Applied Technology Council (ATC) was founded to perform the research and development necessary to improve the code. This effort culminated in 1978, with publication of ATC3.06, a report titled *Tentative Recommended Provisions for Seismic Regulation of Buildings*. The Structural Engineers Association of California (SEAOC) incorporated many of the recommendations in that report into the 1988 edition of the UBC. Perhaps more important, however, was that publication of this report coincided with the adoption of the National Earthquake Hazards Reduction Program (NEHRP).

Although NEHRP provisions were first published in 1985, they were not formally used as the basis of any model building codes until the early 1990s. Prior to that time, these codes had adopted seismic provisions based on the American National Standards Institute's (ANSI) publication, ANSI A58.1 (later ASCE-7), which had been based on SEAOC recommendations. In 1993, the American Society of Civil Engineers (ASCE) revised its ASCE-7 standard to include seismic provisions that closely mirrored the NEHRP document.

Design for seismic resistance involves providing structures with proper configuration and continuity, adequate strength and stiffness, and structural detailing capable of resisting inelastic cyclic loading. The NEHRP provisions deal with all these aspects of seismic design, which are closely interrelated. A building's configuration and its inelastic toughness, as controlled by its detailing, affects the amount of ductility it can exhibit, and therefore, the amount of strength required.

Currently, two model building codes are developed and maintained in the United States. One is the International Building Code (IBC), developed by the International Code Council

(ICC), a coalition of the Building Code officials and Administrators International, (BOCA), International Conference of Building Officials (ICBO), and Southern Building Code Conference International (SBCCI). The other is the National Fire Protection Agency (NFPA) 5000 Code. Although there are differences in the way these codes are developed, from the perspective of structural engineering design criteria, the requirements are very similar and in many aspects, essentially identical. NFPA adopts seismic design provisions, and indeed nearly all structural design criteria, by reference to industry standards (ASCE-7, ACI-318, etc.). From a seismic perspective, these provisions are all based on and consistent with NEHRP provisions. The seismic provisions in the first edition of the IBC were developed directly from NEHRP provisions. In the most recent (2003) edition, greater reliance has been placed on reference to the same industry standards adopted by the NFPA, and in the future, the ICC may decide to primarily adopt provisions by reference only, as does the NFPA.

Both model building codes—the IBC-03 and the 2003 edition of NFPA 5000—are now in print. Both these codes have incorporated major national standards such as AISC, ACI, and the seismic provisions of ASCE 7-02 "by reference." The ASCE-7 is used as a reference for loads; load combinations; and seismic, wind, and show loads. AISC is a reference for steel design, ACI 318 is a reference for concrete design, ACI 530 is a reference for masonry design, and the National Design Specifications for wood, (NDS) are a reference for wood construction.

It should be noted that some of the sections in these codes are hybrids consisting of provisions stated directly in them, with selected sections of ANSI standards incorporated by reference into other code provisions. Other code provisions, such as for soils and foundations, remain as "stand alone" provisions.

The seismic section of the 2003 IBC is still 42 pages long, reduced from 75 pages in the 2000 IBC. Whereas the 2000 IBC was a stand-alone document, the 2003 IBC has ASCE-7 embedded throughout the seismic section. Therefore, instead of wading through the 42 pages of IBC provisions, the user can go directly to ASCE-7. IBC allows one to do so, in Exception 1, Paragraph 1614.1.NFPA, on the other hand, does not include any seismic provisions within its document. In a total of eight paragraphs it simply refers to ASCE-7.

From the preceding discussion, it is obvious that if designers use the provisions given in ASCE-7, they will be satisfying both the IBC and NFPA, a unique situation that has not occurred before. Since either of these is likely to be adopted by governing agencies throughout the United States in the near future, designers have a unique fortuity of working with just one code. Additionally, ASCE 7-02 is likely to form the basis for most earthquake-resistant design in other nations that base their codes on U.S. practices.

Seismic design using the provisions of ASCE 7-02 is fully permitted by IBC 2003. However, it is important to understand, as stated in the Appendix Section A.9.1. of ASCE 7-02, that the seismic provisions are not directly related to computation of earthquake loads. As in other code provisions, the design loads are based on the assumption that substantial cyclic inelastic strain capacity exists in the structure.

The 2003 IBC is organized such that for structures that can be designed according to the Simplified Analysis Procedure, Section 1617.5, all of the seismic design provisions are contained within the code itself. In reading the IBC, unfortunately, the intent that IBC's applicability is limited to Simplified Analysis Procedure does not come through quite clearly. A certain amount of potential confusion exists within the code.

It is important to know that when using the IBC 2003 Simplified Analysis Procedure, the designer is required to use the redundancy provisions of Section 1617.2.2. This section has some important modifications to ASCE 7-02, most notably in the determination of Seismic Design Categories (SDCs).

Recall that the SDC for a structure is determined twice: once based on the short-period design response acceleration, S_{DS}, and a second time based on the long-period design spectral response acceleration, S_{D1}. The more severe of the two SDCs governs the design of the structure. This is well and good for long-period structures, but is a wasted and unnecessary effort for short-period buildings. The amendments to ASCE 7-02, already incorporated in the IBC-2003, address the issue and permit the SDC to be determined based on short-period spectral acceleration S_{DS} alone when the following conditions are met:

- The approximate fundamental period of the structure T_a, in each orthogonal direction, is less than $0.8T_s$.
- Eq. (9.5.5.2.1-1) of ASCE-7 is used to determine the seismic response coefficient, C_s.
- The diaphragms are rigid.

Thus, IBC 2003 provides the user with two distinct choices for seismic design:

- Use ASCE 7-02 instead of IBC 2003, as permitted by IBC.
- For structures designed using simplified procedures, use Sections 1613 through 1623 of IBC 2003 that adopt many provisions of ASCE 7-02 by reference, but have also incorporated certain amendments to ASCE 7-02, most notably to the determination of SDC.

Although the second option enables the designer to take advantage of the amendments to ASCE 7-02, the first option is easier and hence, more attractive.

The seismic design provisions presented in the following sections are based on ASCE 7-02. Since both IBC 2003 and NFPA 5000 have adapted this document by reference, the design provisions given here apply equally to these model codes. Therefore, although for simplicity, only ASCE-02 is referenced in the following text, it is understood that the provisions are also applicable to IBC 2003 and NFPA 5000.

2.4.1. Seismic Design Highlights: ASCE 7-02, IBC-03, NFPA 5000

ASCE 7-02 utilizes spectral response seismic design maps to quantify seismic hazards on the basis of the contours. These maps were prepared by the U.S. Geological Survey (USGS), along with a companion CD-ROM. This CD-ROM provides mapped spectral values for a specific site based on the site's longitude, latitude, and site soil classification. Examples are shown in Figs. 2.43 and 2.44. Longitude and latitude for a given address can be found at Web sites such as www.Oeocode.com. Use of the CD-ROM is recommended for establishing spectral values for design, since the maps found in ASCE-7 and at Web sites are at too large a scale to provide accurate spectral values for most sites.

The origin of present-day (2004) seismic codes may be traced back to the 1971 San Fernando Valley earthquake, which demonstrated that design rules of that time for seismic resistance had some serious shortcomings. Each subsequent major earthquake has taught new lessons. Seismic codes such as ASCE-7 have endeavored to work to improve each succeeding edition so that it would be based upon the best earthquake engineering research as applicable to design and construction and that they would have nationwide applicability.

The seismic provisions are stated in terms of forces and loads; however, the designer should always bear in mind that there are no external forces applied to the aboveground portion of a structure during an earthquake. The design forces are intended only as approximations to produce the same deformations, when multiplied by the deflection amplification factor C_d, as would occur in the same structure should an earthquake ground motion at the design level occur.

The design limit state for resistance to an earthquake is unlike that for any other load. The earthquake limit state is based upon system performance, not member performance. In

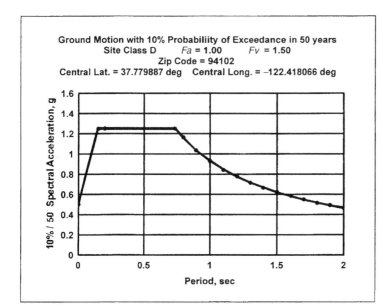

Period, sec	10%/50 Sa, g
0.00	0.500
0.15	1.250
0.20	1.250
0.74	1.250
0.80	1.163
0.90	1.034
1.00	0.930
1.10	0.846
1.20	0.775
1.30	0.716
1.40	0.665
1.50	0.620
1.60	0.582
1.70	0.547
1.80	0.517
1.90	0.490
2.00	0.465

Figure 2.43. Design basis earthquake ground motions; 10% probability of exceedence in 50 years, corresponding to a return period of 474 years, typically rounded to 500 years.

determining design forces, considerable energy dissipation through repeated cycles of inelastic straining is assumed. The reason is the large demand exerted by the earthquake and the associated high cost of providing enough strength to maintain linear elastic response in ordinary buildings. This unusual limit state means that several conveniences of elastic behavior, such as the principle of superposition, are not applicable. This is the reason why seismic provisions contain so may provisions that modify customary requirements for proportioning and detailing structural members and systems. It is also the reason for more stringent construction quality assurance requirements.

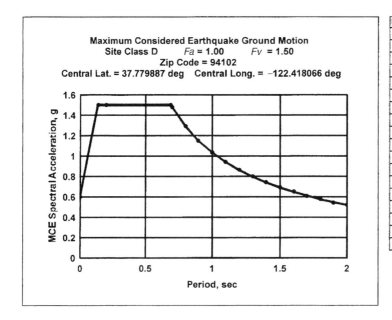

Period, sec	MCE Sa, g
0.00	0.600
0.14	1.500
0.20	1.500
0.69	1.500
0.70	1.479
0.80	1.294
0.90	1.151
1.00	1.036
1.10	0.941
1.20	0.863
1.30	0.797
1.40	0.740
1.50	0.690
1.60	0.647
1.70	0.609
1.80	0.575
1.90	0.545
2.00	0.518

Figure 2.44. Maximum considered earthquake ground motions; 2% probability of exceedence in 50 years, corresponding to a return period of 2475 years, typically rounded to 2500 years.

2.4.1.1. Importance Factor I

The purpose of this factor, I, is to specifically improve the capability of essential facilities and structures containing substantial quantities of hazardous materials to function during and after design earthquakes. This is achieved by introducing the occupancy importance factor of 1.25 for seismic use group (SUG) II structures and 1.5 for SUG III structures. This factor is intended to reduce the ductility demands and result in less damage. When combined with the more stringent drift limits for such hazardous facilities, the result is improved performance of such facilities.

2.4.1.2. Redundancy

This factor applies for structures in seismic design categories (SDC) D, E, and F. The value of this factor varies from 1.0 to 1.5. It has the effect of reducing the R factor for less redundant structures, thereby increasing the seismic demand. The factor recognizes the need to quantify the issue of redundancy in the design. It should be noted that many nonredundant structures have been designed in the past using values of R that were intended for use in designing structures with higher levels of redundancy. The intent of redundancy factor is to prevent such misuse.

2.4.1.3. Elements Supporting Discontinuous Walls or Frames

The purpose of the special load combinations is to protect the gravity load-carrying system against possible overloads caused by overstrength of the lateral force-resisting system. Either columns or beams may be subject to such failure; therefore, both should include this design requirement. Beams may be subject to failure due to overloads in either the downward or upward directions of force. Examples include reinforced concrete beams or unbraced flanges of steel beams or trusses. Hence, the provision has not been limited simply to downward force, but instead to the larger context of vertical load. An issue that has not been fully addressed in the ASCE-7 is clarification of the appropriate load case for the design of the connection between the discontinuous walls or frames and the supporting elements.

2.4.1.4. Special Seismic Load Combinations

Some elements of properly detailed structures are not capable of safely resisting ground shaking demands through inelastic behavior. To ensure safety, these elements must be designed with sufficient strength to remain elastic. The Ω_o coefficient approximates the inherent overstrength in typical structures having different seismic force-resisting systems. The special seismic loads, factored by the Ω_o coefficient, are an approximation of the maximum force these elements are ever likely to experience. ASCE-7 permits the special seismic loads to be taken as less than the amount computed by applying the Ω_o coefficient to the design seismic forces when it can be shown that yielding of other elements in the structure will limit the amount of load that can be delivered to the element. A case in point is the axial load induced in a column of a moment-resisting frame from the shear forces in the beams that connect to this column. The axial loads due to lateral seismic action need never be taken greater than the sum of the shears in these beams at the development of a full structural mechanism, considering the probable strength of the materials and strain-hardening effect. For frames controlled by beam hinge-type mechanisms, this would typically be $2Mp/L$, where for steel frames, M_p is the expected plastic moment capacity of the beam as defined in the AISC Seismic specifications. For concrete frames, M_p is the probable flexural strength of the beams. L is the clear span length for both steel and concrete beams. In the context of seismic design, the term capacity means the expected or median anticipated strength of the element, considering potential variation in material yield strength- and strain-hardening effects. When calculating the capacity of elements for this purpose, material strengths should not be reduced by capacity or resistance factors.

2.4.1.5. Torsion

Where earthquake forces are applied concurrently in two orthogonal directions, the 5% displacement of the center of mass should be applied along a single orthogonal axis chosen to produce the greatest effect, but need not be applied simultaneously along two axes (i.e., in a diagonal direction).

Most diaphragms of light-framed construction are somewhere between rigid and flexible for analysis purpose, i.e., they are semirigid. Such diaphragm behavior is difficult to analyze when considering torsion of the structure. As a result, it is believed that consideration of the amplification of the torsional moment is a refinement that is not warranted for light-framed construction.

The intent is not to amplify the actual, i.e., the calculated torsion component, but only the component due to accidental torsion. There is no theoretical justification to further increase design forces by amplifying both components together.

2.4.1.6. Relative Displacements

The design of some nonstructural components that span vertically in the structure can be complicated when supports for the element do not occur at horizontal diaphragms. Therefore, story drift must be accommodated in the elements that will actually distort. For example, a glazing system supported by precast concrete spandrels must be designed to accommodate the full story drift, even though the height of the glazing system is only a fraction of the floor-to-floor height. The condition arises because the precast spandrels will behave as rigid bodies relative to the glazing system and therefore, all the drift must be accommodated by the joint between the precast spandrel and the glazing unit.

2.4.1.7. Special Requirements for Piles and Grade Beam

Anchorage of the pile into the pile cap should be conservatively designed to allow energy-dissipating mechanisms, such as rocking, to occur in the soil without structural failure of the pile. Precast prestressed concrete piles are exempt from the concrete special moment frame column confinement requirements since these requirements were never intended for slender, precast prestressed concrete elements and will result in unbuildable piles. These piles have been proven through cyclic testing to have adequate performance with substantially less confinement reinforcing than required by ACI 318. Therefore, a transverse steel ratio reduced from that required in frame columns is permitted in concrete piles. It should be noted that confinement provided by the soil improves the behavior of concrete piles.

Batter pile systems that are partially embedded have historically performed poorly under strong ground motions. Difficulties in examining fully embedded batter piles have led to uncertainties as to the extent of damage for this type of foundation. Batter piles are considered as limited ductile systems and should be designed using the special seismic load combinations.

2.4.2. ASCE 7-02: Detail Description of Seismic Provisions

Engineers who design and detail structures for many areas of the United States with low seismic risk have not had to deal with design and detailing requirements that apply to moderate and high seismic zones on the west coast. But that may change due to major revisions of seismic provisions published in ASCE 7-02.

Traditionally, the magnitude of the seismic force and level of seismic detailing were strictly a function of structure location. With the latest seismic design provisions, these are now a function of

1. Structure location.
2. Nature of the structure's use and occupancy.
3. Type of soil the structure rests upon.

Does this affect the design of a structure in a low seismic risk zone? You bet. Consider, for example, the design of an essential facility such as a hospital in Charlotte, North Carolina on a site with a soft soil profile. These two factors—the nature of the building's occupancy and the type of soil it rests upon—could place the structure in an SDC equivalent to that for seismic zones 3 or 4, indicating high seismic risk. This, is turn, triggers a whole host of seismic detailing requirements, as explained later in this section.

IBC earthquake provisions are substantially different in format and content form the 1997 UBC. Design ground motion parameters are determined from mapped values of S_s and S_1 rather than the seismic zone factor Z. The mapped contours of these parameters attain high values in the vicinity of seismic sources that are judged capable of generating large earthquakes. Therefore, near-source factors N_a and N_v of the 1997 UBC are unnecessary, and are not used in the IBC. The spectral response accelerations S_s and S_1 are specified on the seismic hazard maps prepared by the United States Geological Survey (USGS). Two sets of eight (a total of 16) maps give the maximum considered response accelerations at short periods of 0.2 seconds S_s, and at one-second periods S_1. The maps are for 5% of critical damping for site class B, soft rock, commonly found in the U.S. west coast.

Seismic design category (SDC) has replaced the seismic zone factor Z as the trigger for seismic design requirements including the choice of analysis procedure, the required level of strength and detailing, and the permissible irregularities and the height of buildings. A major departure from 1997 UBC is that detailing and other seismic restrictions are now dependent on the soil characteristics at the site of the structure. The type and usage of the building establishes the seismic use group (SUG) for the building. The SUG is based on the occupancy of the building and the consequences of severe earthquake damage to the building. Three seismic hazard groups are defined.

Group III is for facilities such as fire and police stations, hospitals, medical facilities with emergency treatment facilities, and structures containing toxic or explosive substances. Group II is for high-occupancy buildings and utilities not required for emergency backup. Group I applies to all other buildings. Based on the SUGs, an occupancy importance factor I is assigned. This factor, equal to 1.5 for Group III and 1.25 for Group II, is used to increase the design base shear for structures in SUG II and III. No increase is required for Group I structures. Hence, the importance factor is equal to 1.0 for Group I.

The mapped spectral accelerations S_s and S_1 for site class B are modified to other site conditions by using site coefficients F_a and F_v. The modified values denoted as S_{MS} and S_{M1} are the maximum considered earthquake (MCE), which has a 2% probability of occurrence in 50 years corresponding approximately to a 2500-year recurrence interval. The design response spectral accelerations S_{DS} and S_{D1} are simply the two-third values of S_{MS} and S_{M1}.

Traditionally, for seismic design, engineers on the U.S. west coast have used a ground acceleration, with a 10% probability of occurrence in 50 years corresponding to a 475-year recurrence interval. In coastal California, a 2500-year earthquake is considered the largest possible earthquake and, it is the considered opinion of the engineering community that a building with proper seismic details designed for an earthquake of 475-year recurrence interval, has a margin of safety of 1.5 against collapse in an MCE event. In other parts of the United States, however, notably in the New Madrid fault area, a 2500-year earthquake may be as much as four-to-five times the 475-year earthquake. Therefore, a building designed in California for a 475-year earthquake has a good chance of not collapsing under a 2500-year earthquake, whereas its counterpart in the New Madrid area may not have this chance. To keep a uniform margin against collapse, the IBC uses a 2500-year earthquake spectral response acceleration for all the areas of the United States. To bring the design up to par with the current practice of designing with a 1.5 margin against collapse, a 2/3 value (the reciprocal of 1.5) of the MCE is used in design. This is the rationale for taking the 2/3 values of S_{MS} and S_{M1} to arrive at the design response accelerations S_{DS} and S_{D1}.

2.4.2.1. Seismic Use Group

The expected performance of buildings under earthquakes is controlled by assignment of each building to SUGs I, II, or III, shown in Table 2.12. These SUGs are categorized based on the occupancy of the buildings within the group and the relative consequences of severe earthquake damage to these buildings. Progressively more conservative strength, drift control, system selection, and detailing requirements are specified for buildings in the three groups, in order to attain minimum levels of earthquake performance suitable to the individual occupancies.

Specific consideration is given to Group III, essential facilities required for post-earthquake recovery. Also included are structures housing substances deemed to be hazardous to the public if they are released. Group II structures are those where the occupants' ability to exit is restrained. Group I contains all uses other than those in Group III or II.

2.4.2.2. Occupancy Importance Factor I

ASCE-7 assigns seismic importance factors I to the various SUGs, as shown in Table 2.13. These factors are used to increase the design base shear for structures in SUGs II and III with the idea that the design for increased loads results in a relatively lower demand on ductility. The values of occupancy importance factor are

$I = 1.0$ for SUG I

$I = 1.25$ for SUG II

$I = 1.50$ for SUG III

TABLE 2.12 Seismic Use Groups

		Seismic use group		
		I	II	III
Occupancy category	I	X		
(Table 1.1)	II	X		
	III		X	
	IV			X

(From ASCE 7-02, Table 9.1.3.)

TABLE 2.13 Occupancy Importance Factors

Seismic use group	I
I	1.0
II	1.25
III	1.5

(From ASCE 7-02, Table 9.1.4.)

ASCE-7 uses the same seismic importance factors for structures as well as structural components, as opposed to two separate values used in the 1997 UBC. Additionally, a component importance factor I_p is given for architectural, electrical, and mechanical components in ASCE 7-02.

2.4.2.3. Maximum Considered Earthquake Ground Motion

The basis of ASCE-7 provisions is spectral accelerations resulting from an earthquake corresponding to a return period of 2500 years. This earthquake is variously termed maximum credible earthquake, maximum capable earthquake, and most recently, maximum considered earthquake (MCE). The design ground motion given in ASCE-7 is taken as two-thirds of the MCE ground motion. Such ground motion may have a return period varying from a few hundred years to a few thousand years, depending upon the regional seismicity. Group I buildings designed for this ground motion are expected to achieve the life-safe or better performance. Buildings in Group III should be able to achieve immediate occupancy or better performance for this ground motion. Group II buildings would be expected to achieve performance better than the life-safe level, but perhaps less than the immediate occupancy level for this ground motion.

The design approach is to provide for a uniform margin against collapse at the design ground motion. To accomplish this, ground motion hazards are defined in terms of the MCE ground motions. These are based on a set of rules that depend on the seismicity of an individual region. The design ground motions are based on a lower bound estimate of the margin against collapse believed to be inherent in structures designed to ASCE-7 provisions. This lower bound has been considered, based on experience, to be about a factor of 1.5 in ground motion. Consequently, the design earthquake ground motion has been selected at a ground motion shaking level that is 1/1.5, which is equal to the 2/3 of the MCE ground motion.

For most regions of the United States, the MCE ground motion is defined with a uniform likelihood of exceedence of 2% in 50 years (return period of about 2500 years). In regions of high seismicity, such as coastal California, the seismic hazard is typically controlled by large-magnitude events occurring on a limited number of well-defined fault systems. Ground shaking calculated at a 2%-in-50-years likelihood would be much larger near active faults than what would be expected based on the characteristic magnitudes of earthquakes on these known active faults. For these regions, it is considered more appropriate to directly determine maximum considered earthquake ground motions based on the characteristic earthquakes of these defined faults. To provide for an appropriate level of conservatism in the design process, when this approach to calculation of the maximum considered earthquake ground motion is used, the median estimate of ground motion resulting for the characteristics event is multiplied by 1.5.

ASCE-7 defines the maximum considered earthquake ground motion in terms of the mapped values of the spectral response acceleration at short periods, S_s, and at 1 second,

TABLE 2.14 Site Classification

Site class	$\bar{v}_a$	$\bar{N}$ or $\bar{N}_{ch}$	$\bar{s}_U$
A Hard rock	>5,000 ft/s (>1,500 m/s)	Not applicable	Not applicable
B Rock	2,500 to 5,000 ft/s (760 to 1,500 m/s)	Not applicable	Not applicable
C Very dense soil and soft rock	1,200 to 2,500 ft/s (370 to 760 m/s)	>50	>2,000 psf (>100 kPa)
D Stiff soil	600 to 1,200 ft/s (180 to 370 m/s)	15 to 50	1,000 to 2,000 psf (50 to 100 kPa)
E Soil	<600 ft/s (<180 m/s)	<15	<1000 psf (<50 kPa)

Any profile with more than 10 ft of soil having the following characteristics:
– Plasticity index $PI > 20$,
– Moisture content $\omega \geq 40\%$, and
– Undrained shear strength $\bar{s}_u < 500$ psf

F Soils requiring site-specific evaluation	1. Soils vulnerable to potential failure or collapse 2. Peats and/or highly organic clays 3. Very high plasticity clays 4. Very thick soft/medium clays

(From ASCE 7-02, Table 9.4.1.2.)

S_1, for site class B, soft rock. These values may be obtained directly from the map published by USGS. The maps specify contours of random horizontal acceleration values, and locations of faults using both the deterministic and probabilistic procedures.

2.4.2.4. Site Class

A set of six site classifications, S_A through S_F, based on the average properties of the upper 100 feet of soil profile are defined in ASCE-7 and shown in Table 2.14. Since, in practice, geotechnical investigations are seldom conducted to depths of 100 feet, ASCE-7 allows the geotechnical engineers to determine site class based on site-specific data and professional judgment.

2.4.2.5. Site Coefficients F_a and F_v

To obtain acceleration response parameters that are appropriate for sites with characteristics, other than those for S_B sites, it is necessary to modify the S_s and S_1 values. This modification is performed with the use of two coefficients, F_a and F_v, which respectively scale the S_s and S_1 values for other site conditions. The MCE spectral response accelerations adjusted for site class effects are designated, respectively, S_{MS} and S_{M1}, for short-period and 1-second period response (see Tables 2.15 and 2.16).

$$S_{MS} = F_a S_s$$

$$S_{M1} = F_v S_1$$

TABLE 2.15 Values of F_a as a Function of Site Class and Mapped Short-Period MCE Spectral Acceleration

| Site class | Mapped maximum considered earthquake spectral response acceleration at short periods | | | | |
	$S_S \leq 0.25$	$S_S = 0.5$	$S_S = 0.75$	$S_S = 1.0$	$S_S \geq 1.25$
A	0.8	0.8	0.8	0.8	0.8
B	1.0	1.0	1.0	1.0	1.0
C	1.2	1.2	1.1	1.0	1.0
D	1.6	1.4	1.2	1.1	1.0
E	2.5	1.7	1.2	0.9	0.9
F	a	a	a	a	a

Note: Use straight-line interpolation for intermediate values of S_S.

[a] Site-specific geotechnical investigation and dynamic site response analyses shall be performed except that for structures with periods of vibration equal to or less than 0.5 seconds, values of F_a for liquefiable soils may be assumed equal to the values for the site class determined without regard to liquefaction in Step 3 of Section 9.4.1.2.2.

(From ASCE 7-02, Table 9.4.1.2.4a.)

2.4.2.6. Design Spectral Response Accelerations

Design spectral response accelerations, denoted by S_{DS} and S_{D1} are given by:

$$S_{DS} = \frac{2}{3} S_{MS}$$
$$S_{D1} = \frac{2}{3} S_{M1}$$

For buildings and structures designed using the equivalent lateral force technique, the design spectral response acceleration parameters, S_{DS} and S_{D1}, which are directly used in the design, may be derived from S_S and S_1 response acceleration. For structures designed

TABLE 2.16 Values of F_v as a Function of Site Class and Mapped 1-Second Period MCE Spectral Acceleration

| Site class | Mapped maximum considered earthquake spectral response acceleration at 1-second periods | | | | |
	$S_1 \leq 0.1$	$S_1 = 0.2$	$S_1 = 0.3$	$S_1 = 0.4$	$S_1 \geq 0.5$
A	0.8	0.8	0.8	0.8	0.8
B	1.0	1.0	1.0	1.0	1.0
C	1.7	1.6	1.5	1.4	1.3
D	2.4	2.0	1.8	1.6	1.5
E	3.5	3.2	2.8	2.4	2.4
F	a	a	a	a	a

Note: Use straight-line interpolation for intermediate values of S_1.

[a] Site-specific geotechnical investigation and dynamic site response analyses shall be performed except that for structures with periods of vibration equal to or less than 0.5 seconds, values of F_v for liquefiable soils may be assumed equal to the values for the site class determined without regard to liquefaction in Step 3 of Section 9.4.1.2.2.

(From ASCE 7-02, Table 9.4.1.2.4b.)

using modal analysis procedures, a general response spectrum may be developed from the design spectral response acceleration parameters, S_{DS} and S_{D1}.

However, for some sites with special soil conditions or for some buildings with special design requirements, it may be more appropriate to determine a site-specific estimate of the MCE ground shaking response accelerations.

The mapped values S_s for the short-period acceleration has been determined at a period of 0.2 seconds. This is because 0.2 seconds is reasonably representative of the shortest effective period of the buildings and structures.

The spectral response acceleration at periods other than 1 second can typically be derived from S_1, the acceleration at 1 second. Consequently, these two response acceleration parameters, S_s and S_1, are sufficient to define an entire response spectrum for the period range of importance for most buildings and structures.

2.4.2.7. Seismic Design Category

The intent of the IBC is to provide uniform levels of performance for structures, depending on their occupancy and use and the risk to society inherent in their failure. To this end, the IBC establishes a series of SUGs that are used to categorize structures based on the specific SDC. The intent is that a uniform margin of failure to meet the seismic design criteria be provided for all structures within a given SUG.

ASCE-7 establishes five design categories that are the keys for establishing design requirements for any building based on its use (SUG) and on the level of expected seismic ground motion. Once the SDC (A, B, C, D, E, or F) for the building is established, many other requirements are related to it.

Parameters S_{DS} and S_{D1}, which include site soil effects, are used for the purpose of determining the SUG. Tables 2.17 and 2.18 are provided, relating respectively to short-period and long-period structures.

SDC A represents structures in regions where anticipated ground motions are minor, even for very long return periods. For such structures, ASCE-7 requires only that a complete lateral force-resisting system be provided and that all elements of the structure be tied together. A nominal design base shear equal to 1% of the weight of the structure is used to proportion the lateral system.

SDC B includes SUG I and II structures in regions of seismicity where only moderately destructive ground shaking is anticipated. In addition to the requirements for SDC A, structures in SDC B must be designed for forces determined using ASCE 7-02 Seismic maps.

TABLE 2.17 Seismic Design Category Based on Short-Period Response Accelerations

	Seismic use group		
Value of S_{DS}	I	II	III
$S_{DS} < 0.167g$	A	A	A
$0.167g \leq S_{DS} < 0.33g$	B	B	C
$0.33g \leq S_{DS} < 0.50g$	C	C	D
$0.50g \leq S_{DS}$	D[a]	D[a]	D[a]

[a]SUG I and II structures located on sites with mapped maximum considered earthquake spectral response acceleration at 1-second period, S_1, equal to or greater than 0.75g shall be assigned to SDC E, and SUG III structures located on such sites shall be assigned to SDC F.
(From ASCE 7-02, Table 9.4.2.1a.)

TABLE 2.18 Seismic Design Category Based on 1-Second Period Response Accelerations

Value of S_{D1}	Seismic use group		
	I	II	III
$S_{D1} < 0.067g$	A	A	A
$0.067g \leq S_{D1} < 0.133g$	B	B	C
$0.133g \leq S_{D1} < 0.20g$	C	C	D
$0.20g \leq S_{D1}$	D[a]	D[a]	D[a]

[a]SUG I and II structures located on sites with mapped maximum considered earthquake spectral response acceleration at 1-second period, S_1, equal to or greater than $0.75g$ shall be assigned to SDC E, and SUG III structures located on such sites shall be assigned to SDC F. (From ASCE 7-02, Table 9.4.2.1b.)

SDC C includes SUG III structures in regions where moderately destructive ground shaking may occur as well as SUG I and II structures in regions with somewhat more severe ground shaking potential. In SDC C, the use of some structural systems is limited and some nonstructural components must be specifically designed for seismic resistance.

SDC D includes structures of SUG I, II, or III located in regions expected to experience destructive ground shaking, but not located very near major active faults. In SDC D, severe limits are placed on the use of some structural systems and irregular structures must be subjected to dynamic analysis techniques as part of the design process.

SDC E includes SUG I and II structures in regions located very close to major active faults and SDC F includes Seismic Use Group III structures in these locations. Very severe limitations on systems, irregularities, and design methods are specified for SDC E and F. For the purpose of determining if a structure is located in a region that is very close to a major active fault, ASCE-7 uses a trigger of mapped MCE spectral response acceleration at 1-second periods, S_1 of 0.75 g or more regardless of the structure's fundamental period. The mapped short-period acceleration, S_s, is not used for this purpose because short-period response accelerations do not tend to be affected by near-source conditions as strongly as do response accelerations at longer periods.

2.4.2.8. *Development of Response Spectrum*

To proceed with an equivalent static analysis of a structure, we need to determine only the two values of the design acceleration response parameters, S_{D1} and S_{Ds}. This is because the base shear equations, discussed presently, are directly related to these parameters. However, for buildings and structures requiring modal analysis procedures, it is necessary to develop an acceleration graph, commonly referred to as an acceleration spectrum, because design acceleration values are required for an entire range of building periods. In a modal analysis, we attempt to capture the multimodal response of a building by statistically combining its individual modal responses. Therefore, accelerations corresponding to an entire range of building periods are typically required in performing the dynamic analysis.

The characteristic features of an acceleration response spectrum are as follows:

1. For very stiff buildings, the acceleration response approaches the maximum ground acceleration. Buildings in this period, with a range of 0.3 seconds or less, behave as rigid bodies attached to the ground.

2. For moderately short periods of the order of 0.1 to 0.3 seconds, the maximum response accelerations are about two to three times the maximum ground acceleration, and remain constant over this period range.
3. For long-period buildings, the maximum response velocity is the same as the maximum ground velocity.
4. For the very long-period buildings, the maximum displacement response is the same as the maximum ground displacement.

However, in the development of a generalized acceleration response curve, the constant displacement domain is not included because relatively few buildings have a period long enough to fall into this range. Thus, a generalized response curve for all practical purposes may be developed by three curves. The procedure is as follows:

1. Determine the period $T = S_{D1}/S_{DS}$, which defines the period at which the constant spectral acceleration and constant velocity portions of the spectra meet.
2. Determine the spectral acceleration at zero period, T_0, by the relation: $T_0 = 0.4\, S_{D1}$, i.e., the spectral acceleration at zero period is equal to 40% of the spectral acceleration corresponding to the flat top, S_{DS}.
3. For periods greater than or equal to T_0 and less than or equal to T_s, determine S_a by: $S_a = S_{DS}$.
4. For periods less than or equal to T_0, determine the spectral response acceleration, S_a, by: $S_a = S_{DS}(0.6\, T/T_0 + 0.4)$. This region, referred to as the upramp, is used in computer analyses to capture the modal response in the very short-period range of the building.
5. For periods greater than T_s, determine S_a by: $S_a = S_{DS}/T$, where T is the desired range of building periods corresponding to the acceleration input in the computer analysis.

Design Example

Given. A building on site class D, near the city of Memphis, which is close to the New Madrid fault. Partial regionalization maps of the MCE ground motion contours for 0.2-sec and 1.0-sec spectral response accelerations, S_s and S_1, are given in terms of percentage of g in Figs. 2.45a and 2.45b. The maps are for site class B. A site-specific response spectrum is not required for the building.

Required. Develop a general design response spectra for the building site.

Solution.

1. Read the maximum considered earthquake spectral response accelerations S_S and S_1, from the given maps (Fig. 2.45a and b). It is perhaps obvious that Figs. 2.45a and b are too small to read the values of S_S and S_1. In practice, the designer would be using the large maps developed by USGS. However, for purposes of this example, we will assign the following values for S_S and S_1:

$$S_S = 150\% \text{ of } g = 1.50g$$
$$S_1 = 40\% \text{ of } g = 0.4g$$

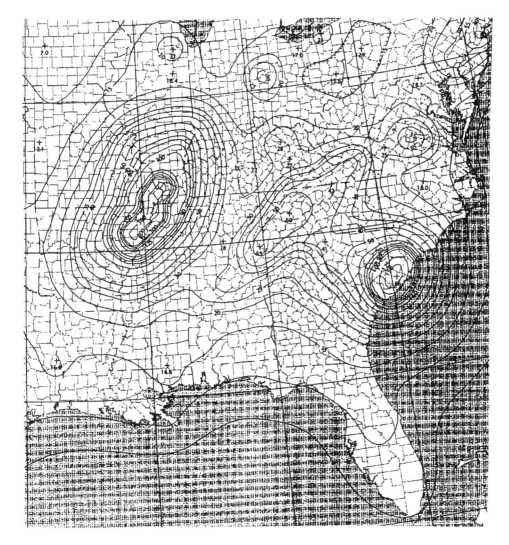

Figure 2.45a.   Maximum considered earthquake ground motion map of 0.2 sec (short-period) spectral response acceleration S_S (5% of critical damping), site class B.

2. Find site coefficients F_a and F_v from Tables 2.15 and 2.16.

    For the given site class D, and $S_S = 1.5g$ $F_a = 1.0$

    For the given site class D, and $S_1 = 0.4$, $F_v = 1.6$

3. Calculate adjusted MCE spectral response accelerations for short period as $S_{MS} = F_a S_S$, and for 1-sec period as $S_{M1} = F_v S_1$.

    $S_{MS} = 1.0 \times 1.5 = 1.5$
    $S_{M1} = 1.6 \times 0.4 = 0.64$

4. Determine design spectral response accelerations as

    $S_{DS} = {}^2/_3 \times S_{MS} = {}^2/_3 \times 1.5 = 1.0$
    $S_{D1} = {}^2/_3 \times S_{M1} = {}^2/_3 \times 0.64 = 0.43$

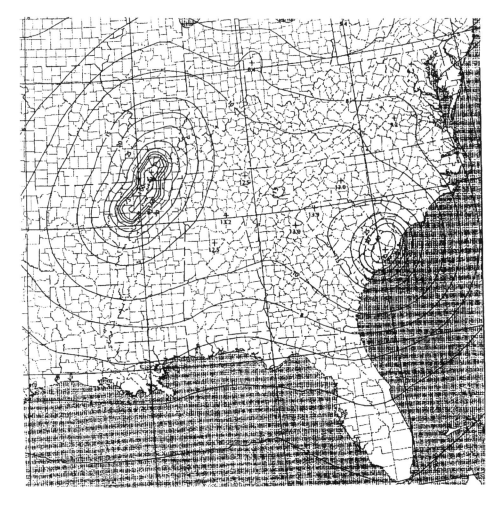

Figure 2.45b.   Maximum considered earthquake ground motion map of 1-sec spectral response acceleration, S_1 (5% of critical damping), site class B.

5. Formulate the general design acceleration response spectrum, S_a. Use the following relations:

$$T_s = S_{D1}/S_{DS} \text{ and } T_0 = 0.2\ T_s$$

For $T \leq T_0$ $S_a = S_{DS}\left(0.4 + 0.6\dfrac{T}{T_0}\right)$

For $T_0 \leq T \leq T_s,$ $S_a = S_{DS}$

For $T \geq T_s,$ $S_a = \dfrac{S_{D1}}{T}$

For the example,

$$T_s = \frac{S_{D1}}{S_{DS}} = \frac{0.43}{1.0} = 0.43 \text{ sec}$$

$$T_0 = 0.2 \times T_s = 0.2 \times 0.43 = 0.086 \text{ sec}$$

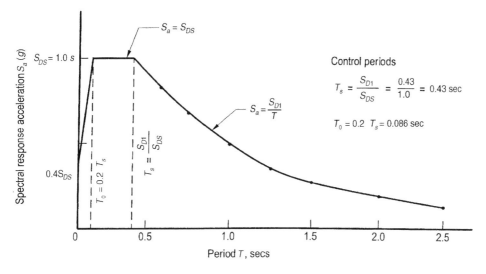

Figure 2.45c. General design response spectra; design example.

$$\text{For } T \leq 0.086 \text{ sec, } S_a = 1.0 \left(0.4 + 0.6 \frac{T}{0.086} \right)$$

For $T = 0$, $S_a = 0.4$

For $T_0 \leq T \leq T_s$, $S_a = S_{DS} = 1.0$

$$\text{For } T \geq T_s, \qquad S_a = \frac{S_{D1}}{T} \qquad \text{For } T = 1 \text{ sec, } S_a = 0.43 \text{ sec}$$

$$\text{For } T = 2 \text{ sec, } S_a = \frac{0.43}{2} = 0.215 \text{ sec}$$

$$\text{For } T = 3 \text{ sec, } S_a = \frac{0.43}{3} = 0.14 \text{ sec}$$

The values of S_a are graphically shown in the generalized response spectrum (Fig. 2.45c).

2.4.2.9. Summary of Design Ground Motions

S_s = Maximum capable earthquake (MCE) spectral acceleration in the short-period range for site class B.

S_1 = MCE spectral acceleration at 1-second period for site class B.

$S_{MS} = F_a S_s$, MCE spectral acceleration in the short-period range adjusted for site class effects.

$S_{M1} = F_v S_1$, MCE spectral acceleration at 1-second period adjusted for site class effects.

$S_{DS} = 2/3 \, S_{MS}$, spectral acceleration in the short-period range for the design ground motion.

$S_{D1} = 2/3 \, S_{M1}$, spectral acceleration at 1-second period for the design ground motion.

2.4.2.10. Building Irregularities

Prior to the 1988 UBC, building codes published a list of irregularities defining the conditions, but provided no quantitative basis for determining the relative significance of a given irregularity. However, starting in 1988, seismic codes have attempted to quantify irregularities by establishing geometrically or by use of building dimensions the points at which the specific irregularity becomes an issue as to require extra analysis and design considerations over and above those of the equivalent lateral procedure. The code requirements for determining the presence of irregularity, and the required methods to compensate for it, have now become complex, as can be seen in a graphic interpretation of the irregularities given in Table 2.19 and 2.20. Observe that the remedial measures range from a simple requirement of a dynamic distribution of lateral forces (e.g., mass irregularity), to special load combination of gravity and seismic forces (e.g., out-of-plane offset irregularity).

2.4.2.11. Load Combinations

ASCE-7 allows for both allowable stress design (ASD), and alternatively, strength design (SD), also referred to as load resistance factor design (LRFD). The emphasis is, however, on the latter, using one set of factored load combinations for all materials. Because the use of ASD is also prevalent among design professionals, ASCE continues to provide ASD combinations.

> **Factored Loads for Strength Design, SD (LRFD); Basic Combinations One-third Increase Not Permitted.**
> 1. $U = 1.4(D + F)$
> 2. $U = 1.2(D + F + T) + 1.6(L + H) + 0.5(L_r$ or S or $R)$
> 3. $U = 1.2D + 1.6 (L_r$ or S or $R) + (L$ or $0.8W)$
> 4. $U = 1.2D + 1.6W + L + 0.5(L_r$ or S or $R)$ (2.25)
> 5. $U = 1.2D + 1.0E + L + 0.2S$
> 6. $U = 0.9D + 1.6W + 1.6H$
> 7. $U = 0.9D + 1.0E + 1.6H$

The notation U in Eqs. (2.25) stands for ultimate load combinations.
Exceptions:
1. The load factor on L in combinations (3), (4), and (5) is permitted to equal 0.5 for all occupancies in which L_0 is less than or equal to 100 psf, with the exception of garages or areas occupied as places of public assembly.
2. The load factor on H shall be set equal to zero in combinations (6) and (7) if the structural action due to H counteracts that due to W or E. Where lateral earth pressure provides resistance to structural actions from other forces, it shall not be included in H, but shall be included in the design resistance.

A significant difference the designer will notice is the jump in the load factor for wind, from 1.3 in UBC 1997 to 1.6 in ASCE 7-02. This is due to the fact that the wind directionality factor K_d is now explicitly furnished instead of being combined with the load factor for wind.

The factor K_d, typically equal to 0.85 for buildings (see Table 1.8 of this text, ASCE 7-02 Table 6.4), when multiplied by the 1.6 load factor, gives an equivalent load factor equal to $0.85 \times 1.6 = 1.36 \cong 1.30$, given in the 1997 UBC.

TABLE 2.19 Vertical Irregularities; ASCE 7-02 Seismic Provisions

Type of irregularity	Graphic interpretation		Remedial measures	Seismic design category application
1a. Stiffness irregularity (soft story)		Stiffness $A < 70\% B$ or $A < 80\% \dfrac{(B + C + D)}{3}$	1	D, E, F
1b. Stiffness irregularity (extreme soft story)		Stiffness $A < 60\% B$ or $A < 70\% \dfrac{(B + C + D)}{3}$	1 2 (NP)	D E, F
2. Weight (mass) irregularity		Mass $B > 150\%$ Mass A	1	D, E, F
3. Vertical geometric irregularity		Dimension $X > 130\% Y$	1	D, E, F
4. In-plane discontinuity in vertical lateral-force-resisting systems		Dimension $L_1 > L$	1, 3	B, C, D, E, F
5. Discontinuity in capacity (weak story)		Shear strength $A < 80\% B$	4 5 2 (NP)	B, C, D, E, F D, E, F E, F

Remedial measures

1. Use modal analysis or more rigorous procedure.
2. Not permitted (NP).
3. Columns or other elements supporting discontinuous walls or frames shall have the design strength to resist special seismic load combination: $1.2D + f_1 L + 1.0E_m$ and $0.9D \pm 1.0E_m$ where $E_m = \Omega_o E_h$, the estimated maximum earthquake force that can be developed in the structure.
4. Where openings occur in walls or diaphragms, extend edge reinforcement to develop the design forces.
5. Multiply the sum of actual and accidental torsion by torsional amplification factor A_x.

TABLE 2.20　　Plan Irregularities; ASCE 7-02 Seismic Provisions

Types of irregularity	Graphic interpretation		Remedial measures	Seismic design category application
1a. Torsional irregularity		Story drift $\Delta_1 > 1.2 \dfrac{(\Delta_1+\Delta_2)}{2}$	6 5	D, E, F C, D, E, F
1b. Extreme torsion		Story drift $\Delta_1 > 1.4 \dfrac{(\Delta_1+\Delta_2)}{2}$	6 5 2 (NP)	D C, D E, F
2. Reentrant corners		Projection beyond reentrant corners $B > 15\%\ A$ $D > 15\%\ C$ $E > 15\%\ C$	6	D, E, F
3. Diaphragm discontinuity		Area $X\,Y > 50\%\ AB$	6	D, E, F
4. Out-of-plane offsets		Out-of-plane offset	6 3	D, E, F B, C, D, E, F
5. Nonparallel system		Nonparallel system	7	C, D, E, F

Remedial measures

2. Not permitted (NP).

3. Columns or other elements supporting discontinuous walls or frames shall have the design strength to resist special seismic load combination: $1.2D + f_1L + 1.0E_m$ and $0.9D \pm 1.0E_m$ where $E_m = \Omega_o E_h$, the estimated maximum earthquake force that can be developed in the structure.

5. Multiply the sum of actual and accidental torsion by torsional amplification factor A_x.

6. Increase forces determined by static procedure by 25% for connection of diaphragms to vertical elements and to collectors, and for connection of collectors to the vertical elements.

7. Design for orthogonal seismic effects. Analyze for 100% of seismic forces in one direction plus 30% of forces in the perpendicular direction. Alternatively, combine orthogonal effects by using square root of the sum of squares (SRSS) procedure.

Nominal Loads for Allowable Stress Design (ASD)
One-Third Increase Not Permitted.
1. $D + F$
2. $D + H + F + L + T$
3. $D + H + F + (L_r \text{ or } S \text{ or } R)$
4. $D + H + F + 0.75(L + T) + 0.75(L_r \text{ or } S \text{ or } R)$ (2.26)
5. $D + H + F + (W \text{ or } 0.7E)$
6. $D + H + F + 0.75(W \text{ or } 0.7E) + 0.75L + 0.75(L_r \text{ or } S \text{ or } R)$
7. $0.6D + W + H$
8. $0.6D + 0.7E + H$

Notations used in the load combinations are as follows:

D = dead load
E = earthquake load
F = load due to fluids with well-defined pressures and maximum heights
H = load due to lateral earth pressure, ground water pressure, or pressure of bulk
       materials
L = live load
L_r = roof live load
R = rain load
S = snow load
T = self-straining force
W = wind load

The designer is referred to AISC 7-02 for load combinations that include ice, flood, and wind on ice.

Special Seismic Load Combinations.   In seismic design, certain elements such as those supporting discontinuous systems, collectors, and transfer diaphragms require special consideration. ASCE-7 requires the design of these elements for a maximum seismic load effect given by

$$E = \Omega_o Q_E \pm 0.2 S_{DS} D$$ (2.27)

where

E = estimated maximum earthquake force that can be developed in the structure.
Ω_o = system overstrength factor
Q_E = horizontal seismic force due to base shear V
S_{DS} = short-period spectral acceleration for design earthquake = 2/3 S_{MS}
D = dead load

Special load combinations apply to both ASD and LRFD.

2.4.2.12.   Redundancy Factor ρ

This factor recognizes multiple paths of resistance that may be present in a lateral-force-resisting system. It takes into account: 1) plan area of the building; 2) number of lateral-force-resisting elements; and 3) magnitude of shear force resisted by the lateral-force-resisting elements. ρ is given by the relation

$$\rho_i = 2 - \frac{20}{V_{max,i}\sqrt{A_i}}$$ (2.27a)

where

ρ_i = redundancy at each level i of building.

Ω_o = maximum value of element story shear ratio at level i.

A_i = area of the floor above level i, in square feet.

The IBC requirements for ρ are:

1. ρ may be taken = 1.0 when calculating building drift.
2. Since ρ may have two different values, one for each original direction. IBC allows the use of these values in two two-dimensional analyses of a building. In a three-dimensional analysis, it may be prudent to use a weighted average of the two ρ values.
3. ρ for the entire building is the maximum value of ρ_i, calculated at each level. It is not a story-by-story nor an element-by-element factor. It is a factor for the entire building in one direction. ρ_i is calculated at all floor levels and not just at the lower $^2/_3$ height of the building, as in the 1997 UBC.
4. $\rho = 1.0$ for SDC A, B, and C.
5. For SDC D, E, and F, $1.0 < \rho < 1.5$.
6. For SMRF, ρ is limited to 1.25 for SDC D, and 1.1 for SDC E and F.

2.4.2.13. *Effect of Vertical Acceleration*

This design concept is related to the load effects of the vertical component of the earthquake ground motion, particularly in SDC D, E, and F. The earthquake load effect, E, must be considered as a combination of horizontal effect ρQ_E and a vertical component equal to $0.2\,S_{DS}D$. Thus

$$E = \rho Q_E + 0.2S_{DS}D \tag{2.28}$$

where:

Q_E = earthquake effect due to design base shear V

ρ = redundancy of the lateral force-resisting systems

$0.2S_{DS}D$ = vertical acceleration effect

2.4.3.   IBC 2003, NFPA 5000 (ASCE 7-02) Equivalent Lateral-Force Procedure

2.4.3.1. *Design Base Shear, V*

The design base shear, as set forth by

$$V = C_s W \tag{2.29}$$

is the starting point. It is given as a seismic response coefficient, C_s, times the effective seismic weight of the structure, W. The effective seismic weight is the total weight of the building and other gravity loads that might reasonably be expected to be acting on the building at the time of an earthquake. It includes permanent and movable partitions and permanent equipment such as mechanical and electrical equipment, piping, and ceilings. The human live load is taken to be negligibly small in its contribution to the seismic lateral forces. Building intended for storage or warehouse occupancy must have at least 25% of the design floor live load included in the calculation of W.

Freshly fallen snow has little effect on the lateral force, but ice firmly attached to the roof of a building would contribute significantly to the inertia force. For this reason, effective snow load is taken as the full design snow load for those regions where the snow load exceeds 30 psf, with the provision that the local authority having jurisdiction may allow a reduction of up to 80%. The magnitude of snow load to be included in the calculation of W depends on how much ice buildup or snow entrapment is expected for the roof configuration and site topography. IBC-03 requires inclusion of a fixed 20% of the flat roof snow load in W, where the flat roof snow load exceeds 30 psf. When the flat roof snow load is lower, no portion of it needs to be included in W.

2.4.3.2. Seismic Response Coefficient C_s

The seismic response coefficient, C_s, as defined by

$$C_s = \frac{S_{DS}}{\left[\frac{R}{I}\right]} \tag{2.30}$$

and

$$C_s = \frac{S_{D1}}{\left[\frac{R}{I}\right]T} \tag{2.31}$$

describes the general design response spectrum with two modifications. First, the ramp building up to the flat top of the design spectrum is excluded from consideration of C_s. This is because the equivalent lateral-force procedure is based on only the fundamental response of the structure. The period T of the first mode of a practical building is unlikely to be in the very short period range, i.e., T is unlikely to be less than 0.2 seconds. Second, the seismic response coefficient C_s, defined by Eqs. (2.30) and (2.31), is the spectral response acceleration amplified by the occupancy importance factor, I_E, and reduced by the response modification factor, R.

Equation (2.30) represents the constant acceleration portion of the spectrum, whereas Eq. (2.31) represents the constant velocity portion. The design force level defined in the static procedure is based on the assumption that a structure will undergo several cycles of inelastic deformation during major earthquake ground motions and, therefore, the force level is related to the type of structural system and its estimated ability to sustain these inelastic deformations without collapse. This force level is used not only in static lateral-force procedure, but also as a lower bound in dynamic analysis procedure.

2.4.3.3. Response Modification Factor, R

The factor R in the denominator of base shear equations is an empirical response reduction factor intended to account for both the damping and ductility (or inelastic deformability) presumed to exist in a structural system at displacements surpassing initial yield, and approaching maximum inelastic response displacements. The factor R is also intended to account for overstrength, which is partly material-dependent and partly system-dependent. Since the design force levels are based on the onset of first yield of the highest stressed element of a system, the maximum force level that the system can resist after the formation of successive hinges, bracing yield, or shear wall yield or cracking, is significantly higher than the initial yield value. Designs are also based on minimum expected yield or strength values, whereas the average strength of a material could be significantly higher. See Table 2.21 for R values.

TABLE 2.21 Design Coefficients and Factors for Basic Seismic Force-Resisting Systems

Basic seismic force-resisting system	Response modification coefficient, R^a	System over-strength factor, Ω_o^g	Deflection amplification factor, C_d^b	Structural system limitations and building height (ft) limitations^c Seismic design category				
				A&B	C	D^d	E^c	F^e
Bearing wall systems								
Ordinary steel concentrically braced frames	4	2	$3\frac{1}{2}$	NL	NL	35^k	35^k	NP^k
Special reinforced concrete shear walls	5	$2\frac{1}{2}$	5	NL	NL	160	160	100
Ordinary reinforced concrete shear walls	4	$2\frac{1}{2}$	4	NL	NL	NP	NP	NP
Detailed plain concrete shear walls	$2\frac{1}{2}$	$2\frac{1}{2}$	2	NL	NP	NP	NP	NP
Ordinary plain concrete shear walls	$1\frac{1}{2}$	$2\frac{1}{2}$	$1\frac{1}{2}$	NL	NP	NP	NP	NP
Building frame systems								
Steel eccentrically braced frames, moment-resisting connections, at columns away from links	8	2	4	NL	NL	160	160	100
Steel eccentrically braced frames, nonmoment-resisting connections, at columns away from links	7	2	4	NL	NL	160	160	100
Special steel concentrically braced frames	6	2	5	NL	NL	160	160	100
Ordinary steel concentrically braced frames	5	2	$4\frac{1}{2}$	NL	NL	35^k	35^k	NP^k
Special reinforced concrete shear walls	6	$2\frac{1}{2}$	5	NL	NL	160	160	100
Ordinary reinforced concrete shear walls	5	$2\frac{1}{2}$	$4\frac{1}{2}$	NL	NL	NP	NP	NP
Detailed plain concrete shear walls	3	$2\frac{1}{2}$	$2\frac{1}{2}$	NL	NP	NP	NP	NP
Ordinary plain concrete shear walls	2	$2\frac{1}{2}$	2	NL	NP	NP	NP	NP
Composite eccentrically braced frames	8	2	4	NL	NL	160	160	100
Composite concentrically braced frames	5	2	$4\frac{1}{2}$	NL	NL	160	160	100
Ordinary composite braced frames	3	2	3	NL	NL	NP	NP	NP
Composite steel plate shear walls	$6\frac{1}{2}$	$2\frac{1}{2}$	$5\frac{1}{2}$	NL	NL	160	160	100

(*Continued*)

TABLE 2.21 (Continued)

Basic seismic force-resisting system	Response modification coefficient, R^a	System over-strength factor, Ω_o^g	Deflection amplification factor, C_d^b	Structural system limitations and building height (ft) limitations^c Seismic design category				
				A&B	C	D^d	E^c	F^e
Special composite reinforced concrete shear walls with steel elements	6	$2\frac{1}{2}$	5	NL	NL	160	160	100
Ordinary composite reinforced concrete shear walls with steel elements	5	$2\frac{1}{2}$	$4\frac{1}{4}$	NL	NL	NP	NP	NP
Moment-resisting frame systems								
Special steel moment frames	8	3	$5\frac{1}{2}$	NL	NL	NL	NL	NL
Special steel truss moment frames	7	3	$5\frac{1}{2}$	NL	NL	160	100	NP
Intermediate steel moment frames	4.5	3	4	NL	NL	35^h	NP^{h,i}	NP^{h,i}
Ordinary steel moment frames	3.5	3	3	NL	NL	NP^{h,i}	NP^{h,i}	NP^{h,i}
Special reinforced concrete moment frames	8	3	$5\frac{1}{2}$	NL	NL	NL	NL	NL
Intermediate reinforced concrete moment frames	5	3	$4\frac{1}{2}$	NL	NL	NP	NP	NP
Ordinary reinforced concrete moment frames	3	3	$2\frac{1}{2}$	NL	NP	NP	NP	NP
Special composite moment frames	8	3	$5\frac{1}{2}$	NL	NL	NL	NL	NL
Intermediate composite moment frames	5	3	$4\frac{1}{2}$	NL	NL	NP	NP	NP
Composite partially restrained moment frames	6	3	$5\frac{1}{2}$	160	160	100	NP	NP
Ordinary composite moment frames	3	3	$2\frac{1}{2}$	NL	NP	NP	NP	NP
Dual systems with special moment frames capable of resisting at least 25% of prescribed seismic forces								
Steel eccentrically braced frames, moment-resisting connections, at columns away from links	8	$2\frac{1}{2}$	4	NL	NL	NL	NL	NL
Steel eccentrically braced frames, nonmoment-resisting connections, at columns away from links	7	$2\frac{1}{2}$	4	NL	NL	NL	NL	NL

(Continued)

TABLE 2.21 (Continued)

Basic seismic force-resisting system	Response modification coefficient, R[a]	System over-strength factor, Ω_o[g]	Deflection amplification factor, C_d[b]	Structural system limitations and building height (ft) limitations[c] Seismic design category				
				A&B	C	D[d]	E[c]	F[e]
Special steel concentrically braced frames	8	$2\frac{1}{2}$	$6\frac{1}{2}$	NL	NL	NL	NL	NL
Special reinforced concrete shear walls	8	$2\frac{1}{2}$	$6\frac{1}{2}$	NL	NL	NL	NL	NL
Ordinary reinforced concrete shear walls	7	$2\frac{1}{2}$	6	NL	NL	NP	NP	NP
Composite eccentrically braced frames	8	$2\frac{1}{2}$	4	NL	NL	NL	NL	NL
Composite concentrically braced frames	6	$2\frac{1}{2}$	5	NL	NL	NL	NL	NL
Composite steel plate shear walls	8	$2\frac{1}{2}$	$6\frac{1}{2}$	NL	NL	NL	NL	NL
Special composite reinforced concrete shear walls with steel elements	8	$2\frac{1}{2}$	$6\frac{1}{2}$	NL	NL	NL	NL	NL
Ordinary composite reinforced concrete shear walls with steel elements	7	$2\frac{1}{2}$	6	NL	NL	NP	NP	NP
Ordinary steel concentrically braced frames	6	$2\frac{1}{2}$	5	NL	NL	NL	NL	NL
Dual systems with intermediate moment frames capable of resisting at least 25% of prescribed seismic forces								
Special steel concentrically braced frames[f]	$4\frac{1}{2}$	$2\frac{1}{2}$	$4\frac{1}{2}$	NL	NL	35	NP	NP[h,i]
Special reinforced concrete shear walls	6	$2\frac{1}{2}$	5	NL	NL	160	100	100
Composite concentrically braced frames	5	$2\frac{1}{2}$	$4\frac{1}{2}$	NL	NL	160	100	NP
Ordinary composite braced frames	4	$2\frac{1}{2}$	3	NL	NL	NP	NP	NP
Ordinary composite reinforced concrete shear walls with steel elements	5	3	$4\frac{1}{2}$	NL	NL	NP	NP	NP
Ordinary steel concentrically braced frames	5	$2\frac{1}{2}$	$4\frac{1}{2}$	NL	NL	160	100	NP
Ordinary reinforced concrete shear walls	$5\frac{1}{2}$	$2\frac{1}{2}$	$4\frac{1}{2}$	NL	NL	NP	NP	NP

(*Continued*)

TABLE 2.21   (Continued)

Basic seismic force-resisting system	Response modification coefficient, R^a	System over-strength factor, Ω_o^g	Deflection amplification factor, C_d^b	Structural system limitations and building height (ft) limitations^c Seismic design category				
				A&B	C	D^d	E^c	F^e
Inverted pendulum systems and cantilevered column systems								
Special steel moment frames	$2\frac{1}{2}$	2	$2\frac{1}{2}$	NL	NL	NL	NL	NL
Ordinary steel moment frames	$1\frac{1}{4}$	2	$2\frac{1}{2}$	NL	NL	NP	NP	NP
Special reinforced concrete moment frames	$2\frac{1}{2}$	2	$1\frac{1}{4}$	NL	NL	NL	NL	NL
Structural steel systems not specifically detailed for seismic resistance	3	3	3	NL	NL	NP	NP	NP

^{a} Response modification coefficient, R, for use throughout the standard. Note R reduces forces to a strength level, not an allowable stress level.

^{b} Deflection amplification factor, C_d, for use in Sections 9.5.7.1 and 9.5.3.7.2.

^{c} NL = Not Limited and NP = Not Permitted. For metric units use 30 m for 100 ft and use 50 m for 160 ft. Heights are measured from the base of the structure as defined in Section 9.2.1.

^{d} See Section 9.5.2.2.4.1 for a description of building systems limited to buildings with a height of 240 ft (75 m) or less.

^{e} See Sections 9.5.2.2.4 and 9.5.2.2.4.5 for building systems limited to buildings with a height of 160 ft (50 m) or less.

^{f} Ordinary moment frame is permitted to be used in lieu of intermediate moment frame in SDC B and C.

^{g} The tabulated value of the overstrength factor, Ω_o, may be reduced by subtracting 1/2 for structures with flexible diaphragms, but shall not be taken as less than 2.0 for any structure.

^{h} Steel ordinary moment frames and intermediate moment frames are permitted in single-story buildings up to a height of 60 ft, when the moment joints of field connections are constructed of bolted end plates and the dead load of the roof does not exceed 15 psf.

^{i} Steel ordinary moment frames are permitted in buildings up to a height of 35 ft where the dead load of the walls, floors, and roofs does not exceed 15 psf.

^{k} Steel ordinary concentrically braced frames are permitted in single-story buildings up to a height of 60 ft when the dead load of the roof does not exceed 15 psf and in penthouse structures.

(Condensed from ASCE 7-02, Table 9.5.2.2.)

For a slightly damped building of brittle material unable to tolerate any appreciable deformation beyond the elastic range, the factor R would be close to 1.0. There is no reduction from the force level corresponding to linear elastic response. At the other extreme, a heavily damped building with a very ductile structural system is likely to withstand deformations considerably in excess of initial yield and would, therefore, merit the assignment of a relatively large response modification factor, R. The coefficient R ranges in value from a minimum of 1.5 for a bearing wall system consisting of ordinary plain masonry shear walls to a maximum of 8.0 for a special moment frame system or a dual system consisting of special moment frames. It should be noted that the numerical values assigned to the coefficient R are based on historical experience, and not on rigorous analysis or experimentation.

For a given fundamental period T, the design base shear is given as a product of the acceleration response spectrum ordinate, S_{DS}, for $T \leq T_S$ and S_{D1}/T for $T > T_S$, and the total structure weight W, multiplied by the importance factor, I_E, and divided by the structure coefficient response R. While only the fundamental mode period is employed, the additional response due to higher modes is approximated by use of the total weight, W, and not just the weight corresponding to the first mode response. The fundamental mode base shear is the effective mass of the first mode times the first mode spectral ordinate, S_a. This effective mass for regular buildings is about 0.7 times the total mass, W/g. Therefore, the use of W, the total structure weight in the base shear equations, results in an upper bound value of the statistically combined mode (SRSS or CQC) base shear response, either by the SRSS, Square Root of Sum of the Squares method, or the CQC, the Complete Quadratic Combination method.

The response spectrum for earthquake ground motions has a descending branch for longer values of T. It varies as $1/T$, as shown in Fig. 2.45 a and b. Because the total weight, W, is used in the design base shear equations, along with appropriate S_{DS} and S_{D1} values, the base shear calculated in the region of long-period plateau provides a reasonable representation of multimode response.

2.4.3.4. *Importance Factor, I*

The value of R in the base shear equations is adjusted by the occupancy importance factor, I, which ranges between 1.0 and 1.5, as shown in Table 2.13. A value of I greater than unity has the effect of reducing the ductility expected of a structure. However, added strength due to higher design forces by itself is not sufficient to ensure superior seismic performance. Connection details that assure ductility, quality assurance procedures, and limitations on building deformation are also important to improve the functionality and safety in critical facilities and those with high-density occupancy. Consequently, the reduction in the damage potential of critical facilities is also addressed by using more conservative drift controls and by providing special design and detailing requirements and materials limitations.

2.4.3.5. *Minimum Base Shear*

The minimum design base shear is $0.044 S_{DS}I_E W$. This is included in view of the uncertainty and the lack of knowledge of actual structural response of long-period buildings subject to earthquake ground motions.

Following the Northridge earthquake of 1994, a second lower bound on the design base shear, applicable in seismic zone 4 only, was added to 1997 UBC. This second minimum is in terms of Z, N_V, I_E and R, where N_V is the velocity-dependent near-source factor, and is specifically intended to account for the large displacement and velocity pulses that were observed in near-fault ground motion in the Northridge earthquake. A corresponding minimum has been adopted into the IBC, and is represented by Eq. 2.10. The minimum is applicable to all structures located where the mapped spectral response acceleration at 1-second period, S_{D1}, equals or exceeds 0.6g.

2.4.3.6. *Period Determination*

In the denominator of the base shear equation, T is the fundamental period of vibration of the building. It is preferable that this be determined using the structural properties and deformational characteristics of the resisting elements in a properly substantiated analysis, i.e., by a dynamic analysis using computers. However, a dynamic analysis is useful to calculate the period of vibration only after the building has been designed. Therefore, an approximate method is necessary to estimate building period, with minimal information available on the building characteristics. Hence, the simple formula of IBC involves only a general description of the building type (such as steel moment frame, concrete moment frame, etc.) and the overall height or number of stories.

Building periods, computed even with the use of very sophisticated software, are only as good as the modeling assumptions used in the analysis and, to a great extent, are dependent on stiffness assumptions. The smaller the assumed stiffness, the longer the computed period, which translates directly into a lower design base shear. The computed period is thus open to possible abuse. Therefore, IBC, just as UBC, imposes a limit on the computed period. For design purposes, it may not be taken any larger than a coefficient C_u times the approximate period calculated. Reasonable mathematical rules should be followed such that the increase in period allowed by the C_u coefficient is not taken advantage of when the structure does not merit it. Note that for purposes of drift analysis only, the upper bound limitation on the computed fundamental period T of the building does not apply.

It may be noted that larger values of C_u are permitted as the soil-dependent seismic risk of a location decreases. This is because buildings in areas with lower lateral-force requirements are thought likely to be more flexible. Higher values of C_u for lower values of S_{D1} also result in less dramatic changes from prior practice in lower-risk areas. It is generally accepted that the equations for T_a are tailored to fit the types of construction common in areas with high lateral-force requirements. It is unlikely that buildings in lower seismic risk areas would be designed to produce as high a drift level as allowed by IBC, due to stability ($P\Delta$) considerations and wind requirements. For buildings with design controlled by wind, the use of a large T will not really result in a lower design force.

Using the assumptions that: 1) the seismic base shear varies as $1/T$; 2) the lateral forces are distributed linearly over the building height; and 3) deflections are controlled by drift limitations, it can be shown that the period of a moment-resisting frame varies roughly with h_n, where h_n is the total height of the building. It is recognized for quite some time that the periods calculated by using the values for C_t are lower than the measured values in the elastic range. However, these estimated periods provide design values that are judged to be appropriate and consistent with past design experience. For the usual case of a descending spectrum, the decrease in demand due to the increase in period as the structure deforms into the inelastic range is already included in the R value of a given structural system.

Observe that for the 0.035 or the 0.030 coefficient to be applicable, 100% of the required seismic force must be resisted by moment frames. Such frames must not be enclosed or adjoined by more rigid elements that would keep them from deforming freely under seismic excitation. If either condition is violated, the designer must use the 0.02 coefficient specified for other building systems. The optional use of $T_a = 0.1N$ that has long been in use is an approximation for low-to-moderate-height frames.

2.4.3.7.　Vertical Distribution of Seismic Forces

The distribution of forces over the height of a building is complex because these forces are the result of superposition of a number of modes of vibration. The relative contribution of these vibration modes to the overall distribution of lateral forces over the height of the building depends on a number of factors including the shape of the earthquake response spectrum, the period of vibration of the building, and the characteristic shapes of the vibration modes which, in turn, depend on the magnitude and distribution of mass and stiffness over the height of the structure. Taking this into consideration, IBC-03 prescribes three types of distribution of the entire base shear:

- A triangular distribution for buildings having a fundamental period not exceeding 0.5 seconds.
- A parabolic distribution for building having an elastic fundamental period in excess of 2.5 seconds.
- A linear interpolation between linear and parabolic distribution for buildings with periods between 0.5 and 2.5 seconds.

2.4.3.8.　Horizontal Shear Distribution

Rigid Diaphragm.　When the deformation of a diaphragm is less than or equal to twice the associated story drift, the diaphragm is considered rigid. In most buildings, the diaphragm may be modeled as fully rigid without in-plane deformability. However, the effects of diaphragm deformability must be investigated for buildings with vertical and plan irregularities. The use of the most critical results obtained from the fully rigid and the flexible diaphragm models is generally considered acceptable in building design.

Flexible Diaphragm.　When its deformation is more than two times the associated story drift, a diaphragm is considered flexible. The term signifies that a diaphragm segment between two vertical lateral-force-resisting elements may be modeled as a simple beam spanning between these elements. In a flexible diaphragm, an out-of-plane offset in a vertical lateral-force-resisting element is allowed to be ignored, provided the offset does not exceed 5% of the plan dimension perpendicular to the direction of lateral load.

2.4.3.9.　Overturning

The design overturning moment must be statically consistent with the design story shears, except for the reduction, factor, I. At any level, the incremental changes of the design overturning moment are to be distributed to the various resisting elements in the same proportion as the distribution of the horizontal shears to these elements. Following are the reasons for reducing the statically computed overturning moments:

1. The distribution of design story shears over the height of the building computed from lateral forces is intended to provide an envelope, recognizing that the shears in all stories do not attain their maximum values simultaneously. If the shear in a specific story is close to the computed value, the shears in almost all other stories are almost necessarily overestimated. Hence, the overturning moments statically consistent with the design story shears are overestimated.

2. Under the action of overturning moments, one edge of the foundation may lift off the ground for a short duration of time. Such behavior leads to substantial reduction in the seismic forces and, consequently, in the overturning moments.

The overturning moments computed statically from the envelope of story shears may be reduced by no more than 20%. This value is similar to those obtained from the results of dynamic analysis. No reduction is permitted in the uppermost 10 stories, primarily because studies have shown that the statically computed overturning moments in these stories may not be on the conservative side. There is hardly any benefit anyway in reducing overturning moments in stories near the top of a structure, since the design of vertical elements in these stories is rarely governed by overturning moments.

For the eleventh to the twentieth stories from the top, linear variation of T provides the simplest transition between the minimum and maximum values of 0.8 and 1.0. Many older building codes used to allow more reduction of overturning moments. These reductions were judged to be excessive because of the damage observed during the 1967 Caracas, Venezuela earthquake where a number of column failures were due primarily to the effect of overturning moments. The 1976 and subsequent editions of the UBC have not permitted any reduction in overturning moments. IBC-03 has chosen the middle ground by allowing moderate reductions.

2.4.3.10. Story Drift

Determination of design story drift, as shown in Table 2.22, involves the following steps:

1. Determine the lateral deflections at the various floor levels by an elastic analysis of the building under the design base shear. The lateral deflection at floor level x, obtained from this analysis, is δ_{xe}.
2. Amplify δ_{xe} by the deflection amplification factor, C_d The quantity $C_d\delta_{xe}$ is an estimated design earthquake displacement at floor level x. IBC-03 requires this quantity to be divided by the importance factor, I_E, because the forces under which the δ_{xe}, displacement is computed are already amplified by I_E. Since IBC drift limits are tighter for buildings in higher occupancy categories, this division by I_E is important. Without it, there would be a double tightening

TABLE 2.22 Allowable Story Drift, $\Delta_a{}^a$

| | Seismic use group | | |
Structure	I	II	III
Structures, other than masonry shear wall or masonry wall frame structures, four stories or less with interior walls, partitions, ceilings, and exterior wall systems that have been designed to accommodate the story drifts.	$0.025h_{sx}{}^b$	$0.020h_{sx}$	$0.015h_{sx}$
Masonry cantilever shear wall structures[c]	$0.010h_{sx}$	$0.010h_{sx}$	$0.010h_{sx}$
Other masonry shear wall structures	$0.007h_{sx}$	$0.007h_{sx}$	$0.007h_{sx}$
Masonry wall frame structures	$0.013h_{sx}$	$0.013h_{sx}$	$0.010h_{sx}$
All other structures	$0.020h_{sx}$	$0.015h_{sx}$	$0.010h_{sx}$

[a] h_{sx} is the story height below level x.

[b] There shall be no drift limit for single-story structures with interior walls, partitions, ceilings, and exterior wall systems that have been designed to accommodate the story drifts. The structure separation requirement of Section 9.5.2.8 is not waived.

[c] Structures in which the basic structural system consists of masonry shear walls designed as vertical elements cantilevered from their base of foundation support, which, are so constructed that moment transfer between shear walls (coupling) is negligible.

(From ASCE 7-02, Table 9.5.2.8.)

of drift limitations for buildings with seismic importance factors greater than one. The quantity $C_d \delta_{xe}/I_E$ at floor level x is δ_x, the adjusted design earthquake displacement.

3. Calculate the design story drift Δ_x for story x (the story below floor level x) by deducting the adjusted design earthquake displacement at the bottom of story x (floor level $x - 1$) from the adjusted design earthquake displacement at the top of story x

$$\Delta_x = \delta_x - \delta_{x-1}$$

The Δ_x values must be kept within limits, as given in ASCE-7 Table 9.5.2.8 (Table 2.22). Three items are worth noting:

1. The design story drift must be computed under the strength-level design earthquake forces, irrespective of whether member design is done using the strength design or the allowable stress design load combinations.
2. The redundancy coefficient, ρ, is equal to 1.0 for computation of the design story drift.
3. For determining compliance with the story drift limitations, the deflections, Δ_x, may be calculated as indicated previously for the seismic force-resisting system, using design forces corresponding to the fundamental period of the structure, T, calculated without the limit, $T < C_u T_a$. The same model of the seismic force-resisting system used in determining the deflections must be used for determining T. The waiver does not pertain to the calculation of drifts for determining $P\Delta$ effects on member forces, overturning moments, etc. If $P\Delta$ effects are significant, the design story drift must be increased by the resulting incremental factor.

The $P\Delta$ effects in a given story are due to the eccentricity of the gravity load above the story. If the design story drift due to the lateral forces is Δ, the bending moments in the story are augmented by an amount equal to Δ times the gravity load above the story. The ratio of the $P\Delta$ moment to the lateral-force story moment is designated as the stability coefficient. If the stability coefficient, θ, is less than 0.10 for every story, then the $P\Delta$ effects on story shears and moments and member forces may be ignored. If not, the $P\Delta$ effects on story drifts, shears, member forces, etc., must be determined by a rational analysis. However, with the availability of computer programs that take into consideration $P\Delta$ effects automatically within the analysis, hand calculations of θ, for determining whether $P\Delta$ is significant are rarely necessary.

$P\Delta$ effects are much more significant in buildings assigned to low-seismic design categories than in buildings assigned to high-seismic design categories. This is because lateral stiffness of buildings is typically greater for higher seismic design categories.

2.4.3.11. Seismic Force-Resisting Systems

Moment-Resisting Frame System. This is a structural system with an essentially complete space frame providing support for gravity loads. For a building to qualify as a moment-resisting frame system, it must have a substantially complete vertical load-carrying frame. For those portions of the space frame that are not part of the designated lateral-force-resisting system, the deformation compatibility requirements must be complied with.

Bearing Wall System. Buildings with this system do not have an essentially complete space frame providing support for gravity loads. Bearing walls provide support for all or most gravity loads. Resistance to lateral load is provided by the same bearing walls acting as shear walls.

Dual System.   A dual system must have three features:

1. An essentially complete space frame provides support for gravity loads.
2. Resistance to lateral loads is provided by moment-resisting frames capable of resisting at least 25% of the design base shear, and by shear walls or braced frames.
3. The two systems (moment frames and shear walls or braced frames) are designed to resist the design base shear in proportion to their relative rigidities.

Building Frame System.   In this system, an essentially complete space frame provides support for gravity loads. Resistance to lateral loads is by shear walls or braced frames or moment frames. The seismic safety of a building frame system is dependent on satisfying the deformation compatibility requirements. These recognize that when the designated lateral-force-resisting system of a structure deforms laterally, the subsystems that have been arbitrarily designated as gravity systems will have no choice but to deform together with the lateral systems, because they are connected at every floor level. If, in the course of that earthquake-induced lateral displacement, the subsystems designed for gravity loads only are unable to sustain their gravity load-carrying capacity, then life-safety is compromised. It is thus a specific requirement of all seismic codes, including the IBC-03, that structural elements or subsystems designated not to be part of the lateral-force-resisting system be able to sustain their gravity load-carrying capacity at a lateral displacement equal to a multiple times the computed elastic displacement of the lateral-force-resisting system under code-prescribed design seismic forces. The amplified elastic displacement of the lateral-force-resisting system is an estimate of the actual displacement of the entire structure caused by an earthquake of intensity anticipated by the code. If, under the estimated earthquake-induced displacements, the gravity loads would cause inelasticity in any structural element initially designed for gravity only, that structural element should also be detailed for inelastic deformability.

Shear Wall Frame-Interactive System.   Shear walls or braced frames used in conjunction with moment frames in buildings assigned to SDC C, D, E, or F must be designed as either building frame systems or as dual systems. Central to the concept of the dual system is the backup frame capable of independently resisting at least 25% of the design lateral forces. A building frame system, on the other hand, has shear walls or braced frames designed to resist 100% of the lateral forces. The attraction of this system is that the moment frames, because they are not part of the designated lateral-force-resisting system, require only ordinary detailing.

Inverted Pendulum System.   This type of structure, with a large portion of its mass concentrated near the top, has very little redundancy and overstrength and all its inelastic behavior is concentrated at the base. As a result, it has substantially less energy dissipation capacity than other systems. Included under inverted pendulum systems is the cantilevered column system that relies solely on column elements for resistance to lateral forces. The columns cantilever from a fixed base and have minimal moment capacity at the top. The lateral forces are applied essentially at the top. This type of structural system is common for multifamily residential occupancies over carports, strip shopping center storefronts, and single-family dwellings on oceanside or hillside lots. In cantilevered column systems, the column elements acting in cantilever action often provide support for the gravity loads in addition to resisting all lateral forces. Hence, there is no independent vertical load-carrying system and the failure of the primary lateral system compromises the ability of the structure to carry gravity loads. Overstrength in the cantilevered column system is minimal because the ability to form a progression of plastic hinges is limited. Hence. design for higher strength and stiffness by use of a low R value is necessary to reduce the high ductility demands.

Interaction Effects. This relates to the interaction of elements of the seismic-force-resisting system with elements designated not to be part of that system. An example is that of infill masonry walls used as architectural elements in between a seismic-force-resisting system consisting of moment-resisting frames. Although not intended to resist seismic forces, the masonry walls at low levels of deformation are substantially more rigid than the moment-resisting frames and thus participate in lateral-force resistance. Such walls often create shear-critical conditions in the columns of the moment frames by reducing the effective flexural height of these columns to the height of the openings in the walls. If these walls are not uniformly distributed throughout the building, or not effectively isolated, they can also create torsional irregularities and soft-story irregularities in structures that would otherwise have a regular configuration. Another example is the presence of ramps in parking garages which can act as effective bracing elements and resist a large portion of the seismic forces. They can induce large axial forces in the diaphragms and large vertical forces on adjacent columns and beams. Additionally if not symmetrically placed, they can cause torsional irregularities.

2.4.3.12. *Deformation Compatibility*

The IBC-03 requires that all structural framing elements and their connections, not required by design to be part of the lateral-force-resisting system, must be designed and/or detailed to be adequate to maintain support of design dead plus live loads when subjected to the expected deformations caused by seismic forces. Important features of deformation compatibility requirements are

1. Expected deformations must be the greater of the maximum inelastic response displacement. Δ_m, considering $P\Delta$ effects an deformation induced by a story drift of 0.0025 times the story height.
2. When computing expected deformations, stiffening effects of those elements not part of the lateral-force-resisting system must be neglected.
3. Forces induced by expected deformations may be considered factored forces.
4. In computing the preceding forces, restraining effect of adjoining rigid structures and nonstructural elements must be considered.
5. For concrete elements that are not part of the lateral-force-resisting system, assigned flexural and shear stiffness properties must not exceed one-half of gross section properties. unless a rational cracked section analysis is performed.
6. Additional deformations that may result from foundation flexibility and diaphragm deflection must be considered (Figs. 2.40 and 2.41). The deformation compatibility requirements of IBC-03 are essentially the same as those of 1997 UBC, except that the expected deformation is taken equal to the design story drift, Δ, times the deflection amplification factor C_d. And when allowable stress design is used, Δ is required to be computed without dividing the specified earthquake forces by 1.4.

2.4.4. Dynamic Analysis Procedure

Dynamic analysis is always acceptable for design. Static procedures are allowed for structures assigned to the higher seismic design categories only under certain conditions of regularity and height. ASCE 7-02 recognizes three dynamic analysis procedures: modal analysis, elastic time-history analysis, and inelastic time-history analysis.

Modal analysis is used for calculating the linear response of multi-degree-of-freedom systems. It is based on the idea that the response of a building is the superposition of the responses of individual modes of vibration, each mode responding with its own particular deformed shape, its own frequency, and with its own modal damping. The response of the structure is therefore determined from the responses of a number of single-degree-of-freedom systems with properties chosen to be representative of the modes and the degree to which the modes are excited by the earthquake motion.

The equivalent lateral force procedure is simply a first mode application of this technique that assumes all the mass of the structure to be active in the first mode. The purpose of modal analysis is to obtain the maximum response of the structure in each of its important modes, which are then summed in an appropriate manner. The results of the analysis are required to be scaled up to, and are permitted to be scaled down to the base shear calculated with the equivalent lateral force procedure. The building period may be taken as 1.2 times the upper limit coefficients for period calculation, C_u times the period calculated using approximate period formulas. This scaling is primarily to ensure that the design forces are not underestimated through the use of a structural model that is excessively flexible.

For buildings with T rationally determined, subject to a maximum of $C_u T_a \geq 0.7$ second, located on site class E or F where $S_{D1} > 0.2g$, scaling must be done on the basis of elastic lateral force base shear calculated using the aforementioned period.

2.4.5. Design and Detailing Requirements

The seismic design and detailing requirements are "cascading" meaning that requirements pertaining to a lower category also apply to a higher category. Therefore, SDC A requirements also apply to SDC B, SDC B to SDC C, and so on.

2.4.5.1. *Seismic Design Category A (ASCE 7-02 Sect. 9.5.2.6.1)*

Design and detailing of structures with a complete and identifiable load path for seismic forces is mandatory. Other requirements are

1. Tie smaller portions of structure to the main structure using a design force $F_p = 0.05$ times the weight of smaller element w_p. Alternatively, F_p may be taken equal to $0.133S_{DS}w_p$. Design connections at the support of each end of beam, girder, or truss for a horizontal force equal to 5% of the vertical dead and live load reaction.
2. Design anchorage of concrete and masonry walls for a minimum out-of-plane lateral force equal to 280 plf of wall.

2.4.5.2. *Seismic Design Category B (ASCE 7-02 Sect. 9.5.2.6.2)*

1. Include $P\Delta$ effects.
2. Provide reinforcement at the edges of wall and diaphragm openings. Extend the reinforcement into the wall or diaphragm to develop the force in the reinforcement.
3. Limit structures with weak stories (type 5 vertical irregularity) to a maximum of two stories or 30 ft when the weak story strength is less than 65% of the story above. This restriction is waived if the weak story is capable of resisting a seismic force equal to Ω_o times the equivalent static force.

4. Design diaphragms for a force equal to

    $$F_p = 0.2 \, IS_{DS} \, w_p + V_{px} \qquad \text{[ASCE 7-02 Eq. (9.5.2.6.2.7)]}$$

    where
    w_p = the weight of the diaphragm and other elements attached to it
    V_{px} = design seismic force transferred through the diaphragm due to offset in the placement of vertical seismic elements, or due to changes in their stiffness
    S_{DS} = the short-period spectral response coefficient
    I = occupancy importance factor
    F_p = diaphragm seismic force

5. Design columns or other elements supporting discontinuous walls or frames in structures having in-plane discontinuity or out-of-plane offset of vertical seismic elements to resist the following special seismic load combinations:

    $$E = \Omega_o Q_E + 0.2 S_{DS} D$$

    $$E = \Omega_o Q_E - 0.2 S_{DS} D$$

    where
    Q_E = the effect of horizontal seismic forces

2.4.5.3 Seismic Design Category C (ASCE 7-02, Sect. 9.5.2.6.3)

1. Anchorage of concrete or masonry walls to flexible diaphragms shall be designed for an out-of-plane force equal to

    $$F_p = 0.8 \, S_{DS} I_E w_p$$

2. Design collector elements, splices, and their connections to resisting elements for the special seismic loads given in item 5, SDC B.

2.4.5.4 Seismic Design Category D (ASCE 7-02, Sect. 9.5.2.6.4)

1. The familiar formula for the design of diaphragms

    $$F_{px} = \frac{\sum\limits_{i=x}^{n} F_i}{\sum\limits_{i=x}^{n} w_i} \, w_{px} \qquad (2.32a)$$

    is now applicable only to SDC D, E, and F structures. The nonmandatory upper and the mandatory lower limits are

    $$F_{px} = 0.4 \, S_{DS} I w_{px} \quad \text{(max)} \qquad (2.32b)$$

    $$F_{px} = 0.2 \, S_{DS} I w_{px} \quad \text{(min)} \qquad (2.32c)$$

2. For structures having plan irregularity of type 1, 2, 3, or 4, or a vertical irregularity of type 4, increase by 25% the design forces determined by analysis for connections of diaphragm elements to vertical elements and to collectors and their connections. As an alternate, collectors and their connections may be designed for special seismic loads given in item 5, SDC B.

2.4.5.5 Seismic Design Categories E and F (ASCE 7-02, Sect. 9.5.2.6.5)

Do not even think of designing category E and F structures having

1. Extreme torsional irregularity. (Table 2.20, Item 1b)
2. Extreme soft story. (Table 2.19, Item 1b)
3. Discontinuity in capacity; weak story. (Table 2.19, Item 5)

They are simply not permitted by ASCE 7-02.

2.4.6   Seismic Design Example: Static Procedure, IBC 2003 (ASCE 7-02, NFPA 5000)

Given.   A 12-story building located in downtown, Los Angeles, California. The building properties summarized in Fig. 2.46 are the same as those used in the 1997 UBC example, Section 2.13.19.1.

Occupancy group = II   (Table 2.13 ASCE Table 1.1)

SUG = 1   (Table 2.12; ASCE Table 9.1.3)

SDC = D   (Tables 2.17 and 2.18; ASCE Tables 9.4.2.1a and b)

Site class as determined by project geotechnical engineer = D
                    (Table 2.14, ASCE Table 9.4.1.2)

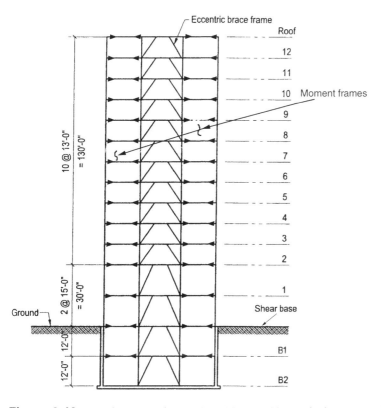

Figure 2.46.   Design example; ASCE 7-02 (IBC-03), static force procedure.

Building's lateral load system = SMRF with EBF

(Table 2.21 ASCE Table 9.5.2.2)

Elastic fundamental period, T_B from computer analysis = 2 secs

Total seismic weight, W = 22,680 kips

Building height h_n, above shear base = 160 ft

Mapped MCE, 5% damped,

Spectral acceleration at short periods, S_S = 1.5

Mapped MCE, 5% damped,

Spectral response acceleration at a period of 1 second, S_1 = 0.6.

Required. Using the equivalent lateral force procedure of ASCE 7-02 determines the following:

- Seismic base shear, V
- Vertical distribution of base shear V
- Seismic story shear
- Overturning moment
- Diaphragm design forces
- Allowable story drifts

Solution. Seismic design coefficients

S_S = 1.5   (Fig. 2.45a; ASCE Fig. 9.4.1.1a)

S_1 = 0.6   (Fig. 2.45b; ASCE Fig. 9.4.1.1.b)

Acceleration-based site coefficient (at 0.2 sec period), F_a = 1

(Table 2.15; ASCE Table 9.4.1.2.4.a)

Velocity-based site coefficient (at 1.0 sec period), F_v = 1.5

(Table 2.16; ASCE Table 9.4.1.2.4b)

MCE spectral response acceleration at short periods for site class effects, $S_{MS} = F_a S_s$
      = $1 \times 1.5 = 1.5$ ASCE Eq. (9.4.1.2.4.1)

MCE spectral response acceleration at a 1-sec period adjusted for site class effects,
      $S_{M1} = F_v \times S_1 = 1.5 \times 0.6 = 0.9$ ASCE Eq. (9.4.1.2.4)

Design, 5% damped, spectral response acceleration at short periods, $S_{DS} = \frac{2}{3}S_{MS}$ =
      $\frac{2}{3} \times 1.5 = 1.0$ ASCE Eq. (9.4.1.2.5.2)

Design, 5% damped, spectral response acceleration at a 1-sec period, $S_{D1} = \frac{2}{3}S_{MS}$ =
      $\frac{2}{3} \times 0.9 = 0.6$ ASCE Eq. (9.4.1.2.5.2)

Occupancy importance factor, I = 1   (Table 2.13; ASCE Table 9.1.4)

Response modification coefficient, R = 8.0   (Table 2.21; ASCE Table 9.5.2.2)

System overstrength factor, Ω_o = 2.5   (Table 2.21; ASCE Table 9.5.2.2)

Coefficient for upper limit on calculated period, C_u = 1.4

(Table 2.23; ASCE Table 9.5.5.3.1)

TABLE 2.23 Coefficient for Upper Limit on Calculated Period

Design spectral response acceleration at 1 Second, S_{D1}	Coefficient C_u
≥ 0.4	1.4
0.3	1.4
0.2	1.5
0.15	1.6
0.1	1.7
≤ 0.05	1.7

(From ASCE 7-02, Table 9.5.5.3.1.)

Deflection amplification factor $C_d = 4$ (Table 2.21; ASCE Table 9.5.2.2)

Vertical load distribution exponent, $k = 1.73$ (as will be shown presently)
    (ASCE Section 9.5.5.4)

Note: For buildings having a period between 0.5 and 2.5 seconds, k may be determined by linear interpolation or may be taken equal to 2.0.

Building period parameters C_t and x (Table 2.24; ASCE Table 9.5.5.3.2)

For moment frames: $C_t = 0.028$, $x = 0.8$

For ECB: $C_t = 0.03$, $x = 0.75$

Approximate Fundamental Period, T_a. Calculate T_a using the three formulas given in ASCE 7-02, Sect. 9.5.5.3.2

1. $T_a = 0.1N$ (for a steel or concrete moment frame building)
   $= 0.1 \times 12 = 1.2$ sec

TABLE 2.24 Values of Approximate Period Parameters C_t and x

Structure type	C_t	x
Moment-resisting frame systems of steel in which the frames resist 100% of the required seismic force and are not enclosed or adjoined by more rigid components that will prevent the frames from deflecting when subjected to seismic forces	0.028	0.8
Moment-resisting frame systems of reinforced concrete in which the frames resist 100% of the required seismic force and are not enclosed or adjoined by more rigid components that will prevent the frame from deflecting when subjected to seismic forces	0.016	0.9
Eccentrically braced steel frames	0.03	0.75
All other structural systems	0.02	0.75

(From ASCE 7-02, Table 9.5.5.3.2.)

2. $T_a = C_t h_n^x$ (for an SMRF building)
    $= 0.028 \times 160^{0.8}$
    $= 1.62$ secs

3. $T_a = 0.03 \times 160^{.75}$ (for an EBF building)
    $= 1.35$ secs

None of the periods calculated previously is explicity applicable to the example problem, which has a dual system of SMRF and EBF. A reasonable approach for a preliminary design would be to take an average of the three.

Thus,

$$T_{a(ave)} = \frac{1.2 + 1.62 + 1.35}{3} = 1.39 \,\text{secs}$$

$T_b = 2$ secs (from computer analysis)   (given)

$T_{max} = C_u \times T_a$ $C_u = 1.4$ (Table 2.23; ASCE Table 9.5.5.3.1)
$\quad\quad = 1.4 \times 1.39$
$\quad\quad = 1.95$

Seismic Base Shear.

$V = C_s W$

$$C_s = \frac{S_{DS}}{R/I}$$

$$= \frac{1}{8/1}$$

$$= 0.125$$

Maximum $C_s = \dfrac{S_{D1}}{T\,(R/I)}$ [ASCE Eq. (9.5.5.2.1.2)]

$$= \frac{0.6}{1.95\,(8)}$$

$$= 0.0385$$

Minimum $C_s = 0.044\, S_{DS}/I$ [ASCE Eq. (9.5.5.2.1.3)]
$\quad\quad = 0.44 \times 1/1$
$\quad\quad = 0.044$ (controls)

If the example building was in SDC E or F,

Minimum $C_s = \dfrac{0.5\, S_1}{R/I}$ [ASCE Eq. (9.5.5.2.1.4)]

$$= \frac{0.5 \times 0.6}{8}$$

$$= 0.0375$$

$$V = 0.044\, W$$

$$= 0.044 \times 22{,}680$$

$$= 998 \text{ kips}$$

Vertical Distribution of Base Shear. Observe that ASCE 7-02 does not give a separate formula for calculating the concentrated force F_t at top. Its effect is automatically included in the manner in which the base shear, V, is distributed vertically over the building height. For a structure with n levels, the force at diaphragm level x is given by the equation:

$$F_x = C_{vx}V \qquad\qquad\qquad\text{[ASCE Eq. (9.5.5.4-1)]}$$

and

$$C_{vx} = \frac{w_x h_x^k}{\sum_{i=1}^{n} w_i h_i^k} \qquad\qquad\qquad\text{[ASCE Eq. (9.5.5.4-2)]}$$

where

$\quad C_{vx}$ = vertical distribution factor

$\quad V$ = total design lateral force or shear at the base of the structure, (kip or kN)

$\quad w_i$ and w_x = the portion of the total gravity load of the structure (W) located or assigned the leverl I or x

$\quad h_I$ and h_2 = the height (ft or m) from the base to level I or x

$\quad k$ = an exponent related to the structure period as follows:

for structures having a period of 0.5 sec or less, $k = 1$

for structures having a period of 2.5 sec or more, $k = 2$

for structures having a period between 0.5 and 2.5 seconds, k shall be 2 or shall be determined by linear interpolation between 1 and 2.

For the example problem, the exponent k by interpolation is equal to

$$k = 1 + \left(\frac{1}{2.5 - 0.5}\right)(1.95 - 0.5) = 1.73 \text{ sec}$$

The vertical distribution of F_x forces computed from the above formula is shown in column 7 of Table 2.25.

Seismic Story Shear. For a given story x, the seismic, story shear is the summation F_x forces above that level. For the example, story shear at level 10 is equal to

$$\sum_{10}^{12} F_x = 123 + 146 + 172 + 199$$

$$= 640 \text{ kips}$$

The values for story shears are shown in column 8 of Table 2.25.

Overturning Moment. This is given by the relation

$$M_x = \sum_{i=x}^{n} F_i(h_i - h_x)$$

TABLE 2.25   ASCE 7-02, IBC-03, NFPA 5000: Static Procedure; Design Example

Level (1)	Height h above shear base, h (ft) (2)	Weight W at height h (kips) (3)	Σ weight = ΣW (kips) (4) = Σ (3)	$W_x h_x^k$ (5)	Vertical distribution factor $\dfrac{W_x h_x^k}{\sum_{i=1}^{n} W_i h_i^k}$ (6)	Lateral seismic force F_x (kips) (7) = (3) × (6)	Lateral seismic story shear (kips) $\sum F_x = \sum(7)$ (8)	Overturning moment (kip-ft) (9) = (7) × (2)	Diaphragm design seismic coefficient (10)	Allowable inelastic story drift Δ_a (in) (11)
Roof	160.0	1,890	1,890	12,057,793	0.199	199	199	–	0.1053	3.12
12	147.0	1,890	3,780	10,416,892	0.172	172	371	2,587	0.0981	3.12
11	134.0	1,890	5,670	8,878,147	0.147	146	517	7,410	0.0912	3.12
10	121.0	1,890	7,560	7,444,183	0.123	123	640	14,131	0.0847	3.12
9	108.0	1,890	9,450	6,117,976	0.101	101	740	22,451	0.0783	3.12
8	95.0	1,890	11,340	4,902,943	0.081	81	821	32,071	0.0724	3.12
7	82.0	1,890	13,230	3,803,069	0.063	63	884	42,744	0.0668	3.12
6	69.0	1,890	15,120	2,823,113	0.047	47	930	54,236	0.0547	3.12
5	56.0	1,890	17,010	1,968,915	0.033	32	963	66,326	0.0510	3.12
4	43.0	1,890	18,900	1,247,944	0.021	21	984	78,845	0.0521	3.12
3	30.0	1,890	20,790	670,355	0.011	11	995	91,637	0.0479	3.6
2	15.0	1,890	22,680	202,609	0.003	3	998	10,652	0.0440	3.6
1	–	–	–	–	–	–	998	121,532	–	–
Σ		22,680	80,533,938		1.00	998	–	121,532		

Note: Δ_a = the allowable inelastic drift = $\dfrac{C_d \delta_{xe}}{I}$ [ASCE 7-02 Eq. (9.5.5.7.1)]

C_d = deflection amplification factor (Table 2.21, ASCE 7-02 Table 9.5.2.2)

I = importance factor (Table 2.13, ASCE 7-02 Table 9.1.4)

δ_{xe} = deflection determined by an elastic analysis for code level forces

For the example, M_x at level 11 is equal to

$$M_{11} = 296 \times 13 + 427 \times 13 = 9399 \text{ kips}$$

Values for M_x are shown in column 9 of Table 2.11.

Diaphragm Design Forces. Forces on diaphragm are computed using an equation different from the one used for determining story shears F_x. This is because higher-mode participation can result in significantly larger forces at individual diaphragm levels than predicted by the relation $F_x = C_{vx}V$. The diaphragm design forces are computed by the equation:

$$F_{px} = \frac{\sum_{i=x}^{n} F_i}{\sum_{i=x}^{n} w_i} w_{px}$$

where

F_{px} = the design force applied to the diaphragm at level x
F_i = the force computed from ASCE 7-02 Eq. (9.5.2.6.4.4) at level i
w_{px} = the effective seismic weight at level x
w_i = the effective seismic weight at level i

Returning to the problem, the diaphragm design force at level 10, for example, is given by

$$F_{p11} = \left(\frac{119 + 131 + 296}{3 \times 1890} \right) w_{px}$$
$$= 0.0963 \, w_{px}$$

The coefficient = 0.0963 is designated the diaphragm design seismic coefficient. Its values for various levels are tabulated in column 10 of Table 2.11.

Allowable Story Drift Δ_a. This is given in Table 2.22 (ASCE 7-02, Table 9.5.2.8). Because the example building is seismic use group (SUG) 1,

$$\Delta_a = 0.02 \, h_{sx}$$

where

h_{sx} = story height below level x

The allowable story drift at level 1 and 2, with h_x = 15 ft is equal to

$$\Delta_a = 0.02 h_x$$
$$= 0.02 \times 15 \times 12$$
$$= 3.6 \text{ in.}$$

For the typical floors and roof

$$\Delta_a = 0.02 \times 13 \times 12$$
$$= 3.12 \text{ in.}$$

These are shown in column 11 of Table 2.25.

2.4.7. Seismic Design Example: Dynamic Analysis Procedure (Response Spectrum Analysis), Hand Calculations

Illustration of dynamic analysis procedure using hand calculations for buildings taller than, say, two or three stories becomes unwieldy. Therefore, in the following example, a planar frame of a two-story building shown in Fig. 2.47 is selected. To keep the explanation simple, infinitely large values are assumed for the flexural stiffness of the beams and the axial stiffness of the columns. Thus, lateral deflection of the frame results from column flexure only.

   Given. A two-story, 30 ft-tall concrete building with a floor-to-floor height of 15 ft

    Structural System: special moment frame system (SMRF)
    I_{cr} = cracked moment of inertia of columns = 12,000 in.^{4} each column
    W = seismic dead load = 580 kips/floor
     = $2 \times 580 = 1160$ kips for the entire building
    E = modules of elasticity of concrete = 4000 ksi

The procedure consists of determining

- Modal periods, T_1 and T_2
- Mode shapes corresponding to T_1 and T_2
- Modal mass and participation factors for each mode
- Modal base shears

To help us understand how static base shear is used to scale dynamic shear, the remainder of this solution consists of determining

- Static base shear using equivalent lateral force procedure
- Scaling of dynamic results
- Distribution of modal base shear in each mode

  Seismic design data. The maximum considered earthquake spectral response acceleration at short period,

     $S_s = 1.5$, and that at 1-second period, $S_1 = 0.6$
     Seismic Use Group = I (standard occupancy)
     Seismic importance factor, $I = 1.0$
     Soil type = S_D
     Site coefficient $F_a = 1.0$
     Site coefficient $F_v = 1.5$

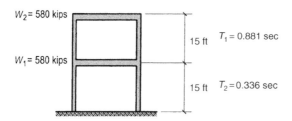

Figure 2.47. Two-story example; dynamic analysis hand calculations.

Modified short period response, $S_{MS} = F_a S_s = 1 \times 1.5 = 1.5$

Modified 1-second period response, $S_{M1} = F_v S_1 = 1.5 \times 0.6 = 0.9$

Design spectral response acceleration parameters at 5% damping:

At short period: $\quad\quad\quad\quad\quad S_{DS} = 2/3 S_{MS} = 2/3 \times 1.5 = 1.0$
At 1-second period: $\quad\quad\quad S_{D1} = 2/3 S_{M1} = 2/3 \times 0.9 = 0.6$

For a special moment frame system (SMRF), $R = 8$, $C_d = 6.5$, where R and C_d are response modification and deflection factor, respectively.

Seismic Design Category based on both S_{Ds} and S_D is D for the example building.

Determine Mass Matrix [m].

$m = W/g = 580/386.4 = 1.5$ kip-sec²/in

$$[m] = \begin{bmatrix} 1.5 & 0 \\ 0 & 1.5 \end{bmatrix} \tag{2.33}$$

Determine Stiffness Matrix. Stiffness K of each column is given by

$$K = \frac{12\,EI}{h_S^3}$$

where

$\quad I$ = total moment of inertia of all columns at level i,
$\quad h_s$ = story height, and
$\quad E$ = modulus of elasticity of concrete

$$K = \frac{12\,EI}{h_S^3} = \frac{12 \times 4000 \times 12000}{(12 \times 15)^3}$$
$$= 98.76 \text{ kips/in. for each column, use } 100 \text{ kips/in}$$

$$\therefore \text{ Stiffness matrix } [K] = 200 \begin{bmatrix} 2 & -1 \\ -1 & 1 \end{bmatrix} \tag{2.34}$$

Find the Determinant of the Matrix

$$[K] - \omega^2[m] \tag{2.34a}$$

$$[K] - \omega^2[m] = \begin{bmatrix} 400 & -200 \\ -200 & 200 \end{bmatrix} - \omega^2 \begin{bmatrix} 1.5 & 0 \\ 0 & 1.5 \end{bmatrix} \tag{2.35}$$

$$= \begin{bmatrix} 400 - 1.5\,\omega^2 & -200 \\ -200 & 200 - 1.5\,\omega^2 \end{bmatrix} \tag{2.36}$$

This matrix is of the form

$$\begin{bmatrix} a_{11} & a_{12} \\ a_{21} & a_{22} \end{bmatrix} \tag{2.37}$$

The determinant of Eq. (2.39) is given by

$$a_{11}\,a_{22} - a_{21}\,a_{12} \tag{2.38}$$

Substituting the elements of the matrix in Eq. (2.38) we get the determinant equal to:

$$(400 - 1.5\omega^2)\,(200 - 1.5\omega^2) - (-200)\,(-200) \tag{2.39}$$

Setting the determinant to zero yields a quadratic equation in ω_i^2. Thus

$$2.25\omega^4 - 900\omega^2 + 40,000 = 0 \tag{2.40}$$

$$\omega^4 - 400\omega^2 + 17,777.7 = 0 \tag{2.41}$$

Solve for the two roots of this characteristic equation. Label these roots ω_1^2 and ω_2^2, with ω_1 being the smaller of the two. ω_1 and ω_2 are called the circular natural frequencies of the system. In mathematical terminology, ω_1^2 and ω_2^2 are called the eigenvalues.

$$\omega_i^2 = \frac{400 \mp \sqrt{400^2 - 4 \times 17,777.7}}{2} \tag{2.42}$$

$\omega_i^2 = 50.9, \qquad \omega_1 = 7.134$ radians/sec

$\omega_2^2 = 349, \qquad \omega_2 = 18.68$ radians/sec

The period T is equal to $\dfrac{2\pi}{\omega}$

Determine Periods.

$$T_1 = \frac{2\pi}{7.134} = 0.881 \text{ seconds}$$

$$T_2 = \frac{2\pi}{18.68} = 0.336 \text{ seconds}$$

Find Mode Shapes.   Substitute ω_i^2 back into the first or second of the characteristic equation to obtain the ratio ϕ_{11}/ϕ_{21}. This ratio defines the natural mode or mode shape corresponding to the natural frequency ω_1.

$$\begin{bmatrix} 400 - 1.5(50.9) & -200 \\ -200 & 100 - 1.5(50.9) \end{bmatrix} \begin{bmatrix} \phi_{11} \\ \phi_{21} \end{bmatrix} = \begin{bmatrix} 0 \\ 0 \end{bmatrix} \tag{2.43}$$

$(400 - 1.5 \times 50.9)\, \phi_{21} = 200_{21}\, \phi_{11} = 1.618\, \phi_{11}$

$\phi_{21} = 1.0, \qquad \phi_{11} = 0.618$

Similarly, by substituting ω_2^2 back into either the first or second of the charactristic equation, we obtain the mode shape corresponding to the frequency ω_2.

$$\begin{bmatrix} 400 - 1.5 \times 349 & -200 \\ 200 & 200 - 1.5 \times 349 \end{bmatrix} \begin{bmatrix} \phi_{12} \\ \phi_{22} \end{bmatrix} = \begin{bmatrix} 0 \\ 0 \end{bmatrix} \tag{2.44}$$

$(400 - 1.5 \times 349)\phi_{22} = 200\phi_{12},$
$$\phi_{22} = -0.618\,\phi_{12}$$
$$\phi_{22} = 1, \qquad \phi_{12} = -1.618$$

In mathematical terminology, natural modes $\begin{bmatrix} \phi_{11} & \phi_{12} \\ \phi_{21} & \phi_{22} \end{bmatrix}$ shown in Fig 2.48 are called eigenvectors.

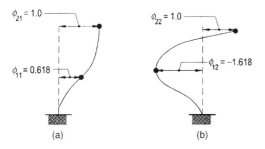

Figure 2.48. Vibration modes; two-story example: (a) first mode; (b) second mode.

ASCE Section 9.5.6.5 states that the portion of the base shear contributed by the mth mode, V_m, shall be determined by the following equations:

$$V_m = C_{sm} W_m \qquad \text{[ASCE Eq. (9.5.6.5-1)]}$$

$$W_m = \frac{\left(\displaystyle\sum_{i=1}^{n} \omega_i \phi_{im} \right)^2}{\displaystyle\sum_{i=1}^{n} \omega_i \phi_{im}^2} \qquad \text{[ASCE Eq. (9.5.6.5-2)]}$$

where

C_{sm} = the modal seismic design coefficient determined below
W_m = the effective modal gravity load
ω_i = the portion of the total gravity load of the structure at level i
ϕ_{im} = the displacement amplitude at the ith level of the structure when vibrating in its mth mode

Determine Modal Mass and Participation Factors for Each Mode. Using the notation

$$L_m = \sum_{i=1}^{n} \frac{w_i}{g} \phi_{im}$$

and

$$M_m = \sum_{i=1}^{n} \frac{w_i}{g} \phi_{im}^2$$

$$L_1 = \sum_{i=1}^{2} \frac{w_i}{g} \phi_{im}$$
$$= 1.5 \text{ kip-sec}^2/\text{in } (\phi_{11} + \phi_{21})$$
$$= 1.5 (0.618 + 1.0)$$
$$= 2.426 \text{ kip-sec}^2/\text{in}$$

$$M_1 = \sum_{i=1}^{2} \frac{w_i \phi_{i1}^2}{g}$$
$$= 1.5 \text{ kip-sec}^2/\text{in} \ (\phi_{11}^2 + \phi_{21}^2)$$
$$= 1.5(0.618^2 + 1.0^2)$$
$$= 2.073 \text{ kip-sec}^2/\text{in}$$

$$L_2 = \sum_{i=1}^{2} \frac{w_i \phi_{im}}{g}$$
$$= 1.5 \text{ kip-sec}^2/\text{in} \ (\phi_{12} + \phi_{22})$$
$$= 1.5(-1.618 + 1)$$
$$= -0.9270 \text{ kip-sec}^2/\text{in}$$

$$M_2 = \sum_{i=1}^{2} \frac{w_i \phi_{12}}{g}$$
$$= 1.5 \text{ kip-sec}^2/\text{in} \ (\phi_{12}^2 + \phi_{22}^2)$$
$$= 1.5 \text{ kip-sec}^2/\text{in} \ (-1.618^2 + 1.0^2)$$
$$= 5.43 \text{ kip-sec}^2/\text{in}$$

Determine Effective Weight and Participating Mass (PM) for Each Mode

$$W_1 = \frac{L_1^2 g}{M_1}$$
$$= \frac{2.426^2 \times 386.4}{2.073}$$
$$= 1098 \text{ kips}$$

$$W_2 = \frac{L_2^2 g}{M_2}$$
$$= \frac{(-0.927)^2 \times 386.4}{5.43}$$
$$= 61.15 \text{ kips} \qquad \text{use 61 kips}$$

$$\sum_{l=1}^{2} W_i = W_1 + W_2$$
$$= 1098 + 61$$
$$= 1159 \text{ kips}$$

$$PM_1 = \frac{1098}{2 \times 580} = 0.95$$

This means that 95% of the total mass participates in the first mode.

$PM_2 = \frac{61}{2 \times 580} = 0.052$ means that 5.2% of the total mass participates in the second mode. Since $PM_1 = 95\%$ is greater than 90% of the total mass, consideration of just the first mode would have been sufficient, per most building codes, to capture the dynamic response of the example building.

Modal Seismic Design Coefficients, C_{sm}.

$$C_{sm} = \frac{S_{am}}{(R/I_E)}$$

where

S_{am} = the modal design spectral response acceleration at period T_m determined from either the general design response spectrum or a site-specific response spectrum.

In the example considered here, the general procedure for determining the spectral acceleration, S_a, will be followed.

For $T \geq T_s$, $\quad S_a = \dfrac{S_{D1}}{T}$

$\quad T_0 < T < T_s, \quad S_a = S_{DS}$

$\quad T \leq T_0, \quad S_a = 0.6 S_{DS} \dfrac{T}{T_0} + 0.4\, S_{DS}$

where

$$T_s = \frac{S_{D1}}{S_{DS}}, \quad \text{and } T_0 = 0.2\, T_s$$

For the example problem,

$$T_s = \frac{0.6}{1.0} = 0.6 \text{ sec}$$

$$T_0 = (0.2) \times (0.6) = 0.12 \text{ sec}$$

Mode 1: $T_1 = 0.881$ sec. This is $> T_s = 0.6$ sec. Therefore,

$$C_s = \frac{S_{D1}}{T\left(\dfrac{R}{I}\right)} = \frac{0.6}{0.881 \times \dfrac{8}{1}} = 0.0851\, g$$

Mode 2: $T_2 = 0.336$ Sec. This is $> T_0$ and $< T_s$. Therefore,

$$C_s = \frac{S_{DS}}{\left(\dfrac{R}{I}\right)} = \frac{1.0}{8} = 0.125\, g$$

Base Shear Using Modal Analysis.

$$V_m = C_{sm}W_m = \frac{L_m^2}{M_m}\, C_{sm}$$

Mode 1: $V_1 = 0.0851 \times 1477 = 125.7$ kips
Mode 2: $V_2 = 0.127 \times 61 = 7.7$ kips

The modal base shear may be combined by taking the square root of the sum of the squares (SRSS) of each of the modal values or by the complete quadratic combination (CQC) technique. The SRSS method is used here.

$$V_t = \left[125.7^2 + 7.7^2\right]^{1/2}$$
$$= 125.9 \text{ kips}, \quad \text{say, } 126 \text{ kips}$$

Design Base Shear Using Equivalent Lateral Force Product.
For the example considered, we have

$$S_{DS} = 1.0$$
$$S_{D1} = 0.6$$
$$S_1 = 0.6$$
$$R = 8$$
$$I = 1.0$$

Approximate fundamental period

$$T_a = C_T(h_n)^x \qquad\qquad \text{[ASCE 7-02, Eq. (9.5.5.3.2-1)]}$$

$C_T = 0.016$ for a moment-resisting concrete frame system
(ASCE 7-02, Table 9.5.5.3.2)

$x = 0.9$ (Table 2.24; ASCE 7-02, Table 9.5.5.3.2)

$h_n =$ total height $= 30$ ft (Table 2.24; ASCE 7-002, Table 9.5.5.3.2)

$T_a = 0.016 \times (30)^{0.9} = 0.34$ sec

$T_B = 0.881$ sec established from modal analysis should not exceed the approximate fundamental period, T_a, by more than a factor C_u. (See ASCE 7-02, Table 9.5.5.3.1.)
For this example problem, $S_{D1} = 0.6 > 0.4$.

Therefore, $C_u = 1.4$ (Table 2.23; ASCE Table 9.5.5.3.1)

$T_{\max} = 1.4 \times 0.34 = 0.48$ sec

Base shear $V = \dfrac{S_{D1}I}{RT} W = \dfrac{0.6 \times 1 \times 1160}{8.0 \times 0.48} = 181.3$ kips

Max $V = \dfrac{S_{DS}I}{R} W = \dfrac{1.0 \times 1.0 \times 1160}{8} = 145$ kips

Min $V = 0.044\, S_{DS}\, I\, W = 0.044 \times 1.0 \times 1160 = 51$ kips

Min V for buildings in SDC E or F $= 0.5 S_1 IW$

$$= \frac{0.52 \times 0.6 \times 1 \times 1160}{8}$$
$$= 43.5 \text{ kips}$$

However, this is not applicable to the example problem since it is in SDC D.

$V = 145$ kips governs.

Scaling of Elastic Response Parameters for Design. The dynamic base shear, V_t, should be scaled up when it is less than 85% of the static base shear V (see ASCE section 9.5.6.8). However, it is permissible to use a fundamental period $T = C_u C_u T_a$ in the calculation of base shear using the equivalent static procedure, instead of $T = C_u T_a$ (ASCE Section 9.5.6.8).

The new period $T = 1.4 \times 1.4 \times 0.34 = 0.67$ sec. The revised base shear for $T = 0.67$ is calculated as follows:

$$V = \frac{S_{D1}I_E}{RT} W = \frac{0.6 \times 1 \times 1160}{8 \times 0.67} = 130 \text{ kips (controls)}$$

$$\text{Max } V = \frac{S_{DS}I_E}{R} W = \frac{1 \times 1 \times 1160}{8} = 145 \text{ kips}$$

Min $V = 0.044 \, S_{DS}I_E W = 0.044 \times 1 \times 1160 = 51$ kips

For buildings in SDC E or F

$$\text{Min } V = \frac{0.5 \times S_1 \times I_E}{R} = \frac{0.5 \times 0.6 \times 1 \times 1160}{8} = 43.5 \text{ kips}$$

This is not applicable to the design example, since it is in SDC D.

Use $V = 130$ kips

The modal base shear $V_t = 126$ kips is not less than 85% of the static base shear $V = 130$ kips. Therefore, modal base shear need not be sealed up by a factor equal to

$$0.85 \frac{V}{V_t} \qquad \text{[ASCE Eq. (9.5.6.8)]}$$

Therefore, use the following shear values derived earlier for modal distribution:

$V_1 = 125.7$ kips
$V_2 = 7.7$ kips

Distribution of Base Shear. Lateral force at level x (levels 1 and 2, in our example), for mode m (modes 1 and 2,) is calculated as follows:

$$F_{xm} = C_{xm}/V_m \qquad \text{[ASCE Eq. (9.5.6.6-1)]}$$

$$C_m = \frac{W_x \phi_{xm}}{\sum_{i=1}^{n} W_1 \phi_{im}} \qquad \text{[ASCE Eq. (9.5.6.6-2)]}$$

where
    C_{vzm} = the vertical distribution factor in the mth mode
    V_m = the total design lateral force or shear at the base in the mth mode
  W_i, W_x = the portion of the gravity load of the building at level i or x
    ϕ_{im} = the displacement amplitude at the ith level of the building when vibrating in its mth mode
    ϕ_{xm} = the desplacement amplitude at the xth level of the building when vibrating in its mth mode

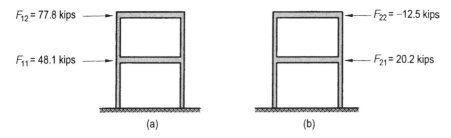

Figure 2.49. Distribution of modal shears: (a) first mode; (b) second mode.

The distribution of modal base shear shown in Fig. 2.49 is calculated as follows:

Mode 1			$V_m = V_1 = 125.7$ kips	
Level	Weight w_i	ϕ_i	$w_i\phi_i$	$F_i = \dfrac{w_i\phi_{im}}{\Sigma w_i\phi_{im}} \times V_m$
2	580	$\phi_{12} = 1.0$	580	$F_{12} = 77.8$ kips
1	580	$\phi_{11} = 0.618$	358.4	$F_{11} = 48.1$ kips
			$\Sigma = 938.4$	$\Sigma = 125.9$ kips
Mode 2			$V_m = V_2 = 7.7$ kips	
2	580	$\phi_{22} = 1.0$	580	$F_{22} = -12.5$ kips
1	580	$\phi_{21} = -1.618$	-938.4	$F_{21} = 20.2$ kips
			$\Sigma = -358.4$	$\Sigma = 7.7$ kips

2.4.8. Anatomy of Computer Response Spectrum Analyses (In Other Words, What Goes on in the Black Box)

Now that we have learned the fundamentals of dynamic analysis, perhaps it is instructive to study a couple of computer dynamic analysis results. This will enhance our understanding of the modal superposition process that takes place in the computer, in the black box.

The examples presented illustrate the modal analysis method. In the first part of each example, the analysis is performed to determine the base shear for each mode using given building characteristics and ground motion spectra. In the second part, the story forces, accelerations, and displacements are calculated for each mode, and are combined statistically using the SRSS combination. The following equations are used in the analysis procedure.

The base shear is determined from

$$V_m = \alpha_m S_{am} W \qquad (2.45)$$

where

V_m = bease shear contributed by the mth mode
α_m = modal base shear participation factor for the mth mode
S_{am} = spectral acceleration for the mth mode determined from the response spectrum
W = total weight of the building including dead loads and applied portions of other loads

The modal base shear participation factor, α_m, for the mth mode is determined from

$$\alpha_m = \frac{\left(\displaystyle\sum_{i=1}^{n} \frac{w_i}{g} \phi_{im}\right)^2}{\displaystyle\sum_{i=1}^{n} \frac{w_i}{g} \sum_{i=1}^{n} \frac{w_i}{g} \phi_{im}^2} \tag{2.46}$$

The story modal participation, PF_{xm}, for the mth mode is determined from

$$PF_{xm} = \left(\frac{\displaystyle\sum_{i=1}^{n} \frac{w_i}{g} \phi_{im}}{\displaystyle\sum_{i=1}^{n} \frac{w_i}{g} \phi_{im}^2}\right) \phi_{xm} \tag{2.47}$$

where

$\quad PF_{xm}$ = modal participation factor at level x for the mth mode
$\quad w_i/g$ = mass assigned to level i
$\quad \phi_{im}$ = amplitude of the mth mode at level i
$\quad \phi_{xm}$ = amplitude of the mth mode at level x
$\quad n$ = level n under consideration

The modal story lateral displacement, δ_{xm}, is determined from

$$\delta_{xm} = PF_{xm} S_{am} \left(\frac{T_m}{2\pi}\right)^2 g \tag{2.48}$$

where

$\quad \delta_{xm}$ = lateral displacement at level x for the mth mode
$\quad S_{am}$ = spectral acceleration for the mth mode determined from the response spectrum
$\quad T_m$ = the period of vibration at the mth mode

2.4.8.1.　*Example 1: Three-Story Building*

Given.　The example is illustrated in Fig. 2.50.
Weights and Masses

$W_R = 187$ kips

$$m_R = \frac{187}{32.2} = 5.81 \text{ kip sec}^2/\text{ft}$$

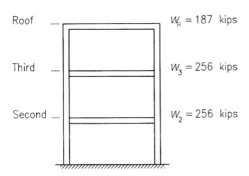

Figure 2.50.　Three-story building example; dynamic analysis.

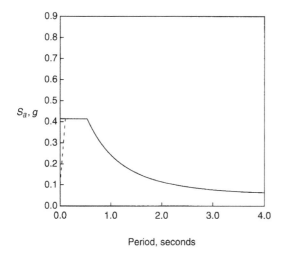

Figure 2.51.   Three-story building; response spectrum.

	Period							
	0.0	.586	.80	1.0	1.5	2.0	3.0	4.0
S_a, g	.14	.41	.300	.240	.160	.120	.080	.060

$W_2 = W_3 = 236$ kips

$$m_2 = m_3 = \frac{236}{32.2} = 7.33 \text{ kip-sec}^2/\text{ft}$$

Periods.

$T_1 = 0.964$ sec

$T_2 = 0.356$ sec

$T_3 = 0.182$ sec

Spectral Acceleration.   From the response spectrum of Fig. 2.51, the spectral accelerations are

$S_{a_1} = 0.251\ g$ for mode 1

$S_{a_2} = 0.41\ g$ for mode 2

$S_{a_3} = 0.41\ g$ for mode 3

Required.

1. Modal analysis to determine base shears.
2. Story forces, overturning moments, accelerations, and displacements for each mode.
3. Square-root-of-the-sum (SRSS) combinations.

T_{m_1}, sec	0.964	0.356	0.182

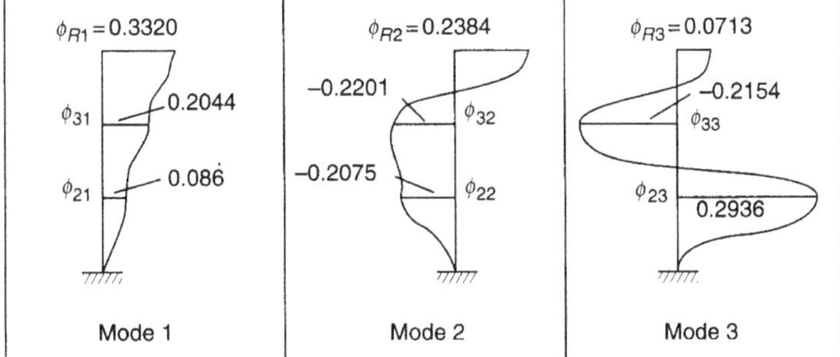

Mode 1   Mode 2   Mode 3

Level	Mass $\left(\dfrac{k \cdot sec^2}{ft}\right)$	Mode 1 ϕ_{x1}	$m_x\phi_{x1}$	$m_x\phi_{x1}^2$	Mode 2 ϕ_{x2}	$m_x\phi_{x2}$	$m_x\phi_{x2}^2$	Mode 3 ϕ_{x3}	$m_x\phi_{x3}$	$m_x\phi_{x3}^2$
R	5.81	0.3320	1.929	0.640	0.2384	1.385	0.330	0.0713	0.4143	0.030
3	7.32	0.2044	1.496	0.306	−0.2201	−1.611	0.355	−0.2154	−1.577	0.340
2	7.32	0.0860	0.630	0.054	−0.2075	−1.519	0.315	0.2936	2.149	0.631
Σ	20.45		4.055	1.000**		−1.745	1.000		0.9863	1.001
PF^*_{Rm}		$\dfrac{\Sigma m\phi}{\Sigma m\phi^2}$ $\phi_{R1}=$ 1.346			−0.416			0.070		
PF_{3m}		0.829			0.384			−0.212		
PF_{2m}		0.349			0.362			0.289		
α_m		$\dfrac{(\Sigma m\phi)^2}{\Sigma m(\Sigma m\phi^2)}$ = 0.8040			0.149			0.048		
S_a		0.251 g			0.41 g			0.41 g		
$v = \alpha_m S_a W$		132.7 kips			40.2 kips			13.0 kips		

* Note that the sum of the modal participation factors $\sum_{m=1}^{3} PF_{xm} = 1.0$ and the sum of modal base shear participation factors $\sum_{m=1}^{3} \alpha_m = 1.0$.

** The mode shapes have been normalized by the computer program so that $\Sigma m\phi^2 = 1.0$.

Figure 2.52. Three-story building; modal analysis to determine base shears.

Solution. The results of the modal analysis are shown in Figs. 2.52, 2.53, and 2.54. It should be noted that higher modes of response become increasingly important for taller or irregular buildings. For the regular 3-story building, the first mode dominates the lateral response as shown in the comparison of the modal story shears and the SRSS story shears in Fig. 2.53. For example, if only the first mode shears had been used for analysis, we would have obtained 89% of the SRSS shear at the roof, 99% at the third floor, and 95% at the second floor. While the second mode shear at the roof is 50% of the first mode shear, when combined on SRSS basis, the first mode accounts for 79% of the SRSS response, with 20% for the second mode and 0.6% for the third mode. These percentages are 91%, 8%, and 1% at the base. The effective modal weight factor, α_m, also shows the relative importance of each mode. In this example, with $\alpha_1 = 0.804$, $\alpha_2 = 0.149$, and $\alpha_3 = 0.048$. This indicate that 80.4% of the building mass participation is in the first mode, 14.9% in the second, and 4.8% in the third.

Level	PF_{xm}	$\dfrac{m_x\phi_{xm}}{\sum m_x\phi_{xm}}$	F_{xm} (k)	V_{xm} (k)	ΔOTM_{xm} (ft·k)	OTM_{xm} (ft·k)	$a_{xm}=\dfrac{F_{xm}}{w_x}$	δ_{xm} (in.)	Δ_{xm} (in.)
R	1.346	0.476	63.2	63.2	772	0	0.337	3.065	1.182
3	0.829	0.369	48.9	112.1	1233	772	0.208	1.892	1.101
2	0.349	0.155	20.6	132.7	1416	2005	0.087	0.791	0.791
		1.000				3421			

(a) Mode 1

Level	PF_{xm}	$\dfrac{m_x\phi_{xm}}{\sum m_x\phi_{xm}}$	F_{xm} (k)	V_{xm} (k)	ΔOTM_{xm} (ft·k)	OTM_{xm} (ft·k)	$a_{xm}=\dfrac{F_{xm}}{w_x}$	δ_{xm} (in.)	Δ_{xm} (in.)
R	−0.416	−0.793	−31.9	−31.9	−389	0	−0.171	−0.212	0.407
3	0.384	0.923	37.1	5.2	57	−389	−0.157	0.195	0.011
2	0.362	0.870	35.0	40.2	429	−332	−0.148	0.184	0.184
		1.000				97			

(b) Mode 2

Level	PF_{xm}	$\dfrac{m_x\phi_{xm}}{\sum m_x\phi_{xm}}$	F_{xm} (k)	V_{xm} (k)	ΔOTM_{xm} (ft·k)	OTM_{xm} (ft·k)	$a_{xm}=\dfrac{F_{xm}}{w_x}$	δ_{xm} (in.)	Δ_{xm} (in.)
R	0.070	0.420	5.5	5.5	67	0	−0.029	0.0094	0.037
3	−0.212	−1.599	−20.8	−15.3	−168	67	−0.087	−0.028	0.066
2	0.289	2.179	28.3	13.0	139	−101	0.118	0.038	0.038
		1.000				38			

(c) Mode 3

Level			F_{xm} (k)	V_{xm} (k)	ΔOTM_{xm} (ft·k)	OTM_{xm} (ft·k)	$a_{xm}=\dfrac{F_{xm}}{w_x}$	δ_{xm} (in.)	Δ_{xm} (in.)
R			71.0	71.0	867	0	0.379	3.072	1.251
3			64.8	113.3	1246	867	0.275	1.893	1.094
2			49.5	139.3	1486	2035	0.208	0.812	0.813
						3423			

(d) SRSS combination

Figure 2.53. Three-story building: modal analysis to determine story forces, accelerations, and displacements.

2.4.8.2. Example 2: Seven-Story Building

Given. See the seven-story building illustration in Fig. 2.55.
Weights and Masses

$W_R = 1410$ kips

$$m_R = \frac{W_R}{g} = \frac{1410}{32.2} = 43.79 \text{ kip-sec}^2/\text{ft}$$

$W_7 = W_6 = W_5 = W_4 = W_3 = 1460$ kips

$$m_7 = m_6 = m_5 = m_4 = m_3 = \frac{1460}{32.2} = 45.34 \text{ kip-sec}^2/\text{ft}$$

$W_2 = 1830$ kips

$$m_2 = \frac{1830}{32.2} = 56.83 \text{ kip-sec}^2/\text{ft}$$

Level	V_{SRSS}	Mode 1			Mode 2		Mode 3	
		V_1	V_1/V_{SRSS}	$(V_1/V_{SRSS})^2$	V_2	$(V_2/V_{SRSS})^2$	V_1	$(V_3/V_{SRSS})^2$
R	71.0	63.2	0.89	0.79	−31.9	0.202	5.5	0.006
3	119.3	112.1	0.989	0.98	5.2	0.002	−15.3	0.018
2	139.3	132.7	0.953	0.91	40.2	0.083	13.0	0.009

Figure 2.54. Three-story building: comparison of modal story shears and the SRSS story.

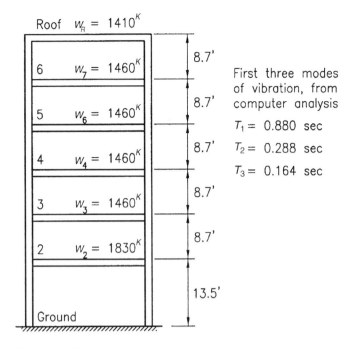

Figure 2.55. Seven-story building example; dynamic analysis.

Periods.

$T_1 = 0.880$ sec

$T_2 = 0.288$ sec

$T_3 = 0.164$ sec

Spectral Accelerations. From the response spectrum of Fig. 2.56 a, b, and c, the spectral accelerations are

$S_{a_1} = 0.276g$

$S_{a_2} = 0.500g$

$S_{a_3} = 0.500g$

Observe that all three parts of Fig. 2.56 contain the same information related to the acceleration response, S_a. Only the format is different. Figure 2.56a shows the building periods and spectral accelerations in a format similar to that in 1997 UBC and IBC-03. Figure 2.56b is a tripartite response spectrum with additional values for displacements and velocities. Figure 2.56c shows the building periods and response accelerations in tabular format.

It should be noted that in the computer program used for calculation of the Eigen values, each mode is normalized for a value of $\Sigma \dfrac{W}{g} \phi^2 = 1.0$. In some programs, ϕ is normalized to 1.0 at the uppermost level.

Required.

1. Modal analysis to determine base shears.
2. First, second, and third mode forces and displacements.
3. Modal analysis summary.

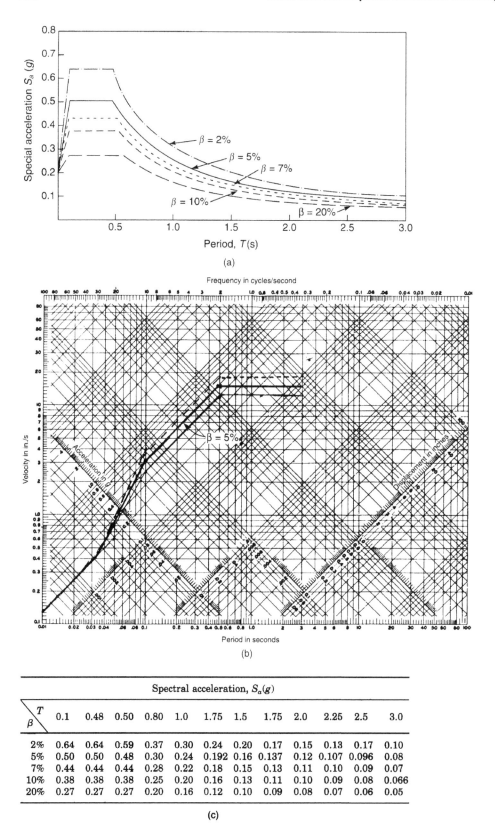

T / β	Spectral acceleration, $S_a(g)$										
	0.1	0.48	0.50	0.80	1.0	1.75	1.5	1.75	2.0	2.25	2.5	3.0
2%	0.64	0.64	0.59	0.37	0.30	0.24	0.20	0.17	0.15	0.13	0.17	0.10
5%	0.50	0.50	0.48	0.30	0.24	0.192	0.16	0.137	0.12	0.107	0.096	0.08
7%	0.44	0.44	0.44	0.28	0.22	0.18	0.15	0.13	0.11	0.10	0.09	0.07
10%	0.38	0.38	0.38	0.25	0.20	0.16	0.13	0.11	0.10	0.09	0.08	0.066
20%	0.27	0.27	0.27	0.20	0.16	0.12	0.10	0.09	0.08	0.07	0.06	0.05

(c)

Figure 2.56. Response spectrum for seven-story building example: (a) acceleration spectrum; (b) tripartite diagram; (c) response spectra numerical representation.

Level	$\frac{w}{g}$ $\left(\frac{\text{k-s}^2}{\text{ft}}\right)$	Mode 1 ϕ_1	$\frac{w}{g}\phi_1$	$\frac{w}{g}\phi_1^2$	a_1 (g)	Mode 2 ϕ_2	$\frac{w}{g}\phi_2$	$\frac{w}{g}\phi_2^2$	a_2 (g)	Mode 3 ϕ_3	$\frac{w}{g}\phi_3$	$\frac{w}{g}\phi_3^2$	a_3 (g)	SRSS a_x (g)
Roof	43.78	0.0794	3.48	0.276	0.362	0.0747	3.27	0.744	−0.235	0.0684	2.99	0.205	0.120	0.448
7	45.34	0.0745	3.38	0.252	0.340	0.0411	1.86	0.076	−0.129	−0.0040	−0.18	0.001	−0.007	0.364
6	45.34	0.0666	3.02	0.201	0.304	−0.0042	−0.19	0.001	0.013	−0.0644	−2.92	0.188	−0.113	0.325
5	45.34	0.0558	2.53	0.141	0.254	−0.0471	−2.14	0.101	0.148	−0.0630	−2.86	0.180	−0.111	0.314
4	45.34	0.0425	1.93	0.082	0.194	−0.0718	−3.26	0.234	0.226	−0.0023	−0.10	0.000	−0.004	0.298
3	45.34	0.0279	1.27	0.035	0.127	−0.0697	−3.16	0.220	0.219	0.0604	2.74	0.166	0.106	0.275
2	56.83	0.0149	0.85	0.013	0.068	−0.0467	−2.65	0.124	0.147	0.0677	3.85	0.261	0.119	0.201
1	—	0	0	0	0	0	0	0	0	0	0	0	0	0
Σ	327.31		16.46	1.000			−6.27	1.000			3.52	1.001		

		Mode 1	Mode 2	Mode 3	
PF_{roof}	Eq. (3.8)	$\frac{16.46}{1.000}(0.0794) = 1.31$	$\frac{-6.37}{1.000}(0.0747) = -0.47$	$\frac{3.52}{1.001}(0.0684) \approx 0.24$ $\ \Sigma = 1.08$	
α	Eq. (3.9)	$\frac{(16.46)^2}{(327.31)(1.000)} = 0.828$	$\frac{(-6.27)^2}{(327.31)(1.000)} = 0.120$	$\frac{(3.52)^2}{(927.31)(1.001)} = 0.038$ $\ \Sigma = 0.986$	
T		0.880 sec	0.288 sec	0.164 sec	
S_a		0.276 g	0.500 g	0.500 g	
a_{roof}	Eq. (3.10)	$(1.31)(0.276) = 0.362\,g$	$(-0.47)(0.500) = -0.235\,g$	$(0.24)(0.500) = 0.120\,g$ $\quad 0.448$	
V	Eq. (3.11)	$(0.828)(0.276)(10,539) = 2408$ kips	$(0.12)(0.500)(10,539) = 632$ kips	$(0.038)(0.500)(10,539) = 200$ kips $\ 2498$ kips (SRSS)	
V/W		0.229	0.060	0.019 $\quad 0.237$	

$W = \Sigma \left(\frac{w}{g}\right) \times g = 327.31 \times 32.2 = 10,539$ kips = Building Weight.
$A_G = 0.20\,g$ Site PGA.
$\beta = 0.05$ Damping Factor.

Figure 2.57. Seven-story building; modal analysis to determine base shears.

Solution. From the modal analysis results shown in Fig. 2.57, the sum of the participation factors, PF_{xm} and α_m, add up to 1.08 and 0.986, respectively. These values being close to 1.0 indicates that most of the modal participation is included in the three modes considered in the example. The story accelerations and the base shears are combined by the square-root-of-the-sum-of-the-squares (SRSS). The modal base shears are 2408 kips, 632 kips, and 200 kips for the first, second, and third modes, respectively. These are used in Fig. 2.61 to determine story forces. The SRSS base shear is 2498 kips.

Story Forces, Accelerations, and Displacements. Figures 2.57–2.60 are set up in a manner similar to the static design procedure described previously. In the static lateral procedure, $Wh/\Sigma Wh$ is used to distribute the force on the assumption of a straight line mode shape. In the dynamic analysis, the more representative $W\phi/\Sigma W\phi$ distribution is used to distribute the forces. The story shears and overturning moments are determined in the same manner for each method. Modal story accelerations are determined by dividing the story force by the story weight. Modal story displacements are calculated from the accelerations and the period by using the following equations:

$$\delta_{xm} = PF_{xm}S_{am}\left(\frac{T_m}{2\pi}\right)^2 g \tag{2.49}$$

where

δ_{xm} = lateral displacement at level x for mode m
S_{am} = spectral displacement for mode m calculated from response spectrum
T_m = modal period of vibration

Modal interstory drifts $\Delta\delta$ are calculated by taking the difference between the δ values of adjacent stories. The values shown in Figs. 2.58–2.60 are summarized in Fig. 2.61.

The fundamental period of vibration as determined from a computer analysis is 0.88 sec. The periods of the second and third modes of vibration are 0.288 sec and 0.164 sec, respectively. From Figs. 2.56, using a response curve with 5% of critical damping

$$T_1 = 0.880 \text{ sec}$$

Modal base shear $V_1 = 2408$ kips

(1)	(2)	(3)	(4)	(5)	(6)	(7)	(8)	(9)	(10)	(11)		
						F						
						kips	V	ΔOTM	OTM	Accel.		
		h	Δh	w	$w\phi$	(V_1)	kips	K-ft	K-ft	g	δ^*	$\Delta\delta$
Story	ϕ	ft	ft	kips	$\dfrac{w\phi}{\Sigma w\phi}$	$\times$(6)	Σ(7)	(4)–(8)	Σ(9)	(7)÷(5)	ft	ft
Roof	0.0794	65.7		1410	0.211	508			0	0.360	0.228	
			8.7				508	4420				0.014
7	0.7450	57.0		1460	0.205	494			4420	0.338	0.214	
			8.7				1002	8717				0.022
6	0.0666	48.3		1460	0.184	443			13,137	0.303	0.192	
			8.7				1445	12,572				0.031
5	0.0558	59.6		1460	0.154	371			25,709	0.254	0.161	
			8.7				1816	15,799				0.039
4	0.0425	30.9		1460	0.117	282			41,508	0.193	0.122	
			8.7				2098	10,253				0.042
3	0.0279	22.2		1460	0.077	185			59,761	0.127	0.080	
			8.7				2283	19,862				0.057
2	0.0149	13.5		1830	0.052	125			79,623	0.068	0.043	
			13.5				2408	32,508				0.043
Grd.	0	0		0	0	0			112,131	0	0	
				Σ	1.000	2408			112,191			

$$\text{*Displacement } \delta_{x1} = \frac{g}{4\pi^2} \times T_1^2 \times \frac{F}{W}$$

$$= \frac{32}{4\pi^2} \times 0.88^2 \times \text{acceleration}$$

$$= 0.632 \times \text{acceleration}$$

Figure 2.58. Seven-story building; first-mode forces and displacements.

$$T_2 = 0.288 \text{ sec}$$

Modal base shear $V_2 = 632$ kips

(1)	(2)	(3)	(4)	(5)	(6)	(7)	(8)	(9)	(10)	(11)		
						F		ΔOTM				
						kips	V	K-ft	OTM	Accel.		
		h	Δh	w	$w\phi$	(V_2)	kips	(4)	k-ft	g	δ^*	$\Delta\delta$
Story	ϕ	ft	ft	kips	$\dfrac{w\phi}{\Sigma w\phi}$	$\times$(6)	Σ(7)	$\times$(8)	Z(9)	(7)÷(5)	ft	ft
Roof	0.0747	65.7		1410	0.522	−330			0	−0.234	−0.016	
			8.7				−330	−2871				0.007
7	0.0411	57.0		1460	0.297	−188			−2871	−0.129	−0.009	
			8.7				−518	−4507				0.010
6	−0.0042	48.3		1460	0.030	19			−7378	0.013	0.001	
			8.7				−499	−4341				0.009
5	−0.0471	39.6		1460	0.341	216			−11,719	0.148	0.010	
			8.7				−283	−2462				0.005
4	−0.0718	30.9		1460	0.520	329			−14,181	0.225	0.015	
			8.7				46	400				0.000
3	−0.0697	22.2		1460	0.504	319			−13,781	0.219	0.015	
			8.7				365	3176				0.005
2	−0.0467	13.5		1830	0.423	267			−10,605	0.146	0.010	
			13.5				632	8532				0.010
Grd.	0	0							−2073	0	0	
				Σ	0.999	632			−2073			

$$\text{*Displacement } \delta_{x2} = \frac{g}{4\pi^2} \times T_2^2 \times \frac{F}{w}$$

$$= \frac{32}{4\pi^2} \times 0.288^2 \times \text{acceleration}$$

$$= 0.068 \times \text{acceleration}$$

Figure 2.59. Seven-story building; second-mode forces and displacements.

$$T_3 = 0.164 \text{ sec}$$

Modal base shear $V_3 = 200$ kips

(1)	(2)	(3)	(4)	(5)	(6)	(7)	(8)	(9)	(10)	(11)		
						F	V	ΔOTM	OTM	Accel.		
						kips	kips	K-ft	K-ft	g	δ^*	$\Delta\delta$
		h	Δh	w	$\dfrac{w\phi}{\sum w\phi}$	(V_3)	$\sum(7)$	$(4)\times(8)$	$\sum(9)$	$(7)\div(5)$		
Story	ϕ	ft	ft	kips		$\times(6)$					ft	ft
Roof	0.0684	65.7		1410	0.849	170			0	0.121	0.003	
			8.7				170	1479				0.003
7	−0.0040	57.0		1460	−0.051	−10			1479	−0.007	0.000	
			8.7				160	1392				0.003
6	−0.0644	48.3		1460	−0.830	−166			2871	−0.114	−0.003	
			8.7				−6	−52				0.000
5	−0.0630	39.6		1460	−0.813	−163			2819	−0.112	−0.003	
			8.7				−169	−1470				0.003
4	−0.0023	30.9		1460	−0.028	−6			1349	−0.004	0.000	
			8.7				−175	−1523				0.003
3	0.0604	22.2		1460	0.778	156			−174	0.107	0.002	
			8.7				−19	−165				0.002
2	0.0677	13.5		1830	1.094	219			−339	0.120	0.003	
			13.5				200	2700				0.001
Grd.	0	0							2361	0	0	
												0.003
				Σ	0.999	200		2361				

$$^*\text{Displacement } \delta_{x3} = \frac{g}{4\pi^2} \times T_3^2 \times \frac{F}{W}$$

$$= \frac{32}{4\pi^2} \times 0.64^2 \times \text{acceleration}$$

$$= 0.022 \times \text{acceleration}$$

Figure 2.60. Seven-story building; third-mode forces and displacements.

($\beta = 0.05$), it is determined that the second and third mode spectral accelerations ($0.500g$) are 80% greater than the first mode spectral acceleration ($0.276g$). On the basis of mode shapes and modal participation factors, modal story forces, shears, overturning moments, acceleration, and displacements are determined.

Figure 2.61a shows story forces obtained by multiplying the story acceleration by the story mass. The shapes of story force curves (Fig. 2.61a) are quite similar to the shapes of the acceleration curves (Fig. 2.61d), because the building mass is essentially uniform.

Figure 2.61b shows story shears that are a summation of the modal story forces in Fig. 2.61a. The higher modes become less significant in relation to the first mode because the forces tend to cancel each other due to the reversal of direction. The SRSS values do not differ substantially from the first mode values.

Figure 2.61c shows the building overturning moments. Again, the higher modes become somewhat less significant because of the reversal of force direction. The SRSS curve is essentially equal to the first mode curve.

Figure 2.61d shows story accelerations. Observe that the second and third modes do play a significant role in the structure's maximum response. While the shape of an individual mode is the same for displacements and accelerations, accelerations are proportional to displacements divided by the squared value of the modal period, which accounts for the greater accelerations in the higher modes. The shape of the SRSS combination of the accelerations is substantially different from the shapes of any of the individual modes because it accounts for the predominance of the various modes at different story levels.

Figure 2.61e shows the modal displacements. Observe that the fundamental mode predominates, while the second and third mode displacements are relatively insignificant. The SRSS combination does not differ greatly from the fundamental mode. It should be noted that for taller and irregular buildings, the influence of the higher modes becomes larger.

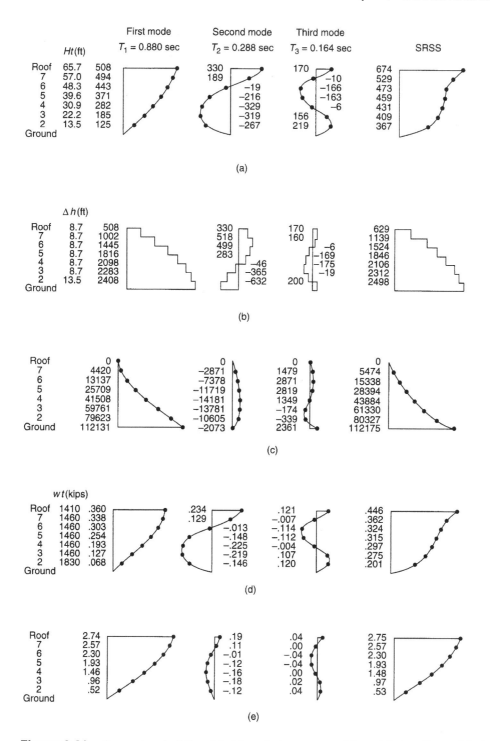

Figure 2.61.   Seven-story building. Modal analysis summary: (a) modal story forces (kips); (b) modal story shears (kips); (c) modal story overturning moments (kip-ft); (d) modal story acceleration (*g*'s); (e) modal lateral displacements (in.).

2.5. SEISMIC DESIGN OF STRUCTURAL ELEMENTS, NONSTRUCTURAL COMPONENTS, AND EQUIPMENT; 1997 UBC PROVISIONS

The seismic demands imposed on equipment are a function of the ground shaking, the dynamic response of the supporting structure, the type of equipment attachment, and the behavior of the equipment. Determination of these demands is quite complex, requiring complicated dynamic analysis of the structure. However, in building engineering, a more simplified procedure is used where the demand is determined more or less independently of the structural characteristics. Typically it is assumed that elements, components, and equipment attached to structures do not substantially alter the structure's seismic response. However, the design lateral-force coefficients used for the design of the items themselves are generally higher than the force coefficients used for the structure's design. This is because

- Absolute accelerations acting on items supported by the structure above ground level are generally greater than at ground level.
- Additional amplified response can occur within items unless they are themselves quite rigid.
- The items may lack the redundancy or energy absorption properties that allow the rational reduction of force levels used for design of the items.

Usually, friction resulting from gravity loads should not be used to resist seismic forces. However, friction achieved by clamping and friction caused by seismic overturning forces may be used. If they are used, the structural system must be checked to assure that sufficient strength and rigidity exists to transmit the overturning forces. Additionally, the supports and foundations in the friction load path must be designed to resist the apportioned seismic shear transferred by friction.

Not all components and equipment are required to have seismic restraints. Unrestrained components and equipment may be permitted, particularly temporary and moveable items, such as heavy furnishings, provided that the safety of occupants and the public is not comprised. Items with low ratios of height-to-base width or length that are inherently stable against overturning may be unrestrained. In such cases, lateral movement may be limited by friction forces based on gravity forces with reductions due to vertical earthquake accelerations. The maximum lateral movement of components and equipment that may occur under maximum expected ground motions should be estimated and incorporated into the design of utility connections and seismic restraints. Space should be provided to accommodate component or equipment movement.

Elimination or reduction of the threat to life is the primary consideration in seismic design. Therefore, a nonessential air handler package unit that is bolted to a mechanical room floor and stands less than 3 feet above the floor is not a threat to life safety as long as adequate anchorage is provided. Hence, the air handler itself need not be designed for seismic forces. Only the attachments such as anchor bolts need to be designed to restrain the movement of the air handler during an earthquake so as to not harm any occupant of the mechanical room. On the other hand, a tall tank mounted on the roof or near a building exit way does pose a hazard. It is the intent of UBC that the tank legs, and the connections between the roof and the tank legs, and the connections between the legs and the tank be designed to resist seismic forces. Alternatively, restraint of the tank by guys or additional bracing could be acceptable.

Nonbuilding structures such as cooling towers, industrial storage rack systems, pressure vessels, and tanks supported on grade do not depend on another structure for lateral stability. Certain categories of these components such as pressure vessels, boilers, and chillers are often rigid, massively constructed with little or no inherent ductility. Seismic response of these components is typically characterized by sliding or overturning at the level of connection to the ground. When damage occurs to these components, it is often concentrated in the connections. At the opposite end of the spectrum are structures such as cooling towers, which are flexible and highly redundant, with behavior quite similar to that of buildings.

Essential structures that are required to provide continuous uninterrupted operation during and after an earthquake will require nonstructural component design that exceed the levels specified in most building codes. Seismic design for the component is generally limited to anchorage and bracing. The integrity of the component and its internal contents is not expressly considered. For example, seismic design of an electrical transformer is typically limited to the design of anchorage of the unit of the structure. Although the internal component may be acceleration-sensitive and vulnerable to damage at acceleration levels significantly lower than the design anchorage force, in North American practice, their design is not part of building design.

The basic objective of seismic design is to provide an adequate level of life safety to protect occupants from life-threatening injury or death. Beyond this basic level of safety, higher levels of performance may be demanded, to limit damage or protect against loss of function. Extensive descriptions of damage states to various systems and components at different performance levels may be found in Tables C1-5, C1-6, and C1-7 of FEMA 356 Publication. These descriptions depict the condition of the component or system following a design-level earthquake. For example, the damage state of cladding of a building for life safety performance is described as "severe distortions in connections with distributed cracking, bending, crushing, and spalling of cladding elements. Some fracturing of cladding, but panels do not fall." On the other hand, a higher level of performance is expected of the same component for immediate occupancy performance by limiting the damage to "mere yielding of connections and minor cracks or bending in cladding."

For new construction, the minimum design objective is to provide life safety. Nonstructural components constructed to this performance objective do not pose a significant threat to life, although the building may close for repairs following a strong earthquake. The emphasis is in elimination of falling hazards, but the nonstructural elements may not be functional or repairable following a strong earthquake.

Essential facilities, such as hospitals, police and fire stations, and emergency command centers are typically designed for higher performance objective with the expectation that they be functional during or shortly after an earthquake.

Seismic design of nonstructural components is a balance between the potential losses versus the cost of damage mitigation measures. However, there are many cases where significant damage can be prevented by simply anchoring components to the floor or walls, at little cost.

2.5.1. Architectural Components

Architectural components include items such as nonload-bearing partitions, exterior curtain walls, and cladding. For life safety, the design objective should be to limit the severity of damage to the components so that they do not topple, or detach themselves from the structure and fall. For higher performance objectives, it may be necessary to control damage to the components so that functionality is not impaired.

In curtain wall terminology, adhered veneer refers to thin surface materials, such as tile, thin-set brick, or stone, which rely on adhesive attachment to a backing or substrate for support. This includes tile, masonry, stone, terra cotta, and similar materials not over 1 inch in thickness. These materials are glued by using adhesive without mechanical attachments to a supporting substrate, which may be masonry, concrete, cement plaster, or a structural framework. Adhered veneers are inherently brittle, sensitive to deformation, and their seismic performance depends on the performance of the supporting substrate. Deformation of the substrate leads to cracking, which can result in the veneer separating from the substrate. The key to good seismic performance is to detail the substrate so as to isolate it from the effects of story drifts.

The threat to life safety posed by adhered veneers depends on the height of the veneer, and the size and weight of the fragments likely to become dislodged. Falling of individual units such as thin tiles, typically would not be considered a life-safety issue as opposed to the tumbling of large areas of the veneer.

Anchored veneer consists of masonry units that are attached to the supporting structure by mechanical means. This type of veneer is both acceleration-and deformation-sensitive. The masonry units can be dislodged by the distortion or failure of the mechanical connectors. Deformations of the supporting structure may displace or dislodge the units by racking. Damage to anchored veneers can be controlled by limiting the drift ratios of the supporting structure, isolating units from story distortions through slip connections or joints, and by anchoring the veneers for an adequate force level that includes consideration of the vertical component of ground shaking. Special attention should be paid at locations likely to experience large deformations, especially at corners and around openings.

Masonry veneer facades on steel frame buildings should be avoided unless the veneer is securely tied to a separate wall or framework that is independent of the primary steel frame. Otherwise, adequate provisions for the large expected lateral deformation of the steel frame must be made.

2.5.2. Exterior Ornaments and Appendages

Exterior ornaments and appendages are nonstructural components that project above or away from the building. They include marquees, canopies, sings, sculptures, and ornaments, as well as concrete and masonry parapets. These components are acceleration-sensitive, and if not properly braced or anchored, can become disengaged from the structure and topple. Building codes require consideration of vertical accelerations for cantilever components. Features such as balconies are typically an extension of the floor structure, and should be designed as part of the structure. Parapets and cornices, unless well braced, are flexible components and design forces for these components and should amplified accordingly.

2.5.3. Component Behavior

Nonstructural components can be classified as deformation-or acceleration-sensitive. If the performance of a component is controlled by the supporting structure's deformation, such as the interstory drift, it is deformation-sensitive. Curtain walls and piping systems running floor-to-floor are some examples of deformation-sensitive components. These components spanning from floor-to-floor are often rigidly connected to the structure. They are thus deformation-sensitive and are susceptible to damage due to a building's interstory drifts.

When a component is not vulnerable to damage from the interstory displacements, it is generally acceleration-sensitive. A mechanical unit anchored to the floor or a roof of a building is a good example. Acceleration-sensitive components are vulnerable to shifting and overturning and as such their anchorage or bracing is of prime concern. The force provisions of building codes generally predicate design forces high enough to prevent sliding, toppling, or collapse of acceleration-sensitive components. Many components are both deformation-and acceleration-sensitive, although a primary mode of behavior can generally be identified. For example, the exterior skin of a building such as anchored veneer or prefabricated panels are both deformation-and acceleration-sensitive. However, their design is primarily controlled by deformation.

Acceleration-sensitive components should be anchored or braced to the structure to limit their movement during seismic events. However, these components should not be anchored in such a way as to inadvertently affect the seismic behavior of the structural system. For example, if the base of a component is anchored to the floor with its top rigidly braced to the floor above, it can have the unintended effect of altering the response of the structural system. An example is a nonstructural masonry partition rigidly connected at the top and bottom to the building floors. The wall acts as a shear wall, leading to an

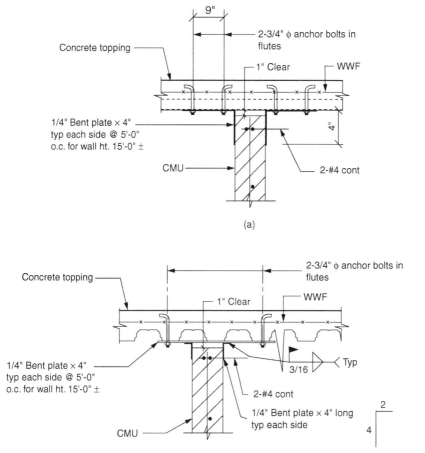

Figure 2.62. Slip joint in nonstructural masonry partition; connection at top provides out-of-plane support without restricting in-plane movement of wall. (a) Wall perpendicular to metal deck span; (b) wall parallel to metal deck span.

unintended redistribution of lateral load. A solution to prevent this condition is to provide isolation joints between the masonry wall and the structural columns wide enough to prevent interaction between the two elements. A sliding connection at the top of the wall should be designed to provide out-of-plane support allowing in-plane movement of the wall (see Fig. 2.62a and b).

The dynamic behavior of components mounted at or below grade is similar to that of buildings. On the other hand, the behavior of components attached to the upper floors of buildings is quite complicated. Its response not only depends on the mass and stiffness of the component, and the characteristics of the ground motion, but also on the dynamic characteristics of the structure itself.

Mechanical components are often fitted with vibration isolation mounts to prevent transmission of vibrations to the structure. By increasing their flexibility, the vibration isolation mounts can alter the dynamic properties of the components, resulting in a dramatic increase in seismic inertial forces. Isolation mounts must be specifically designed to resist these increased seismic effects. For example, 1997 UBC requires the design forces for equipment mounted on a vibration isolator to be based on a dynamic amplification factor, $a_p = 2.5$, and a component response modification factor, $R_p = 1.5$. Comparable values for a rigid equipment with supports fabricated of ductile materials attached to rigid mounts are $a_p = 1.0$ and $R_p = 3.0$. Since the seismic design force is a function of the ratio a_p/R_p, all other things being equal, the design force for an equipment with vibration isolation mounts would be five times larger than the design force when it is mounted on rigid supports.

2.5.4. 1997 UBC Provisions

Building codes may exempt nonstructural components and equipment supported by structures from anchorage and bracing requirements, depending on the level of seismic risk at the site, the occupancy of the structure, and the importance of the components. In regions of low seismicity, all components are typically exempt from seismic bracing requirements. In regions of moderate seismicity, bracing requirements are often limited to critical systems or hazardous components, such as cantilever parapets. In areas of high seismicity, furniture and components that are floor mounted and weight less than 400 pounds are generally exempt from anchorage and bracing requirements. Items that are suspended from walls or ceilings and weigh less than 20 pounds are also typically exempt.

All other components not exempt from seismic design should be designed for seismic forces that are dependent on the following factors.

- Component's weight
- Flexibility or stiffness of the component and its supports including that of the vibration isolator if present
- Acceleration imparted to the component at the point of attachment to the structure
- Redundancy and energy absorption capacity of the component and its attachment to the structures
- An importance factor based on the hazard imposed by the item or the functionality requirement of the building

Although UBC 1997 exempts the attachments for floor- or roof-mounted equipment weighing less than 400 pounds (181 kg) from design, an item of special safety-related equipment should not be exempt from being seismically designed or restrained just because its weight is less than 400 pounds. Also, although it is not specifically required that equipment and furniture weighing less than 400 pounds be attached to the structure, an evaluation of the inherent fragility or ruggedness of equipment or furniture, or the hazard created if the item slides, topples, or otherwise moves as a result of earthquake effects, may require the design of seismic restraints.

Miscellaneous elements of structures consist of elements such as

- Walls, including parapets, exterior walls, and interior bearing and nonbearing wall.
- Penthouses, except where framed by an extension of the structural frame.
- Prefabricated structural elements other than walls.

Components are permanent assemblies not having a structural function and include items such as

- Exterior and interior ornamentation and appendages.
- Chimneys, stacks, and towers.
- Signs and billboards.
- Storage racks, cabinets, and bookshelves, including contents, over 6 feet tall.
- Suspended ceilings and light fixtures.
- Access floor systems.
- Masonry or concrete fences over 6 feet tall.
- Interior partitions.

Equipment consists of mechanical, plumbing, and electrical assemblies attached to the structure and includes items such as

- Tanks and vessels.
- Boilers, heat exchangers, tanks, turbines, chillers, pumps, motors, air handling units, cooling towers, transformers, and switchgear and control panels.
- Piping, conduits, ducts, and cable trays.
- Emergency power supply systems and essential communication equipment.
- Temporary containers of flammable or hazardous materials.

1997 UBC addresses equipment supports that transfer seismic forces from the equipment through attachments to the seismic resisting system of the structure. Equipment supports, even when supplied by the manufacturer, should be verified to ascertain sufficient strength is present to adequately transfer the combined forces. An equipment is considered rigid when the equipment, its supports, and its attachments considered as a single dynamic system, has a fundamental mode of vibration with a natural period less than 0.06 seconds (natural frequency greater than 16.7 Hz). A rigid piece of equipment supported on vibration isolation devices or other flexible supports is considered flexible.

One special category is where loss of structural integrity causes a loss in physical connectivity or restraint under seismic motions, resulting in a direct life hazard. An example is where the entire item or a part of the item breaks off and falls, slides, or otherwise moves, posing a threat to occupants, or blocks a means of egress or an exit way. In these cases, the use of bumpers, braces, guys, or gapped restraints may protect the occupant, even if the item itself is damaged. A similar consideration is required for special safety-related items where the failure of equipment to perform a required function could cause a more indirect life hazard. Examples include fire protection piping or a standby power system in a hospital. It is a good idea to locate such equipment at or below ground level, where it can be easily maintained to ensure its operation during an emergency.

Special safety-related equipment consist of items needed after an earthquake, fire, or other emergencies. It also includes equipment that contains a sufficient quantity of explosive or toxic substances which, if released, would threaten the life safety of the general public. Special safety-related equipment does not include equipment that threatens life safety solely through structural failure.

Other than the design of this anchorage, UBC requirements need not be applied to the design of most air handlers; compressors; pumps; motors; engines; generators; valves; pneumatic, hydraulic, or motor operators; chillers; small horizontal vessels or heat exchangers' evaporators; heaters; condensers; motor control centers; low-or medium-voltage switchgear; transformer with anchored internal coils; small factory-manufactured boilers; inverters; batteries; battery chargers; and distribution panels. In general, such equipment has been found to be rugged. Nonstructural components or systems in facilities with critical functions (e.g., computer centers, hospitals, manufacturing plants with especially hazardous materials, museums with tragile/valuable collection items) are of special interest.

2.5.4.1. *Design Force, F_p*

Equivalent lateral-force procedures, where the component is designed for a lateral seismic force that is expressed as a fraction of the component weight, is the most used method for acceleration-controlled components. Deformation-sensitive components are designed to accommodate the design story drifts, amplified to the levels expected in the design earthquake. The objective of these approaches is to design the anchorage or bracing system that can withstand the accelerations generated by the earthquake, without allowing the component to shift or topple. In addition, the component must be able to tolerate the actual deformations of the primary structure without becoming dislodged or adversely affecting the primary structure's dynamic response.

The 1997 UBC introduced significant changes in the design procedures for non-structural components. The most notable change is in the design acceleration at the upper levels of the building, which could be as much as three or four times the ground acceleration. This change was driven by the analysis of instrument records obtained from buildings that experienced earthquake shaking during the 1994 Northridge earthquake.

In addition, the design force equations are calibrated to a strength design level, which translates to an approximate 1.4 increase in force levels compared to those from 1994 UBC, in which allowable stress design was used.

New equations for determining the design seismic force, F_p, for elements, components, and equipment are dependent on

- Weight of the system or component, W_p.
- Component amplification factor, a_p.
- Horizontal acceleration of the structure for the design ground motion at the point of component attachment to the structure.
- Component importance factor I_p.
- Component response modification factor, R_p.

The design lateral force for nonstructural components is given by

$$F_p = \frac{a_p C_a I_p}{R_p}\left(1 + 3\frac{h_x}{h_r}\right)W_p \qquad (2.50)$$

TABLE 2.26 Horizontal Force Factor, a_p and R_p

	a_p	R_p
1. Elements of structure		
a. Cantilevered parapets	2.5	3.0
b. All interior bearing and nonbearing walls	1.0	3.0
c. Penthouse (not an extension of structural frame)	2.5	4.0
d. Cladding connections	1.0	3.0
2. Nonstructural components		
a. Ornamentations and appendages	2.5	3.0
b. Floor-supported cabinets and book stacks more than 6 feet in height	1.0	3.0
c. Partitions	1.0	3.0
3. Equipment		
a. Emergency power supply systems	1.0	3.0
b. Tanks and vessels	1.0	3.0
4. Other components		
a. Rigid components with ductile material and attachments	1.0	3.0
b. Rigid components with nonductile material and attachments	1.0	1.5
c. Flexible components with ductile material and attachments	2.5	3.0
d. Flexible components with nonductile material or attachments	2.5	1.5

(Condensed from 1997 UBC, Table 16-O.)

where

F_p = lateral force applied to the center of mass of the component

a_p = in-structure amplification factor that varies from 1.0 to 2.5 (Table 2.26)

C_a = seismic coefficient that varies depending on the seismic zone in which the structure is located and the proximity to active earthquake faults. C_a varies from 0.075 to 0.66.

I_p = component importance factor, which depends on the occupancy of the structure and varies from 1.0 to 1.5

R_p = component response modification factor, which varies from 1.5 to 3.0 (Table 2.26)

h_x = element or attachment elevation with respect to grade. h_x shall not be taken as less than 0.

h_r = the structure roof elevation, with respect to grade

W_p = weight of the component

Upper-and lower-bound limits for F_p are defined as follows:

F_p need not exceed $4C_a I_p W_p$

F_p shall not be less than $0.7\ C_a I_p W_p$

The a_p factor accounts for the dynamic amplification of force levels for flexible equipment. Rigid components, defined as components including attachments that have a period less than 0.06 seconds, are assigned an $a_p = 1.0$. Flexible components have a period greater than 0.06 seconds, and are assigned an $a_p = 2.5$. The component response modification factor, R_p, represents the energy absorption capability of the component's structure and attachments. Conceptually, the value considers both the overstrength and ductility of the component's structure and attachments. It is believed that in the absence of research, these separate considerations can be adequately combined into a single factor for nonstructural components. In general, the following benchmark values were used:

$R_p = 1.5$ brittle or buckling failure mode expected

$R_p = 3.0$ moderately ductile materials and detailing

$R_p = 4.0$ highly ductile materials and details

Where connection of the component to concrete or masonry is made with shallow expansion, chemical, or cast-in-place anchors, R_p is taken as 1.5. Shallow anchors are defined as those anchors with an embedment length-to-diameter ratio of less than 8. If the anchors are constructed of brittle materials (such as ceramic elements in electrical components), or when anchorage is provided by an adhesive, R_p is taken as 1.0. The term adhesive in this case refers to connections made by using surface application of a bonding agent, and not anchor bolts embedded using expoxy or other adhesives. An example, of anchorage made with adhesive would be base plates for posts glued to the surface of the structural floor in a raised access floor system.

The reduced R_p values, 1.5 for shallow embedment (post installed and cast) anchors and 1.0 for adhesive anchors, are intended to account for poor anchor performance observed after the Northridge earthquake. When anchors are installed into "housekeeping" pads, these pads should be adequately reinforced and positively anchored to the supporting structural system.

The design forces for equipment mounted on vibration isolation mounts must be computed using an a_p of 2.5 and an R_p of 1.5. If the isolation mount is attached to the structure using shallow or expansion-type anchors, the design forces for the anchors must be doubled.

Equation (2.50) represents a trapezoidal distribution of floor accelerations within the structure, linearly varying from C_a, at the ground, 4.0 C_a at the roof. The ground acceleration, C_a, is intended to be the same acceleration used as design input for the structure itself and will include site effects.

To meet the need for a simpler formulation, a conservative maximum value for $F_p = 4I_pW_p$ has been set.

A lower limit for $F_p = 0.7C_aI_pW_p$ is prescribed to ensure a minimal seismic design force. The redundancy factor R has been set equal to unity since the limiting redundancy of nonstructural components has already been accommodated in the selection R_p factors.

The out-of-plane design loads for exterior walls or wall panels that have points of attachment at two or more different elevations may be determined as follows. For the vertical span of a wall between two successive attachment elevations, h_x and h_{x+1}, evaluate the seismic force coefficients F_p/W_p at each of the two points, observing the minimum and maximum limits, and compute the average of the two values. The average seismic coefficient times the unit weight of the wall provides the distributed load for the span between the given attachment points, and it should extend to the top of any wall parapet above the roof attachment point at h_r.

For a single-story exterior wall, the seismic force coefficient at the base is $0.7C_aI_p$, and at the roof is $1.33C_aI_p$. An average value of $1.02C_aI_p$ applies to the unit weight of the wall for the distributed load over the entire wall.

In addition to lateral force requirements, the 1997 UBC specifies that for essential or hazardous facilities, components must be designed for the effects of relative motion, if the component is attached to the structure at several points. An example would be vertical riser in a piping system that runs from floor to floor. The component must accommodate the maximum inelastic response displacement, defined as

$$\Delta_M = 0.7R\Delta_S$$

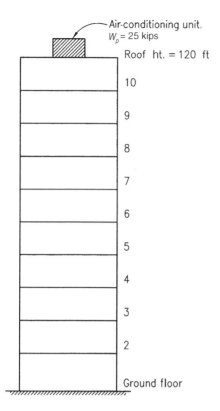

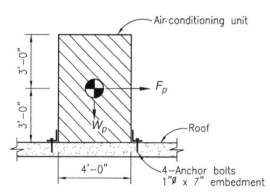

Figure 2.63. Equipment anchorage design; air-conditioning unit example.

2.5.4.2. Design Examples

Example 1.

Given. An air-conditioning unit weighing 25 kips will be installed in the mechanical penthouse on the roof of a 10-story building. The dimensions of the unit are shown in Fig. 2.63. The fundamental period of the air-conditioning unit is 0.05 seconds. There are four 1-inch diameter anchor bolts, one at each corner of the unit, embedded in the roof concrete slab. The bolts have an embedment length of 7 inches. The building is in seismic zone 4 and the building site is within 5 kilometers of a type-B seismic source, and located on soil profile type S_D.

Required. Using the 1997 UBC provisions, determine the shear and tension demands on the anchor bolts, assuming the bolts will be designed using ASD, allowable stress design.

Solution

h_x = 120 feet (attachment height of element with respect to grade)

h_r = 120 feet (roof elevation with respect to grade)

I_p = 1.0 (standard occupancy)

a_p = 1.0 (in-structure amplification factor, values range from 1.0 to 2.5)
(In our case, this is equal to 1.0.)

R_p = 1.25 (component response modification factor varies from 1.25 to 3) In our case, R_p = 1.25, calculated as follows:

The ratio of anchor bolt embedment length to bolt diameter $^{le}/_d = ^7/_1 = 7$. This is less than 8. Therefore, from footnotes of 1997 UBC, Table 16-N, R_p = 1.25. If the ratio $^{le}/_d$ was more than 8, we would have used a higher value of R_p.

C_a = 0.66 (seismic coefficient that is dependent on the seismic zone in which the structure is located and its proximity to active faults) C_a varies from a low of 0.075 to a high of 0.66.

In our case, C_a = 0.66, calculated as follows:

For soil profile type S_D (stiff soil), $C_a = 0.44N_a$, where N_a is the near-source factor. Our site is within 2 kilometers of a type A seismic source. Therefore, N_a = 1.5 and $C_a = 0.44N_a = 0.44 \times 1.5 = 0.66$.

The design lateral force for the equipment using Eq. (2.50) is

$$
\begin{aligned}
F_p &= \frac{a_p C_a I_p \left(1 + 3\dfrac{h_x}{h_r}\right) W_p}{R_p} \\
&= \frac{110 \times 0.66 \times 1}{1.5}\left(1 + 3 \times \frac{120}{120}\right)25 \\
&= 144 \text{ kips} \leftarrow \text{controls}
\end{aligned}
$$

$$
\begin{aligned}
F_{p\,min} &= 0.7 C_a I_p W_p \\
&= 1.7 \times 0.66 \times 1 \times 25 \\
&= 11.55 \text{ kips}
\end{aligned}
$$

$$
\begin{aligned}
F_{p\,max} &= 4 C_a I_p W_p \\
&= 4 \times 0.66 \times 1 \times 25 \\
&= 66 \text{ kips} > 44 \text{ kips}
\end{aligned}
$$

Therefore, the design lateral force for the equipment is

F_p = 44 kips

The ultimate shear per bolt, $V_u = \dfrac{44}{4} = 11$ kips-bolt

The ultimate overturning moment M_{OT} is

$M_{OT} = 44 \times 3 = 132$ kip-ft

The net weight resisting the overturning moment is equal to $0.9\,W_p$, which accounts for the uplift effects due to vertical accelerations. The uplift is typically 10% of the weight W_p.

The resisting movement is

$$M_R = 0.9 \times 25 \times 2 = 45 \text{ kip-ft}$$

Taking the sum of the overturning and resisting moments about a corner of the base of the equipment, the ultimate uplift force F_t in the anchors equals.

$$F_{t\,\text{ult}} = \frac{132 - 45}{4 \times 2} = 10.88 \text{ kips}$$

To convert the ultimate shear and tension forces to allowable stress design (ASD) levels, we divide by a factor of 1.4 to obtain

$$V_{\text{ASD}} = \frac{11}{1.4} = 7.86 \text{ kips}$$

$$F_{t\text{ASD}} = \frac{10.88}{1.4} = 7.77 \text{ kips}$$

Example 2

Given. An emergency generator weighing 20 kips is installed on the fifth floor of a 7-story command center. The dimensions of the unit are shown in Fig. 2.64. The generator is mounted on six vibration isolation mounts with a lateral stiffness of 6 kips/inch. The building floor-to-floor height is 14 feet and the fundamental period of the building is 0.8 seconds. The building is in UBC seismic zone 4, and is in the proximity of an active fault. The site is within 5 kilometers of a type B seismic source and located on soil profile type S_C.

Required. Using the 1997 UBC provisions, determine the shear and tension demands on the vibration isolation mounts, for seismic forces in the east-west direction. Assume the design is by ASD.

Solution

$h_x = 5$ floors @ 14 feet = 70 feet

$h_r = 7$ floors @ 14 feet = 98 feet

$I_p = 1.5$ (essential occupancy structure) (1997 UBC, Table 16-K)

$a_p = 2.5$ (flexible component) (1997 UBC, Table 16-O)

$R_p = 1.5$ (vibration isolated component) (1997 UBC, Table 16-O)

$W_p = 20$ kips (given)

For soil profile type S_C (soft rock, very dense soil), $C_a = 0.40\,N_a$ where N_a is the near-source factor. Our building is within 5 km of a type-B seismic source. Therefore, $N_a = 1.2$, and $C_a = 0.4N_a = 0.4 \times 1.2 = 0.48$.

The design lateral force for the component, F_p, is given by

$$
\begin{aligned}
F_p &= \frac{a_p C_a I_p \left(1 + \dfrac{3\,h_x}{h_r}\right) W_p}{R_p} \\[2mm]
&= \frac{2.5 \times 0.48 \times 1.5}{1.5}\left(1 + 3 \times \frac{70}{98}\right) \times 20 \\[2mm]
&= 75.43 \text{ kips}
\end{aligned}
$$

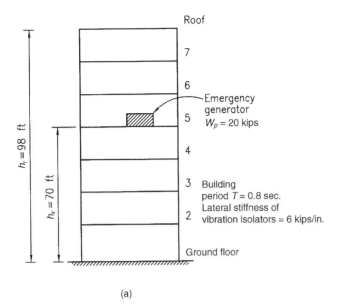

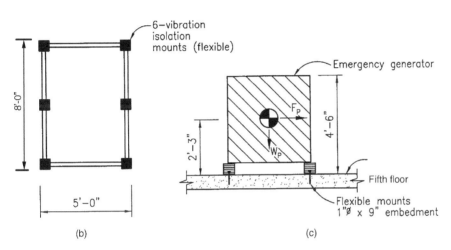

Figure 2.64. Example 2: Emergency generator: (a) building elevation; (b) equipment plan dimensions; (c) equipment elevation.

$$F_{p\,min} = 0.7\,C_a I_p W_p$$
$$= 0.7 \times 0.48 \times 1.5 \times 20$$
$$= 10.08 \text{ kips}$$
$$F_{p\,max} = 4\,C_a I_p W_p$$
$$= 4 \times 0.48 \times 1.5 \times 20$$
$$= 57.00 \text{ kips} \leftarrow \text{controls}$$

Therefore, the design lateral force for the equipment is

$F_p = 57.60$ kips

The ultimate shear, V_u, per isolater mount $= \dfrac{57.60}{6} = 9.60$ kips-mount.

The ultimate overturning moment M_{OT} is

$$M_{OT} = 57.60 \times 2.25 = 129.60 \text{ kip-ft}$$

Allow for a reduction of 10% of the weight W_p to account for the uplift effects due to vertical accelerations. Therefore, the reduced weight is equal to $0.9W_p$ for calculating the resisting moment M_R.

$$M_R = 0.9 \times 20 \times 2.5 = 45 \text{ kip-ft}$$

Noting that there are three mounts on each side of the unit, the ultimate uplift force, i.e., the tension in each mount, is

$$F_t = \frac{129.50 - 45}{5 \times 3}$$
$$= 5.63 \text{ kips (ultimate)}$$

The ASD level forces are determined by dividing the ultimate values by a factor of 1.4.

$$V_{\text{ASD}} = \frac{9.60}{1.4} = 6.86 \text{ kips-mount}$$

$$F_{t,\text{ASD}} = \frac{5.63}{1.4} = 4.02 \text{ kips-mount}$$

2.6.   DYNAMIC ANALYSIS THEORY

A good portion of the loads that occur in buildings can be considered static, requiring static analysis only. Although almost all loads except dead loads are transient, meaning that they change with time, it is customary to treat them as static. For example, lateral loads imposed by transient wind pulses are usually treated as static loads and even in earthquake design, one of the acceptable methods of design, particularly for buildings with regular configuration, is to use an equivalent static force procedure. Under these circumstances, the analysis of a structure reduces to a single solution for a given set of static loads. Although the equivalent static method is a recognized method, most building codes typically mandate dynamic analysis for certain types of buildings such as those with irregular configurations (see ASCE 7-03, Table 9.5.2.5.1). It is therefore necessary, particularly in seismic design, to have a thorough understanding of dynamic analysis concept.

Consider a building subjected to lateral wind loads. Although wind loads are dynamic, in typical design practice, except in the case of slender buildings, wind loads are considered as equivalent static loads. The variation of wind velocity with time is taken into account by including a gust factor in the determination of wind loads. Therefore, for a given set of wind loads, there is but one unique solution.

Now consider the same building, instead of being buffeted by wind, subjected to ground motions due to an earthquake. The input shaking causes the foundation of the building to oscillate back and forth in a more or less horizontal plane. The building would follow the movement of the ground without experiencing lateral loads if the ground oscillation took place very slowly over a long period of time. The building would simply ride to the new displaced position. On the other hand, when the ground moves suddenly as in an earthquake, building mass, which has inertia, attempts to prevent the displacement of the structure.

Therefore, lateral forces are exerted on the mass in order to bring it along with the foundation. This dynamic action maybe visualized as a group of horizontal forces applied to the structure in proportion to its mass, and to the height of the mass above the ground.

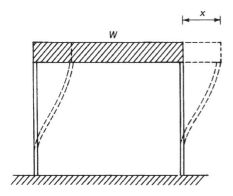

Figure 2.65. Single-bay single-story portal frame.

These earthquake forces are considered dynamic, because they vary with time. Since the load is time-varying, the response of the structure, including deflections, axial and shear forces, and bending moments is also time-dependent. Therefore, instead of a single solution, a separate solution is required to capture the response of the building at each instant of time for the entire duration of an earthquake. Because the resulting inertia forces are a function of building accelerations, which are themselves related to the inertia forces, it is necessary to formulate the dynamic problem in terms of differential equations.

2.6.1.   Single-Degree-of-Freedom Systems

Consider a portal frame, shown in Fig. 2.65, consisting of an infinitely stiff beam supported by flexible columns that have negligible mass as compared to that of the beam. For horizontal motions, the structure can be visualized as a spring-supported mass, as shown in Fig. 2.66a, or as a weight W suspended from a spring, as shown in Fig. 2 66b. Under the action of gravity force on W, the spring will extend by a certain amount x. If the spring is very stiff, x is small, and vice versa. The extension x can be related to the stiffness of the spring k by the relation

$$x = \frac{W}{k} \qquad (2.51)$$

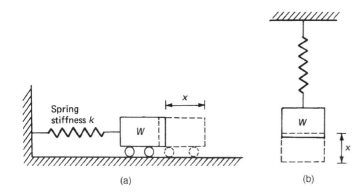

Figure 2.66. Analytical models for single-degree-of-freedom stem: (a) model in horizontal position; (b) model in vertical position.

The spring constant or spring stiffness k denotes the load required to produce unit extension of the spring. If W is measured in kips and the extension in inches, the spring stiffness will have a dimension of kips per inch. The weight W comes to rest after the spring has extended by the length x. Equation (2.51) expresses the familiar static equilibrium condition between the internal force in the spring and the externally applied force W.

If a vertical force is applied or removed suddenly, vibrations of the system are produced. Such vibrations, maintained by the elastic force in the spring alone, are called free or natural vibrations. The weight moves up and down, and therefore is subjected to an acceleration $\ddot{x}$ given by the second derivative of displacement x, with respect to time t. At any instant t, there are three forces acting on the body: the dynamic force equal to the product of the body mass and its acceleration, the gravity force W acting downward, and the force in the spring equal to $W + kx$ for the position of weight shown in Fig. 2.67. These are in a state of dynamic equilibrium given by the relation

$$\frac{W}{g}\ddot{x} = W - (W + kx) = -kx \tag{2.52}$$

The preceding equation of motion is called Newton's law of motion and is governed by the equilibrium of inertia force that is a product of the mass W/g, and acceleration $\ddot{x}$, and the resisting forces that are a function of the stiffness of the spring.

The principle of virtual work can be used as an alternative to derive Newton's law of motion. Although the method was first developed for static problems, it can readily be applied to dynamic problems by using D'Alembert's principle. The method establishes dynamic equilibrium by including inertial forces in the system.

The principle of virtual work can be stated as follows: For a system in equilibrium, the work done by all the forces during a virtual displacement is equal to zero. Consider

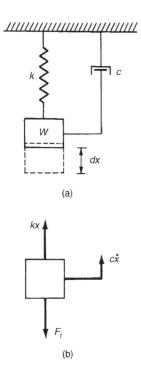

(a)

(b)

Figure 2.67. Damped oscillator: (a) analytical model; (b) forces in equilibrium.

a damped oscillator subjected to a time-dependent force $F_{(t)}$, as shown in Fig. 2.67. The free-body diagram of the oscillator subjected to various forces is shown in Fig. 2.67b.

Let δx be the virtual displacement. The total work done by the system is zero and is given by

$$m\ddot{x}\,\delta x + c\dot{x}\,\delta x + kx\,\delta x - F_{(t)}\,\delta x = 0 \tag{2.53}$$

$$(m\ddot{x} + c\dot{x} + kx - F_{(t)})\,\delta x = 0 \tag{2.54}$$

Since δx is arbitrarily selected,

$$m\ddot{x} + c\dot{x} + kx - F_{(t)} = 0 \tag{2.55}$$

This is the differential equation of motion of the damped oscillator.

The equation of motion for an undamped system can also be obtained from the principle of conservation of energy. It states that if no external forces are acting on the system, and there is no dissipation of energy due to damping, then the total energy of the system must remain constant during motion and consequently, its derivative with respect to time must be equal to zero.s

Consider again the oscillator shown in Fig. 2.67 without the damper. The two energies associated with this system are the kinetic energy of the mass and the potential energy of the spring.

The kinetic energy of the spring

$$T = \frac{1}{2}m\dot{x}^2 \tag{2.56}$$

where $\dot{x}$ is the instantaneous velocity of the mass.

The force in the spring is kx; work done by the spring is $kx\,\delta x$. The potential energy is the work done by this force and is given by

$$V = \int_0^x kx\,\delta x = \frac{1}{2}kx^2 \tag{2.57}$$

The total energy in the system is a constant. Thus

$$\frac{1}{2}m\dot{x}^2 + \frac{1}{2}kx^2 = \text{constant } c_0 \tag{2.58}$$

Differentiating with respect to x, we get

$$m\dot{x}\ddot{x} + kx\dot{x} = 0 \tag{2.59}$$

Since $\dot{x}$ cannot be zero for all values of t, we get

$$m\ddot{x} + kx = 0 \tag{2.60}$$

which has the same form as Eq. (2.52). This differential equation has a solution of the form

$$x = A\sin(\omega t + \alpha) \tag{2.61}$$

$$\dot{x} = \omega A\cos(\omega t + \alpha) \tag{2.62}$$

where A is the maximum displacement and ωA is the maximum velocity. Maximum kinetic energy is given by

$$T_{max} = \frac{1}{2}m(\omega A)^2 \tag{2.63}$$

Maximum potential energy is

$$V_{max} = \frac{1}{2}kA^2 \tag{2.64}$$

Since $T = V$,

$$\frac{1}{2}m(\omega A)^2 = \frac{1}{2}kA^2$$

or

$$\omega = \sqrt{\frac{k}{m}} \tag{2.65}$$

which is the natural frequency of the simple oscillator. This method, in which the natural frequency is obtained by equating maximum kinetic energy and maximum potential energy, is known as Rayleigh's method.

2.6.2.   Multidegree-of-Freedom Systems

In these systems, the displacement configuration is determined by a finite number of displacement coordinates. The true response of a multidegree system can be determined only by evaluating the inertia effects at each mass particle because structures are continuous systems with an infinite number of degrees-of-freedom. Although analytical methods are available to describe the behavior of such systems, these are limited to structures with uniform material properties and regular geometry. The methods are complex, requiring formulation of partial differential equations. However, the analysis is greatly simplified by replacing the entire displacement of the structure by a limited number of displacement components, and assuming the entire mass of the structure is concentrated in a number of discrete points.

Consider a multistory building with n degrees-of-freedom, as shown in Fig. 2.68. The dynamic equilibrium equations for undamped free vibration can be written in the general form

$$\begin{bmatrix} m_{11} & m_{12} & m_{13} & \cdots & m_{1m} \\ m_{21} & m_{22} & m_{23} & \cdots & m_{2m} \\ m_{31} & m_{32} & m_{33} & \cdots & m_{3m} \\ \cdots & \cdots & \cdots & \cdots & \cdots \\ m_{n1} & m_{n2} & m_{n3} & \cdots & m_{nm} \end{bmatrix} \begin{bmatrix} \ddot{x}_1 \\ \ddot{x}_2 \\ \ddot{x}_3 \\ \vdots \\ \ddot{x}_n \end{bmatrix}$$

$$+ \begin{bmatrix} k_{11} & k_{12} & k_{13} & \cdots & k_{1n} \\ k_{21} & k_{22} & k_{23} & \cdots & k_{2n} \\ k_{31} & k_{32} & k_{33} & \cdots & k_{3n} \\ \cdots & \cdots & \cdots & \cdots & \cdots \\ k_{n1} & k_{42} & k_{n3} & \cdots & k_{nm} \end{bmatrix} \begin{bmatrix} x_1 \\ x_2 \\ x_3 \\ \vdots \\ x_n \end{bmatrix} = 0$$

Writing the equations in matrix form

$$[M]\{\ddot{x}\} + [K]\{x\} = 0 \tag{2.66}$$

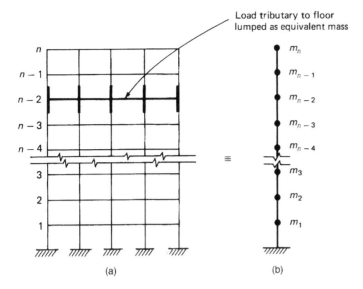

Load tributary to floor
lumped as equivalent mass

Figure 2.68. Multidegree-of-freedom system: (a) multistory analytical model with lumped masses.

where

$[M]$ = the mass or inertia matrix
$\{\ddot{x}\}$ = the column vector of accelerations
$[K]$ = the structure stiffness matrix
$\{x\}$ = the column vector of displacements of the structure

If the effect of damping is included, the equations of motion would be of the form

$$[M]\{\ddot{x}\} + [C]\{\dot{x}\} + [K]\{x\} = \{P\} \tag{2.67}$$

where

$[C]$ = the damping matrix
$\{\dot{x}\}$ = the column vector of velocity
$\{P\}$ = the column vector of external forces

General methods of solutions of these equations are available, but tend to be cumbersome. Therefore, in solving seismic problems, simplified methods are used; the problem is first solved by neglecting damping. Its effects are later included by modifying the design spectrum to account for damping. The absence of precise data on damping does not usually justify a more rigorous treatment. Neglecting damping results in dropping the second term, and limiting the problem to free-vibrations results in dropping the right-hand side of Eq. (2.67). The resulting equations of motion will become identical to Eq. (2.66).

During free vibrations, the motions of the system are simple harmonic, which means that the system oscillates about the stationary position in a sinusoidal manner; all masses follow the same harmonic function, having similar angular frequency, ω. Thus.

$$x_1 = a_1 \sin \omega_1 t$$
$$x_2 = a_2 \sin \omega_2 t$$
$$\vdots$$
$$x_n = a_n \sin \omega_n t$$

or in matrix notation

$$\{x\} = \{a_n\} \sin \omega_n t$$

where $\{a_n\}$ represents the column vector of modal amplitudes for the nth mode, and ω_n the corresponding frequency. Substituting for $\{x\}$ and its second derivative $\{\ddot{x}\}$ in Eq. (2.66) results in a set of algebraic expressions.

$$-\omega_n^2[M]\{a_n\} + [K]\{a_n\} = 0 \tag{2.68}$$

Using a procedure known as Cramer's rule, the preceding expressions can be solved for determining the frequencies of vibrations and relative values of amplitudes of motion a_{11}, $a_{12}, \ldots, a_n$. The rule states that nontrivial values of amplitudes exist only if the determinant of the coefficients of a is equal to zero because the equations are homogeneous, meaning that the right-hand side of Eq. (2.68) is zero. Setting the determinant of Eq. (2.68) equal to zero, we get

$$\begin{bmatrix} k_{11} - \omega_1^2 m_{11} & k_{12} - \omega_1^2 m_{12} & k_{13} - \omega_1^2 m_{13} & \cdots & k_{1n} - \omega_n m_{1n} \\ k_{21} - \omega_2^2 m_{21} & k_{22} - \omega_2^2 m_{12} & k_{23} - \omega_2^2 m_{23} & \cdots & k_{2n} - \omega_n m_{2n} \\ k_{31} - \omega_3^2 m_{31} & k_{32} - \omega_3^2 m_{32} & k_{33} - \omega_3^2 m_{33} & \cdots & k_{3n} - \omega_n m_{3n} \\ \cdots - \cdots & \cdots - \cdots & \cdots - \cdots & \cdots & \cdots - \cdots \\ k_{n1} - \omega_n^2 m_{n1} & k_{n2} - \omega_n^2 m_{n2} & k_{n3} - \omega_n^2 m_{n3} & \cdots & k_{nn} - \omega_n m_{nn} \end{bmatrix} = 0 \tag{2.69}$$

With the understanding that the values for all the stiffness coefficients k_{11}, k_{12}, etc., and the masses m_1, m_2, etc., are known, the determinant of the equation can be expanded, leading to a polynomial expression in ω^2. Solution of the polynomial gives one real root for each mode of vibration. Hence, for a system with n degrees of freedom, n natural frequencies are obtained. The smallest of the values obtained is called the fundamental frequency and the corresponding mode, the fundamental or first mode.

In mathematical terms, the vibration problem is similar to those encountered in stability analyses. The determination of frequency of vibrations can be considered similar to the determination of critical loads, while the modes of vibration can be likened to evaluation of buckling modes. Such types of problems are known as eigenvalue, or characteristic value, problems. The quantities ω^2, which are analogous to critical loads, are called eigenvalues or characteristic values, and in a broad sense can be looked upon as unique properties of the structure similar to geometric properties such as area or moment of inertia of individual elements.

Unique values for characteristic shapes, on the other hand, cannot be determined because substitution of ω^2 for a particular mode into the dynamic equilibrium equation [Eq. (2.68)] results in exactly n unknowns for the characteristic amplitudes $x_1 \ldots x_n$ for that mode. However, it is possible to obtain relative values for all amplitudes in terms of any particular amplitude. We are, therefore, able to obtain the pattern or the shape of the vibrating mode, but not its absolute magnitude. The set of modal amplitudes that describe the vibrating pattern is called eigenvector or characteristic vector.

2.6.3. Modal Superposition Method

In this method, the equations of motions are transformed from a set of n simultaneous differential equations to a set of n independent equations by the use of normal coordinates. The equations are solved for the response of each mode, and the total response of the system is obtained by superposing individual solutions. Two concepts are necessary for the understanding of the modal superposition method: (1) the normal coordinates; and (2) the property of orthogonality.

2.6.3.1. *Normal Coordinates*

In a static analysis, it is common to represent structural displacements by a Cartesian system of coordinates. For example, in a planar system, coordinates x and y and rotation θ are used to describe the position of a displaced structure with respect to its static position. If the structure is restrained to move only in the horizontal direction and if rotations are of no consequence, only one coordinate x is sufficient to describe the displacement. The displacements can also be identified by using any other independent system of coordinates. The only stipulation is that a sufficient number of coordinates are included to capture the deflected shape of the structure. These coordinates are commonly referred to as generalized coordinates and their number equal the number of degrees-of-freedom of the system. In dynamic analysis, however, it is advantageous to use free-vibration mode shapes known as normal modes to represent the displacements. While a mathematical description of normal modes and their properties may be intriguing, there is nothing complicated about their concept. Let us indulge in some analogies to bring home the idea. For example, normal modes may be considered as being similar to the primary colors red, blue, and yellow. None of these primary colors can be obtained as a combination of the others, but any secondary color such as green, pink, or orange can be created by combining the primary colors, each with a distinct proportion of the primary colors. The proportions of the primary colors can be looked upon as scale factors, while the primary colors themselves can be considered similar to normal modes. To further reinforce the concept of generalized coordinates, recall beam bending problems in which the deflection curve of a beam is represented in the form of trigonometric series. Considering the case of a simply supported beam subjected to vertical loads, as shown in Fig. 2.69, the deflection y, at any point can be represented by the following series:

$$y = a_1 \frac{\sin \pi x}{l} + a_2 \frac{\sin 2\pi x}{l} + a_3 \frac{\sin 3\pi x}{l} \tag{2.70}$$

Geometrically, this means that the deflection curve can be obtained by superposing simple sinusoidal shown in Fig. 2.69.

The first term in Eq. (2.70) represents the full-sine curve, the second term, the half-sine, etc. The coefficients $a_1, a_2, a_3, \ldots$, represent the maximum ordinates of the curves, while the numbers 1, 2, and 3, the number of waves or mode shapes. By determining the coefficients $a_1, a_2, a_3, \ldots$, the series can represent the deflection curve to any desired degree of accuracy, depending on the number of terms considered in the series.

2.6.3.2. *Orthogonality*

This force-displacement relationship is rarely used in static problems, but is of great significance in structural dynamics. This is best explained with an example shown in Fig. 2.70.

Consider a two-story, lumped-mass system subjected to free vibrations. The system's two modes of vibrations can be considered as elastic displacements due to two different loading conditions, as shown in Fig. 2.70b and c. We will use a theorem known as Betti's reciprocal theorem to demonstrate the derivation of orthogonality conditions. This theorem states that the work done by one set of loads on the deflections due to a second set of

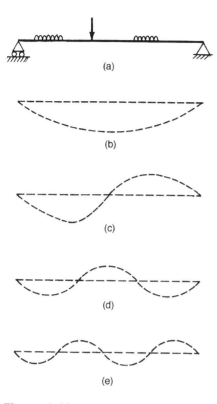

Figure 2.69. Generalized displacement of a simply supported beam (a) loading; (b) full-sine curve; (c) half-sine curve; (d) one-third sine curve; (e) one-fourth sine curve.

loads is equal to the work done by the second set of loads acting on the deflections due to the first. Using this theorem with reference to Fig. 2.70, we get

$$\omega_1^2 m_1 x_{1b} + \omega_1^2 m_2 x_{2b} = \omega_2^2 m_1 x_{1a} + \omega_1^2 m_2 x_{2a} \tag{2.71}$$

This can be written in matrix form

$$\omega_1^2 \begin{bmatrix} m_1 & 0 \\ 0 & m_2 \end{bmatrix} \begin{bmatrix} x_{1b} \\ x_{2b} \end{bmatrix} = \omega_2^2 \begin{bmatrix} m_1 & 0 \\ 0 & m_2 \end{bmatrix} \begin{bmatrix} x_{1a} \\ x_{2a} \end{bmatrix}$$

or

$$(\omega_1^2 - \omega_2^2 \{x_b\}^T [M]\{x_a\} = 0 \tag{2.72}$$

If the two frequencies are not the same, i.e., $\omega_1 \neq \omega_2$, we get

$$\{x_b\}^T [M]\{x_a\} = 0 \tag{2.73}$$

This condition is called the orthogonality condition, and the vibrating shapes, $\{x_a\}$ and $\{x_b\}$, are said to be orthogonal with respect to the mass matrix, $[M]$. By using a similar procedure, it can be shown that

$$\{x_a\}^T [k]\{x_b\} = 0 \tag{2.74}$$

The vibrating shapes are therefore orthogonal with respect to the stiffness matrix as they are with respect to the mass matrix. In the general case of the structures with damping, it is necessary to make a further assumption in the modal analysis that the orthogonality

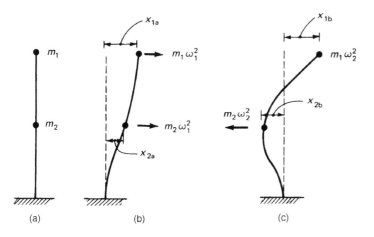

Figure 2.70. Two-story lumped-mass system illustrating Betti's reciprocal theorem: (a) lumped model; (b) forces acting during first mode of vibration; (c) forces acting during second mode of vibration.

condition also applies for the damping matrix. This is for mathematical convenience only and has no theoretical basis. Therefore, in addition to the two orthogonality conditions mentioned previously, a third orthogonality condition of the form

$$\left\{x\frac{T}{a}\right\}c\{x_b\} = 0 \tag{2.75}$$

is used in the modal analysis.

    To bring out the essentials of the normal mode method, it is convenient to consider the dynamic analysis of a two-degree-of-freedom system. We will first analyze the system by a direct method and then show how the analysis can be simplified by the modal superposition method.

    Consider a 2-story dynamic model of a shear building shown in Fig. 2.71a, b, and c, subject to free vibrations. The masses m_1 and m_2 at levels 1 and 2 can be considered connected to each other and to the ground by two springs having stiffnesses k_1 and k_2. The stiffness coefficients are mathematically equivalent to the forces required at levels 1 and 2 to produce unit horizontal displacements relative to each level.

    It is assumed that the floors, and therefore the masses m_1 and m_2, are restrained to move in the direction x and that there is no damping in the system. Using Newton's second law of motion, the equations of dynamic equilibrium for masses m_1 and m_2 are given by

$$m_1\ddot{x}_1 = -k_1x + k_2(x_2 - x_1) \tag{2.76}$$

$$m_2\ddot{x}_2 = -k_2(x_2 - x_1) \tag{2.77}$$

Rearranging terms in these equations gives

$$m_1\ddot{x}_1 + (k_1 + k_2)x_1 - k_2x_2 = 0 \tag{2.78}$$

$$m_2\ddot{x}_2 - k_2x_1 + k_2x_2 = 0 \tag{2.79}$$

The solutions for the displacements x_1 and x_2 can be assumed to be of the form

$$x_1 = A\sin(\omega t + \alpha) \tag{2.80}$$

$$x_2 = B\sin(\omega t + \alpha) \tag{2.81}$$

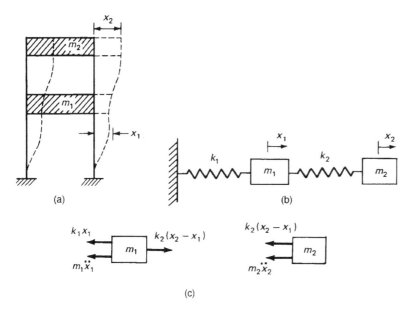

Figure 2.71. Two-story shear building; free vibrations: (a) building with d masses; (b) mathematical model; (c) free-body diagram with d masses.

where ω represents the angular frequency and α represents the phase angle of the harmonic motion of the two masses. A and B represent the maximum amplitudes of the vibratory motion. Substitution of Eqs. (2.80) and (2.81) into Eqs. (2.78) and (2.79) gives the following equations:

$$(k_1 + k_2 - \omega^2 m_1)A - k_2 B = 0 \tag{2.82}$$

$$k_2 A + (k_2 - \omega^2 m_2)B = 0 \tag{2.83}$$

To obtain the solution for the nontrivial case of A and $B \neq 0$, the determinant of the coefficients of A and B must be equal to zero, thus

$$\begin{bmatrix} (k_1 + k_2 - \omega^2 m_1) & -k_2 \\ -k_2 & (k_2 - \omega^2 m_2) \end{bmatrix} = 0 \tag{2.84}$$

Expansion of the determinant gives the relation

$$(k_1 + k_2 - \omega^2 m_1)(k_2 - \omega^2 m_2) - k_2^2 = 0 \tag{2.85}$$

or

$$m_1 m_2 \omega^4 - m_1 k_2 + m_2(k_1 + k_2)\omega^2 + k_1 k_2 = 0 \tag{2.86}$$

Solution of this quadratic equation yields two values for ω^2 of the form

$$\omega_1^2 = \frac{-b + \sqrt{b^2 - 4ac}}{2a} \tag{2.87}$$

$$\omega_2^2 = \frac{-b - \sqrt{b^2 - 4ac}}{2a} \tag{2.88}$$

where

$$a = m_1 m_2$$
$$b = - [m_1 k_2 + m_2(k_1 + k_2)]$$
$$c = k_1 k_2$$

As mentioned previously, the two frequencies ω_1 and ω_2, which can be considered intrinsic properties of the system, are uniquely determined.

The magnitudes of the amplitudes A and B cannot be determined uniquely, but can be obtained in terms of ratios $r_1 = A_1/B_1$ and $r_2 = A_2/B_2$ corresponding to ω_1^2 and ω_2^2, respectively. Thus

$$r_1 = \frac{A_1}{B_1} = \frac{k_2}{k_1 + k_2 - \omega_1^2 m_1} \tag{2.89}$$

$$r_2 = \frac{A_2}{B_2} = \frac{k_2}{k_1 + k_2 - \omega_2^2 m_1} \tag{2.90}$$

The ratios r_1 and r_2 are called the amplitude ratios and represent the shapes of the two natural modes of vibration of the system.

Substituting the angular frequency ω_1 and the corresponding ratio r_1 in Eqs. (2.80) and (2.81), we get

$$x_1' = r_1 B_1 \sin(\omega_1 t + \alpha_1) \tag{2.90}$$

$$x_2' = B_1 \sin(\omega_1 t + \alpha_1) \tag{2.91}$$

These expressions describe the first mode of vibration, also called the fundamental mode. Substituting the larger angular frequency ω_2 and the corresponding ratio r_2 in Eqs. (2.80) and (2.81), we get

$$x_1'' = r_2 B_2 \sin(\omega_2 t + \alpha_2) \tag{2.92}$$

$$x_2'' = B_2 \sin(\omega_2 t + \alpha_2) \tag{2.93}$$

The displacements x_1'' and x_2'' describe the second mode of vibration. The general displacement of the system is obtained by summing the modal displacements, thus

$$x_1 = x_1' + x_1''$$
$$x_2 = x_2' + x_2''$$

Thus, for systems having two degrees of freedom, we are able to determine the frequencies and mode shapes without undue mathematical difficulties. Although the equations of motions for multidegree systems have similar mathematical form, solutions for modal amplitudes in terms of geometrical coordinates become unwieldy. Use of orthogonal properties of mode shapes makes this laborious process unnecessary. We will demonstrate how the analysis can be simplified by using the modal superposition method. Consider again the equations of motion for the idealized two-story building discussed in the previous section. As before, damping is neglected, but instead of free vibrations, we will consider the analysis of the system subject to time-varying force functions F_1 and F_2 at levels 1 and 2. The dynamic equilibrium for masses m_1 and m_2 is given by

$$m_1 \ddot{x}_1 + (k_1 + k_2)x_1 - k_2 x_2 = F_1 \tag{2.94}$$

$$m_2 \ddot{x}_2 - k_2 x_1 + k_2 x_2 = F_2 \tag{2.95}$$

These two equations are interdependent because they contain both the unknowns x_1 and x_2. These can be solved simultaneously to get the response of the system, which was indeed the method used in the previous section to obtain the values for frequencies and mode shapes. Modal superposition method offers an alternate procedure for solving such problems. Instead of requiring simultaneous solution of the equations, we seek to transform the system of interdependent or coupled equations into a system of independent or uncoupled equations. Since the resulting equations contain only one unknown function of time, solutions are greatly simplified. Let us assume that solution for the preceding dynamic equations is of the form

$$x_1 = a_{11}z_1 + a_{12}z_2 \tag{2.96}$$

$$x_2 = a_{21}z_1 + a_{22}z_2 \tag{2.97}$$

What we have done in the preceding equations is to express displacement x_1 and x_2 at levels 1 and 2 as a linear combination of properly scaled values of two independent modes. For example, a_{11} and a_{12}, which are the mode shapes at level 1, are combined linearly to give the displacement x_1, z_1, and z_2 can be looked upon as scaling functions. Substituting for x_1 and x_2 and their derivatives $\ddot{x}_1$ and $\ddot{x}_2$ in the equilibrium Eqs. (2.94) and (2.95), we get

$$m_1 a_{11}\ddot{z}_1 + (k_1 + k_2)a_{11}z_1 - k_2 a_{21}z_1 + m_1 a_{12}\ddot{z}_2 + (k_1 + k_2)a_{12}z_2 - k_2 a_{22}z_2 = F_1 \tag{2.98}$$

$$m_2 a_{21}\ddot{z}_1 - k_2 a_{11}z_1 + k_2 a_{21}z_1 + m_2 a_{24}z_2 - k_2 a_{12}z_2 + k_2 a_{22}z_2 = F_2 \tag{2.99}$$

We seek to uncouple Eqs. (2.98) and (2.99) by using the orthogonality conditions. Multiplying Eqs. (2.98) by a_{11} and Eqs. (2.99) by a_{21}, we get

$$m_1 a_{11}^2 \ddot{z}_1 + (k_1 + k_2)a_{11}^2 z_1 - k_2 a_{11}a_{21}z_1 + m_1 a_{11}a_{12}\ddot{z}_2$$
$$+ (k_1 + k_2)a_{11}a_{12}z_2 - k_2 a_{11}a_{22}z_2 = a_{11}F_1 \tag{2.100}$$

$$m_1 a_{21}^2 \ddot{z}_1 - k_2 a_{11}a_{21}z_1 + k_2 a_{21}^2 z_1 + m_2 a_{21}a_{22}\ddot{z}_2 - k_2 a_{12}a_{21}z_2$$
$$+ k_2 a_{21}a_{22}z_2 = a_{21}F_2 \tag{2.101}$$

Adding the preceding two equations, we get

$$(m_1 a_{11}^2 + m_2 a_{21}^2)\ddot{z}_1 + \omega_1^2(m_1 a_{11}^2 + m_2 a_{21}^2)z_1 = a_{11}F_1 + a_{21}F_2 \tag{2.102}$$

Similarly, multiplying Eqs. (3.66) and (3.67) by a_{12} and a_{22} and adding, we obtain

$$(m_1 a_{12}^2 + m_2 a_{22}^2)\ddot{z}_2 + \omega_2^2(m_1 a_{12}^2 + m_2 a_{22}^2)z_2 = a_{12}F_1 - a_{22}F_2 \tag{2.103}$$

Equations (2.102) and (2.103) are independent of each other and are the uncoupled form of the original system of coupled differential equations. These can be further written in a simplified form by making use of the following abbreviations:

$$M_1 = m_1 a_{11}^2 + m_2 a_{21}^2$$
$$M_2 = m_1 a_{12}^2 + m_2 a_{22}^2 \tag{2.104}$$

$$K_1 = \omega_1^2 M_1$$
$$K_2 = \omega_2^2 M_2 \tag{2.105}$$

$$P_1 = a_{11}F_1 + a_{21}F_2$$
$$P_2 = a_{12}F_1 + a_{22}F_2 \tag{2.106}$$

M_1 and M_2 are called the generalized masses, K_1 and K_2 the generalized stiffnesses, and P_1 and P_2 the generalized forces.

Using these notations, each of the Eqs. (2.102) and (2.103) takes the form similar to the equations of motion of a single-degree-of-freedom system, thus

$$M_1\ddot{z}_1 + k_1 z_1 = P_1 \tag{2.107}$$

$$M_2\ddot{z}_2 + k_2 z_2 = P_2 \tag{2.108}$$

The solution of these uncoupled differential equations can be found by any of the standard procedures given in textbooks on vibration analysis. In particular, Duhamel's integral provides a general method of solving these equations irrespective of the complexity of the loading function. However, in seismic analysis, usually a response spectrum is used instead of a forcing function to obtain the maximum values of the response corresponding to each modal equation. Direct superposition of modal maximum would, however, give only an upper limit for the total system which, in many engineering problems, would be too conservative. To alleviate this problem, approximations based on probability considerations are generally employed. One method employs the so-called root mean square procedure, also called the square root of sum of the squares (SRSS) method. As the name implies, a probable maximum value is obtained by evaluating the square root of the sum of the squares of the modal quantities. Although this method is simple and widely used, it is not always a conservative predictor of earthquake response because more severe combinations of modal quantities can occur, as for example, when two modes have nearly the same natural period. In such cases, it is more appropriate to use the complete quadratic combination (CQC) procedure.

The aim of this section is to bring out the essentials of structural dynamics as related to seismic design of buildings. A certain amount of mathematical presentation has been unavoidable. Lest the reader lose the physical meaning of the various steps, it is worthwhile to summarize the essential features of dynamic analysis.

Dynamic analysis of buildings is performed by idealizing them as multidegree-of-freedom systems. The dead load of the building together with a percentage of live load (estimated to be present during an earthquake) is considered as lumped masses at each floor level. In a planar analysis, each mass has one degree-of-freedom corresponding to lateral displacement in the direction under consideration, while in a three-dimensional analysis, it has three degrees-of-freedom corresponding to two translational and one torsional displacement. Free vibrations of the buildings are evaluated, without including the effect of damping. Damping is taken into account by modifying the design response spectrum. The dynamic model representing a building has the number of mode shapes equal to the number of degrees-of-freedom of the model. Mode shapes have the property of orthogonality, which means that no given mode shape can be constructed as a combination of others, yet any deformation of the dynamic model can be described as a combination of its mode shapes, each multiplied by a scale factor. Each mode shape has a natural frequency of vibration. The mode shapes and frequencies are determined by solving for the eigenvalues. The total response of the building to a given response spectrum is obtained by statistically summing a predetermined number of modal responses. The number of modes required to adequately determine the design forces is a function of the dynamic characteristics of the building. Generally, for regular buildings, six to 10 modes in each direction are considered sufficient. Since each mass responds to earthquakes in more than one mode, it is necessary to evaluate effective modal mass values. These values indicate the percentage of the total mass that is mobilized in each mode. The acceleration experienced by each mass undergoing various modal deformations is determined from the response spectrum, which has been adjusted for damping. The product of the acceleration for a particular mode, multiplied by the effective modal mass for that mode, gives the static equivalent of forces at each discrete level. Since these forces do not reach their maximum values simultaneously, statistical methods such as SRSS or CQC are used for the combinations. The resulting forces are used as design static forces.

2.7. CHAPTER SUMMARY

Since earthquakes can occur almost anywhere, some measure of earthquake resistance in the form of reserve ductility and redundancy should be built into the design of all structures to prevent catastrophic failures. The magnitude of inertial forces induced by earthquakes depends on the building mass, ground acceleration, and the dynamic response of the structure. The shape and proportion of a building have a major effect on the distribution of earthquake forces as they work their way through the building. If irregular features are unavoidable, special design considerations are required to account for load transfer at abrupt changes in structural resistance.

Two approaches are recognized in modern codes for estimating the magnitude of seismic loads. The first approach, termed the equivalent lateral force procedure, uses a simple method to take into account the properties of the structure and the foundation material. The second is a dynamic analysis procedure in which the modal responses are combined in a statistical manner to find the maximum values of the building response. Note that the level of force experienced by a structure during a major earthquake is much larger than the forces usually employed in the design. However, by prescribing detailing requirements, the structure is relied upon to sustain postyield displacements without collapse.

The complex and random nature of ground motion makes it necessary to work with a more general characterization of ground motion. This is achieved by using earthquake response spectra to postulate the intensity and vibration content of ground motion at a given site. Duration of ground motion, although important, is not used explicitly in establishing design criteria at present (2004).

Earthquakes "load" structures indirectly. As the ground displaces, a building will follow and vibrate. The vibration produces deformations with associated strains and stresses in the structure. Computation of dynamic response to earthquake ground shaking is complex. As a simplification. the concept of a response spectrum is used in practice. A response spectrum for specific earthquake ground motion does not reflect the total time history of response, but only approximates the maximum value of response for simple structures to that ground motion. The design response spectrum is a smoothed and normalized approximation for many different ground motions, adjusted at the extremes for characteristics of larger structures.

Multistory buildings are analyzed as multidegree-of-freedom systems. They are represented by lumped masses at story intervals along the height of a vertically cantilevered pole. Each mode of the building system is represented by an equivalent single-degree-of-freedom system using the concept of generalized mass and stiffness. With the known period, mode shape, mass distribution, and acceleration, one can compute the deflected shape, story accelerations, forces, and overturning moments. Each predominant mode is analyzed separately, and by using either the SRSS or CQC method, the peak modal responses are combined to give a reasonable value between an upper bound as the absolute sum of the modes and a lower bound as the maximum value of a single mode.

The time-history analysis technique represents the most sophisticated method of dynamic analysis for buildings. In this method, the mathematical model of the building is subjected to full range of accelerations for the entire duration of earthquake by using earthquake records that represent the expected earthquake at the base of the structure. The equations of motion are integrated by using computers to obtain a complete record of acceleration, velocity, and displacement of each lumped mass. The maximum value is found by scanning the output record. Even with the availability of sophisticated computers, use of this method is restricted to the design of special structures such as nuclear facilities, military installations, and base-isolated structures.

In seismic design, nearly elastic behavior is interpreted as allowing some structural elements to slightly exceed specified yield stress on the condition that the elastic linear behavior of the overall structure is not substantially altered. For a structure with a multiplicity of structural elements forming the lateral-force-resisting system, the yielding of a small number of elements will generally not affect the overall elastic behavior of the structure if excess load can be distributed to other structural elements that have not exceeded their yield strength.

Although for new buildings, the ductile design approach is quite routine, seismic retrofitting of existing nonductile buildings with poor confinement details is generally extremely expensive. Therefore, it is necessary to formulate an alternative method that attempts a realistic assessment of damage resistance of the building. One method, discussed in Chapter 6, is based on the concept of trade-off between ductility and strength. In other words, structural systems of limited ductility may be considered valid in seismic design, provided they can resist correspondingly higher forces. In this method, the concept of the inelastic demand ratio is used to describe the ability of the structural elements to resist stresses beyond yield stress.

Shanghai World Plaza
Shanghai, China

John A. Martin & Associates, Inc.
Structural Engineers
Los Angeles, CA

Chang Ping Stadium
Beijing, China

John A. Martin & Associates, Inc.
Structural Engineers
Los Angeles, CA

3
Steel Buildings

The development of structural steel systems for tall buildings can be traced back to William LeBaron Jenny, who in 1885 used metal framework for the construction of the Home Insurance Building, an eight-story structure in Chicago. This, combined with the invention of a safe passenger elevator by Otis in 1854 led to an explosion of high-rise buildings. In the ensuing 28-year period from 1885 to 1913, the design of steel frame evolved from an eight-story building to the 800-ft tall Woolworth Building in New York City. The first generation of skyscrapers culminated with the erection of the Chrysler Building in New York in 1930, immediately followed by the Empire State Building in 1931, which held the record as the world's tallest building for 41 years.

The second wave of tall buildings began in 1956 based on new building technology and new concepts in structural design, climaxing in 1974 with the completion of the Sears Tower, a 110-story, 1450-ft tall building in Chicago. Following the Sears Tower, the post-second generation of supertall buildings has included only "mixed" construction, consisting of both steel and reinforced concrete. The 1476-ft Petronas Towers, built in Kuala Lampur, Malaysia, in 1997, and the 1667-ft tall Taipei 101 building, which attained its full height on Oct. 17, 2003, unseating Malaysia's Towers as the world's tallest building, are two examples.

Although today's building systems have evolved from an entirely different structural concept than those of the first generation of skyscrapers, it is of interest to group the systems into specific categories, each with an applicable height range, as shown in Fig. 3.1.

At the top of the list is the rigid frame with an economical height range of about 20 stories. In its simplest form it is composed of orthogonally arranged bents consisting of columns and beams with the beams rigidly connected to columns. At the other end of the list is the bundled tube system used for the Sears Tower, consisting of an exterior framed tube stiffened by interior frames to reduce the effect of shear lag in the exterior columns.

The height range for structural arrangements shown in Fig. 3.1 is particularly suitable for prismatically shaped buildings without serious plan or vertical irregularities. They can be structured by a single identifiable system. For example, a 50-story regular prismatic building can be executed with a single system such as a tubular frame consisting of closely spaced exterior columns rigidly connected to a deep spandrels. However, buildings that are emphatically irregular in shape, with an intricate configuration such as large cutouts, vertical step-backs, etc., are less amenable to a single structural system. Therefore, the engineer has to improvise in developing an architecturally acceptable and structurally economical solution. In such situations combinations of two or even more of the basic structural arrangements need to be used in the same building, either by direct combination or by adapting different systems in different parts of the building. Therefore, the structural system for a building evolves as a response to a unique set of demands, giving engineers an opportunity to combine known systems or create their own.

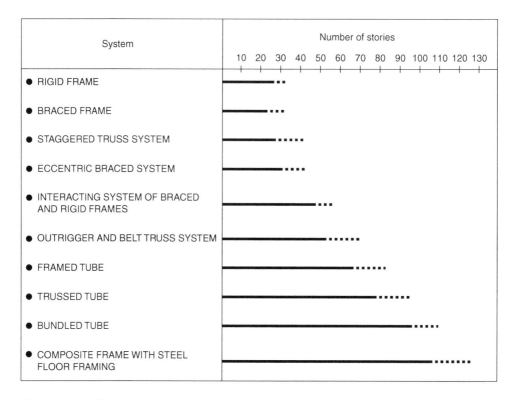

Figure 3.1. Height comparison of steel building systems.

3.1. RIGID FRAMES (MOMENT FRAMES)

A frame is considered rigid when its beam-to-column connections have sufficient rigidity to hold virtually unchanged the original angles between intersecting members. In this system, shown in Fig 3.2, lateral loads are resisted primarily by the rigid frame action; that is, by the development of shear forces and bending moments in the frame members and joints. The continuity at both ends of beams also assists in resisting gravity loads more efficiently by reducing positive moments in beam spans. Moment frames have certain advantages in building applications due to their flexibility in architectural planning. They may be placed at the building exterior without undue restrictions on their depths. They may also be located throughout the interior of the building with certain limitations on beam depths to allow for passage of mechanical and air conditioning ducts. Because there are no bracing elements present to block lease space or window openings, they are considered architecturally more versatile than other systems such as braced frames or shear walls.

The depths of frame members are often controlled by stiffness rather than strength to limit story drift under lateral loads. The story drift is defined as the lateral displacement of one level relative to the level below. It is of concern in serviceability checking arising primarily from the effects of wind. Drift limits in common usage for wind designs of buildings are of the order of 1/400–1/500 of the story height. These limits are believed to be generally sufficient to minimize damage to cladding and nonstructural walls and partitions.

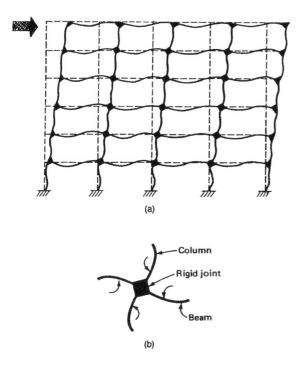

(a)

(b)

Figure 3.2. (a) Response of rigid frame to lateral loads; (b) flexural deformation of beams and columns due to nondeformability of connections.

The inherent flexibility of moment frames may also result in greater drift-induced nonstructural damage under seismic loading than in other systems. It should be noted that seismic drift Δ_M, including inelastic response of buildings, is typically limited to 1/50 of the story height, about 10 times larger than the allowable wind drift.

The strength and ductility of the connections between beams and columns are also important considerations, particularly for frames designed to resist seismic loads. Take, for example, the welded moment connection details used in North American seismic regions during the 25 years preceding the Northridge earthquake, which are shown in Fig. 3.3. The connection typically consisted of full-penetration field-welded top and bottom beam flanges, and a high-strength bolted shear tab connection. It was believed that this type of connection was capable of developing large inelastic rotations.

The Northridge, Richter magnitude 6.7 earthquake of January 17, 1994 in California, which caused damage to over 200 steel moment-resisting frame buildings, and the January 18, 1995, Richter magnitude 6.8 earthquake in Kobe, Japan, have shaken engineers' confidence in the use of the moment frame for seismic design. In both of these earthquakes, steel moment frames did not perform as well as expected. Almost without exception, the connections that failed were of the type shown in Fig. 3.3. The majority of the damage consisted of fractures of the bottom flange weld between the column and girder flanges. There were also many instances where top flange fractures occurred. In view of the observed brittle fracture at the beam-to-column intersections, new connection strategies have been developed, and most building codes are being revised. The new game plan typically consists of designing beams such that the plastic hinges form away from the column face. Current moment frame design practice in regions of high seismicity is addressed later in this chapter.

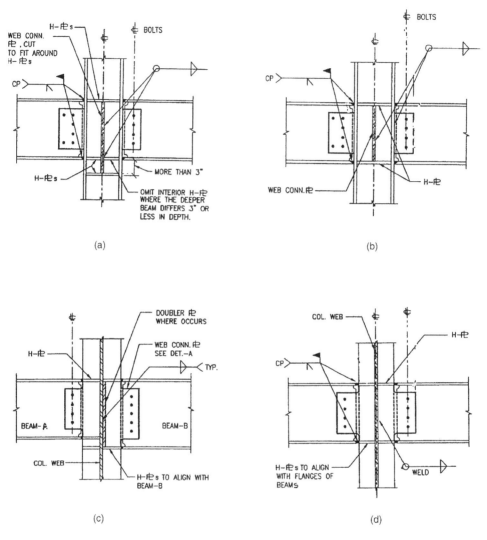

Figure 3.3. Pre-Northridge moment connections: (a) unequal-depth beams to column flange connection; (b) equal-depth beams to column flange connection; (c) unequal-depth beams to column web connection; (d) equal-depth beams to column web connection.

3.1.1. Deflection Characteristics

Because of the rigid beam-to-column connections, a moment frame cannot displace laterally without bending the beams and columns. The primary source of lateral stiffness of the entire frame is therefore dependent on the bending rigidity of the frame members. To understand its lateral deflection characteristics, it is convenient to consider the deflections of a rigid frame as composed of two components similar to the deflection components of a structural element such as a vertical cantilever column. The primary deflection of the cantilever column is due to bending and the secondary component is due to shear. Unless the column is relatively short, the shear component may be ignored in deflection computations. However, in a moment frame, both of these components are equally important. The bending and the shear deflection components of a rigid frame are usually referred to as the cantilever bending and frame racking.

3.1.2. Cantilever Bending Component

In resisting overturning moments, a moment frame responds as a vertical cantilever, resulting in axial deformation of the columns. The columns on the windward face lengthen while those on the leeward face shorten. This change in column lengths causes the building to rotate about a horizontal axis. The resulting lateral deflection, as shown in Fig. 3.4a, is analogous to the bending deflection component of the cantilever.

3.1.3. Shear Racking Component

This response in a rigid frame, shown in Fig. 3.4b, is similar to the shear deflection component of the cantilever column. As the frame displaces laterally, by virtue of the rigid beam-to-column connections, bending moments and shears are developed in the beams and columns. The horizontal shear above a given level due to lateral loads is resisted by shear in each of the columns of that story (Fig. 3.4b). This shear in turn causes the story-height columns to bend in double curvature with points of contraflexure at approximately midstory levels. To satisfy equilibrium, the sum of column moments above and below a joint must equal the sum of beam moments on either side of the column. In resisting the bending, the beams also bend in a double curvature, with points of contraflexure at approximately midspan. The cumulative bending of the columns and beams results in the overall shear racking of the frame. The deflected shape due to this component has a shear deflection configuration, as shown in Fig. 3.4b.

The shear mode of deformation accounts for about 70% of the total sway of a moment frame, with the beam flexure contributing about 10 to 15%, and the column

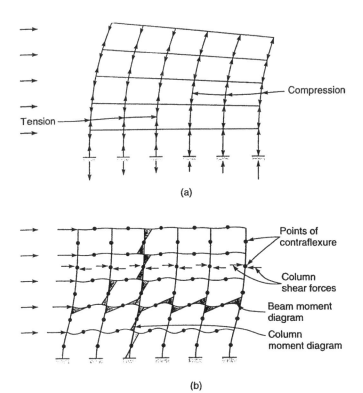

Figure 3.4. Rigid frame deflections: (a) forces and deformations caused by external overturning moment; (b) forces and deformations caused by external shear.

bending furnishing the remainder. This is because in a rigid frame, typically, the column stiffness, as measured by the I_c/L_c ratio, is substantially greater than the beam stiffness ratio, I_b/L_b,

where

I_b = moment of inertia of beam
I_c = moment of inertia of column
L_b = length of beam
L_c = length of column

Therefore, in general, to reduce lateral deflection, the place to start adding stiffness is in the beams. However, in nontypical frames, such as for those in framed tubes with column spacing approaching floor-to-floor height, it is prudent to study the relative beam and column stiffnesses before making adjustments to the member stiffnesses.

Because of the cumulative effect of building rotation up the height, the story drift increases with height, while that due to shear racking tends to stay the same up the height. The contribution to story drift due to cantilever bending in the uppermost stories exceeds that from shear racking. However, the bending effect usually does not exceed 10 to 20% of that due to shear racking, except in very tall and slender rigid frames. Therefore, the overall deflected shape of a medium-rise frame usually has a shear deflection configuration. Thus, the total lateral deflection of a rigid frame may be considered a combination of the following factors:

- Cantilever deflection due to axial deformation of columns (15 to 20%).
- Frame shear racking due to bending of beams (50 to 60%).
- Frame racking due to bending of columns (15 to 20%).

In addition to the preceding factors, the deformations of the panel zone of a beam-column joint, defined as the rectangular segment of the column web within the column flanges and beam continuity plates, also contributes to the total lateral deflection of the frame. Its effect, however, rarely exceeds 5% of the total deflection.

3.2. BRACED FRAMES

Rigid frame systems are not efficient for buildings taller than about 20 stories because the shear racking component of deflection due to the bending of columns and beams causes the story drift to be too great. Addition of diagonal or V-braces within the frame transforms the system into a vertical truss, virtually eliminating the bending of columns and beams. High stiffness is achieved because the horizontal shear is now primarily absorbed by the web members and not by the columns. The webs resist lateral forces by developing internal axial actions and relatively small flexural actions. The braces can be configured by using any number of steel shapes such as I-shaped sections, rectangular or circular tubes, single or double angles stitched together, T-shape sections, or channels. Brace connections to the framing systems commonly consist of gusset plates with bolted or welded connections to the braces.

In simple terms, braced frames may be considered cantilevered vertical trusses resisting lateral loads primarily through the axial stiffness of columns and braces. The columns act as the chords in resisting the overturning moment, with tension in the windward column and compression in the leeward column. The diagonals work as the web members resisting the horizontal shear in axial compression or tension, depending on the direction of inclination. The beams act axially, when the system is a fully triangulated truss. They undergo bending only when the braces are eccentrically connected to them. Because the

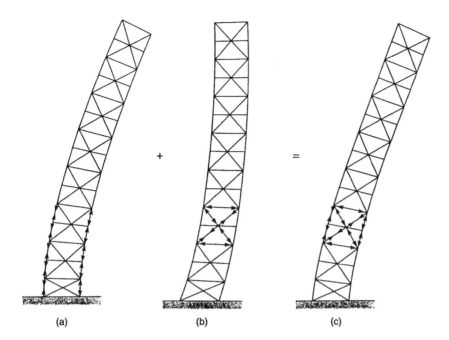

Figure 3.5. Braced frame deformation: (a) flexural deformation; (b) shear deformation; (c) combined configuration.

lateral loads are reversible, braces are subjected to both compression and tension; they are most often designed for the more stringent case of compression.

The effect of axial deformation of the columns results in a "flexural" configuration of the deflection with concavity downwind and a maximum slope at the top (Fig. 3.5a). The axial deformations of the web members, on the other hand, cause a "shear" configuration of deflection with concavity upwind, a maximum slope at the base, and a zero slope at the top (Fig. 3.5b). The resulting deflected shape of the frame (Fig. 3.5c) is a combination of the effects of the flexural and shear curves, with a resultant configuration depending on their relative magnitudes, as determined mainly by the type of bracing. Nevertheless, it is the flexural deflection that most often dominates the deflection characteristics.

The role of web members in resisting shear can be demonstrated by following the path of the horizontal shear down the braced bent. Consider the braced frames shown in Fig. 3.6a–e, subjected to an external shear force at the top level. In Fig. 3.6a, the diagonal in each story is in compression, causing the beams to be in axial tension; therefore, the shortening of the diagonal and extension of the beams gives rise to the shear deformation of the bent. In Fig. 3.6b, the forces in the braces connecting to each beam-end are in equilibrium horizontally with the beam carrying insignificant axial load. In Fig. 3.6c, half of each beam is in compression while the other half is in tension. In Fig. 3.6d, the braces are alternately in compression and tension while the beams remain basically unstressed. Finally, in Fig. 3.6e, the end parts of the beam are in compression and tension with the entire beam subjected to double curvature bending. Observe that with a reversal in the direction of horizontal load, all actions and deformations in each member will also be reversed.

In a braced frame, the principal function of web members is to resist the horizontal shear forces. However, depending on the configuration of the bracing, the web members may pick up substantial compressive forces as the columns shorten vertically under gravity loads. Consider, for example, the typical bracing configurations shown in Fig. 3.7a through d. As the columns in Fig. 3.7a and b shorten, the diagonals are subjected to

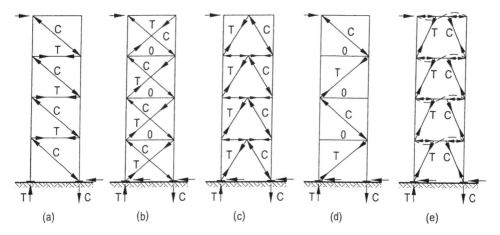

Figure 3.6. Load path for horizontal shear through web numbers: (a) single diagonal bracing; (b) X-bracing; (c) chevron bracing; (d) single-diagonal, alternate direction bracing; (e) knee bracing.

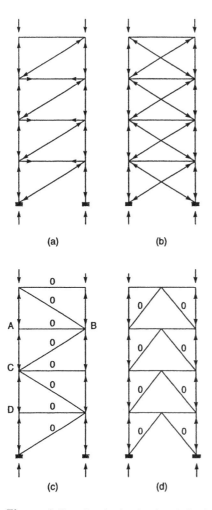

Figure 3.7. Gravity load path: (a) single diagonal single direction bracing; (b) X-bracing; (c) single diagonal alternate direction bracing; (d) chevron bracing.

compression forces because the beams at each end of the braces are effective in resisting the horizontal component of the compressive forces in the diagonal. At first glance, this may appear to be the case for the frame shown in Fig. 3.7c. However, the diagonal shown in Fig. 3.7c will not attract significant gravity forces because there is no triangulation at the ends of beams where the diagonals are not connected (nodes A and D, in Fig. 3.7c). The only horizontal restraint at the end is by the bending resistance of columns, which usually is of minor significance in the overall behavior. Similarly, in Fig. 3.7d, the vertical restraint from the bending stiffness of the beam is not large; therefore, as in the previous case, the braces experience only negligible gravity forces.

3.2.1. Types of Braces

Braced frames may be grouped into two categories, as either concentric braced frames (CBF) or eccentric braced frames (EBF), depending on their geometric characteristics. In CBFs, the axes of all members—i.e., columns, beams, and braces—intersect at a common point such that the member forces are axial. EBFs utilize axis offsets to deliberately introduce flexure and shear into framing beams. The primary goal is to increase ductility, as discussed later in this chapter.

The CBFs can be configured in various forms, some of which are shown in Fig. 3.8. Depending on the magnitude of force, length, required stiffness, and clearances, the diagonal

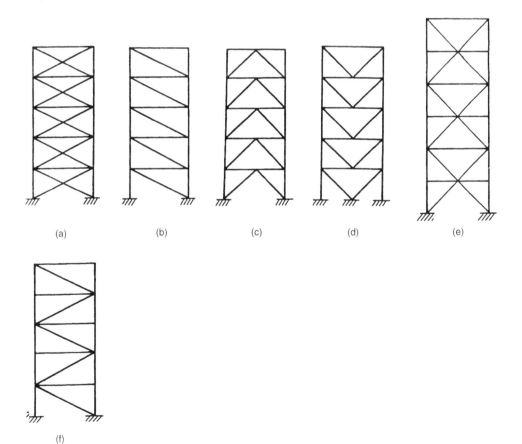

Figure 3.8. Typical concentric braced frame (CBF) configurations: (a) one-story X-bracing; (b) single-diagonal bracing; (c) and (d) chevron bracing; (e) two-story X-bracing; (f) single-diagonal, alternate-direction bracing.

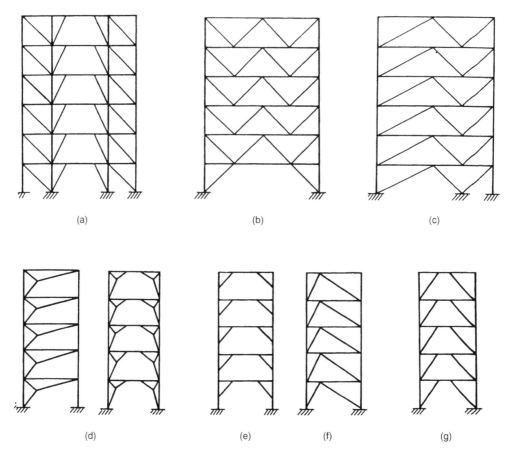

Figure 3.9a–g. Brace configurations that allow for door-size openings in interior space layouts. Note: Some configurations are not permitted in areas of high seismicity.

member can be made of double angles, channels, T-sections, tubes, or wide flange shapes. Besides performance, the shape of the diagonal is often based on connection considerations. The least objectionable locations for braces are around service cores and elevators, where frame diagonals may be enclosed within permanent walls. The braces can be joined together to form a closed or partially closed three-dimensional cell for effectively resisting torsional loads.

The most efficient (but also the most obstructive) types of bracing are those that form a fully triangulated vertical truss. Figure 3.9 shows other types of braced bents that pose fewer problems in the architectural organization of internal space as well as in locating door openings, but may cause bending in columns and girders. Historically, bracing has been used to provide lateral resistance to the majority of the world's tallest buildings, from the earliest examples at the end of the 19th century to the present. An outstanding example is the 1250-ft-high Empire State Building (Fig. 3.9h), completed in 1931.

3.3. STAGGERED TRUSS SYSTEM

In this system, story-high trusses span in the transverse direction between the columns at the exterior of the building. Flexibility in architectural planning is achieved by arranging the trusses in a staggered pattern at alternate floors, as shown in Fig. 3.10. The floor system

acts as a diaphragm transferring lateral loads in the short
direction to the trusses. Lateral loads are thereby resisted
by truss diagonals and are transferred into direct loads in
the columns. The columns therefore receive no bending
moments. The truss diagonals are eliminated at the cor-
ridor locations to allow for openings. Since the diagonal
is eliminated, the shear is carried by the bending action
of the top and bottom chord members at these locations.

Because the staggered truss system resists a majority of gravity and lateral loads in direct stresses, it is quite
stiff. In general, additional steel tonnage required for
controlling drift is quite small. Therefore, high-strength
steels may be used throughout the entire frame. In regions
of low seismicity, the system has been used for buildings
in the 35- to 40-story range. Transverse spans must be
long enough to make the trusses efficient, with 45ft (13.72
m) considered the minimum practical limit. Since the
trusses are supported only at the peri-meter, the need for
interior columns and associated foundations is eliminated, contributing to the economy of the system. The
system is not limited to simple rectangular plans, and can
be used for curvilinear plan shapes, as shown in Fig. 3.11.

The essential structural action in a staggered truss
system is the transfer of lateral loads across the floor to
the truss on the adjacent column line. This action continues down on the truss line across the next floor down the
next truss, etc., as shown schematically in Figs. 3.12
and 3.13. Thus, between the floors, lateral forces are
resisted by the truss diagonals, and at each floor, these
forces are transferred to the truss below with the floor
system acting as a diaphragm. The columns between the
floors receive no bending moments, resulting in a very
efficient and stiff structure. Since the trusses are placed
at alternate levels on adjacent column lines, a two-bay-
wide column-free interior floor space is created in the
longitudinal direction.

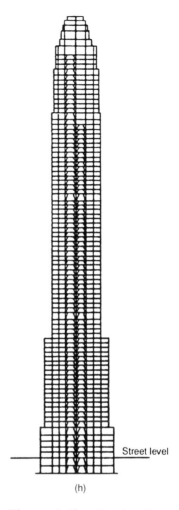

Street level

(h)

Figure 3.9h. Empire State
Building bracing system; riveted
structural steel frame encased in
cinder concrete.

3.3.1. Floor System

The floor system has two primary functions:

- Collect and transmit gravity loads to the vertical elements.
- Resist lateral loads through diaphragm action by providing a continuous path
  for transferring lateral loads from the bottom chord of one truss to the top
  chord of the adjacent truss down through the structure.

Thus, one could use precast concrete planks, long-span composite steel decks, open-
web joists, or any other system consistent with the structural and architectural requirements. Precast planks and flat-bottomed steel decks are often used as exposed ceilings
requiring a minimum of finish. For spans up to 30 ft (9.15 m), 8-in. (203-mm) thick
planks are required, whereas for spans less than 24 ft (7.3 m), 6-in. (152-mm) thick planks

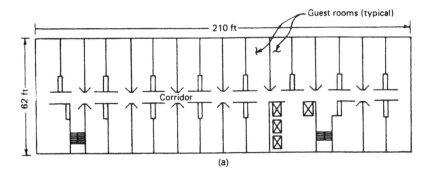

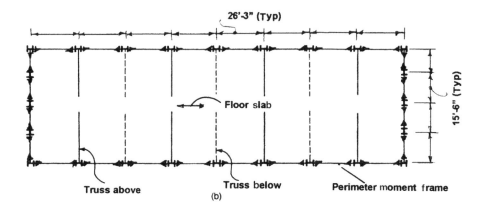

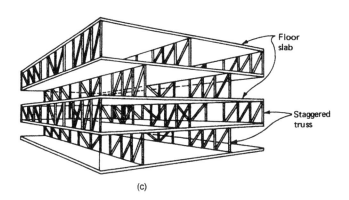

Figure 3.10. Staggered truss system: (a) hotel plan showing guest rooms; (b) arrangement of staggered trusses; (c) perspective view of truss arrangement.

are adequate. In a steel deck system, a 7-$\frac{1}{2}$-in. (190-mm) deep deck is required for spans up to 30 ft (9.15 m), and for spans up to 24 ft (7.3 m), a 6-in. (152-mm) deep steel deck is adequate. When precast planks are used, shear transfer is achieved by using welded plates cast in the planks or by welding shear connectors to the truss chords.

For metal deck systems, generally adequate shear transfer is achieved by welding the steel deck to the trusses. Planks used for erection purposes should have connection weld plates, even when shear connectors are provided. The choice of the floor system

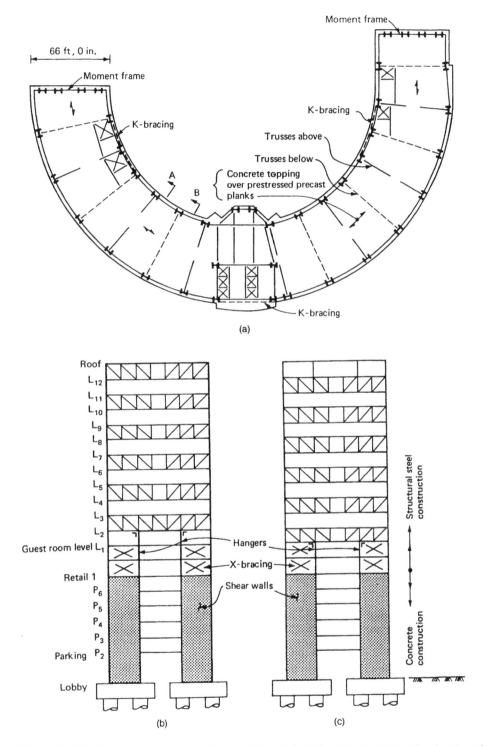

Figure 3.11. Staggered truss system for a semicircular building: (a) plan; (b) section A; (c) section B.

depends on the geographical location as well as local conditions. In cold climates, the cost of grouting between precast planks in winter is increased by the necessity for heating.

The floor system may consist of either a series of simple or continuous spans over the chords of the trusses. Because of the large spacing between the trusses, continuous

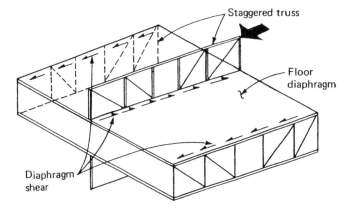

Figure 3.12.   Load path in staggered truss system.

spans are usually limited to a maximum of two bays. Generally, one end of each span is supported on the lower chord, while the other end is made continuous by simply running the floor system across the top chord of the trusses.

Since the trusses are staggered at alternate floors, the equivalent lateral load on each truss is equal to the lateral load acting on two bays. Hence, floor panels on each side of the truss must transmit half that load to the adjacent truss in the story immediately below. The floor system is designed as a deep beam to resist both the in-plane shears and bending moments. Gravity load design is identical to that of a conventional system. Because the integrity of the system depends on the ability of the diaphragm to transfer lateral loads from one truss to another, it must be detailed with chord elements at the boundaries to resist axial compression and tension.

3.3.2. Columns

Since the lateral loads are resisted by the truss diagonals, there is no bending of columns in the transverse direction. Thus, the columns are typically oriented with their weak axis perpendicular to the longitudinal direction of the building. Consequently, the weak-axis bending of columns due to gravity deflection of trusses must be considered in designing columns. This effect can, however, be minimized by introducing a camber in the truss by

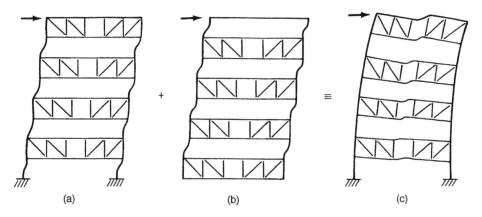

Figure 3.13.   Conceptual two-dimensional model for staggered truss system: (a, b) lateral deformation of adjacent bays; (c) overall behavior. Note the absence of local bending of columns.

deliberately making the truss bottom chord shorter than the top chord. An alternate solution is to design the connection between the truss bottom chord and column as a slip connection for dead loads. The bolts may then be torqued after the application of dead loads, thus limiting the bending moments in the columns due to truss deflections. If neither of these two methods is applicable, then the moments in columns due to the deflection of truss must be provided for in the design.

3.3.3. Trusses

The design of trusses is quite conventional. The large floor area supported by the truss allows for maximum live load reduction. Because the gravity and lateral loads are primarily resisted by axial compression and tension of the truss members, the system is very efficient. The only bending that occurs in the truss chords is at locations where the diagonals are eliminated, as at the openings for corridors.

The span-to-depth ratio of trusses is typically in the range of 6:1, giving adequate depth for the efficient design of top and bottom chords. Usually, the panel width of trusses is not a governing criterion. Larger panel lengths with fewer web members decrease fabrication costs and may work out to be more economical than for a shorter panel length.

For maximum efficiency, as in any other structural system, it is preferable to maintain a uniform spacing of trusses. This allows for maximization of typical truss units and reduces fabrication costs. However, when required by architectural arrangement, it is possible to vary the column and thus, the truss spacing. Truss design is based on assuming continuous top and bottom chords with pin-ended diagonals. Generally, wide flange shapes are selected for the chord members since angles are not efficient in resisting the secondary bending. Also, when planks are used, wide flanges offer good bearing areas. Since the truss system resists loads primarily by direct stresses, deflections are generally not a problem and therefore, high-strength steels can be economically employed.

The simplest method of stacking trusses is a configuration called the checkerboard pattern, in which the trusses are placed at alternate columns and floors. It is possible, however, to obtain greater variety of spaces by using different layouts on alternate levels.

Longitudinally, the lateral forces can be resisted by any conventional bracing system such as perimeter rigid frames, braced frames, or core shear walls. Rigid frames may be particularly attractive because: 1) columns along the longitudinal faces of the building are typically oriented with webs parallel to the spandrels; and 2) a large number of columns is generally available to participate in frame action.

It should be noted that particular attention is required for transferring lateral loads from the lowest staggered truss level to the foundations.

3.4. ECCENTRIC BRACED FRAME (EBF)

An eccentric bracing system attempts to combine the strength and stiffness of a braced frame with the inelastic behavior and energy dissipation characteristics of a moment frame. The system is called eccentric because deliberate eccentricities are inserted between beam-to-column or beam-to-brace connections. The eccentric beam element acts as a fuse by limiting large forces from entering into and causing buckling of braces. The eccentric segment of the beam, called the link, undergoes flexural or shear yielding prior to formation of plastic hinges in other bending members and well before buckling of any compression members. Thus, the system maintains stability even under large inelastic deformations. The required stiffness during wind or minor earthquakes is maintained because no plastic hinges are formed under these loads and all behavior is entirely elastic. Although the lateral deformation is larger than in a concentrically braced frame due to bending and shear deformation of the "fuse," its contribution to deflection is not significant because of its relatively small length. Thus, the elastic stiffness of eccentrically braced frames can be considered to be about the same as that of a concentrically braced frame.

3.4.1. Ductility

A ductile behavior is highly desirable when a structure is called upon to absorb energy well in excess of its elastic capacity, as when it is subjected to strong seismic ground motions. Steel's capacity for deformation without fracture when connections between structural elements are properly detailed, combined with its high strength, makes steel an ideal material for use in eccentric bracing systems. In a properly designed and executed connection, steel continues to resist loads even after the maximum load is reached. This property, by virtue of which steel sustains the load without fracture, is called ductility. A brittle material, on the other hand, does not undergo large deformations at the onset of yielding. It fractures prior to or just when it reaches the maximum load.

3.4.2. Behavior

Eccentrically braced frames can be configured in various forms as long as the brace is connected to at least one link (see Fig. 3.14). The underlying principle is to prevent buckling of the brace from large overloads that may occur during major earthquakes. This is achieved by designing the link to yield prior to distress in other structural elements.

The shear yielding of beams is a relatively well-defined phenomenon; the load required for shear yielding of a beam of given dimensions can be calculated fairly accurately. The corresponding axial load and moments in columns and braces connected to the link can also be assessed with reasonable accuracy. Using certain overload factors, the braces and columns are designed to carry more load than could be imposed on them by the shear yielding of link. This assures that in the event of a large earthquake, it is the link that takes the hit, and not the columns or braces connected to it.

Consider the bracing shown in Fig. 3.14e subjected to cyclic horizontal loads caused by an earthquake. The axial force in the brace is transmitted to the beam as a horizontal force inducing axial stresses, and as a vertical force inducing shear stresses in the beam web. Of more concern in the design of the link are the cyclic shear forces induced in the beam. Assuming the link and its moment connection to the column are adequate in bending, the failure mechanism of the braced frame is by shear yielding of the beam web, provided that the buckling of the web itself is prevented. This is typically achieved by designing adequate vertical web stiffeners between the beam flanges in the link region (Fig 3.14f).

3.4.3. Essential Features of Link

Depending upon its length, the link may dissipate seismic energy either by developing plastic hinges or by yielding of the web in shear. Links longer than twice their depths tend to develop plastic hinges whereas shorter links tend to yield in shear. Thus, the links can be identified either as short or long, with the former experiencing moderate rotations and the latter, relatively larger rotations.

The shear yielding is an excellent energy dissipation mechanism because large cyclic deflections can take place without failure or deterioration of the link. This is because yielding occurs over a large segment of the beam web and is followed by a cyclic diagonal field.

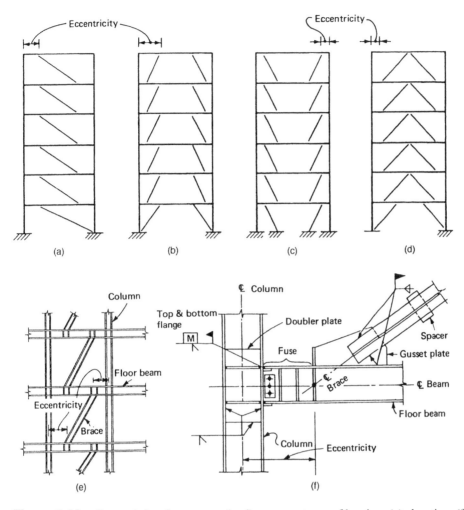

Figure 3.14. Eccentric bracing system: (a–d) common types of bracing; (e) elevation; (f) detail.

The web buckles after yielding in shear, but the tension field takes over the load-carrying mechanism to prevent failure, resulting in good energy dissipation. Thus the link may bend and twist, but will not lose its gravity load-carrying capacity.

3.4.4. Analysis and Design Considerations

To force the formation of a hinge in the beam web, the plastic moment capacity of the beam should exceed the beam shear yield capacity. In calculating the plastic moment capacity of the beam, the contribution of the web is neglected because the web is assumed to have yielded. The beam is first selected for the required shear capacity and then the plastic moment capacity is checked to assure that it is slightly larger than the shear yield capacity. As in a ductile frame design, the column is selected by adhering to the weak beam–strong column concept. This is to prevent the so-called story collapse mechanism by assuring that plastic hinges are formed in beams and not in columns. If the plastic moment of the beam selected is larger than that required by design, the column is designed in an equally conservative manner. To assure that the braces are prevented from buckling, they are designed to withstand forces somewhat larger than those given by the analysis.

This conservatism is necessary to take into account the fact that the actual beam designed is likely to have additional capacity due to factors such as

- Beam strain hardening.
- Actual yield stress of beam steel greater than the theoretical value.
- Interaction of floor slabs with link beam with attendant increase in its plastic moment capacity.

The brace-to-beam connection can be designed as either a welded or bolted connection. The bolts are designed as slip-resistant bolts and checked for bearing capacity because of the likelihood of slippage in the event of a large earthquake. The beam-to-column connection is designed as a moment connection by welding the beam flanges to the column with full-penetration welds. A single-side shear plate connection with fillet welds is used to develop the high-shear forces in the link. Lateral support is provided at the top and bottom flanges of the beam to prevent lateral torsional buckling and weak axis bending.

3.4.5. Deflection Considerations

The lateral deflection of an eccentrically braced frame can be estimated as the sum of the following three components:

- Deflection due to axial strain in the brace.
- Deflection due to axial strain in the columns.
- Deflection due to bending deformation of the link.

Because the braces and columns are designed to remain elastic even under a severe earthquake, their deflection contributions are very nearly constant even after the shear yielding of the link. And, since the beams in an eccentric bracing system are much heavier than those in a concentrically braced frame, they are likely to contribute little to the deflection under elastic conditions. Therefore, even with the bending of link beams, an eccentrically braced frame is not an unreasonably flexible system.

3.4.6. Conclusions

The structural characteristics of an eccentrically braced frame may be summarized as follows:

- It provides a stiff structural system without imposing undue penalty on the steel tonnage.
- Eccentric beam elements yielding either in shear or in bending act as fuses to dissipate excess energy during severe earthquakes.
- The yielding of the link does not cause the structure to collapse because the structure continues to retain its vertical load-carrying capacity.

3.5. INTERACTING SYSTEM OF BRACED AND RIGID FRAMES

Even for buildings in the of 10- to 15-story range, unreasonably heavy columns may result if the lateral bracing is confined to the building's service core because the depth available for bracing is usually limited. Additionally, high uplift forces may occur at the bottom of core columns, presenting foundation problems. For such buildings, an economical structural system can be devised, using a combination of rigid frames with a core bracing system. Although relatively deep girders are required for a substantial frame action, rigid frames are often architecturally preferred because they are least objectionable from the interior space planning considerations. When used on the building exterior, deep spandrels and closely spaced columns may be permissible because columns usually will not interfere with the space planning, and the depth of spandrels need not be shallow as for interior beams, for the passage of air conditioning and other utility ducts. A schematic floor plan of a building using this concept is shown in Fig. 3.15a.

As an alternative to perimeter frames, a set of interior frames can be used with the core bracing, as shown in Fig. 3.15b, in which frames on grid lines 1, 2, 6, and 7 interact with core bracing on lines 3, 4, and 5. Yet another option is to moment-connect the girders between the braced core and perimeter columns, as shown in Fig. 3.15c. In this example, the moment-connected girders act as outriggers connecting the exterior columns to the braced core.

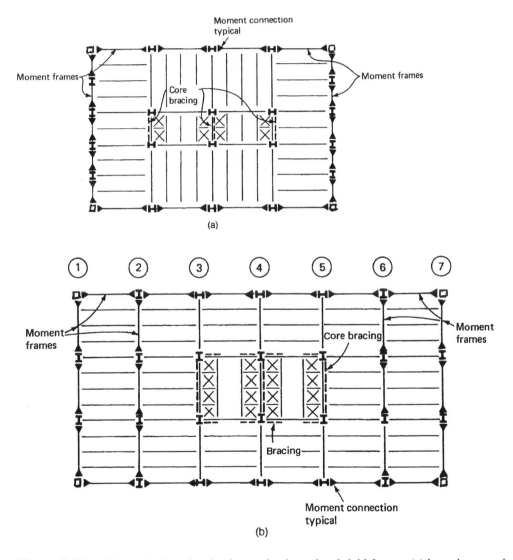

Figure 3.15. Schematic plans showing interacting braced and rigid frames: (a) braced core and perimeter frames; (b) braced core and interior and exterior frames; (c) braced core and interior frames; (d) full-depth interior bracing and exterior frames; (e) transverse cross section showing primary interior bracing, secondary bracing, and basement construction.

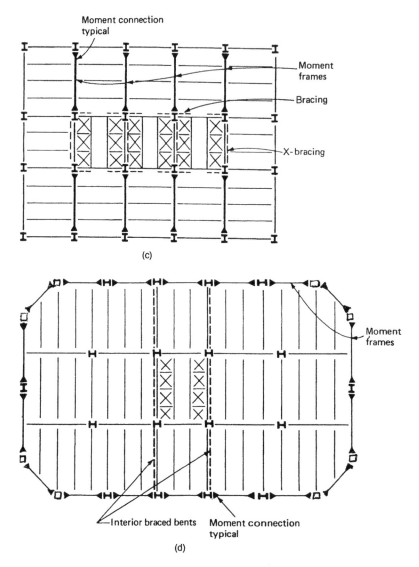

Figure 3.15. (*Continued*).

For slender buildings with height-to-width ratios in excess of 5, an interacting system of moment frames and braces becomes uneconomical if braces are placed only within the building core. A good structural solution, if architecturally acceptable, is to spread the bracing for the full width of the building along the facades.

Another possibility is to move the full-depth bracing to the interior of the building, as shown in Fig. 3.15d. The braces stretched out for the full width of the building form giant K-braces, resisting overturning and shear forces by developing predominantly axial forces. A transverse cross section of the building is shown in Fig. 3.15e, which identifies a secondary system of braces required to transfer the lateral loads to the panel points of the K-braces. The diagonals of the K-braces running through the interior of the building result in sloping columns whose presence, must be architecturally acknowledged as a trade-off for structural efficiency.

All of the aforementioned bracing systems or any number of their variations can be used singly or in combination with one another, depending on the layout of the building

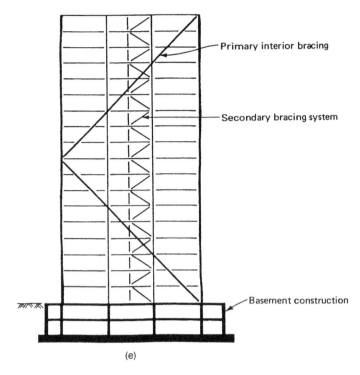

(e)

Figure 3.15. (*Continued*).

and architectural requirements. The lateral resisting system can be turned by varying the relative stiffness of braces and frames to achieve an economical and sound structural system.

3.5.1. Behavior

If the lateral deflection patterns of braced and unbraced frames were similar, the lateral loads would be distributed between the two systems according to their relative stiffness. However, in general, the unbraced and braced frames deform with their own characteristic shapes, resulting in a heavy interaction between the two, particularly at the upper levels of the buildings.

Insofar as the lateral-load-resistance is concerned, rigid and braced frames can be considered as two distinct units. The basis of classification is the mode of deformation of the two when subjected to lateral loading.

The deflection characteristics of a braced frame are similar to those of a cantilever beam. Near the bottom, the braced frame is relatively stiff, and therefore, the floor-to-floor deflections will be less than half the values near the top. Near the top, the floor-to-floor deflections increase rather rapidly, resulting mainly from the cumulative effect of braced frame rotation due to axial deformations of the columns. Since this effect occurs at every floor, the resulting deflection at the top is cumulative. This type of deflection due to axial strains in columns—often referred to as chord drift—is difficult to control unless the column areas are increased far above those required for gravity needs.

Rigid frames deform predominantly in a shear mode. The relative story deflections depend primarily on the magnitude of shear applied at each story level. Although the deflections are larger near the bottom and smaller near the top as compared to the braced frames, the floor-to-floor deflections can be considered more nearly uniform throughout the

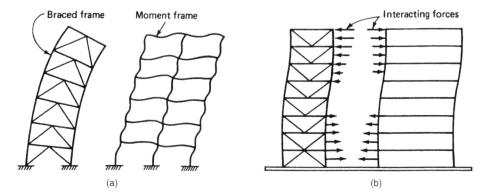

Figure 3.16. Interaction between braced and unbraced frames: (a) characteristic deformation shapes; (b) variation of shear forces resulting from interaction.

height. When the two systems—the braced and rigid frames—are connected by rigid floor diaphragms, a nonuniform shear force develops between the two. The resulting interaction typically results in a more economical system.

Figure 3.16 shows the deformation patterns of a braced and unbraced frame subjected to lateral loads. Also shown are the horizontal shear forces between the two, the length of arrows schematically representing the level of interaction between them. Observe that the braced frame acts as a vertical cantilever beam, with the slope of the deflection greatest at the top, indicating that in this region the braced frame contributes the least to the lateral stiffness.

The rigid frame, on the other hand, deforms in a shear mode, with the slope greater at the base of the structure where the shear is maximum. Since the lateral deflection characteristics of the two frames are entirely different, the rigid frame tends to pull back the brace frame in the upper portion of the building while pushing it forward in the lower portion. As a result, the rigid frame participates more effectively in the upper portion of the building where lateral shears are relatively weaker, while the braced frame carries most of the shear in the lower portion of the building. Because of the distinct difference in the deflection characteristics, the two systems tend to help each other a great deal. The rigid frame tends to reduce the lateral deflection of the brace frame at the top, while the braced frame supports the rigid frame near the base. A typical variation of horizontal shear carried by each frame is shown in Fig. 3.16b, in which the length of arrows conceptually indicates the magnitude of interacting shear forces.

3.6. OUTRIGGER AND BELT TRUSS SYSTEMS

The structural arrangement for an outrigger system consists of a main core connected to the exterior columns by relatively stiff horizontal members commonly referred to as outriggers. The main core may consist of a steel braced frame or reinforced concrete shear walls, and may be centrally located with outriggers extending on both sides (Fig. 3.17a). Alternatively, it may be located on one side of the building with outriggers extending to the building columns on one side (Fig. 3.17b).

The basic structural response of the system is quite simple. When subjected to lateral loads, the column-restrained outriggers resist the rotation of the core, causing the lateral deflections and moments in the core to be smaller than if the freestanding core alone

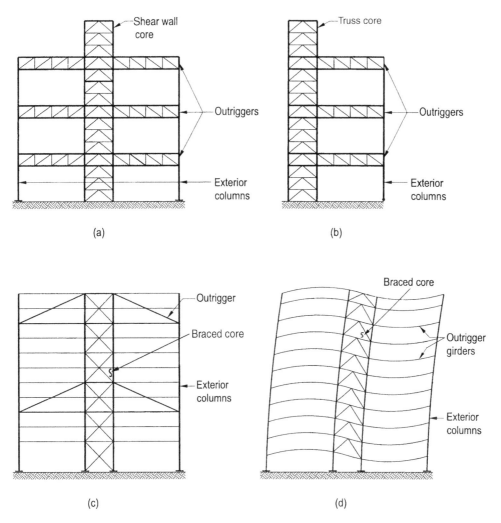

Figure 3.17. (a) Outrigger system with a central core: (b) outrigger system with offset core; (c) diagonals acting as outriggers; (d) floor girders acting as outriggers.

resisted the loading. The external moment is now resisted not by bending of the core alone, but also by the axial tension and compression of the exterior columns connected to the outriggers. As a result, the effective depth of the structure is increased when it flexes as a vertical cantilever, by the development of tension in the windward columns, and by compression in the leeward columns.

In addition to those columns located at the ends of the outriggers, it is usual to also mobilize other peripheral columns to assist in restraining the outriggers. This is achieved by including a deep spandrel girder, or a truss commonly referred to as a "belt truss," around the structure at the levels of the outriggers.

To make the outriggers and belt truss adequately stiff in flexure and shear, they are made at least one—and often two—stories deep. It is also possible to use diagonals extending through several floors to act as outriggers, as shown in Fig. 3.17c. Yet another option is to moment-connect the girders at each floor to the core (Fig. 3.17d). It should be noted that whereas the outrigger system is effective in increasing the structure's flexural stiffness, it does not increase its resistance to shear, which must be carried mainly by the core.

3.6.1. Behavior

To understand the behavior of an outrigger system, consider a building stiffened by a story-high outrigger at top, as shown in Fig. 3.18. Because the outrigger is at the top, the system is often referred to as a cap or hat truss system. The tie-down action of the cap truss generates a restoring couple at the building top, resulting in a point of contraflexure in its deflection curve. This reversal in curvature reduces the bending moment in the core and hence, the building drift.

Although the belt truss shown in Fig. 3.18 functions as a horizontal fascia stiffener mobilizing other exterior columns, for analytical simplicity we will assume that the cumulative effect of the exterior columns may be represented by two equivalent columns, one at each end of the outrigger (Fig. 3.18c). This idealization is not necessary in developing the theory, but keeps the explanation simple.

The core may be considered as a single-redundant cantilever with the rotation restrained at the top by the stretching and shortening of windward and leeward columns. The result of the tensile and compressive forces is equivalent to a restoring couple opposing the rotation of the core. Therefore, the cap truss may be conceptualized as a restraining spring located at the top of the cantilever. Its rotational stiffness may be defined as the restoring couple due to a unit rotation of the core at the top.

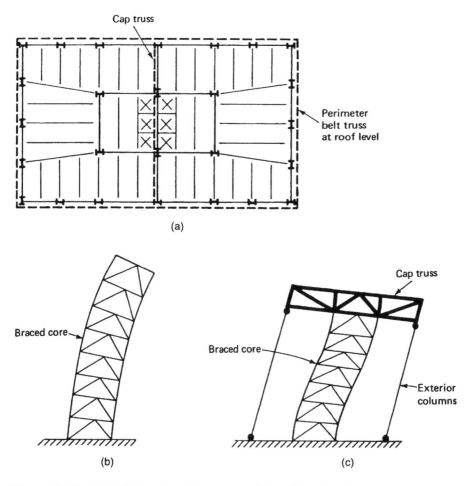

Figure 3.18. (a) Building plan with cap truss; (b) cantilever bending of core; (c) tie-down action of cap truss.

Assuming the cap truss is infinitely rigid, the axial elongation and shortening of columns is equal to the rotation of the core multiplied by their respective distances from the center of the core. If the distance of the equivalent column is $d/2$ from the center of the core, the axial deformation of the columns is then equal to $\theta d/2$, where θ is the rotation of the core. Since the equivalent spring stiffness is calculated for unit rotation of the core (that is, $\theta = 1$), the axial deformation of the equivalent columns is equal to $1 \times d/2 = d/2$ units.

The corresponding axial load is given by

$$P = \frac{AEd}{2L}$$

where

P = axial load in the columns
A = area of columns
E = modulus of elasticity
d = distance between the exterior columns
L = height of the building.

The restoring couple, that is, the rotational stiffness of the cap truss, is given by the axial load in the equivalent columns multiplied by their distance from the center of the core. Using the notation K for the rotational stiffness, and noting that there are two equivalent columns, each located at a distance $d/2$ from the core, we get

$$K = P \times \frac{d}{2} \times 2$$
$$= Pd$$

The reduction in drift depends on the stiffness K and the magnitude of rotation θ at the top.

Several questions arise if we consider the interaction of the core with the outriggers located not at the top, but somewhere up the height. How does the location influence the building drift and moment in the core? Is the top location the best? What if the outrigger is moved toward the bottom, say, to the midheight of the building? Is there an optimum location that reduces the drift to a minimum?

Before answering these rather intriguing questions, it is perhaps instructive to study the behavior of the system with an outrigger located at specific heights of the building, say, at the top, three-quarters height, midheight, and one-quarter height.

3.6.2. Deflection Calculations

Case 1: Outrigger located at top, $z = L$
The rotation compatibility condition at $z = L$ (see Fig. 3.19) can be written as

$$\theta_W - \theta_S = \theta_L \tag{3.1}$$

where

θ_W = rotation of the cantilever at $z = L$ due to a uniform lateral load W, in radians.
θ_S = rotation due to spring restraint located at $z = L$, in radians. The negative sign for θ_S in Eq. 3.1 indicates that the rotation of the cantilever due to the spring stiffness is in a direction opposite to the rotation due to external load.
θ_L = final rotation of the cantilever at $z = L$, in radians.

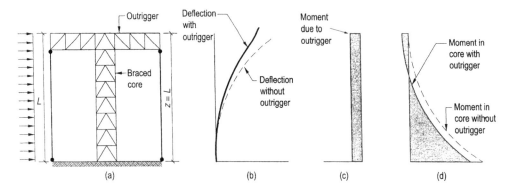

Figure 3.19. Outrigger at top, $z = L$.

For a cantilever with uniform moment of inertia I and modulus of elasticity E subjected to uniform horizontal load W,

$$\theta_W = \frac{WL^3}{6EI} \tag{3.2}$$

If M_1 and K_1 represent the moment and stiffness of the spring located at $z = L$, Eq. (3.1) can be rewritten as

$$\frac{WL^3}{6EI} - \frac{M_1 L}{EI} = \frac{M_1}{K_1} \tag{3.3}$$

and

$$M_1 = \frac{WL^3/6EI}{1/K_1 + L/EI} \tag{3.4}$$

The resulting deflection Δ_1 at the building top can be obtained by superposing the deflection of the cantilever due to external uniform load W, and the deflection due to the moment induced by the spring, thus

$$\Delta_1 = \Delta_{\text{load}} - \Delta_{\text{spring}}$$
$$= \frac{WL^4}{8EI} - \frac{M_1 L^2}{2EI} \tag{3.5}$$
$$= \frac{L^2}{2EI}\left(\frac{WL^2}{4} - M_1\right) \tag{3.6}$$

Case 2: Outrigger located at $z = 3L/4$

The general expression for lateral deflection y, at distance x measured from the top, for a cantilever subjected to a uniform lateral load (see Fig. 3.20) is given by

$$y = \frac{W}{24EI}(x^4 - 4L^3 x + 3L^4) \tag{3.7}$$

Note that x is measured from the top and is equal to $(L - z)$.

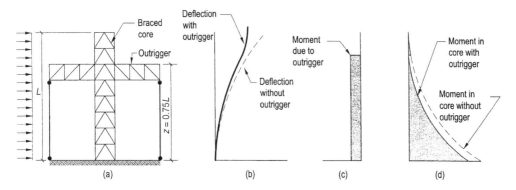

Figure 3.20. Outrigger, three-quarters up the height, $z = 0.75L$.

Differentiating Eq. (3.7) with respect to x, the general expression for slope of the cantilever is given by

$$\frac{dy}{dx} = \frac{W}{6EI}(x^3 - L^3) \tag{3.8}$$

The slope at the spring location is given by substituting $z = 3L/4$, i.e., $x = L/4$ in Eq. (3.8). Thus

$$\frac{dy}{dx}\left(\text{at } z = \frac{3L}{4}\right) = \frac{W}{6EI}\left(\frac{L^3}{64} - L^3\right)$$

$$= \frac{WL^3}{6EI} \times \frac{63}{64} \tag{3.9}$$

Using the notation M_2 and K_2 to represent the moment and stiffness of spring at $z = 3L/4$, the compatibility equation at location 2 can be written as

$$\frac{WL^3}{6EI}\left(\frac{63}{64}\right) - \frac{M_2}{EI}\left(\frac{3L}{4}\right) = \frac{M_2}{K_2} \tag{3.10}$$

Noting that $K_2 = 4K_1/3$, the expression for M_2 can be written as

$$M_2 = \left(\frac{WL^3/6EI}{1/K_1 + L/EI}\right)\frac{63/64}{3/4} = \left(\frac{WL^3/6EI}{1/K_1 + L/EI}\right)1.31 \tag{3.11}$$

Noting that the terms in the parentheses represent M_1, Eq. (3.11) can be expressed in terms of M_1:

$$M_2 = 1.31M_1$$

The drift is given by the relation

$$\Delta_2 = \frac{WL^4}{8EI} - \frac{M_2 3L}{4EI}\left(L - \frac{3L}{8}\right) \tag{3.12}$$

or

$$\Delta_2 = \frac{L^2}{2EI}\left(\frac{WL^2}{4} - 1.23M_1\right) \tag{3.13}$$

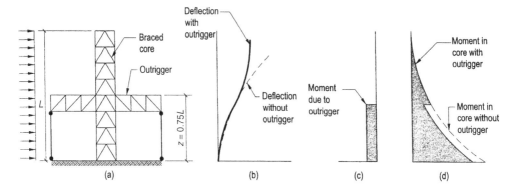

Figure 3.21. Outrigger at midheight, $z = 0.5L$.

Case 3: Outrigger at midheight, $z = L/2$

The rotation at $z = L/2$ due to external load W (see Fig. 3.21) can be shown to be equal to $7WL^3/48EI$, giving the rotation compatibility equation

$$\frac{7WL^3}{48EI} - \frac{M_3 L}{2EI} = \frac{M_3}{K_3} \tag{3.14}$$

where M_3 and K_3 represent the moment and stiffness of the spring at $z = L/2$. Noting that $K_3 = 2K_1$, the expression for M_3 works out as

$$M_3 = \left(\frac{WL^3/6EI}{1/K_1 + L/EI}\right) \times \frac{7}{4} \tag{3.15}$$

Since the expression in the parentheses is equal to M_1, M_3 can be expressed in terms of M_1:

$$M_3 = 1.75M_1 \tag{3.16}$$

The drift is given by the equation

$$\Delta_3 = \frac{WL^4}{8EI} - \frac{M_3 L}{2EI}\left(L - \frac{L}{4}\right) \tag{3.17}$$

or

$$\Delta_3 = \frac{L^2}{2EI}\left(\frac{WL^2}{4} - 1.31M_1\right) \tag{3.18}$$

Case 4: Outriggers at quarter-height, $z = L/4$

The rotation at $z = L/4$ due to uniform lateral load (see Fig. 3.22) can be shown to be equal to $WL^3/6EI[(37/64)]$, giving the rotation compatibility equation

$$\frac{WL^3}{6EI}\left(\frac{37}{64}\right) - \frac{M_4 L}{4EI} = \frac{M_4}{K_4} \tag{3.19}$$

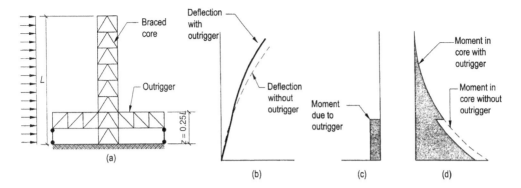

Figure 3.22. Outrigger at quarter-height, $z = 0.25L$.

where M_4 and K_4 represent the moment and stiffness of the spring at $z = L/4$. Noting that $K_4 = 4K_1$, M_4 in Eq. (3.19) can be expressed in terms of M_1:

$$M_4 = 2.3M_1 \tag{3.20}$$

The drift for this case is given by the expression

$$\Delta_4 = \frac{WL^4}{8EI} - \frac{M_4L}{4EI}\left(L - \frac{L}{8}\right) \tag{3.21}$$

or

$$\Delta_4 = \frac{L^2}{2EI}\left(\frac{WL^2}{4} - M_1\right) \tag{3.22}$$

Equations (3.6), (3.13), (3.18), and (3.22) give the building drift for four different locations of the belt and outrigger trusses.

The value of K_1, which corresponds to stiffness of the spring when it is located at $z = L$, can be derived as follows.

A unit rotation given to the core at the top results in extension and compression of all perimeter columns, the magnitudes of which are given by their respective distances from the center of gravity of the core. The resulting force multiplied by the lever arm gives the value for stiffness K_1. Thus, if p_1 is measured in kips and the distance in feet, K_1 has units of kip feet. The force p in each exterior column is given by the relation $p = AE\,\delta/L$ since, by definition, δ corresponds to column extension or compression due to unit rotation of the core, $\delta = d/2$, where d is the distance between the exterior columns. Therefore,

$$p = \frac{AE}{L}\left(\frac{d}{2}\right) \tag{3.23}$$

and its contribution to the stiffness K_1 is given by the relation

$$K_i = p_i d$$
$$= \frac{A_i E}{L}\frac{d^2}{2} \tag{3.24}$$

3.6.3. Optimum Location of a Single Outrigger

The preceding analysis has indicated that the beneficial action of the outrigger is a function of two distinct characteristics: 1) the stiffness of the equivalent spring; and 2) the magnitude of the rotation of the cantilever at the spring location due to lateral loads. The spring stiffness varies inversely as the distance of the outrigger from the base. For example, it is at a minimum when located at the top and a maximum when at the bottom. The rotation, θ, of the free cantilever subjected to a uniformly distributed horizontal load varies parabolically with a maximum value at the top to zero at the bottom. Therefore, from the point of view of spring stiffness, it is desirable to locate the outrigger at the bottom, whereas from consideration of its rotation, the converse is true. It must therefore be obvious that the optimum location is somewhere in between.

We start with the following assumptions:

1. The building is prismatic and vertically is uniform; that is, the perimeter columns have a constant area and the core has a constant moment of inertia for the full height.
2. The outrigger and the belt trusses are flexurally rigid.
3. The lateral resistance is provided only by the bending resistance of the core and the tie-down action of the exterior columns.
4. The core is rigidly fixed at the base.
5. The rotation of the core due to its shear deformation is negligible; that is, the core structure is heavily braced, so that its rotation due to bracing deformations may be assumed negligible.
6. The lateral load is constant for the full height.
7. The exterior columns are pin-connected at the base.

Consider Fig. 3.23 with a single outrigger located at a distance x from the building top. To evaluate the optimum location, first the restoring moment M_x of the outrigger located at x is evaluated. Next, an algebraic equation for the deflection of the core at the top due to M_x is derived. Differentiating this equation and equating to zero yields a third-degree polynomial, the solution of which yields the outrigger optimum location corresponding to the minimum deflection of the building due to external load. The details are as follows.

The rotation θ of the cantilever at a distance x from the top, due to a uniformly distributed load w, is given by the relation

$$\theta = \frac{W}{EI}(x^3 - L^3) \tag{3.25}$$

The rotation at the top due to the restoring couple M_x is given by the relation

$$\theta = \frac{M_x}{EI}(L - x) \tag{3.26}$$

The compatibility relation at x is given by

$$\frac{W}{6EI}(x^3 - L^3) - \frac{M_x}{EI}(L - x) = \frac{M_x}{K_x} \tag{3.27}$$

where

W = intensity of the wind load per unit height of the structure
M_x = the restoring moment due to outrigger restraint
K_x = spring stiffness at x equal to $AE/(L - x) \times (d^2/2)$

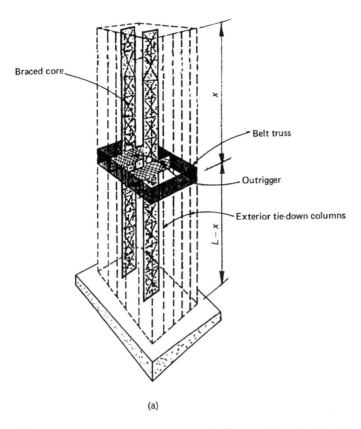

(a)

Figure 3.23a. Single outrigger and belt truss schematic elevation.

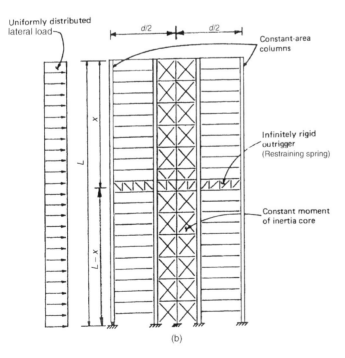

(b)

Figure 3.23b. Conceptual model for a single outrigger and belt truss model. Restraining "spring" occurs at a distance x from top.

E = modulus of elasticity of the core
I = moment of inertia of the core
A = area of the perimeter columns
L = height of the building
x = location of truss measured from the top
d = distance out-to-out of columns

Next, obtain the deflection at the top due to M_x:

$$Y_M = \frac{M_x(L - x)(L + x)}{2EI} \tag{3.28}$$

From our definition, the optimum location of the outrigger is that location for which the deflection Y_M is a maximum. This is obtained by substituting for M_x from Eq. (3.27) into Eq. (3.28) and differentiating it with respect to x and equating to zero. Thus, dy/dx of

$$\left[\frac{W(x^3 - L^3)(L + x)}{12(EI)^2(1/AE + 1/EI)} \right] = 0 \tag{3.29}$$

Simplifying this equation, we get a cubic equation in x.

$$4x^3 + 3x^2L - L^3 = 0 \tag{3.30}$$

This cubic equation has a single positive root, $x = 0.455L$.

Therefore, to minimize drift, a single outrigger must be located at a distance x = 0.455 L from the top or, say, approximately at midheight of the building.

In the discussion, several assumptions were made to simplify the problem for hand calculations. However, in a practical building, many of these assumptions are rarely satisfied. For example:

- The lateral load does not remain constant up the building height. It varies in a trapezoidal or a triangular manner, the former representative of wind loads and the latter, seismic loads.
- The cross-sectional areas of both the exterior and interior columns typically reduce up the building height. A linear variation is perhaps more representative of a practical building column.
- As the areas of core columns decrease up the height, so does the moment of inertia of the core. Therefore, a linear variation of the moment of inertia of the core, up the height is more appropriate.

Incorporating the aforementioned modifications aligns the analytical model closer to a practical structure, but renders the hand calculations all but impossible. Therefore, a computer-assisted analysis has been performed on a representative 46-story building using the modified assumptions previously mentioned. A schematic plan of the building, and an elevation of the idealized structural system and lateral loading are shown in Figs. 3.24 and 3.25. The lateral deflections at the building top are shown in a graphical format in Fig. 3.26 for various outrigger locations.

The deflections shown in a nondimensional format in Fig. 3.26 are relative to that of the core without the outrigger. Thus, the vertical ordinate with a value of unity at the extreme right of Fig. 3.26 is the deflection of the building without the restraining effect of the outrigger. The deflections including the effect of the outriggers are shown in curve 'S.' It is obtained by successively varying the outrigger location starting at the very top and progressively lowering its location in single-story increments, down through the building height.

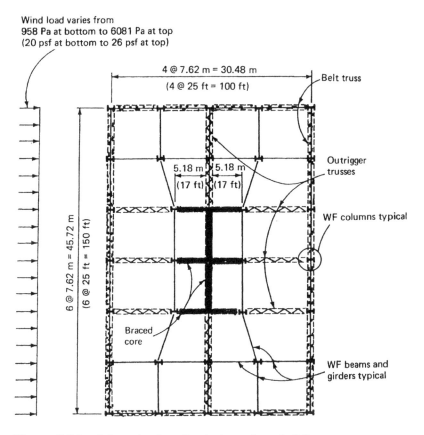

Wind load varies from
958 Pa at bottom to 6081 Pa at top
(20 psf at bottom to 26 psf at top)

4 @ 7.62 m = 30.48 m
(4 @ 25 ft = 100 ft)

Belt truss

5.18 m 5.18 m
(17 ft) (17 ft)

6 @ 7.62 m = 45.72 m
(6 @ 25 ft = 150 ft)

Outrigger
trusses

WF columns typical

Braced
core

WF beams and
girders typical

Figure 3.24. Schematic plan of a single outrigger building.

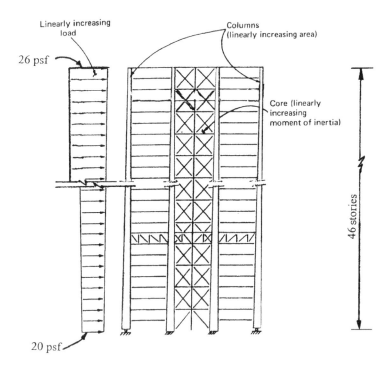

Linearly increasing
load

Columns
(linearly increasing area)

26 psf

Core (linearly
increasing
moment of inertia)

46 stories

20 psf

Figure 3.25. Single-outrigger building, schematic structural system.

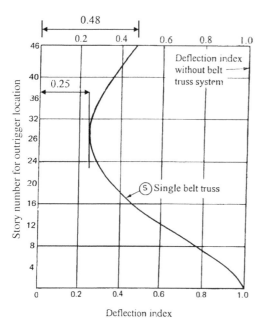

Figure 3.26. Deflection index versus level of outrigger location.

Note: Deflection index = $\dfrac{\text{Deflection at top w/o outrigger}}{\text{Deflection at top with outrigger}}$.

It is seen that lowering the outrigger down from its top location decreases the building drift progressively until the outrigger reaches level 26. Moving it either above or below this "optimum location" only reduces its efficiency. Observe that this level is at distance $(46 - 26/46)L = 0.435L$ from the top, very close to the optimum location of $x = 0.455L$ for the building with uniform characteristics. Furthermore, it can be seen from Fig. 3.26 that the efficiency of the outrigger placed at midheight; that is, at level 23, is very close to that when it is at the optimum location. Therefore, as a rule of thumb, the optimum location for a single outrigger may be considered at midheight.

Observe that when the outrigger is at the top, the building drift is reduced to nearly half the deflection of the unrestrained core. Thus, for example, if the drift of the unrestrained core is, say, $20''$ at the top, the corresponding deflection with an outrigger at level 46 is reduced to $0.48 \times 20 = 9.6$ in. A rather impressive reduction indeed, but what is more important is that the deflection continues to reduce as the outrigger is lowered from level 46 downward. It reaches a minimum value of $0.25 \times 20 = 5$ in. as shown in Fig. 3.26 when the outrigger is placed at the optimum location, level 26. Further lowering of the outrigger will not reduce the drift, but increase it. Its beneficial effect vanishes to nearly nothing when placed very close to the bottom of the building, say, at level 2 of the example problem.

Using the results of the example problem, the following conclusions can be drawn:

- Given a choice, the best location for a single outrigger is about midheight of the building.
- An outrigger placed at the top, acting as a cap or hat truss, is about 50% less efficient than that placed at midheight. However, in many practical situations, it may be more permissible to locate the outrigger at the building top. Therefore, although not as efficient as when at midheight, the benefits of a cap truss are nevertheless quite impressive, resulting in up to a 50% reduction in building drift.

3.6.4. Optimum Location of Two Outriggers

In the preceding analyses, only one compatibility equation was necessary because the one-outrigger structure is once redundant. On the other hand, a two-outrigger structure is twice redundant, requiring a solution of two compatibility equations.

As before, the sectional areas of the exterior columns and the moment of inertia of the core are assumed to decrease up the height. A trapezoidal distribution is assumed as before, for the lateral load. Schematic elevation and a conceptual analytical model used in the analysis are shown in Figs. 3.27 and 3.28.

The method of analysis for calculating the deflections at the top is similar to that used for the previous example. The moments at the outrigger locations are chosen as the unknown arbitrary constants M_1 and M_2, and the structure is rendered statically determinate by removing the rotational restraints at the outrigger locations. Next, the compatibility equations for the rotations at the truss locations are set up and solved simultaneously to obtain the values to M_1 and M_2. The final deflection at the top is obtained by a superposition of the deflection due to the external load and a counteracting deflection due to the moments M_1 and M_2.

The magnitude of the deflection at the top is given for three types of buildings by assuming that the lateral loads are resisted by: 1) the core alone; 2) the core acting together with a single outrigger; or 3) the core acting in conjunction with two outriggers. (See Fig. 3.29.)

As before, the vertical ordinate shown with a value of unity is the deflection index at the top derived by neglecting the restraining effect of outriggers. The resistance is provided by the cantilever action of the braced core alone. Curve S represents the top

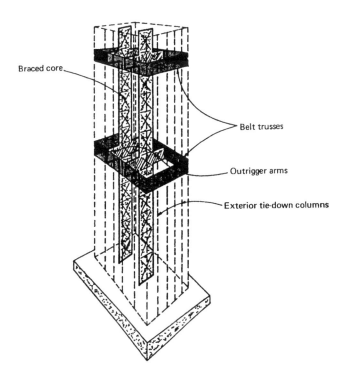

Braced core

Belt trusses

Outrigger arms

Exterior tie-down columns

Figure 3.27. Schematic structural system for a building with two outriggers and belt trusses.

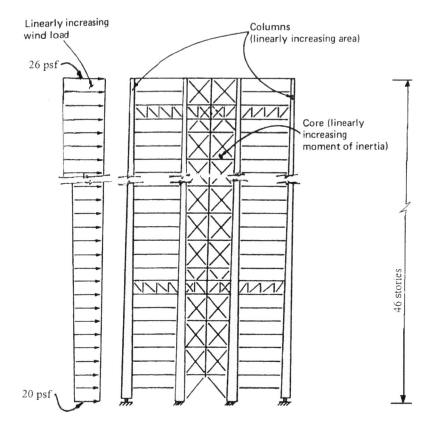

Figure 3.28. Conceptual analytical model and loading diagram for a building with two outriggers and belt trusses.

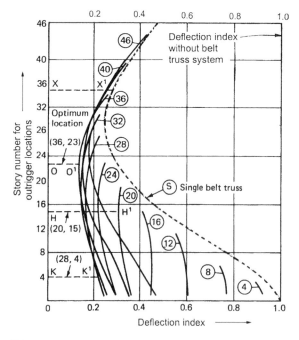

Figure 3.29. Deflection index versus level of outrigger locations.

Note: Deflection index = $\dfrac{\text{Deflection at top w/o outriggers}}{\text{Deflection at top with outriggers}}$.

deflection of the core restrained by a single outrigger located anywhere up the height of the structure.

The curves designated as 4, 8,…, 46 represent the deflections at the top for two outriggers located anywhere up the height of the structure. To plot each curve, the location of the upper outrigger was considered fixed in relation to the building height, while the location of the lower outrigger was moved in single-story increments, starting from the floor immediately below the top outrigger.

The number designations of the curves represent the floor number at which the upper outrigger is located. The second outrigger location is shown by story levels on the vertical axis. The horizontal distance between the curves and the vertical axis is the relative building drift for the particular combination of truss locations given by the curve designation and the story level. For example, let us assume that the relative deflection at the top is desired for a combination (20, 15), the numbers 20 and 15 representing the floors at which the upper and lower outriggers are located. To find the deflection index for this particular combination, the procedure is to select the curve with the designation 20, go down the vertical axis to level 15, and draw a horizontal line from this level to curve 20. The required relative top deflection is the horizontal distance between level 15 and curve 20 (distance HH¹ in Fig. 3.29). Similarly, the length KK¹ gives the relative deflection at the top for the combination (28, 4). It is seen from Fig. 3.29 that the relative location of the trusses has a significant effect on controlling the drift. Furthermore, it is evident that a deflection very nearly equal to the minimum can be achieved by placing the trusses at levels other than at their optimum locations. For the example building, a relative deflection of 0.15, which differs negligibly from the optimum value of 0.13, is achieved by placing the outriggers at (40, 23), (32, 33), etc.

3.6.5. Recommendations for Optimum Locations of Belt and Outrigger Trusses

Based on the approximate method presented thus for, the following recommendations are made for the general arrangement of the structure, particularly for estimating the optimum levels of the outriggers for minimizing the drift:

- The optimum location for a single outrigger is, perhaps unexpectedly, not at the top. The reduction in the drift is about 50%, as compared to a maximum of 75% achievable by placing it at approximately half-height. However, since other architectural requirements take precedence in a structural layout, the benefits of placing a truss at the top are still worth pursuing.
- A two-outrigger structure appears to offer more options in the placements of outriggers. Reductions in building deflections close to the optimum results may be achieved with outriggers placed at levels entirely different from optimum locations. Thus, the engineer and architect have some leeway in choosing the outrigger locations. However, as a rule of thumb, the optimum location for a two-outrigger structure is at one-third and two-third heights. And for a three-outrigger system, they should be at the one-quarter, one-half, and three-quarter heights, and so on. Therefore, for the optimum performance of an n-outrigger structure, the outriggers should be placed at $1/n + 1$, $2/n + 1$, $3/n + 1$, $4/n + 1$,…, $n/n + 1$ height locations. For example, in an 80-story building with four outriggers (i.e., $n = 4$), the optimum locations are at the 16th, 32nd, 48th, and 64th levels. A summary of the recommendations is shown in Fig. 3.30.

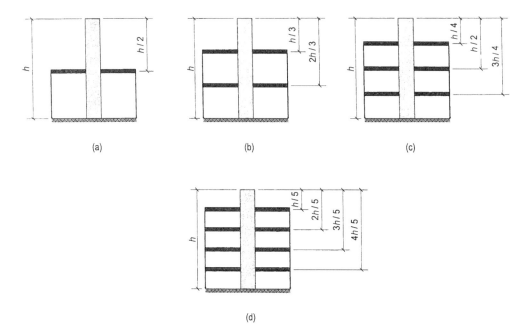

Figure 3.30.   Optimum locations of outriggers: (a) single outrigger; (b) two outriggers; (c) three outriggers; (d) four outriggers.

3.7.   FRAMED TUBE SYSTEM

In the simplest terms, a framed tube can be defined as a three-dimensional system that utilizes the entire building perimeter to resist lateral loads. A necessary requirement to create the wall-like three-dimensional structure is to place columns on the building exterior relatively close to each other, joined by deep spandrel girders. In practice, columns are placed 10 ft (4 m) to as much as 20 ft (6.1 m) apart, with spandrel depths varying from about 3 to 5 ft (0.90 to 1.52 in.). A somewhat different type of tube, often referred to as a braced tube, permits greater spacing of columns. As the name implies, the tube has diagonal or K-type of bracing at the building exterior. Yet another type of tube system called the bundled tube uses two or more tubes tied together to form a single, multicell tube. However, the framed tube is by far the most popular system because rectangular windows can be accommodated in this design.

3.7.1.   Behavior

To understand the behavior of a framed tube, consider a building shown in Fig. 3.31 in which the entire lateral resistance is provided by closely spaced exterior columns and deep spandrel beams. The floor system, typically considered rigid in its own plane, distributes the lateral load to various elements according to their stiffness. Its contribution to lateral resistance in terms of out-of-plane stiffness is considered negligible as in other systems. The lateral load-resisting system thus comprises four orthogonally oriented, rigidly jointed frame panels forming a tube in plan, as shown in Fig. 3.32.

The "strong" bending direction of the columns is typically aligned along the face of the building, in contrast to a typical transverse rigid frame where it is aligned perpendicular to the face. The frames parallel to the lateral load act as webs of the perforated tube, while

Moment Deep Closely spaced
connection spandrels columns

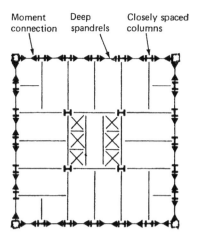

Figure 3.31. Schematic plan of framed tube.

the frames normal to the load act as the flanges. Gravity loads are resisted partly by the exterior frames and partly by interior columns. When subjected to bending, the primary mode of action is that of a vertical cantilever tube, in which the columns on opposite sides of the neutral axis are subjected to tensile and compressive forces. In addition, the frames parallel to the direction of the lateral load are subjected to the in-plane bending and the shearing forces associated with an independent rigid frame. The discrete columns and spandrels may be considered, in a conceptual sense, equivalent to a hollow tube cantilevering from the ground with a linear axial stress distribution, as shown in Fig. 3.33.

Although the structure has a tubelike form, its behavior is much more complex than that of a solid tube. Unlike a solid tube, it is subjected to the effects of shear lag, which

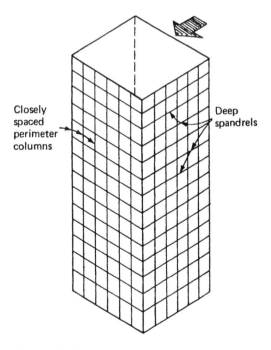

Closely
spaced
perimeter
columns

Deep
spandrels

Figure 3.32. Isometric view of framed tube.

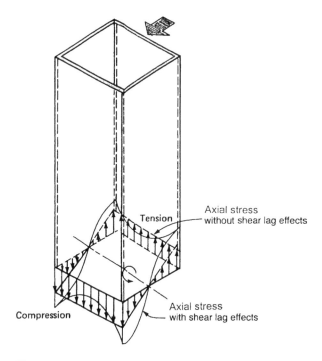

Figure 3.33.   Axial stress distribution in square hollow tube with and without shear lag.

have a tendency to modify the axial distribution in the columns. The influence of shear lag, considered in the following section, is to increase the axial stresses in the corner columns and reduce those in the inner columns of both the flange and the web panels, as shown in Fig. 3.33.

Figure 3.34 shows examples of free-form tubular configurations. Although in simplistic terms, the tube is similar to a hollow cantilever, in reality its response to lateral loads is in a combined bending and shear mode. The overall bending of the tube is due to axial shortening and elongation of the columns, whereas the shear deformation is due to bending of individual columns and spandrels. The underlying principle for an efficient design is to eliminate or minimize the shear deformation.

3.7.2.   Shear Lag Phenomenon

Consider Fig 3.35, in which columns of a tubular building are noted as T and C. T denotes a column in tension while C denotes a column in compression. The primary resistance to lateral loads comes from the web frames with the T columns in tension and the C columns in compression (Fig. 3.35). The web frames are subjected to the usual in-plane bending and racking action associated with an independent rigid frame. The primary action is modified by the flexibility of the spandrel beams, which causes the axial stresses in the corner columns to increase and those in the interior columns to decrease.

The principal interaction between the web and flange frames occurs through the axial displacements of the corner columns. When column C, for example, is under compression, it will tend to compress the adjacent column C_1 (Fig. 3.35) because the two are connected by the spandrel beams. The compressive deformations of C_1 will not be identical to that of corner column C since the connecting spandrel beam will bend. The axial deformation of C_1 will be less, by an amount depending on the stiffness of the connecting beam. The deformation of column C_1 will, in turn, induce compressive deformations of

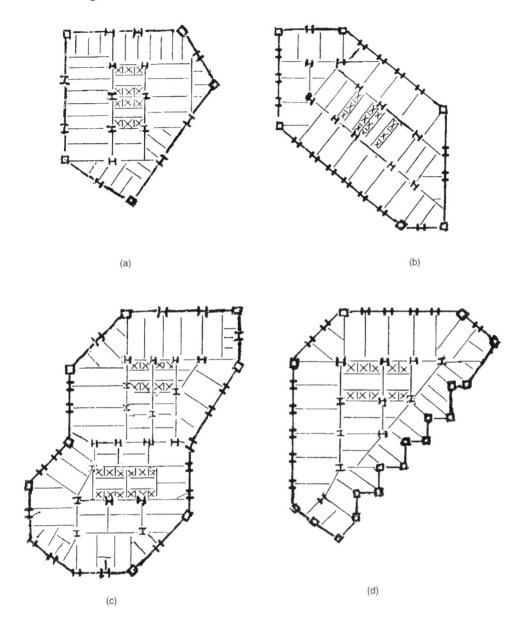

Figure 3.34. Free-form tubular configurations.

the next inner column C_2, but the deformation will again be less. Thus, each successive interior column will experience a smaller deformation and hence a lower stress than the outer ones. The stresses in the corner column will be greater than those from pure tubular action, and those in the inner columns will be less. The stresses in the inner columns lag behind those in the corner columns, hence the term shear lag.

The difference between stress distribution as predicted by ordinary beam theory, which assumes that plane sections remain plane, and the actual distribution due to shear lag is illustrated in Fig. 3.35. Because the column stresses are distributed less effectively than in an ideal tube, the moment resistance and the flexural rigidity are reduced. Thus, although a framed tube is highly efficient, it does not fully utilize the potential stiffness and strength of the structure because of the effects of shear lag.

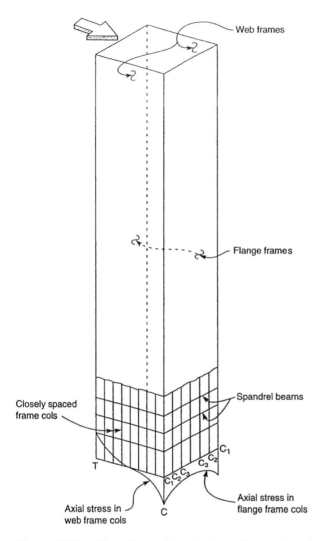

Figure 3.35. Shear lag in framed tube. Observe that the axial stresses are distributed quite differently from those predicted by engineer's bending theory.

3.8. IRREGULAR TUBE

The framed tube concept can be executed with any reasonable arrangement of column and spandrels around the building perimeter. However, noncompact plans and plans with reentrant corners considerably reduce the efficiency of the system. For framed tubes, a compact plan may be defined as one with an aspect ratio not greater than 1.5 or so. Elongated plans with larger aspects ratios impose considerable premium on the system because: 1) in wind-controlled design, the elongated building elevation acts like a sail collecting large wind loads; 2) the resulting shear forces most usually require closer spacing and/or larger columns and spandrels parallel to the wind; and 3) shear lag effects are more pronounced, especially for columns oriented perpendicular to the direction of wind.

In similar manner, a sharp change in the tubular form results in a less efficient system because the shear flow must pass around the corners solely through axial shortening of the columns. Also, a secondary frame action at these locations alters the load distribution in the framed tube columns.

3.9. TRUSSED TUBE

A trussed tube system improves on the efficiency of the framed tube by increasing its potential for use in taller buildings and allowing greater spacing between the columns. This is achieved by adding diagonal bracing at the faces of the tube to virtually eliminate the shear lag in both the flange and web frames.

The framed tube, as discussed previously, even with its close spacing of columns is somewhat flexible because the high axial stresses in the columns parallel to the lateral loads cannot be transferred effectively around the corners. For maximum efficiency, the tube should respond to lateral loads with the purity of a cantilever, with compression and tension forces spread uniformly across the windward and leeward faces. The framed tube, however, behaves more like a thin-walled tube with openings. The axial forces tend to diminish as they travel around the corners, with the result that the columns in the middle

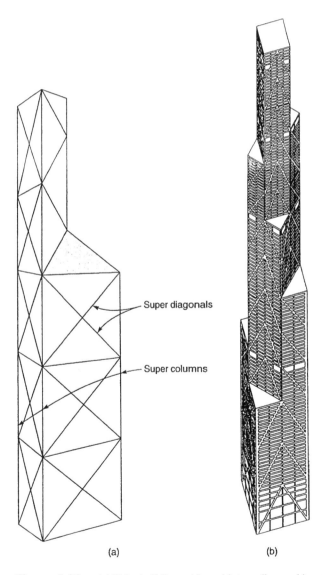

(a)   (b)

Figure 3.36.   (a) Tube building with multistory diagonal bracing; (b) rotated square tube with super diagonals. (Adapted from an article by Mahjoub Elnimeiri, published in *Civil Engineering Journal*.)

of the windward and leeward faces may not sustain their fair share of compressive and tensile forces. This effect, referred to previously as the shear lag, limits the framed tube application to 50- or 60-story buildings unless the column spacing is very close, as was the case with the 109-story World Trade Center Towers, New York, which had columns at 3.8 ft (1.0 m).

Addition of diagonal braces, as shown in Fig. 3.36 is by far the most usual method of increasing the efficiency of a framed tube. The fascia diagonals interact with the trusses on the perpendicular faces to achieve a three-dimensional behavior, virtually eliminating the effects of shear lag in both the flange and web frames. Consequently, the spacing of the columns can be greater and the size of the columns and spandrels less, thereby allowing larger windows than in a conventional tube structure. The bracing also contributes to the improved performance of the tube in carrying gravity loading. Differences between gravity load stresses in the columns are evened out by the braces transferring axial loading from the more highly to the less stressed columns.

An example of an exterior braced steel building is shown in Fig. 3.37. The building has six eight-story deep chevron braces on each facade that collect about half the gravity loads and resist the entire lateral load above the transfer level.

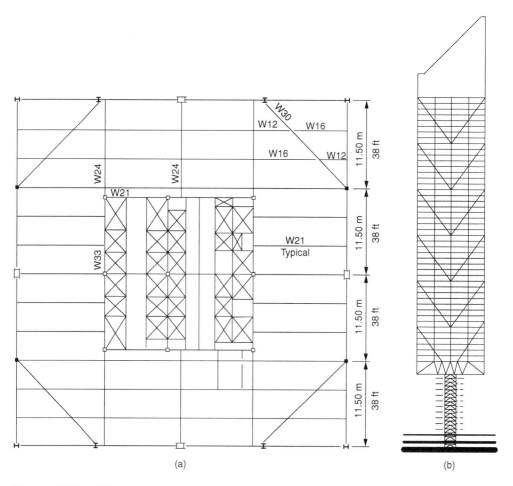

Figure 3.37. Citicorp Center (Structural Engineers, LeMessurier Consultants, Inc.): (a) typical floor framing plan; (b) elevation; (c) lateral bracing system.

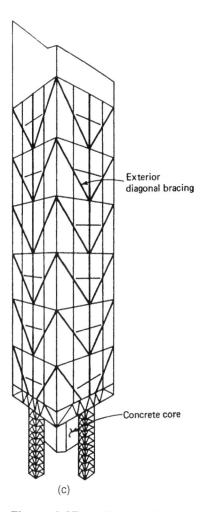

Exterior
diagonal bracing

Concrete core

(c)

Figure 3.37c. *(Continued).*

3.10. BUNDLED TUBE

A bundled tube typically consists of a number of individual tubes interconnected to form a multicell tube, in which the frames in the lateral load direction resist the shears, while the flange frames carry most of the overturning moments. The cells can be curtailed at different heights without diminishing structural integrity. The torsional loads are readily resisted by the closed form of the modules. The greater spacing of the columns and shallower spandrels permitted by the more efficient bundled tube structure provide for larger window openings than are allowed in a single-tube structure.

The shear lag experienced by conventional framed tubes is greatly reduced by the addition of interior framed web panels across the entire width of the building. When the building is subjected to bending under the action of lateral forces, the high in-plane rigidity of the floor slabs constrains the interior web frames to deflect equally with the exterior web frames. Thus, the shear carried by each web frame is proportional to its lateral stiffness. Since the end columns of the interior webs are activated directly by the webs, they are more highly stressed than in a single tube where they are activated indirectly by the exterior web through the flange frame spandrels. Consequently, the presence of the interior webs reduces substantially the nonuniformity of column forces caused by the shear lag. The vertical

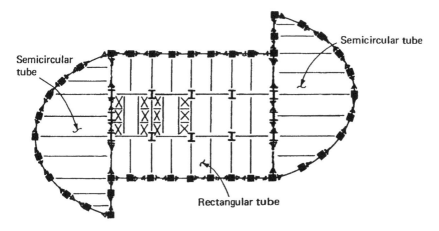

Figure 3.38A. Bundled tube, schematic plan.

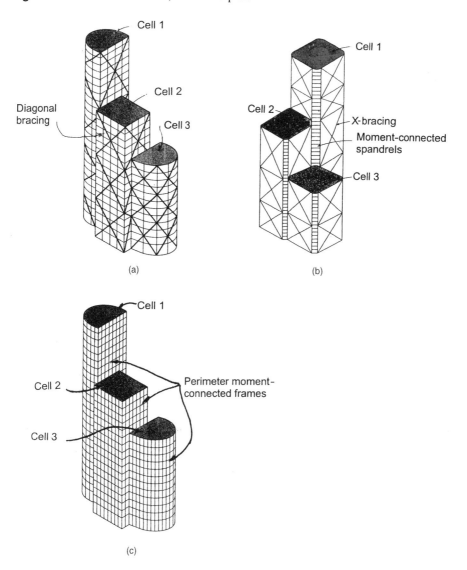

Figure 3.38B. Bundled tube concept: (a) perimeter diagonal bracing; (b) X-bracing with moment-connected spandrels; (c) perimeter moment frames.

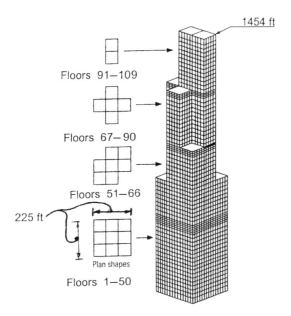

Figure 3.39. Bundled tube structural system; Sears Tower, Chicago. Building height: 1454 ft.

stresses in the normal panels are more nearly uniform, and the structural behavior is much closer to that of a braced tube than a framed tube.

Because a bundled tube is configured from a layout of individual tubes, it is possible to achieve a variety of floor configurations by simply terminating a given tube at any desired level. See Fig. 3.38A. Schematic elevations of structural systems consisting of perimeter diagonal bracing and moment-connected frames are shown in Figs. 3.38Ba, b, and c. Figure 3.39 shows the bundled tube system for the 1454-ft tall Sears Tower consisting of nine tubes at the bottom, with only two of the constituent tubes rising to the top. The building designed by the Chicago office of Skidmore, Owings, and Merrill was completed in 1974.

3.11. SEISMIC DESIGN

In the event of a large earthquake, structural steel buildings, similar to their counterpart concrete buildings, are expected to dissipate seismic energy through inelastic deformations of certain designated structural elements. Their design is governed by the seismic provisions given in AISC 341-02 (hereafter referred to as AISC-Seismic), and the general design provisions are given in AISC Load and Resistance Factor Design (LRFD) 1999 Specifications.

The AISC-Seismic provisions are intended to be compulsory for buildings in seismic design category (SDC) D and above. For buildings in SDC A, B, or C, the designer has a choice to either exclusively use AISC LRFD with a typical R factor of 3, or may choose to assign a higher R factor and detail the system following the requirements of AISC-Seismic.

When designing buildings to resist earthquake motions, each building is categorized according to its use or occupancy to establish its potential earthquake hazard, and then assigned to one of three seismic use groups, depending upon occupancy or use. Next, based on the seismic use group and seismicity of the site, buildings are assigned to one of seven seismic design categories, A through F. Those in areas of low-to-moderate seismicity are generally assigned to SDC A, B, or C. AISC-Seismic provisions are non-mandatory for these buildings. However, special seismic provisions are mandatory for buildings in areas of high seismicity, assigned typically to SDC D, E, or F.

3.11.1. Concentric Braced Frames

Just about any system that has identifiable load paths for gravity and lateral loads is permitted without any height limits for buildings in regions of low seismicity or assigned to SDC A, B, or C. These may be designed by using the provisions of American Institute of Steel Construction (AISC), Load and Resistance Factor Design (LRFD), Allowable Stress Design (ASD), and Hollow Structural Section (HSS) specifications, all of which have been adapted by provisions of the American Society of Civil Engineers (ASCE) 7-02, National Fire Protection Association (NFPA) 5000, and International Building Code (IBC) 2003.

The design of braced frames in regions of high seismicity, or those assigned to SDC D, E, or F, is performed according to the provisions of AISC 341-02, "Seismic Provisions for Structural Steel Buildings," commonly referred to as AISC-Seismic. A brief discussion of the seismic provisions of this publication, including the salient characteristics of braced frame design, are as follows:

- A variety of braced frame configurations are permitted by AISC-Seismic. Some of these are shown in Fig. 3.40a.
- AISC-Seismic permits seismic design of braced frames either as an ordinary concentric braced frame (OCBF) or as a special concentric braced frame (SCBF). The only difference between the two is in detailing of the connections and some prescriptive requirements for SCBF intended to enable them to respond to seismic forces with greater ductility.
- Both the V- and inverted V-braced frames, often referred to as chevron braces, have been poor performers during past earthquakes because of buckling of braces and excessive flexure of beam at midspan where the braces intersect the beam. Buildings with single or multistory X-braces or V-braces with zipper columns are deemed better performers and hence, should be considered for braced frame configurations in high seismic zones.
- Braced frames with single diagonals are also permitted by AISC-Seismic. However, there is a heavy penalty since the braces must be designed to resist 100% of the seismic force in compression, unless multiple single-diagonal braces are provided along a given brace frame line.
- A preferred but difficult-to-achieve behavior in an SCBF is the in-plane buckling of the brace. Given a choice, a brace would buckle out-of-plane rather than buckling in the plane of the braced frame. This is so because the in-plane buckling is inhibited because: 1) placement of braces in a flat position is generally not permitted for architectural reasons; and 2) the presence of infill metal studs above and below the braces adds considerable in-plane stiffness to the braces. Recognizing these features, both UBC 1997 and AISC-Seismic permit out-of-plane buckling of braces, provided an uninterrupted yield line can develop in gusset plates at each end of the brace connection. This is achieved by prescribing the following detailing requirements:
  1. Provide a minimum of $2t$ and maximum of $4t$ offset from the end of brace to the yield line, where t is the thickness of gusset plate.
  2. Provide a 1-in. minimum offset from the brace to the edge of gusset plate.
  3. Isolate gusset plate yield line from the floor slab.

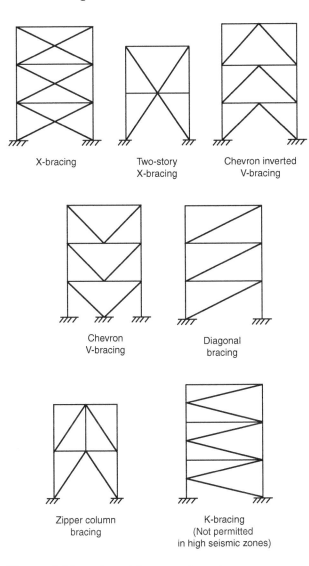

Figure 3.40a. Braced frame configurations.

- In detailing the SCBF's gusset plate, the potential restraint that occurs due to the floor slab must be considered. To keep the gusset plate as small as possible, it may be isolated from the slab to allow the yield line to extend below the concrete surface. Note that the entire gusset plate does not have to be isolated, just the area where the yield line occurs. A compressible material 2- to 3-in. thick on each side may be used to isolate the plate, as shown in Fig. 3.40b.
- Beams or columns of braced frames should not be interrupted at the brace intersections. This is to ensure out-of-plane stability of the bracing system at those locations. However, mere continuity of columns or beams at the brace intersections may not be sufficient to provide the required stability. Typical practice is to provide perpendicular framing that engages a diaphragm to provide out-of-plane strength and stiffness, and resistance to lateral torsional buckling of beams.

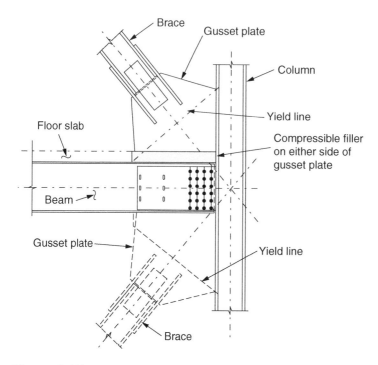

Figure 3.40b. Gusset plate isolation from floor slab.

- SCBFs are expected to achieve trilinear hysteretic behavior in a large earthquake by going through three ranges of displacement of its brace: the elastic range, the postbuckling range, and the tensile yielding range.
- Basic design concept for an OCBF is to limit its response to an elastic behavior. Therefore, a higher seismic force (using a lower value of $R = 5$ versus $R = 6$ for an SCBF) and lower brace capacity are used in the design. Increase in capacity is achieved by limiting $Kl/r \leq 720/\sqrt{F_y}$ versus $1000/\sqrt{F_y}$ for an SCBF.
- Compactness requirements for braces are the same for OCBFs and SCBFs.
- Basic design concept for an SCBF is to mitigate brittle modes of failure by controlling its behavior through better detailing. Therefore, connections are designed to develop yield capacity of brace.
- In a chevron-configured SCBF, instead of increasing earthquake load by 1.5, the beam is designed for unbalanced load requirements by assuming one brace at the beam intersection fails in compression. This is not a design requirement for an OCBF.
- Special requirements apply to the design of chevron-braced frames. Because braces meet at the midspan of beams, the vertical force resulting from the unequal compression and tension strengths of these braces can have a considerable impact on the cyclic behavior of the frame. That vertical force introduces flexure in the beam, and possibly a plastic hinge, producing a plastic collapse mechanism. Therefore, beams in chevron-braced frames must be continuous between columns.
- Seismic provisions require that beams in chevron-braced frames be capable of resisting their tributary gravity loads, neglecting the presence of the braces, and that each beam in an SCBF be designed to resist a maximum unbalanced

vertical load calculated using full-yield strength or the brace in tension, and 30% of the brace buckling strength in compression. In an OCBF, this latter provision need not be considered. However, braces in the OCBF must be designed to have 1.5 times the strength required by load combinations that include seismic forces, which is equivalent to designing chevron-braced frames for a smaller value of R to compensate for their lower ductility.

- To prevent instability of a beam bottom flange at the intersection of the braces in a chevron-braced frame, the top and bottom flanges of beams in both SCBF and OCBF must be designed to resist a lateral force equal to 2% of the nominal beam flange strength (i.e., $0.02A_f F_y$). This requirement is best met by the addition of a beam perpendicular to the chevron-braced frame.

- The preceding concepts explain why a K-type braced frame configuration is prohibited in high seismic regions. The unequal buckling and tension-yielding strengths of the braces would create an unbalanced horizontal load at the midheight of the columns, jeopardizing the ability of the column to carry gravity loads if a plastic hinge forms at the midheight of the column.

- Concentrically braced frames are expected to undergo inelastic response during large earthquakes. Specially designed diagonal braces in these frames can sustain plastic deformations and dissipate hysteretic energy in a stable manner through successive cycles of buckling in compression and yielding in tension. The preferred design strategy is, therefore, to ensure that plastic deformations only occur in the braces, leaving the columns, beams, and connections undamaged, thus allowing the structure to survive strong earthquakes without losing gravity-load resistance.

- The plastic hinge that forms at midspan of a buckled brace may develop large plastic rotations that could lead to local buckling and rapid loss of compressive capacity and energy dissipation characteristic during repeated cycles of inelastic deformations. Locally buckled braces can also suffer low-cycle fatigue and fracture after a few cycles of severe inelastic deformations, especially when braces are cold-formed rectangular hollow sections. For these reasons, braces in SCBFs must satisfy the width-to-thickness ratio limits for compact sections. For OCBFs, braces can be compact or noncompact, but not slender. In particular, the width-to-thickness ratio of angles (b/t), the outside-diameter–to–wall-thickness ratio of unstiffened circular hollow sections (D/t), and the outside-width–to–wall-thickness ratio of unstiffened rectangular sections must not exceed $52/\sqrt{F_y}$, $1300/\sqrt{F_y}$, and $110/\sqrt{F_y}$, respectively. Note that the AISC- Seismic provisions define b for rectangular hollow sections as the "out-to-out width," not the flat-width equal to $b - 3t$, as defined in the AISC Allowable Stress Design Specifications (AISC 1989).

- When a brace is in tension, net section fracture and block shear rupture at the end of the brace must be avoided. Likewise, the brace connections to beams and columns must be stronger than the braces themselves. Using capacity design, calculation of brace strength must recognize that the expected yield strength of the brace, F_{ye}, will typically exceed its specified minimum yield strength, F_y. Thus, connections must be designed to resist an axial force equal to $R_y F_y A_g$, where R_y is the ratio of expected yield strength to specified yield strength F_y. See Table 3.1. Connections must also be able to resist the forces due to buckling of the brace. If strong connections permit the development of a plastic hinge at each end of a brace, they should be designed to resist a moment equal to $1.1R_y M_p$ of the brace in the direction of buckling. Otherwise, the connecting elements will themselves yield in flexure (such as gussets out of their plane); these must then be designed to resist the maximum brace compression force in a stable manner while undergoing the large plastic rotations that result from brace buckling. Providing a clear distance of twice the plate thickness between the end of the brace and the assumed line of restraint for the gusset plate permits plastic rotations and precludes plate buckling.

- Beams and columns in braced frames must be designed to remain elastic when braces have reached their maximum tension or compression capacity ($1.1R_y$ times the nominal strength) to preclude inelastic response in all components except the braces. This requirement could be too severe for columns of a multistory frame because the braces along the height of the frame do not necessarily reach their capacity simultaneously during an earthquake. AISC-Seismic provisions address this issue using special load conditions with the further specification that the maximum axial tension forces in columns need not be taken larger than the value corresponding to foundation uplift. For SCBFs, the provisions also require that columns satisfy the same width-to-thickness ratio limits as braces.

- Partial penetration groove welds in column splices have been observed to fail in a brittle manner. When a welded column splice is expected to be in tension under the loading combination, the AISC-Seismic provisions mandate that the partial joint penetration groove welded joints in SCBFs be designed to resist 200% of the strength required by elastic analysis using code-specified forces. Column splices also need to be designed to develop at least the nominal shear strength of the smaller connected member and 50% of the nominal flexural strength of the smaller connected section.

3.11.1.1. *Ordinary Concentric Braced Frame (OCBF)*

ASCE 7-02, Seismic Coefficients
OCBF
   $R = 5, \Omega_o = 2.0$ No height limit for SDC A, B, or C
                              Maximum height limit for SDC D or E = 35 ft
                              Not permitted for SDC F

OCBF in dual systems with special moment-resisting space frames; OCBF + SMRF
   $R = 7, \Omega_o = 2.5$ No height limit for SDC A, B, or C
                              Not permitted for SDC D, E, or F

OCBF in dual systems with intermediate moment frames and with bearing wall systems: Not allowed.

Ordinary concentric braced frames (OCBFs) are designed for the maximum anticipated force so as to remain essentially elastic under a seismic event. Therefore, the need for significant ductility in the design of members and their connections is not anticipated.

When the effects of gravity loads are additive, the ultimate load combination is given by

$$U = 1.2D + f_1L + 0.2S + E$$

where

D = dead load

L = floor live load

S = snow load

f_1 = 1.0 for floors in garages and places of public assembly and for floor loads in excess of 100 lb/ft²

= 0.5 for other live loads

$E = \rho Q_E + 0.2\,S_{DS}D$

Ω_o = structure overstrength factor

Q_E = effect of horizontal seismic forces

ρ = redundancy factor

S_{DS} = short-period spectral response acceleration

When the effects of gravity and seismic loads counteract, the ultimate load combination is given by

$$U = 0.9D + 1.6H + E$$

where H = hydrostatic load.

The design strength of brace connections shall not be less than the expected tensile strength of the brace as determined by

$$P_{ut} = R_y F_y A_g$$

where

R_y = ratio of the expected yield stress to the minimum specified yield strength

F_y = specified minimum yield stress of the type of steel used

A_g = gross area of section

Values of R_y given in AISC-Seismic Table I-6-1 are reproduced in Table 3.1.

To reduce the likelihood of buckling causing large deflections in the floor beam, bracing members in a chevron configuration should be designed with a slenderness ratio not exceeding

$$Kl/r = 4.23 \sqrt{\frac{E_s}{F_y}}$$

TABLE 3.1 Values of R_y

Application	Grade	R_y[a]
Hot-rolled structural W-sections, angles, bars	ASTM A36	1.5
	ASTM A572 Grade 42	1.3
	ASTM A992	1.1
	All other grades	1.1
Hollow structural sections	ASTM A500, A501, A618, and A847	1.3
	All other grades	1.3
Pipes	ASTM A53	1.4
Plates and all other products		1.1

[a] R_y is the ratio of expected yield strength to minimum specified yield strength F_y.
(From AISC-Seismic Table I-6-1.)

where

    l = length of the bracing member

    r = governing radius of gyration

    K = effective length factor

    E_s = modulus of elasticity of steel = 29,000 ksi (200,000 Mpa)

    F_y = specified minimum yield stress of the type of steel to be used, ksi. Yield stress in LRFD denotes either the minimum specified yield point for those steels that have a yield point, or the specified yield strength for those steels that do not have a yield point.

Thus, the limiting slenderness ratio is expressed as

$$Kl/r = \frac{4.23 \times \sqrt{29,000}}{\sqrt{F_y}} = \frac{720}{\sqrt{F_y}}$$

Design Example

Given. A two-story OCBF shown in Fig. 3.40c that forms part of the building frame system in SDC E. The axial loads on the ground floor brace B1 are as follows:

Dead load D = 30 kips

Live load L = 15 kips

Seismic force Q_E = ±80 kips

Snow load S = 0 kips

Hydrostatic load H = 0

The redundancy coefficient ρ = 1.1.

Mapped two-second såpectral acceleration, S_{DS} = 0.826 g.

Required. Determine a pipe section for brace B1, ASTM A53 Grade B steel; F_y = 35 ksi, F_u = 60 ksi

Solution. When gravity loads are additive, the basic load combination including earthquake effects E, is given by

$$U = 1.2D + 0.5L + 0.2S + E$$

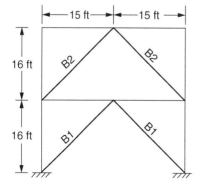

Figure 3.40c.   Ordinary concentric brace frame (OCBF) example.

where

$$E = \rho Q_E + 0.2\,S_{DS}D$$
$$U = 1.2D + 0.5L + 0.2S + \rho Q_E + 0.2S_{DS}D$$
$$= 1.2 \times 30 + 0.5 \times 15 + 0.2 \times 0 + 1.1 \times 80 + 0.2 \times 0.826 \times 30$$
$$= 136.5 \text{ kips, compression governs}$$

When gravity loads counteract seismic loads, the basic load combination is given by

$$U = 0.9D + 1.6H + 1.0E$$

where

$$E = -\rho Q_E - 0.2S_{DS}D$$
$$U = 0.9D + 1.6H - \rho Q_E - 0.2S_{DS}D$$
$$= 0.9 \times 30 + 1.6 \times 0 - 1.1 \times 80 - 0.2 \times 0.826 \times 30$$
$$= -66 \text{ kips, tension}$$

The given frame in SDC E has a height of 32 ft, which is less than the maximum permitted height of 35 ft. Therefore OK.

The unbraced length of the brace B1, using centerline dimensions, is

$$l = \sqrt{16^2 + 15^2} = 21.93 \text{ ft, use 22 ft for the design.}$$

The effective length factor k for the brace, assuming hinged ends, is equal to 1.0.

The effective length of the brace is

$$Kl = 1.0 \times 22 = 22 \text{ ft}$$

The design strength in axial compression is given by

$$\phi_c P_n = \phi_c A_g F_{cr} \qquad\qquad \text{(LRFD E 2-1)}$$

where

ϕ_c = resistance factor for compression
  $= 0.85$
A_g = gross area of member
F_{cr} = critical stress
P_n = nominal axial strength

From LRFD Table 4.8, select an 8-in.-diameter standard steel pipe that has a design strength in axial compression of 165 kips for an effective length of 22 feet.

$$\phi_c P_n = 165 \text{ kips}$$
$$> P_{uc} = 136.5 \text{ kips} \qquad \text{OK}$$

The section properties of an 8-in.-diameter standard pipe are given in LRFD Table 4.8 as

$A_g = 8.6 \text{ in.}^2$
$r = 2.94 \text{ in.}$
$t = 0.332 \text{ in.}$
$D = 8.625 \text{ in.}$
$F_y = 35 \text{ ksi}$
$F_u = 60 \text{ ksi}$

The diameter-to-thickness ratio of a circular hollow section is limited by AISC-Seismic to a maximum value of

$$D/t = \lambda_p = 0.044 \frac{E_s}{F_y} = \frac{0.044 \times 29,000}{F_y}$$
$$= 1300/F_y$$
$$= 1300/35$$
$$= 37.1 \qquad \qquad \text{(AISC-Seismic Table I-8-1)}$$

The actual diameter-to-thickness ratio is

$$D/t = \frac{8.63}{0.332} = 26$$
$$< \lambda_p \qquad \text{OK}$$

AISC-Seismic requires bracing members in a chevron configuration of V or inverted V to be designed with a slenderness ratio not exceeding

$$\frac{Kl}{r} = 4.23 \sqrt{\frac{E_s}{F_y}}$$
$$= 4.23 \sqrt{\frac{29,000}{F_y}}$$
$$= \frac{720}{\sqrt{F_y}}$$
$$= \frac{720}{\sqrt{46}}$$
$$= 106 \qquad \qquad \text{(AISC-Seismic Part I, Sec. 14.2)}$$

The actual slenderness ratio is

$$\frac{Kl}{r} = \frac{22 \times 12}{2.94}$$
$$= 89.8 < 106. \qquad \text{OK.}$$

Hence, 8-in.-diameter. Standard weight pipe satisfies design requirements.

3.11.1.2. *Special Concentric Braced Frame (SCBF)*

ASCE 7-02, Seismic Coefficients
SCBF
$R = 6$, $\Omega_o = 2.0$ No height limit for SDC A, B, or C
                           Maximum height for SDC D or E = 160 ft
                           Maximum height limit for SDC F = 100 ft

SCBF in dual systems with SMRF; SCBF + SMRF
$R = 8$, $\Omega_o = 2.5$ No height limit for any SDC

SCBF in dual systems with IMF: SCBF + IMF

$R = 6$, $\Omega_o = 2.5$ No height limit for SDC A, B, or C

Maximum height for SDC D or E = 160 ft

Maximum height for SDC E = 100 ft

Not permitted for SDC F

An SCBF has increased ductility over OCBF due to lesser strength degradation when compression braces buckle. Hence, it is expected to withstand substantial inelastic excursions when subjected to forces resulting from ground motions of large earthquakes. The design requirements are as follows:

- To reduce the possibility of global buckling, bracing members are to have

$$\frac{Kl}{r} \le 5.87\sqrt{\frac{E_s}{F_y}} = 5.87\frac{\sqrt{29,000}}{\sqrt{F_y}} = \frac{1000}{\sqrt{F_y}}$$

  and a limiting width-thickness ratio as given in Table 3.2 (AISC-Seismic Table 1.8.1).
- The required compressive strength of a brace must not exceed $\phi_c P_n$.
- To forestall a cumulation of inelastic deformation in one direction, and to preclude the tension-only diagonal bracing, AISC-Seismic stipulates that neither the sum of the horizontal components of the compressive member forces nor the sum of the horizontal components of tensile member forces along a given line of bracing shall exceed 70% of the lateral force. This provision attempts to balance the compressive and tensile resistance across the width and the breadth of a building. An exception is provided for cases where an essentially elastic response is expected because of oversized bracing members.
- The tensile strength of the connection shall be based upon tensile rupture on the effective net section and block shear. The flexural strength in the anticipated direction of brace buckling, shall be equal to $R_y M_p$ of the brace about the critical buckling axis. The design of the gusset plate must include consideration of buckling.
- Beams receiving V-type or inverted V-type bracing must be
  1. Continuous between columns.
  2. Designed for gravity loads, assuming that bracing is not present.
  3. Designed for the maximum unbalanced vertical load equal to $(R_y P_y - 0.3\ \phi_c P_n)$.
  4. Braced at the top and bottom flanges or designed to support a lateral force that is equal to 2% of the nominal beam strength $F_y b_f t_{bf}$.
- In concentrically braced frames, particularly in those not used as part of a dual system, the bracing members carry most of the seismic story shear. Therefore, bracing connections must be designed to prevent failure by out-of-plane buckling of gusset plate and brittle fracture of the brace. Although connections with stiffness in two directions, such as crossed gusset plates, can be detailed, satisfactory performance can be ensured by allowing the gusset plate to develop restraint-free plastic rotations.
- In a major seismic event, columns in braced frames can experience significant bending beyond the elastic range after buckling and yielding of the braces. Although their bending strength is not used in design, columns in SCBFs are required to have adequate compactness and flexural strength in order to maintain their lateral strength during large cyclic deformations of the frame. Two requirements are given in AISC 341-02 (AISC-Seismic).

TABLE 3.2 Limiting Width-Thickness Ratios λ_{ps} for Compression Elements

Description of element	Width-thickness ratio	Limiting width-thickness ratios λ_{ps} (seismically compact)
Unstiffened elements		
Flanges of I-shaped rolled, hybrid, or welded beams[a,b,f,h]	b/t	$0.30\sqrt{E_s/F_y}$
Flanges of I-shaped rolled, hybrid or welded columns[a,c]	b/t	$0.30\sqrt{E_s/F_y}$
Flanges of channels, angles, and I-shaped rolled, hybrid, or welded beams and braces[a,d,h]	b/t	$0.30\sqrt{E_s/F_y}$
Flanges of I-shaped rolled, hybrid, or welded columns[a,e]	b/t	$0.38\sqrt{E_s/F_y}$
Flanges of H-pile sections	b/t	$0.45\sqrt{E_s/F_y}$
Flat bars[g]	b/t	2.5
Legs of single angle, legs of double-angle members with separators, or flanges of tees[h]	b/t	$0.30\sqrt{E_s/F_y}$
Webs of tees[h]	d/t	$0.30\sqrt{E_s/F_y}$
Stiffened elements		
Webs in flexural compression in beams in SMF, Section 9, unless noted otherwise[a]	h/t_w	$2.45\sqrt{E_s/F_y}$
Other webs in flexural compression[a]	h/t_w	$3.14\sqrt{E_s/F_y}$
Webs in combined flexure and axial compression[a–f,g]	h/t_w	For $Pu/\phi_b P_y \le 0.125$ $$3.14\sqrt{\frac{E_s}{F_y}}\left(1-1.54\frac{P_u}{\phi_b P_y}\right)$$ For $Pu/\phi_b P_y > 0.125$ $$1.12\sqrt{\frac{E_s}{F_y}}\left(2.33-\frac{P_u}{\phi_b P_y}\right)$$
Round HSS in axial and/or flexural compression[d,h]	D/t	$0.044\,E_s/F_y$
Rectangular HSS in axial and/or flexural compression[d,h]	b/t or h/t_w	$0.64\sqrt{E_s/F_y}$
Webs of H-pile sections	h/t_w	$0.94\sqrt{E_s/F_y}$

[a] For hybrid beams, use the yield strength of the flange F_{yf} instead of F_y.

[b] Required for beams in SMF, Section 9.

[c] Required for columns in SMF, Section 9, unless the ratios from Eq. (9.3) are greater than 2.0 where it is permitted to use λ_p in LRFD Specification Table B5.1

[d] Required for beams and braces in SCBF, Section 13.

[e] It is permitted to use λ_p in LRFD Specification Table B5.1 for columns in STMF, Section 12 and EBF, Section 15.

[f] Required for link in EBF, Section 15.

[g] Diagonal web members within the special segment of STMF, Section 12.

[h] Chord members of STMF, Section 12.

(From AISC 341-02 Table I-8-1.)

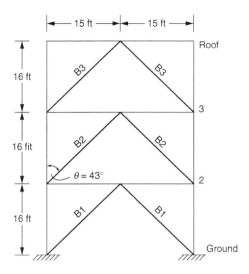

Figure 3.41. Special concentric brace frame (SCBF) example.

- Width-thickness ratios of columns shall meet the requirements specified for bracing members of an SCBF.
- Column splices must develop
  1. At least the nominal shear strength of the smaller column.
  2. At least 50% of the nominal flexural strength of the smaller column.

Design Example

Given. A three-story, two-bay SCBF shown in Fig. 3.41. It is part of a building frame system of a structure in seismic design category (SDC) D with a redundancy coefficient $\rho = 1.20$. The building is on a site with a short period mapped acceleration response $S_{DS} = 0.90$. The axial loads acting on the second story brace B2 are

Dead load $D = 40$ kips

Live load $L = 12$ kips

Snow load $S = 0$ kips

Hydrostatic load $H = 0$ kips

Seismic force $Q_E = \pm 90$ kips

Required. Select an appropriate tube section for the second-floor brace B2. Use ASTM A 500, Grade B, $F_y = 42$ ksi, F_u (minimum tensile stress) = 58 ksi, steel.

Solution. The factored compression load P_{uc} in the brace is

$$P_{uc} = 1.2D + 0.5L + 0.2S + \rho Q_E + 0.2S_{DS}D$$
$$= 1.2 \times 40 + 0.5 \times 12 + 0 + 1.2 \times 90 + 0.2 \times 0.9 \times 40$$
$$= 169.2 \text{ kips, compression} \leftarrow \text{governs.}$$

The factored tensile load P_{ut} on the brace is

$$P_{ut} = 0.9D + 1.6H - \rho Q_E - 0.2S_{DS}D$$
$$= 0.9 \times 40 + 0 - 1.2 \times 90 - 0.2 \times 0.9 \times 40$$
$$= -144 \text{ kips, tension}$$

The unbraced length of the brace, using centerline dimensions, is

$$l = H/\sin \theta$$
$$= H/\sin 43°$$
$$= 21.9 \text{ ft, use 22 ft}$$

The effective length factor for the brace, assuming hinged ends, is given by LRFD Table C–C2.1, item (d), as

$$K = 1.0$$

The effective length of the brace is

$$Kl = 1 \times 22 = 22 \text{ ft}$$

From LRFD Table 4.7, select a round HSS 8.625 × 0.322 in. with a design strength in axial compression of 171 kips for an effective length of 22 feet.

$$\phi_c P_n = 171 \text{ kips}$$
$$> P_{uc} = 162.2 \text{ kips} \quad \text{OK}$$

The section properties of an HSS 8.625 × 0.332 in. are given in LRFD Table 1.13 as

$$A = 7.85 \text{ in.}^2$$
$$r = 2.95 \text{ in.}$$
$$t = 0.3 \text{ in. (Note: Design thickness} = 0.93 \times \text{ nominal wall thickness)}$$
$$D = 8.625 \text{ in.}$$

The diameter-to-thickness ratio of a hollow section is limited by Table 3.2 to a maximum value of

$$= 1300/F_y$$
$$= 1300/35$$
$$= 37.1$$

The actual diameter-to-thickness ratio is

$$D/t = 8.625/0.3$$
$$= 28.75$$
$$< 37.1 \quad \text{OK}$$

AISC-Seismic requires bracing members in a chevron configuration to be designed with a slenderness ratio net exceeding

$$l/r = 1000/(F_y)^{0.5}$$
$$= 1000/(35)^{0.5}$$
$$= 169$$

The actual slenderness ratio is

$$l/r = (22 \times 12)/2.95$$
$$= 89.5$$
$$< 169 \quad \text{OK}$$

Hence, an HSS 8.625 × 0.332-in. brace satisfies all bracing requirements.

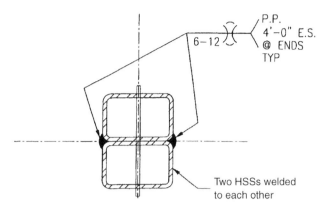

Figure 3.42. Built-up brace detail for use in an SCBF. (Courtesy of Louis Choi, S.E., John A. Martin & Assoc. Structural Engineers, Los Angeles, CA.)

In some practical applications, the ratio D/t of available hollow sections for a given load may not be able to satisfy the design criteria. In such cases, doubling of hollow structural sections, HSS, as shown in Fig. 3.42, may provide the required section properties.

Figures 3.43 through 3.50 show some typical details for braced frames.

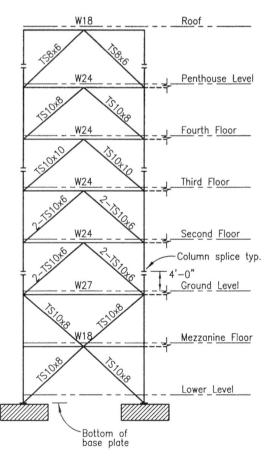

Figure 3.43. Braced frame with story-high chevron inverted-V braces. (Courtesy of Louis Choi, S.E., John A. Martin & Assoc. Structural Engineers, Los Angeles, CA.)

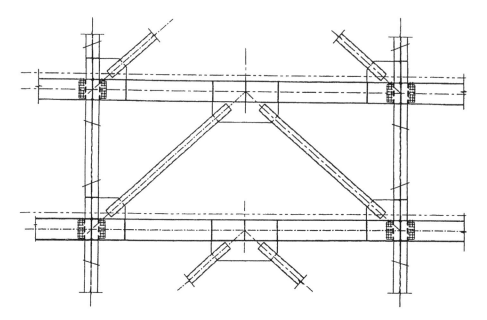

Figure 3.44. Schematic elevation of an OCBF showing gusset plates. (Courtesy of Louis Choi, S.E., John A. Martin & Associates Structural Engineers, Los Angeles, CA.)

Typical procedure for connection design is to use the Uniform Force Method (UFM, AISC, LRFD, Vol. II). The gusset dimensions are configured such that there are no moments at the connection interfaces: gusset-to-column and gusset-to-beam. Then, the plate and connection capacities are calculated and compared with the required capacities.

The gusset plate is designed to carry the compressive strength of the brace without buckling. Using the Whitmore method (UFM, AISC, LRFD, Vol. II), the effective plate width W_w at the critical section is calculated. The unsupported plate length L_u is taken as centerline length from the end of the brace to the edge of beam or column. An effective length factor, K, representative of the plate boundary conditions, is used to calculate KL_u/r.

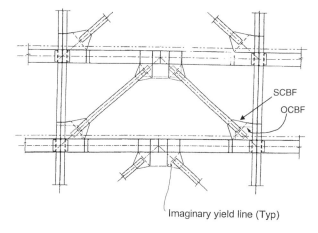

Figure 3.45. Schematic elevation of an SCBF showing imaginary yield lines in gusset plates. Also shown are the gusset plates for an OCBF. Observe the substantial increase in gusset plate sizes required for the SCBF. (Courtesy of Louis Choi, S.E., John A. Martin & Assoc. Structural Engineers, Los Angeles, CA.)

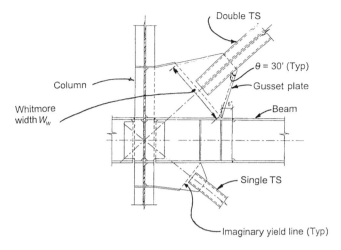

Figure 3.46. An SCBF connection detail at column. (Courtesy of Louis Choi, S.E., John A. Martin & Associates Structural Engineers, Los Angeles, CA.)

Based on the yield strength F_y of the plate, the allowable axial stress F_a is obtained from AISC tables. The axial capacity is calculated as

$$P_{\text{plate}} = 1.7 \times t \times W_w \times F_a$$

and compared with the lesser of

- The strength of the brace in axial tension, P_{st}.
- The maximum force that can be transferred to the brace by the system.

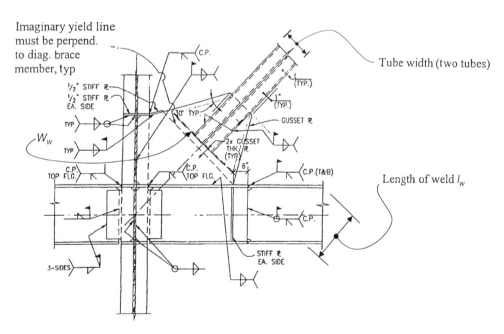

Figure 3.47. Typical bracing-to-column connection, SCBF.
Notes: 1. Yield line intersects free edge of gusset plate and not beam flange.
    2. Whitmore effective width W_w = tube width + $2l_w$ (tan 30).
    3. Welding of beam flanges is for resisting drag forces.
(Courtesy of Louis Choi, S.E., John A. Martin & Associates Structural Engineers, Los Angeles, CA.)

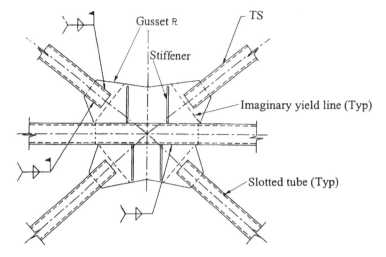

Figure 3.48. X-brace–to–beam connection. (Courtesy of Louis Choi, S.E., John A. Martin & Assoc. Structural Engineers, Los Angeles, CA.)

3.11.2. Eccentric Braced Frame (EBF)

The design principles of an eccentric braced frame are perhaps best understood by considering the tensile strength of the chain shown in Fig. 3.51a. Using the well-known adage that the strength of a chain is controlled by the strength of its weakest link, the ductility of the entire chain may be controlled by the ductility of a single link, L (Fig. 3.51a). The nominal, or ideal, tensile strength of this ductile link is T_1, while the other links, presumed to be brittle, are designed to have a strength in excess of the maximum feasible strength of the weak link. Observe that if the other links were designed to have the same nominal strength as the ductile link, the randomness of strength variation between all links, including the ductile link, would imply a high probability that failure would occur in a brittle link and the chain would not have the intended ductility.

In an eccentric braced system, the segment e of the beam is our ductile link. The segment outside of e, the brace, and the columns are the other links presumed to be brittle and designed to have a strength in excess of the strength of the weak link to account for the normal uncertainties of material strength and strain-hardening effects at high strains.

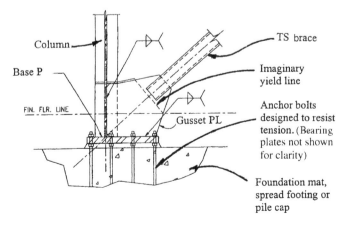

Figure 3.49. Column and brace connection at foundation. (Courtesy of Louis Choi, S.E., John A. Martin & Assoc. Structural Engineers, Los Angeles, CA.)

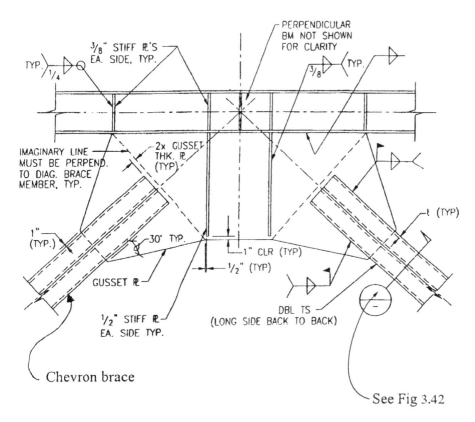

Figure 3.50. Chevron brace-to-beam connection.
Notes: 1. Chevron braces designed to SCBF requirements are not subject to the load amplification
factor of 1.5 imposed on chevron braces in OCBF systems.
    2. Beam depth shown is not to scale.
(Courtesy of Louis Choi, S.E., John A. Martin & Assoc. Structural Engineers, Los Angeles, CA.)

The link e is to be designed to carry an earthquake-induced force $P_U = P_E$. Hence, the ideal strength of the link P_i needs to be greater than $P_i = P_E$. Having chosen an appropriate segment of the beam as the eccentric link e, its overstrength, which becomes the design force, can be readily established and hence, the required strength for the strong and presumed brittle braces and columns. This illustrates the important relationship between the ductility potential of the entire bracing system and the corresponding ductility demand of a single ductile link.

Essential design features of the EBF are as follows:

- An eccentrically braced frame dissipates energy by controlled yielding of its link. It is a framing system in which the axial forces induced in the braces are transferred either to a column or another brace through shear and bending in a small segment of the bem called the link. The links in EBFs act as

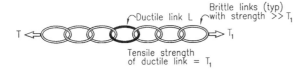

Figure 3.51a. Chain with ductile and brittle links.

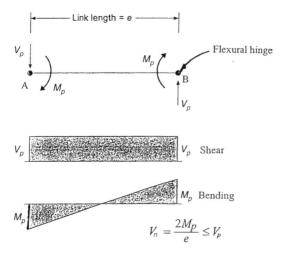

Figure 3.51b.   Bending moments and shear forces in link beam. Flexural hinges form at A and B when both M_A and M_B reach the plastic moment M_p.

structural fuses to dissipate earthquake-induced energy in a stable manner. To do so, a link needs to be properly detailed such that it has adequate strength and stable energy dissipation characteristics. All other structural components such as beam segments outside of the link, braces, columns, and connections are proportioned following capacity design provisions to remain essentially elastic during the design earthquake. They are designed for the forces generated by the actual (or expected) capacity of the links rather than the code-specified design seismic forces. The capacity design concept thus requires that the computation of the link strength be based not only on the expected yield strength of the steel, but also on considerations of strain-hardening and overstrength due to composite action of the slab.

- If we ignore the effects of axial force and the interaction between moment and shear in the link, flexural hinges always form at two ends of the link when both M_A and M_B reach the plastic moment, M_p. A shear hinge is said to form when the shear reaches the plastic shear capacity, V_p (see Fig. 3.51).
- The presence of an axial force in a link reduces not only the flexural and shear capacities, but also its inelastic deformation capacity. When the axial force P_n exceeds 15% of the yield force, $P_y = A_g F_y$, the P-M interaction formula for plastic design can be used to compute the reduced plastic moment. $M_{pa} = 1.18\, M_p\, (1 - P_a/P_y)$.
- Composite action due to the presence of slabs can significantly increase the link shear capacity during the first few cycles of large inelastic deformations. However, composite action deteriorates rapidly in subsequent cycles due to local concrete floor damage at both ends of the link. For design purposes, it is conservative to ignore the contribution of composite action for calculating the link shear strength. But the overstrength produced by the composite slab effect must be considered when estimating the maximum forces imposed by the link on other structural components.
- When detailing a link, full-depth web stiffeners must be placed symmetrically on both sides of the link web at the diagonal brace ends of the link. These end stiffeners are required to have a combined width not less than $(b_f - 2t_w)$

and a thickness not less than $0.75t_w$ or 3/8 in., whichever is larger. The link section must satisfy the same compactness requirement as the beam section for special moment frames. Further, the link must be stiffened in order to delay the onset of web buckling and to prevent flange local buckling. The stiffening requirement is dependent on the length of link.

- Intermediate link web stiffeners must be full depth. Whereas two-sided stiffeners are required at the end of the link where the diagonal brace intersects the link, intermediate stiffeners placed on one side of the link web are sufficient for links less than 25 in. in depth. Fillet welds connecting a link stiffener to the link web are to have a design strength to resist a force of $A_{st} F_y$, where A_{st} is the stiffener area. The design strength of fillet welds fastening the stiffener to the flanges shall be adequate to resist a force of $A_{st}F_y/4$.

- To ensure stable hysteresis, a link must be laterally braced at each end to avoid out-of-plane twisting. Lateral bracing also stabilizes the eccentric bracing and the beam segment outside the link. The concrete slab alone cannot be relied upon to provide lateral bracing. Therefore, both top and bottom flanges of the link beam must be braced. The bracing should be designed for 2% of the expected link flange strength. See Figs. 3.52 and 3.53 for a graphic representation of stiffener requirements.

- The required axial and flexural strength of the diagonal brace shall be those generated by the expected shear strength of the link increased by 125% to account for strain-hardening. Although braces are not expected to experience buckling, the AISC-Siesmic provisions take a conservative approach by requiring that a compact section be used for the brace.

- At the connection between the diagonal brace and the beam, the intersection of the brace and beam centerlines shall be at the end of the link or within the length of the link. If the intersection point lies outside of the link length, the eccentricity together with the brace axial force produces additional moments in the beam and brace, which should be accounted for in the design. The diagonal brace-to-beam connection at the link end of the brace is also to be designed to develop the expected strength of the brace. No part of this

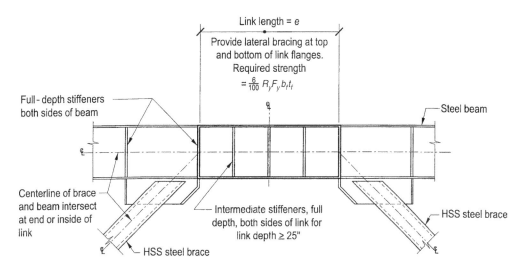

Figure 3.52. An EBF with HSS bracing; stiffener requirements.

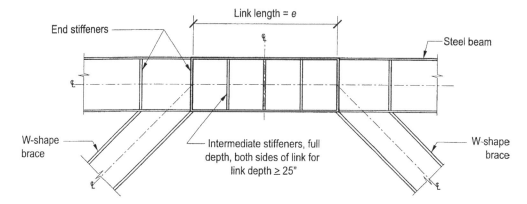

Figure 3.53. An EBF with W-shape bracing; stiffener requirements.
Notes: 1. See Fig. 3.52 for items not noted.
2. Refer to AISC-Seismic, Sec. 15.3, and Sec. 3.11.2.1 of this text for required spacing of stiffeners.

connection shall extend over the link length to reduce the link length, e. If the connection is designed as a pin, the gusset plate must be properly stiffened at the free edge to avoid local buckling.

- It is highly desirable to use a split V-braced EBF to avoid a moment connection between the link and column. Test results have shown that a fully-restrained welded connection between a column and a link, particularly if the link is relatively long, is vulnerable to brittle fracture similar to those found in the beam-to-column moment connections after the Northridge earthquake. Therefore, AISC-Seismic provisions require that the deformation capacity of the link-to-column connections be verified by qualifying cyclic tests to demonstrate that the link inelastic rotation capacity is at least 20% greater than the calculated values.

- When cover plates are used to reinforce a link-to-column connection, the link over the reinforced length must be designed such that no yielding takes place in this region. In this context, the link is defined as the segment between the end of the reinforcement and the brace connection. Cyclic testing is not needed when: 1) the shortened link length does not exceed $e_0 = 2M_p/V_p$; and 2) the design strength of the reinforced connection is equal to or greater than the force produced by a shear force of $1.25\ R_y V_n$ in the link.

- For the preferred EBF configuration, where the link is not adjacent to a column, a simple connection between the beam and column is considered adequate if it provides some restraint against torsion in the beam. Provisions of AISC-Seismic stipulate that the magnitude of this torsion be calculated by considering perpendicular forces equal to 2% of the beam flange nominal strength, $F_y b_f t_f$, applied in opposite directions on each flange.

- Although the link end moment is distributed between the brace and the beam outside of the link according to their relative stiffness, in preliminary design, it is conservative to assume that all the link end moment is resisted by the beam. Because a single member is generally used for both the link and the beam outside the link, it is too conservative to use the expected yield strength, $R_y F_y$, for estimating the force demand produced by the link while the beam strength is based on the nominal yield strength, F_y. Therefore, AISC-Seismic provisions allow designers to increase the design strength of the beam by a factor R_y.

- The horizontal component of the brace produces a significant axial force in the beam, particularly if the angle between the diagonal brace and the beam is small. Therefore, the beam outside the link must be designed as a beam-column. When lateral bracing is used to increase the capacity of the beam-column, this bracing must be designed to resist 2% of the beam flange nominal strength, $F_y b_f t_f$.
- Using a capacity design approach, columns in braced bays are designed to have sufficient strength to resist the gravity-load actions, moments, and axial forces generated by 1.1 times the expected nominal strength, $R_y V_n$, of the link.

  Based on the results of limited tests, this design procedure may be appropriate for low-rise buildings and the upper stories of medium- and high-rise buildings, but may be too conservative in other instances. Therefore, an alternative design procedure is permitted by AISC-Seismic provisions. The method consists of amplifying the design seismic axial forces and moments in columns by the overstrength factor, $\Omega_o = 2.0$. The computed column forces need not exceed those computed by the first procedure. Therefore, the first design procedure will generally produce a more conservative design for columns.

3.11.2.1. Link Design

- To ensure stability of the link during inelastic deformations, compact sections shall be used, complying with the flange width-thickness ratios of

$$b_f / 2t_f = 52/(F_y)^{0.5}$$

  where
  b_f = flange width
  t_f = flange thickness

- Doubler plates on the web of the link are not allowed because they are ineffective during inelastic deformation.
- Holes are not allowed in the web of the link because they affect the inelastic deformation of the link web.
- To ensure ductile behavior, the specified minimum yield stress of steel used for links shall not exceed 50 ksi.
- The effect of axial force on the link design shear capacity need not be considered when

$$P_u \leq 0.15 P_y$$

  where
  P_u = required axial strength
  $P_y = F_y A_g$
  A_g = gross area of section

- If plastic hinges form at the ends of the link (see Fig. 3.51), a point of inflection occurs at the center of the link and the nominal required shear capacity is given by

$$V_n = 2M_p / e$$
$$\leq V_p$$

where

M_p = nominal plastic flexural strength

$= ZF_y$

Z = plastic section modulus

e = length of link

V_p = nominal shear strength of link

$= 0.60F_y A_w$

A_w = web area

$= (d_b - 2t_f)t_w$

d_b = depth of link

t_f = flange thickness

t_w = web thickness

and

ϕV_p = design shear capacity of link

ϕ = resistance factor

$= 0.9$

A balanced shear condition exists when flexural and shear hinges occur simultaneously for a link length of $e_y = 2M_p/V_p$. For lengths less than e_y, a shear mode predominates and for lengths greater than e_y, a flexural mode predominates.

- When $P_u > 0.15P_y$

  1. The nominal required shear capacity of the link is given by

$$V_{na} = 2M_{pa}/e$$
$$\leq V_{pa}$$

  where

M_{pa} = reduced nominal plastic flexural capacity

$= 1.18M_p(1 - P_u/P_y)$

V_{pa} = reduced nominal shear capacity of link

$= V_p[1 - (P_u/P_y)^2]^{0.5}$

  and

ϕV_{na} = reduced design shear capacity of link

ϕ = resistance factor

$= 0.9$

  2. From AISC-Seismic, Eqs. (15.3) and (15.4), the length of the link is limited to the lesser of

$$e = [1.15 - 0.5\rho'(A_w/A_g)]1.6M_p/V_p \ldots \text{when } \rho'(A_w/A_g) \geq 0.3 \text{ or}$$
$$= 1.6M_p/V_p \ldots \text{when } \rho'(A_w/A_g) < 0.3$$

  where

$\rho' = P_u/V_u$

V_u = required shear strength

A_g = gross area of link

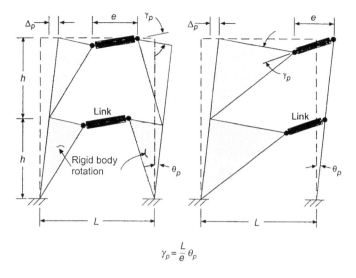

$$\gamma_p = \frac{L}{e}\,\theta_p$$

Figure 3.54. Link rotation angle γ_p. The link rotation angle γ_p is estimated by assuming the EBF bay rotates as a rigid body. By geometry, γ_p is related to plastic story drift angle θ_p, which in turn is related to plastic story drift $= \Delta_p/h$. Conservatively, Δ_p may be taken to equal design story drift. Refer to AISC 341-02, C15.2 for additional information.

- For the maximum inelastic story drift, the elements of the frame may be considered rigid and the link rotation angle γ_p is derived as shown in Fig. 3.54 and is given by

$$\gamma_p = L\Delta/he$$
$$= L\theta_p/e$$

where

L = beam length between column centers
Δ = maximum inelastic story drift
h = story height
e = length of link
θ_p = story drift angle
γ_p = link rotation angle

To limit the inelastic deformation of the frame, the link rotation angle is limited to the following values:

$$\gamma_p \leq 0.080 \text{ radian} \ldots \text{for short links of length } e \leq 1.6 M_p/V_p$$
$$\gamma_p \leq 0.020 \text{ radian} \ldots \text{for long links of length } e \geq 2.6 M_p/V_p$$

These limits are illustrated in Fig. 3.55 and linear interpolation may be used for intermediate link lengths. To ensure stable behavior of the link under cyclic loading, AISC-Seismic specifies the following detailing requirements:

- To prevent web instability under cyclic loading, full-depth web stiffeners shall be provided on both sides of the link web at the brace end of the link. As shown in Fig. 3.56, the stiffeners shall have a combined width of

$$2b_{st} \geq b_f - 2t_w$$

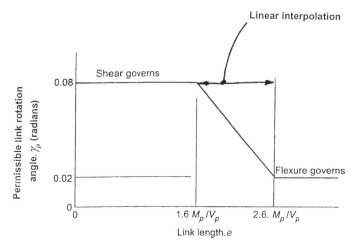

Figure 3.55. Permissible link rotation angle γ_p versus link length, e. The permissible link plastic rotation angle γ_p is the primary variable that describes the inelastic link deformation. It is strongly influenced by the length of the link, e, and its M_p/V_p ratio. When $e \leq 2.6M_p/V_p$, shear yielding dominates the inelastic response, and when $e \geq 2.6M_p/V_p$, flexural yielding governs the inelastic response.

and a thickness of

$$t_{st} = 0.75t_w$$
$$\geq 3/8 \text{ in.}$$

where

b_f = link flange width
t_w = link web thickness

- The weld between the stiffener and the web is required to develop the full strength of the stiffener, as shown in Fig. 3.56. The weld must be adequate to resist the force as given by

$$P_w \geq A_{st} F_y$$

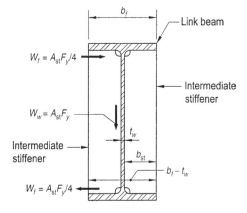

Figure 3.56. Intermediate web stiffener details for link beams. The required strength of welds, W_f, connecting the stiffener to link flanges is $A_{st} F_y$, whereas the strength W_w of welds connecting the stiffener to the link web is $A_{st} F_y$. A_{st} is the area of the stiffener.

where

A_{st} = area of stiffener

$= b_{st}t_{st}$

The weld between the stiffener and the flange is necessary to develop the rigidity of the stiffener and restrain flange buckling. The weld force is given by

$P_w \geq A_{st}F_y/4$

- For a shear link with $e \leq 1.6M_p/V_p$, intermediate stiffeners are required, at a spacing of

  $s \leq 30t_w - d/5 \dots$ for $\gamma_p = 0.08$ radian

  $s \leq 52t_w - d/5 \dots$ for $\gamma_p \leq 0.02$ radian

- Linear interpolation may be used for intermediate link rotations.

  For $\quad 2.6M_p/V_p \leq e \leq 5M_p/V_p$,

  intermediate stiffeners are required at a distance of $1.5b_f$ from each end of the link.

  For $\quad 1.6M_p/V_p \leq e \leq 2.6M_p/V_p$,

  intermediate stiffeners are required to satisfy both the aforementioned requirements.

  For $\quad e > 5M_p/V_p$,

  intermediate stiffeners are not required.

- Single-sided, full-depth web intermediate stiffeners are permitted, provided the link depth is less than 24 in. The required width is given by

  $b_{st} \geq b_f/2 - t_w$

  and the thickness by

  $t_{st} = t_w$

  $\geq \frac{3}{8}$ in.

- As specified in AISC-Seismic Section 15.5, lateral bracing to the top and bottom flanges is necessary at each end of the link to prevent instability and restrain the link from twisting out-of-plane. Lateral support shall be provided at the ends of the link with a design strength of

  $P_{bl} = 0.06R_yF_yb_ft_f$

where

R_y = ratio of the expected yield stress to the minimum specified yield strength as given in Table 3.1.

3.11.2.2. Beam Design

AISC-Seismic specifies the following design requirements for the beam outside the link:

- The nominal required axial and flexural capacity of the beam shall be determined from the forces generated by 1.1 times the nominal shear capacity of the link defined as

  $1.1R_yV_n$

where

R_y = ratio of the expected yield stress to the minimum specified yield strength of the link

V_n = nominal required shear capacity of the link

In determining the design capacity of the beam, the design capacity determined using the procedures from LRFD Sections C, E, F, and H may be multiplied by R_y. For Grade 50 steel, $R_y = 1.1$ and the enhanced design capacity becomes

$$R_y \phi R_n = 1.1 \times \phi R_n$$
$$= \text{nominal capacity (for } \phi = 0.9)$$
$$= R_n$$

- Where required, the beam shall be provided with lateral support at both the top and bottom flanges. Each support shall have a design capacity of

$$P_b = 0.02 F_y B_f t_f$$

where

F_y = specified yield strength of the beam

b_f = width-of-beam flange

t_f = thickness-of-beam flange

- In accordance with AISC-Seismic, Section 15.7, beam-to-column connections away from the link may be designed as pinned in the plane of the web. The connection shall have a design capacity to resist a torsional moment about the longitudinal axis of the beam, with a magnitude of

$$M_T = 0.02 F_y b_f t_f d$$

3.11.2.3. Brace Design

AISC-Seismic specifies the following design requirements for the diagonal brace:

- To allow for strain hardening in the link, the nominal required axial and flexural capacity of the brace shall be determined from the forces generated by the amplified nominal shear capacity of the link defined as

$$1.25 R_y V_n$$

where

R_y = ratio of the expected yield stress to the minimum specified yield strength of the link

V_n = nominal required shear capacity of the link

- The width-thickness ratios of the brace shall satisfy the requirements of LRFD Table B5.1.
- As shown in Fig. 3.14, the intersection of the brace and beam centerlines shall be at the end of the link or within the link. In accordance with AISC-Seismic, Commentary Section C15.6c, the intersection of the brace and beam centerlines should not be located outside the link because the eccentricity creates additional moment in the beam.

- The required strength of the brace-to-beam connection shall not be less than the nominal strength of the brace. No part of the connection shall extend over the length of the link. If the brace resists a portion of the link end moment, the connection shall be designed as a fully restrained moment connection.

3.11.2.4 Column Design

To ensure that link yielding is the predominant inelastic behavior, AISC-Seismic specifies the following loading combinations for the design of the column:

$$1.2D + 0.5L + 0.2S + Q_L$$
$$0.9D - Q_L$$

where

D = dead load
L = floor live load
S = snow load
Q_L = forces generated by 1.1 times the nominal strength of the link. This is equal to $1.1R_yV_n$.
R_y = ratio of the expected yield stress to the minimum specified yield strength of the link
V_n = nominal required shear capacity of the link

3.11.3. Moment Frames

The Northridge earthquake demonstrated that the pre-1994 prescriptive connection shown in Fig. 3.3 was inadequate for anticipated seismic demands. Following that earthquake, a number of steel moment-frame buildings were found to have experienced brittle fractures of beam-to-column connections, shattering the belief that steel moment-frame buildings were essentially invulnerable to earthquake-induced structural damage. It was also thought that should such damage occur, it would be limited to ductile yielding of members and connections.

The Northridge earthquake changed all that. Moreover, it showed that brittle fracture was initiated within connections at very low levels of plastic demand and, in some cases, while the structures remained essentially elastic. Fractures at the complete joint penetration (CJP) weld, between the beam bottom flange and column bottom flange, once initiated, progressed along a number of paths, depending on individual joint conditions.

Based on test results of more than 150 connection assemblies, the Federal Emergency Management Agency (FEMA) published a July 2000 document titled "Recommended Seismic Design Criteria for New Steel Moment-Frame Buildings"—FEMA 350. This publication allows new prequalifications for connection details believed to be capable of providing reliable service when subjected to large earthquake demands. The criteria given in the publication allow the design of steel moment-frame structures to be performed in a straightforward, select-design-detail method, and are believed to provide the reliability incorrectly assumed to exist in pre-Northridge moment-frame connections. For the majority of structures and conditions of use, the designer is now able to select, design, and detail prequalified moment-frame connections using FEMA 350 criteria without the need to perform project-specific prototype qualification testing. For connection details other than those included in FEMA 350, qualification tests must still be performed.

As many as eight types of prequalified connections (including two proprietary types) for use in special moment frames (SMFs) are given in FEMA 350. Of these, only two—the welded flange plate (WFP) and reduced beam section (RBS), shown in Figs. 3.57 and 3.58—are considered here.

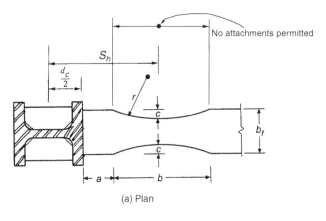

(a) Plan

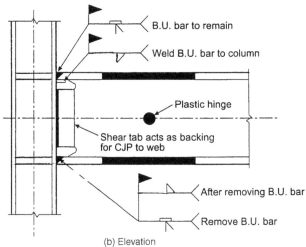

(b) Elevation

Figure 3.57.   Prequalified reduced beam section (RBS) connection.
Note: For SMFs, this connection is limited to use with W12 and W14 columns and with W36 and
    shallower beams, maximum weight of 300 lbs/ft. See FEMA 350 for exceptions.

3.11.3.1.   Reduced Beam Section (RBS) Connection

This type of connection, shown in Fig. 3.57, utilizes circular radius cuts in both top and
bottom flanges of the beam to reduce the flange area over a length of the beam near the
ends of the beam span. Welds of beam flanges to column flanges are complete joint
penetration (CJP) groove welds. In this connection, no reinforcement other than weld metal
is used to joint the flanges of the beam to the column. Web joints may be either CJP groove
welds, or bolted or welded shear tabs. When this type of connection is used, the elastic
drift calculations should consider the effect of flange reduction. In lieu of specific calculations, a drift increase of 9% may be applied for flange reductions of up to 50% of the beam
flange width, with linear interpolation for lesser values of flange reduction.

The flange reduction referred to as the RBS cut is normally made by thermal cutting.
The requirements for minimizing the notch effects are described in FEMA 353.

3.11.3.2.   Welded Flange Plate (WSP) Connection

Figure 3.58 shows a typical detail for this type of connection. Observe that there is no direct
connection between the beam and column flanges. Instead, the flange plates are used to
connect the beam flanges to the column flanges. The flange-plate–to–column-flange-joint is

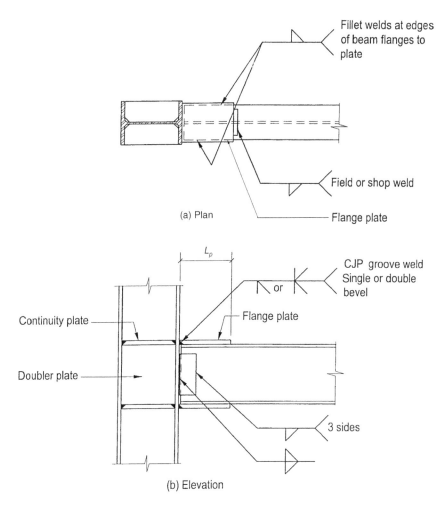

Figure 3.58. Prequalified welded flange plate (WFP) connection.
Note: For SMFs this connection is limited to use with W12 and W14 columns and with W36 and
shallower beams. FEMA 350 specifies no weight limit for beams.

a CJP groove weld. The flange plates are fillet welded to the top and bottom of beam
flanges. This connection, rather than the cover-plated connection, is recommended by
FEMA because the welding of a single plate is considered more reliable than the welding
of the combination of beam flange and cover-plate.

3.11.3.3. Connections Designed to Induce Plastic Hinges Within Beam Span: Design Principles

The formation of plastic hinges at the beam-column interface during a seismic event results
in large inelastic strain demands at the connection leading to brittle failure. To prevent this
occurrence, the prequalified connections are designed to produce the plastic hinges within
the beam span, as shown in Fig. 3.59. This condition may be achieved by reducing the
section of the beam at the desired location of the plastic hinge or by reinforcing the beam
at the connection to prevent the formation of a hinge in this region. By this means, the
connection at the beam-column interface remains nominally elastic and the inelastic deformation occurs away from the connection. The hinge location distances given are valid for
beams in which gravity loading represents only a small portion of the flexural demand.

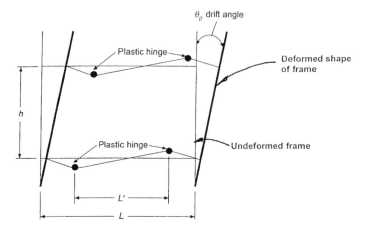

Figure 3.59. Inelastic drift of special moment frames with plastic hinges within beam span. A frame in which inelastic excursion occurs through the formation of plastic hinges within the beam span is capable of dissipating large amounts of energy. Such a behavior may be obtained by: 1) locally stiffening and strengthening fully restrained connections; or 2) locally reducing the beam section at desired locations.

The probable beam plastic moment, allowing for overstrength of the steel, the difference in yield strengths of the beam flanges and web materials, and the estimated strain-hardening is given by

$$M_{pr} = C_{pr}R_y Z_{be}F_y$$

where

R_y = overstrength coefficient
　　= ratio of the expected yield stress to the minimum specified yield strength of the material
F_y = minimum specified yield stress of the beam
Z_{be} = effective plastic section modulus of the beam at the zone of plastic hinging
C_{pr} = peak connection strength coefficient given as
　　= $(F_y + F_u)/2F_y$
　　= 1.15... for reduced beam section connections
　　= 1.2... for other connections
F_u = minimum specified tensile strength of the beam

The shear force at the plastic hinge is given by

$$V_p = 2M_{pr}/L' + w_u L'/2$$

where

w_u = factored gravity load on the beam
L' = length between plastic hinges

Neglecting the gravity load on the length x (as shown in Fig. 3.60), the resulting bending moment at the face of the column is

$$M_f = M_{pr} + V_p x$$

For reduced beam section connections, the bending moment at the face of the column is limited to

$$M_f < R_y Z_b F_y$$

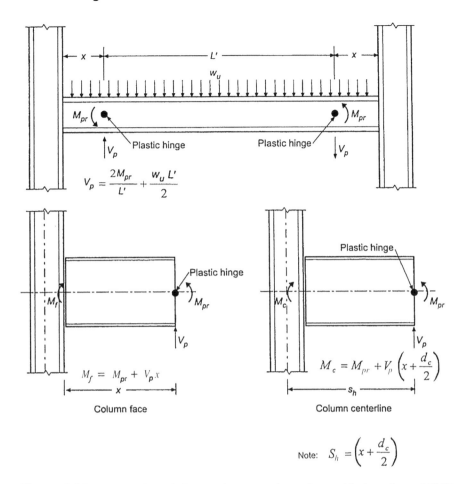

Figure 3.60. Calculation of shear and moment demands at critical sections of SMF.

where

Z_b = plastic section modulus of the beam at the column face

The resulting shear force at the face of the column is

$V_f = 2M_f/(L - d_c) + w_u L/2$

The resulting bending moment at the center of the column (as shown in Fig. 3.60) is given as

$M_c = M_{pr} + V_p S_h$

For the general case, with beams framing into both sides of the column

$\Sigma M_c = \Sigma (M_{pr} + V_p S_h)$

3.11.3.4 Strong Column–Weak Beam

A strong column–weak beam concept should be adopted to ensure frame stability, as the formation of plastic hinges in the columns of a story may cause a weak story condition. In addition, large inelastic displacements produced in the columns increase the P-delta effect and may lead to column failure. The strong column–weak beam concept may be achieved in accordance with the requirement

$\Sigma M_{pc}^* / \Sigma M_c > 1.0$

where

ΣM_{pc}^* = sum of the nominal flexural strengths of the column above and below the joint at the beam centerline with a reduction for the factored axial force in the column as given by

$\quad = \Sigma Z_c(F_{yc} - P_{uc}/A_g)$

P_{uc} = required axial compression strength in the column

Z_c = plastic section modulus of the column

F_{yc} = minimum specified yield stress of the column

A_g = gross area of the column

ΣM_c = sum of the bending moments at the center of the column resulting from the development of the probable beam plastic moments

Provided that a column complies with the width-thickness ratio provisions of Table 3.2, AISC-Seismic relaxes the strong column–weak beam requirement. In addition, for this relaxation to be allowed, the column is also required to have an axial stress less than $0.3F_y$ and

1. Be located in a one-story building or in the top story of a multistory building.
2. Be located in a column line in which the design shear strength of all exempted columns is less than 33% of the required shear strength of the column line, and the design shear strength of all exempted columns in the story is less than 20% of the required shear strength of the story.

AISC-Seismic also provides an exemption for a column located in a story with a design shear strength 50% greater than that of the story above.

3.11.3.5. *Beam Buckling*

To limit local flange buckling, AISC-Seismic specifies the use of sections with a maximum flange width-to-thickness ratio of

$$b_f/2t_f = 52/(F_y)^{0.5}$$

This ratio may be determined, in reduced beam section connections, at the ends of the center two-thirds of the reduced section of the beam, unless gravity loading moves the hinge point significantly from the center of the reduced section.

To prevent stress concentrations resulting in a brittle mode of failure, abrupt changes of flange area are not permitted in the hinging area. The hinging area is defined as the distance from the face of the column to one-half the beam depth beyond the theoretical hinge point. Connections, shear studs, or other attachments are not permitted in the hinging area.

To provide adequate web stability, the height-to-thickness ratio of the web shall not exceed

$$h_c/t_w = 418/(F_y)^{0.5}$$

Lateral bracing is necessary on the top and bottom flanges of the beam to prevent instability. Bracing is required near all concentrated loads, at changes in cross section, where a hinge may form, and at a maximum spacing of

$$l_{cr} = 2500r_y/F_y$$

When the beam supports a concrete slab along its whole length, lateral bracing is not required.

3.11.3.6. Column Design

When the ratio of column moments to beam moments is

$$\Sigma M_{pc}^* / \Sigma M_c \leq 2.0$$

columns shall comply with the slenderness requirements of Table 3.2. Otherwise, columns shall comply with the slenderness requirements of LRFD Table B5.1.

When the ratio of column moments to beam moments is

$$\Sigma M_{pc}^* / \Sigma M_c < 2.0$$

lateral bracing of column flanges at beam column connections shall be provided.

3.11.3.7. Continuity Plates

Continuity plates, as shown in Fig. 3.61, are required when the column flange thickness is less than the value given by either of the following two expressions:

$$t_{cf} = \sqrt{1.8\, b_f t_f\, \frac{F_{yb} R_{yb}}{F_{yc} R_{yc}}} \qquad \text{(AISC-Seismic C 9.3)}$$

or

$$t_{cf} = b_f / 6 \qquad \text{(AISC-Seismic C 9.4)}$$

where

t_{cf} = minimum required thickness of column flange when no continuity plates are provided

b_f = beam flange width

t_f = beam flange thickness

F_{yb} = minimum specified yield stress of the beam flange

F_{yc} = minimum specified yield stress of the column flange

R_{yb} = ratio of the expected yield strength of the beam material to the minimum specified yield strength

R_{yc} = ratio of the expected yield strength of the column material to the minimum specified yield strength

The minimum continuity plate thickness is

$$t_{st} = t_f \qquad \text{(for two-sided (interior) connections)}$$

and

$$t_{st} = t_f / 2 \qquad \text{(for one-sided (exterior) connections)}$$

The minimum width of a continuity plate is required to match the beam flange. The maximum width-thickness ratio is defined as

$$b_{st} / t_{st} = 1.79 / (F_{yst} / E)^{0.5}$$

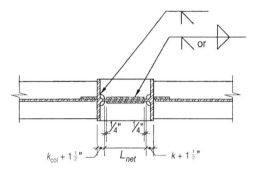

(a) Plan

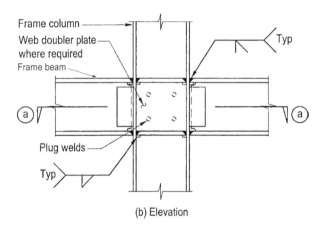

(b) Elevation

Figure 3.61.   Typical continuity plates and doubler plates.

When continuity plates are required, they are to be designed as axially loaded columns to support the beam flange force. The effective length is taken as

$$l_e = 0.75h$$

where

$\quad\quad h$ = clear distance between flanges, less the corner radii
$\quad\quad\quad = d_c - 2k$
$\quad\quad k$ = distance from outer face of column flange to web toe of fillet
$\quad\quad d_c$ = depth of column

The cross section of the column may be considered to consist of the stiffener and a strip of column web having a width of $25t_w$.

Continuity plates are welded to the column flange using complete joint penetration groove welds, as shown in Fig. 3.61. Continuity plates are clipped to avoid the column k-area and are welded to the column web to develop the shear capacity of the net length of the continuity plate, which is

$$P_w = 0.6t_{st}L_{net}F_{yst}$$

where

L_{net} = net length of continuity plate
= $d_c - 2(k + 1.5)$
k = distance from outer face of column flange to web toe of fillet

3.11.3.8. Panel Zone

The thickness of the panel zone to ensure simultaneous yielding of the beam and panel zone is given as

$$t = C_y M_c (h - d_b)/[0.9 \times 0.6 F_{yc} R_{yc} d_c (d_b - t_{fb}) h]$$

where

C_y = $S_b/C_{pr}Z_{be}$
S_b = elastic section modulus of the beam at the zone of plastic hinging
Z_{be} = effective plastic section modulus of the beam at the zone of plastic hinging
C_{pr} = peak connection strength coefficient defined
= $(F_y + F_u)/2F_y$
= 1.15 (for reduced beam section connections)
= 1.2 (for other connections)
M_c = moment at center of column
= $M_{pr} + V_p s_h$
M_{pr} = probable beam plastic moment
= $C_{pr} R_{yb} Z_{be} F_{yb}$
R_{yb} = overstrength coefficient
= ratio of the expected yield stress to the minimum specified yield strength of the beam
F_{yb} = minimum specified yield stress of the beam
V_p = $2M_{pr}/L' + w_u L'/2$
w_u = factored gravity load on the beam
L' = length between plastic hinges
s_h = hinge location distance
R_{yc} = overstrength coefficient
= ratio of the expected yield stress to the minimum specified yield strength of the column
F_{yc} = minimum specified yield stress of the column
d_b = depth of beam
d_c = depth of column
t_{fb} = thickness of the beam flange
h = average story height of the stories above and below the panel zone

The thickness of the column web must at least equal t, otherwise doubler plates are requred. The thickness of a doubler plate may be included in t, provided it is connected to the column web with plug welds, as shown in Fig. 3.61, adequate to prevent local buckling of the plate. When the doubler plate is placed against the column web, it shall be welded at top and bottom to develop the proportion of the total force that is transmitted to the doubler plate. The doubler plate shall be either butt- or fillet-welded to the column flanges to develop its shear strength. Doubler plates may be placed between continuity plates or may extend above and below the continuity plates. When the doubler plates are placed away from the column web, they shall be placed symmetrically in pairs and welded to continuity plates, to develop their share of the total force transmitted to the doubler plate.

To prevent shear buckling during cyclic loading, the individual thicknesses of column webs and doubler plates shall not be less than the value given as

$$t = (d_z + w_z)/90$$

where

d_z = panel zone depth between continuity plates
w_z = panel zone width between column flanges.

The thickness of any doubler plate may be included in t, provided it is connected to the column web with plug welds adequate to prevent local buckling of the plate.

3.11.3.9. *Reduced Beam Section (RBS) Connection: Design Example*

Given.   A two-bay frame with fully restrained moment connections. It forms part of a lateral resisting system of a building in a high seismic zone. A structural analysis using center-to-center framing dimensions (without explicitly modeling panel zones, as permitted in FEMA 350, Section 2.8.2.3) has been performed for earthquake loading to verify the adequacy of the lateral system using the applicable R, C_d, and Ω_0 values, and redundancy limits. Strength criteria and drift limits have been verified using the sizes shown in Fig. 3.10, and have been found to be adequate. The engineer has selected RBS as an appropriate connection for the project. Using the provisions of the FEMA 350 publication, "Recommended Seismic Design Criteria for New Steel Moment-Frame Buildings" as a guide, it is initially assumed that flange reductions are in the range of 50% of beam flange width.

Because the selected connection type with the presumed 50% reduction in flange width has appreciable effects on frame stiffness, the calculated drift is increased by 9% as suggested in the FEMA 350 recommendations to verify drift limits. The frame satisfies the drift requirements. It is assumed that gravity loads represent a small portion of the total flexural demand, and do not influence the location of plastic hinges.

Typical column and beam member sizes, specified strengths, and gravity loads are as follows:

Frame column: W14 × 342, ASTM A913 Grade 65, F_{yc} = 65 ksi, F_u = 80 ksi
Frame beam: W33 × 141, ASTM A572 Grade 50, F_{yb} = 50 ksi, F_u = 65 ksi
Center-to-center span of frame beam = 32.5 ft
Factored axial load on the column = 818 kips
Dead load D = 2.29 kip/ft
Live load L = 1.04 kip/ft

Required.   Check critical parameters for a typical beam and column for compliance with FEMA 350 Prequalification Data for RBS Connection (FEMA Table 3-6). Design a sample connection in accordance with the criteria recommended in FEMA 350.

Solution.   The solution consists of verifying critical parameters for beams and columns, strong column–weak beam criteria, and designing the RBS connection using the procedure given in FEMA 350.

Beam Critical Parameters.

1. Depth: Given beam depth = 33″, which is less than the maximum allowed depth of 36″. OK
2. Span-to-depth ratio = 390/33.3 = 11.71 > 7 permitted minimum for SMRF. OK
3. Given weight of 141 lbs/ft < 200 lbs/ft, less than the permitted maximum. OK

4. $b_f/2t_f = 11.50/(2 \times 0.96) = 5.99$, maximum permitted $= 52/\sqrt{F_y} = 52/\sqrt{50} = 7.35$.   OK
5. Thickness of flange $t_f = 0.96$ in. < 1.75 in.   OK
6. Beam material is A572, Grade 50, permitted by FEMA 350.   OK
7. Flange reduction will be within FEMA guidelines [see FEMA 350, Eqs. (3.15) and (3.16)].   OK

Column Critical Parameters.

1. Depth: Given size W14. W12 and W14 are permitted for SMRF.   OK
2. Material: ASTM A913, Grade 65, permitted by FEMA 350.   OK

Strong Column–Weak Beam (SC–WB Concept). Building frames in high seismic zones are typically designed with a strong column–weak beam configuration The main reason being that when subjected to strong ground shaking, frames with columns that are weaker in flexure than the framing beams can form single-story mechanisms, in which plastic hinges form at the base and top of all columns in a story. This can lead to very large local drifts, $P\Delta$ instability, and possible collapse.

To prevent this undesirable behavior, FEMA 350 recommends the following relationship:

$\Sigma M^*_{PC}/\Sigma M_c > 1.0$

Observe this is the same as Eq. (9.3) of AISC 341-02 except that M_c is substituted for the quantity M^*_{pb}.

In the preceding equations,

ΣM^*_{PC} = the sum of the moments in the column above and below the joint at the intersection of the beam and column centerlines with a reduction for the axial force in the column. It is permitted to take

$\Sigma M^*_{PC} = \Sigma Z_C(F_{yc} - P_{uc}/A_g)$

Substituting the given values,

$$M^*_{pc} = \Sigma Z_c \left(F_{yc} - \frac{P_{uc}}{A_g} \right)$$

$$= 2 \times 672 \left(65 - \frac{818}{101} \right)$$

$$= 76{,}454 \text{ kip-in.} = 6373 \text{ kip-ft}$$

ΣM_c, as will be shown shortly, is equal to $2 \times 2373 = 4746$ kip-ft

$$\therefore \frac{M_{pc}}{M_c} = \frac{6373}{4746} = 1.34 > 1.0 \qquad \text{OK}$$

Connection Design.

Step 1. Determine the length and location of the beam flanges' reduction, based on the following:

$a \cong (0.5 \text{ to } 0.75) \, b_f$ $\qquad\qquad$ [FEMA 350, Eq. (3.15)]

$b \cong (0.65 \text{ to } 0.85) \, d_b$ $\qquad\qquad$ [FEMA 350, Eq. (3.16)]

where a and b are shown in Fig. 3.57, and b_f and d_b are the beam flange width and depth, respectively.

In our case the properties for the given beam W33 $\times$ 141 are as follows:

d_b = 33.30 in., b_{fb} = 11.50 in., t_{fb} = 0.96 in, Z_{xb} = 514.0 in.³, S_{xb} = 448.0 in.³, F_{yb} = 50.0 ksi

R_{yb} = 1.1 [An overstrength coefficient specified in AISC-Seismic (AISC 341-02) Table 1.6.1]

F_{yc} = $R_{yb} \times F_{yb}$ = 1.1 $\times$ 50 = 55 ksi

C_{pr} = a factor to account for peak connection strength, including strain hardening, local restraint, additional reinforcement, and other connection conditions. C_{pr} is given by the formula

$$c_{pr} = \frac{F_y + F_u}{2F_y}$$
[FEMA 350 Eq. (3.2)]

FEMA recommends a value of 1.5 for RBS connections.

For our RBS, assume

$a = 0.5\, b_f = 0.5 \times 11.50 = 5.75$ in.

$b = 0.78\, d_b = 0.778 \times 33.50 = 26.05$ in.

Step 2.   Determine the depth of flange reduction, c, using a trial and error procedure, according to the following guidelines:

1. Assume $c = 0.20 b_f$.
2. Calculate Z_{RBS}.
3. Calculate $M_f = M_{pr} + V_p x$.
4. If $M_f < C_{pr} Z_b F_y$, the design is acceptable. If not, increase c. Limit the value of c to a maximum of 0.25 b_f.

Notations of additional terms used above are as follows:

Z_{RBS} = section modules of the beam at the reduced section equal to

$Z_{RBS} = Z_{zb} - 2ct_{fp}(d_b - t_{fb})$ (Note: Z_{RBS} is denoted as Z_c in FEMA 350)

In our case,

1. $c = 0.20\, b_{fb} = 0.2 \times 11.50 = 2.33$ in.
2. $Z_{RBS} = 514 - 2 \times 2.33 \times 0.96\,(33.60 - 0.96)$
   $= 369.62$ in.³
3. $M_f = M_{pr} + V_p x$ (see FEMA 350, Fig. 3.4)
   M_{pr} = probable peak plastic moment
   $= C_{pr}\, R_y\, Z_e\, F_y$
   $= 1.15 \times 1.1 \times 369.62 \times 50$
   $= 23318$ kip-in. $= 1942.2$ kip-ft
   $= 1.15 \times 1.1 \times 369.62 \times 50$
   $= 23318$ kip-in. $= 1942.2$ kip-ft

$$V_{\text{Gravity}} = (1.2D + 0.5L) \times \frac{L'}{2}$$

$$= (1.2 \times 2.29 + 0.5 \times 1.04) \times \frac{27.91}{2}$$

$$= 45.63 \text{ kips}$$

$$V_{\text{Seismic}} = \frac{2M_{pr}}{L'} = \frac{2 \times 1942.2}{27.91} = 139.61 \text{ kips}$$

$$V_p = V_{\text{Gravity}} + V_{\text{Seismic}}$$

$$= 45.63 + 139.61 = 185.24 \text{ kips.}$$

$$M_f = 1942.2 + 185.24 \times \frac{98.78}{12}$$

$$= 2238 \text{ kip-ft}$$

4. $M_{pc} = C_{pr}R_yZ_bF_y$

$$= 1.15 \times 1.1 \times 514 \times 50$$

$$= 32510.5 \text{ kip-in} = 2709 \text{ kip-ft} \qquad \text{[FEMA 350, Eq. (3.1)]}$$

Since $M_f = 2238$ kip-ft is less than $M_{pc} = 2709$ kip-ft, the design is acceptable.

Step 3. Calculate M_f and M_c based on final RBS dimensions. In our case, $M_f = 2238$ kip-ft, the same as calculated earlier because we did not revise the RBS dimensions. M_c is the plastic moment at the centerline of the column given by

$$M_c = M_{pr} + V_p \left(x + \frac{d_c}{2} \right) = M_{pr} + V_pS_h$$

$$= 1948.2 + 185.24 \left(\frac{27.53}{12} \right)$$

$$= 2373 \text{ kip-ft} \qquad \text{(see Fig. 3.60)}$$

This value will be used to verify thickness of panel zone in Step 5.

Stpe 4. Calculate the shear at the column face according to the equation:

$$V_f = \frac{2M_f}{(L - d_c)} + V_g$$

$$= \frac{2 \times 2238}{(32.5 - 17.5)} = 298.40 \text{ kips} \qquad \text{[FEMA 350, Eq. (3.8)]}$$

If we were to use a complete joint penetration (CJP) weld between the column flange and the beam web, no further calculations would be necessary. If a bolted shear tab is used, as is the case for the example, the tab and bolts should be designed for the shear of 298.40 kips calculated above. Bolts should be designed for bearing, using a resistance factor ϕ of unity. For a 1-in. nominal bolt diameter,

$$\phi R_n = 43.5^k \ (F_u = 58 \text{ ksi}) \text{ where } \phi = 0.75$$

Therefore, for $\phi = 1.0$, the bolt capacity is $43.5/0.75 = 58$ kips.
No. of bolts = 298.50/58 = 5.14; use six 1-in.-diameter A325 bolts. Fully tighten the bolts.

Step 5. Design of Panel zone. To design a moment-resisting connection that has a desirable seismic behavior, we basically have two choices. Proportion the joint such that: 1) shear yielding of the panel zone initiates at the same time as flexural yielding of beam elements or; 2) design the joint such that all yielding occurs in the beam. The best performance is likely to be achieved when there is a good balance between beam bending and panel zone distortion.

This is achieved when t, the thickness of the panel zone, satisfies the following relationship:

$$t = \frac{C_y M_c \dfrac{(h - d_b)}{h}}{(0.9)0.6 F_{yc} R_{ye} d_c (d_b - t_{fb})}$$ [FEMA 350, Eq. (3.7)]

where

h = the average height of the stories above and below the panel zone.

Other terms are as defined earlier.

If t calculated from the above equation is greater than the thickness of the column web, typically the solution is to provide doubler plates. Another option is to select a heavier column with a thicker web. The designer is referred to AISC 341-02 for guidance on welding of doubler plates.

For the example problem

$$C_y = \frac{1}{C_{pr} \dfrac{Z_{be}}{S_b}}$$ [FEMA 350, Eq. (3.4)]

$$= \frac{1}{1.15 \times 369.62/307.76} = 0.72$$

$\Sigma M_c = 2 \times 2373 = 4746$ kip-ft = 56952 kip-in.

$$t = \frac{0.72 \times 56952 \left(\dfrac{13.5 \times 12 - 33.30}{13.5 \times 12}\right)}{0.9 \times 0.6 \times 65 \times 1.1 \times 17.50(33.30 - 0.96)}$$
$$= 1.49 \text{ in.}$$

The thickness of W14 × 342 column, t_{wc} = 1.54 in. This is greater than t = 1.49 in. calculated above. Therefore, doubler plates are not required.

Step 6.   Check Continuity Plate Requirements. Beam flange continuity plates are required across the column web when t_{cf}, the thickness of the column flange, is less than the value given by:

$$t_{cf} < 0.4 \sqrt{\frac{1.8 b_{fb} t_{fb} F_{yb} R_{yb}}{F_{yc} R_{yc}}}$$ [FEMA 350, Eq. (3.5)]

$$t_{cf} < \frac{b_f}{6}$$ [FEMA 350, Eq. (3.6)]

In our case

$$t_{cf} = 0.4 \sqrt{\frac{1.8 \times 11.5 \times 0.96 \times 50 \times 1.1}{65 \times 1.1}}$$
$$= 1.56 \text{ in.}$$

$$\frac{b_f}{6} = \frac{11.5}{6} = 1.92 \text{ in.}$$

The column flange thickness t_{fc} of W14 × 132 is 2.5 in. This is greater than 1.92 in. calculated above. Therefore, continuity plates are not required.

Walt Disney Concert Hall
Los Angeles, California

John A. Martin & Associates, Inc.
Structural Engineers
Los Angeles, CA

4

Concrete Buildings

4.1. STRUCTURAL SYSTEMS

For buildings in regions of low seismic risk Uniform Building Code (UBC) zones 0 and 1, and for those assigned to International Building Code (IBC) seismic design categories (SDC) A or B, just about any structural system that has a recognizable load path for gravity and lateral loads is permitted. The load path should be continuous and have adequate strength and stiffness to transfer the forces from the point of application to the point of resistance. This is one of the most fundamental considerations in structural design that applies across the board to all engineered structures. The identification, selection, design, and detailing of load path using the provisions of governing building codes constitutes the principal structural engineering task.

Reinforced structural concrete, known to humans since the 19th century, offers a wide range of structural systems that may be grouped into distinct categories, each with an applicable height range, as shown in Fig. 4.1. The height range given for each group, although logical for normally proportioned buildings, should be verified for a specific application by considering such factors as building geometry, severity of wind exposure, seismicity of the region, seismic design category assigned to the building, and the height limitations imposed by the governing codes. The systems shown in Fig. 4.1 are for buildings in areas of low seismic risk and are permitted with no height limit. Structural system limitations and maximum heights permitted in areas of higher seismic risk are discussed later.

4.1.1. Flat Slab–Beam System

The structural system list shown in Fig. 4.1 starts with a simple system consisting of floor slabs and columns designed to carry both gravity and lateral loads. This system—referred to as flat slab–frame—is not permitted in SDC D, E, and F, and has stringent detailing requirements for buildings in UBC zones 2A and 2B. In areas of low seismicity they may be designed without any limitations on height. However, lateral drift requirements limit their economical height to about 10 stories, as shown in Fig. 4.1. The nonductile detailing requirements given in the first 20 chapters of the American Concrete Institute's ACI-318-99 are presumed to be sufficient to provide the necessary strength and nominal ductility for buildings in regions of low seismicity.

Floors in concrete buildings often are of a two-way system such as a flat plate, flat slab, or waffle slab (Fig. 4.2.) A flat plate system consists of a concrete slab that frames directly into columns, whereas a flat slab has column capitals, drop panels, or both to increase the shear and moment resistance of the system at the columns where the shears and moment are greatest. A drop panel is considered as part of a slab and its design is accounted for along with the slab design, whereas a column capital is deemed part of a column and its design is considered column design. The waffle slab consists of orthogonal

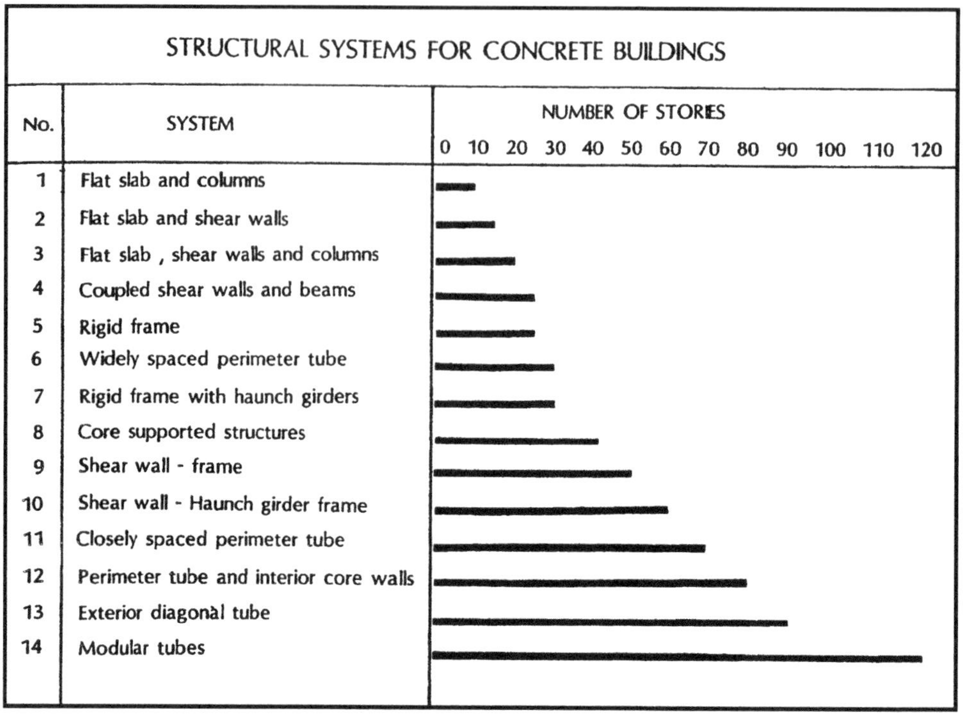

STRUCTURAL SYSTEMS FOR CONCRETE BUILDINGS

No.	SYSTEM	NUMBER OF STORIES
		0 10 20 30 40 50 60 70 80 90 100 110 120
1	Flat slab and columns	
2	Flat slab and shear walls	
3	Flat slab , shear walls and columns	
4	Coupled shear walls and beams	
5	Rigid frame	
6	Widely spaced perimeter tube	
7	Rigid frame with haunch girders	
8	Core supported structures	
9	Shear wall - frame	
10	Shear wall - Haunch girder frame	
11	Closely spaced perimeter tube	
12	Perimeter tube and interior core walls	
13	Exterior diagonal tube	
14	Modular tubes	

Figure 4.1. Structural systems for concrete buildings.

rows of joists commonly formed by using square domes. The domes are omitted around the columns to increase the moment and shear capacity of the slab. Any of the three systems may be used as an integral part of a lateral resisting system and all are popular for apartments and hotels in areas of low seismicity.

The slab system shown in Fig. 4.2 has two distinct actions in resisting lateral loads. First, because of its high in-plane stiffness, it distributes the lateral loads to various vertical elements in proportion to their stiffness. Second, because of its significant out-of-plane stiffness, it restrains the vertical displacements and rotations of columns as if they were interconnected by a shallow wide beam.

The concept of effective width can be used to determine the equivalent width of a flat slab–beam. Although physically no beam exists between the columns, for analytical purposes a certain width of slab may be considered as a beam framing between the columns. The effective width is, however, dependent on various parameters, such as column aspect ratios, distance between the columns, thickness of the slab, etc. Research has shown that values less than, equal to, and greater than full width are all valid depending upon the parameters mentioned above.

The American Concrete Institute (ACI) permits a full width of slab between adjacent panel center lines for both gravity and lateral load analysis with the stipulation that the effect of slab cracking be considered in evaluating stiffness of frame members. Use of a full width is explicit for gravity analysis, and implicit (because it is not specifically prohibited) for the lateral loads. However, engineers generally agree that use of a full width is unconservative for lateral analysis. It overestimates the column stiffness, compounding the error in the distribution of moments due to lateral loads.

Of particular concern in the design of a flat slab–frame is the problem of shear stress concentration at the column–slab joint. Shear reinforcement is almost always necessary to

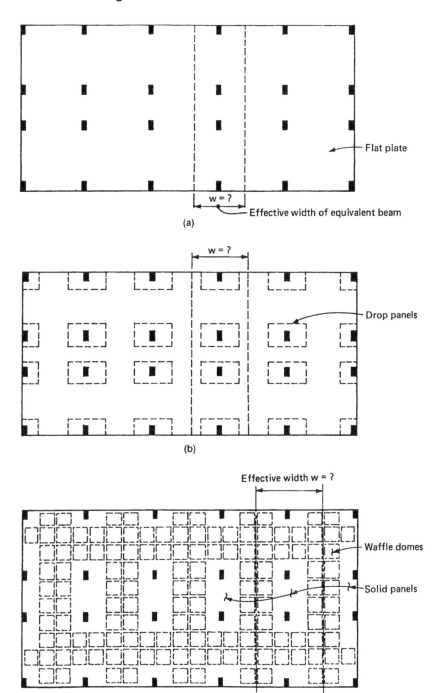

Figure 4.2. Lateral systems using slab and columns: (a) flat plate; (b) flat slab with drop panels; (c) two-way waffle system.

improve joint behavior and avoid early stiffness deterioration under lateral cyclic loading. This is one of the primary reasons that two-way slab systems are not permitted by the ACI code in regions of high seismic risk (UBC zones 3 and 4). Their use in regions of moderate seismic risk (UBC zones 2 and 2B) is permitted, subject to certain requirements, mainly relating to reinforcement placement in the column strip.

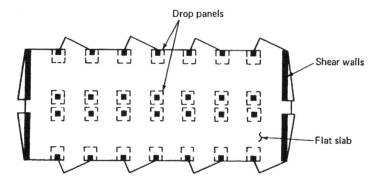

Figure 4.3. Shear wall–flat slab system.

4.1.2. Flat Slab–Frame with Shear Walls

Frame action provided by flat slab–beam and column interaction is generally insufficient for buildings taller than about 10 stories. A system consisting of shear walls and flat slab–frames may provide an appropriate lateral bracing system. Figure 4.3 shows an example.

Coupling of walls and columns solely by slabs is a relatively weak source of energy dissipation. When sufficiently large rotations occur in the walls during an earthquake, shear transmission from slab into wall occurs mainly around the inner edges of the wall. Because of torsional cracking of the slab and shear distortions around the columns, the system hysteretic response is poor. Therefore, seismic codes discourage the use of slab–beam frames by limiting the width of slab that can be considered as an equivalent beam. For buildings in high seismic zones (UBC zones 3 and 4) the width of the equivalent beam is limited to the width of the supporting column plus 1.5 times the thickness of the slab. Only in this limited width are we allowed to place the top and bottom flexural reinforcement. This requirement precludes the use of flat slab–beams as part of a seismic system in zones of high seismicity.

It should be noted that deformation compatibility requirements impose severe punching stress demands in the flat slabs of buildings in regions of high seismic risk.

4.1.3. Coupled Shear Walls

A system of interconnected shear walls exhibits a stiffness that far exceeds the summation of the individual wall stiffnesses. This is because the interconnecting slab or beam restrains the cantilever bending of individual walls by forcing the system to work as a composite unit.

The system is economical for buildings in the 40-story range. Since planar shear walls carry loads only in their plane, walls in two orthogonal directions are generally required to resist lateral loads in two directions. Placement of walls around elevators, stairs, and utility shafts is common because they do not interfere with interior architectural layout. However, resistance to torsional loads must be considered in determining their location.

4.1.4. Rigid Frame

Cast-in-place concrete has an inherent advantage of continuity at joints. The design and detailing of joints at the intersection of beams and columns is of concern particularly in seismic design because the column height within the depth of the girder is subjected to large shear forces. Horizontal seismic ties at very close spacing may be required to avoid uncontrolled diagonal cracking and disintegration of concrete and to promote ductile behavior. The design intent in high seismic zones is to have a system that can respond to earthquake loads without loss in gravity-load carrying capacity.

A rigid frame is characterized by flexure of beams and columns and rotation at the joints. Interior rigid frames for office buildings are generally inefficient because: 1) the number of columns in any given frame is limited due to leasing considerations; and 2) the beam depths are often limited by the floor-to-floor height. However, frames located at the building exterior do not necessarily have these limitations. An efficient frame action can thus be developed by providing closely spaced columns and deep spandrels at the building exterior.

4.1.5. Tube System with Widely Spaced Columns

The term tube, in usual building terminology, suggests a system of closely spaced columns say, 8 to 15 ft on center (2.43 to 4.57 m) tied together with a relatively deep spandrel. However, for buildings with compact plans it is possible to achieve tube action with relatively widely spaced columns interconnected with deep spandrels. As an example, the plan of a 28-story building constructed in New Orleans, LA, is shown in Fig. 4.4. Lateral resistance is provided by a perimeter frame consisting of columns 5 ft (1.5 m) wide, spaced at 25-ft (7.62-m) centers, and tied together with a spandrel 5 ft (1.53 m) deep.

4.1.6. Rigid Frame with Haunch Girders

Office buildings usually have a lease depth of about 40 ft (12.19 m) without interior columns. A girder about 2 ft-6 in. (0.76 m) in depth is required to carry gravity loads for

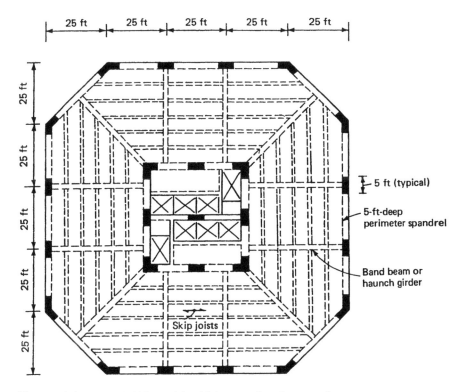

Figure 4.4. Tube building with widely spaced perimeter columns.

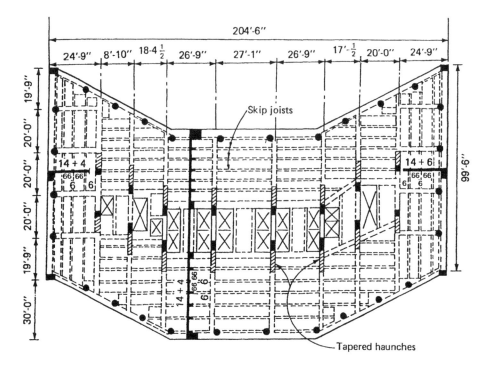

Figure 4.5. A 28-story haunch girder building: typical floor framing plan.

a 40-ft (12.19-m) span unless the girder is post-tensioned. The beam depth has an impact on the floor–floor height and is often limited because of the additional cost for the increased height of interior partitions, a curtain wall, and the added heating and cooling loads due to the increased volume of the building. A variable-depth haunch girder, as shown in Fig. 4.5, is often the solution for resisting both gravity and lateral loads. No increase in floor-to-floor height is required because the depth of girder at the midsection is flush with the floor system, thus providing ample beamless space for passage of mechanical ducts.

4.1.7.   Core-Supported Structures

Shear walls around building cores can be considered as a spatial system capable of transmitting lateral loads in both directions. The advantage of shear walls around the elevator and staircases is that, being spatial structures, they are able to resist gravity loads, shear forces, bending moments, and torsion in two directions, especially when adequate stiffness and strength are provided between the openings. The shape of the core is governed by the elevator and stair requirements, and can vary from a single rectangular core to multiple cores. Structural floor framing surrounding the core may consist of any type of common system such as cast-in-place concrete, precast concrete, or structural steel (Fig. 4.6).

4.1.8.   Shear Wall–Frame Interaction

Without question, this system is one of the most—if not the most—popular systems for resisting lateral loads in medium- to high-rise buildings. The system has a broad range of application and has been used for buildings as low as 10 stories to as high as 50 stories or even taller. With the advent of haunch girders, the applicability of the system can be extended to buildings in the 70- to 80-story range.

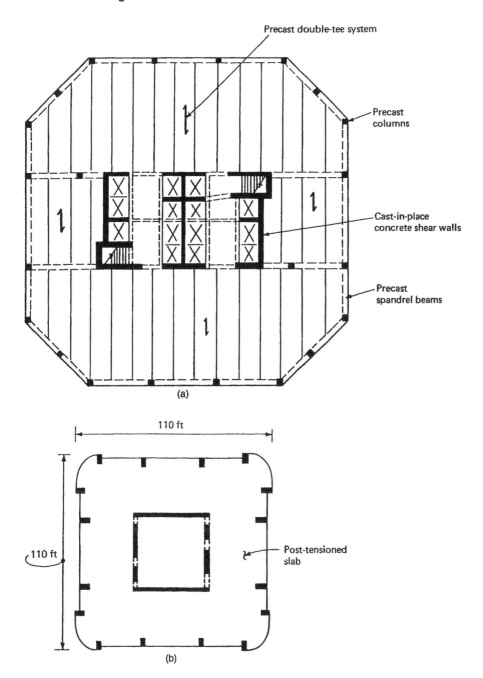

Figure 4.6. Examples of shear core buildings: (a) cast-in-place shear walls with precast surround; (b) shear walls with post-tensioned flat plate; (c) shear walls with one-way joist system.

The classical mode of interaction between a prismatic shear wall and a moment frame is shown in Fig. 4.7; the frame deflects in a so-called shear mode whereas the shear wall predominantly responds by bending as a cantilever. Compatibility of horizontal deflection generates interaction between the two. The linear sway of the moment frame, combined with the parabolic sway of the shear wall, results in enhanced stiffness because the wall is restrained by the frame at the upper levels while at the lower levels the shear wall is restrained by the frame. However, a frame consisting of closely spaced columns

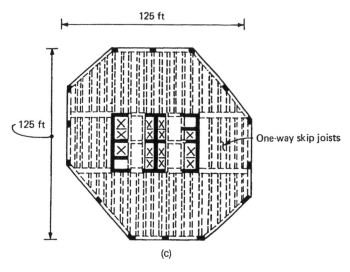

(c)

Figure 4.6. *(Continued)*

and deep beams tends to behave more like a shear wall responding predominantly in a bending mode. And similarly, a shear wall weakened by large openings acts more like a frame by deflecting in a shear mode. The combined structural action, therefore, depends on the relative rigidity of the two, and their modes of deformation.

4.1.9. Frame Tube System

In this system, the perimeter of the building consists of closely spaced columns connected by deep spandrels. The system works as a hollow vertical cantilever and is efficient because of the maximum distance separating the windward and leeward columns. However, lateral drift due to the axial displacement of the columns—commonly referred to as chord drift—and web drift, caused by shear and bending deformations of the spandrels and columns, may be quite large depending upon the tube geometry. For example, if the plan aspect ratio is large, say, much in excess of 1:2.5, it is likely that supplemental lateral bracing may be necessary to satisfy drift limitations. The economy of the tube system

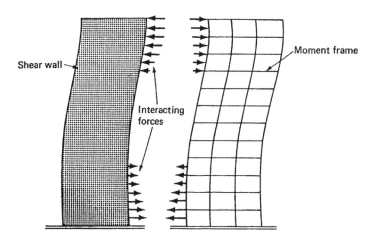

Figure 4.7. Shear wall–frame interaction.

therefore depends on factors such as spacing and size of columns, depth of perimeter spandrels, and the plan aspect ratio of the building. This system should, however, be given consideration for buildings taller than about 40 stories.

4.1.10. Exterior Diagonal Tube

By applying structural principles similar to those of a trussed steel tube, it is possible to visualize a concrete system consisting of closely spaced exterior columns with blocked-out windows at each floor to create a diagonal pattern on the building facade. The diagonals carry lateral shear forces in axial compression and tension, thus eliminating bending in the columns and girders. Currently, two buildings have been built using this approach. The first is a 50-story office building in New York, and the second is a mixed-use building in Chicago. The structural system for the building in New York consists of a combination of a framed and a trussed tube interacting with a system of interior core walls. The building is 570 ft (173.73 m) tall with a height-to-width ratio of 8:1. Schematic elevation and floor plan of the building are shown in Fig. 4.8.

Figure 4.8. Exterior braced tube: (a) schematic elevation; (b) plan.

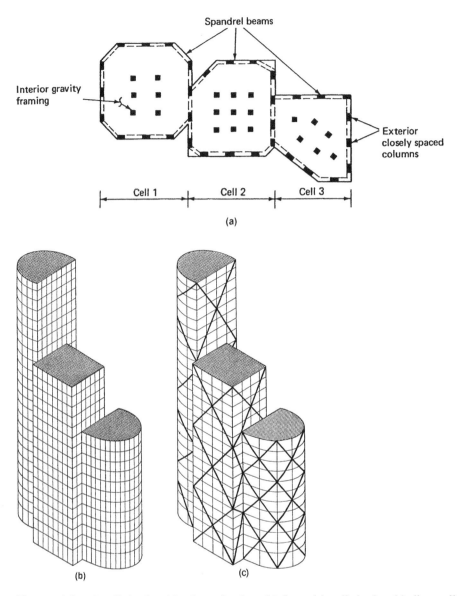

Figure 4.9. Bundled tube: (a) schematic plan; (b) framed bundled tube; (c) diagonally braced bundled tube.

4.1.11. Bundled Tube

The underlying principle is to connect two or more individual tubes in a bundle with the object of decreasing shear lag effects (Fig. 4.9a). Two basic versions are possible using either framed or diagonally braced tubes as shown in Fig. 4.9b and c. A mixture of the two is, of course, feasible.

4.1.12. Miscellaneous Systems

Figure 4.10 shows a schematic plan of a building with a cap wall consisting of a 1- or 2-story-high outrigger wall that connects the core to the perimeter columns. A 1- or 2-story

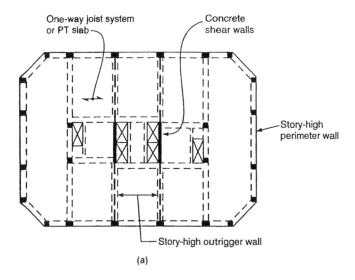

One-way joist system
or PT slab

Concrete
shear walls

Story-high
perimeter wall

Story-high outrigger wall

(a)

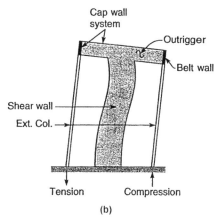

Cap wall
system

Outrigger

Belt wall

Shear wall

Ext. Col.

Tension   Compression

(b)

Figure 4.10.   Building with cap wall: (a) schematic plan; (b) structural behavior.

wall at the perimeter acting as a belt wall is typically used in the system to tie the exterior columns together. The cap wall at the top tends to reverse the bending curvature of the cantilever shear core. A substantial portion of moment in the core is thus transferred to the perimeter columns by inducing tension in the windward columns and compression in the leeward columns (Fig. 4.10b). Optimum locations discussed in Chapter 3, for single and multiple outriggers related to steel systems, are also relevant to concrete systems (Fig. 4.11). In high seismic zones it is prudent to use a 1- or 2-story-deep vierendeel-type ductile frame for outriggers and belt trusses instead of walls. A schematic elevation of a building with a two-story vierendeel outrigger is shown in Fig. 4.12.

Buildings with high plan aspect ratios tend to be inefficient in resisting lateral loads because of shear lag effects. By introducing interior columns (three at every other floor in the example building shown in Fig. 4.13), it is possible to reduce the effect of shear lag, and thus increase the bending efficiency. A 2-story haunch girder vierendeel frame at every other floor effectively ties the building exterior columns to the interior shear walls thus mobilizing the entire flange–frames in resisting overturning moments.

One concept of full-depth interior bracing interacting with the building's perimeter frame is shown in Fig. 4.14. The interior diagonal bracing consists of a series of wall

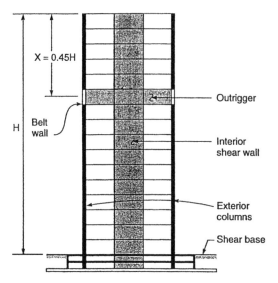

Figure 4.11. Single outrigger system: optimum location.

panels interconnected between interior columns to form a giant K-brace stretched out for the full width of the building.

A system suitable for super-tall buildings—taller than, say, 80 stories—is shown in Fig. 4.15. It consists of a service core located at each corner of the building interconnected by a super diagonal in-fill bracing. The service core at each corner acts as a giant column carrying most of the gravity load and overturning moments. The eccentricity between the super diagonals and exterior columns is a deliberate design strategy to enhance the ductility of the lateral bracing systems. The ductile response of the links helps in dissipating seismic energy, thus assuring the gravity-carrying capacity of the building during and after a large earthquake.

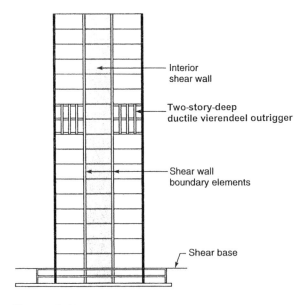

Figure 4.12. Outrigger system: seismic version.

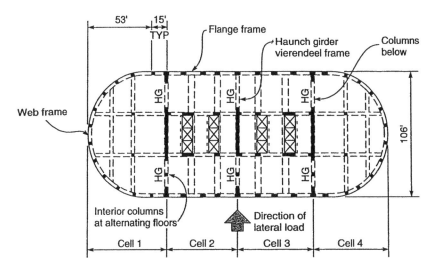

Figure 4.13. Cellular tube with interior vierendeel frames.

4.2. SEISMIC DESIGN

For buildings in regions of low seismic risk—UBC zones 0 and 1—or buildings assigned to SDC A or B, the provisions for the design of elements given in the first 20 chapters of the ACI code are considered sufficient. There are no requirements for special ductile detailing for walls or moment frames. In regions of moderate seismic risk—UBC zones 2A and 2B—or for buildings assigned to SDC C, there is no special ductile detailing required for walls. However, some ductile detailing requirements for moment frames including flat slab–frames are required. In regions of high seismicity—UBC zones 3 and 4 or SDC D, E, and F—almost all structural elements require ductile detailing.

Given the ready availability of computer programs, the analysis of a building is the easy part and, in a broad sense, is the same whether the building is in a high or a low seismic zone. The detailing requirements, particularly at the joints, are the factor that sets the designs apart.

In regions of low seismic risk, it is likely that a building will never experience forces that will result in an inelastic excursion of the building. For these buildings, a safe and economic design is achieved by using an appropriate margin of safety against gravity and lateral overload. This is typically realized in structural steel design by using allowable stress design (ASD), which limits the allowable stress to a percentage of yield stress.

In ultimate strength design, also referred to as strength design, or load resistance factor design (LRFD), the margin of safety is achieved by use of load factors and strength reduction factors. Either of these two methods, ASD or LRFD, results in structures that are believed to have an adequate margin of safety against overloads. Put another way, the probability of yielding of the structure designed by these methods is considered very low. Structural deflections under lateral loads are expected to be elastic and thus fully recoverable. For example, a very tall building, say, at a height of 1400 feet, on a windy day may experience as much as 3 feet of lateral deflection but would not endure any permanent deflection. The elastic design used in the sizing of structural members for these loads assures that after the winds have subsided, the building would come back to its prewind plumbness without any permanent set.

Such is not the case for buildings in moderate-to-high seismic-risk zones. Yes, they too respond elastically under the most severe wind conditions, because the design is meant

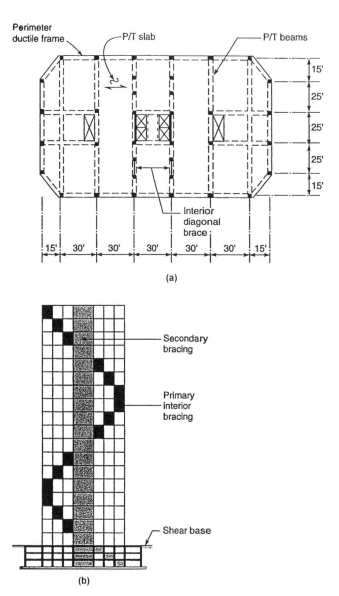

Figure 4.14. Full-depth interior brace: (a) plan; (b) schematic section.

to keep the structure elastic under the generally predictable wind loads. However, the lateral loads that we use in seismic design are highly unpredictable. We know this much, as past earthquakes have taught us: The magnitudes of lateral loads experienced by buildings under large earthquakes are so large that an elastic design under these loads is simply not possible. The building designed to perform elastically in a large seismic event will have structural members so large, and costing so much more, that society has accepted the risk of buildings going beyond their elastic limit, with the stipulation that they do not fall down or collapse. In other words, a building may be utterly damaged beyond repair and may never be occupied again, but if it stays up, providing life safety for the building occupants during and after a large earthquake, it is deemed to have performed adequately under present seismic codes.

The collapse of a building is generally preventable if brittle failure of its members and connections is prevented. In other words, the structural elements may bend and twist

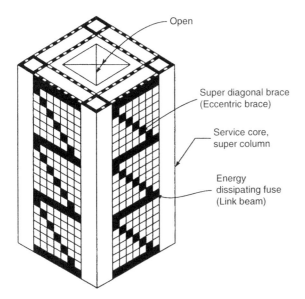

Open

Super diagonal brace
(Eccentric brace)

Service core,
super column

Energy
dissipating fuse
(Link beam)

Figure 4.15. Eccentric bracing system for super-tall buildings.

to their hearts' content, but may not snap. The intent, then, is to build ductility into the structure so that it will absorb energy, and thus prevent sudden breaking up of members that result in collapse.

Therefore, structures in regions of high seismic risk are detailed to have ductility. The degree of detailing is entirely dependent on the severity of seismic risk. This is the reason that a building in seismic zone 3 or 4, or assigned to SDC D, E, or F, is designed to be more ductile than its counterpart in a less severe seismic zone, or assigned to SDC A, B, or C. The vast difference in design requirements may be appreciated by studying Table 4.1, which gives a comparison of nonseismic and seismic design criteria for moment frames and shear walls.

Seismic design coefficients for concrete buildings as specified in UBC 1997 and IBC-03 are given in Tables 4.2 and 4.3. Since both IBC-03 and NFPA 5000 are based on the same resource document—ASCE 7-02—Table 4.3 is also applicable to NFPA 5000.

Seismic design of reinforced concrete buildings entails the following steps:

1. Determination of design earthquake forces, including
   - Calculation of base shear corresponding to the computed or estimated fundamental period of vibration of the structure.
   - Distribution of the base shear over the height of the building.
2. Analysis of the structure for the lateral forces calculated in step 1, as well as under gravity and wind loads, to obtain member design forces and story drift ratios.

   Observe that for certain classes of structures having plan or vertical irregularities, or for structures over 240 feet in height, dynamic analysis is required by most codes. The story shears, moments, drifts, and deflections determined from dynamic analysis are to be adjusted for the static base shear value.
3. Design of members and joints for the most unfavorable combination of gravity and lateral loads, including the design and detailing of members and their connections to ensure their ductile behavior.

TABLE 4.1 Seismic and Nonseismic Design Criteria Comparison, ACI 318-02

Member	Type of checks/design	Ordinary moment-resisting frames (nonseismic)	Intermediate moment-resisting frames (seismic)	Special moment-resisting frames (seismic)
Frame–column	Column design: Flexure and axial loads	Ultimate load combinations $1\% < \rho < 8\%$	Ultimate load combinations $1\% < \rho < 8\%$	Ultimate load combinations. Column capacity $\geq \frac{6}{5}$ beam capacity with $\alpha = 1.0$, $\sum M_c \geq \frac{6}{5} \sum M_g$ $1\% < \rho < 6\%$
	Column design: Shear	Ultimate load combinations	Modified ultimate load combination (earthquake loads doubled). Column capacity $\phi = 1.0$ and $\alpha = 1.0$	Ultimate load beam capacity combinations with $\phi = 1.0$ and $\alpha = 1.25$
Frame–beam	Beam design: Flexure	Ultimate load combinations	Ultimate load combinations	Ultimate load combinations. $\rho_{\max} \leq 0.025$
	Beam min. moment requirements	No requirement	$M_{u\text{END}}^+ \geq \frac{1}{3} M_{u\text{END}}^-$ $M_{u\text{SPAN}}^+ \geq \frac{1}{5} M_{u\text{END}}^+$ $M_{u\text{SPAN}}^- \geq \frac{1}{5} M_{u\text{END}}^-$	$M_{u\text{END}}^+ \geq \frac{1}{2} M_{u\text{END}}^-$ $M_{u\text{SPAN}}^+ \geq \frac{1}{4} M_{u\text{END}}^+$ $M_{u\text{SPAN}}^- \geq \frac{1}{4} M_{u\text{END}}^-$
	Beam design: Shear	Ultimate load combinations	Modified ultimate load combinations (earthquake loads doubled). Beam capacity shear (V_p) with $\alpha = 1.0$ and $\phi = 1.0$ plus V_{D+L}	Beam shear capacity $\geq$ plastic shear (V_p) with $\alpha = 1.25$ and $\phi = 1.0$ plus V_{D+L} (ult). $V_c = 0$, if $V_p \geq \frac{V_{\text{total}}}{2}$, and axial compressive force $< \frac{A_g f_c}{20}$.
Beam–column joint	Shear design	No requirement	No requirement	Shear capacity of joint area, $A_j \geq$ Beam plastic shear capacity (V_p) with $\alpha = 1.25$ and $\phi = 1.0$
	Beam/column ratios	No requirement	No requirement	Column capacity based on uniaxial capacity under axial loads from ultimate load combinations $\geq$ beam capacity with $\alpha = 1.0$.
Shear wall	Flexure design	Ultimate load combinations; no special ductile detailing	Ultimate load combinations; no special ductile detailing	Boundary elements as required by displacement-based or stress-based design.
	Shear design	Ultimate load combinations; no special requirement	Ultimate load combinations; no special requirement	$\phi = 0.6$ (shear controlled) $\phi = 0.75$ (flexure controlled) $\phi_{\text{diaphragm}} \leq \phi_{\text{shear wall}}$

TABLE 4.2 1997 UBC Seismic Coefficients; Concrete Systems

System	Zone 3 or 4			Zone 2A or 2B			Zone 1		
	R	Ω_o	H	R	Ω_o	H	R	Ω_o	H
Bearing wall systems									
• Shear walls	4.5	2.8	160	4.5	2.8	NL	4.5	2.8	NL
• Braced frames	Not allowed			2.8	2.2	NL	2.8	2.2	NL
Building frame systems									
• Shear walls	5.5	2.8	240	5.6	2.2	NL	5.5	2.8	NL
• Braced frames	Not allowed			5.6	2.2	NL	5.5	2.8	NL
Moment-resisting frames									
• SMRF	8.5	2.8	NL	8.5	2.8	NL	8.5	2.8	NL
• IMRF	Not allowed			5.5	2.8	NL	5.5	2.8	NL
• OMRF	Not allowed			Not allowed			3.5	2.8	NL
Dual systems									
• Shear walls + SMRF	8.5	2.8	NL	8.5	2.8	NL	8.5	2.8	NL
• Shear walls + IMRF	Not allowed			6.5	2.8	NL	6.5	2.8	NL
• Braced frames + SMRF	Not allowed			6.5	2.8	NL	6.5	2.8	NL
• Braced frames + IMRF	Not allowed			4.2	2.8	NL	4.2	2.8	NL

SMRF: Special moment-resisting frame (ductile frame).
IMRF: Intermediate moment-resisting frame (semiductile frame).
OMRF: Ordinary moment-resisting frame (nonductile frame).
NL: No height limit.
H: Height in feet.

The above steps are carried out in each principal direction of the buildings, assuming that the design lateral forces act nonconcurrently in each principal direction. However, for certain building categories that may be sensitive to torsional oscillations or characterized by significant irregularities, and for elements forming part of two or more intersecting lateral force-resisting systems, orthogonal effects need to be considered. The orthogonal effects requirement is deemed to be satisfied if the design is based on the more severe combination of 100% of the prescribed seismic forces in one direction plus 30% of the forces in the perpendicular direction.

Experience has shown that reinforced concrete members achieve ductility when certain limits are placed on steel in tension and on concrete in compression. Reinforced concrete beams with common proportions can possess ductility under monotonic loading even greater than common steel beams in which buckling is usually a limiting factor. However, providing stability or resistance to reversed inelastic strains requires special detailing. Thus, there is a wide range of reduction factors from elastic response to design response, depending on the degree of detailing provided for achieving stable and assured resistance. The essence of seismic detailing is to prevent premature shear failures in members and joints, buckling of compression bars, and crushing of concrete. It is not sufficient to have only strength capability; there must also be special details to actualize the inelastic behavior of the seismic-resisting elements to ensure that the system remains stable at deformations corresponding to maximum expected ground motion. Vertical loads must be supported even when maximum elastic deformations are exceeded. In other words, inelastic yielding is allowed in resisting seismic loads as long as yielding does not impair the vertical load capacity of the structure.

Why shear reinforcement in a beam–column joint? Because the mechanism of shear failure in a joint is different from shear-flexure failure in beams, the nominal shear

TABLE 4.3 Seismic Design Coefficients for Concrete Systems: ASCE 7-02

Basic seismic force-resisting system	Response modification coefficient, R	System over strength factor, Ω_o	Deflection amplification factor, C_d	System limitations and building height limitations (feet) by seismic design category				
				A or B	C	D	E	F
1. Bearing wall systems								
Special reinforced concrete shear walls	5	2½	5	NL	NL	160	160	160
Ordinary reinforced concrete shear walls	4	2½	4	NL	NL	NP	NP	NP
2. Building frame systems								
Special reinforced concrete shear walls	6	2½	5	NL	NL	160	160	100
Ordinary reinforced concrete shear walls	5	2½	4½	NL	NL	NP	NP	NP
3. Moment-resisting frame systems								
Special reinforced concrete moment frames	8	3	5½	NL	NL	NL	NL	NL
Intermediate reinforced concrete moment frames	5	3	4½	NL	NL	NP	NP	NP
Ordinary reinforced concrete moment frames	3	3	2½	NL	NP	NP	NP	NP
4. Dual systems with special moment frames capable of resisting at least 25% of prescribed seismic forces								
Special reinforced concrete shear walls	8	2½	6½	NL	NL	NL	NL	NL
Ordinary reinforced concrete shear walls	7	2½	6	NL	NL	NP	NP	NP
5. Dual systems with intermediate moment frames capable of resisting at least 25% of prescribed seismic forces								
Special reinforced concrete shear walls	6	2½	5	NL	NL	160	100	100
Ordinary reinforced concrete shear walls	5½	2½	4½	NL	NL	NP	NP	NP
6. Inverted pendulum systems and cantilevered column systems								
Special reinforced concrete moment frames	2½	2	1¼	NL	NL	NL	NL	NL

NL = No height limit
NP = Not permitted
(Adapted from ASCE 7-02, Table 9.5.2.2)

capacities are considerably higher than the values the designers in the nonseismic areas are accustomed to. For example, ACI 318-02 permits the ultimate shear stress in joints as listed:

- $\phi\, 20\sqrt{f_c'}$ for joints confined on all four sides.
- $\phi\, 15\sqrt{f_c'}$ for joints confined on three sides or two opposite sides.
- $\phi\, 12\sqrt{f_c'}$ for all other cases.

Compare this to the value of $\phi 4\sqrt{f_c'}$ allowed for punching shear in flat slabs without shear reinforcement, and $\phi 6\sqrt{f_c'}$ with shear reinforcement. The relatively high values for joint shear allowed in seismic design may give the wrong impression that joint shear will not be a problem in sizing of columns in high seismic zones. This is not the case. Even with the very high shear stresses permitted, joint shear most often controls the size of frame–columns. Also note that shear reinforcement extending through the beam–column joint is required even though no increase in shear capacity is credited for its presence.

Why a strong column–weak beam? The reason is to prevent a story mechanism. This is achieved by assuring that, at each beam–column joint, the flexural resistance of columns is substantially (20%) more than the flexural strength of beams. In calculating the nominal flexural strength of columns, the effect of column axial loads should be included.

Why minimum positive reinforcement? The reason for minimum positive moment at beam ends is because actual seismic loads are much larger than what we calculate, and also reverse in direction. The bending moment and shear in a beam, therefore, can be positive or negative at different points in time. Simple elastic analysis and load combinations cannot possibly give reliable results, hence, the necessity to provide a minimum capacity for positive moments at the ends as well as negative moments at the midspans of beams.

Why closely spaced ties? To ensure that a plastic hinge develops in frame beams, it is necessary: 1) to attain yielding of reinforcement well before concrete fails in compression; 2) and to provide transverse reinforcement at close intervals to confine the concrete core within the longitudinal reinforcement.

Closely spaced ties enhance the ductility of concrete by allowing large compression strains to develop in concrete without spalling. The ties prevent buckling of longitudinal bars. A buckled or kinked bar has a tendency to fracture when the bar straightens in tension under load reversals. Therefore, in seismic detailing it is necessary to use seismic hooks in the ties. This is because when the concrete cover spalls, the hoops may themselves be exposed and lose their confining capacity. The ties must be anchored into the confined zone of concrete. A structure designed in accordance with seismic provisions is expected to survive even after responding inelastically in strong earthquakes.

The members must be designed and detailed with prior realization of the inevitability of inelastic response. The detailing provisions promote a relatively benign ductile response rather than an undesirable brittle response. This is achieved by ensuring that members have inelastic energy-dissipation characteristics, through yielding of reinforcement as opposed to the shearing or crushing failure of concrete.

The vertical elements designed to partake in energy dissipation should have proper confinement such that the vertical load-carrying capacity is not compromised. The seismic provisions encourage formation of beam hinges rather than column hinges to prevent story mechanisms. To ensure adequate flexural ductility in critical regions of beams, most codes specify a factor of 1.25 as the ratio of ultimate tensile strength to actual yield strength of the reinforcement. Also, the amount by which the actual yield strength can exceed the minimum specified value is limited to 18,000 psi. The joints of frames are designed for shears corresponding to the development of maximum beam moments, assuming the longitudinal reinforcement is stressed to 1.25 times the specified yield strength. This is to allow for the effects of strain hardening and for the possibility of actual yield strengths exceeding the specified minimum values.

Why diagonal reinforcement in deep coupling beams? Coupling beams designed as conventional flexural members with stirrups, and with some shear resistance allocated to concrete, are unsuitable for energy dissipation by formation of plastic hinges at the beam ends, as implied for typical frame beams. The relatively short beam between the walls has a tendency to divide itself into two triangular parts if the shear force associated with the flexural overstrength of the beam cannot be effectively transmitted by the vertical stirrups.

This consideration has led to the use of diagonal reinforcement in relatively deep coupling beams. The shear resistance is provided by the diagonal tension and compression in the reinforcement. This results in a very ductile behavior that can then sustain large deformations imposed on the beams during seismic inelastic excursions.

Why Boundary Elements in special reinforced concrete walls? Boundary elements and corresponding detailing requirements are required at the vertical edges of walls to provide proper confinement of concrete at these locations. These are required to confine the concrete where the strain at the extreme fiber of the wall exceeds a threshold value when the wall is subjected to a lateral displacement corresponding to a displacement likely to occur in a large earthquake. A similar provision is given based on calculated compressive stress at the extreme fibers. In either case, transverse reinforcement similar to that required for a frame column is required to prevent buckling of the longitudinal reinforcement due to cyclic load reversals.

Why heavy transverse reinforcement in frame–columns? The amount of transverse reinforcement provided in columns is controlled by four design requirements: 1) shear strength; 2) lateral support of compression reinforcement to prevent buckling; 3) confinement of highly stressed compression zones, both in potential plastic hinge regions and over the full height of columns; and 4) prevention of bond strength loss within column vertical bar splices.

1. Shear resistance. Some or all of the design force V_u must be resisted by the transverse reinforcement in the form of spiral or circular hoops and column ties. The approach to shear design in potential plastic hinge regions is different from that for other parts of column, as in beam shear design.
2. Lateral support for compression reinforcement. Antibuckling reinforcement should be provided in the plastic hinge regions of frame columns in the same manner as for the end regions of frame beams. The design of transverse reinforcement in between the end regions is as for nonseismic design. However, most usually the minimum spacing requirements for shear strength or confinement of compression reinforcement generally govern the spacing.
3. Confinement of concrete. Confinement is essential to preserve adequate rotational ductility in potential plastic hinge regions of columns. Lengths of potential plastic hinge regions in columns are generally smaller than beams partly because column moments vary along the story height with a relatively large gradient. Therefore the region of a frame column subjected to tension yielding of reinforcement is limited. The full amount of confining reinforcement is required for the entire plastic region, with only one-half of this required in between.

4. Transverse reinforcement at lapped splices. Splicing of reinforcement in structural members is not a requirement but a necessity for building practical structures. This is commonly achieved by overlapping parallel bars. Force transmission from one bar to the next occurs through the response of surrounding concrete. However, when large forces are to be transmitted by bond, splitting of concrete resulting in cracks may develop. To mobilize a load path for the force transmission through the cracked concrete, a shear friction reinforcement, in the form of transverse reinforcement, is required at lapped splices.

It is most important to design and detail the reinforcement in members and their connections to ensure their ductile behavior and thus allow the structure to sustain, without collapse, the severe distortions that may occur during a major earthquake. This requirement—intended to ensure adequate ductility in structural elements—represents the major difference between the design requirements for conventional, nonearthquake-resistant structures and those located in regions of high seismic risk.

4.2.1. Load Factors, Strength Reduction Factors, and Load Combinations

Concrete structures are commonly designed in the United States using the ultimate strength method. Since the American Concrete Institute published ACI 318-71, the term "ultimate" has been dropped, so that what used to be referred to as ultimate-strength design is now simply called strength design. In this approach, structures are proportioned such that their ultimate capacity is equal to or greater than the required ultimate strength. The required strength is based on the most critical combination of factored loads, obtained by multiplying specified service loads by appropriate load factors. The capacity of an element, on the other hand, is obtained by applying a strength reduction factor ϕ to the nominal resistance of the element. Load factors are intended to take into account the variability in the magnitude of the specified loads. Lower load factors are used for types of loads that are less likely to vary significantly from the specified values. To allow for the lesser likelihood of certain types of loads occurring simultaneously, reduced load factors are specified for some loads when considered in combination with other loads.

For the most common dead load D, live load L, roof live load L_r, wind load W, and earthquake load E, the simplified load combinations of ACI 318-02 are

$$U = 1.4D$$
$$U = 1.2D + 1.6L + 0.5L_r$$
$$U = 1.2D + 1.6L_r + (1.0L \text{ or } 0.8W)$$
$$U = 1.2D + 1.6W + 1.0L + 0.5L_r$$
$$U = 1.2D + 1.0E + 1.0L$$
$$U = 0.9D + 1.6W$$
$$U = 0.9D + 1.0E$$

The designer is referred to ACI 318-02, Section 9.2, for load combinations that include loads due to:

1. H = weight and pressure of soil, water in soil, or other materials.
2. F = weight and pressures of fluids.
3. T = temperature, creep, shrinkage, differential settlement.
4. R = rain load.
5. S = snow load.

ACI 318-02 permits a reduction of 50% on the load factor for L, except for garages, areas occupied as places of public assembly, and all areas where the live load L is greater than 100 lb/ft^2.

The load factor of 1.6 for wind is based on the premise that the designers will be using wind loads determined by the provisions of ASCE 7-02 which includes a factor for directionality that is equal to 0.85 for buildings. Therefore, the corresponding load factor for wind is increased accordingly in the ACI 318-02 (1.3/0.85 = 1.53 rounded up to 1.6). Use of a previous wind load factor of 1.3 is permitted when wind load is obtained from other sources that do not include the directionality factor.

A reduced load factor of 1.0 for earthquake forces is used because model building codes such as ASCE 7-02 have converted earthquake forces to strength level.

ASCE 7-02 and IBC-03 require the same combinations except the effect of seismic load E is defined as follows:

$$E = \rho Q_E + 0.2 S_{DS} D$$
$$E = \rho Q_E - 0.2 S_{DS} D$$

where

$\quad\quad E$ = the effect of horizontal and vertical earthquake-induced forces
$\quad\quad S_{DS}$ = the design spectral response acceleration at short periods
$\quad\quad D$ = the effect of dead load
$\quad\quad \rho$ = the reliability or the penalty factor for buildings in which the lateral resistance is limited to only few members in the structure. The maximum value of ρ is limited to 1.5.
$\quad\quad Q_E$ = the effect of horizontal seismic forces

The factor $0.2 S_{DS}$ placed on the dead load in the above equations is to account for the effects of vertical acceleration.

For situations where failure of an isolated, individual, brittle element can result in the loss of a complete lateral force-resisting system or in instability and collapse, ASCE 7-02 has a specific requirement to determine the seismic design forces. These elements are referred to as collector elements. Columns supporting discontinuous lateral load-resisting elements such as walls also fall under this category.

Seismic loads for such elements are as follows:

$$E = \Omega_o Q_E + 0.2 S_{DS} D$$
$$E = \Omega_o Q_E - 0.2 S_{DS} D$$

where Ω_o is the system overstrength factor, defined as the ratio of the ultimate lateral force the structure is capable of resisting to the design strength. The value of Ω_o varies between 2 to 3 depending on the type of lateral force-resisting system (See Table 4.3).

In concrete buildings, the capacity of a structural element is calculated by applying a strength reduction factor, ϕ, to the nominal strength of the element. The factor ϕ is intended to take account of variations in material strength and uncertainties in the estimation of the nominal member strength, the nature of the expected failure mode, and the importance of a member to the overall safety of the structure.

The values of the strength reduction factor ϕ are

$\phi = 0.90$ for tension-controlled sections (no change from the previous edition)

$\phi = 0.70$ for spirally reinforced compression members

$\phi = 0.65$ for other reinforced members

$\phi = 0.75$ for shear and torsion

$\phi = 0.65$ for bearing on concrete (except for post-tensioned anchorage zones and strut-and-tie models)

$\phi = 0.85$ for post-tensioned anchorage zones

However, an exception to the value of $\phi = 0.75$ in shear is specified for structures designed in high seismic zones. For shear capacity calculations of structural members other than joints, a value $\phi = 0.60$ is used when the nominal shear strength of a member is less than the shear corresponding to the development of the nominal flexural strength of the member. For shear in joints and diagonally reinforced coupling beams, ϕ is equal to 0.85. The above exception applies mainly to brittle members such as low-rise walls, portions of walls between openings, or diaphragms that are impractical to reinforce to raise their nominal shear strength above nominal flexural strength for the pertinent loading conditions.

Reference is made in the remainder of this chapter to various equations and sections given in ACI 318. Unless specifically stated otherwise, it is understood, that these refer to ACI 318-02.

4.2.2. Integrity Reinforcement

The goal of structural integrity is: If a structure or part of a structure is subjected to an abnormal loading condition, or if a primary element sustains damage from an unanticipated event, tying the members together should result in confining the resulting damage to a relatively small area. Requirements for structural integrity included in Sections 7.13 and 13.3.8.5, focus on the structural detailing of cast-in-place concrete. Basically, prescribed amounts of longitudinal reinforcement must be continuous over the support or reinforcing bars that terminate at discontinuous ends of a member and must be anchored with hooks.

Since accidents and misuse are normally unforeseeable events, they cannot be defined precisely. Similarly, providing general structural integrity to a structure is a requirement that cannot be stated in simple terms. The Code's performance provision—"a structure shall be effectively tied together to improve integrity of the overall structure"—requires considerable judgment on the part of the design engineer. Opinions among engineers differ on the effectiveness of a general structural integrity solution for a particular framing system. However, the Code does set forth specific examples of certain reinforcing details for cast-in-place joists, beams, and two-way slab construction.

With damage to a support, top reinforcement that is continuous over the support will tend to tear out of the concrete. It will not provide the catenary action needed to bridge the damaged support unless it is confined by stirrups. By making a portion of the bottom reinforcement in beams continuous over supports, some catenary action can be provided. By providing some continuous top and bottom reinforcement in edge or perimeter beams, an entire structure can be tied together. Also, continuous ties provided to perimeter beams of a structure will toughen the exterior portion of a structure, should an exterior column be severely damaged.

Provisions for integrity reinforcement, first introduced in ACI 318-89, require continuous reinforcement in beams around the perimeter of the structure. The required amount is at minimum one-sixth of the tension reinforcement for negative moment at the support and one-fourth of the tension reinforcement for positive moment at the midspan. In either case a minimum of two bars is required. Continuity in rebars is achieved by providing class A tension lap splicers, mechanical or welded splices in cast-in-place joists and beams.

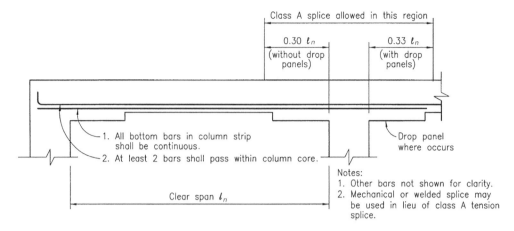

Figure 4.16. Structural integrity reinfothtrcement in flat slabs without beams.

Two-way slabs. In a two-way slab construction, all bottom bars within the column strip in each direction must be lap-spliced with class A tension laps. Figure 13.3.8 given in the ACI 318-02 (Fig. 4.16 of this text) shows the locations where the lap splices are permitted. At least two of the bottom bars in the column strip must pass within the core of the columns and be anchored at exterior supports.

Joists. At least one reinforcing bar in the bottom of a rib is to be continuous over supports or the bar must be spliced with a class A tension lap splice to a bar in the adjacent span. At discontinuous ends of joists, anchorage of at least one bottom bar must be provided with a standard hook (Fig. 4.17).

Beams. Beams are categorized as either perimeter or nonperimeter beams. A spandrel beam would be a perimeter beam. The detailing of top and bottom bars and of stirrups in perimeter beams is impacted by the structural integrity provisions. At least one-sixth of the $-A_s$ required for negative-factored moment at the face of supports and one-quarter of the $+A_s$ required for positive-factored moment at midspan are to be made continuous around the perimeter of the structure. Closed stirrups are also required in perimeter beams. It is not necessary to place closed stirrups within the joints. It is permissible to provide continuity of the top and bottom bars by splicing the top bars at midspan and the bottom bars at or near the supports. Lap-splicing with class A tension lap splices is required (Fig. 4.18).

For nonperimeter beams, the engineer has two choices to satisfy the structural integrity requirements: 1) provide closed stirrups; or 2) make at least one-quarter of the $+A_s$

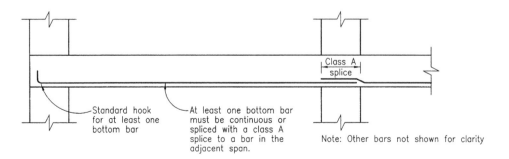

Figure 4.17. Structural integrity reinforcement in joists.

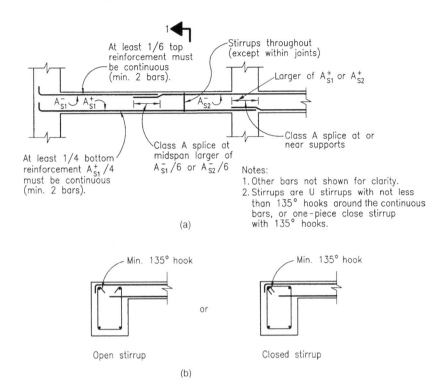

(a)

(b)

Figure 4.18.   Integrity reinforcement in perimeter beams: (a) perimeter beam elevation; (b) Section 1.

required for positive-factored moment at midspan continuous. Splicing the prescribed number of bottom bars over the supports with class A tension lap spices is acceptable. At discontinuous ends of nonperimeter beams, the bottom bars must be anchored with standard hooks (Fig. 4.19). In all cases, mechanical or welded splices may be used instead of class A tension lap splices.

4.2.3.   Intermediate Moment-Resisting Frames

4.2.3.1.   General Requirements: Frame Beams

General requirements for frame beams of intermediate moment frames given in ACI 318-02 Sections 21.12.2 and 21.12.3 are as follows:

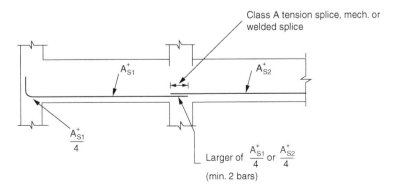

Figure 4.19.   Integrity reinforcement in beams other than perimeter beams.

- Reinforcement details in a frame member shall satisfy Section 21.12.4 if the factored compressive axial load $\leq A_g f_c'/10$.
- If the factored compressive axial load $> A_g f_c'/10$, frame reinforcement details shall satisfy Section 21.12.5, unless the member has spiral reinforcement in accordance with Eq. (10-5).
- If a two-way slab system without beams is treated as part of the lateral force-resisting system, reinforcement details in any span-resisting moments caused by lateral forces shall satisfy Section 21.12.6.
- Design shear strength of beams, columns, and two-way slabs resisting earthquake effects shall not be less than either
    - The sum of the shear forces associated with development of nominal moment strengths of the member at each restrained end of the clear span and the shear force calculated for factored gravity loads; or
    - The maximum shear force obtained from design load combinations that include earthquake effects, with the shear force from earthquake effects assumed to be twice that prescribed by the governing code for earthquake-resistant design.

4.2.3.2. *Flexural and Transverse Reinforcement: Frame Beams*

Flexural and transverse reinforcement requirements for frame beams given in Sections 21.12.4.1, 21.12.4.2, and 21.12.4.3 are as follows:

- Positive moment strength at joint face $\geq$ one-third negative moment strength provided at that face of the joint.
- Neither the negative nor the positive moment strength at any section along the member length shall be less than one-fifth the maximum moment strength provided at the face of either joint.
- Stirrups shall be provided at both ends of a member over a length equal to $2h$ from the face of the supporting member toward midspan.
- The first stirrup shall be located no more than 2 in. from the face of the supporting member.
- Maximum stirrup spacing shall not exceed
    - $d/4$.
    - $8 \times$ diameter of smallest longitudinal bar.
    - $24 \times$ diameter of stirrup bar.
    - 12 in.
- Stirrups shall be spaced at no more than $d/2$ throughout the length of the member.

Refer to Figs. 4.20 and 4.21 for schematic flexural and transverse reinforcement details for frame beams.

4.2.3.3. *Transverse Reinforcement: Frame Columns*

Transverse reinforcement requirements for frame columns given in Sections 20.12.5.1 through 20.12.5.4 are as follows:

- Maximum tie spacing shall not exceed s_o over a length ℓ_o measured from each joint face. Spacing s_o shall not exceed the smallest of:

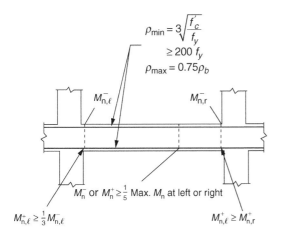

$$\rho_{\min} = 3\sqrt{\frac{f'_c}{f_y}}$$
$$\geq 200\, f_y$$
$$\rho_{\max} = 0.75\rho_b$$

$M_{n,\ell}^-$ $M_{n,r}^-$

M_n^- or $M_n^+ \geq \frac{1}{5}$ Max. M_n at left or right

$M_{n,\ell}^+ \geq \frac{1}{3}M_{n,\ell}^-$ $M_{n,\ell}^+ \geq M_{n,r}^+$

Figure 4.20. Intermediate moment-resisting frame (IMRF); flexural reinforcement requirements for frame beams.

- 8 × diameter of smallest longitudinal bar.
- 24 × diameter of tie bar.
- Minimum member dimension/2.
- 12 in.
- The length ℓ_o shall not be less than the largest of
  - Clear span/6.
  - Maximum cross-sectional dimension of member.
  - 18 in.
- The first tie shall be located no farther than $s_o/2$ from the joint face.
- Joint reinforcement shall conform to Section 11.11.2.
- Tie spacing outside of the length ℓ_o shall not exceed $2s_o$.

Figure 4.22 provides a schematic interpretation of these requirements.

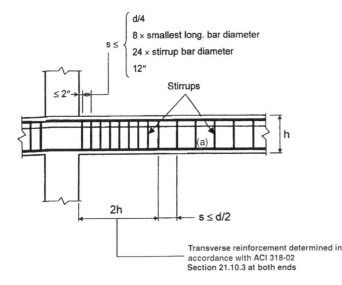

$$s \leq \begin{cases} d/4 \\ 8 \times \text{smallest long. bar diameter} \\ 24 \times \text{stirrup bar diameter} \\ 12'' \end{cases}$$

≤ 2″

Stirrups

(a)

h

2h

$s \leq d/2$

Transverse reinforcement determined in accordance with ACI 318-02 Section 21.10.3 at both ends

Figure 4.21. Intermediate moment-resisting frame (IMRF); transverse reinforcement requirements for frame beams.

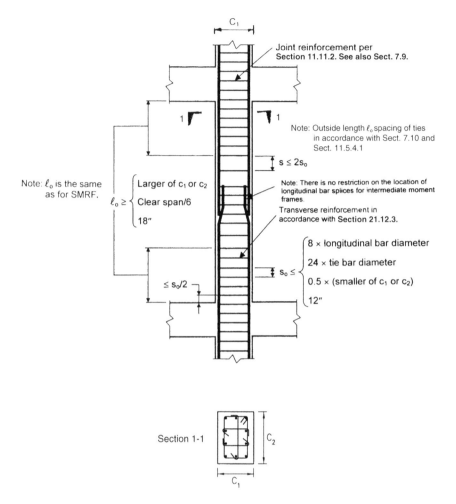

Figure 4.22. Intermediate moment-resisting frame; transverse reinforcement requirements for frame columns.

4.2.3.4. Detailing Requirements for Two-Way Slab Systems Without Beams

Detailing requirements given in Sections 21.12.6.1 through 21.12.6.7 may be summarized as follows:

- All reinforcement provided to resist M_s shall be placed within the column strip defined in Section 13.2.1.
- Reinforcement to resist $\gamma_f M_s$ shall be placed within the effective slab width defined in Section 13.5.3.2.
- Not less than one-half of column strip reinforcement at the support shall be placed within the effective slab width defined in Section 13.5.3.2.
- Not less than one-quarter of top reinforcement at the support in the column strip shall be continuous throughout the span.
- All bottom reinforcement in the column strip shall be continuous or spliced with class A splices. At least two of the column strip bottom bars shall pass within the column core and shall be anchored at exterior supports.
- Not less than one-half of all bottom reinforcement at midspan shall be continuous and shall develop its yield strength at the face of the support as defined in Section 13.6.2.5.

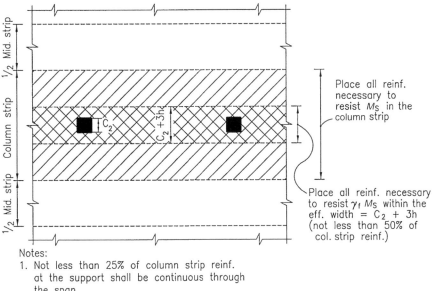

Notes:
1. Not less than 25% of column strip reinf. at the support shall be continuous through the span.
2. See Fig. 4.16 for integrity reinf.

Figure 4.23. Seismic detailing requirements for two-way slabs in areas of moderate seismic risk; flat slab–beams not permitted in UBC zones 3 and 4, or for buildings assigned to SDC C, D, E, or F.

• At discontinuous edges of the slab, all top and bottom reinforcement at the support shall be developed at the face of the support as defined in Section 13.6.2.5.

Refer to Figs. 4.23, 4.24, and 4.25 for pictorial representations of these items.

4.2.4. Special Moment-Resisting Frames

4.2.4.1. *General Requirements: Frame Beams*

General requirements for the design and detailing of special moment-resisting frames (SMRF) given in Sections 21.3.1.1 through 21.3.1.4 are summarized as follows:

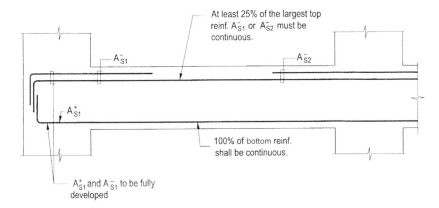

Figure 4.24. Seismic detailing requirements for two-way slabs in areas of moderate seismic risk; column strip.

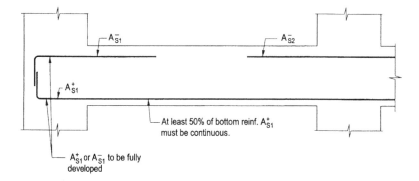

Figure 4.25. Seismic detailing requirements for two-way slabs in areas of moderate seismic risk; middle strip.

- Factored axial compressive force $\leq A_g f'_c / 10$.
- Clear span $\geq 4 \times$ effective depth.
- Width-to-depth ratio ≥ 0.3.
- Width ≥ 10 in.
- Width $\leq$ width of supporting member (measured on a plane perpendicular to the longitudinal axis of the flexural member) + distances on each side of the supporting member not exceeding three-fourths of the depth of the flexural member.

See Fig. 4.26 for schematics of general requirements.

4.2.4.2. *Flexural Reinforcement: Frame Beams*

This last requirement, referring to the width limitation, effectively eliminates the use of flat slabs as frame beams in areas of high seismicity or for buildings assigned to SDC D, E, or F.

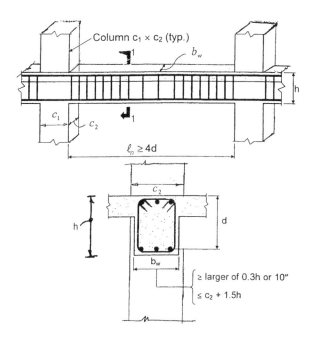

Figure 4.26. Frame beam; general requirements, special moment frame.

Structural requirements for flexural reinforcements and their splices given in Sections 21.3.2.1 through 21.3.2.4 for frame beams are as follows:

- Minimum reinforcement shall not be less than

$$\frac{3\sqrt{f_c'}}{f_y} \times b_w d \quad \text{and} \quad \frac{200\, b_w d}{f_y}$$

  at any section, top and bottom, unless provisions of Section 10.5.3 are satisfied.
- The reinforcement ratio ρ shall not exceed 0.025.
- At least two bars must be provided continuously at both top and bottom of section.
- Positive moment strength at joint face shall be $\geq 1/2$ negative moment strength provided at that face of the joint.
- Neither the negative nor the positive moment strength at any section along the member length shall be less than $1/4$ the maximum moment strength provided at the face of either joint.
- Lap splices of flexural reinforcement are permitted only if hoop or spiral reinforcement is provided over the lap length. Hoop and spiral reinforcement spacing shall not exceed
  - $d/4$.
  - 4 in.
- Lap splices are not permitted
  - Within joints.
  - Within a distance of $2h$ from the face of the joint.
  - At locations where analysis indicates flexural yielding caused by inelastic lateral displacements of the frame.
- Mechanical splices shall conform to Section 21.2.6 and welded splices shall conform to Section 21.2.7.1.

4.2.4.3. *Transverse Reinforcement: Frame Beams*

Requirements for transverse reinforcement (hoops and stirrups) in frame beams given in Sections 21.3.3.1 through 21.3.3.6 and Section 21.3.4 are summarized as follows:

- Hoops are required in the following regions of frame members:
  - Over a length equal to $2h$ from the face of the supporting member toward midspan at both ends of the flexural member.
  - Over lengths equal to $2h$ on both sides of a section where flexural yielding may occur in connection with inelastic lateral displacements of the frame.
- Where hoops are required, the spacing shall not exceed:
  - $d/4$.
  - $8 \times$ diameter of smallest longitudinal bar.
  - $24 \times$ diameter of hoop bars.
  - 12 in.
- The first hoop shall be located no more than 2 in. from the face of the supporting member.
- Where hoops are required, longitudinal bars on the perimeter shall have lateral support conforming to Section 7.10.5.3.
- Where hoops are not required, stirrups with seismic hooks at both ends shall be spaced at a distance not more than $d/2$ throughout the length of the member.

- Stirrups or ties required to resist shear shall be hoops over lengths of members in Sections 21.3.3, 21.4.4, and 21.5.2.
- Hoops in flexural members shall be permitted to be made up of two pieces of reinforcement: a stirrup having seismic hooks at both ends and closed by a crosstie. Consecutive crossties engaging the same longitudinal bar shall have their 90-degree hooks at opposite sides of the flexural member. If the longitudinal bars secured by the crossties are confined by a slab on only one side of the flexural frame member, the 90-degree hooks of the crossties shall be placed on that side.
- Transverse reinforcement must also be proportioned to resist the design shear forces.

Figures 4.27 and 4.28 show transverse reinforcement schematics for frame beams.

4.2.4.4.   General Requirements: Frame Columns

The requirements given in Section 21.4 are summarized as follows:

- Factored axial compressive force $> A_g f_c'/10$.
- Shortest cross-sectional dimension measured on a straight line passing through the geometric centroid ≥ 12 in.
- Ratio of the shortest cross-sectional dimension to the perpendicular dimension ≥ 0.4.

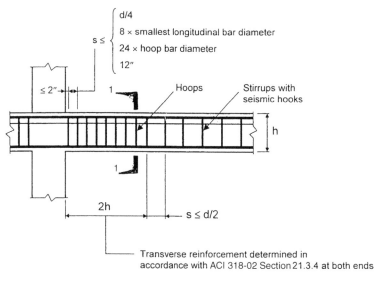

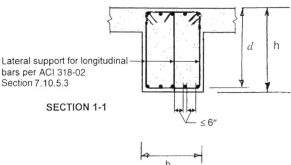

Figure 4.27.   Frame beam; transverse reinforcement requirements, special moment frame.

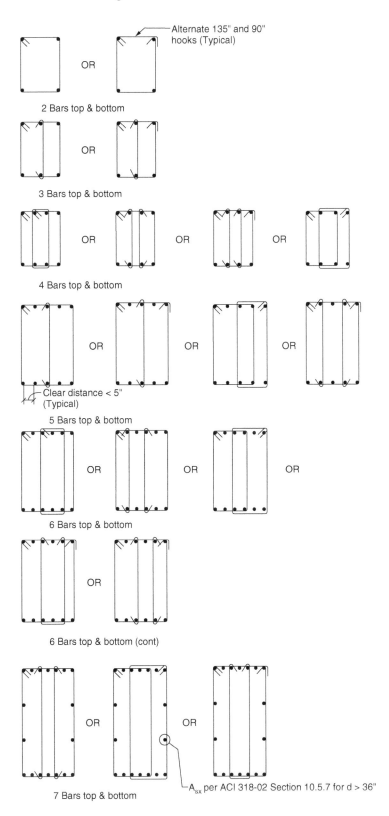

Figure 4.28.   Arrangement of hoops and crossties; frame beams; special moment frame.

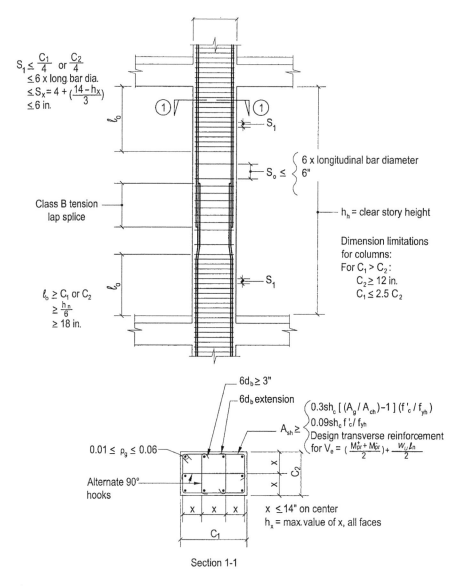

$S_1 \leq \dfrac{C_1}{4}$ or $\dfrac{C_2}{4}$
$\leq 6 \times$ long. bar dia.
$\leq S_x = 4 + \left(\dfrac{14 - h_x}{3}\right)$
≤ 6 in.

$6 \times$ longitudinal bar diameter
$6''$

$S_o \leq \left\{ \begin{array}{l} \end{array}\right.$

Class B tension lap splice

h_n = clear story height

Dimension limitations for columns:
For $C_1 > C_2$:
 $C_2 \geq 12$ in.
 $C_1 \leq 2.5\, C_2$

$\ell_o \geq C_1$ or C_2
$\geq \dfrac{h_n}{6}$
≥ 18 in.

$6d_b \geq 3''$

$6d_b$ extension

$A_{sh} \geq \begin{cases} 0.3sh_c\,[\,(A_g/A_{ch})-1\,]\,(f'_c/f_{yh}) \\ 0.09sh_c\,f'_c/f_{yh} \\ \text{Design transverse reinforcement} \\ \text{for } V_e = \left(\dfrac{M^+_{pr}+M^-_{pr}}{2}\right)+\dfrac{w_u\,l_n}{2} \end{cases}$

$0.01 \leq \rho_g \leq 0.06$

Alternate $90°$ hooks

$x \leq 14''$ on center
h_x = max. value of x, all faces

C_1

Section 1-1

Figure 4.29. Frame column; detailing requirements, special moment frame.

4.2.4.5. Flexural Reinforcement: Frame Columns

Refer to Figs. 4.29 and 4.30 for schematic details and minimum requirements of transverse reinforcement.

- The flexural strengths of columns shall satisfy the following:

$$\sum M_c \geq (6/5) \sum M_g \tag{21.1}$$

where
  $\sum M_c$ = sum of moments at the faces of the joint, corresponding to the nominal flexural strength of the columns framing into that joint. Column flexural strength shall be calculated for the factored axial force, consistent with the direction of the lateral forces considered, resulting in the lowest flexural strength.

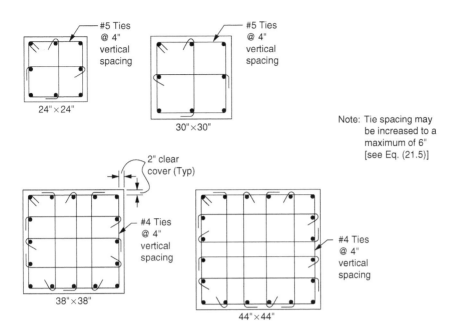

Figure 4.30. Examples of minimum transverse reinforcement in frame columns of SMRF (Eqs. 21.3 and 21.4).

Note: $f_c' = 5$ ksi, $f_y = 60$ ksi

Vertical spacing of ties = 4 in.

Ties #5 for 24" × 24" and 30" × 30" columns

#4 for 38" × 38" and 44" × 44" columns

ΣM_g = sum of moments at the faces of the joint, corresponding to the nominal flexural strength of the girders framing into that joint. In T-beam construction, slab reinforcement within an effective slab width defined in Section 8.10 shall contribute to flexural strength.

- If Eq. (21.1) is not satisfied, the lateral strength and stiffness of the columns shall not be considered when determining the strength and stiffness of the structure, and the columns shall conform to Section 21.11. Also, the columns must have transverse reinforcement over their full height as specified in Sections 21.4.4.1 through 21.4.4.3.
- The reinforcement ratio ρ_g shall not be less than 0.01 and shall not exceed 0.06.
- Mechanical splices shall conform to Section 21.2.6 and welded splices shall conform to Section 21.2.7.1. Lap splices are permitted only within the center half of the member length, must be tension lap splices, and shall be enclosed within transverse reinforcement conforming to Sections 21.4.4.2 and 21.4.4.3.

4.2.4.6. *Transverse Reinforcement: Frame Columns*

- The transverse reinforcement requirements discussed in the following need be provided only over a length ℓ_o from each joint face and on both sides of any section where flexural yielding is likely to occur. The length ℓ_o shall not be less than
  - Depth of member at joint face or at section where flexural yielding is likely to occur.

- Clear span/6.
- 18 in.
- Ratio of spiral or circular hoop reinforcement ρ_s shall not be less than that given by:

$$\rho_s = 0.12 \frac{f_c'}{f_{yh}} \geq 0.45 \left(\frac{A_g}{A_c} - 1 \right) \frac{f_c'}{f_{yh}} . \qquad (21.2) \text{ and } (10.6)$$

- Total cross-sectional area of rectangular hoop reinforcement for confinement A_{sh} shall not be less than that given by the following two equations:

$$A_{sh} = 0.3(sh_c f_c'/f_{yh})[(A_g/A_{ch}) - 1], \qquad (21.3)$$

$$A_{sh} = 0.09 sh_c f_c'/f_{yh} . \qquad (21.4)$$

- Transverse reinforcement shall be provided by either single or overlapping hoops. Crossties of the same bar size and spacing as the hoops are permitted, with each end of the crosstie engaging a peripheral longitudinal reinforcing bar. Consecutive crossties shall be alternated end for end along the longitudinal reinforcement.
- Eqs. (21.3) and (10.6) need not be satisfied if the design strength of the member core satisfies the requirement of the design loading combinations, including the earthquake effects.

    If the thickness of the concrete outside of the confining transverse reinforcement >4 in., additional transverse reinforcement shall be provided at a spacing ≤12 in. Concrete cover on the additional reinforcement ≤4 in.
- Transverse reinforcement shall be spaced at distances not exceeding
  - Minimum member dimension/4.
  - 6 × longitudinal bar diameter.
  - s_x

    where $4 \text{ in.} \leq s_x = 4 + \dfrac{14 - h_x}{3} \leq 6 \text{ in.}$ \qquad (21.5)

- Crossties or legs of overlapping hoops shall not be spaced more than 14 in. on center in the direction perpendicular to the longitudinal axis of the structural member. Vertical bars shall not be farther than 6 in. clear from a laterally supported bar.

    Where transverse reinforcement as required in Sections 21.4.4.1–21.4.4.3 is no longer required, the remainder of the column shall contain spiral or hoop reinforcement spaced at distances not to exceed
  - 6 × longitudinal bar diameter.
  - 6 in.
- Transverse reinforcement must also be proportioned to resist the design shear forces.
- Columns supporting reactions from discontinued stiff members, such as walls, shall have transverse reinforcement as specified in Sections 21.4.4.1–21.4.4.3 over their full height, if the factored axial compressive force related to earthquake effects $>A_g f_c'/10$. This transverse reinforcement shall extend into the discontinued member for at least the development length of the largest longitudinal reinforcement in the column in accordance with Section 21.5.4.

- If the lower end of the column terminates on a wall, transverse reinforcement per Sections 21.4.4.1–21.4.4.3 shall extend into the wall for at least the development length of the largest longitudinal bar in the column at the point of termination.
- If the column terminates on a footing or mat, transverse reinforcement per Sections 21.4.4.1–21.4.4.3 shall extend at least 12 in. into the footing or mat.

Schematic details of reinforcement for a ductile frame are shown in Fig. 4.31.

4.2.4.7. Transverse Reinforcement: Joints

Transverse reinforcement requirements for joints of SMRFs given in Sections 21.5.2.1 through 21.5.2.3 are as follows:

- Transverse hoop reinforcement required for column ends per Section 21.4.4 shall be provided within a joint, unless structural members confine the joint as specified in Section 21.5.2.2.
- Where members frame into all four sides of a joint and each member width is at least ¾ the column width, the transverse reinforcement within the depth of the shallowest member may be reduced to ½ of the amount required by Section 21.4.4.1. The spacing of the transverse reinforcement required in Section 21.4.4.2(b) shall not exceed 6 in. at these locations.
- Transverse reinforcement per Section 21.4.4 shall be provided through the joint to confine longitudinal beam reinforcement outside the column core if a beam framing into the joint does not provide such confinement.

Figure 4.31 shows reinforcement schematics for a ductile frame:

4.2.4.8. Shear Strength of Joint

Shear strength requirements for joints in special moment-resisting frames (SMRFs) given in Sections 21.5.3.1 and 21.5.3.2, are summarized as follows:

- For normal weight concrete, the nominal shear strength of the joint shall not exceed the following forces:
  - For joints confined on all four faces $20\sqrt{f_c'}A_j$
  - For joints confined on three faces or on two opposite faces $15\sqrt{f_c'}A_j$
  - For other joints $12\sqrt{f_c'}A_j$

  where A_j = effective cross-sectional area within a joint in a plane parallel to the plane of the reinforcement generating shear in the joint. The overall depth shall be the overall depth of the column. Where a beam frames into a support of larger width, the effective width of the joint shall not exceed the smaller of
  1. Beam width plus the joint depth.
  2. Twice the smaller perpendicular distance from the longitudinal axis of the beam to the column side.
- A joint is considered confined if the confining members frame into all faces of the joint. A member is considered to provide confinement at the joint if the framing member covers at least ¾ of the joint face.
- For lightweight aggregate concrete, the nominal shear strength of the joint shall not exceed ¾ of the limits given in Section 21.5.3.1.

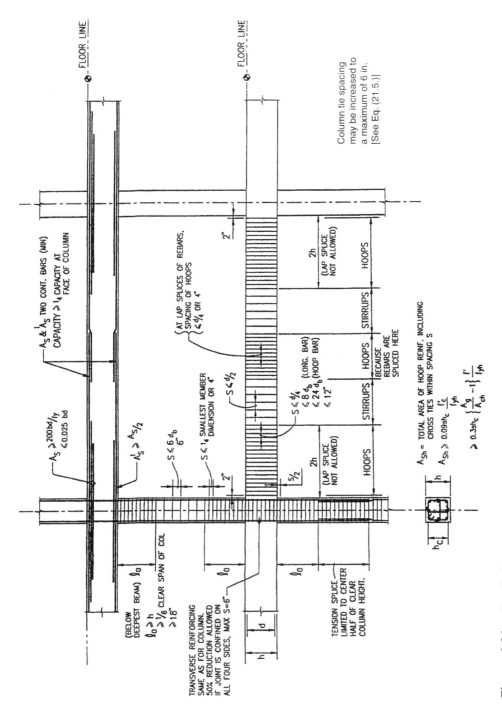

Figure 4.31. Ductile frame, special moment-resisting frame (SMRF); schematic reinforcement detail.

4.2.4.9. Development of Bars in Tension

Criteria for development of bars in tension, given in Sections 21.5.4.1 through 21.5.4.4, are as follows:

- For normal weight concrete, the development length ℓ_{dh} for a bar with a standard 90-degree hook shall not be less than the largest of
  - $8 \times$ diameter of the bar.
  - 6 in.
  - $f_y d_b/(65\sqrt{f_c'})$ (21.6)

  for bar sizes No. 3 through No. 11. The 90-degree hook shall be located within the confined core of a column or boundary element.
- For lightweight aggregate concrete, the development length ℓ_{dh} for a bar with a standard 90-degree hook shall not be less than the largest of:
  - $10 \times$ diameter of the bar.
  - 7.5 in.
  - $1.25 f_y d_b/(65\sqrt{f_c'})$

  for bar sizes No. 3 through No. 11. The 90-degree hook shall be located within the confined core of a column or boundary element.
- For bar sizes No. 3 through No. 11, the development length ℓ_d for a straight bar shall not be less than:
  - $2.5\,\ell_{dh}$ if the depth of the concrete cast in one lift beneath the bar ≤ 12 in.
  - $3.5\,\ell_{dh}$ if the depth of the concrete cast in one lift beneath the bar >12 in.
- Straight bars terminated at a joint shall pass through the confined core of a column or boundary element. Any portion of the straight embedment length not within the confined core shall be increased by a factor of 1.6.
- For epoxy-coated reinforcement, the development lengths in Sections 21.5.4.1–21.5.4.3 shall be multiplied by
  - 1.5 for straight bars with cover less than $3d_b$ or clear spacing less than $6d_b$.
  - 1.2 for all other straight bars.
  - 1.2 for bars terminating in a standard hook.

4.2.5. Shear Walls

4.2.5.1. Minimum Web Reinforcement: Design for Shear

Requirements for minimum web reinforcement and design for shear strength of shear walls are given in Sections 21.7.2.1 through 21.7.4.5. A summary follows:

- The required amounts of vertical and horizontal web reinforcement depend on the magnitude of the design shear force V_u:
  - For $V_u \leq A_{cv}\sqrt{f_c'}$:

    Vertical reinf. ratio ≥ 0.0012 for No. 5 bars or smaller,
    ≥ 0.0015 for No. 6 bars or larger.
    Horizontal reinf. ratio ≥ 0.0020 for No. 5 bars or smaller,
    ≥ 0.0025 for No. 6 bars or larger.
  - For $V_u > A_{cv}\sqrt{f_c'}$:
    $\rho_v \geq 0.0025$,
    $\rho_n \geq 0.0025$.

- Reinforcement spacing each way shall not exceed 18 in.
- Reinforcement provided for shear strength shall be continuous and shall be distributed across the shear plane.

- For $V_u > 2A_{cv}\sqrt{f'_c}$, two curtains of reinforcement must be provided.
- All continuous reinforcement in structural walls shall be anchored or spliced in accordance with the provisions for reinforcement in tension in Section 21.5.4.
- The nominal shear strength V_n of structural walls shall not exceed:

$$V_n = A_{cv}\left(\alpha_c\sqrt{f'_c} + \rho_n f_y\right)$$

where $\alpha_c = 3.0$ for $h_w/l_w \leq 1.5$
$\qquad = 2.0$ for $h_w/l_w \geq 2.0$.
α_c varies linearly between 3.0 and 2.0 for h_w/l_w between 1.5 and 2.0.

- The value of h_w/l_w used for determining V_n for segments of a wall shall be the larger of the ratios for the entire wall and the segment of wall considered.
- Walls shall have distributed shear reinforcement in two orthogonal directions in the plane of the wall. If $h_w/l_w \leq 2.0$, $\rho_v \geq \rho_n$.
- Nominal shear strength of all wall piers sharing a common lateral force shall not be assumed to exceed $8A_{cv}\sqrt{f'_c}$, where A_{cv} is the total cross-sectional area, and the nominal shear strength of any one of the individual wall piers shall not be assumed to exceed $10A_{cp}\sqrt{f'_c}$, where A_{cp} is the cross-sectional area of the pier considered.
- Nominal shear strength of horizontal wall segments and coupling beams shall be assumed not to exceed $10A_{cp}\sqrt{f'_c}$, where A_{cp} is the cross-sectional area of a horizontal wall segment or coupling beam.

4.2.5.2. Boundary Elements

Boundary element requirements for shear walls given in Sections 21.7.6.2 through 21.7.6.4 are as follows:

- Compression zones of walls or wall piers that are effectively continuous over their entire height and designed to have a single critical section for flexure and axial loads shall be reinforced with special boundary elements where:

$$c \geq \frac{\ell_w}{600(\delta_u/h_w)} \tag{21.8}$$

where $\delta_u/h_w \geq 0.007$.
- Special boundary element reinforcement shall extend vertically from the critical section a distance not less than the larger of l_w or $M_u/4V_u$.
- Structural walls not designed by the provisions of Section 21.7.6.2 shall have special boundary elements at boundaries and around openings of structural walls where the maximum extreme fiber compressive stress, corresponding to factored forces including earthquake effects, exceeds $0.2f'_c$.
- Special boundary elements may be discontinued where the calculated compressive strength is less than $0.15f'_c$.
- Stresses shall be calculated using a linearly elastic model and gross section properties.
- Where special boundary elements are required by Sections 21.7.6.2 or 21.7.6.3, the following shall be satisfied:

- The boundary element shall extend horizontally from the extreme compression fiber a distance not less than the larger of $c - 0.1l_w$ and $c/2$.
- In flanged sections, the boundary element shall include the effective flange width in compression and shall extend at least 12 in. into the web.
- Special boundary element transverse reinforcement shall satisfy the requirements of Sections 21.4.4.1 through 21.4.4.3, except Eq. (21.3) need not be satisfied.
- Special boundary element transverse reinforcement at the base of the wall shall extend into the support at least the development length of the largest longitudinal bar in the special boundary element. If the special boundary element terminates on a footing or mat, the special boundary element transverse reinforcement shall extend at least 12 in. into the footing or mat.
- Horizontal reinforcement in the web shall be anchored to develop the specified yield strength f_y within the confined core of the boundary element.
- Mechanical splices and welded splices of longitudinal reinforcement of boundary elements shall conform to Sections 21.2.6 and 21.2.7, respectively.

Although boundary elements may not be required by calculations, Section 21.6.6.5 stipulates certain requirements as follows:

- Where special boundary elements are not required by Sections 21.7.6.2 or 21.7.6.3, the following shall be satisfied:
  - Boundary transverse reinforcement shall satisfy Sections 21.4.4.1(c), 21.4.4.3, and 21.7.6.4(c) if the longitudinal reinforcement ratio at the wall boundary is greater than $400/f_y$. The maximum longitudinal spacing of transverse reinforcement in the boundary shall not exceed 8 in.
  - Horizontal wall reinforcement terminating at the ends of structural walls without boundary elements shall have a standard hook engaging the edge reinforcement or the edge reinforcement shall be enclosed in U-stirrups having the same size and spacing as, and spliced to, the horizontal reinforcement when $V_u \geq A_{cv}\sqrt{f_c'}$.

4.2.5.3. *Coupling Beams*

Design requirements for coupling beams given in Sections 21.7.7.1 through 21.7.7.4 are as follows:

- Coupling beams with aspect ratio $l_n/d \geq 4$ shall satisfy the requirements of Eq. (21.3), except the provisions of Sections 21.3.1.3 and 21.3.1.4(a) shall not be required if it can be shown by analysis that the beam has adequate lateral stability.
- Coupling beams with aspect ratio $l_n/d < 4$ shall be permitted to be reinforced with two intersecting groups of diagonally placed bars symmetrical about the midspan.
- Coupling beams with aspect ratio $l_n/d < 2$ and $V_u > 4\sqrt{f_c'}b_w d$ shall be reinforced with two intersecting groups of diagonally placed bars symmetrical about the midspan, unless it can be shown that loss of stiffness and strength of the coupling beams will not impair the vertical load carrying capacity of the structure, or the egress from the structure, or the integrity of nonstructural components and their connections to the structure.

- Coupling beams reinforced with two intersecting groups of diagonally placed bars symmetrical about the midspan shall satisfy the following:
  - A minimum of four bars is required in each group of diagonally placed bars. Each diagonal group of bars is assembled in a core having sides measured to the outside of transverse reinforcement greater than or equal to $b_w/2$ perpendicular to the plane of the beam and $b_w/5$ in the plane of the beam and perpendicular to the diagonal bars.
  - The nominal shear strength V_n is determined from the following:

$$V_n = 2A_{vd}f_y \sin \alpha \le 10\sqrt{f_c'}b_w d. \tag{21.9}$$

  - Each group of diagonally reinforced bars shall be enclosed in transverse reinforcement satisfying Sections 21.4.4.1 through 21.4.4.3. The minimum concrete cover required in Section 7.7 shall be assumed on all four sides of each group of diagonally placed reinforcing bars for purposes of computing A_g in Eqs. (10.6) and (21.3).
  - The diagonally placed bars shall be developed for tension in the wall.
  - The diagonally placed bars shall be considered to contribute to the nominal flexural strength of the coupling beam.
  - Reinforcement conforming to Sections 11.8.9 and 11.8.10 shall be provided as a minimum parallel and transverse to the longitudinal axis of the beam.

4.2.6. Frame Members Not Designed to Resist Earthquake Forces

Detailing requirements for frame members not designed to resist earthquake forces are given in Sections 21.11.2 and 21.11.3. Requirements of Section 21.11.2 are for frame members expected to experience only moderate excursions into inelastic range during design earthquake motions. Those given in Section 21.11.3 are for members expected to experience nearly the same magnitude of inelastic deformations as members designed to resist earthquake motions. If $M_u \le \phi M_n$ and $V_u \le \phi V_n$, the members are designed according to Section 21.11.2 (case 1). If $M_u \ge \phi M_n$ and $V_u \ge \phi_2 V_n$, the detailing requirements are more stringent, i.e., nearly the same as those specified for members proportioned to resist forces induced by earthquake motions (case 2).

Case 1. $M_u \le \phi M_n$ and $V_u \le \phi V_n$

- Factored gravity axial force $\le A_g f_c'/10$
  - Satisfy detailing requirements of Section 21.3.2.1.
  - Provide stirrups spaced not more than $d/2$ throughout the length of the member.
- Factored gravity axial force $> A_g f_c'/10$.
  - Satisfy detailing requirements of Sections 21.4.3, 21.4.4.1(c), 21.4.4.3, and 21.4.5.
  - Maximum longitudinal spacing of ties shall be s_o for the full column height.
  - Spacing s_o shall not be more than the smaller of 6 diameters of the smallest longitudinal bar enclosed or 6 in.
- Factored gravity axial force $> 0.35 P_o$.
  - Satisfy detailing requirements of Section 21.11.2.2.

- Provide transverse reinforcement ≥ one-half of that required by Section 21.4.4.1.
- Maximum longitudinal spacing of ties shall be s_o for the full column height.
- Spacing s_o shall not be more than the smaller of 6 diameters of the smallest longitudinal bar or 6 in.

Case 2: $M_u > \phi M_n$ or $V_u > \phi V_m$ or induced moments not calculated

- Materials shall satisfy Sections 21.2.4 and 21.2.5. Mechanical and welded splices shall satisfy Sections 21.2.6 and 21.2.7.1, respectively.
- Factored gravity axial force $\leq A_g f_c'/10$.
  - Satisfy detailing requirements of Sections 21.3.2.1 and 21.3.4.
  - Provide stirrups spaced not more than $d/2$ throughout the length of the member.
- Factored gravity axial force $> A_g f_c'/10$.
  - Satisfy detailing requirements of Sections 21.4.4, 21.4.5, and 21.5.2.1.

4.2.7. Diaphragms

4.2.7.1. *Minimum Thickness and Reinforcement*

Minimum thickness and reinforcement requirements for diaphragms as given in Sections 21.9.4 and 21.9.5.1 through 21.9.5.5 are as follows:

- Concrete slabs and composite topping slabs serving as structural diaphragms to transmit earthquake forces shall not be less than 2 in. thick.
- Topping slabs over precast floor or roof elements, acting as structural diaphragms and not relying on composite action with the precast elements to resist earthquake forces, shall not be less than $2^1/2$ in. thick.
- For structural diaphragms:
  - Minimum reinforcement shall be in conformance with Section 7.12.
  - Spacing of nonprestressed reinforcement shall not exceed 18 in.
  - Where welded wire fabric is utilized to resist shear forces in topping slabs over precast floor and roof elements, the wires parallel to the span of the precast elements shall be spaced not less than 10 in. on center.
  - Reinforcement provided for shear strength shall be continuous and shall be distributed uniformly across the shear plane.
  - In diaphragm chords or collectors utilizing bonded prestressing tendons as primary reinforcement, the stress due to design seismic forces shall not exceed 60,000 psi.
  - Precompression from unbonded tendons shall be permitted to resist diaphragm design forces if a complete load path is provided.
- Structural truss elements, struts, ties, diaphragm chords, and collector elements shall have transverse reinforcement in accordance with Sections 21.4.4.1 through 21.4.4.3 over the length of the element where compressive stresses exceed $0.2 f_c'$. Special transverse reinforcement may be discontinued where the compressive stress is less than $0.15 f_c'$. Stresses shall be calculated for the factored forces using a linearly elastic model and gross section properties.
- All continuous reinforcement in diaphragms, trusses, struts, ties, chords, and collector elements shall be anchored or spliced in accordance with the provisions for reinforcement in tension as specified in Section 21.5.4.

- Type 2 splices are required where mechanical splices are used to transfer forces between the diaphragm and the vertical components of the lateral force-resisting system.

4.2.7.2. Shear Strength

Shear strength requirements for diaphragms given in Section 21.9 are summarized as follows:

- The nominal shear strength V_n of structural diaphragms shall not exceed:

$$V_n = A_{cv}\left(2\sqrt{f_c'} + \rho_n f_y\right).$$

- The nominal shear strength of cast-in-place composite-topping slab diaphragms and cast-in-place noncomposite topping slab diaphragms on a precast floor or roof shall not exceed:

$$V_n = A_{cv}\rho_n f_y$$

where A_{cv} is based on the thickness of the topping slab. The required web reinforcement shall be distributed uniformly in both directions.
- Nominal shear strength shall not exceed $8A_{cv}\sqrt{f_c'}$, where A_{cv} is the gross cross-sectional area of the diaphragm.

4.2.7.3. Boundary Elements

A summary of boundary element requirements for diaphragms given in Sections 21.9.8.1 through 21.9.8.3 are given as follows:

- Boundary elements of structural diaphragms shall be proportioned to resist the sum of the factored axial forces acting in the plane of the diaphragm and the force obtained by dividing the factored moment at the section by the distance between the boundary elements of the diaphragm at that section.
- Splices of tensile reinforcement in chords and collector elements of diaphragms shall develop f_y of the reinforcement. Mechanical and welded splices shall conform to Sections 21.2.6 and 21.2.7, respectively.
- Reinforcement for chords and collectors at splices and anchorage zones shall have either
  - A minimum spacing of 3 longitudinal bar diameters, but not less than $1^1/2$ in., and a minimum concrete cover of $2^1/2$ longitudinal bar diameters, but not less than 2 in.; or
  - Transverse reinforcement per Section 11.5.5.3, except as required in Section 21.9.5.3.

4.2.8. Foundations

4.2.8.1. Footings, Mats, and Piles

Structural requirements for footings, foundation mats, and piles are given in Sections 21.10.2.1 through 21.10.2.5 and in Section 22.10. They are summarized as follows:

- Longitudinal reinforcement of columns and structural walls resisting eartquake-induced forces shall extend into the footing, mat, or pile cap, and shall be fully developed for tension at the interface.

- Columns designed assuming fixed end conditions at the foundation shall comply with Section 21.10.2.1.
- If longitudinal reinforcement of a column requires hooks, the hooks shall have a 90-degree bend and shall be located near the bottom of the foundation with the free end of the bars oriented towards the center of the column.
- Transverse reinforcement in accordance with Section 21.4.4 shall be provided below the top of a footing when columns or boundary elements of special reinforced concrete structural walls have an edge located within one-half the footing depth from an edge of a footing. The transverse reinforcement shall extend into the footing a distance greater than or equal to the smaller of
  - Depth of the footing, mat, or pile cap; or
  - Development length in tension of the longitudinal reinforcement.
- Flexural reinforcement shall be provided in the top of a footing, mat, or pile cap supporting columns or boundary elements of special reinforced concrete structural walls subjected to uplift forces from earthquake effects. Flexural reinforcement shall not be less than that required by Section 10.5.
- The use of structural plain concrete in footings and basement walls is prohibited, except for specific cases cited in Section 22.10.

4.2.8.2. Grade Beams and Slabs-on-Grade

Requirements for grade beams and slabs-on-grade given in Sections 21.10.3.1 through 21.10.3.4 are summarized as follows:

- Grade beams acting as horizontal ties between pile caps or footings shall have continuous longitudinal reinforcement that shall be developed within or beyond the supported column. At all discontinuities, the longitudinal reinforcement must be anchored within the pile cap or footing.
- Grade beams acting as horizontal ties between pile caps or footings shall be proportioned such that the smallest cross-section dimension is greater than or equal to the clear spacing between connected columns divided by 20, but need not be greater than 18 in.
- Closed ties shall be provided at a spacing not to exceed the lesser of one-half the smallest orthogonal cross-section dimension or 12 in.
- Grade beams and beams that are part of a mat foundation subjected to flexure from columns that are part of the lateral-force-resisting system shall conform to Section 21.3.
- Slabs on grade that resist seismic forces from columns or walls that are part of the lateral-force-resisting system shall be designed as structural diaphragms per Section 21.9.
- The design drawings shall clearly state that the slab on grade is a structural diaphragm and is part of the lateral-force-resisting system.

4.2.8.3. Piles, Piers, and Caissons

Requirements for piles, piers, and caissons are given in Sections 21.10.4.2 through 21.10.4.7. They may be summarized as follows:

- Piles, piers, or caissons resisting tension loads shall have continuous longitudinal reinforcement over the length resisting the design tension forces. The longitudinal reinforcement shall be detailed to transfer tensile forces between the pile cap and the supported structural members.

- Where tension forces induced by earthquake effects are transferred between a pile cap or mat foundation and a precast pile by reinforcing bars that are grouted or post-installed in the top of the pile, the grouting system shall demonstrate by test that it can develop at least $1.25f_y$ of the bar.
- Piles, piers, or caissons shall have transverse reinforcement in accordance with Section 21.4.4 at the following locations:
  - At the top of the member for at least 5 times the member cross-section dimension, but not less than 6 ft below the bottom of the pile cap.
  - Along the entire unsupported length plus the length required in Section 21.10.4.4(a) for portions of piles in soil that is not capable of providing lateral support, or in air or water.
- For precast concrete driven piles, the provided length of transverse reinforcement shall be sufficient to account for potential variations in the elevation in pile tips.
- Concrete piles, piers, or caissons in foundations supporting one- and two-story stud bearing wall construction are exempt from the transverse reinforcement requirements of Sections 21.10.4.4 and 21.5.4.5.
- Pile caps incorporating batter piles shall be designed to resist the full compressive strength of the batter piles acting as short columns. For portions of piles in soil that is not capable of providing lateral support, or in air or water, the slenderness effects of batter piles shall be considered.

4.2.9. Design Examples*

Several design examples are given in the following sections to explain the provisions of the ACI 318 Building Code. The examples range from ordinary moment-resisting frames, OMRFs (some times referred to as nonseismic frames) to coupled shear walls with diagonal beams, applicable to designs in high seismic zones.

The first eight examples are worked out using the provisions of ACI 318-99, although the reference to design equations and seismic provisions are to ACI 318-02. The last two examples of specially reinforced concrete shear walls (often referred to as California walls) are executed using the provisions of ACI 318-02. Note that the use of ACI 318-99 load factors and ϕ factors is still permitted in ACI 318-02, Appendix C.

An attempt is made to keep the numerical work simple. For example, the tension-controlled flexural reinforcement A_s is calculated by using the relation

$$A_s = \frac{M_u}{a_u d}$$

with a_u typically taken at 4.0 or 4.1, for $f_c' = 4000$ psi and $f_y = 60,000$ psi. Other similar shortcuts are used throughout. The designer is referred to standard reinforced concrete design handbooks for more precise design calculations.

4.2.9.1. Frame Beam Example: Ordinary Reinforced Concrete Moment Frame

Given. Figure 4.32 shows frame beam B3 of an ordinary moment frame of a building located in an area of low seismicity corresponding to UBC 1997 seismic zone 0 or 1. The seismic characteristics of the building site are: $S_s = 0.14g$ and $S_1 = 0.03g$. The

* The author wishes to acknowledge gratitude to Filbert B. Apanay for checking the design examples for numerical accuracy.

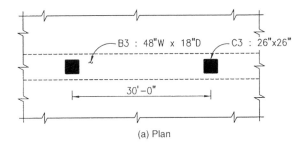

(a) Plan

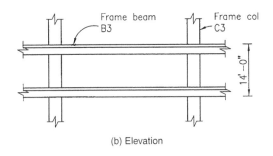

(b) Elevation

Figure 4.32. Frame beam and column example; ordinary moment frame: (a) plan; (b) elevation.

building has been analyzed using a commercially avaible three-dimensional analysis program. Cracked section properties have been input for the members; for beams, $I_{eff} = 0.5I_g$; for columns, $I_{eff} = 0.7I_g$; and for shear walls, $I_{eff} = 0.5I_g$. Rigid diaphragms and rigid-end offsets have been assumed, consistent with the assumptions commonly used in practice. The analysis automatically has taken the effects of $P\Delta$ into consideration. The analysis results for beam B3 are as follows:

Dead load D
 At supports: $M = -150$ kip-ft, $V = 40$ kips
 At midspan: $M = 90$ kip-ft, $V = 0$

Live load L
 At supports: $M = -20$ kip-ft, $V = 12$ kips
 At midspan: $M = 15$ kip-ft, $V = 0$

Wind W
 At supports: $M = \pm 95$ kip-ft, $V = \pm 18$ kips

Required. Design and a schematic reinforcement detail for B3 using the provisions of ACI 318-99. (Note: The design procedure is essentially the same for ACI 318-02 except for the load and ϕ factors)

Solution. The ultimate design load combinations consisting of dead, live, and wind loads are shown in Table 4.4.

Check Limitations on Beam Section Dimensions. According to ACI 318-02 Section 21.2.1.2, the provisions of Chapters 1 through 18 and 22 are adequate to provide a threshold of toughness expected of structures assigned to ordinary categories. These are structures in regions of low seismic risk, corresponding approximately to UBC zones 0 and 1, and assigned to SDC A or B.

No dimensional limitations are specified for frame beams of buildings assigned to SDC A or B. Thus, the given dimensions of 48 in. wide × 18 in. deep for the example

TABLE 4.4   Design Bending Moments and Shear Forces for Frame Beam B3; Ordinary Moment Frame

Load case	Location	Bending moment (kip-ft)	Shear force (kips)
Dead load D	Support	−150	40
	Midspan	90	0
Live load L	Support	−20	12
	Midspan	15	
Wind W	Support	±95	±18
Load combinations (ACI 318-99)			
1. $1.4D + 1.7L$	Support	244	76
	Midspan	152	0
2. $0.75(1.4D + 1.7L + 1.7W)$	Support	−304	80
		−62	34
	Midspan	114	0
3. $0.9D + 1.3W$	Support	−259	59
		12	9
	Midspan	81	0

beam is OK. Note that beam depth of 18 in. satisfies minimum requirements specified in ACI 318-02 Table 9.5(a) for non-prestressed beams and slabs.

Calculate Required Flexural Reinforcement.   At support: $-M_u = 304$ kip-ft (load combination 2, see Table 4.4).

$$-A_s = \frac{304}{4 \times 15.25} = 4.98 \text{ in.}^2$$

Use five #9 at top, giving $-A_s = 5.0$ in.^{2}
At midspan: $+M_u = 152$ kip-ft (load combination 1)

$$+A_s = \frac{152}{4 \times 15.25} = 2.49 \text{ in.}^2$$

Use five #7 at bottom, giving $+A_s = 3.0$ in.^{2}

The reinforcement ratios provided are

$$\rho_{\text{top}} = \frac{5 \times 1}{48 \times 15.25} = 0.0068 \quad \text{and}$$

$$\rho_{\text{bot}} = \frac{5 \times 0.6}{48 \times 15.25} = 0.0041$$

These are more than the minimum required by Section 10.5 and, by inspection, less than the maximum permitted by Section 10.3.3.

Shear Design.   The maximum factored shear force V_u is 80 kips, as calculated in load combination 2, Table 4.5. Assume an equivalent factored uniform load w_u equal to 4.3 kip/ft, the same as calculated for the frame beam B2 of SMRF.

TABLE 4.5 Design Axial Forces, Bending Moments, and Shear Forces for Frame Column C3; Ordinary Moment Frame

Load case	Axial force (kips)	Bending moment (kip-ft)	Shear force (kips)
Dead load D	1500	0	0
Live load L	200	0	0
Wind W	0	±210	±40
Load combinations (ACI 318-99)			
1. $1.4D + 1.7L$	2440	0	0
2. $0.75 (1.4D + 1.7L + 1.7W)$	1830	±268	±30
3. $0.9D + 1.3W$	1350	±273	±52

At critical section distance d from the face of columns, $V_u = 80 - 15.25/12 \times 4.3 = 75$ kips.

$$V_c = 2\sqrt{f_c'}\, b_w d$$

$$= 2\sqrt{4000} \times \frac{48 \times 15.25}{1000}$$

$$= 93 \text{ kips} \tag{11.3}$$

Since V_u is greater than $\phi V_c/2 = 0.85 \times 93/2 = 39.5$ kips, and is less than $\phi V_c = 0.85 \times 93 = 79$ kips, the required shear reinforcement is governed by the minimum specified in Section 11.5.5. Assuming #3 stirrups with four vertical legs, the required spacing s is:

$$s = \frac{A_v f_y}{50\, b_w}$$

$$= \frac{4 \times 0.11 \times 60,000}{50 \times 48} = 11 \text{ in.} \tag{11.5.5}$$

The maximum spacing of shear reinforcement, according to Section 11.5.4, is $d/2 = 15.25/2 = 7.6$ in. or 24 in. Thus, the governing spacing of stirrups is 7.6 in. According to Section 11.5.5.1, stirrups may be discontinued at sections where $V_u \leq V_c/2$. For the example beam, this occurs at 10.2 ft from the face of the column. Provide 18 #3, four-legged stirrups at 7-in. spacing at each end. Place first stirrup 2 in. from the face of support. See Fig. 4.33 for a schematic reinforcement layout.

4.2.9.2. Frame Column Example: Ordinary Reinforced Concrete Moment Frame

Given. Values of axial loads, bending moments, and shear forces obtained from an analysis for column C3 are given in Table 4.5. To keep the calculations simple, the values of bending moments and shear forces due to dead and live loads and the axial load due to wind are assumed negligible.

Required. Design and a schematic reinforcement detail for column C3 using provisions of ACI 318-99.

Solution. Similar to frame beams of OMF, frame columns must satisfy the design provisions of ACI Chapters 1 through 18 and 22. Chapter 22 refers to structural plane concrete and has limited impact on the design.

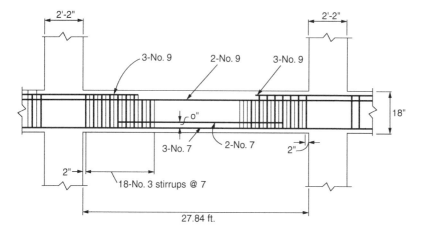

Figure 4.33. Design example, frame beam; ordinary moment frame. Although by calculations no shear reinforcement is required in the midsection of the beam, it is good practice to provide #3 four-legged stirrups at 15 in. spacing.

Since there are no dimensional limitations specified for frame columns of OMF, the given column dimensions of 26 in. × 26 in. are OK.

From an interaction diagram not shown here, a 26 in. × 26 in. column with 12 #11 vertical bars has been found to be adequate for the ultimate load combinations given in Table 4.5. The reinforcement ratio of $(12 \times 1.56)/(26 \times 26) \times 100 = 2.7\%$ is within the maximum and minimum limits of 1 and 8%. Thus the design of column C3 is OK.

Design for Shear.   The shear design of a frame column of OMF is no different from that of a nonframe column. The shear strength of column is verified using ACI Eq. (11.4) for members subject to axial compression:

$$
V_c = 2\left(1 + \frac{N_u}{2000\,A_g}\right)\sqrt{f_c'}\,b_w d
$$

$$
= 2\left(1 + \frac{1,350,000}{2000 \times 26 \times 26}\right)\sqrt{4000} \times \frac{26 \times 22}{1000}
$$

$$
= 145 \text{ kips} \tag{11.4}
$$

Observe that N_u is the smallest axial force corresponding to the largest shear force $V_u = \pm 52$ kips. (See Table 4.5.)

Since $V_u = 52$ kips $< \phi V_c/2 = 0.85 \times 145/2 = 62$ kips, column tie requirements must satisfy Section 7.10.5. Using #4 ties, the minimum vertical spacing of ties is given by the smallest of

- 16 times the diameter of vertical bars = $16 \times 1.41 = 22.5$ in.
- 48 times the diameter of tie bars = $48 \times 0.5 = 24$ in.
- The least column dimension = 26 in.

Use #4 ties at 22 in. Observe that at least #4 ties are required for vertical bars of sizes #11, 14, and 18, and for bundled vertical bars. See Fig. 4.34 for column reinforcement.

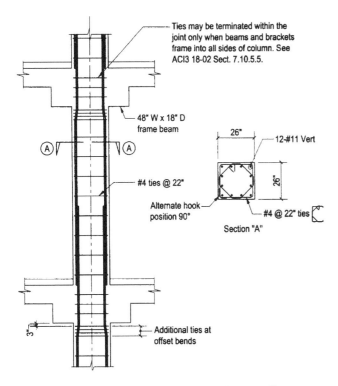

Figure 4.34. Design example, frame column; ordinary moment frame.

4.2.9.3. *Frame Beam Example: Intermediate Reinforced Concrete Moment Frame*

Given. A beam 24 in. wide × 26 in. deep as shown in Fig. 4.35. The beam is part of the lateral resisting system that consists of an intermediate reinforced concrete moment frame.

Ultimate design values are as follows: Nominal moments ($\phi = 1$) are as follows:

Support moment $-M_{ul} = 376$ kip-ft   At supports: $-M_{nl} = 418$ kip-ft

$\qquad -M_{ur} = 188$ kip-ft $\qquad -M_{nr} = 209$ kip-ft

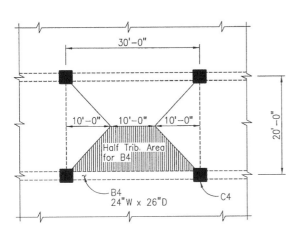

Figure 4.35. Frame beam and column example; intermediate moment frame.

Midspan moment $+M_u = 157$ kip-ft

Clear span $= 30 - \dfrac{30}{12} = 27.50$ ft

Shear force due to seismic $= 50$ kips

Shear force due to gravity loads $= 63$ kips

$f'_c = 4000$ psi, $f_y = 60,000$ psi

Required. Design and schematic of reinforcement for beam B4 using the provisions of ACI318-99,

Solution.

At left support, $-A_{s,\text{left}} = \dfrac{376}{4 \times 23.5}$

$$= 4.0 \text{ in.}^2$$

Use 4 #9 @ top giving $-A_{s,\text{left}} = 4.0$ in.^{2}

At right support, $-A_{s,\text{right}} = \dfrac{188}{4 \times 23.5}$

$$= 2.01 \text{ in.}^2$$

Use 2 #9 @ top, giving $-A_{s,\text{right}} = 2.0$ in.^{2}

At center span, $+A_s = \dfrac{157}{4 \times 23.5} = 1.67$ in.^{2}

Use 4 #6 @ bottom, giving $+A_s = 4 \times 0.44 = 1.76$ in.^{2}

Capacity $= \dfrac{1.76}{1.67} \times 157 = 165$ kip-ft

Verify Minimum Strength Requirements.

1. At joint face, positive moment $\geq \phi M_n^- / 3$.

2. At any section along the beam, both positive and negative moments $\geq \phi M_n^- / 5$.
    In our case item 1, positive moment strength criteria are satisfied because
    165 kip-ft > 376/3 = 125 kip-ft   OK
    Referring to item 2, both positive and negative moment criteria are satisfied along the beam, because two #9 bars are continued at top and four #6 are continued at bottom. The flexural capacity of 188 kip-ft provided at top and 165 kip-ft at bottom are greater than 376/5 = 75 kip-ft.   OK

Shear Design. The designer is given the following two options for determining the factored design shear force (ACI 318-02 Section 21.12.3):

1. Use the nominal moment strength of the member and the gravity load on it to determine the design shear force. Assume that nominal moment strengths ($\phi = 1.0$) are developed at both ends of its clear span. Use statics to evaluate the shear associated with this condition. Add the effect of the factored gravity loads W_D and W_L to obtain the total design shear.

Observe that the procedure is the same as for frame beams of SMRF. The only difference is that for an intermediate moment frame, nominal moment M_n, and not probable moment M_{pr}, is used at the beam ends.

2. Use a factored design shear V_u based on load combinations that include earthquake effects E, where E is taken to be twice that prescribed by the governing code.

For the example problem, we use the first option. Shear force associated with nominal moments M_{nl} and M_{nr} is equal to $\frac{M_{nl}+M_{nr}}{l_n}$.

$$V = \frac{418+219}{\left(30 - \dfrac{30}{12}\right)} = 28\,\text{kips}$$

Shear force due to factored gravity load = 63 kips.
Design shear force = 28 + 63 = 91 kips.

$$V_c = 2\sqrt{f_c'}\,b_w d = \frac{2\sqrt{4000} \times 26 \times 23.5}{1000} = 77.3\ \text{kips}$$

Assuming #3 stirrups, the required spacing s is

$$s = \frac{A_v f_y d}{V_s} = \frac{2 \times 0.11 \times 60 \times 23.5}{\left(\dfrac{91}{0.85} - 77.3\right)} = 10.4\,\text{in.}$$

Maximum spacing of stirrups over a length equal to $2h = 2 \times 26 = 52$ in. from the face of the supports is the smallest of

- $\frac{d}{4} = \frac{23.5}{4} = 5.9$ in.   (Controls)
- $8 \times$ diameter of smallest longitudinal bar $8 \times 1 = 8.0$ in.
- $24 \times$ diameter of stirrup bar = $24 \times 0.375 = 9.0$ in.
- 12 in.

Observe that the allowable maximum spacing is the same as for the frame beams of SMRFs. However, hoops and crossties with seismic hooks are not required for frame beams of IMFs.

Provide 12 #3 stirrups at each end spaced at 5 in. on centers. Place the first stirrup 2 in. from the face of each column. For the remainder of the beam, the maximum spacing of stirrups is $d/2 = 23.5/2 = 11.8$ in. Use 11-in. spacing. Figure 4.36 provides a schematic reinforcement layout.

4.2.9.4. Frame Column Example: Intermediate Reinforced Concrete Moment Frame

Given. A 30 in. $\times$ 30 in. frame column of an intermediate reinforced concrete moment frame. The column has been designed with 10 #11 longitudinal reinforcement to satisfy the ultimate axial load and moment combinations.

The ultimate design shear force due to earthquake loads $E = 35$ kips. The smallest axial load, N_u, corresponding to the shear force, = 1040 kips.

$$f_c' = 4000\ \text{psi} \qquad f_y = 60,000\ \text{psi}$$

Clear height of the column = 11.84 ft.

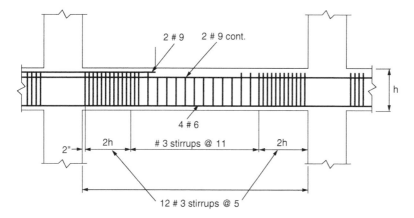

Figure 4.36. Design example, frame beam; intermediate moment frame.

Required. Seismic design and a schematic reinforcement detail for column C4 using the provisions given in Section 21.10 of ACI 318-99.

Check Limitations on Column Cross-Sectional Dimensions. No limitations are specified in ACI 318-99. Therefore, the given dimensions of 30 in. × 30 in. for the column are OK.

Design for Bending and Axial Loads. The statement of the problem acknowledges that the column has been designed for the governing load combinations with 10 # 11 vertical reinforcement. The reinforcement ratio (equal to 15.6/(30 × 30) × 100 = 1.74%) is within the allowable range of 1% and 8%.   OK.

Design for Shear. Similar to that for beams, the shear design of columns in intermediate moment frames is based on providing a threshold of toughness. The design shear in columns may be determined by using either of the two options similar to those given earlier for beams. The first choice is to use the shear associated with development of nominal moment strengths of column at each end of the clear span. The second is to double the earthquake effect E when calculating ultimate design load combinations that include the earthquake effect E.

The ultimate shear force E due to earthquake = 35 kips, as given in the statement of the problem. Using the second option, the design shear force V_u is equal to $2 \times E = 2 \times 35 = 70$ kips. The shear capacity of the column is:

$$V_c = 2\sqrt{f_c'}\, b_w d$$

$$= \frac{2\sqrt{4000}}{1000} \times 30 \times 23.5 = 89 \text{ kips,} \tag{11.3}$$

The shear capacity may also be calculated by taking advantage of the axial compression present in the column by using the equation

$$V_c = 2\left(1 + \frac{N_u}{2000\,A_g}\right)\sqrt{f_c'}\, b_w d$$

$$= 2\left(1 + \frac{1040,000}{2000 \times 30 \times 30}\right)\frac{\sqrt{4000} \times 30 \times 23.5}{1000}$$

$$= 141 \text{ kips} \tag{11.4}$$

In the above equation, N_u is the smallest axial load = 1040 kips corresponding to the largest shear force on the column (given).

Since $V_u = 70$ kips < $141/2 = 70.5$ kips, column tie requirements given in Section 7.10.5 would have sufficed: However, frame columns of intermediate moment frames are required to have a minimum threshold of toughness. Hence the requirements of Section 21.10.4.

To properly confine the concrete core in the plastic hinge length region l_o, and to maintain lateral support of column vertical bars, transverse reinforcement requirements for frame columns of intermediate moment frame are as follows:

- 8 × the diameter of the smallest
  vertical bar of column = 8 × 1.41 = 11.3 in.   ← controls
- 24 × the diameter of the tie bar
  = 24 × 0.5 = 12 in.   (Section 21.10.4.2)
- one-half the least column dimension = 30/2 = 15 in.
- 12 in.

The plastic hinge length l_o is the largest of

- one-sixth of the column clear height
  $= \frac{11.84}{6} \times 12 = 24$ in.
- maximum cross-sectional dimension of the column
  = 30 in.   ← controls
- 18 in.

Use #4 ties and crossties at 10 in. spacing within the l_o region. In between l_o, provide ties @ 20 in. spacing. Figure 4.37 provides a schematic reinforcement layout of column vertical bar and ties.

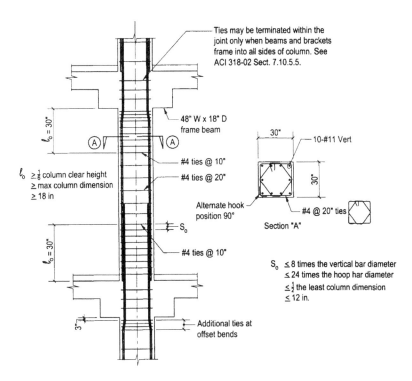

Figure 4.37. Design example, frame column; ordinary moment frame. Note: ℓ_o is the same as for columns of SMRF. There is no requirement to splice the column bars at mid-height.

4.2.9.5.   *Shear Wall Example: Seismic Design Category (SDC) A, B, or C*

Although ACI 318-99 specifies certain seismic design and detailing requirements for intermediate moment frames (IMFs) there are no requirements for shear walls in buildings assigned to SDC A, B, or C. For these buildings, ACI considers that the requirements given in Chapters 1 through 18 and 22 are sufficient to provide a degree of toughness that is consistent with the seismic risk associated with zone 2 or for buildings assigned to SDC C.

The design procedure for a reinforced shear wall subject to bending and axial loads is a two-step process. First, generate an axial load-moment interaction diagram for the shear wall of given dimensions and concrete strength, with various percentages of reinforcement. This is done by taking successive choices of neutral axis distance measured from one face of the wall, and then calculating the axial force P_u and the corresponding moment M_u. Each sequence of calculations is repeated until the complete interaction diagram is obtained. The next step is the selection of reinforcement that satisfies the design requirement under loads and moments equal to or larger than the factored loads and moments. The formulation is based on the principles of ultimate strength design with a linear strain diagram that limits the concrete strain at the extremity of the section to 0.003. With the general availability of computers it is no longer tedious to establish axial load-moment interaction diagrams, Therefore, design for axial loads and moments is not discussed further in this section.

Given.   A shear wall 24 ft. long and 12 in. thick with a floor-to-floor height of 14 ft.

Compressive strength of concrete $f'_c = 4000$ psi

Yield strength of reinforcing bars $f_y = 60$ ksi

Maximum factored shear force $V_u = 500$ kips

Vertical reinforcement as determined for bending and axial loads = four # 7 @ 9 in. vertical at each end, and # 6 @ 15 vertical in between each face.

Required.   Shear design using the provisions of ACI 318-99.

Solution.

Shear Design.

$$V_c = 2\sqrt{f'_c}\, hd$$
$$= 2\sqrt{4000} \times \frac{12 \times 0.8 \times 24 \times 12}{1000} = 350 \text{ kips} \tag{11.10.5}$$

Observe that Eq. (11.10.4) permits d to be taken equal to $0.8 l_w = 0.8 \times 24 \times 12 = 230$ in. for the design of horizontal shear forces in the plane of the wall. A larger value of d, equal to the distance from the extreme compression fiber to the centerline of tension force in reinforcement determined by a strain compatibility analysis, is also permitted.

The maximum factored shear force V_u is 500 kips, as given in the statement of problem. Since $V_u = 500$ kips $> \phi V_c /2 = 0.85 \times 350/2 = 149$ kips, provide horizontal reinforcement given by

$$V_s = \frac{A_v f_y d}{s_2} \tag{11.33}$$

Assuming two layers of #4 horizontal reinforcement, one layer at each face

$$s_2 = \frac{2 \times 0.22 \times 60 \times 230}{\left(\dfrac{500}{0.85} - 350\right)} = 25 \text{ in.}$$

However, maximum spacing of horizontal reinforcement must not exceed

- $\dfrac{l_w}{5} = \dfrac{24 \times 12}{5} = 57.6$ in.
- $3h = 3 \times 12 = 36$ in.   or
- 18 in.   (Controls)

(11.10.9.3)

Section 11.10.9.2 requires ratio ρ_h of horizontal reinforcement to be not less than 0.0025.

$$\rho_h = \frac{2 \times 0.20}{12 \times 18} = 0.0019 < 0.0025 \qquad \text{Not OK.}$$

Therefore, use two layers of #4 horizontal bars at 12 in. spacing, giving $\rho_h = 0.0028 > 0.0025$ OK

At any horizontal section, the shear strength V_n must not exceed $10\sqrt{f_c'}hd = 10\sqrt{4000} \times (12 \times 230)/1000 = 1746$ kips

$$V_n = V_c + V_s$$

$$= 350 + \frac{2 \times 0.20 \times 60 \times 230}{12}$$

$$= 810 \text{ kips} < 1746 \text{ kips} \qquad \text{OK} \qquad\qquad (10.11.3)$$

Section 11.10.9.4 requires the area of vertical shear reinforcement to gross concrete area of horizontal section, denoted by ρ_n, to be not less than

- $\rho_n = 0.0025 + 0.5\left(2.5 - \dfrac{h_w}{l_w}\right)(\rho_h - 0.0025)$

  $= 0.0025 + 0.5\left(2.5 - \dfrac{14 \times 12}{24 \times 12}\right)(0.0028 - 0.0025)$

  $= 0.0028.$ (Controls)   (11.34)

  Note: h_w = wall height = 14 ft; l_w = wall length = 24 ft.

- $\rho_n \geq 0.0025.$
- ρ_n need not be greater than the required horizontal shear reinforcement.

Thus, $\rho_n = 0.0028$.

The spacing of vertical shear reinforcement, according to Section 11.10.9.5, must not exceed

- $\dfrac{l_w}{3} = \dfrac{24 \times 12}{3} = 96$ in.
- $3h = 3 \times 12 = 36$ in.
- 18 in.

For two curtains of #6 vertical bars spaced at 15 in. centers, $\rho_n = \frac{2 \times 0.44}{12 \times 15} = 0.0049$ OK.

The provided horizontal and vertical reinforcements satisfy the minimum reinforcement ratios ρ_h and ρ_n given in Sections 14.3.2 and 14.3.3.

Figure 4.38 provides a schematic reinforcement detail.

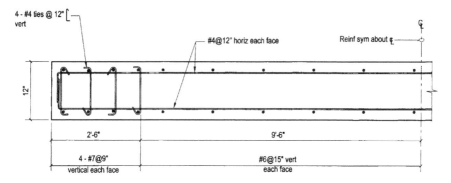

Figure 4.38. Shear wall; low to moderate seismic zones (SDC A, B, or C). Note: Vertical reinforcement of #7 @ 9" at each end is enclosed by lateral ties, since the reinforcement area of eight #7 vertical bars equal to $8 \times 0.6 = 4.8$ in.^{2} is greater than 0.01 times the area of concrete $= 12 \times 30 = 360$ in.^{2} (See ACI 318-02 Sect. 14.3.6.)

4.2.9.6. Frame Beam Example: Special Reinforced Concrete Moment Frame

Given. Values for bending moments and shear forces for beam B2 at the supports and at midspan are shown in Table 4.6.

Beam B2: 28 in. wide $\times$ 33 in. deep clear span $l_n = 25.17$ ft

$f'_c = 4000$ psi; $f_y = 60,000$ psi

Axial force in B2 is negligible.

Required. Schematic design and detail of frame beam B2 using the provisions of ACI 318-99.

Solution. The first step is to calculate the design bending moments and shear forces for the beam using the load combinations specified in governing codes (see Table 4.6).

TABLE 4.6 Design Bending Moments and Shear Forces for B2; Special Moment Frame

Load case	Location	Bending moment (kip-ft)	Shear force (kips)
Dead load, D	Support	−120	35
	Midspan	+90	0
Live load, L	Support	−20	10
	Midspan	9	0
Seismic, Q_E	Support	±600	±60
Load combinations			
1. $U = 1.4D + 1.7L$	Support`	−202	66
	Midspan	+141.3	0
2. $U = 1.2D + F_1L + 1.0E$, where $E = \rho Q_E + 0.2S_{DS}D$ Since $F_1 = 0.5$, $\rho = 1.0$, and $S_{DS} = 1.0$,			
$U = 1.2D + 0.5L + Q_E + 0.2D$	Support	−778	114
$= 1.4D + 0.5L + Q_E$	Midspan	130.5	60
3. $U = 0.9D − 1.0E$ $= 0.9D + \rho Q_E − 0.2S_{DS}D$			
$U = 0.7D + Q_E$	Support	516	−35.5
	Midspan	63	−60

Check Limitations on Section Dimensions.
- Factored axial compressive force on B2 is negligible. Therefore B2 may be designed as a flexural member.

- $\dfrac{l_n}{d} = \dfrac{(28 \times 12) - 34}{30.5} = 10.0 > 4.$ OK   (Section 21.3.1.2)

- $\dfrac{\text{width}}{\text{depth}} = \dfrac{28}{33} = 0.85 > 0.3.$ OK   (Section 21.3.1.3)

- width = 28 in. > 10 in.   OK   (Section 21.3.1.4)
  $$< \text{width of supporting col}$$
  $$+ (1.5 \times \text{beam depth})$$
  $$< 34 + 1.5 \times 33 = 83.5 \text{ in.}$$ OK   (Section 21.3.1.5)

Calculate Required Flexural Reinforcement.

At support: $-M_u = 778$ kip-ft.

$$-A_s = \frac{778}{4.24 \times 30.5} = 6.02 \text{ in.}^2$$

Use 6 #9 top

$$\phi M_n = -778 \text{ kip-ft}$$
$$+M_u = 516 \text{ kip-ft}$$

$$+A_s = \frac{516}{4.1 \times 30.5} = 4.12 \text{ in.}^2$$

Use six #8 bottom, $A_s = 4.74$ in.^{2}

$$\phi M_n = 516 \times \frac{4.74}{4.12} = 594 \text{ kip-ft}$$

At midspan: $M_u = +141.3$ kip-ft

$$A_s = \frac{141.3}{4.1 \times 30.5} = 1.13 \text{ in.}^2$$

$$A_{s(\min)} = \frac{3\sqrt{f_c'} b_w d}{f_y}$$

$$= \frac{3\sqrt{4000} \times 28 \times 30.5}{60,000} = 2.70 \text{ in.}^2 \qquad (10.3)$$

$$A_{s(\min)} = \frac{200 b_w d}{f_y}$$

$$= \frac{200 \times 28 \times 30.5}{60,000} = 2.84 \text{ in}^2. \quad \text{(Controls)} \qquad \text{(Section 21.3.2.1)}$$

$$A_{s(\max)} = \rho_{\max} b_w d$$
$$= 0.025 \times 28 \times 30.5$$
$$= 21.3 \text{ in.}^2$$
$$> 2.84 \text{ in.}^2 \quad \text{OK} \qquad \text{(Section 21.3.2.1)}$$

Use three #9 giving $A_s = 3.0$ in.^{2}

Verify Minimum Strength Requirements.

1. At joint face, for positive moment $\geq \frac{-M_u}{2}$.
2. At any section along the beam, for both positive and negative moments $\geq \frac{-M_u}{4}$ (21.3.2.2).
   1. Positive moment strength criteria are satisfied because 594 kip-ft $> \frac{778}{2}$ = 389 kip-ft.
   2. $-M_u/4 = 778/4 = 194.5$ kip-ft. This can be satisfied by providing two #9 top bars and two #8 bottom bars. However, the minimum reinforcement requirement is 2.84 in.² Therefore, provide continuous three #9 top bars and four #8 bottom bars, giving $A_s = 3.00$ in.² and 3.16 in.², respectively, which are greater than $A_{s(\min)} = 2.84$ in.²

      Observe that this also automatically fulfills the requirement that at least two bars be continuous at both the top and bottom of the beam (Section 21.3.2.1).

Shear Design.   It is worthwhile to mention again that the values for shear obtained from lateral analysis at the beam ends do not play a primary role in determining the shear reinforcement. This is because the method of determining shear forces in beams of special moment frames is based on the premise that plastic hinges may form at regions near the supports. The shear forces are thus computed using statics, based on the assumption that moments of opposite sign corresponding to the probable moment strength M_{pr} act at the beam ends. Additionally, a shear force corresponding to the factored gravity load is added to the shear derived from the probable moment to determine the design shear forces.

The probable moment M_{pr} is determined by using: 1) a stress of $1.25 f_y$ in the tensile reinforcement; and 2) a strength reduction factor ϕ equal to 1.0. In determining the shear strength of a frame beam, both contributions provided by concrete, V_c, and reinforcing steel, V_s, are taken into account. However, V_c is to be taken as zero when both of the following conditions are met: 1) The earthquake induced shear force (calculated using the probable moment M_{pr} and $\phi = 1$) is greater than or equal to 50% of the maximum required shear strength; and 2) The factored axial compressive force in the beam, including earthquake effects, is less than $A_g f_c'/20$.

The second condition reflects the neccessity of increasing the shear reinforcement in the case of no axial load.

The following equation may be used to compute M_{pr}:

$$M_{pr} = A_s(1.25 f_y)\left(d - \frac{a}{2}\right)$$

where

$$a = \frac{A_s(1.25 f_y)}{0.85 f_c' b}$$

Returning to the example problem,
for six #9 top bars, $A_s = 6$ in.²

$$a = \frac{6 \times 1.25 \times 60}{0.85 \times 4 \times 28} = 4.73 \text{ in.}$$

$$M_{pr} = 6 \times 1.25 \times 60\left(30.5 - \frac{4.73}{2}\right)$$

$$= 12{,}661 \text{ kip-in.} = 1055 \text{ kip-ft}$$

For six #8 bottom bars, $A_s = 4.74$ in.^{2}

$$a = \frac{4.74 \times 1.25 \times 60}{0.85 \times 4 \times 28} = 3.73 \text{ in.}$$

$$M_{pr} = 4.74 \times 1.25 \times 60 \left(30.5 - \frac{3.73}{2} \right)$$

$$= 10{,}179 \text{ kip-in.} = 848 \text{ kip-ft}$$

The shear forces corresponding to M_{pr} at each end (positive at one end and negative at the other) are computed from a free body diagram of the beam. Added to these are the shear forces due to factored gravity loads to obtain the design shear force V_e at each end of the beam.

The following data will be used to determine the uniformly distributed gravity load on the frame beam.

Area of trapezoid tributary to B2 (see Fig. 4.35)

$$= 8 \times 10 \times 2 + 4 \times \frac{1}{2} \times 10 \times 10 = 360 \text{ ft}^2$$

Dead load of slab assuming a thickness of 7.5 in.

$$= \frac{7.5}{12} \times 0.15 \times 360 = 33.75 \text{ kips}$$

Dead load of B2 $= \dfrac{28 \times 33}{144} \times 0.15 \times 25.17 = 24.22$ kips

Superimposed dead load @ 20 psf
for partitions and 15 psf for

ceiling, mechanical, and floor finishes $= \dfrac{35 \times 360}{100} = 12.60$ kips

Total $D_L = 70.57$ kips

Live load at 50 psf $= \dfrac{50 \times 360}{1000} = 18.0$ kips

Equivalent dead load $= \dfrac{70.57}{25.17} = 2.8$ kip-ft

Equivalent live load $= \dfrac{18}{25.17} = 0.72$ kip-ft

Factored gravity load $= 1.4D + 0.5L$ (load combination 2)
$$= 1.4 \times 2.8 + 0.5 \times 0.72$$
$$= 4.3 \text{ kip-ft}$$

Therefore,

$w_u = 4.3$ kip-ft

The maximum combined designed shear force V_e equal to 129.6 kips, as will be computed shortly, is larger than the shear force value of 114 kips obtained from load combination 2, based on structural analysis. To determine whether the shear strength V_c provided by the concrete can be used in calculating the shear resistance, two checks are performed:

- Determine whether earthquake-induced shear force based on M_{pr} is larger than 50% of the total shear V_c.
- Determine whether the compressive axial force in the beam is less than $A_g f_c'/20$. (For the example, it is given in the statement of the problem that axial compressive force in the beam is negligible.)

If both of these criteria are satisfied, then V_c must be taken equal to zero (Section 21.3.4.2). For example,

Shear force due to plastic moments at each end of beam = $\dfrac{848 + 1055}{25.17}$ = 75.6 kips

Design shear force V_e = shear due to M_{pr} + gravity shear
$$= 75.6 + 4.3 \times 25.17/2 = 129.6 \text{ kips}$$

Since 74.6 kips is greater than 50% of 129.6 = 64.8 kips, and the axial compressive force is negligible, $V_c = 0$.

Design shear V_u (ACI 318-02 uses notation V_e) is equal to

$$V_u = 129.6 = \phi V_c + \phi V_s$$

Since $V_c = 0$, $V_u = \phi V_s$, and $V_s = V_u/\phi = 129.6/0.85 = 152.5$ kips, the spacing s of #4 hoops (closed stirrups) with four legs is given by

$$s = \frac{A_v f_y d}{V_s} = \frac{0.2 \times 4 \times 60 \times 30.5}{152.5}$$
$$= 9.6 \text{ in.} \tag{11.15}$$

Observe that four legs are required because longitudinal beam bars on the perimeter are to have lateral confinement conforming to Section 7.10.5.3; every corner and alternate bar must have lateral support provided by the corner of a tie with an included angle of not more than 135 degrees, and no bar shall be farther than 6 in. clear on each side along the tie from such a laterally supported bar.

Additionally, 135° hooks are required for hoops and ties. Maximum spacing of hoops within the plastic hinge length, equal to a distance of 2 times the beam depth, $2h = 2 \times 33 = 66$ in., is the smaller of

- $\dfrac{d}{4} = \dfrac{30.5}{4} = 7.6$ in.   (governs)
- 8 times the diameter of smallest longitudinal bar = $8 \times 1 = 8$ in.

(Section 21.3.3.2)

- 24 times the diameter of hoop bar = $24 \times 0.5 = 12$ in.

Use nine #4 hoops at each end of the beam spaced 7.5 in. apart. Place the first loop 2 in. from the face of support, as required by Section 21.3.3.2.

Hoops are required only in the plastic hinge length; stirrups with seismic hooks at both ends may be used elsewhere along the beam length. Additionally, the shear strength contribution V_c of the beam concrete may be used in calculating the shear resistance.

At a distance 6.6 in. from the face of support

$$V_u = 129.6 - \left[\frac{66}{12} \times 4.3\right] = 106 \text{ kips}$$

$$V_c = 2\sqrt{f_c'} b_w d$$

$$= 2 \times \sqrt{4000} \times \frac{28 \times 30.5}{1000} = 108 \text{ kips}$$

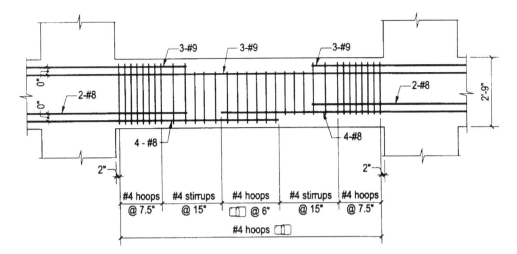

Figure 4.39. Design example, frame beam; special moment frame.

$$V_s = \frac{V_u}{\phi} - V_c$$

$$= \frac{106}{0.85} - 108 = 16.7 \text{ kips}$$

Required stirrup spacing of two-legged #4 stirrups is

$$s = \frac{A_v f_y d}{V_s}$$

$$= \frac{0.2 \times 2 \times 60 \times 30.5}{16.7} = 43.8$$

Maximum allowable spacing is $d/2 = 30.5/2 = 15.25$ in.   (Section 21.3.3.4.)

Use 15-in. spacing for the portion of the beam bounded between the plastic hinge-length and the bottom bar splice at the center. Use 6 in. spacing for the length of splice (Fig. 4.39).

4.2.9.7. Frame Column Example: Special Reinforced Concrete Moment Frame

Given. A 34 in. × 34 in. column (column C2) with 10 # 11 vertical reinforcement. See Fig. 4.40. The column has been verified for the axial loads and bending moments resulting from the following ultimate load combinations.

Load combination	P_u (kips)	M_u (kip-ft)	V_u (kips)
1	2372	0	0
2	2180	400	80
3	1050	−400	−80

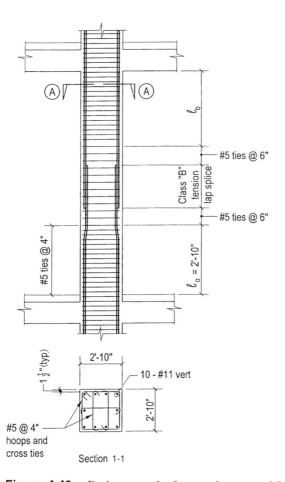

Figure 4.40.   Design example, frame column; special moment frame.

Beam moments framing into the column: $M_{g,\text{left}} = 848$ kip-ft

$$-M_{g,\text{right}} = -1055 \text{ kip-ft}$$

Nominal flexural strength of column at the beam–column joint.

Above the joint = 1769 kip-ft

Below the joint = 1819 kip-ft

Clear height of column = 13 ft – beam depth

$$= 13 - \frac{33}{12} = 10.25 \text{ ft}$$

$f_c' = 5000$ psi, $f_y = 60,000$ psi

Required.   Design and a schematic reinforcement detail for C2 using the provisions of ACI 318-99.

Solution.

Check Limitations on Section Dimensions.

- The index axial force that delineates a frame column from a frame beam is given by

$$\frac{A_g f_c'}{10} = \frac{34 \times 34 \times 5}{10} = 578 \text{ kips.} \qquad \text{(Section 21.4.1)}$$

Since the factored axial loads given in the load combinations exceed the index value, C2 may be designed as a column.

- Column cross-sectional dimension of 34 in. > 12 in.   OK   (Section 21.4.1.1)
- Ratio of column cross-sectional dimensions = 34/34 = 1.0 > 0.4.   OK
(Section 21.4.1.2)

Design for Bending and Axial Loads.   The 34×34 in. column with ten #11 vertical bars is adequate for the combined bending and axial loads as stated in the problem. Reinforcement ratio $\rho_g = (10 \times 156)/(34 \times 34) \times 100 = 1.35\%$ is within the allowable range of 1.0% and 6.0% (Section 21.4.3.1).   OK

Minimum Flexural Strength of Columns.   The sum of the nominal flexural strengths of columns at a joint must be greater than or equal to 6/5 the sum of the nominal flexural strengths of girders framing into that joint.

$$\Sigma M_c \geq \frac{6}{5} \Sigma M_g \tag{21.1}$$

When computing the nominal flexural strengths of T-beams, top slab reinforcement with in an effective width of beam as defined in Section 8.10 must be included if the slab reinforcement is developed at the critical section for flexure. For the example problem, it is assumed that the top slab reinforcement is not developed at the critical bending region. Therefore its contribution may be ignored in computing M_g.

$\Sigma M_g = 1055 + 848 = 1903$ kip-ft, and $\Sigma M_c = 1815 + 1769 = 3584$ kip-ft.

Checking Eq. (21.1):

$$\Sigma M_c = 3584 > \frac{6}{5}(1903) = 2283 \text{ kip-ft.}   \text{OK}$$

Therefore, the lateral strength and stiffness of a column can be considered when evaluating the strength and stiffness of the structure. If these strong column–weak beam criteria are not satisfied, then any positive contribution of the column to the strength and stiffness is to be ignored. Negative impacts of ignoring the stiffness and strength of the column must, however, be taken into account. For example, if ignoring the strength and stiffness of the column results in a decrease in torsional effects, the decrease should not be considered in the analysis.

Design for Shear.   The method of determining design shear forces in columns is similar to that for beams. It takes into consideration the likelihood of formation of plastic hinges in regions near the ends of columns. This region, denoted l_o, is the largest of

- Depth of member = 34 in.   (Controls)
- Clear span/6 = $(13 \times 12 - 33)/6 = 20.5$ in.   (Section 21.4.4)
- 18 in.

To maintain lateral support of column vertical bars and to confine the concrete core in the region l_o, transverse reinforcement requirements are as follows:

- One-forth of the minimum member dimension = 34/4 = 8.5 in.
- Six times the diameter of the longitudinal reinforcement = $6 \times 1.41 = 8.46$ in.
- $s_x = 4 + \left(\frac{14 - h_x}{3}\right)$ but not greater than 6 in. or less than 4 in.

    $= 4 + \left(\frac{14 - 10}{3}\right) = 5.33$ in. but use $s_x = 4$ in.   (Controls)

From Fig. 4.40, it is seen that $h_x = 10$ in. < 14 in.   OK   (Section 21.4.4.3)

Determine Area of Hoops and Crossties.   Assuming 4-in. vertical spacing,

$$A_{sh} = \frac{0.3 s h_c f_c'}{f_{yh}}\left[\left(\frac{A_g}{A_{ch}}\right) - 1\right]$$

$$= \frac{0.3 \times 4 \times 30.5 \times 5}{60}\left(\frac{34 \times 34}{961}\right) - 1 = 0.62 \text{ in.}^2 \qquad\qquad (21.3)$$

$$A_{sh} = \frac{0.09 s h_c f_c'}{f_{yh}}$$

$$= \frac{0.09 \times 4 \times 30.5 \times 5}{60} = 0.92 \text{ in.}^2 \quad \text{(Controls)} \qquad\qquad (21.4)$$

Using #5 hoops with two crossties in the longitudinal and one in the transverse direction,

$A_{sh} = 4 \times 0.31 = 1.24$ in.^{2} in the longitudinal direction and,
$A_{sh} = 3 \times 0.31 = 0.93$ in.^{2} in the transverse direction.

This is larger than 0.92 required by Eq. (21.4). Use #5 hoops and crossties at 4-in. vertical spacing as shown in Fig. 4.40.

Verify Confining Reinforcement for Shear.   In the previous step, we determined transverse reinforcement required for confining column concrete and for providing lateral support to column vertical bars. In this step, we check if this reinforcement is adequate to resist shear forces resulting from the probable flexural strengths M_{pr} at each end of a column.

The positive probable flexural strength of the beam framing to the left face of column at third level is 848 kip-ft. The negative probable strength on the right face is 1055 kip-ft. Assuming that the flexural reinforcement for the beam below the level under consideration is the same, the design strength V_e is given by

$$V_e = \frac{\dfrac{2(848 + 1055)}{2}}{\left(13 - \dfrac{33}{12}\right)} = 186 \text{ kips}$$

$$V_u = V_e = \phi\,(V_c + V_s)$$

Since the factored axial forces are greater than $A_g f_c'/20$, the shear strength V_c of concrete may be included in calculating the column shear capacity. For simplicity, we use $V_c = 2\sqrt{f_c'}\,bd$, although for members subjected to axial compression (as is the case for the example column), Eq. (11.4) permits higher shear values in concrete.

$$V_c = \frac{2\sqrt{5000} \times 34 \times (34 - 3)}{1000}$$

$$= 149 \text{ kips}$$

$$V_s = \frac{A_v f_y d}{s} = \frac{1.24 \times 60 \times 31}{4} = 577 \text{ kips}$$

Shear capacity $\phi V_n = \phi\,(V_c + V_s)$
$$= 0.85\,(149 + 577)$$
$$= 617 \text{ kips} > 186 \text{ kips}$$

Therefore, #5 hoops and crossties provided at a spacing of 4 in. for confinement over a length of $l_o = 34$ in. at column ends is also adequate for design shear.

The midlength of the column between the plastic hinging lengths must be provided with hoop reinforcement not exceeding a spacing of 6 times the diameter of the longitudinal bar $= 6 \times 1.56 = 9.36$ in. or 6 in. In our case, the spacing of 6 in. governs. Therefore, provide #5 hoops and crossties at 6 in. for the midlength of the column. See Fig. 4.40 for a schematic layout of reinforcement.

4.2.9.8. *Beam Column Joint Example: Special Reinforced Concrete Frame*

To ensure that the beam–column joint of special moment-resisting frames have adequate shear strength, first, an analysis of the beam–column panel zone is performed to determine the shear forces generated in the joint. The next step is to check this against allowable shear stress.

The joint analysis is done in the major and the minor directions of the column. The procedure involves the following steps:

- Determination of panel zone design shear force.
- Determination of effective area of the joint.
- Verification of panel zone shear stress.

Determination of panel zone shear force. Consider the free body stress condition of a typical beam–column intersection showing the forces P_u, V_u, M_u^L and M_u^R (Fig. 4.40a). The force V_u^h, the horizontal panel zone shear force, is to be calculated.

The forces P_u and V_u are the axial force and shear force, respectively, from the column framing into the top of the joint. The moment M_u^L and M_u^R are the beam moments framing into the joint. The joint shear force V_u^h is calculated by resolving the moments into compression C and tension T forces. The location of C or T is determined by the direction of the moment using basic principles of ultimate strength design. Noting that $T_L = C_L$ and $T_R = C_R$, $V_u^h = T_L + T_R - V_u$.

The moments and the C and T forces from beams that frame into the joint in a direction that is not parallel to the major or minor directions of the column are resolved along the direction that is being investigated.

In the design of special moment-resisting concrete frames, the evaluation of the design shear force is based upon the moment capacities (with reinforcing steel overstrength factor α and no ϕ factors) of the beams framing into the joint. The C and T forces are based upon these moment capacities. The column shear force V_u is calculated from the beam moment capacities as follows:

$$V_u = \frac{M_u^L + M_u^R}{H}$$

It should be noted that the points of inflection shown on Fig. 4.40a are taken as midway between actual lateral support points for the columns.

The effects of load reversals, as illustrated in cases 1 and 2 of Fig. 4.40b, are investigated and the design is based upon the maximum of joint shears obtained from the two cases.

Determine the effective area of joint. The joint area that resists the shear forces is assumed always to be rectangular. The dimensions of the rectangle correspond to the major and minor dimensions of the column below the joint, except that if the beam framing into the joint is very narrow, the width of the joint is limited to the depth of the joint plus the width of the beam. The area of the joint is assumed not to exceed the area of the

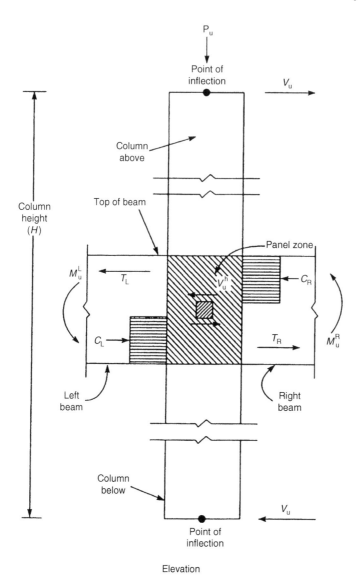

Figure 4.40a. Column panel shear forces.

column below. It should be noted that if the beam frames into the joint eccentrically, the above assumptions may be unconservative.

Given. A frame column joint.

Beam: 28 in. wide × 33 in. deep

Column: 34 in. × 34 in.   Floor-to-floor height = 10.23 ft

Beam top reinforcement, six #9 top

Beam bottom reinforcement, eight #8 bottom

Beam moment $M_u^L = 1055$ kip-ft

Beam moment $M_u^R = 186$ kip-ft

Beam confined on two faces.

$f_c' = 5000$ psi, $f_y = 60,000$ psi

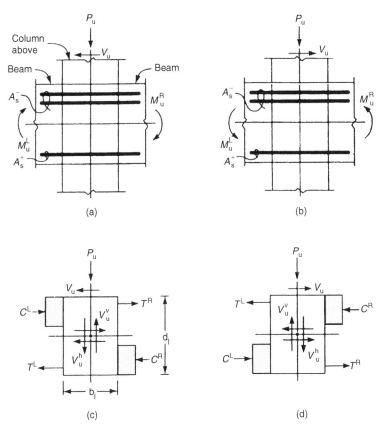

Figure 4.40b. Beam–column joint analysis: (a) forces and moments, case 1; (b) forces and moments, case 2; (c) resolved forces, case 1; (d) resolved forces, case 2.

Confining reinforcement through the joint of a frame column is required no matter how low the calculated shear force is. This is to ensure ductile behavior of the joint and to allow it to maintain its load carrying capacity even after possible spalling of concrete outside of transverse reinforcement.

The design shear force is determined by subtracting the column shear force from the tensile force in the top beam reinforcement and the compressive force at the bottom of the beam on the opposite face of the column. The stress in the beam reinforcement is taken as $1.25 f_y$ (Section 21.5.1.1).

T_L due to six #9 $= A_s \times 1.25 f_y = 6 \times 1.0 \times 1.25 \times 60 = 450$ kips

C_R due to six #8 $= 6 \times 0.79 \times 1.25 \times 60 = 356$ kips.

Column horizontal shear force, V_h is obtained by assuming a point of contraflexure at midheight of column and by moment equilibrium condition at the frame joint.

$$
\begin{aligned}
V_h &= \frac{M_u^L + M_u^R}{H} \\
&= \frac{1055 + 186}{13} = 95.46 \text{ kips}, \quad \text{use 96 kips}
\end{aligned}
$$

The net shear force is $T_L + C_R - V_h = 450 + 356 - 96 = 710$ kips.

The example column joint is confined on two opposite faces as given in the statement of the problem. Therefore,

$$\phi V_c = 15\sqrt{f_c'}A_j$$

$$= \frac{15\sqrt{5000} \times 34 \times 34}{1000} = 1226 \text{ kips} > 620 \text{ kips} \quad \text{OK} \qquad \text{(Section 21.5.3)}$$

Note: A_j = effective cross-sectional area within the joint equal to the joint depth times an effective width. The effective width is the smaller of
- Beam width + joint depth = 28 + 34 = 62 in.
- Beam width + twice the smaller distance from beam edge to the column edge equal to 28 + 2 × 3 = 34 in.

Observe that joint shear is a function of effective cross-sectional area A_j of the joint and the square root of the concrete compressive strength $\sqrt{f_c'}$ only. If the net shear exceeds the nominal shear strength ϕV_c (equal to $20\sqrt{f_c'}A_j, 15\sqrt{f_c'}A_j$, or $12\sqrt{f_c'}A_j$, depending upon the confinement provided at the joint), then the designer has no choice but to increase f_c' of concrete and/or the size of columns.

A column face is considered confined by a beam if the beam width is equal to at least 75% of the column width. (No mention is made in ACI 318-02 for the required depth of beam.) When joints are confined on all four sides, transverse reinforcement within the joint required by Section 21.4.4 may be reduced by 50%. Hoop spacing is permitted to a maximum of 6 in. See Fig. 4.41.

4.2.9.9.   *Special Reinforced Concrete Shear Wall*

Given.   A shear wall that is part of a lateral load resisting system of a 10-story building located in a high seismic zone that has the following seismic characteristics.

S_1 = maximum considered earthquake, 5% damped, spectral response acceleration at a period of 1 sec = 0.85g.

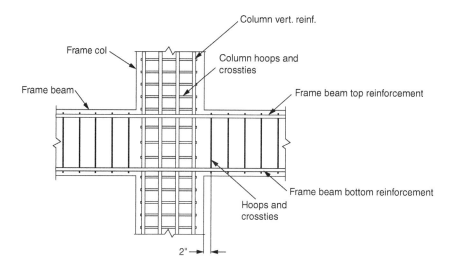

Figure 4.41.   Beam–column joint; special moment frame. Transverse reinforcing in the joint is the same as for the frame column. A 50% reduction is allowed if the joint is confined on all the four faces. Maximum spacing of transverse reinyforcement = 6 in.

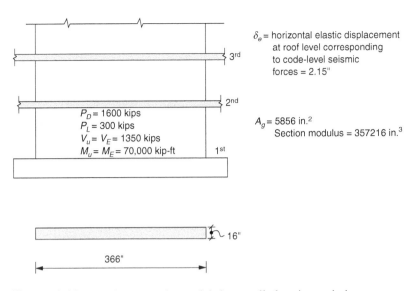

Figure 4.42. Design example; partial shear wall elevation and plan.

S_s = maximum considered earthquake, 5% damped, spectral response acceleration at short periods = 1.80g.

Site class = D (as determined by project geotechnical engineer).

Seismic use group, SUG = 1

Seismic design category, SDC = D

Reliability/redundancy factor, $\rho = 1.0$

Seismic importance factor, $I_E = 1.0$

Specified compressive strength of concrete $f_c' = 5000$ psi

Specified yield strength of reinforcement $f_y = 60$ ksi

Figure 4.42 shows a partial elevation and plan of the wall along with the ultimate axial forces and moments due to gravity and lateral loads. The dead load P_D includes the self-weight of the wall. P_L is the reduced live load. Also shown therein are the section properties of the wall and the horizontal displacement δ_e equal to 2.15 inches at the roof level. The displacement is the lateral elastic deflection due to design basis code level earthquake loads. As will be seen presently, this displacement multiplied by the C_d factor is used to determine the requirements for detailing boundary elements.

The wall has been analyzed using the following assumptions:

- Base of the wall is fixed.
- Effective section properties of the wall are based on a cracked section.
- Flexural rigidity = $0.5E_cI_g$.
- Shear rigidity = $0.4\,E_cA_w$.
- Actual rigidity = E_cA_g.

It should be noted that a computer analysis is almost always necessary to determine the building's response. This is because it is mandated in recent seismic codes to consider variables such as uncracked and cracked concrete section properties and some soil or foundation deformation beneath the structure's base.

Required.

- Calculation of ultimate design loads and moments using ASCE 7-02 load combinations.
- Preliminary sizing of the wall using a rule-of-thumb approach.

- Design of wall for shear.
- Design of wall for combined axial load and bending moment.
- Determination of boundary element requirements using both stress index and displacement-based methods.
- Design of boundary elements.
- Schematics showing reinforcement layout.
- The design shall be in accordance with ACI 318-02.

Solution.
Load Combinations.
1. $1.2D + 1.0E + f_1L + f_2S$
2. $0.9D + 1.0E$

For compression check, $E = \rho Q_E + 0.2S_{DS}D$

For tension check, $E = \rho Q_E - 0.2S_{DS}D$

$\rho = 1$ and $S_s = 1.80$ as given in the statement of the problem.

$S_{MS} = F_a S_s, \quad F_a = 1.0$ for site class D with $S \geq 1.25$

$\quad = 1.0\, S_s = 1.80$

$$S_{DS} = \frac{2}{3} S_{MS} = \frac{2}{3} \times 1.80 = 1.20$$

Factored axial load P_u for compression check

$P_u = 1.2(1600) + 1.0(1 \times 0 + 0.2 \times 1.20 \times 1600) + 300 + 0$

$\quad = 2604$ kips

Factored axial load P_u for tension check

$P_u = 0.9 \times 1600 - 1 \times 0 - 0.2 \times 1.2 \times 1600$

$\quad = 1056$ kips

The two sets of design forces and moments for the example are

$P_u = 2604$ kips	$P_u = 1056$ kips
$M_u = 71{,}000$ kip-ft	$M_u = 71{,}000$ kip-ft
$V_u = 1400$ kips	$V_u = 1400$ kips

4.2.9.9.1. Preliminary Size Determination. Since the length of the wall has been set at 30.5 ft, only the thickness t is adjusted to limit shear stress. The maximum shear stress allowed by Section 21.7.4.4. is $8\sqrt{f_c'}$, but experience has shown that limiting shear stress between $3\sqrt{f_c'}$ and $5\sqrt{f_c'}$ usually results in an economical wall design. For the example wall, using $4\sqrt{f_c'} = 4\sqrt{5000} = 283$ psi as the limiting shear stress, the required wall thickness equals $t = 1{,}350{,}00/(30 \times 12 \times 283) = 13.25$ in.

However, because of boundary element considerations we will use 16 in. as the wall thickness.

A few thoughts of about preliminary sizing of shear walls. An estimate of wall length and thickness based on a reasonable shear stress using only the base shear may not be adequate for resisting design moments. The resulting area of vertical boundary reinforcement may be too high, quickly leading to unworkable details. Thus it is prudent to verify that the wall thickness determined on the basis of shear stress is also thick enough to allow room for placement of reinforcing steel and concrete.

4.2.9.9.2. Shear Design. Shear design using ACI 318-02 requirements is quite straight forward. Typically, the shear demand is taken directly from the lateral analysis without having to go through load combinations because, most often, horizontal shear resulting from gravity loads is negligible unless, of course, the building is highly irregular with built-in $P\Delta$ effects. For the example wall, $V_u = V_E = 1350$ kips as obtained from a lateral analysis performed by using the ultimate earthquake loads.

Next the required horizontal reinforcement is calculated from the usable shear capacity equation,

$$\phi V_n = \phi A_{CV}\left(\alpha_c \sqrt{f_c'} + P_n f_y\right) \tag{21.7}$$

where

V_n = nominal shear capacity
ϕ = strength reduction factor = 0.6 (see Section 9.3.4)
A_{CV} = gross area of wall equal to its length times the thickness.
α_c = coefficient defining the relative contribution of concrete strength to wall strength, typically taken as equal to 2.0 (Note: Section 21.7.4.1 permits $\alpha_c = 3.0$ for squat walls with $h_w/l_w \leq 1.5$, 2.0 for $h_w/l_w \geq 2.0$, and a linear variation between 3.0 and 2.0 for intermediate values of h_w/l_w. The controlling ratio for the design of wall pier is based on the larger of overall dimensions of the wall or a segment of the wall. It is permitted to use $\alpha_c = 2.0$ in all cases.
P_n = ratio of area horizontal reinforcement to gross concreate area perpendicular to it.
f_c' = specified compressive strength of concrete, psi
f_y = specified yield strength of reinforcement, psi

For the example wall, the shear demand

$V_u = V_E = 1350$ kips

Assuming #5 @ 15 horizontal reinforcement, each face

$$P_n = \frac{0.31 \times 2 \times 12}{16 \times 12 \times 15} = 0.0026$$

$$\phi V_n = \frac{0.6 \times 16 \times 366}{1000}(2\sqrt{5000} + 0.0026 \times 60,000)$$

$$= 1045 \text{ kips} < 1350 \text{ kips} \quad \text{NG}$$

Try #6 @ 12 horizontal, each face

$$P_n = \frac{0.44 \times 12}{16 \times 12} = 0.0046$$

$$\phi V_n = \frac{0.6 \times 16 \times 366}{1000}(2\sqrt{5000} + 0.0046 \times 60,000)$$

$$= 1467 > V_u = 1350 \text{ kips} \quad \text{OK}$$

Use #6 @ 12 horizontal, each face.

Check for minimum horizontal reinforcement

$$\rho_n \geq 0.0025$$

ρ_n provided $= 0.0046 > 0.0025$ (Section 21.7.2)

Check for maximum allowable nominal shear strength

$$V_n \not> 8A_{CV}\sqrt{f_c'}$$

$$8A_{CV}\sqrt{f_c'} = \frac{8 \times 16 \times 366}{1000}\sqrt{5000}$$

$$= 3312 \text{ kips} > \frac{1350}{0.6} = 2250 \text{ kips} \qquad \text{OK} \qquad \text{(Section 21.7.4.4)}$$

4.2.9.9.3.   Shear Friction (Sliding Shear).   The shear design performed in the previous section is intended to prevent diagonal tension failures rather than direct shear transfer failures. Direct shear transfer failure, also referred to as sliding shear failure, can occur by the sliding of two vertical segments of a wall at weak sections such as at construction joints. The shear resistance is verified by using the equation

$$V_n = A_{vf}f_y\mu \tag{11.25}$$

where

A_{vf} = area of shear friction reinforcement, in.^{2}, that crosses the potential sliding plane

μ = coefficient of friction = 1.0 for a normal weight concrete surface roughened to $^1/_4$-inch amplitude.

Additionally, ACI 318-02 permits permanent net compression across the shear plane as additive to the resistance provided by shear friction reinforcement. For the example shear wall we will conservatively ignore the beneficial effect of compression.

As will be seen presently, the vertical reinforcement A_{vf} required to satisfy the governing axial load and moment combination is equal to

A_{vf} = 32 # 11 plus 36 # 7

$\quad\;\; = 32 \times 1.56 + 36 \times 0.60$

$\quad\;\; = 71.5 \text{ in.}^2$

The sliding shear resistance $V_n = 71.5 \times 60 \times 1 = 4290$ kips

$\phi V_n = 0.65 \times 4290 = 2788 \text{ kips} > 1350 \text{ kips}$

Therefore the wall is OK for sliding shear.

Section 11.7.5 limits the shear friction strength to $0.2 f_c' A_c$ or $800 A_c$ inch-lb, where A_c is the area of concrete resisting shear transfer.

For the example wall

$$V_n = 0.2 f_c' A_c = \frac{0.2 \times 5000 \times 16 \times 366}{1000} = 5836 \text{ kips}$$

$$V_n = 800 A_c = \frac{800 \times 16 \times 366}{1000} = 4685 \text{ kips} \qquad \leftarrow \text{controls}$$

$\phi V_n = 0.65 \times 4685 = 3045 \text{ kips} > 1350 \text{ kips} \qquad \text{OK}$

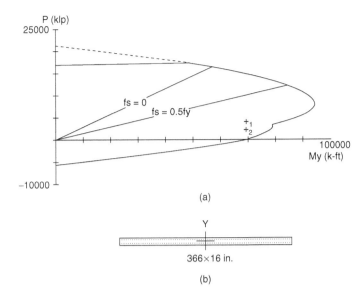

Figure 4.43. (a) Shear wall interaction diagram; (b) cross section of wall.

4.2.9.9.4. Longitudinal Reinforcement. Design of vertical reinforcement to resist a given set of axial loads and bending moments is typically a trial and error procedure. Given a wall section and an assumed reinforcement layout, the section is checked for the governing axial load and bending moment combinations. Although hand calculations and spreadsheet approaches are possible, the most desirable and expedient method is to use a computer program such as PCACOL developed by Portland Cement Association.

Figure 4.43 shows an interaction diagram for the wall with 16 #11 placed near the wall boundaries and #7 @ 9, each face, in between the boundaries for a total $A_{vf} = 71.5$ in.[2] The figure is a printed screen output of the PCACOL run. The points 1 and 2 that lie within the interaction curve represent the governing loads. Point 1 is for $P_u = 2604$ kips and $M_u = 71,000$ kip-ft, and 2 is for $P_u = 1056$ kips and $M_u = 71,000$ kip-ft. Since both 1 and 2 lie within the interaction curve, the example wall is OK for the ultimate axial load and moment combinations.

4.2.9.9.5. Web Reinforcement. Section 21.7.2.1 requires a uniform distribution of both horizontal shear reinforcement ρ_n and vertical reinforcement ρ_v. Further, to control width of inclined cracks due to shear, a minimum reinforcement ratio equal to 0.0025 and a maximum spacing of 18 inches is specified for both ρ_n and ρ_u. However, a reduction in the reinforcement ratio is permitted if the design shear force V_u is less than $A_c \sqrt{f_c'}$.

Minimum ratios of ρ_u if $V_u \le A_c \sqrt{f_c'}$ (see Section 14.3) are

- 0.0020 for #5 and smaller bars, with $f_y \ge 60,000$ psi.
- 0.0025 for other bars.
- 0.0020 for welded fabric not larger than W31 or D31.

The minimum ratios of ρ_v (vertical reinforcement) for the same condition are

- 0.0012 for #5 and smaller bars with $f_y \ge 60,000$ psi.
- 0.0015 for other bars.
- 0.0012 for welded wire fabric not larger than W31 or D31.

In seismic design, the vertical reinforcement at the bottom few stories of a shear wall is typically controlled by bending requirements. The upper levels are likely to be controlled by the ACI 318-02 minimum reinforcement ratio of 0.0025.

For the example wall,

$$A_{CV}\sqrt{f'_c} = \frac{366 \times 16\sqrt{5000}}{1000} = 414 \text{ kips} < V_u = 1400 \text{ kips}$$

The minimum horizontal reinforcement

$$= 0.0025 \times b \times t$$
$$= 0.0025 \times 16 \times 12$$
$$= 0.48 \text{ in.}^2$$

Use #5 @ 15 giving a steel area $= \dfrac{2 \times 0.31 \times 12}{15} = 0.496 \text{ in.}^2$

$$> 0.48 \text{ in.}^2 \qquad \text{OK}$$

Section 21.7.2.2 requires at least two curtains of reinforcement if the factored shear force V_u exceeds $2A_{CV}\sqrt{f'_c}$.

For the example wall,

$$2A_{CV}\sqrt{f'_c} = \frac{2 \times 366 \times 16}{1000}\sqrt{5000}$$
$$= 828 \text{ kips}$$

Since $V_u = 1350$ kips is greater than 828 kips, we use two layers of #5 @ 15. The reason for two layers of reinforcement is to place web reinforcement close to the wall surface to inhibit fragmentation of concrete in the event of severe cracking of concrete during an earthquake.

4.2.9.9.6 Boundary Elements.

4.2.9.9.6.a. Stress Index Procedure.   This method is quite straightforward (Section 21.7.6.3). A stress index of $0.2 f'_c$ is used as a benchmark for the maximum extreme fiber compressive stress corresponding to factored forces that include gravity and earthquake effects. If the calculated compressive stress is less than the index value, special boundary elements are not required. If not, detailing of boundary elements in accordance with Section 21.7.6.4 is required. The compressive stresses are calculated for the factored axial forces and bending moments using a linear elastic model and gross section properties.

For the example wall,

$$A_g = 366 \times 16 = 5856 \text{ in.}^2$$

$$I_g = 16 \times \frac{366^3}{12} = 65{,}370{,}528 \text{ in.}^4$$

$$S_{YY} = \frac{65{,}370{,}528}{183} = 357{,}216 \text{ in.}^3$$

$$\frac{P_u}{A_g} + \frac{M_u}{S_{YY}} = \frac{2604}{5856} + \frac{71{,}000 \times 12}{357{,}216}$$

$$= 0.445 + 2.385 = 2.83 \text{ ksi} > 0.2 f'_c$$
$$= 0.2 \times 5000$$
$$= 1.0 \text{ ksi}$$

Therefore, boundary elements are required by the stress–index procedure.

4.2.9.9.6.b. Displacement-Based Procedure. In this procedure (Section 21.7.6.2), the neutral axis depth c, which is directly related to the strain at the extreme compression fiber, is used as an index to determine whether or not boundary elements are required. Boundary zone detailing is required if

$$c > \frac{l_w}{600\left(\dfrac{\delta_u}{h_w}\right)} \qquad (21.8)$$

where

c = distance from the extreme compression fiber to the neutral axis

l_w = length of entire wall or wall–pier (segment)

δ_u = design displacement at top of wall or segment equal to elastic displacement δ_c due to code level seismic forces multiplied by C_d, the deflection amplification factor given in governing codes.

h_w = height of entire wall or wall segment.

The displacement-based approach is founded on the assumption that the inelastic response of the wall is due to flexural yielding at a critical section, typically at its base. Given this proviso, the method of determining whether or not boundary elements are required follows:

- Analytically displace the wall at top equal to the design displacement δ_u. This displacement is equal to the elastic displacement δ_c calculated for code seismic loads, multiplied by a deflection amplification factor C_d. Thus, $\delta_u = \delta_c \times C_d$.
- Calculate the strain in the extreme compression fiber of the wall corresponding to the horizontal displacement of δ_u. Since the strain is related to the depth of neutral axis c, it is used indirectly for evaluating the strain. Equation (21.8) of ACI 318-02 is used to calculate c. The depth c may be considered, in a conceptual sense, as an index–depth of neutral axis for comparing against the actual depth calculated for the largest ultimate load P_u and corresponding moment M_n.
- Next, compute the neutral axis depth c, using a linear strain distribution (Section 10.2), or by assuming yielding of all vertical reinforcement in compression or tension. The latter is recommended by the 1999 Blue Book of the Structural Engineers Association of California (SEAOC). The depth c is calculated for the factored axial force and nominal moment strength consistent with the displacement δ_u at the top of the wall resulting in the largest neutral axis depth.
- If the calculated value of c is greater than the index value, then special boundary elements detailed are similar to those of a ductile column.

For the example wall, we have the following two load combinations:

P_u = 2470 kips	P_u = 1056 kips
M_u = 70,000 kip-ft	M_u = 70,000 kip-ft
V_u = 1350 kips	V_u = 1350 kips

Using the PCACOL column design program, the depth of the neutral axis was found to be 108 inches.

The term δ_u is design displacement defined as the lateral displacement expected for the design-basis earthquake. It is invariably larger than the elastic displacement δ_e calculated for code-level forces applied to a linear elastic model. Although the analysis may consider the effects of cracked sections, torsion, $P\Delta$ forces, and foundation flexibility, it does not account for the expected inelastic response. Thus δ_u is calculated by multiplying δ_e by a deflection amplification factor C_d given in the governing codes or standards. For example, ASCE 7-02 and IBC-03 specify $C_d = 5.5$ and 6.5 for special reinforced concrete moment frames and dual systems consisting of SMRF and special reinforced concrete walls. For the example problem having a building system of special reinforced concrete wall, $C_d = 5.0$, by both ASCE and IBC.

The elastic deflection δ_e of the shear wall at the roof level = 2.15 in., as obtained from a linear elastic analysis of the building under code-prescribed seismic forces. This is given in the statement of the problem.

Therefore

$$\frac{\delta_u}{h_w} = \frac{C_d \delta_e}{I_E} = \frac{5 \times 2.15}{1} = 10.75 \text{ in.}$$

$$\frac{\delta_u}{h_w} = \frac{10.75}{118 \times 12} = 0.0076 > 0.007 \text{ (min)}$$

$$c = \frac{l_w}{600\left(\dfrac{h_w}{l_w}\right)} = \frac{366}{600 \times 0.0076} = 80.26 \text{ in.} < 108 \text{ in.}$$

Special boundary elements are therefore required. It is interesting to note that for the example wall, both the stress index and the strain index methods lead to the same conclusion, namely, that boundary elements are required. Thus may not be the case in all designs. A more likely scenario would be for the stress index method to show that boundary elements are required, while the strain method does not. Although ACI 318-02 does not require both criteria to be satisfied, many engineers choose to detail the boundary zones as required by the stress index method. Keep in mind, in seismic design, more is less!

4.2.9.9.6.c.   Reinforcement Details.   Irrespective of the method used to determine whether or not special boundary elements are required, the detailing is performed according to Sections 21.6.6.4 through 21.6.6.6, summarized as follows:

- The required width of boundary element is given by the larger of $c - 0.1l_w$ and $c/2$.
- Where required, special boundary elements are extended from the critical section a distance not less than l_w or $M_u/4V_u$.

For the example wall, the width of boundary element is the larger of

$$c - 0.1 \, l_w = 108 - 0.1 \times 366 = 71.4 \text{ in.} \qquad \leftarrow \text{controls}$$

$$\frac{c}{2} = \frac{108}{2} = 54 \text{ in.}$$

Considering the placement of vertical bars, detail a boundary element for a width of 75 in. (Fig. 4.44).

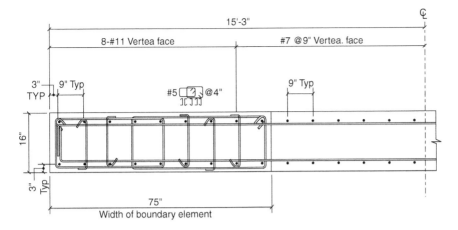

Figure 4.44. Shear wall example; schematic reinforcement.

The vertical extension of the boundary element must not be less than

$$l_w = 366 \text{ in.} \quad \text{or} \quad \leftarrow \text{controls}$$

$$\frac{M_u}{4V_u} = \frac{71000 \times 12}{1400} = 608.6 \text{ in.}$$

Confinement of 16 × 75 in. Boundary Elements.
Confinement Perpendicular to the Wall.   Maximum allowable spacing of hoops and crossties, assuming #5 bars,

$$\begin{aligned}
s_{\max} &= 0.25 \times \text{minimum member dimensions} \\
&= 0.25 \times 16 = 4 \text{ in.} \quad \text{(Controls)} \\
&= 6 \times \text{diameter of longitudinal bar} \\
&= 6 \times 1.41 = 8.5 \text{ in.} \\
&= s_x = 4 + \left(\frac{14 - b_x}{3}\right) = 4 + \left(\frac{14 - 10}{3}\right) = 5.33 \text{ in.}
\end{aligned}$$

The required cross-sectional area of confining reinforcement A_{sh}, in the 16×75 in. boundary elements, using $s = 4$ in., is given by

$$A_{sh} = 0.09 sh_c \frac{f'_c}{f_y} \tag{21.9}$$

where
   h_c = cross-sectional dimension of boundary element measured center-to-center of confining reinforcement.

In our case, $h_c = 16 - (3 + 3) + 1.41 + 0.625 = 12$ in.

$$A_{sh} = 0.09 \times 4 \times 12 \times \frac{5}{60} = 0.36 \text{ in.}^2$$

No. 5 hoops with two legs provide $A_{sh} = 2 \times 0.31 = 0.62$ in.^{2} > 0.36 in.^{2}

Confinement Parallel to the Wall.

$$h_c = 75 - (3 + 3) + 1.41 + 0.625 = 71 \text{ in.}$$

$$A_{sh} = 0.09 \times 4 \times 71 \times \frac{5}{60} = 2.13 \text{ in.}^2$$

With two hoops consisting of two legs each, and five crossties,

$$A_{sh} \text{ provided} = 9 \times 0.31 = 2.79 \text{ in.}^2 > 2.13 \text{ in.}^2 \qquad \text{OK}$$

In most designs, special boundary elements may not be required by calculations for the entire height of walls. However, to prevent buckling of boundary longitudinal elements even in cases where they are not needed by design, Section 21.7.6.5 requires transverse ties not exceeding a vertical spacing of 8 in., if the vertical reinforcement ratio is greater than $400/f_y$. The transverse reinforcement shall consist of either single or overlapping hooks. As in ductile columns, crossties are permitted. For calculating the ratio $400/f_y$, only the reinforcement within the wall boundary element is included.

Using the most common value of $f_y = 60,000$ psi, the ratio $400/f_y = 400/60,000 = 0.0067$. If the ratio of vertical reinforcement is greater than this value, then hoops supplemented with crossties are required. What if the ratio of vertical bars placed in between the boundary zones is greater than 0.0067? Do they also need to be tied? Yes, but only if the vertical reinforcement ratio is greater than 0.01, or where the vertical reinforcement is required as a compression reinforcement. See Section 14.3.6. A schematic placement of reinforcement is shown in Fig. 4.45. A construction photograph of a shear wall in zone 4 is shown in Fig. 4.45a.

4.2.9.10.   Special Reinforced Concrete Coupled Shear Walls

Given.   A 40-ft-long by 16-inch-thick shear wall with openings as shown in Fig. 4.46. The shear wall forms part of a lateral load-resisting system of a 10-story concrete building located in a high seismic zone. A computer analysis has been performed for the building using code-prescribed lateral forces and gravity loads. The analysis typically has provided moment and shear forces for each coupling beam, and moments, shear forces, and axial forces for each wall segment commonly referred to as wall pier. In modeling the shear walls, effective section properties, rather than gross properties, are used as required by most current codes.

The first step in design is the determination of ultimate design values, generally the P_u, M_u, and V_u using code-specified load combinations. Typically, the design of an element such as a wall pier or a coupling beam is verified for a number of design load combinations. This is because several lateral load analyses are performed to account for changes in load directions, minimum eccentricities in each direction, uplift and downward effects of seismic loads, etc. Computation of design values using different load combinations that includes several lateral load analyses is indeed a major task invariably necessitating use of computers. Without dwelling on this further, we will proceed with the design of coupling beam CB1 and wall pier W1 by presupposing the following ultimate design values.

CB1	$V_u = 300$ kips
	$M_u = 12,000$ kip-ft, left end
	$M_u = 8000$ kip-ft, right end
Wall pier W1	$P_u = 1500$ kips
	$V_u = 210$ kips
	$M_u = 45,000$ kip-ft.

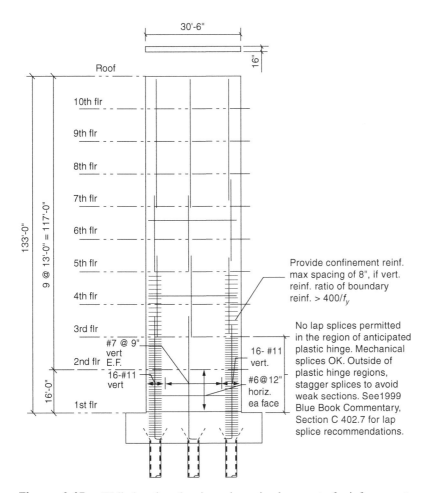

Figure 4.45. Wall elevation showing schematic placement of reinforcement.

Figure 4.45a. Photograph of a shear wall in seismic zone 4 showing wall-pier and boundary-element reinforcement. (Photo courtesy of Mr. Walter Steimle, John A. Martin and Assoc., Los Angeles, CA.)

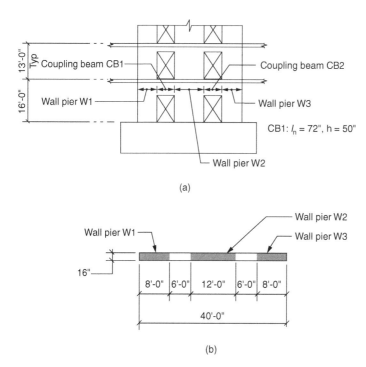

Figure 4.46. Coupled shear walls: (a) partial elevation; (b) plan.

Required
- Coupling beam design.
    1. Design of diagonal reinforcement.
    2. Design of transverse reinforcement.
    3. Schematic section through coupling beam.

- Design of wall pier W1.
    1. Design for shear.
    2. Design for combined flexure and axial loads.
    3. Determine boundary elements requirements using
    - Stress-index procedure.
    - Displacement-based procedure.
    - Schematic layout of reinforcement.

The design shall be in accordance with ACI 318-02.
Solution.
4.2.9.10.1. Coupling Beams.
4.2.9.10.1.a. Diagonal Reinforcement.   Two simultaneous criteria establish whether diagonal reinforcement is required in a coupling beam.

1. Clear length-to-span ratio, often referred to as the aspect ratio of the beam, is less than 2, i.e., $l_n/h < 2.0$.
2. The factored shear force V_u is greater than $4\sqrt{f_c'}A_{cp}$.

For the example coupling beam CB1, we have

$f'_c = 4000$ psi, $f_y = 60,000$ psi, $A_{cp} = b \times h = 16 \times 50 = 800$ in.^{2}

$V_u = \phi V_n = 210$ kips, $l_n = 72$ in.

$\dfrac{l_n}{h} = \dfrac{72}{50} = 1.44 < 2$ (Aspect ratio criterion)

$V_u = 210$ kips $> \dfrac{4\sqrt{4000} \times 800}{1000} = 203$ kips (V_u criterion)

Therefore, because of both the aspect ratio and the V_u criteria, diagonal reinforcement must be provided.

Observe that if either of the criteria was not satisfied, we would have had the option of designing the beam CB1 without the diagonal reinforcement. We could have used conventional horizontal reinforcement to resist flexure and vertical stirrups to resist shear. However, research has shown that diagonal reinforcement improves coupling beam performance, even at lower shear stress level. See SEAOC's 1999 Blue Book Commentary, Section C 407.7.

In some buildings it may be impractical to use diagonal reinforcement. Do the designers have any fallback position? Yes, they do. The requirements of Section 21.6.7.3 for diagonal reinforcement may be waived if coupling beams are not used as part of the lateral force resisting system. Such beams are permitted at locations where damage to these elements does not impair vertical load-carrying capacity or egress of the structure, or integrity of nonstructural components and their connections to the structure.

Returning to the example problem, the equation that determines the area of diagonal reinforcement A_{vd} is given by

$$V_n = 2A_{vd}f_y \sin \alpha \leq 10\sqrt{f'_c}A_{cp} \tag{21.9}$$

This can be written as

$$A_{vd} = \frac{\phi V_n}{2\phi f_y \sin \alpha} = \frac{V_u}{2\phi f_y \sin \alpha}$$

where
  α = angle between the diagonal reinforcement and the longitudinal axis of the coupling beam (Fig. 4.47).
  A_{vd} = area of diagonal reinforcement in each diagonally reinforced beam.

It should be noted that diagonally oriented reinforcement is effective only if the bars are placed with a reasonably large inclination angle α. So, diagonally reinforced coupling beams are restricted to beams having an aspects ratio $l_n/h < 4.0$. This ratio approximately corresponds to $\alpha = 13°$. Therefore, for beams with a geometry that results in α less than about 13°, ACI 318-02 does not permit diagonal reinforcement.

Each diagonal element is reinforced similar to a column consisting of longitudinal and transverse reinforcement. The column cage must consist of at least four longitudinal

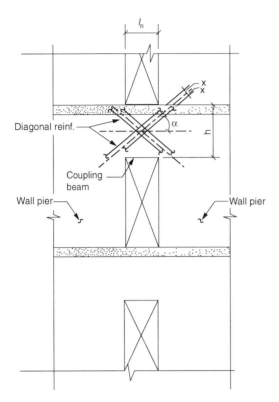

Figure 4.47. Geometry for calculating α, the angle between diagonal reinforcement and the longitudinal axis of the coupling beam.

Note: $\tan \alpha = \dfrac{\dfrac{h}{2} - \dfrac{x}{\cos \alpha}}{\dfrac{\ell_n}{2}}$

(solve for α by trial and error.)

bars, with core dimensions measured to the outside of the transverse reinforcement not less than $b_w/2$ and $b_w/5$. It should be noted that, in practice, minimum required reinforcement clearance often controls the thickness of walls. Typically, a wall thickness of 16 in. or larger is required for the detailing of diagonally reinforced coupling beams.

The required area of longitudinal reinforcement A_{vd} is calculated as follows:

$$A_{vd} = \frac{V_u}{2\phi f_y \sin \alpha}$$

This is the same as Eq. 21.9, written in a different form. An upper limit of $10\sqrt{f_c'} A_{cp}$ is imposed for the nominal capacity $V_n = \dfrac{V_u}{\phi}$. For the example problem, the upper limit is equal to

$$\frac{10\sqrt{f_c'}}{\phi} A_{cp} = \frac{10\sqrt{4000} \times 800}{1000 \times 0.75} = 675 \text{ kips}$$

This is greater than the design value of $V_u = 210$ kips.   OK

Referring to Fig. 4.47, we have $\tan \alpha = h - \frac{2x}{\cos \alpha}/l_n$. This is a transcendental equation, best solved by trial and error.

Try $\alpha = 30°$. $\tan \alpha = \tan 30° = 0.577$, $\cos 30° = 0.866$

$$\frac{h - \dfrac{2x}{\cos \alpha}}{l_n} = \frac{50 - \dfrac{2 \times 6}{0.866}}{72} = 0.50$$

$\tan \alpha = \tan 30° = 0.577$. Compared to 0.50, this is not close enough.

Try $\alpha = 28°$, $\tan 28° = 0.552$, $\cos 28° = 0.883$

$$\frac{h - \dfrac{2x}{\cos \alpha}}{l_n} = \frac{50 - \dfrac{2 \times 6}{0.883}}{72} = 0.506   \tan \alpha = 0.552$$

Again, not close enough.

Try $\alpha = 27°$, $\tan 27° = 0.509$, $\cos 27° = 0.891$

$$\frac{h - \dfrac{2x}{\cos \alpha}}{l_n} = \frac{50 - \dfrac{2 \times 6}{0.891}}{72} = 0.507   \tan \alpha = 0.509   \text{OK}$$

Use $\alpha = 27°$, $\sin \alpha = 0.459$

$$A_{vd} = \frac{V_u}{2 \phi f_y \sin \alpha}$$

$$= \frac{210}{2 \times 0.75 \times 60 \times 0.459} = 5.0 \text{ in.}^2$$

Use four #10 diagonal reinforcement giving $A_{vd} = 4 \times 1.27 = 5.08$ in.^{2}

4.2.9.10.1.b.   Transverse Reinforcement. The requirements of transverse reinforcement given below are the same as for frame columns of SMRFs.

$$A_{sh} = 0.3 \left(s h_c \frac{f'_c}{f_{yh}} \right) \left[\frac{A_g}{A_{ch}} - 1 \right] \tag{21.3}$$

$$A_{sh} = 0.09 s h_c \frac{f'_c}{f_y} \tag{21.4}$$

The maximum spacing limits of transverse reinforcement, also referred to as ties, are once again the same as for frame columns. According to Section 21.4.4.2, the limits are

1.   One-quarter the minimum member dimensions.
2.   Six times the diameter of diagonal reinforcement.
3. $s_x = 4 + \left(\dfrac{14 - h_x}{3} \right)$.

s_x need not be less than 4 in. nor can it exceed 6 in., i.e., 4 in. $\leq s_x \leq$ 6 in.

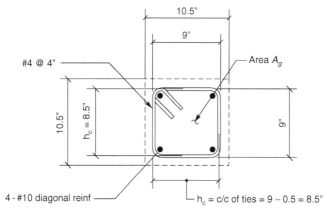

A_g is calculated by assuming a $\frac{3}{4}''$ cover around #4 ties.

$A_g = (9 + 2 \times 0.75)^2 = 110.25$ in.^{2}

$A_{ch} = 9 \times 9 = 81$ in.^{2}

Figure 4.48. Parameters for calculating diagonal beam reinforcement.

For the diagonally reinforced CB1 we have

1. $\dfrac{b_w}{4} = \dfrac{16}{4} = 4$ in.   ←   controls
2. $6d_b = 6 \times 1.27 = 7.62$ in.
3. $s_x = 4 + \left(\dfrac{14 - 14}{3}\right) = 4$ in.

Substituting the controlling value of 4.0 in. for the spacing s in Eqs. (21.3) and (21.4), we get

$$A_g = (9 + 2 \times 0.75)(9 + 2 \times 0.75) = 110.25 \text{ in.}^2$$

Note that A_g is calculated assuming a minimum cover of $^3/_4$ in. around the diagonal core (Fig. 4.48).

By Eq. (21.3),

$$A_{sh} = 0.3\left(4 \times 8.5 \frac{4}{60}\right)\left[\frac{110.25}{81} - 1\right]$$

$$= 0.246 \text{ in.}^2 \quad \leftarrow \text{ controls}$$

By Eq. (21.4),

$$A_{sh} = 0.09 \times 4 \times 8.5 \times \frac{4}{60} = 0.204 \text{ in.}^2$$

A single #4 loop around four diagonal bars with two legs gives $A_{sh} = 0.40$ in.^{2} Hence, #4 ties at 4-inch spacing are OK for the bursting steel requirements.

Note in our example, the core dimensions of diagonal reinforcement are the same in both directions. In a general case, with differing cross-section dimensions, A_{sh} is calculated for each direction.

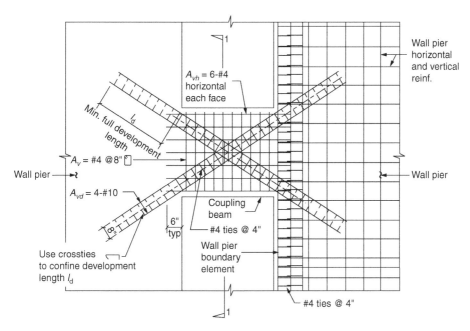

Figure 4.49. Coupling beam with diagonal reinforcement. Each diagonal reinforcement must consist of at least four bars with closely spaced ties. Use wider closed ties or crossties at central intersection. Use crossties to confine development length ℓ_d.

By Section 21.7.7.4 (d), diagonal bars are required to be developed for tension into the wall piers. This is shown in Fig. 4.49 where the diagonal bars extend a distance of l_d beyond the face of the wall pier. Instead of loops, crossties are used along the development length and at the intersection of diagonal bars at the center of diagonal beam.

In addition to the reinforcement calculated thus far, supplemental horizontal and vertical reinforcements are required by Sections 11.8.4 and 11.8.5. The intent of additional reinforcement is to contain the concrete outside the diagonal cores, in case the concrete is damaged by earthquake loading. Since the diagonal reinforcement is designed to resist the entire shear and flexure in the coupling beam, additional transverse and longitudinal reinforcement acts primarily as a basketing reinforcement to contain concrete that may spall. It is not necessary to develop the horizontal bars into the wall piers.

The minimum reinforcement, A_v, perpendicular to the longitudinal axis of the coupling beam (meaning vertical reinforcement) shall not be less than $A_v \geq 0.0025\, b_w s$ (Section 11.8.4). The area of horizontal (longitudinal) reinforcement, A_{vh}, shall not be less than $0.0015\, b_w s_2$, and s_2 shall not exceed $d/5$ or 12 in. (Section 11.8.5).

For the example beam, assuming #4 @ eight loops as vertical reinforcement,

$$A_v = \frac{0.2 \times 2 \times 12}{8} = 0.60 \text{ in.}^2 > 0.0025 b_w s = 0.0025 \times 16 \times 8 = 0.32 \text{ in.}^2 \quad \text{OK}$$

Assuming six #4 horizontal bars at each face,

$$A_{vh} \geq 0.0025\, b_w \times h = 0.0025 \times 16 \times 50 = 2 \text{ in.}^2 < 6 \times 2 \times 0.2 = 2.4 \text{ in.}^2 \quad \text{OK}$$

A schematic section of the coupling beam is shown in Fig 4.50.

4.2.9.10.2. Wall Piers.

4.2.9.10.2.a. Shear Design. For the example pier W1, $V_u = 300$ kips. The parameter α_c, the coefficient defining the relative contribution of concrete shear strength to the total

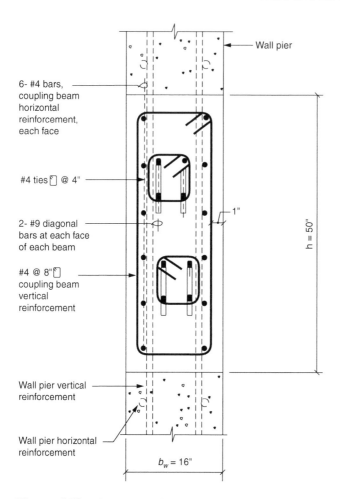

Figure 4.50. Section 1-1. Schematic section through coupling beam. The purpose of this sketch is to ensure that the wall is thick enough for the proper placement of wall and diagonal beam reinforcement and concrete.

shear strength of wall, may be conservatively taken to equal 2.0. However, if the designer chooses to calculate α_c, it should be based on the ratio h_w/l_w, taken as the larger for the individual wall pier and for the entire wall (see Section 21.7.4.2).

For the example wall pier, the overall $h_w/l_w = 133/40 = 3.32.5$ and the individual wall pier $h_w/l_w = 15/8 = 1.875$.

Thus the ratio 3.325 controls the determination of α, giving $\alpha = 2.0$.

$$\phi V_n = \phi A_{cv}\left(\alpha_c \sqrt{f_c'} + \rho_n f_y\right) \tag{21.7}$$

Try #5 @ 15 horizontal, each face

$$\rho_n = \frac{0.31 \times 2 \times 12}{16 \times 12 \times 15} = 0.0026$$

$$A_{CV} = 8 \times 12 \times 16 = 1536 \text{ in.}^2$$

$$\phi V_n = 0.6 \times 1536\left(2\sqrt{4000} + 0.0026 \times 60{,}000\right)$$

$$= 260 \text{ kips} < V_u = 300 \text{ kips} \quad \text{NG}$$

Try #6 @ 15 horizontal, each face.

$$\rho_n = \frac{0.44 \times 2 \times 12}{16 \times 12 \times 15} = 0.0037$$

$$\phi V_n = 0.6 \times 1536\left(2\sqrt{4000} + 0.0026 \times 60{,}000\right)$$

$$= 320 \text{ kips} > 300 \text{ kips} \qquad \text{OK}$$

Use #6 @ 15 horizontal, each face.

$$\rho_n = 0.0037 > \rho_{\min} = 0.0025 \qquad \text{OK}$$

Check for maximum allowable nominal strength.

$$V_n \ngtr 10\, A_{CV}\sqrt{f_c'} = \frac{10 \times 1536}{1000}\sqrt{4000} = 971 \text{ kips}$$

$$V_n = \frac{320}{0.6} = 533 \text{ kips} < 971 \text{ kips} \qquad \text{OK}$$

4.2.9.10.2.b.   Shear Friction (Sliding Shear).   To determine sliding shear resistance, we need to know the area of vertical reinforcement A_{vf} crossing the assumed shear plane. For the example problem, we have not yet determined the vertical reinforcement in wall pier W1 across this plane which typically occurs at construction joints. A_{vf} is calculated presently in the combined axial and flexural design part of this problem. It is equal to 12 #11 plus 8 #6, giving $A_{vf} = 22.24$ in.², which will be used to check the sliding shear.

The sliding shear resistance is given by

$$V_n = A_{vf} f_y \mu$$

Using $\mu = 1.0\,\lambda$, where $\lambda = 1$ for normal weight concrete, and $A_{vf} = 22.4$ in.²,

$$V_n = 22.4 \times 60 \times 1 = 1334.4 \text{ kips} > 300 \text{ kips} \quad \text{OK}$$

Section 11.7.5 limits shear friction strength to $0.2 f_c' A_c$ or $800\, A_c$. For the example wall pier,

$$V_n = 0.2 f_c' A_c = \frac{0.2 \times 4000 \times 16 \times 8 \times 12}{1000}$$

$$= 1228 \text{ kips} > 300 \text{ kips}$$

$$V_n = 800 A_c = \frac{800 \times 16 \times 8 \times 12}{1000} = 1228 \text{ kips} > 300 \text{ kips}$$

Therefore, wall pier W1 OK for sliding shear

4.2.9.10.2.c.   Longitudinal Reinforcement.   Factored axial forces and moments for the design of W1 are as follows:

$$P_u = 1200 \text{ kips}$$

$$M_u = 25{,}200 \text{ kip-in.} = 2100 \text{ kip-ft}$$

Figure 4.51a shows the arrangement of vertical reinforcement in wall pier W1. Six #11 are placed near the wall boundary zones, with #6 @ 10 at each face in between the boundary elements. The printed output of the PCACOL screen is shown in Fig. 4.51b. The interaction point A, corresponding to $P_u = 1200$ kips, $M_u = 2100$ kips, is well within the interaction curve, justifying the design of W1 for the combined axial load and building moments.

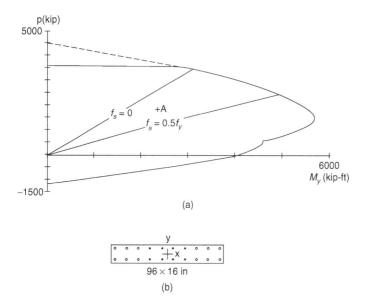

Figure 4.51.   (a) Wall–pier W1 interaction diagram; (b) cross section of wall pier W1.

4.2.9.10.2.d.   Web Reinforcement.   The minimum vertical reinforcement ratio ρ_v, by Section 21.7.2.1 is 0.0025. However, Section 14.3 permits a reduction in ρ_v if $V_u \leq A_c\sqrt{f_c'}$. In our case,

$$V_u = 210 \text{ kips} > \frac{96 \times 16\sqrt{4000}}{1000} = 97 \text{ kips}$$

Therefore, minimum $\rho_v = 0.0025$

$$\rho_v \text{ provided at \#6 @ eight, each face} = \frac{0.44 \times 2 \times 12}{8 \times 16 \times 12}$$
$$= 0.0069$$
$$> 0.0025 \quad \text{OK}$$

Section 21.7.2.2 requires at least two curtains of reinforcement, for both ρ_v and ρ_n, if $V_u > 2A_{CV}\sqrt{f_c'}$.
For the example wall pier W1,

$$V_u = 210 \text{ kips} > 2A_{CV}\sqrt{f_c'} = \frac{2 \times 96 \times 16\sqrt{4000}}{1000} = 194.3 \text{ kips}$$

Therefore, two curtains of #6 @ eight at each face is OK.

4.2.9.10.2.e.   Boundary Elements.   The design of wall segments for flexure is identical to that for a conventional solid wall. However, in designing boundary elements, a question comes up, whether to use the displacement-based approach or the stress-index method. Section 21.7.6.2 limits the use of displacement-based approach to walls that are continuous from the base of the structure to the top of wall and designed to have a single critical section for flexure and axial loads. A coupled shear wall as a whole is typically not designed to have a single critical section for flexure and axial loads because plastic hinges may form in the coupling beams as well as at the base of each pier. Therefore, by this interpretation, displacement-based design is not permitted for wall piers.

However, if the makeup of the wall is considered as an assemblage of independent wall piers, then it can be argued that each wall pier is continuous and is designed to have a single critical section at its base for flexure and axial loads. Using this interpretation, the evaluation of special boundary elements may be based on the displacement-based method of comparing neutral axis depths.

Faced with this uncertainty, what is the best way to tackle wall pier designs? Keeping in mind that "more is less" in seismic design, the author recommends use of the more conservative stress-index method. However, for illustration purposes, the example wall pier W1 will be designed using both methods.

Stress-Index Procedure. For the example wall pier W1, $A_g = 96 \times 16 = 1356$ in.[2], and the combined compressive stress for the factored axial load and bending moments are:

$$\frac{P_u}{A_g} + \frac{M_u}{S_{y-y}} = \frac{1200}{1536} + \frac{2100 \times 12}{24576}$$

$$= 0.78 + 1.02 = 1.8 \text{ ksi}$$

This is greater than $0.2 f'_c = \frac{0.2 \times 4000}{1000} = 0.80$ ksi.
Therefore, boundary elements are required by the stress-index procedure.

Displacement-Based Procedure. Boundary zone detailing is required if the depth of the neutral axis c from the extreme compression fiber is greater than an index depth as given by

$$c > \frac{l_w}{600 \left(\dfrac{\delta_u}{h_w} \right)} \tag{21.8}$$

For the example wall W1, the value of c calculated by using the PCACOL program is equal to 40.63 in.

The elastic deflection δ_c is equal to 1.97 in. at the roof, as given in the statement of the problem.

The design displacement $\quad \delta_u = C_d \delta_e = 5.5 \times 1.97 = 10.85$ in.

$$\frac{\delta_u}{h_w} = \frac{10.85}{133 \times 12} = 0.0068 < 0.007$$

Therefore $\quad c = \dfrac{96}{600 \times 0.007} = 22.86$ in.

The value of $c = 40.63$ in. calculated using the PCACOL program is greater than the index value of 22.86 in. Therefore, boundary elements are required by the displacement-based procedure.

Reinforcement Details. The required width of boundary element is the larger of

$c - 0.1\, l_w = 40.63 - 0.1 \times 96 = 31$ in. $\qquad \leftarrow$ controls

$$\frac{c}{2} = \frac{40.63}{2} = 20.31 \text{ in.}$$

Considering the placement of the vertical bars, confine 36 in. width of wall at both ends of the wall pier.

The vertical extension must not be less than

$$l_w = 96 \text{ in.} \qquad \text{(Controls)}$$

$$\frac{M_n}{4V_n} = \frac{2100 \times 12}{4 \times 210} = 30 \text{ in.}$$

Although the boundary element need not extend more than 96 in., we choose to extend it to the full height of the first floor.

Confinement of 16×36 in. Boundary Elements.

Confinement Perpendicular to the Wall.   Minimum allowable spacing of hoops and crossties is given by

$$s_{max} = 0.25 \text{ times minimum member dimension}$$
$$= 0.25 \times 16 = 4 \text{ in.} \qquad \text{(Controls)}$$
$$= 6 \times \text{diameter of longitudinal bar}$$
$$= 6 \times 1.41 = 8.5 \text{ in.}$$

$$= 4 + \left(\frac{14 - h_x}{3} \right) = 4 + \left(\frac{14 - 11.91}{3} \right) = 4.7 \text{ in.}$$

Notes: Crossties are not used in this example because
$$h_x = 16 - (3 + 3) + 1.41 + 0.5 = 11.91 < 14 \text{ in.} \qquad \text{OK}$$
The required cross-sectional area of confining reinforcement using $s = 4$ in. is given by

$$A_{sh} = 0.09sh_c \frac{f'_c}{f_y} \tag{21.9}$$

where
h_c = cross-sectional dimension of boundary element measured center-to-center of confining reinforcement.

In our case, $h_c = 16 - (3 + 3) + 1.41 + 0.5 = 11.91$ in.

$$A_{sh} \text{ required} = 0.09 \times 4 \times 12 \times \frac{4}{60} = 0.29 \text{ in.}^2$$

No. 4 hoops with two legs provide $A_{sh} = 2 \times 0.2 = 0.4 \text{ in.}^2 > 0.29 \text{ in.}^2 \qquad$ OK

Confinement parallel to the wall.

$$h_c = 33 - (3 + 3) + 1.41 + 0.625 = 29 \text{ in.}$$

$$A_{sh} = 0.09 \times 4 \times 29 \times \frac{4}{60} = 0.7 \text{ in.}^2$$

With two hoops, A_{sh} provided $= 4 \times 0.2 = 0.8 \text{ in}^2 > 0.7 \text{ in.}^2 \qquad$ OK

Figure 4.52 shows a schematic layout of reinforcement in wall pier W1.

The analysis and design performed thus far does not consider post-elastic behavior of coupled walls, nor does it explain how a plastic analysis may be performed for seismic forces when the elements of the wall are yielding. This type of post-yield analysis is not required by ACI 318-02 but is recommended in the 1999 Blue Book, *Recommended Lateral force Requirements and Commentary*, published by the Structural Engineers Association of California (SEAOC). The designer is referred to Chapter 4 of this reference for further details.

A construction photograph of a diagonally reinforced coupling beam is shown in Fig. 4.52a.

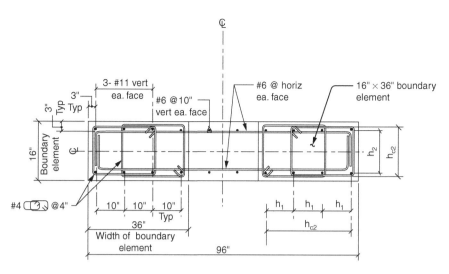

Figure 4.52. Schematic reinforcement layout for wall pier, example 2.

Figure 4.52a. Construction photograph of a diagonally reinforced coupling beam. (Photo courtesy of Mr. Walter Steimle, John A. Martin and Assoc., Los Angeles, CA.)

Walt Disney Concert Hall
Los Angeles, California

John A. Martin & Associates, Inc.
Structural Engineers
Los Angeles, CA

5
Composite Buildings

It is worth noting, for the sake of humility, that the basic ideas of composite behavior were conceived and attempts at composite construction made long before anyone reading this work was born. In the United States, composite construction first appeared in 1894. In that year, Joseph Milan, an engineer from Vienna, built a composite bridge in Rock Rapids, IA, using steel I-beams bent into an arch shape, and then encasing concrete around them; he claimed that steel and concrete acted together. So confident was he of this system that he applied for and eventually got an American patent by submitting deflection calculations to prove his theory. Replacement of the curved beams with straight beams was the next step. By the turn of the century, steel beam encasement in concrete was a regular practice in both buildings and bridges. However, it took some 30 more years to codify the composite design. The design rules first appeared in the 1930 New York City Building Code when fully encased steel beams were permitted a bending stress of 20 ksi, an 11% increase over the 18 ksi allowed for noncomposite beams at that time.

In an ideal world, a logical sequence of events that occurs before codifying a unique structural system would be as follows: First, an uncommon system is conceived not because it is unique, not because it is clever, but because it is the market economy that motivates the conception of novel systems. Ideally one would make conceptual sketches of the proposed component and its connections, and get an opinion of its constructability from contractors familiar with the construction practices in that area.

After establishing the theory, the basic behavior of the new system is verified by near full-scale tests to develop codes based on the test results. And only then are the code regulations for actual design and construction released. In composite construction, this order of priority appears to have occurred in a reverse direction, especially in areas of low seismicity. For example, in Houston, TX, and New Orleans, LA, some 30 years ago, tall buildings using composite systems were first built using a design procedure not prescribed in building codes or backed by research, attesting to the adage that it is the economy that motivates new construction techniques and not just research.

Another example is the use of shear studs in composite beams. Mechanical shear connectors were being used as early as 1903, but it was not until 1956 that a formula for the use of stud connectors was published after tests at the University of Illinois. Since then, of course, it has become virtually the only system used in steel building floor construction.

The success of combining steel and concrete in composite floor systems motivated the development of composite building systems, particularly composite columns some 30 years ago. Economic studies in the United States have consistently shown that a composite column designed for buildings in areas of low seismicity is about 4 to 6 times more economical than an all-steel column.

Structural steel is well suited for providing the column-free span of about 40 feet that is desirable for leasing spaces in contemporary office buildings. Because of its light weight, steel imposes less severe foundation requirements. It goes up faster. Its lightness is often a major consideration in seismic design. Also, a steel frame is simpler to modify to fit the changing needs of building tenants. And, in a steel frame, it is less expensive to increase the load capacity of an existing floor, or to cut holes to install new stairways, atriums, etc., that may be required by changes in tenancy. Therefore, the ability to renovate and rehabilitate the building plays an important role in the structural system selection process.

Similarly, concrete buildings have their own advantages. With the use of high-strength concrete well in excess of 10 ksi, it is possible to have relatively small-size columns. Floor-framing techniques have progressed from typical flat slab construction to skip-joist and haunch girder systems that can provide spans in excess of 40 feet economically. In addition to the improvements in gravity systems, lateral bracings that rival those of steel have been developed. Other advantages of concrete framing are low material costs, moldability, insulating and fire-resisting quality, and, most of all, inherent stiffness. However, in relation to steel, concrete construction is generally slower.

The two building systems, concrete and steel, evolved independently of each other until the 1960s. Until then engineers were trained to think of buildings either in steel or concrete. This belief was overcome in 1969 when blending of steel and concrete occured in a relatively short 20-story building, in which the exterior columns and spandrels were encased in concrete to provide the required lateral resistance. The system was basically a steel frame stabilized by reinforced concrete. The building consisted of an exterior frame tube with composite columns spaced at 10 ft on centers. A light steel column section, a W8 × 35, served as an erection column for the steel frame prior to concrete encasement.

The term composite system has taken on numerous meanings in recent years to describe many combinations of steel and concrete. As used here, the term means any and all combinations of steel and reinforced concrete elements and systems, and is considered synonymous with other definitions such as mixed systems, hybrid systems, etc.

5.1. COMPOSITE ELEMENTS

The primary structural components used in composite building construction consist of the following horizontal and vertical elements:

1. Composite slabs/diaphragms.
2. Composite beams.
3. Composite columns.
4. Composite diagonals.
5. Composite shear walls.

5.1.1. Composite Slabs

In steel buildings, the use of high-strength, light-gauge (16- to 20-gauge) metal deck with concrete topping has become the standard floor-framing method. The metal deck has embossments pressed into the sheet metal to achieve composite action with the concrete topping. Once the concrete hardens, the metal deck acts as the bottom tension reinforcement while the concrete acts as the compression component. The resulting composite slab acts as a diaphragm providing for the horizontal transfer of shear forces to vertical bracing elements. Furthermore, it acts as a stability bracing for the compression flange of steel beams.

In a concrete-filled steel deck, concrete is the stiffest part of the system in the horizontal plane. Therefore the shear stresses are primarily resisted by concrete. Thus, for transmission of force between a steel column and the diaphragm, forces must first transfer to the beam through the beam-to-column connection and then to the concrete fill either through welded studs or through puddle welds to the steel deck and finally through the bond and the embossments of the decking to the concrete fill. Each of these transfers must be adequate for the intended forces. When the concrete fill connects directly to concrete shear walls or steel-encased composite concrete beams, reinforcing dowels can be used for direct shear transfer.

Special construction considerations are necessary for studs welded to steel beams through galvanized steel decking because the zinc used for galvanizing can result in poor-quality welds.

When the published values for diaphragm capacities, based on test data, are less than those required by design, the concrete fill can be increased in thickness and adequately reinforced to resist the entire horizontal diaphragm shear. In this case, the metal deck serves only as a form for concrete placement and as gravity tension reinforcement of the composite floor slab as it spans between adjacent floor beams.

As with any diaphragm, the seismic load path including the chord and collector requirements, can best be identified by visualizing the slab as a horizontal beam in each direction. It is usually possible to utilize members already present in the floor system to serve as chords and collectors. For example, the perimeter beams may be designed for chord forces, so the only issues are ensuring that their splices and connections are adequate to resist the resulting forces.

5.1.2. Composite Frame Beams

The gravity design of composite beams is discussed in Chapter 7. Here the focus is on the design of frame beams subject to lateral loads.

For a medium-rise building in the 20- to 30-story range, the typical frame column spacing is usually 25 to 35 ft (7.6 to 10.67 m) and the floor-to-floor height is $12^1/_2$ to $13^1/_2$ ft (3.81 to 4.12 m). This geometry results in columns that are much stiffer in bending than the beams. Therefore, to limit the deflection of the frame under lateral loads, it is generally more economical to increase the beam stiffness than the column stiffness. The frame beams are typically designed noncomposite because they are subject to a reversal of curvature under lateral loads. However, the shear connectors provided for the transfer of diaphragm shear from slab to the beams also increase their moment of inertia. Nevertheless, the moment of inertia does not increase for the full length of the beam because its response to lateral loads is by bending in a reversed curvature with resultant tension in the top flange. Since concrete is ineffective in resisting tension, the increase in moment of inertia due to composite action can be counted on only in the positive moment region. Although design rules are not well established, a rational method may be used to take advantage of the increased moment of inertia. Occasionally engineers have used a dual approach in wind design by using bare steel beam properties for strength calculations, and composite properties in the positive regions for drift calculations.

5.1.3. Composite Columns

Two types of composite columns are used in building construction. The first consists of a steel section encased in a reinforced concrete envelope (Fig. 5.1). The second consists of a steel pipe or tube filled with structural concrete, as shown in Fig. 5.2. Conceptually, the behavior of a composite column may be considered similar to that of a reinforced

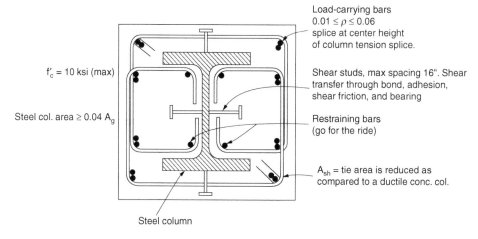

Load-carrying bars
$0.01 \leq \rho \leq 0.06$
splice at center height
of column tension splice.

$f'_c = 10$ ksi (max)

Shear studs, max spacing 16". Shear
transfer through bond, adhesion,
shear friction, and bearing

Steel col. area ≥ 0.04 A_g

Restraining bars
(go for the ride)

A_{sh} = tie area is reduced as
compared to a ductile conc. col.

Steel column

Note: Bond and adhesion must be ignored in calculating shear transfer

Figure 5.1.   Concrete-encased composite column; design considerations.

concrete column: The only difference is that in generating the axial load versus moment interaction diagram, the steel section is analytically replaced with an equivalent mild steel reinforcement.

Compositing of building exterior columns by encasing steel sections with concrete is by far the most frequent application. The reasons are entirely economic because form work around interior columns does not lend itself to jump forms. The exterior columns, on the other hand, are relatively easy to form using jump forms because they are open-faced: The form work can be "folded" around the steel columns for placement of concrete, and then unfolded and jumped to the next floor, repeating the cycle without having to dismantle the entire form work. However, in Japanese construction it is a common practice to composite the interior columns as well. Their construction makes extensive use of welding for vertical as well as transverse reinforcement (Fig. 5.3).

In the second type of composite column consisting of a steel pipe or tube filled with concrete, typically neither vertical nor transverse reinforcement is used (see Fig. 5.2).

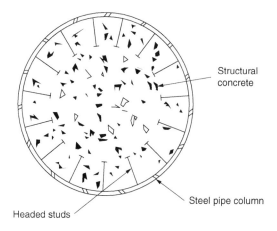

Structural
concrete

Steel pipe column

Headed studs

Figure 5.2.   Concrete-filled composite pipe column.

(a)

(b)

Figure 5.3. Japanese composite construction details: (a) I-beam column intersection; (b,c) composite column with welded ties; (d) general view.

However, shear connectors welded to the inner face of the structural steel section provide for the interaction between the concrete and outer shell. Since the placement of concrete does not require form work, this type of composite column may be used in both the interior and the exterior of buildings.

A method of attaching frame beams to pipe columns is to weld the beams directly to the pipes, as shown in Fig. 5.4.

(c)

(d)

Figure 5.3. (*Continued*).

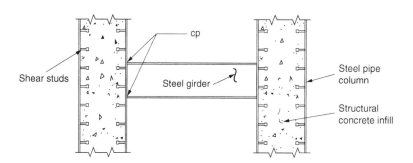

Figure 5.4. Composite column-to-steel girder moment connection.

Figure 5.5. Bank of China Tower, Hong Kong.

5.1.4. Composite Diagonals

Braced frame buildings are mostly of structural steel construction. However, braced frames using composite diagonals and composite columns have been used in a few buildings since the mid-1980s. The majority of these are multistoried braces working in concert with "super composite columns." An outstanding example is the 76-story Bank of China Tower, in Hong Kong, shown in Fig. 5.5.

5.1.5. Composite Shear Walls

A schematic plan of a composite shear wall system is shown in Fig. 5.6. This is similar to a reinforced concrete shear wall system with the exception that a structural steel frame placed within the walls speeds up the construction process (see Fig. 5.6a).

Generally, in an all-concrete system, the walls are interconnected with concrete beams, commonly referred to as link beams, to increase their bending stiffness. If the link beams are relatively short, the resulting shear forces due to lateral loads may be quite large. This may lead to a brittle fracture of the beam unless the beam is detailed with diagonal reinforcement as mandated in most seismic provisions. The resulting detail often leads to congestion of reinforcement. A method of overcoming the problem is to use

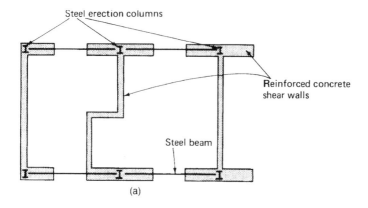

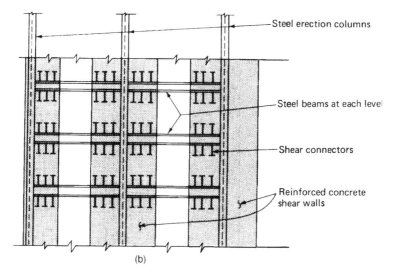

Figure 5.6. Composite shear wall with steel beams: (a) plan; (b) elevation.

structural steel beams as link beams between the shear walls, as shown in Fig. 5.6b. The moment capacity of the steel beam is developed in the wall by welding shear connectors to the top and bottom flanges of the beam, as shown in Fig. 5.7.

For resisting large in-plane shear forces, a full-length steel web plate attached to a concrete shear wall may be used. An example of such a construction is the core wall of the Bank of China Building in Hong Kong. In this building all the lateral forces are transferred to the core at the base. To resist the high shear forces, steel plates are attached to the concrete core through shear studs welded to the steel plates, as shown in Fig. 5.8.

5.2. COMPOSITE BUILDING SYSTEMS

Composite building systems may be classified into the following categories:

1. Composite shear wall system.
2. Composite shear wall–frame interacting system.
3. Composite tube sytem.
4. Composite vertically mixed system.
5. Composite mega frames with super columns.

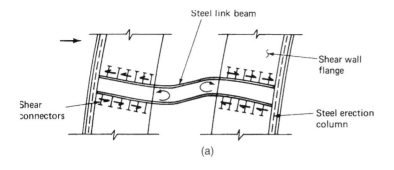

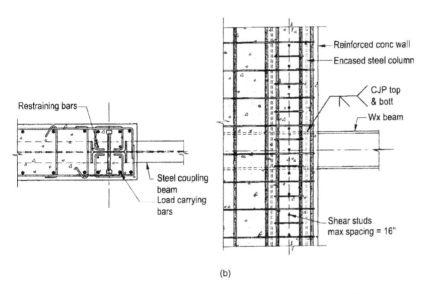

Figure 5.7. Moment transfer between steel beam and concrete wall.

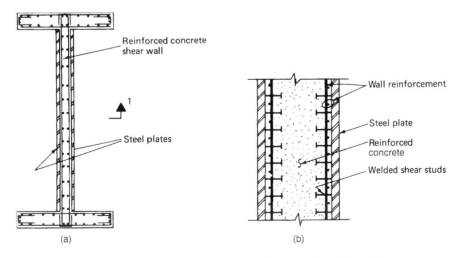

Figure 5.8. Composite shear walls with steel plates: (a) plan; (b) section.

(a)   (b)

Figure 5.9.   Core-supported composite building: (a) concrete core; (b) concrete core encompassed with steel frame.

5.2.1. Composite Shear Wall Systems

In this system a central core consisting of concrete shear walls is designed to resist the total lateral load while the remainder of the construction surrounding the core is designed for gravity loads using structural steel. The construction sequence, whether the concrete core or the steel surround goes up first, is often project specific. In one version, concrete core is built first, using jump or slip forms, followed by erection of the steel surround, as shown in Fig. 5.9. Although the structural steel framing may not proceed as fast as in a conventional steel building, the overall construction time is likely to be less because the building's vertical transportion, consisting of stairs and elevators, and the mechanical and electrical services can be installed in the core while erection of steel outside· of the core is still in progress. In another version, steel erection columns within the shear walls serve as erection columns, and erection of steel for the entire building proceeds as in a conventional steel building. After the steel erection has reached a predetermined level, concreting of the core takes place using conventional forming techniques. To facilitate jumping of forms from one level to the next through the floor system that is already in place, temporary openings are provided in the floor framing around the shear walls.

The behavior of a core-supported composite shear wall building is no different from that of a concrete building designed to resist all the lateral forces in the core. However, the absence of torsional stiffness due to lack of bracing at the building perimeter must be recognized in design. It is advisable to provide at least some reasonable lateral resistance, for example, one bay of lateral bracing at each building face.

If the entire lateral load, including torsional effects, is resisted by concrete shear walls, the steel surround may be designed as a simple framing for gravity loads only. Since there are no moment connections in the steel frame requiring welding or heavy bolting,

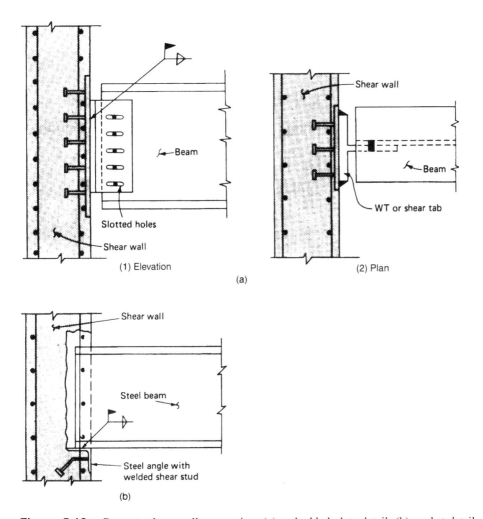

Figure 5.10. Beam-to-shear wall connection: (a) embedded plate detail; (b) pocket detail.

the erection proceeds much faster. The only nonstandard connection is between the shear walls and the floor beams. Various techniques have been developed for this connection, chief among them the embedded plate and pocket details shown in Fig. 5.10. The floor construction invariably consists of a composite metal deck with a structural concrete topping. The composite shear wall system has the advantage of keeping the steel fabrication and erection simple. Since columns carry only gravity loads, high-strength steel can be used with the attendant savings.

The construction of the floor within the concrete core can be of cast-in-place concrete or of structural steel consisting of steel decking and concrete topping. The connection between the floor slab and the core walls should provide for the transmission of diaphragm shear forces from the floor system to the core. The weld plate detail shown in Fig. 5.10a is the most popular, particularly in a slip-formed construction. During the slip-form operation, weld plates are set at the required locations, with the outer surface of the plate set flush with the wall surface. The plate is anchored to the wall by shear connectors welded to the plate. The bending capacity of the connection is often supplemented with a bent steel bar welded to the plate at top. Experience in slip-form construction indicates that it is prudent to overdesign these connections to compensate for misalignment. Subsequent to the installation of weld plates, structural tee or shear tab connections with slotted holes are field-welded to the plate. Slotted holes provide for additional tolerance in the erection of floor beams.

Slip forming is a special construction technique that uses a mechanized moving platform system. The process of slip forming is similar to an extrusion process. The difference is that, whereas in an extrusion process the extrusion moves, in a slip-forming process the die moves while the extrusion remains fixed.

An important consideration in the design of core-supported buildings is the resistance (or lack of it) to overturning forces. Generally the vertical load resisted by the core due to gravity effects is limited because the floor area supported by the core is relatively small. For tall buildings, this can result in an unfavorable stability condition due to large tensile forces at the base. A method of counteracting the tensile forces is to apply an external prestressing force to the core. Similarly, an equivalent passive prestressing effect can be achieved by increasing the vertical load resisted by the core by manupulating the layout of floor beams. Extending this idea to its limit results in the concept of a building entirely supported on a single central core. Depending upon the floor area and the number of levels supported, several options present themselves for the support of the floor system from a central core. For example, 1) floors can be hung from the top of the center core; 2) they may be hung from story-deep cantilever trusses located at one or two intermediate levels, such as at the top and midheight of the building; or 3) the floor system can be cantilevered at each level. The second scheme has certain advantages primarily due to reduced length of hangers resulting in fewer floor-leveling problems. The advantages of a core-only supported building are 1) it offers views unobstructed by exterior columns at each floor; 2) the absence of exterior columns provides for the commonly sought column-free entrances at the street level; and 3) the undulations on the building exterior common in today's architecture are easy to accommodate.

Galvanized bridge-strand cables can be used as hangers to support the structural steel framing consisting of composite beams, metal deck, and concrete topping. The floor beams are attached to the hanger with simple supports, whereas at the core, pockets or anchor plates cast into the core walls provide for the support of floor beams.

It is common practice to slip-form the center core with an average concrete growth rate of 6 to 18 in./hr (152 to 457 mm/hr). After completion of the core, the second stage of construction in the hung-floor system is the erection of roof girders and draping of the floor-supporting cables. Erection of floor members between the core and the perimeter cables proceeds in a manner similar to that in typical steel building construction. Placement of steel floor decks and welding of shear studs for composite action is followed by placement of concrete topping. Because elongation of the cable due to cumulative floor loads can be substantial, it is necessary to compensate for this effect, during the design and possibly during the construction.

5.2.2. Shear Wall–Frame Interacting Systems

This system has applications in buildings that do not have a sufficiently large core to resist the entire lateral loads. Supplementing the resistance of the core with steel or composite moment frames located at the building exterior is perhaps the most common method of increasing the lateral stiffness. In North American practice, use of interior composite frames is not popular because the cost of form work, placing of reinforcement, and encasing of steel columns and beams with structural concrete far outstrips the advantages of additional strength and stiffness. Therefore, use of composite construction is typically confined to the exterior components. If the erection of steel members within the composite core precedes concrete encasement, it is usually more cost-effective to use

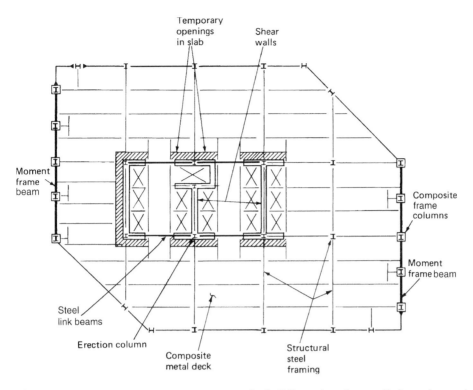

Figure 5.11.   Typical floor plan of a composite building using shear wall–frame interaction.

steel link beams between the shear walls. A schematic plan of this system is shown in Fig. 5.11.

5.2.3. Tube Systems

A framing system often referred to as a composite tube, used extensively in Louisiana and Texas, makes use of the well-known virtues of a concrete tube system along with the speed of steel construction. As in a concrete or steel tube system, closely spaced columns around the building's perimeter connected to deep spandrels form the backbone of the system. Two versions, both using composite columns are popular: One system uses cast-in-place concrete spandrels and the other structural steel spandrels. A relatively small steel beam is often used in the first system to stabilize the steel columns prior to casting concrete for the spandrels. However, in the design of the concrete spandrel, its strength and stiffness contribution is generally neglected because of its relatively small size. Schematic plan and sections for the two versions of tubular system are shown in Figs. 5.12 and 5.13.

   In either of these systems, the speed of construction rivaling that of an all-steel building is achieved by erecting steel columns for the perimeter tube along with interior steel columns. Usually steel is erected some 10 to 12 stories ahead of encasing the perimeter columns with concrete. The key to the success of this type of construction for high-rise buildings lies in the rigidity of closely spaced exterior columns which, together with deep spandrels, results in an exterior facade that behaves more like a bearing wall with punched windows than as a moment frame.

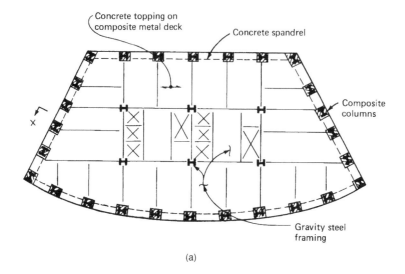

(a)

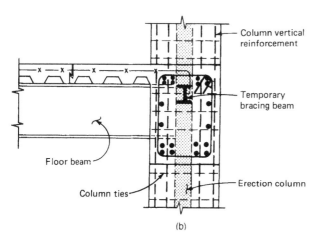

(b)

(c)

Figure 5.12. Composite tube with concrete spandrels: (a) typical floor plan; (b) typical cross section through spandrel; (c) detail at perimeter column and spandrel intersection.

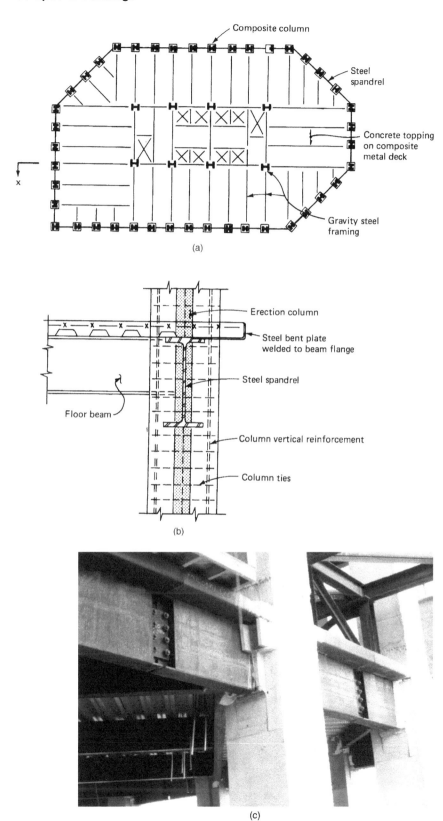

Figure 5.13. Composite tube with steel spandrels: (a) typical floor plan; (b) typical cross section through spandrel; (c) detail at perimeter column and spandrel intersection.

5.2.4. Vertically Mixed Systems

Mixed-use buildings typically provide for two or more types of occupancies. This is often achieved by vertically stacking different amenities in a single building. For example, the lower levels may house parking, midlevels, office floors; and the top levels, residential units. Since different types of occupancies economically favor different types of construction, it is logical to mix construction vertically up the building height. For example, beamless flat ceilings with a relatively short span of about 25 ft are preferred in residential occupancies, whereas large spans of the order of 40 ft (12.2 m) are required in office buildings for optimum lease space. These spans are, however, too large for apartments, condominiums, and hotel suites. Therefore, it is possible to introduce additional columns without adversely affecting the architectural layout. The relatively short spans combined with the requirement of a beamless ceiling points toward concrete construction for the floors dedicated to residential occupancies.

In certain types of buildings, use of concrete for the lower levels and structural steel for the upper levels may provide an optimum solution. As an example, a 26-story building constructed in Houston, TX, is shown in Fig. 5.14. The bracing for the lower 13 floors is

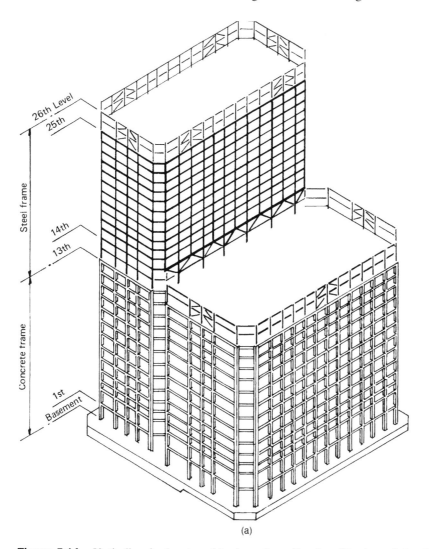

(a)

Figure 5.14.   Vertically mixed system: (a) schematic periframing; (b) schematic bracing concept.

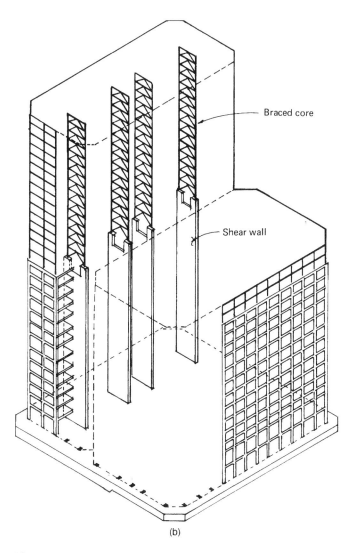

Braced core

Shear wall

(b)

Figure 5.14. (*Continued*).

provided by a combination of moment frame and shear walls, while a braced steel core and steel moment frame interacting system provides lateral resistance for the upper levels. The steel columns are transferred onto concrete elements by embedding them in concrete for two levels below the transfer level. Shear studs, shop-welded to the embedded steel columns, provide for the transfer of axial loads from steel to concrete.

5.2.5. Mega Frames with Super Columns

An efficient method of resisting lateral loads for buildings in the 60-plus-story range is to position columns farthest from the building center with shear-resisting elements in-between. This idea has given rise to a whole new category of composite systems characterized by their use of super columns interconnected across the building with a shear-resisting web-like framing.

The construction of super columns can take on many forms. One system uses large-diameter pipes or tubes filled with high-strength concrete in the range of 6 to 20 ksi

(41 to 138 MPa). Generally, neither longitudinal nor transverse reinforcement is used within the steel pipe or tube. Another method is to encase the steel column with reinforced concrete using conventional forming techniques.

5.3. EXAMPLE PROJECTS

5.3.1. Buildings with Composite Steel Pipe Columns

Examples of buildings with large-diameter composite columns are shown in Figs. 5.15, 5.16 and 5.17. The 44-story Pacific First Center has eight 7.5-ft (2.3-m)-diameter pipe columns in the core and several 2.5-ft (0.76-m)-diameter columns at the perimeter, each filled with 19-ksi (131-Mpa) concrete. The second example is a 62-story tower with 9-ft (2.7-m)-diameter pipe columns tied together with 10-story-high X-braces. The third is a 58-story building with four 10-ft (3-m)-diameter pipe columns filled with 19,000-psi (131-Mpa) concrete. To achieve composite action, steel studs welded to the pipes' interior surfaces are used. All three buildings were designed by the Seattle-based structural engineering firm, Skilling Ward Magnusson Barkshire, Inc.

An example of non-high-rise application of composite columns is shown in Fig. 5.18. Called the Fremont Street Experience, the space frame has an overall dimension of 1387 ft (422.75 m) by 100 ft (30.5 m) with a 50-ft (15.25-m) radius. The space frame is 5.77 ft (1.76 m) deep and is supported on composite columns spaced longitudinally at 180 and 200 ft (54.87 and 61 m). The space frame's top and bottom chords and web members consist of a 3-in. (76-mm)-diameter steel tubing, with a typical wall thickness of 0.120 in. (3 mm).

The composite columns consist of 42-in.-diameter by 0.75-in.-thick (1067 mm × 19 mm) steel pipes with 8000-psi (55.16-MPa) concrete. Headed studs 1/2 in. (12.7 mm)

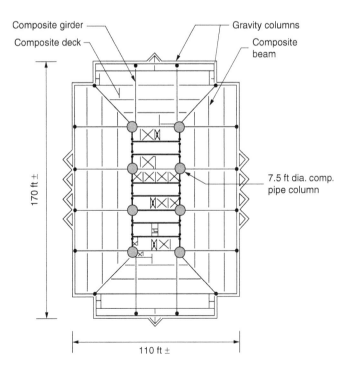

Figure 5.15.   Pacific First Center; 44-story building.

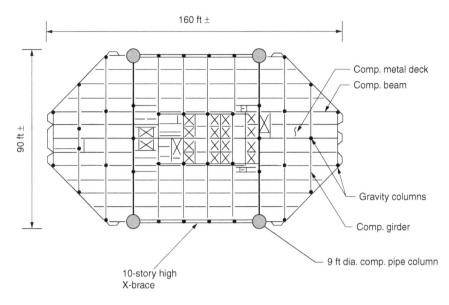

Figure 5.16. Gateway Tower; 62-story building.

in diameter by 8 in. (203 mm) long are welded to the inside face of the tube at a vertical spacing of 12 in. (305 mm) and a radial spacing of 9 in. (228 mm). The bending capacity of the pipe column is developed into the foundation by 1) welding shear studs around the outer surface of the column embedded in the foundation; and 2) by extending the mild steel reinforcement inside the column into the foundation. Figure 5.18b shows a

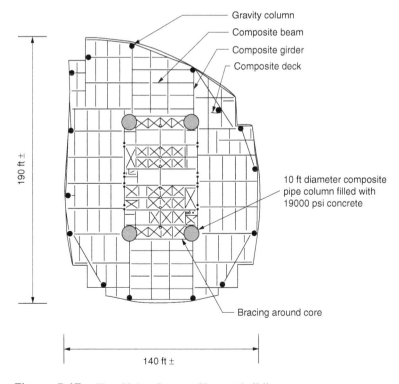

Figure 5.17. Two Union Square; 58-story building.

(a)

(b)

Figure 5.18. Fremont Street Experience: (a) general view; (b) detail at column; (c) typical section; (d) concrete-filled composite pipe column.

schematic section through the space frame. The architectural design is by the Jerde Partnership, Inc., Venice, CA; the structural engineering of the vault support is by John A. Martin & Associates, Inc., Los Angeles; and the space frame design is by Pearce Systems International, Inc.

5.3.2. Buildings with Formed Composite Columns

5.3.2.1. InterFirst Plaza, Dallas

An example of a building that uses formed composite super columns is shown in Fig. 5.19. In this 73-story, 921-ft (281-m)-tall building, the entire weight of the building is supported on 16 composite columns located up to 20 ft (6 m) inboard from the building exterior. The lateral loads are resisted by the composite columns interconnected to a system of 7-story two-way vierendeel trusses. The composite columns vary in size from 7 ft × 7 ft (2.1 m × 2.1 m)

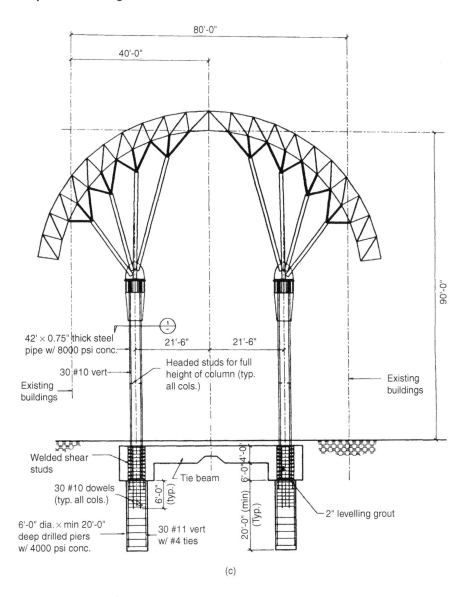

80'-0"

40'-0"

90'-0"

42' × 0.75" thick steel pipe w/ 8000 psi conc.

21'-6"   21'-6"

Headed studs for full height of column (typ. all cols.)

30 #10 vert

Existing buildings

Existing buildings

Welded shear studs

Tie beam

30 #10 dowels (typ. all cols.)

6'-0" (typ.)

6'-0"   4'-0"

20'-0" (min) (Typ.)

6'-0" dia. × min 20'-0" deep drilled piers w/ 4000 psi conc.

30 #11 vert w/ #4 ties

2" levelling grout

(c)

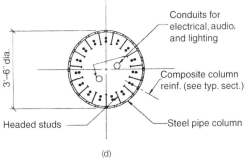

Conduits for electrical, audio, and lighting

3'-6" dia.

Composite column reinf. (see typ. sect.)

Headed studs

Steel pipe column

(d)

Figure 5.18. (*Continued*).

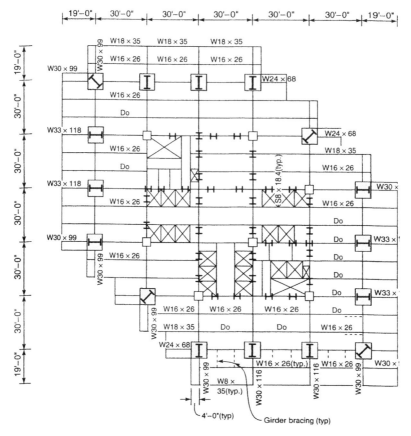

Figure 5.19. Dallas Main Center: InterFirst Plaza, 26th–43rd floor framing plan.

at the base to 5 ft × 5 ft (1.5 m × 1.5 m) at the top; 36-in. (0.30-m) deep steel shapes are encased in 10-ksi (69-Mpa) concrete columns, reinforced with 75 ksi (517 Mpa) mild steel reinforcement.

5.3.2.2. Bank of China Tower, Hong Kong

This prism-shaped building (shown in Fig. 5.20), designed by the architectural firm of I. M. Pei and Partners and structural engineer Leslie E. Robertson, is a 76-story building consisting of four quadrants. Each of the quadrants rises to a different height, and only one reaches the full 76 stories. The lateral bracing consists of a space truss spanning between the four corner columns. From the top quadrant down, the gravity loads are systematically transferred out to the building corner columns by truss action. Transverse trusses wrap around the building at selected levels. At the 25th floor, the center column is transferred to the corners by the space truss, providing for an uninterrupted 158-ft (48-m) clear span at the lobby. At the fourth floor, the horizontal shear forces are transferred from the space truss to the interior composite core walls through 1/2-in. (12-mm)-thick steel plate diaphragms acting compositely with the floor slab. The foundation for the building consists of caissons as large as 30 ft (9.1 m) in diameter hand-dug to bedrock.

5.3.2.3. Bank of Southwest Tower, Houston, TX

The Bank of Southwest Tower, an 82-story, 1220-ft (372-m) building proposed for, and not built in, downtown Houston, TX, uses the unique concept of composite columns with

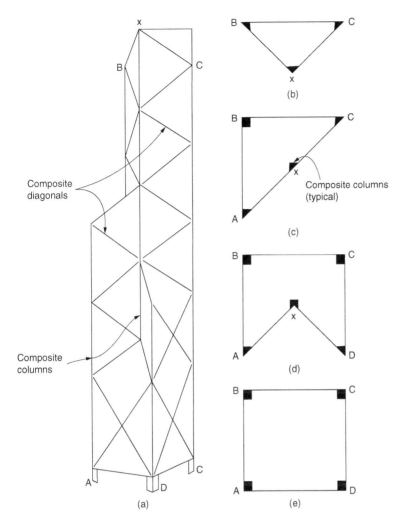

Figure 5.20. Bank of China Tower, Hong Kong.

interior steel diagonal bracings. The diagonals transfer both the gravity and the lateral loads into eight composite super columns. The building has a base of only 165 ft (50.32 in.), giving it a height-to-width ratio of 7.4. The characteristic feature of the design consists of a system of internal braces that extend through the service core and span the entire width of the building in two directions. A typical bracing consists of an inverted K-type brace rising for nine floors, there being two such braces in each direction. Eight of these 9-story trusses are assembled one on top of another within the tower. All of the gravity loads are transferred to eight composite columns located at the building perimeter. The structural engineering is by LeMessurier Consultants, Inc., and Walter P. Moore & Associates, Inc. A schematic bracing diagram is shown in Fig. 5.21.

5.3.3. Buildings with Composite Shear Walls and Frames

5.3.3.1. First City Tower

Designed by structural engineers Walter P. Moore & Associates, Inc., this 49-story tower comprises a number of distinctly different composite elements (Figs. 5.22 and 5.23).

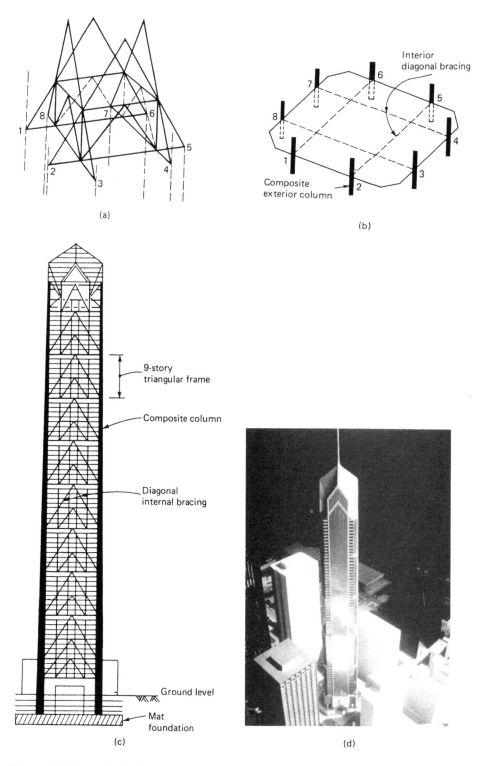

Figure 5.21. Bank of Southwest Tower (structural engineers: LeMessurier Associates and Walter P. Moore Associates; architects: Murphy/Jahn and Lloyd Jones and Fillpot). (a) Schematic representation of interior diagonal bracing; (b) schematic plan; (c) schematic section; (d) photograph of model. (Photo courtesy of Malcolm Stewart, Century Development Corporation.)

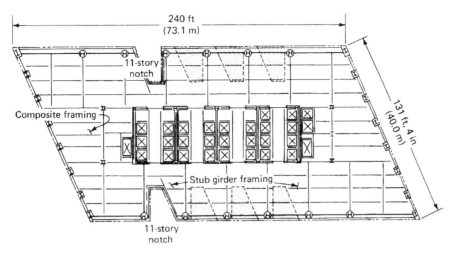

Figure 5.22. Composite floor-framing plan.

Shown in Figs. 5.24 and 5.25 are the details of the composite columns. Typiclly, the embedded steel columns vary from a W14 × 370 (W360 × 551) at the bottom to a W14 × 68 (W360 × 101) at the top. The vertical reinforcement in the columns varies from #18 bars (57 mm) at the bottom to #7 bars (22 mm) at the top. Open ties permitted in low seismic zones are used throughout.

Figure 5.24 shows the arrangement of reinforcement around a W10 × 72 (W250 × 107) erection column embedded in the shear walls. Ties are used where the vertical reinforcement ratio is more than 1% or where the reinforcement is required for resisting compression. This requirement has been in the ACI 318 code for the past 25 years. See Section 14.3.6 of ACI 318-02.

Figure 5.25 shows the connection detail between the concrete shear wall and a typical steel link beam. The moment capacity of the beam is developed through a shear transfer mechanism by steel studs welded to the top and bottom flanges of the beam. The stiffener plate, set flush with the wall face, has no structural purpose but helps in simplifying the form work around the beam. The construction sequence generally used in a composite section is shown in Fig. 5.26.

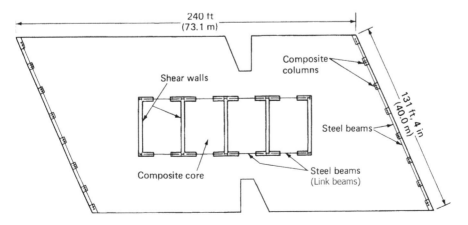

Figure 5.23. Composite elements.

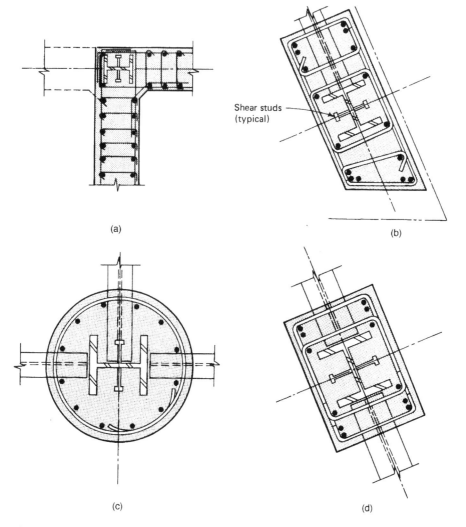

Figure 5.24. Composite vertical elements: (a) composite shear wall; (b) composite corner column; (c) typical circular column on long faces; (d) typical exterior column on short faces.

5.3.4. Building with Composite Tube System

5.3.4.1. America Tower, Houston, TX

Shown Figs. 5.27, 5.28, and 5.29 are the details for a 42-story office building called America Tower, designed by structural engineers Walter P. Moore and Associates, Houston, TX. This building uses a hybrid tubular perimeter frame consisting of composite columns above the third floor and structural steel columns below (see Fig. 5.29). This integration of steel columns and composite columns eliminated form work for columns at nontypical lower levels.

5.4. SUPER-TALL BUILDINGS: STRUCTURAL CONCEPT

A super-tall building is generally referred to as a skyscraper when it is taller than some 80 stories or so. Its silhouette has a slender form with a height-to-width ratio well in

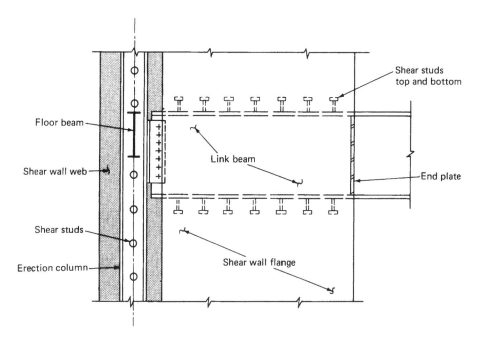

Figure 5.25.   Arrangement of link beam in shear wall.

excess of 8. An ideal structural system for such a slim building is one that can at once resist the effect of bending, torsion, shear, and vibration in a unified manner. A perfect form is a chimney with its walls located at the farthest extremity from the horizontal center, but as an architectural form, it is less than inspiring as a building model. A practical interpretation presents itself in a skeletal structure with its lateral stiffness located at the farthest extremity from the building center. Two additional requirements need to be incorporated within this basic concept to achieve high efficiency: 1) transfer as much of the gravity load, preferably all of the gravity load, into these columns to

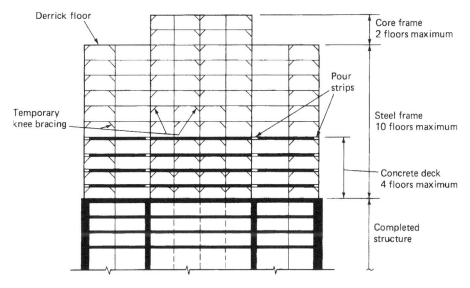

Figure 5.26.   General construction sequence in composite structures.

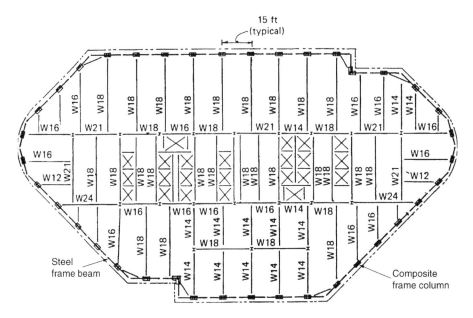

Figure 5.27. Structural floor framing plan; 42-story America Tower, Houston, TX.

enhance their capacity for resisting overturning moment; and 2) connect the columns with a system capable of resisting the shear forces.

The ultimate structure for a rectangular building, then, will have just four corner columns interconnected with a shear-resisting system. Such a concept, proposed by the author for a super-tall building, is shown in Fig. 5.30. The columns are deliberately located inboard from the building corners to allow for architectural freedom in modulating the short faces of the building. The shear in the transverse direction is resisted by a system of 12-story-high braces, while in the longitudinal direction the shear resistance is provided primarily by the full-height vierendeel frames located on the long faces. The story-high longitudinal trusses located at every 12th floor permit cantilevering of the floor system. The primary function of the interior vierendeel frame is to transfer the gravity loads of the interior columns to the composite columns via chevron braces. However, because of the geometry, it also resists external shear forces in the long direction.

The scheme shown in Fig. 5.30 can be modified to fit a variety of architectural shapes. Any desired slicing and dicing of the building on the short faces may be accommodated without inflicting an undue penalty on systems efficiency.

5.5. SEISMIC COMPOSITE SYSTEMS

The progress of composite systems in regions of high seismicity has been the least because of the difficulties associated with the detailing of joints. Until recently, joints for composite members were designed as for steel structures without regard to the structural concrete encasing the steel sections. However, studies reported over the years from Japan, concerned principally with the resistance of joints to earthquake forces, have given new impetus to composite construction.

With proper detailing, the components of a composite structure can be tied together well. This major characteristic, particularly beneficial in seismic design, has made composite design popular in Japan. The acceptance of composite construction in California, another

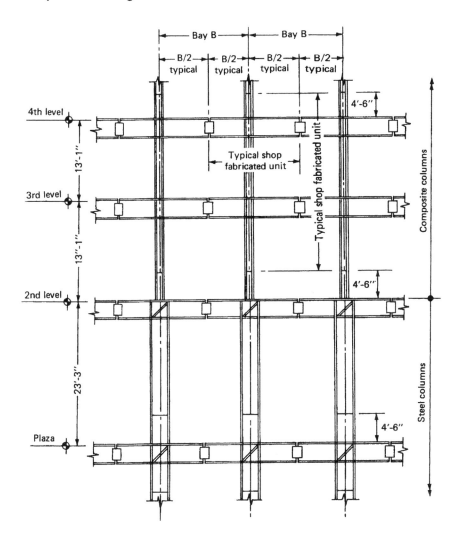

Figure 5.28.   Typical frame fabrication unit.

earthquake area, has been less enthusiastic. However, in the rest of the United States and around the world, composite construction has in the past two decades gotten a strong foothold. The two tallest buildings in the world are of composite construction.

Because composite systems are assemblies of steel and concrete components, their design is governed by both AISC and ACI specifications. The available research and

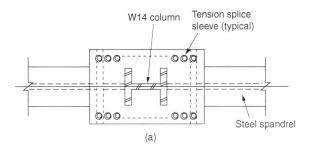

Figure 5.29.   Composite column-to-steel column connection detail: (a) plan; (b) elevation; (c) America Tower, Houston, TX. Photograph of building under construction.

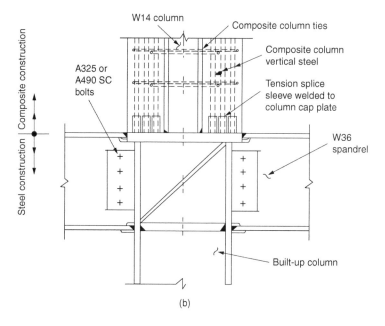

(b)

(c)

Figure 5.29. (*Continued*).

limited experience has demonstrated that properly detailed composite members and connections can perform reliably when subjected to seismic ground motions. Because there is at present limited experience with composite building systems subjected to extreme seismic forces, careful attention to all design aspects is necessary, particularly

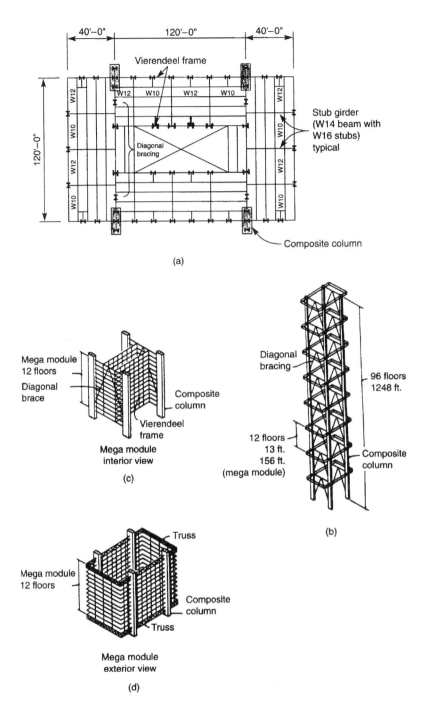

Figure 5.30. Structural concept for a super-tall building: (a) plan; (b) schematic elevation; (c) interior view of mega module; (d) exterior view of mega module.

to the detailing of members and connections. It is generally recognized that overall behavior of seismic composite systems will be similar to that for counterpart steel or reinforced concrete systems. For example, it is anticipated that inelastic deformations such as flexural yielding occurs in beams of moment frames, and in braced frames axial yielding and/or bulking occurs in braces. However, in composite systems, differential stiffness between steel and concrete components is more pronounced in the calculations of internal forces and deformations than for structural-steel-only or reinforced-concrete-only systems. This is because stiffness of reinforced concrete elements can vary considerably due to effects of cracking.

The seismic response modification factors such as R, Ω_o, and C_d for composite systems are similar to those for comparable systems of steel and concrete. The current ASCE 7-02, NFPA 5000, and IBC-03 include these factors along with design criteria for composite structures. The UBC 1997 does not explicitly make reference to composite construction, but by invoking Section 1605.2 of the code, one would tacitly presume that composite systems are also permitted. This section permits any system that is based on a rational analysis in accordance with well-established principles of mechanics. The expectation is that, when carefully designed and detailed, the overall inelastic response of composite systems should be similar to comparable steel and reinforced concrete systems.

Before discussing seismic design of composite systems, it is worthwhile to revisit the frequently used phrase "design for wind and seismic forces." A clear understanding of this phrase is crucial for proper implementation of a design. The design is made for the greater of wind or seismic forces according to the applicable building code, often supplemented by site-specific studies accepted by the building official. Whereas wind forces are based on wind exposure category, those due to seismic activity are based on a number of design parameters such as seismic design category (SDC) of the building, seismicity of the region, the mass of the building, and the type of lateral-force-resisting system. The greater of the two sets of forces calculated for wind or seismicity is used for the design of the lateral-force-resisting system. However, in seismic design it is recognized that actual seismic forces can be significantly greater than the code-prescribed values. Thus, seismic design includes not only strength requirements but also material and system limitations and special provision for member proportioning and detailing. The purpose of these additional provisions is to ensure that the members and joints do not snap in a large seismic event, but have the necessary ductility to ride out the forces. Therefore, when designing a building located in a high-seismic zone, even when wind forces govern the design, the detailing and proportioning requirements for seismic resistance must also be satisfied.

Bracing systems for buildings consist of structural components in both vertical and horizontal planes. Vertical bracing is provided by the primary elements of the building such as moment-resisting frames, diagonally braced frames, or shear walls. Horizontal bracing typically includes floor and roof diaphragms. Both the horizontal and the vertical bracing should be properly interconnected in order to transfer all lateral forces from their point of origin through the horizontal bracing to the vertical bracing and into the base of the structure. A complete load path throughout the structure interconnecting all elements of the bracing system is an essential ingredient of a properly designed bracing system.

5.5.1. Moment-Resisting Frames

Seismic code provisions distinguish between "special" and "ordinary" moment frames of both steel and reinforced concrete construction. Special moment frames, which must meet additional detailing requirements to provide ductile inelastic response, are designed for lower force levels than ordinary moment-resisting frames.

Early composite designs focused on combining perimeter steel beams with composite steel and concrete columns, with the lateral force design generally controlled by wind forces. Often the steel column section was used solely for erection purposes, with the concrete section designed to provide the required stiffness and strength. Although other possible combinations exist for providing interior moment-resisting frames, few such buildings have been constructed in high seismic regions in the United States. The practice of compositing interior frames is, however, more popular in Japan. In the United States no code provisions were available prior to 1993 addressing design of composite systems and the attendant ductility requirements in areas of high seismicity. During that year, the Building Seismic Safety Council (BSSC) developed recommendations for seismic design of composite steel and concrete construction. These provisions with their subsequent modifications have served as the basis for the design of seismic lateral-force-resisting systems, which are now included in the ASCE 7-02. The two model codes, IBC-2003 and NFPA 5000, by adaptation of the ASCE 7-02, now provide a firm basis for design and detailing of composite building systems. As many as 18 types of composite systems, listed in Table 5.1, are recognized.

Three potential classes of composite moment-resisting framing systems are identified in the model codes: 1) partially restrained moment-resisting frames; 2) ordinary moment-resisting frames; and 3) special moment-resisting frames. These three systems are similar to the moment-resisting frame systems presently identified for use in steel construction. Only the ordinary and special moment-resisting frames are discussed in this work. The designer is referred to seismic provisions for Structural Steel Buildings, ANSI/AISC 341-02 for additional information.

5.5.1.1. *Ordinary Moment Frames*

The term "ordinary" refers to systems in which the elements are not designed or detailed to provide the maximum potential ductility during inelastic cyclic response. However, to provide acceptable performance and to reduce the potential ductility demand, the lateral design forces are increased significantly over those required of "special" moment-resisting frame systems. Because of their limited ductility, seismic codes have imposed certain restrictions on the use of ordinary moment frames in areas of high seismicity. Where permitted in areas of low seismicity, these systems are often economical because the expense of providing the ductile elements and connections required for special moment-resisting frames far exceeds the cost of providing for increased lateral loads.

A composite ordinary moment-resisting frame may be developed by combining steel and concrete components in a number of ways. These include steel or composite beams combined with steel, reinforced concrete, or composite columns. The most commonly used system to date in areas of low seismicity has included steel beams and composite columns. The columns may consist of encased or filled composite columns. The connections in composite ordinary moment-resisting frames are generally designed to develop the full moment capacity of the steel beams.

The design requirements for ordinary composite moment-resisting frames and the limit imposed on their use are similar to those specified for steel or concrete ordinary moment-resisting frames. The beams and columns of composite ordinary moment-resisting frames may consist of one of a number of possible combinations of structural steel, reinforced concrete, and composite sections. The analysis, design, and detailing of the frame members is quite similar to that required for steel or concrete moment-resisting frames. Force transfer between the elements of a composite frame is somewhat unique, since in general the connections are designed to be stronger than the beams framing into the joint.

The analytical procedures used in the design of composite moment-resisting frames are identical to those used in the design of structural steel or reinforced concrete frames.

TABLE 5.1 ASCE 7-02 (IBC 2003 & NFPA 5000) Design Coefficients and Factors for Seismic-Force-Resisting Systems of Composite Buildings

Basic seismic-force-resisting system	AISC Seismic Part II Section Number	Response modification coefficient, R	System over-strength factor, Ω_o	Deflection amplification factor, C_d	System limitations and building height limitations (feet) by seismic design category				
					A or B	C	D	E	F
Building Frame Systems									
Composite eccentrically braced frames (C-EBF)	(14)	8	2	4	NL[a]	NL	160	160	100
Composite concentrically braced frames (C-CBF)	(13)	5	2	$4\frac{1}{2}$	NL	NL	160	160	100
Ordinary composite braced frames (C-OBF)	(12)	3	2	3	NL	NL	NP[b]	NP	NP
Composite steel plate shear walls (C-SPW)	(17)	$6\frac{1}{2}$	$2\frac{1}{2}$	$5\frac{1}{2}$	NL	NL	160	160	100
Special composite reinforced concrete shear walls with steel elements (C-SRCW)	(16)	6	$2\frac{1}{2}$	5	NL	NL	160	160	100
Ordinary composite reinforced concrete shear walls with steel elements (C-ORCW)	(15)	5	$2\frac{1}{2}$	$4\frac{1}{4}$	NL	NL	NP	NP	NP
Moment-Resisting Frame Systems									
Special composite moment frames (C-SMF)	(9)	8	3	$5\frac{1}{2}$	NL	NL	NL	NL	NL
Intermediate composite moment frames (C-IMF)	(10)	5	3	$4\frac{1}{2}$	NL	NL	NP	NP	NP
Composite partially restrained moment frames (C-PRMF)	(8)	6	3	$5\frac{1}{2}$	160	160	100	NP	NP
Ordinary composite moment frames (C-OMF)	(11)	3	3	$2\frac{1}{2}$	NL	NP	NP	NP	NP

TABLE 5.1 (Continued)

Basic seismic-force-resisting system	AISC Seismic Part II Section Number	Response modification coefficient, R	System over-strength factor, Ω_o	Deflection amplification factor, C_d	System limitations and building height limitations (feet) by seismic design category				
					A or B	C	D	E	F
Dual Systems with Special Moment Frames Capable of Resisting at Least 25% of Prescribed Seismic Forces									
Composite eccentrically braced frames (C-EBF)	(14)	8	$2^1/_2$	4	NL	NL	NL	NL	NL
Composite concentrically braced frames (C-CBF)	(13)	6	$2^1/_2$	5	NL	NL	NL	NL	NL
Composite steel plate shear walls (C-SPW)	(17)	8	$2^1/_2$	$6^1/_2$	NL	NL	NL	NL	NL
Special composite reinforced concrete shear walls with steel elements (C-SRCW)	(16)	8	$2^1/_2$	$6^1/_2$	NL	NL	NL	NL	NL
Ordinary composite reinforced concrete shear walls with steel elements (C-ORCW)	(15)	7	$2^1/_2$	6	NL	NL	NP	NP	NP
Dual Systems with Intermediate Moment Frames Capable of Resisting at Least 25% of Prescribed Seismic Forces									
Composite concentrically braced frames (C-CBF)	(13)	5	$2^1/_2$	$4^1/_2$	NL	NL	160	100	NP
Ordinary composite braced frames (C-OBF)	(12)	4	$2^1/_2$	3	NL	NL	NP	NP	NP
Ordinary composite reinforced concrete shear walls with steel elements (C-ORCW)	(15)	5	3	$4^1/_2$	NL	NL	NP	NP	NP

[a]NL = No limit.
[b]NP = Not permitted.
(Data from ASCE 7-02 [IBC 2003 and NFPA 5000].)

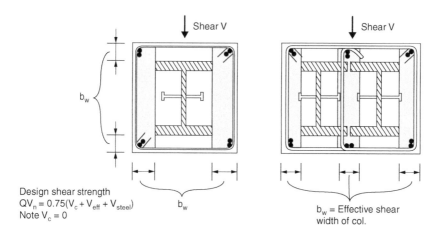

Design shear strength
$QV_n = 0.75(V_c + V_{eff} + V_{steel})$
Note $V_c = 0$

b_w = Effective shear width of col.

Figure 5.31. Encased composite column; design shear strength parameters.

Elastic properties of composite elements can be transformed into equivalent properties of steel for stiffness analyses using standard procedures. As with steel or concrete frames, it may be more accurate to include a finite rigid joint size in the frame model, particularly when the composite columns are quite large.

The design of composite ordinary moment-resisting frames is not significantly different from the procedures for structural steel and reinforced concrete moment frames. Encased composite columns should have a minimum ratio of structural steel to gross column area of 4%. The shear strength of these columns generally ignores the contribution of the concrete. However, the contribution of the shear strength of the reinforcing ties based on an effective shear width b_f of the section, as noted in Fig. 5.31, is permitted. For filled composite columns, it is conservative to neglect the contribution of the concrete to the shear strength of the column. For conditions where shear strength becomes critical, analytically the composite column may be treated as a reinforced concrete column with the steel tube considered as shear reinforcement. Transfer of forces between the structural steel and reinforced concrete should be made through shear connectors, ignoring the contribution of bond or friction.

The design and detailing of reinforced concrete columns in ordinary composite frames is similar to those of intermediate or special moment frames of reinforced concrete. Conservative detailing based on the special moment frame requirements are recommended for these frames in high seismic zones. This is because there is little research on the use of intermediate detailing of concrete columns in these applications.

In ordinary moment-resisting frames, the connections are typically designed to develop the strength of the connected members. However, in seismic design, it is generally desirable to avoid inelastic action in the frame connections. Both structural steel and concrete's contribution to member strength must be considered in the determination of connection strength, including the strengthening effect of the composite action of a steel beam and a concrete slab in the joint regions.

Transfer of loads between structural steel and reinforced concrete elements of a composite moment-resisting frame should be made only through shear friction and direct bearing. Reliance on bond and adhesion forces should not be considered because of the cyclic nature of the lateral loading. In addition, AISC Seismic (i.e., AISC 341-02) recommends that a 25% reduction in the typical shear-friction capacities be imposed for buildings in areas of high seismicity.

Panel zone strength calculations for composite frames with fully encased steel columns may typically be taken as the sum of the steel and the reinforced-concrete capacities. Reinforcing bar development lengths as required by the ACI 318 Provisions should be provided in the detailing of these joints.

Ordinary composite moment frames are permitted for buildings only in seismic design category (SDC) A or B where there is no height restriction. The values for seismic design factors are $R = 3$, $\Omega_o = 3$, and $C_d = 2.5$. These frames are not permitted for buildings in SDC C, D, E, or F.

5.5.1.2. *Special Moment-Resisting Frames*

The term "special" refers to systems where the elements and connections are designed and detailed to provide maximum ductility and toughness, implying excellent energy dissipation and seismic performance during severe earthquake shaking. In recognition of this ductility, seismic codes allow a maximum reduction in the design base shear for special moment-resisting systems. Because of the recognized ductility and the limited interference with architectural planning, special moment-resisting frames are most commonly used for resisting lateral forces.

Composite special moment-resisting framing systems are similar in configuration to ordinary moment-resisting frames. As in the steel or concrete systems, more stringent detailing provisions are required to increase the system ductility and toughness of the composite speical moment-resisting frame. The commensurate reduction in design lateral forces is identical to that in steel or concrete special moment frames. The goal of seismic detailing provisions is to confine inelastic hinging to the beams, whereas the columns and connections remain essentially elastic. The design base shear prescribed for this system is similar to the special moment-resisting frame systems of steel or reinforced concrete. Likewise, no limitations have been placed on their usage in high seismic zones.

Composite speical moment-resisting frames according to current (2004) belief are subject to the same potential failure mechanisms as experienced by special moment-resisting frame steel buildings during the Northridge, CA, earthquake of 1994, and the Kobe, Japan, earthquake of 1995. The design approach for composite special moment-resisting frames attempts to provide the maximum possible frame ductility, toughness, and energy-dissipation capacity. This requirement results in more stringent provisions for element and joint detailing. Generally these frames are designed to limit inelastic action to the beams, with the intent of preventing potential yielding in columns and connections.

The design and detailing provisions for composite special moment-resisting frames should incorporate all the corresponding provisions of steel and concrete special moment frames. The design should include the strong column–weak beam concept. For composite columns, transverse reinforcement requirements should be equivalent to those required for reinforced concrete columns in special moment-resisting frames. Special details are invariably required to meet the intent of closed-hoop and cross-tie requirements for composite columns with a structural steel core. An example of a closed-hoop detail for an encased composite column is shown in Fig. 5.32.

Steel and composite beams should be designed to meet the more restrictive $b_f/2t_f$ and d/t_w compactness limits and the lateral bracing requirements of steel special moment-resisting frames. The additional restrictions are necessary to increase the resistance to local and lateral torsional buckling, allowing the beam elements to develop their fully plastic flexural capacity. However, steel flanges connected to a concrete slab with shear connectors are exempted from this provision since the lateral torsional and local buckling forces are substantially inhibited by the presence of shear connectors and the concrete slab.

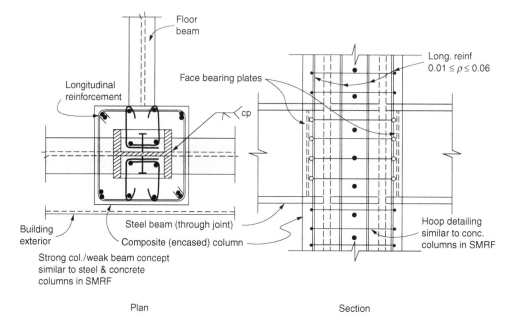

Figure 5.32. Composite special moment-resisting frames; composite perimeter frame column-to-frame beam connection detail.

Special composite moment frames are permitted for buildings in SDC A, B, C, D, E, or F without any height restrictions. The values for seismic design factors are: $R = 8$, $\Omega_o = 3$, and $C_d = 5.5$.

5.5.2. Braced Frames

Most braced frame construction is of structural steel, although there have been some examples of concrete-braced frames in taller buildings designed to resist wind loads. Two types of steel-braced frame construction are recognized in building codes: 1) concentric bracing, where the centerlines of the various members that frame into a joint meet at a single point; and 2) a relatively new form of braced frame called eccentric brace. Developed during the 1970s and 1980s, this system attempts to combine the ductility of moment-resisting frames with the high stiffness of concentrically braced frames. It consists of bracing elements that are deliberately offset from the centerline of beam–column joints.

The short portion of beam between braces or between the brace and the column is referred to as the link. The link of an eccentrically braced frame is designed to act as a ductile fuse to dissipate energy during seismic overloads. As a result, the design of brace elements can be performed so as to preclude the possibility of brace buckling. With proper choice of the brace eccentricity, i.e., of the length of the link beams, the stiffness of this system can approach that of a concentrically braced frame. The ability to combine the ductility of moment frames and the stiffness of concentrically braced frames has led to increasing use of the system in areas of high seismicity. The two types of bracing, namely, concentric and eccentric, are also applicable to composite systems.

5.5.2.1. Concentrically Braced Frames

As shown in Fig. 5.33, composite concentrically braced frames may be configured using a number of possible combinations of steel, reinforced concrete, and composite elements.

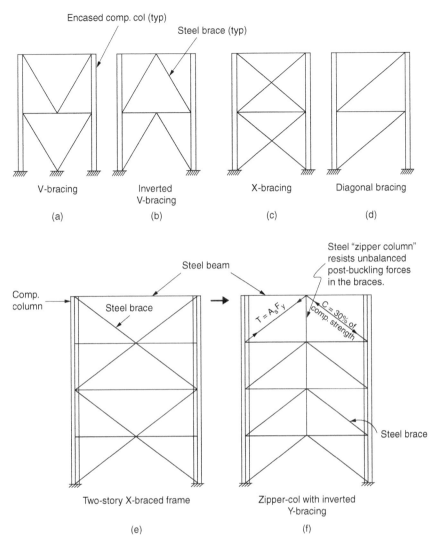

Encased comp. col (typ)

Steel brace (typ)

V-bracing	Inverted V-bracing	X-bracing	Diagonal bracing
(a)	(b)	(c)	(d)

Steel "zipper column" resists unbalanced post-buckling forces in the braces.

Steel beam

Comp. column

Steel brace

$T = A_s F_y$

$C = 30\%$ of comp. strength

Steel brace

Two-story X-braced frame

Zipper-col with inverted Y-bracing

(e) (f)

Figure 5.33. Composite concentrically braced frames: (a) V-bracing; (b) inverted V-bracing; (c) X-bracing; (d) diagonal bracing; (e) two-story X-bracing; (f) zipper column with inverted V-bracing.

Composite braces of either concrete-filled steel tubes or concrete-encased steel braces may be combined with steel frame elements. Composite columns may also be used in conjunction with composite floors and steel bracing members, as used frequently in the design of tall buildings.

The lateral load capacity of a concentrically braced composite frame is somewhat limited for seismic loading. This is because the energy-dissipation capacity of brace elements deteriorates during repeated inelastic cycles. For small or moderate earthquakes where the braced frame elements remain essentially elastic, the response of these frames can be expected to be satisfactory. Certain techniques such as filling steel tubes with concrete may be used to inhibit the onset of local buckling and thereby improve the cyclic response of the brace elements.

Design of connections should be similar to that of steel-braced frames, where the connections are intended to develop the capacity of the brace elements. Where composite elements are used, the connection design must consider the increased capacity caused by the addition of concrete to the steel bracing elements.

The design of elements in composite concentrically braced frames is similar to the design of corresponding elements in steel and concrete systems. Encased composite columns should have a minimum ratio of structural steel to gross column area of 4%. Transfer of forces between the structural steel and reinforced concrete portions of the section should be made through shear connectors, ignoring the contribution of bond or friction. The capacity design of reinforced concrete columns should meet the requirements for columns in ordinary moment-resisting frames. The detailing of both composite and reinforced concrete columns should provide ductility comparable to that of composite ordinary moment-resisting frames.

Composite brace design in concentrically braced frames must recognize that these elements are expected to provide the inelastic action during large seismic overloads. Braces consisting of concrete-encased steel elements should include reinforcing and confinement steel sufficient to provide the intended stiffening effect even after the brace has buckled during multiple cycles of seismic motion. As a result, it is recommended that these elements should meet detailing requirements similar to those for composite columns. Composite braces in tension should be designed considering only the structural steel.

The general intent of the connection design is to provide strength to develop the capacity of the braces in tension or compression. For composite brace sections, the additional strength of the concrete must be considered, since it would be unconservative to consider only the strength of the structural steel section. Brace buckling and the resulting large rotation demands which could result at the brace ends should be considered in connection detailing. Two schematics of composite concentric bracing connections are shown in Figs. 5.34 and 5.35.

Transfer of loads between structural steel and reinforced concrete elements of a composite-braced frame should be made only through shear friction and direct bearing. Reliance on bond and adhesion should not be considered because of the cyclic nature of the lateral loading. In addition, where shear–friction equations are used in the calculation of connection transfer forces, AISC 341-02, i.e., AISC Seismic, recommends that a 25% reduction in the typical shear–friction capacities be imposed for buildings in areas of high seismicity.

Composite concentrically braced frames are permitted similar to composite eccentrically braced frames in SDC A, B, or C without any restrictions on height. The values of

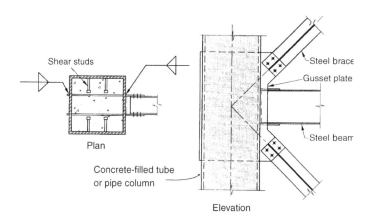

Figure 5.34.   Concrete-filled composite column-to-steel concentric brace connection.

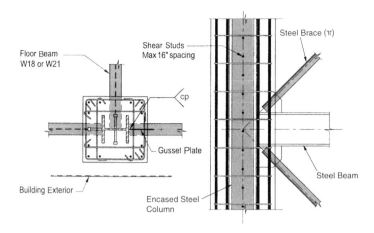

Figure 5.35. Concrete-encased composite column-to-steel concentric brace connection.

seismic design factors are: $R = 5$, $\Omega_o = 2$, and $C_d = 4.5$. Buildings in SDC D or E are permitted with a height limitation of 160 feet. The height limit for buildings in SDC F is 100 ft.

5.5.2.2. Eccentrically Braced Frames

The beam elements of composite eccentrically braced frames will generally consist of structural steel elements. Any concrete encasement of the beam elements should not extend into the link regions where large inelastic action is developed (Fig. 5.36). The column and brace elements of these frames can be of either structural steel or composite construction with structural steel and reinforced concrete. The analysis, design, and detailing of the frames is similar to that required for steel eccentrically braced frames. Since the force transfer mechanisms between the elements of a composite frame rely on bearing and shear friction, special attention must be paid to the design of these connections to realize the intended inelastic action in the ductile link members.

Composite action of the concrete slab with the structural steel link beam may become significant in determining the initial capacity of the link section. This should be considered in sizing the brace and column elements.

The design of composite columns must consider the maximum load that will be generated by yielding and strain hardening of the link beam elements, similar to those required for steel columns. Encased composite columns should have a minimum ratio of structural steel-to-gross column area of 4% unless they are designed as reinforced concrete columns. Transfer of forces between the structural steel and reinforced concrete portions of the section should be made through shear connectors, ignoring the contribution of bond or friction. The capacity design of reinforced concrete and encased composite columns in these frames should meet the requirements for columns in ordinary moment-resisting frames. The detailing of both encased composite and reinforced concrete columns should provide ductility comparable to that of intermediate moment-resisting frames. In addition, for higher-performance categories, these columns should meet the transverse reinforcement requirements for special moment-resisting frames. This requirement is extended to all performance categories when the link element is located adjacent to the column.

Composite brace design in eccentrically braced frames must recognize that these elements are intended to remain essentially elastic during large seismic overloads, since they are designed to be strong enough to yield the link beam elements. The design strength of these braces must consider the yielding and significant strain hardening which can occur

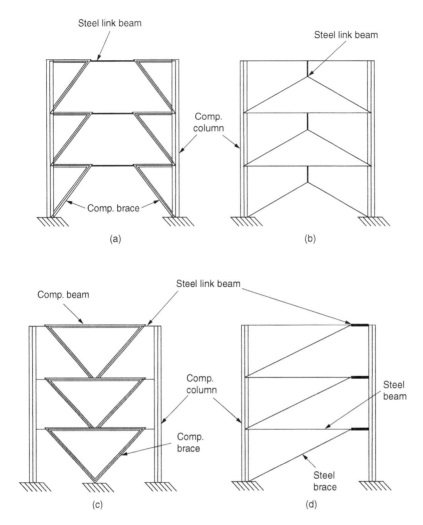

Figure 5.36.   Examples of composite eccentrically braced frames.

in properly designed and detailed link elements. Both axial and bending forces generated in the braces by the strain-hardened link beams must be considered. Braces should therefore be designed to meet detailing requirements similar to those for columns. Composite braces in tension should be designed considering only the structural steel.

The general intent of the connection design is to provide strength to develop the capacity of the link-beam elements. For composite braces, the additional strength of the concrete must be considered, since it would be unconservative to consider only the strength of the structural steel section. Where the shear link is not adjacent to the column, the connections between the braces and columns are similar to those in composite concentrically braced frames. Where the shear link is adjacent to the columns, the connections should be detailed similar to composite beam–column connections in special moment-resisting frames. The large rotation demands that could result at the ends of the link beams should be considered in detailing the connections of composite eccentrically braced frames. Schematic details for two locations of link are shown in Fig. 5.37.

Composite eccentrically braced frames are permitted for buildings in SDC A, B, or C without any height restrictions. The values for seismic design factors are: $R = 8$, $\Omega_o = 2$, and $C_d = 4$. For buildings in SDC D or E, the height limit is 160 feet, and for

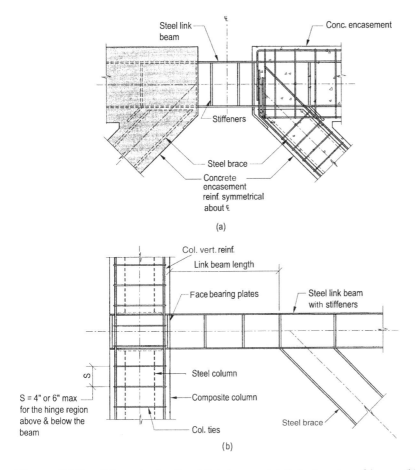

Figure 5.37. Schematic details of link beams: (a) link at center of beam; (b) link adjacent to column.

those in SDC F, the limit is 100 feet. Note that the height limits for both concentric and eccentric composite braced frames are the same.

5.5.3. Composite Shear Walls

One of the most common types of composite shear walls consists of a structural steel frame in which some bays are encased in a reinforced concrete wall. In essence, this results in a reinforced concrete shear wall with structural steel boundary elements and coupling beams. The steel coupling beam is subjected to high shear and moment at each end, requiring a moment-resisting connection to the column. A strong shear connection is also invariably required to resist high shear forces.

If the coupling beams were pin-connected at each end to the boundary elements, they would be ineffective in improving the lateral resistance of the wall; the two wall piers would resist lateral loads independently. On the other hand, if the coupling beams are infinitely stiff, they can fully couple the two piers and make them work as a single unit. If the coupling beam stiffness is in between the two extremes, as is the case in most practical buildings, a portion of the lateral forces will be resisted by the overall system and a portion by the individual elements, typically resulting in an economical structure.

Adding reinforced concrete or structural steel to an existing structural system to achieve composite action of shear walls is a prevalent method of retrofit for resisting lateral loads, particularly in seismic strengthening of buildings.

AISC 7-02, IBC-03, and NFPA 5000 recognize three types of composite shear walls with their attendant seismic factors as follows:

- Composite steel plate shear walls, $R = 6.5$, $\Omega_o = 2.5$, and $C_d = 5.5$.
- Special composite-reinforced concrete shear walls with steel elements, $R = 6$, $\Omega_o = 2.5$, and $C_d = 5$.
- Ordinary composite-reinforced concrete shear walls with steel elements, $R = 5$, $\Omega_o = 2.5$, and $C_d = 4.25$.

The height limitations for the first two types of composite shear walls are 160 ft for buildings in SDC D or E, and 100 ft for buildings in SDC F. There are no height limits

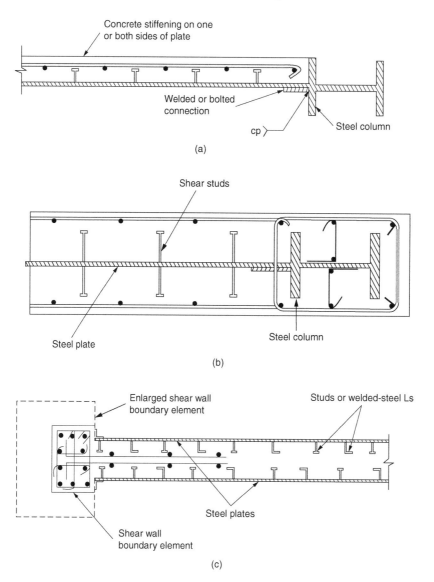

Figure 5.38. Composite steel plate shear walls.

for buildings in SDC A, B, or C using these two types of walls. Ordinary composite shear walls are not permitted for buildings in SDC D, E, or F. They are permitted for buildings in SDC A, B, or C without any height limit.

Composite steel plate shear walls are appropriate when extremely high shear forces must be resisted by a limited length of walls. An example of this use may be found in the 76-story Bank of China Tower, Hong Kong, in which the entire base shear is transferred to the building core at the base.

Possible details for concrete-encased shear plates are shown in Fig. 5.38. In these details, structural steel framing surrounds the steel plates with entire steel assembly encased in reinforced concrete. The steel columns not only resist gravity loads but also act as boundary members resisting overturning forces. The shear-wall web is a steel plate welded to the boundary members. A simple practical detail would be to provide a steel tab continuously fillet-welded in the shop to the beams and columns. The shear-wall steel plate can then be attached to the tabs of the beams and columns with erection bolts. Field fillet welds can then be installed between the steel plate and the tabs. If the plates need to be installed in pieces because of size limitations in shipping or erection, field splices can be of simple fillet welds using a common back-up plate. If there are openings in the wall, additional steel boundary members or flanges must be installed as required.

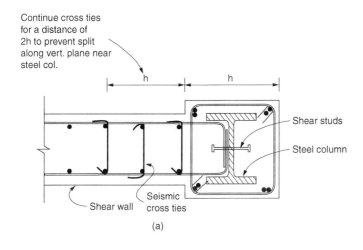

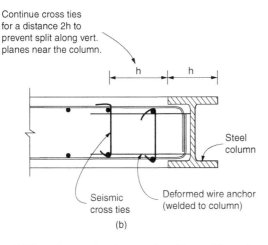

Figure 5.39. Composite shear walls with steel boundary elements.

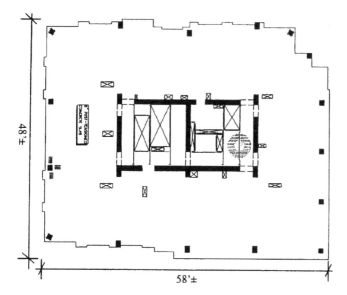

Figure 5.40. The Renaissance project, San Diego, CA; typical floor framing plan.

To prevent buckling of the steel plate, the completed steel assembly is encased in reinforced concrete. This also fireproofs the steel. The encasement should be thick enough to provide the stiffness needed to prevent buckling and should be reinforced for strength. Common details would include a regular pattern of welded studs on each side of the plate or a regular pattern of holes in the plate to pass reinforcing bars hooked at each end. This provides a composite sandwich of steel and concrete with the entire thickness effective in preventing buckling of the composite plate. Schematic details of composite shear walls with structural steel boundary elements are shown in Fig. 5.39.

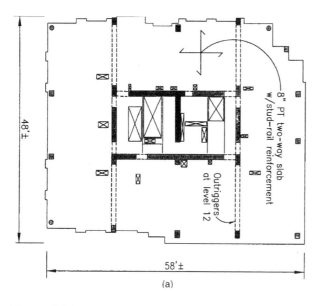

(a)

Figure 5.41. The Renaissance project, San Diego, CA; (a) plan at outrigger level; (b) transverse wall elevation showing composite outriggers; (c) composite outrigger details; (d) section through composite outrigger.

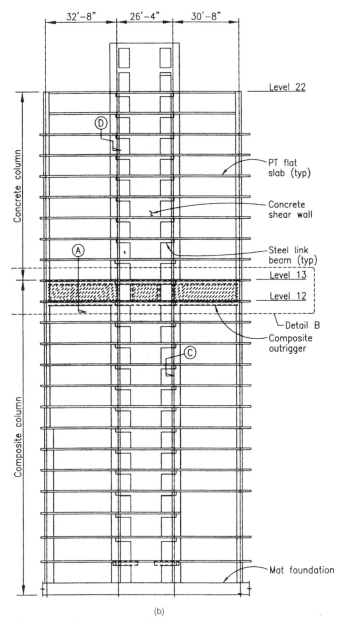

(b)

Figure 5.41. (*Continued*).

5.5.4. Example Projects

5.5.4.1. The Renaissance Project, San Diego, CA*

An example of composite construction in a high-seismic zone is the Renaissance project, a residential development in downtown San Diego, CA. It is constructed primarily of cost-in-place conventional and post-tensioned concrete with some structural steel. It has certain unique design features, including the use of steel link beams embedded in reinforced

---

* Photographs and figures courtesy of Eric Lehmkuhl, S.E., Associate KPFF Consulting Engineers, San Diego, CA.

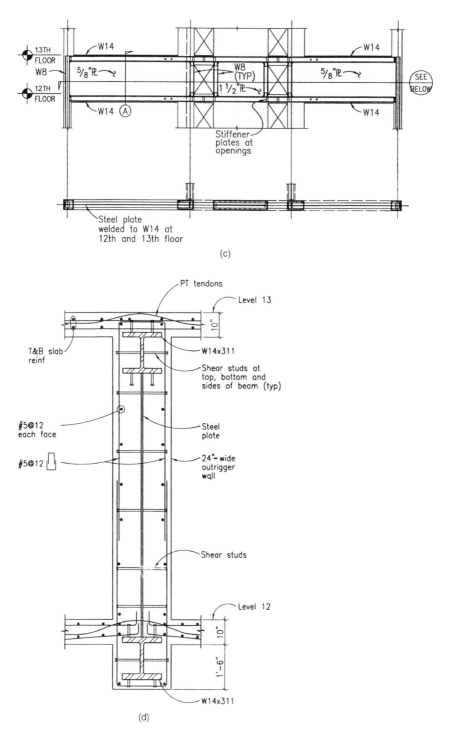

Figure 5.41. (*Continued*).

concrete shear walls and story-high composite outrigger beams coupling the shear walls to exterior composite columns. The development consists of two 24-story towers placed within a three-story-high podium, and houses residential and retail facilities. Additionally, a two-level below-grade parking structure is present under the entire podium. Structural engineering for the project is by KPFF Consulting Engineers, San Diego, CA.

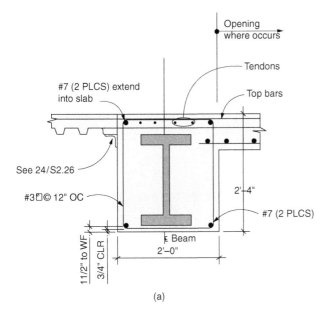

(a)

Note:
1. No P/T tendons allowed perp
   to coupling beams over doors

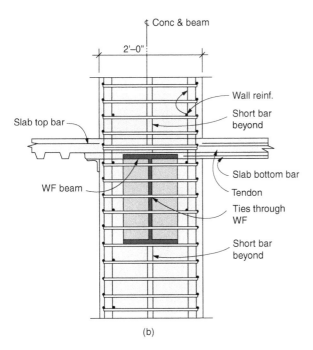

(b)

Figure 5.42. Renaissance project: (a) section through link beam at door opening; (b) coupling beam embedded in shear walls.

Typical framing for residential floors (see Fig. 5.40) consists of an 8-in. (203-mm)-thick two-way flat plate. Stud-rail reinforcement is used to resist punching shear at column heads. A 12-inch (305-mm)-thick slab is used at the third level to resist a relatively heavy landscape loading. The slab also acts as a diaphragm by distributing the lateral loads to the podium shear walls. A 12-in. (305-mm)-thick slab also occurs at the 22nd floor to support a two-story steel structure. The entire structure is founded on a variable thickness mat up to 7 ft (2.13 m) thick under the towers. Structural steel framing is used for a two-story structure atop the 22nd level, and inside the tower core walls.

(a)

(b)

Figure 5.43.   Renaissance Towers, San Diego, CA; (a,b,c,d) construction photographs.

(c)

(d)

Figure 5.43. (*Continued*).

    The lateral system is a composite shear wall core with story-high composite outrigger beams interconnected to exterior composite columns at the mid-height of the building in the transverse, slender direction of the core. Steel beams, W14 × 311 (W 360 × 196), are used as flanges, and plates up to $1^1/_2$-inch (38-mm)-thick are used as webs for the story-high outriggers. See Figs. 5.41a through d.

    A diagonally reinforced concrete coupling beam was judged by the design engineers to be impractical if not unbuildable because the beams were required to link the shear walls in two directions at the corners. Therefore, steel wide flange beams up to W18 × 258 (W460 × 383) are used as link beams by embedding them in the shear walls as shown in Fig. 5.42a and b. The core walls are 24-inch (0.60-m)-thick from the base to the 16th floor, and then step down to 16 in. thick. The embedment length of the link beams is sufficiently long to develop the full plastic capacity of the steel beam. Typically this is the plastic shear capacity, $V_p = 0.6F_y(d - 2t_f)t_w$ (see AISC Seismic, Sect. 15), as the beams are designed to function similar in manner to a ductile link of an eccentric braced system.

Ascertaining satisfactory performance of the gravity system—particularly the flat slab system—subjected to deformations due to seismic lateral loads was a concern. The slab system, together with the columns, behaves as a flat slab-frame and is thus subjected to additional punching shears by virtue of the fact that the slab-frame experiences the same lateral deformations as the lateral-load-resisting elements. To determine the additional shears, a two-dimensional model of an equivalent frame with slabs and columns was analyzed by applying building's drifts to it.

Recommendations given in FEMA 356 were used to determine the equivalent width of the slab and the degree of slab craking. The resulting slab moments were limited to the moment resulting from the yielding of top or bottom slab reinforcement. The punching shears derived from the moments were used to design the slab shear reinforcement.

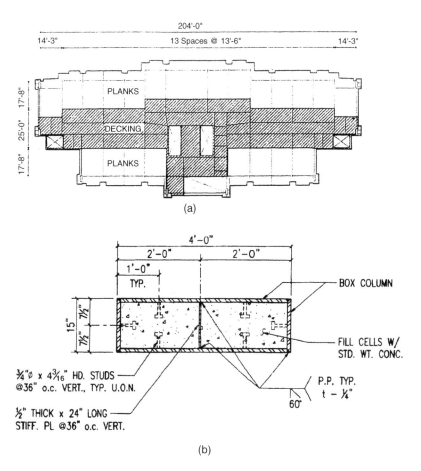

(a)

(b)

Figure 5.44. Kalia Towers, Waikiki, HI; (a) Typical floor plan; (b) box column plan section; (c) brace-to-box column connection; (d) construction photo. (The tower utilized a composite system composed of steel beams with a 6"-thick precast, prestressed concrete plank in the guest rooms, and metal deck with steel beams in the corridors. The lateral system was a braced frame with composite box columns filled with concrete.) Photograph and figures courtesy of Gary Y. K. Chock, S.E., Martin & Chock Inc., Honolulu, HI.

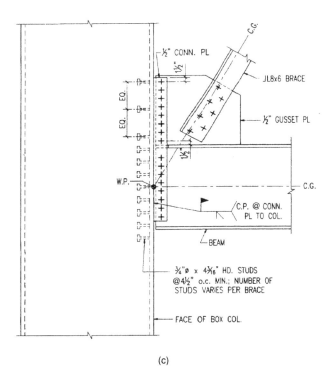

(c)

(d)

Figure 5.44. (*Continued*).

5.5.4.2. *Kalia Towers, Waikiki, HI*

Kalia Towers, constructed in UBC seismic zone 3, Waikiki, HI, is a 24-story, 300 ft-high, 450-room hotel featuring structural steel with precast, prestressed concrete floor system. An $8\frac{1}{2}$-ft (2.60-m) ceiling height is achieved while still keeping the overall floor-to-floor height at 9 ft (2.74 m). The floor system consists of steel beams with 6-in.-thick precast, prestressed concrete planks in the guest rooms, and a metal deck with concrete topping in the corridors. The structural steel framing allowed engineers to lay out the steel system so that most of the beams linedup within the demising walls. As a result, the economics that come with keeping overall building heights at a minimum were realized.

Composite box columns filled with concrete are used in a steel bracing system to resist lateral loads. A typical floor plan, details of composite column, and construction photographs are shown in Fig. 5.44. The structural engineering for the project is by Martin and Chock, Inc., Honolulu, HI.

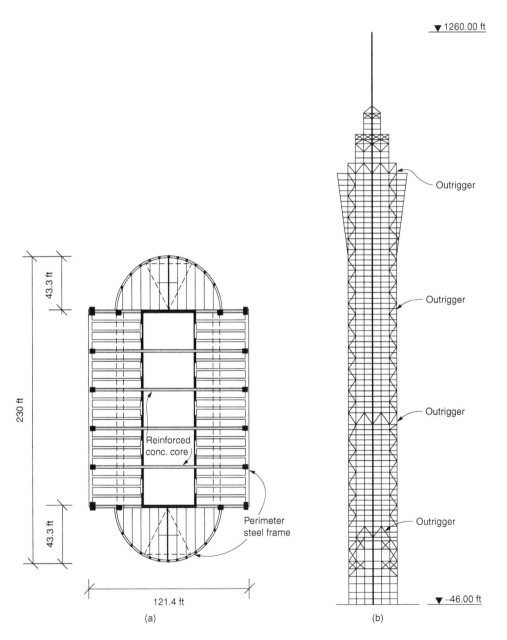

Figure 5.45. Di Wang Building, Shenzhen City, China: (a) schematic plan; (b) schematic elevation; (c) photograph of rendering. (Photograph and sketches courtesy of Chaoying Luo, P.E., John A. Martin & Associates, Structural Engineers, Los Angeles, CA.)

5.5.4.3. Di Wang Building, Shenzhen City, China

An example of one of the world's tallest composite buildings is the Di Wang Building in downtown Shenzhen City, China, about 1.25 miles (2 km) from Hong Kong. At 1260 ft (384 m) to the top of the masts, the building's structural system consists of a reinforced concrete core coupled to the perimeter steel frame by steel outrigger trusses at four levels. The floor plan, composed of a rectangle and two semicircles, measures 300 ft × 121.4 ft (70 m × 37 m). The building is indeed slim, with a height-to-width ratio of 8.78 in the narrow direction. A schematic plan and an elevation are shown in Fig. 5.45a and b. A photograph of the rendering is shown in Fig. 5.45c.

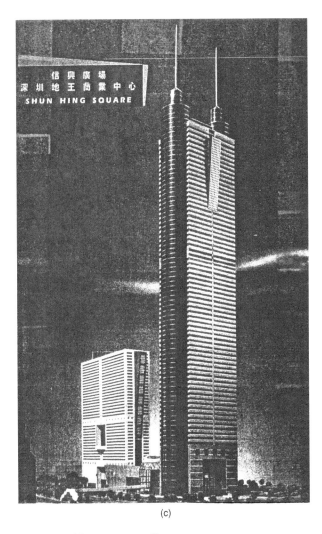

(c)

Figure 5.45. *(Continued)*.

**MIT Stata Complex
Engineering Building**
Boston, Massachusetts

John A. Martin & Associates, Inc.
Structural Engineers
Los Angeles, CA

Photo Credit:
JAMA / Ron S. Lee, PE

6
Seismic Rehabilitation of Existing Buildings

Seismic rehabilitation of a building entails costs as well as disruption of its usage. In fact, the effects of a rehabilitation program are similar to those of an earthquake because strengthening, in terms of cost and the need to vacate the structure while strengthening is underway, is analogous to building repair after an earthquake. The crucial difference is that strengthening occurs at a specified time and no deaths or injuries will occur during the process.

In a seismic rehabilitation study it is convenient to classify the damage within a building in two categories, structural and nonstructural. Structural damage refers to degradation of the building's support system, such as frames and walls, whereas nonstructural damage is any damage that does not affect the integrity of the building's physical support system. Examples of nonstructural damage are chimneys that collapse, broken windows or ornamental features, and collapsed ceilings. The type of damage a building experiences depends on its structural characteristics, age, configuration, construction, materials, site conditions, proximity to neighboring buildings, and the type of nonstructural elements.

An earthquake can cause a building to experience four types of damage:

1. The entire building collapses.
2. Portions of the building collapse.
3. Components of the building fail and fall.
4. Entry-exit routes are blocked, preventing evacuation and rescue.

Any of the above may result in unacceptable risk to human lives. It can also mean loss of property and interruptions of use or normal function.

Another type of damage that should be included in the rehabilitation study is the structural damage from the pounding action that results when two insufficiently separated buildings collide. This condition is particularly severe when the floor levels of the two buildings do not match, because the stiff floor framing of one building can badly damage the more fragile walls or columns of its neighbor.

A rehabilitation objective may be achieved by implementing a variety of measures, including

1. Local modification of deficient components.
2. Removal or partial mitigation of existing irregularities.
3. Global stiffening.
4. Global strengthening.
5. Reduction of mass.
6. Seismic isolation.
7. Installation of supplemental energy dissipation devices.

Failure of nonstructural architectural elements can also create life-threatening hazards. For example, windows may break or architectural cladding such as granite veneer with insufficient anchorage may separate from the building, causing injury to pedestrians. Consequently, a seismic retrofit program should explore techniques for dealing with nonstructural components such as veneers, lighting fixtures, glass doors and windows, raised computer access floors, and ceilings. Similarly, because damage to mechanical and electrical components can impair building functions that may be essential to life safety, seismic strengthening should be considered for components such as mechanical and electrical equipment, ductwork and piping, elevators, emergency power systems, communication systems, and computer equipment.

6.1. CODE-SPONSORED DESIGN

The forces experienced by a structure during a major earthquake are much greater than the design forces. Usually, it is neither practical nor economically feasible to design a building to remain elastic during a major seismic event. Instead, the structure is designed to remain elastic at a reduced force level. By prescribing detailing requirements, engineers can rely upon the structure to sustain post-yield displacements without collapse when subjected to higher levels of ground motion. The rationale for designing with lower forces is based on the premise that the special ductile detailing of the components is adequate to allow for additional deformation without collapse. Historically, this approach has produced buildings with a strength capacity adequate for the scaled-down seismic forces and, more important, with adequate performance characteristics beyond the elastic range. It is the consensus of the structural engineering profession that a building properly designed to both code-specified forces and detailing requirements will have an acceptable level of life safety during a major seismic event.

The ability of a member to undergo large deformations beyond the elastic range is termed ductility. The same property in a building that allows it to absorb earthquake-induced damage and yet remain stable may be considered, in a conceptual sense, similar to ductility. Ductile structures may deform excessively under load, but they remain by and large intact. This characteristic prevents total structural collapse and provides protection to occupants of buildings. Therefore, providing capacity for displacement beyond the elastic range without collapse is a primary goal.

Aside from this implicit philosophy, no explicit earthquake performance objectives are stated in most building codes. However, building structures designed in conformance with modern codes such as the UBC 1997 are expected to

1. Resist low-level earthquakes without damage.
2. Resist moderate-level of earthquakes without structural damage, while possibly experiencing some nonstructural damage.
3. Resist high-level earthquakes of intensity equal to the strongest experienced or forecast for the building site without collapse, while possibly experiencing some structural or nonstructural damage.

It is expected that structural damage, even in a major earthquake, will be limited to a repairable level for structures that meet these requirements. However, conformance to these provisions does not ensure that significant structural damage will not occur in the event of a large earthquake. Therefore, additional requirements are given in the code to provide for structural stability in the event of extreme structural deformations.

The protection of life rather than prevention and repairability of damage is the primary purpose of the code; the protection of life is thus reasonably provided for but not with complete assurance.

6.2. ALTERNATE DESIGN PHILOSOPHY

Although earthquake performance objectives are implicit in building codes, significant questions linger. Is the philosophy of inferring the behavior adequate to define the expected earthquake performance? Can the performance be actually delivered? Should the earthquake response objectives be explicitly stated in building codes? Is it feasible to make an existing nonductile building conform to current detailing and ductility provisions? If not, what level of upgrade will provide for minimum life safety? How much more strengthening is required to achieves an "immediate occupancy rating"?

Explicit answers to these and similar questions cannot be found in current building codes. Although a set of minimum design loads are prescribed, the loads may not be appropriate for seismic performance verification and upgrade design because

1. The code provisions do not provide a dependable or established method to evaluate the performance of noncode compliant structures.
2. They are not readily adaptable to a modified criterion, such as one that attempts to limit damage.
3. Since the primary purpose is protection of life safety, the code does not address some building owners' business concerns such as protection of property, the environment, or business operations.

To overcome these shortcomings, a procedure that uses a two-phase design and analysis approach has been in use for some time. The technique explicitly requires verification of serviceability and survival limit states by using two distinct design earthquakes; one that defines the threshold of damage and the other that defines collapse. The serviceability level earthquake is normally characterized as an earthquake that has a maximum likelihood of occurring once during the life of the structure. The collapse threshold is typically associated with the maximum earthquake that can occur at the building site in the presently known tectonic framework. This characterization can vary, however, to suit the specifics of the project, such as the nature of the facility, associated risk levels, and the threshold of damageability.

The principle behind the two-phase approach may be explained by recalling the primary goal in seismic design, which is to provide capacity for displacement beyond the elastic range. Any combination of elastic and inelastic deformations is possible to attain this goal. For example, we could design a structural system that would remain elastic throughout the displacement range. This system would have a high elastic strength but low ductility. Conversely, it is entirely possible to have a system with relatively low elastic strength but high ductility, meeting the same design objective of remaining stable. It may be easier to understand the methodology if it is recognized that a specific earthquake excitation causes about the same displacement in a structure whether it responds elastically or with any degree of inelasticity.

Figure 6.1 shows the behavior of an idealized structure subjected to three levels of earthquake forces F_L, F_U, and F_C corresponding to lower-level, upper-level, and collapse-level earthquakes. Also shown is an earthquake force F_E experienced by the structure if it were to remain completely elastic. The structure designed using the lower-level earthquake

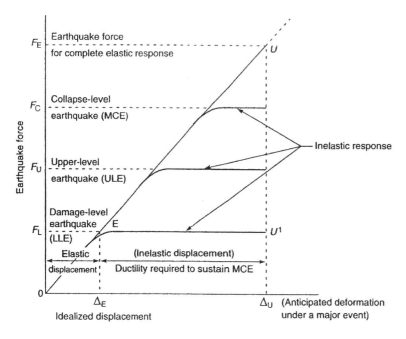

Figure 6.1.   Idealized earthquake force-displacement relationships.

force F_L deforms elastically from 0 to E and inelastically from E to U. The same structure designed using the force F_U needs to deform 0 to U, responding elastically all through the displacement range. Both systems are capable of attaining the anticipated deformation of Δ_U. However, a building designed using the force F_L will require a more ductile system than a building designed for the fully elastic force F_E. More important, it will suffer heavier damage should the postulated event occur. Nevertheless, both systems achieve the primary goal: Both remain stable without collapse under the expected deformation Δ_U. Therefore, it is possible to design the structure using any level of force between F_L and F_E with the understanding that a corresponding ductility is developed by the detailing of the system. For example, a structure designed for the force level F_U requires a higher strength but less ductility than if it were designed for force level F_L. Hence, it is a matter of choice as to how much strength can be traded off for ductility and, conversely, ductility traded for strength. Expressed another way, structural systems of limited ductility may be considered valid, provided they are capable of resisting correspondingly higher seismic forces.

This is the approach used in the seismic retrofit design of existing buildings. Since buildings of pre-1970 vintage do not have the required ductile detailing, the purpose is to establish the strength levels that can be traded off in part for lack of required ductility.

6.3.   CODE PROVISIONS FOR SEISMIC UPGRADE

Most building codes deal primarily with the design of new buildings. For seismic upgrade, the primary use for these documents is determining existing building capacity. They do not, in general, provide guidance for evaluating and upgrading the seismic resistance of an existing building.

Most codes allow existing buildings to use their current lateral-load-resisting systems if only trivial changes to the structure are proposed and the building's use remains unchanged. Codes require upgrading of buildings when major changes or tied-in additions are planned, and when the proposed alterations reduce the existing lateral-load-resisting capacity. A lateral-load upgrade may also be required if the proposed changes move the building into the categories of "essential" or "hazardous" facilities.

The seismic provisions of the IBC attempt to be more specific by quantifying the meaning of "significant change." It requires that the addition itself be compliant with the code for new construction, and requires a seismic upgrade of the existing building if the addition increases the seismic forces in any existing structural member by more than 5% unless that member is already strong enough to comply with the code. Similarly, the addition is not allowed to weaken the seismic capacity of any existing structural member to a level below that specified for new construction. However, there remain some questions as to how to interpret these provisions.

When building codes prescribe full compliance with their current seismic provisions, they are rarely explicit in telling users what measures to take to upgrade the building. There are exceptions, of course. On the U.S. west coast, San Francisco's building code requires upgrading of existing structures to 75% of the strength required by the code for new construction. On the east coast, the *Commonwealth of Massachusetts Building Code* offers an elaborate path for determination of required remedial measures. In some cases it allows lower seismic forces than those used for new construction. In some regions of high seismic activity, state and local codes and ordinances may require a seismic upgrade even for buildings that are not undergoing renovation. Perhaps the best known of these is California's Senate Bill 1953, a seismic retrofit ordinance adopted on Feb. 24, 1994, in the wake of the Northridge earthquake. It requires more than 450 acute care facilities to submit seismic evaluation and compliance plans showing how the facilities will withstand a code-level earthquake, defined as a seismic event with a 10% probability of being exceeded in 100 years. By 2008, all acute care facilities found to be vulnerable to collapse must be removed from service.

In general, the process for seismic upgrade is disorderly. It is not uncommon to have one engineer declare that a building needs a complete seismic upgrade, while another states that none is needed. Some times the owner will "shop" for an engineer in whose opinion an upgrade is not needed, who is willing to justify this interpretation of the code to building officials.

These real-life observations lead to the conclusion that guidance on this issue from an authoritative source is sorely needed. One source—the FEMA 356 publication, discussed shortly—attempts to fill the void.

As compared to seismic upgrade of existing structures, design of a new structure for proper seismic performance is a "cinch". This is because most structural characteristics important to seismic performance including ductility, strength, deformability, continuity, configuration, and construction quality can be designed and, to a certain extent, controlled.

Seismic rehabilitation of existing structures poses a completely different problem. First, until recently (2002), there was no clear professional consensus on appropriate design criteria. That changed substantially with the publication of FEMA 356, *Prestandard and Commentary on the Seismic Rehabilitation of Buildings*. Second, the building codes for new construction are not directly applicable because they incorporate levels of conservatism and performance objectives that may not be appropriate for use on existing structures due to economic limitations. Third, the material strengths and ductility characteristics of an existing structure will, in general, not be well defined. And finally, the details and quality of construction are frequently unknown and, because the structure has been in service for some time, deterioration and damage are often a concern.

The successful seismic upgrade of an existing structure therefore requires a thorough understanding of the existing construction, its limiting strength and deformation characteristics, quantification of the owner's economic and performance objectives, and selection of an appropriate design criterion to meet these objectives, and also be acceptable to the building official. Most of the time it includes the selection of retrofit systems and detailing that can be installed within the existing structure.

6.4. BUILDING DEFORMATIONS

The basic design procedure for new structures consists of the selection of lateral forces appropriate for design purposes, and then providing a complete, appropriately detailed, lateral-force-resisting system to carry these forces from the mass levels to the foundations. Although deformations are checked, experience has shown that new structures with modern materials and ductile detailing can sustain large deformations while experiencing limited damage. Older structures, however, do not have the advantage of this inherent ductility. Therefore, control of deformations becomes an extremely important issue in the design of seismic retrofits.

Determination of the deformations expected in a structure, when subjected to the design earthquake, is the most important task in seismic rehabilitation design. There are three types of deformations that must be considered and controlled in a seismic retrofit design. These are global deformations, elemental deformations, and interstructural deformations. Although they are all interrelated, for purposes of seismic upgrade it is convenient to consider each of these separately.

Global deformations are the only type explicitly controlled by the building codes and are typically considered by reviewing interstory drift. The basic concern is that large interstory drifts can result in $P\Delta$ instabilities. Control of interstory drift can also be used as a means of limiting damage to nonstructural elements of a structure. However, it is less effective than elemental or interstructural deformations in limiting damage to individual structural elements.

Elemental deformation is the amount of seismic distortion experienced by an individual element of a structure such as a beam, column, shear wall, or diaphragm. Building codes have very few provisions that directly control these deformations. They rely on ductility to ensure that individual elements will not fail at the global deformation levels predicted for the structure. In existing structures with questionable ductility, it is therefore critical to evaluate the deformation of each element and to ensure that expected damage to the element is acceptable. This requirement extends to elements not normally considered as participating in the lateral-force-resisting system. A glaring example that is attracting much attention after the Northridge earthquake is the punching shear failure of flat slabs at interior columns, resulting from excessive rotation at the slab–column joint. Often, the slab system is not considered to participate in the lateral-force-resisting system. In fact, building codes indirectly prohibit the use of flat slab–frames in the lateral system of buildings in high seismic zones. However, in relatively flexible buildings such as those without shear walls, when flat slabs "go for a ride," they bend and twist. In doing so, they fail if they do not have adequate ductility. Therefore it is very important to limit the rotational deformation of these joints to prevent a punching shear failure.

Interstructural deformations are those that relate to the differential movement between elements of the structure. Failures that result from lack of such control include failures of masonry walls that have not been anchored to diaphragms and failures resulting from bearing connections slipping off beam seats. Building codes control these deformations, which may cause separation of one element from another, by requiring interconnection of all portions of structures. A similar technique should be considered in the retrofit of an existing structure.

Code methodologies rely on elastic dynamic analysis using base shears that are typically scaled down to base shear values computed on the basis of an equivalent lateral-load procedure. Therefore, design forces are smaller than those likely to be experienced by the building. This reduction factor in the 1994 UBC, which was based on working loads, used to be as large as 12, but in the 1997 UBC, which uses ultimate design values, the corresponding reduction factor is $12/1.4 = 8.57$, rounded to 8. However, it is explicitly recognized that the predicted levels of deformation, termed Δ_s, are substantially smaller than what will be experienced by the building. Hence, amplified deformations $\Delta_M = 0.7R\Delta_s$ are specified in the codes to evaluate the effects of deformation compatibility. It is even more important to use a similar method in evaluating the existing structural elements in a retrofitted structure, because pre-1971 buildings rarely have the required ductility. As with other seismic-design codes and standards, FEMA 356 uses statistical probabilities, not absolute certainty. New or upgraded buildings using the FEMA 356 design approach are not expected to withstand any possible earthquake without a scratch. Instead, they are expected to sustain some damage during strong ground shaking, with a real although very small probability of collapse when a "design" seismic event occurs.

6.5. COMMON DEFICIENCIES AND UPGRADE METHODS

Seismic upgrade of buildings typically involves strengthening of their horizontal and vertical lateral-load-resisting elements. These can be reinforced in-place, or new elements can be added to them. If the existing lateral-load-resisting structure is grossly deficient, it can be replaced. Whenever buildings are upgraded to resist a larger seismic load, their foundations must be checked for the new loading, and be reinforced if necessary.

Prime candidates for renovation and strengthening are

- Buildings with irregular configurations, such as those with abrupt changes in stiffness, large floor openings, very large floor heights, reentrant corners in plan, and soft stories.
- Buildings with walls of unreinforced masonry, which tend to crack and crumble under severe ground motions.
- Buildings with inadequate diaphragms lacking ties between walls and floors or roofs.
- Buildings with nonductile concrete frames, in which shear failures at beam–column joints and column failures are common.
- Concrete buildings with insufficient lengths of bar anchorage and splices.
- Concrete buildings with flat-slab framing, which can be severely affected by large story drifts.
- Buildings with open storefronts.
- Buildings with clear-story conditions.
- Buildings with elements that tend to fail during ground shaking: Examples are unreinforced masonry parapets and chimneys, and nonstructural building elements, which may fall, blocking exits and injuring people.

6.5.1. Diaphragms

A floor deck must act as a diaphragm—a deep horizontal beam capable of lateral-load transfer among the vertical rigid elements. To do so effectively it must have

1. The ability to resist horizontal shear forces, meaning that it must possess a certain degree of strength and rigidity in its plane. This also means that decking elements must be attached to each other and to the supporting floor structures with fasteners capable of transmitting these shear forces. In other words, the decking must be able to function as a web of the beam that does not break and does not deflect excessively under load.
2. Flanges at opposite ends of the diaphragm perpendicular to the applied forces. These flanges, called chords, must be attached to the diaphragm's web with connections capable of transmitting the seismic forces.
3. Drag struts, also called collector elements, to deliver the seismic load from the diaphragm to the vertical lateral-load-resisting elements.

The horizontal distribution of load among the walls or frames depends on the types of floor and roof diaphragms in the building. Flexible systems such as plywood or thin-gauge metal deck diaphragms without structural concrete topping are assumed to distribute lateral loads to the walls or frames in proportion to their tributary areas. In contrast, rigid diaphragms, such as those made of concrete, and concrete topping on composite metal deck distribute lateral loads to the walls or frames in proportion to their relative rigidities. Rigid diaphragms can distribute horizontal forces by developing torsional resistance. This is helpful in buildings with irregular wall layout. Flexible diaphragms are considered too supple to work in torsion. The majority of real-life floor structures fall between the two categories; engineering judgment is required to predict the behavior of these semirigid or semiflexible diaphragms. However, prevailing practice allows the assumption of rigid diaphragms for concrete slabs and concrete-topped composite metal decks, unless diaphragm spans are very large.

The type and function of existing diaphragms must be evaluated prior to making a decision about how to strengthen the vertical lateral-load-resisting elements of the building. For example, it is unwise to add shear walls or braced frames in an asymmetric manner if this introduces torsion into the existing diaphragm and leads to its possible distress. If shear walls or braced frames are placed in the interior of the building, collector elements must be present in the diaphragm to carry the inertial forces to them.

Methods of strengthening diaphragms depend on their composition and the nature of their weaknesses. Deficiencies of existing diaphragms typically fall into two categories: insufficient strength or stiffness, and the absence of chords and collectors or proper connections to them. Replacing a diaphragm, which involves taking out the building floor, is reserved for the most critical condition.

6.5.1.1. Steel Deck Diaphragms

Inadequate diaphragm shear and chord capacities and excessive diaphragm stresses at openings or plan irregularities are common deficiencies in steel deck diaphragms. Steel deck diaphragm shear capacity is limited by the shear capacity of the corrugated sheet steel and the fastener capacity connecting adjacent deck sheets (typically through crimping of the seams or seam welding). Capacity is also controlled by the spacing of deck-to-beam connections, which prevent out-of-plane buckling of the deck.

A modest increase in shear capacity can be achieved by additional welding at sheet seams. This, however, requires the removal of insulation fill on roof decks to provide access for the welding. Should added welding be insufficient or impractical, reducing the demand to below the shear capacity of the diaphragm can be accomplished by adding supplemental vertical lateral-force-resisting elements. New steel braced frames or shear walls can be added to cut down the diaphragm span. Drag struts connecting to the new braced frame or shear wall will be required to distribute the loads into the diaphragm.

Inadequate flexural capacity of steel deck diaphragms may occur due to incomplete or inadequate chord members. Perimeter steel beams or ledgers need to be continuous to act as chords. Beam-to-column connections at the perimeter may have inadequate stiffness or strength in the axial direction of the beams to adequately act as chords.

The following measures may be effective in rehabilitating bare metal diaphragms:

1. Adding shear connectors for transfer of load to chord or collector elements.
2. Strengthening existing chords or connectors by the addition of steel plates to existing frame components.
3. Adding puddle welds or other shear connectors.
4. Adding diagonal steel bracing to form a horizontal truss to supplement diaphragm strength.
5. Replacing nonstructural fill with structural concrete.
6. Adding connections between the deck and supporting members.

6.5.1.2. *Metal Deck Diaphragms with Nonstructural Topping*

Metal deck diaphragms with nonstructural concrete topping are typically evaluated as bare metal deck diaphragms, unless the strength and stiffness of the nonstructural topping is substantiated by approved test data. These diaphragms are commonly used on roofs of buildings where the gravity loads are small. The concrete fill, such as lightweight insulating concrete, usually does not have usable structural properties and is most often unreinforced. Consideration of any composite action must be done with caution after extensive investigation of field conditions. Material properties, force transfer mechanisms, and other factors must be verified in order to include composite action. Typically, decks are composed of corrugated light-gauge sheet steel, with rib depth varying from 9/16 to 3 in. in most cases.

The following measures may be effective in rehabilitating metal deck diaphragms with nonstructural concrete topping:

1. Adding shear connectors to transfer forces to chord or collector elements.
2. Strengthening existing chords or collectors by the addition of steel plates to existing frame components, or attaching plates directly to the slab by embedded bolts or epoxy.
3. Adding puddle welds at the perimeter of diaphragms.
4. Adding diagonal steel bracing to supplement diaphragm strength.
5. Replacing nonstructural fill with structural concrete.

6.5.1.3. *Metal Deck Diaphragms with Structural Concrete Topping*

This system consists of metal deck diaphragms with structural concrete topping, consisting of either a composite deck with indentations or a noncomposite form deck and a concrete topping slab with reinforcement acting together to resist diaphragm loads.

The concrete fill is either normal or lightweight structural concrete, with reinforcing composed of wire mesh or reinforcing steel. Decking units are attached to each other and to structural steel supports by welds or by mechanical fasteners. The steel frame elements to which the topped metal deck diaphragm boundaries are attached are considered to be the chord and collector elements. These types of diaphragms are frequently used on floors and roofs of buildings where typical floor gravity loads are on the order of 100 psf. The resulting concrete slab has structural properties that significantly add to diaphragm stiffness and strength. Concrete reinforcing ranges from light mesh reinforcement to a regular grid of #3 or #4 reinforcing bars. Metal decking is typically composed of corrugated sheet steel from 16 gauge down to 22 gauge. Rib depths vary from $1\frac{1}{2}$ to 3 in. Attachment of the metal deck to the steel frame is usually accomplished using puddle welds at 1 to 2 ft on center. For composite behavior, shear studs are welded to the frame before the concrete is cast.

A relatively recent innovation is to attach the deck to supports with pneumatic shot fasteners. In some cases, self-drilling screws have also been used in these connections. Diaphragms made of concrete fill on steel deck typically fall into the semirigid and semiflexible categories. The flexibility characteristics and shear resistance of a steel deck diaphragm depend on the depth and gauge of the deck, the length of the span between supports, and the method of attachment to the supports.

When the existing steel-deck diaphragm lacks proper attachments to chords or to intermediate beams, attachments can be upgraded. Additional plug welding requires removal of the floor or roof finishes, and a better course of action may be to add overhead fillet welds from below. Attachments to the chords and collectors are usually made in the same manner.

The following measures are effective in rehabilitating metal deck diaphragms with structural concrete topping:

1. Adding shear connectors to transfer forces to chord or collector elements.
2. Strengthening existing chords or collectors by the addition of steel plates to existing frame components, or attaching plates directly to the slab using embedded bolts or epoxy.
3. Adding diagonal steel bracing to supplement diaphragm strength.

6.5.1.4. Cast-in-Place Concrete Diaphragms

Cast-in-place diaphragms are sturdy elements that rarely require major upgrade except at their connections to the chord. However, common deficiencies at diaphragm openings or plan irregularities include inadequate shear capacity, inadequate chord capacity, and excessive shear stresses.

Two alternatives may be effective in correcting the deficiencies: either improve strength and ductility, or reduce demand. Providing additional reinforcement and encasement may be an effective measure to strengthen or improve individual components. Increasing the diaphragm thickness may also be effective, but the added weight may overload the footings and increase the seismic loads. Lowering seismic demand by providing additional lateral-force-resisting elements, introducing additional damping, or base isolating the structure may also be effective rehabilitation measures.

Inadequate shear capacity of concrete diaphragms may be mitigated by reducing the shear demand on the diaphragm by providing additional vertical lateral-force-resisting elements or by increasing the diaphragm capacity by adding a concrete overlay. The addition of a concrete overlay is usually quite expensive, since this requires the removal of existing partitions and floor finishes and may require the strengthening of existing beams and columns to carry the added dead load. Adding supplemental vertical lateral-force-resisting elements will provide additional benefits by reducing demand on other elements that have deficiencies.

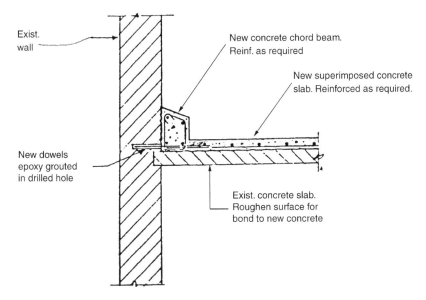

Figure 6.2.   Superimposed diaphragm slab at an existing concrete wall.

Increasing the chord capacity of existing concrete diaphragms can be realized by adding new concrete or steel members or by improving the continuity of existing members. A common method for increasing the chord capacity of a concrete diaphragm with the addition of a new concrete member is shown in Figs. 6.2 and 6.3. This member can be placed above or below the diaphragm. Locating the chord below the diaphragm will typically have less impact on floor space. A common method of increasing the strength and stiffness of an existing simple connection of a steel beam to provide adequate chord capacity is shown in Fig. 6.4. Strengthening of existing concrete and steel deck diaphragms is shown in Figs. 6.5 and 6.6, respectively. Figure 6.7 shows addition of collectors at reentrant corners of a diaphragm.

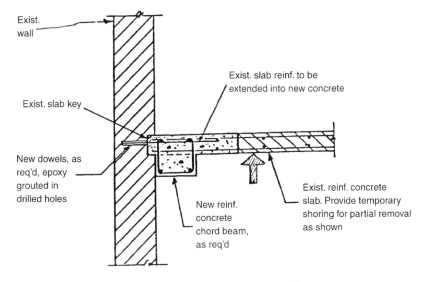

Figure 6.3.   Diaphragm chord for existing concrete slab.

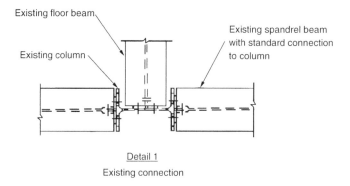

Detail 1
Existing connection

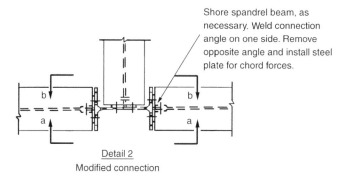

Detail 2
Modified connection

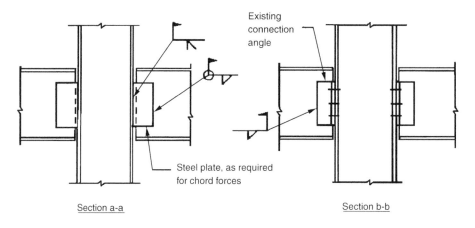

Section a-a   Section b-b

Figure 6.4. Modification of existing steel framing for diaphragm chord forces.

The following measures may be effective in rehabilitating chord and collector elements:

1. Strengthening the connection between diaphragms and chords and collectors.
2. Strengthening steel chords or collectors with steel plates attached directly to the slab with embedded bolts or epoxy, and strengthening slab chord or collectors with added reinforcing bars.
3. Adding chord members.

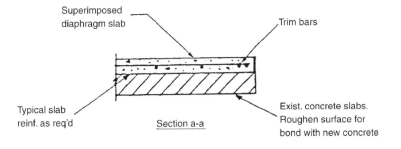

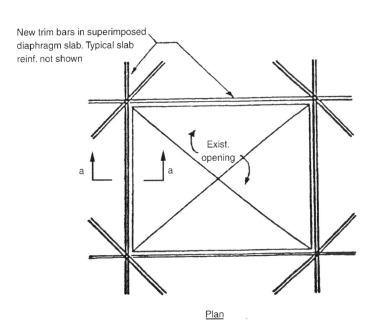

Figure 6.5. Strengthening of openings in a superimposed diaphragm.

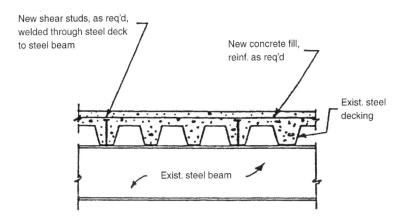

Figure 6.6. Strengthening of existing steel deck diaphragms.

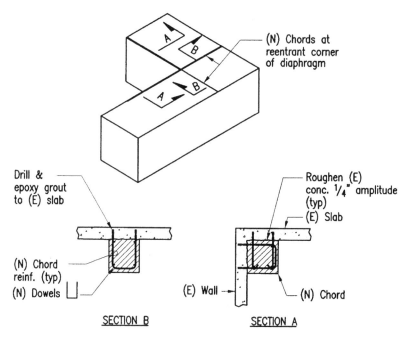

Figure 6.7.   New chords at reentrant corners.

6.5.1.5.   *Precast Concrete Diaphragms*

Common deficiencies of precast concrete diaphragms include inadequate shear capacity, inadequate chord capacity, and excessive shear stresses at diaphragm openings or plan irregularities. Existing precast concrete slabs constructed using precast tees or cored planks commonly have inadequate shear capacity. Frequently, limited shear connectors are provided between adjacent units, and a minimal topping slab with steel mesh reinforcement is placed over the planks to provide an even surface to compensate for irregularities in the precast elements. The composite diaphragm may have limited shear capacity.

Strengthening the existing diaphragm is generally not cost-effective. Adding a reinforced topping slab is generally not feasible because of the added weight. Adding mechanical connectors between units is generally not practical, because the added connectors are unlikely to have sufficient stiffness, compared to the topping slab, to resist an appreciable load. The connectors would therefore need to be designed for the entire shear load assuming the topping slab fails. The number of fasteners, combined with edge distance concerns, typically makes this impractical. The most cost-effective approach is generally to reduce the diaphragm shear forces through the addition of supplemental shear walls or braced frames.

Inadequate chord capacity in a precast concrete deck can be mitigated by adding new concrete or steel members, as discussed earlier for a cast-in-place concrete diaphragm. A new chord member can be added above or below the precast concrete deck. Excessive stresses at diaphragm openings or plan irregularities in precast concrete diaphragms can also be mitigated by introducing drag struts, as described earlier for cast-in-place concrete diaphragms.

6.5.1.6.   *Horizontal Steel Bracing*

Horizontal steel bracing, commonly referred to as steel-truss diaphragms, may be designed to act as dia phragms independently or in conjunction with bare metal deck roofs. Where structural concrete fill is provided over the metal decking, relative rigidities between the steel-truss and concrete systems must be considered in the analysis.

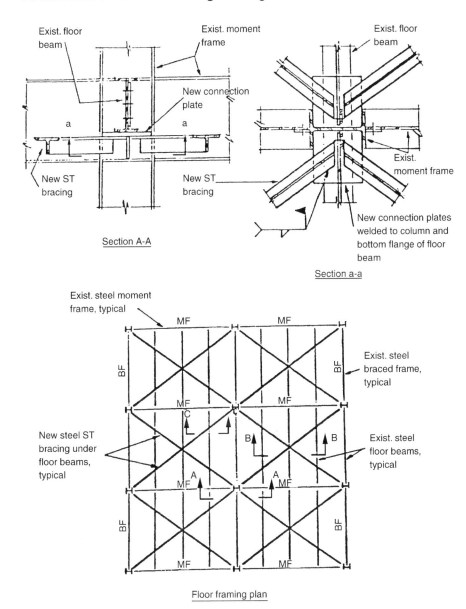

Figure 6.8. Strengthening of an existing steel frame building with horizontal bracing.

Where horizontal steel-truss diaphragms are added as part of a rehabilitation plan, interaction of the new and existing elements in the strengthened diaphragm systems should be evaluated for stiffness compatibility. Also, the load transfer mechanisms between new and existing diaphragm elements must be considered in determining the flexibility of the strengthened diaphragm. Shown in Figs. 6.8 and 6.9 are some common methods of upgrading steel deck diaphragms.

6.5.2. Concrete Shear Walls

The problems that are most difficult to fix are those caused by the irregular configuration of a building (e.g., abrupt changes in stiffness, soft stories, large floor openings, and

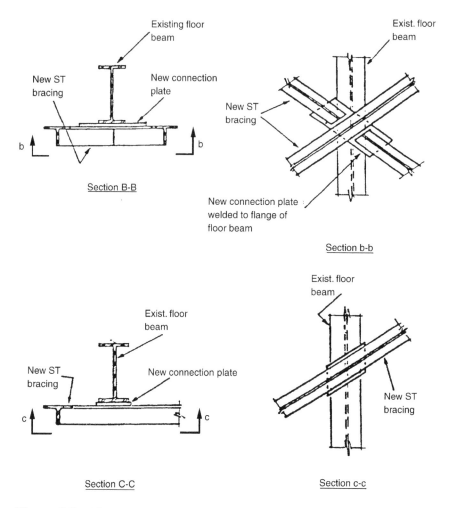

Figure 6.8. (*Continued*)

reentrant floor corners). These cases may require the addition of vertical or horizontal rigid structural elements, as well as strengthening of existing foundations or addition of new ones.

There are several approaches to the reinforcement of existing concrete shear walls, discussed in the following sections.

6.5.2.1. *Increasing Wall Thickness*

Wall thickness is increased by applying reinforced shotcrete to the wall surface. Shotcrete, a mixture of aggregate, cement, and water sprayed by a pneumatic gun at high velocity, is widely used for strengthening walls because it bonds well with concrete. Some prefer application by the dry-mix method (sometimes called gunite) because the slump and stiffness can be better controlled by the nozzle operator and because gunite is applied at higher nozzle velocities, promoting superior bonding.

Concrete shear walls that lack ductility may fail by crushing of their boundary elements, horizontal sliding along construction joints due to shear, or diagonal cracking caused by combined flexure and shear. Among the most common areas of damage are the coupling beams. These can be repaired by through-bolted side plates extending onto the

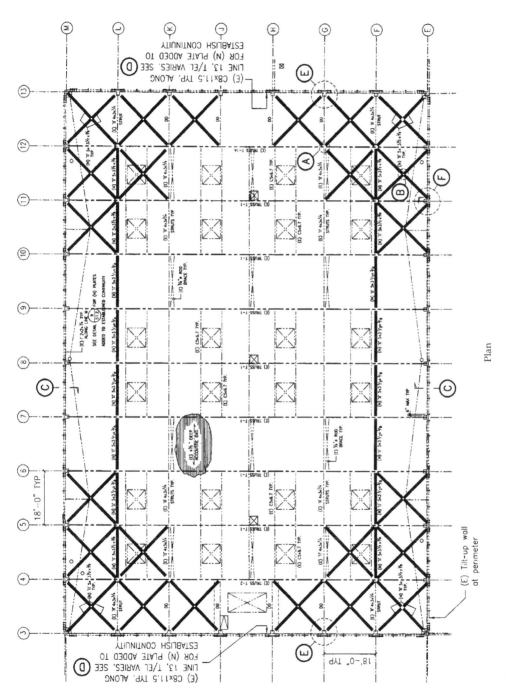

Plan

Figure 6.9. Diaphagm strengthening of an existing gymnasium roof with horizontal bracing.

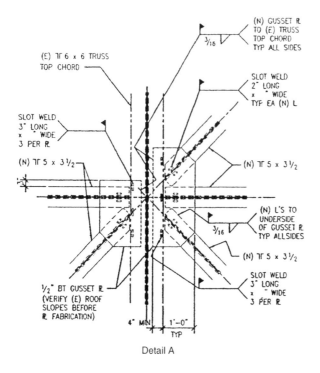

Detail A

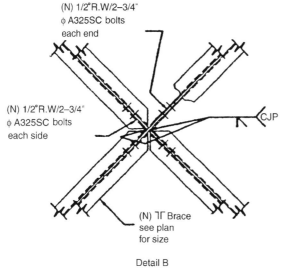

Detail B

Figure 6.9. (*Continued*)

faces of the walls. Short and rigid piers between walls openings also tend to attract an inordinate amount of seismic loading and are therefore prone to damage.

The key to shotcreting walls lies in the surface preparation of the wall because existing concrete may be counted as part of the strengthened wall. All loose and cracked concrete must be removed from the existing wall, and its surface cleaned and roughened by sandblasting or other means. To assure composite action, the overlay is mechanically connected to the wall by closely spaced shear dowels. In addition, steel reinforcement placed in shotcrete is developed at the ends by grouted-in dowels or by continuation into

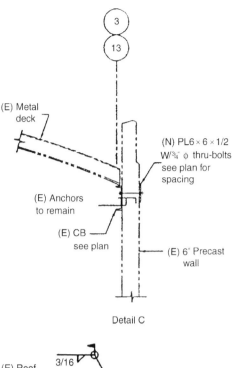

Detail C

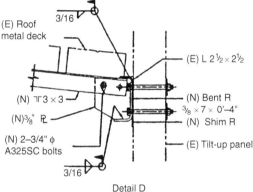

Detail D

Figure 6.9.   (*Continued*)

an adjacent overlay space. This involves drilling through the perimeter beams or columns, filling the drilled openings with epoxy, and splicing the bars with those in the adjoining overlay areas. If the existing wall openings must be filled, the infill should be connected to the roughened edges of the opening with perimeter dowels set in epoxy.

When interior shotcreting is used, attention must be directed toward stabilizing the exterior walls and any exterior ornamental elements of the structure. These may have to be tied back into the new shotcrete by drilled-in dowels set at regular intervals. Dowels placed in exterior elements that are exposed to moisture should be given a measure of corrosion protection, such as galvanizing.

In cases where it is desirable not to increase the wall size, the outer course of bricks can be removed and replaced with shotcrete. The same can be done with interior shotcreting, except that any members framing into the wall may have to be shored during this operation. The added bonus of this approach is that the vertical load on the existing wall foundations changes very little, and they may not require the otherwise necessary enlargement.

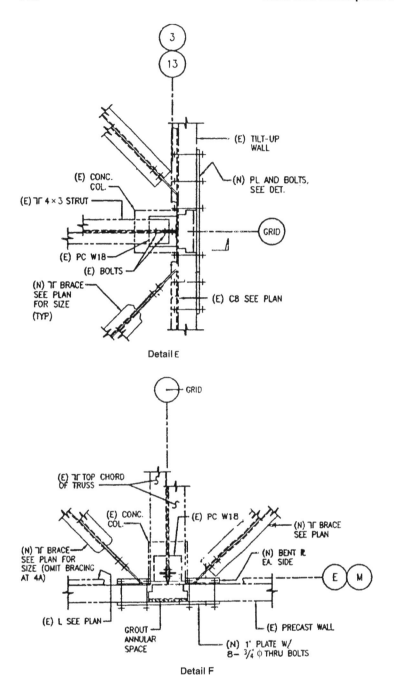

Detail E

Detail F

Figure 6.9. (*Continued*)

6.5.2.2. Increasing Shear Strength of Wall

Increasing the shear strength of the web of a shear wall by casting additional reinforced concrete adjacent to the wall web may be an effective rehabilitation measure. The new concrete should be at least 4-in. thick, contain horizontal and vertical reinforcement, and be properly bonded to the existing web of the shear wall. The use of composite fiber sheets, epoxied to the concrete surface, is another method of increasing the shear capacity of a shear wall. The use of confinement jackets as a rehabilitation measure for wall boundaries may also be effective in increasing both the shear capacity and deformation capacity of coupling beams and columns supporting discontinuous shear walls.

6.5.2.3. *Infilling Between Columns*

Where a discontinuous shear wall is supported on columns that lack either sufficient strength or deformation capacity, making the wall continuous by infilling the opening between these columns may be an effective rehabilitation measure. The infill and existing columns should be designed to satisfy all the requirements for new wall construction, including any strengthening of the existing columns required by adding a composite fiber jacket or a concrete or steel jacket for strength and increased confinement. The opening below a discontinuous shear wall may also be infilled with steel bracing. The bracing members should be sized to satisfy all design requirements for new construction and the columns should be strengthened with a steel or a concrete jacket. All of these rehabilitation measures require an evaluation of the wall foundation, diaphragms, and connections between existing structural elements and any elements added for rehabilitation purposes.

Adding new shear walls or braced frames conforming to current code detailing provisions is among the most common steps taken to strengthen the lateral-load-resisting systems of buildings. The new walls and frames can either: 1) complement the existing elements; or 2) be designed as the sole means of providing vertical rigidity in the building. In the first case, analysis of comparable rigidities must be done to determine what percentage of the total lateral loading the new construction will carry. In the second case, the existing rigid elements that are now considered to be nonstructural must be checked for inelastic deformation compatibility. In any case, new foundations must be provided under the new elements and dowels placed around them for proper transfer of loads.

A common complication of adding shear walls and braced frames is that they tend to interfere with the building layout, circulation, or fenestration. Quite often, shear walls with openings or braced frames of unusual configurations may be needed to accommodate window or door openings. In some rare cases exterior buttresses or counterforts may be considered.

Adding braced frames, usually of structural steel, can be economical in buildings where the existing steel framing will not require strengthening to accommodate the bracing and where the existing framing is readily accessible.

6.5.2.4. *Addition of Boundary Elements*

Addition of boundary members may be an effective measure in strengthening shear walls or wall segments that have insufficient flexural strength. These members may be either cast-in-place reinforced concrete elements or steel sections. In both cases, proper connections should be made between the existing wall and the added members. The shear capacity of the rehabilitated wall should be re-evaluated.

6.5.2.5. *Addition of Confinement Jackets*

Increasing the confinement at the wall boundaries with the addition of a steel or reinforced concrete jacket may be effective in improving the flexural deformation capacity of a shear wall. The minimum thickness for a concrete jacket should be 3 in. A composite fiber jacket may be used to improve the confinement of concrete in compression.

6.5.2.6. *Repair of Cracked Coupling Beams*

These can be repaired by adding side plates extending on the faces of the walls. In this procedure, the plates are attached with both epoxy adhesive and anchor bolts. The plates may be attached to only one face of the wall or can be placed at both faces for extra strength, with the opposite plates through-bolted together. Another possibility for improving coupling beams is by using composite fiber wrapping. This method is least intrusive because the wrapping and the epoxy combined are only 0.25-in. thick.

6.5.2.7. *Adding New Walls*

Adding new shear walls at a few strategic locations can be a very cost-effective approach to a seismic retrofit. The new wall is connected to the adjoining frame by drilled-in dowels. Its foundations are similarly doweled into the existing column footings. To accommodate wall shrinkage, the wall can stop short some distance—2 in., for example—from the existing concrete at the top. The space can be filled later with nonshrink grout.

6.5.2.8. *Precast Concrete Shear Walls*

Precast concrete shear wall systems may suffer from some of the same deficiencies as cast-in-place walls. These may include inadequate flexural capacity, inadequate shear capacity with respect to flexural capacity, lack of confinement at wall boundaries, and inadequate splice lengths for longitudinal reinforcement in wall boundaries. Deficiencies unique to precast wall construction are inadequate connections between panels, to the foundation, and to floor or roof diaphragms.

The rehabilitation measures previously described for concrete buildings may also be effective in rehabilitating precast concrete shear walls. In addition, the following rehabilitation measures may be effective:

- **Enhancement of connections between adjacent or intersecting precast wall panels.**   Mechanical connectors such as steel shapes and various types of drilled-in anchors, cast-in-plane strengthening methods, or a combination of the two may be effective in strengthening connections between precast panels. Cast-in-place strengthening methods include exposing the reinforcing steel at the edges of adjacent panels, adding vertical and transverse reinforcement, and placing new concrete.
- **Enhancement of connections between precast wall panels and foundations.**   Increasing the shear capacity of the wall panel-to-foundation connection by using supplemental mechanical connectors or a cast-in place overlay with new dowels into the foundation may be effective rehabilitation measures. Increasing the overturning moment capacity of the panel-to-foundation connection by using drilled-in dowels within a new cast-in-place connection at the edges of the panel is another effective rehabilitation measure. Adding connections to adjacent panels is also an effective rehabilitation measure, eliminating some of the forces transmitted through the panel-to-foundation connection.

6.5.3. Reinforcing of Steel-Braced Frames

Reinforcement of existing braced frames is relatively straightforward and is often preferable to adding new ones. The work includes adding cover plates, angles, or similar shapes and new welded or bolted connections. For existing bolted connections of the bearing type, new welds can be designed to carry the entire load, or the existing fasteners can be removed and replaced with new, stronger ones. When welded reinforcement is contemplated, it is wise to check the existing steel for weldability, unless some other welding to that steel is already in place.

6.5.4. Infilling of Moment Frames

In many cases, the existing concrete or steel skeleton is stiffened by filling in the space between the beams and columns with masonry or cast-in-place concrete. These infill walls can be a cost-effective method of increasing the lateral strength and rigidity of the building.

Designers should avoid counting on some of the infill walls in structural analysis but not on others, because the stiffness of the frames filled with this nonstructural masonry will increase, whether the designers realize this fact or not. In an earthquake, these panels attract large lateral forces and are damaged, or the perimeter columns, beams, and their connections fail. When a frame, however well designed, is filled with rigid material, however brittle and weak, the fundamental behavior of this structural element is changed from that of a frame to that of a shear wall.

Rehabilitation measures commonly used for concrete frames with masonry infills may also be effective in rehabilitating concrete frames with concrete infills. Additionally, application of shotcrete to the face of an existing wall to increase the thickness and shear strength may be effective. For this purpose, the face of the existing wall should be roughened, a mat of reinforcing steel doweled into the existing structure, and shotcreate applied to the desired thickness.

6.5.5. Reinforced Concrete Moment Frames

Earthquake damage sometimes results in sheared-off columns that formerly were parts of a frame. Typically, the concrete cover is spalled, column bars buckled, and concrete inside broken up. Most problems in concrete frames involve bar splices and failures of beam–column joints that lack confinement and in which reinforcement is stopped prematurely.

Many old buildings with flat-slab and flat-plate floor systems, even those constructed after 1973 (and presumably reflecting the post-San Fernando earthquake code changes), are vulnerable to earthquakes.

Methods available for strengthening traditional concrete frames include encasing the beam–column joints in steel or high-strength fiber jackets. One such design uses jackets consisting of four U-shaped corrugated-metal parts, two around the beam and two around the column. The column jackets are bolted to the end of the beam, the pieces are welded together, and the space between the jackets and the frame is filled with grout.

Frame joints damaged during earthquakes can be repaired with epoxy injection, and badly fractured concrete can be removed and replaced. To minimize shrinkage, the replacement concrete should be made with shrinkage-compensating (type K) cement, or should utilize a shrinkage-reducing admixture. Frame members that have been pushed out of alignment during an earthquake should be jacked back into the proper position before repair. Damaged columns can also be strengthened with fiber-reinforced plastic wraps or other methods of exterior concrete confinement. This is common practice for seismic strengthening of building and bridge columns in California. Another structural issue that requires consideration is the transfer of load from the floor diaphragms to the frames and walls. This may require new drag struts. These elements can be added by attaching new concrete or structural steel sections to the underside of existing floors. They are typically placed against cleaned and roughened concrete surfaces and anchored to the floors and to frames by drilled-in dowels or through-bolts.

Connections between new and existing materials should be designed to transfer the forces anticipated for the design load combinations. Where the existing concrete frame columns and beams act as boundary elements and collectors for the new shear wall or braced frame, these should be checked for adequacy, considering strength, reinforcement development, and deformability. Diaphragms, including drag struts and collectors, should be evaluated and rehabilitated to ensure a complete load path to the new shear wall or braced frame element, if necessary.

Another method of seismic rehabilitation is to jacket existing beams, columns, or joints with new reinforced concrete, steel, or fiber-wrap overlays. The new materials should be designed and constructed to act compositely with the existing concrete. Where reinforced concrete jackets are used, the design should provide detailing to enhance ductility and the jackets should be designed to provide increased connection strength and improved continuity between adjacent components.

Post-tensioning existing beams, columns, or joints using external post-tensioned reinforcement is an effective strategy of seismic rehabilitation. Post-tensioned reinforcement should be unbounded within a distance equal twice the effective depth from sections where inelastic action is expected. Anchors should be located away from regions where inelastic action is anticipated, and be designed considering possible force variations due to earthquake loading.

6.5.6. Steel Moment Frames

The following measures are effective in rehabilitating existing steel moment frames:

1. Adding steel braces to one or more bays of each story to form concentric or eccentric braced frames to increase the stiffness of the frames. The location of added braces should be selected so as not to increase torsion in the system.
2. Adding ductile concrete shear walls or infill walls to one or more bays of each story to increase the stiffness and strength of the structure. The location of added walls should be selected so as not to increase torsion in the system.
3. Attaching new steel frames to the exterior of the building. The rehabilitated structure should be checked for the effects of the change in the distribution of stiffness, the seismic load path, and the connections between the new and existing frames. The rehabilitation scheme of attaching new steel frames to the exterior of the building has been used in the past and has been shown to be very effective under certain conditions. This rehabilitation approach may be structurally efficient, but it changes the architectural appearance of the building. Its advantage is that rehabilitation may take place without disrupting use of building.
4. Adding energy dissipation devices.
5. Increasing the strength and stiffness of existing frames by welding steel plates or shapes to selected members.
6. Reinforcing moment-resisting connections to force plastic hinge locations in the beam away from the joint region. This reduces the stresses in the welded connection, thereby reducing the probability of brittle fractures. This scheme is not recommended if the full-penetration connection of the existing structure does not use weld material of sufficient toughness to avoid fracture at stresses lower than yield or when strain hardening at the new hinge location produces larger stresses than existing at the weld. Rehabilitation measures to reinforce selected moment-resisting connections may consist of providing horizontal

cover plates, vertical stiffeners, or haunches. In regions of high seismicity, pre-Northridge earthquake welded moment connections have typically been found to need strengthening. The upgraded connection must be not only strong enough to resist the stresses resulting from gravity and seismic loading, but also flexible enough to have plastic rotation capacity of at least 0.025 to 0.03 radians.

Repairing connections usually involves, in addition to structural work, removal of wall and ceiling finishes and some disruption of operations, even when the repair is done after working hours. Repair costs can exceed $20,000 per connection (2002 dollars). Further information on seismic upgrade of pre-Northridge welded moment connections is given in the AISC Design Guide 12, *Modification of Existing Welded Steel Moment Frames for Seismic Resistance*. The guide provides information on three designs: reduced beam section, welded haunch, and bolted bracket. In addition to the technical discussion, it also covers practical implementation issues such as reducing tenant disruption in occupied buildings and safety issues.

6.5.7.   Open Storefront

The deficiency in a building with an open storefront is the lack of a vertical line of resistance along one or two sides of a building. This results in a lateral system that is excessively soft at one end of the building, causing significant torsional response and potential instability.

The most effective method of correcting this deficiency is to install a new stiff vertical element in the line of the open-front side or sides. If the open-front appearance is desired, the steel frames may be located directly behind the storefront windows. Shear walls may also be used to provide adequate strength. In both cases collectors are required to adequately distribute the loads from the diaphragm into the vertical lateral-load-resisting element. Adequate anchorage of vertical elements into the foundation is also required to resist overturning forces. Steel moment frames instead of brace frames can also be utilized to provide adequate strength, provided that inelastic deformations of the frame under severe seismic loads are carefully considered to ensure that displacements are controlled. Common methods for upgrading buildings with open storefronts are shown in Fig. 6.10.

6.5.8.   Clerestory

A clerestory, typically designed to produce an open airy feeling, can result in significant discontinuity in a horizontal diaphragm. A common method of correcting the diaphragm discontinuity is to add a horizontal steel truss. Steel members can be designed to transfer diaphragm shears while minimizing the visual obstruction of the clerestory.

An alternate approach is to reduce the demands on the diaphragm through the addition of new vertical lateral-force-resisting elements such as shear walls or braced frames.

6.5.9.   Shallow Foundations

The following rehabilitation measures may be considered for shallow foundations:

1.   Enlarging the existing footing to resist the design loads. Care must be taken to provide adequate shear and moment transfer capacity across the joint between the existing footing and the additions.

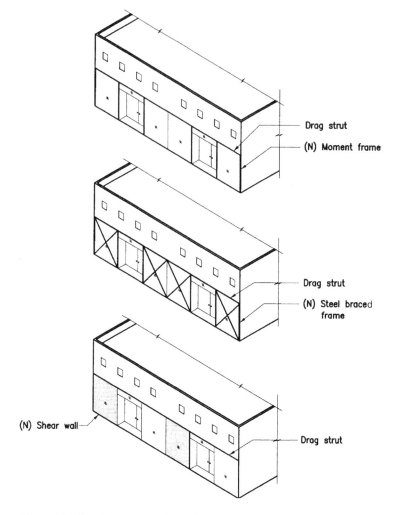

Figure 6.10.   Common methods for upgrading buildings with open storefronts.

2. Underpinning the existing footing, removing of unsuitable soil underneath and replacing it with concrete, soil cement, or another suitable material. Underpinning should be staged in small increments to prevent endangering the stability of the structure. This technique may be used to enlarge an existing footing or to extend it to a more competent soil stratum.

3. Providing tension hold-downs to resist uplift. Tension ties consisting of soil and rock anchors with or without prestress may be drilled and grouted into competent soils and anchored in the existing footing. Piles or drilled piers may also be effective in providing tension hold-downs for existing footings.

4. Increasing the effective depth of the existing footing by placing new concrete to increase shear and moment capacity. The new concrete must be adequately doweled or otherwise connected so that it is integral with the existing footing. New horizontal reinforcement should be provided, if required, to resist increased moments.

5. Increasing the effective depth of a concrete mat foundation with a reinforced concrete overlay. This method involves placing an integral topping slab over the existing mat to increase shear and moment capacity.

6. Providing pile supports for concrete footings or mat foundations. Adding new piles may be effective in providing support for existing concrete footing or mat foundations, provided the pile locations and spacing are designed to avoid overstressing the existing foundations.

7. Changing the building structural characteristics to reduce the demand on the existing elements. This may be accomplished by removing mass or height from the building or adding other elements such as energy dissipation devices to reduce the load transfer at the base. New shear walls or braces may be provided to reduce the demand on foundations.

8. Adding new grade beams to tie existing footings together when soil conditions are poor. This method is useful for providing fixity to column bases, and to distribute lateral loads between individual footings, pile caps, or foundation walls.

9. Grouting techniques to improve existing soil.

6.5.10. Rehabilitation Measures for Deep Foundations

The following rehabilitation measures may be considered for deep foundations:

1. Providing additional piles or piers to increase the load bearing capacity of the existing foundations.

2. Increasing the effective depth of a pile cap by adding concrete and reinforcement to its top. This method is effective in increasing its shear and moment capacity, provided the interface is designed to transfer loads between the existing and new materials.

3. Improving the soil adjacent to an existing pile cap by injection-grouting.

4. Increasing the passive pressure bearing area of a pile cap by addition of new reinforced concrete extensions.

5. Changing the building system to reduce the demands on the existing elements by adding new lateral-load-resisting elements.

6. Adding batter piles or piers to the existing pile or pier foundation to increase resistance to lateral loads. It should be noted that batter piles have performed poorly in recent earthquakes when liquefiable soils were present. This is especially important to consider near wharf structures and in areas with a high water table.

7. Increasing tension tie capacity from a pile or pier to the superstructure.

6.5.11. Nonstructural Elements

6.5.11.1. Nonload-Bearing Walls

The performance of buildings with nonstructural walls that adversely affect the seismic response of a building may be improved by removing and replacing them with walls constructed of relatively flexible materials such as gypsum board sheathing or modifying the wall connections so that they will not resist lateral loads. Removal and replacement of existing hollow clay tile, concrete, or brick masonry partitions is the preferred method of addressing the inadequate out-of-plane capacity of nonstructural partitions. Alternatively, steel strongbacks can provide the out-of-plane support. Steel members are installed at regular intervals and secured to the masonry with drilled and grouted anchors. The masonry spans between the steel members, which span either vertically between floor diaphragms or horizontally between columns. A third method for mitigating masonry walls with inadequate out-of-plane capacity is to provide a structural overlay. The overlay may be constructed of plaster with welded wire mesh reinforcement or concrete with reinforcing steel or welded wire mesh. This approach is used at times merely to provide containment of the masonry. Nonstructural masonry walls are frequently used as firewalls around means of egress. Egress walls with deficient out-of-plane capacity can fail, resulting in rubble blocking the egress. Containment of the masonry with a plaster or concrete overlay can maintain egress, although the walls may need to be replaced following a major seismic event.

6.5.11.2. *Precast Concrete Cladding*

Precast concrete cladding panels with rigid connections may not have the flexibility or ductility to accommodate large building deformations. Failure of the connection may result in heavy panels falling away from the building. Complete correction of this deficiency is likely to be costly, since numerous panel connections would need to be modified to accommodate anticipated building drifts. This may require removal and reinstallation or replacement of the panels. A more economical solution is to install redundant flexible/ductile connections that will keep the panels from falling, should the existing connections fail.

Improper design or installation of precast concrete cladding may also be more than just a connection problem. The cladding may act as an unintended lateral-load-resisting element, should the connections be rigid or insufficient gaps be present between panels. Correcting this deficiency can be accomplished by installing occasional seismic joints in the panels to minimize their stiffness or by stiffening the existing lateral-force-resisting system.

If an entirely new precast cladding system is installed, the connections should be designed to

- Carry gravity loads of precast panels.
- Transfer the in-plane and out-of-plane inertia forces of the panels into the building.
- Isolate the panels from the inelastic drift likely to be experineced by the building in a large earthquake.

6.5.11.3. *Stone or Masonry Veneers*

Stone or masonry veneers may become falling hazards unless their anchorage can accommodate the inelastic deformation of the building. Removal and replacement by veneer with adequate anchorage is one option. A second option is to decrease the deformation of the supporting wall by adding stiffness to the structure.

6.5.11.4. *Building ornamentation*

Building ornamentation such as parapets, cornices, signs, and other appendages are another potential falling hazard during strong ground shaking. Unreinforced masonry parapets with heights greater than $1\frac{1}{2}$ times their width are particularly vulnerable to damage. Parapets are commonly retrofit by providing bracing back to the roof framing.

Cornices and other stone or masonry appendages may be retrofitted by installing drilled and grouted anchors at regular intervals. Sometimes they may be replaced with a lightweight substitute material such as plastic, fiberglass, or metal.

6.5.11.5. *Acoustical Ceiling*

Unbraced suspended acoustical tile ceilings are significantly more flexible than the floors or roofs to which they are attached. The ceilings sway independently from the floor or roof, typically resulting in their connections being broken. This deficiency can be reduced by stiffening the suspended ceiling system with diagonal wires between the ceiling grid and the structural floor or roof members. Vertical compression struts are also required at the location of the diagonal wires to resist the upward component of force caused by the lateral loads. Current code standards can be used for the upgrade of existing ceiling systems.

6.6. FEMA 356: *PRESTANDARD AND COMMENTARY ON THE SEISMIC REHABILITATION OF BUILDINGS*

This standard endorses the use of performance-based design solutions for seismic rehabilitation of buildings. The chosen performance of the building may vary from preventing collapse to a near-perfect building that would survive an expected earthquake without a scratch. The standard allows owners to select their desired performance level and permits designers to choose their own approaches to achieve the desired results rather than strictly adhering to the prescriptive requirements of codes. Instead of dictating how to achieve a given design goal, performance-based design emphasizes the goals that must be met and sets the criteria for acceptance. This way, engineers are free to innovate without running afoul of specific code provisions, within certain limits.

The FEMA documents outline criteria and methods for ensuring the desired performance of buildings at various performance levels selected by the owners with input from their design professionals. The guidelines allow owners to select a level of seismic upgrade that not only protects lives, a goal of all building codes, but also protects their investment.

FEMA 356 is a radical departure from current practice in that it seeks to provide the structural engineering profession with tools to explicitly, rather than implicitly, design for multiple, specifically defined, levels of performance. These performance levels are defined in terms of specifically limiting damage states, against which a structure's performance can be objectively measured. Recommendations are developed as to which performance levels should be attained by buildings of different occupancies and use. This tiered specification of performance levels at predetermined earthquake hazard levels becomes the design performance objective and a basis for design. It recognizes the importance of the performance of all the various component systems to the overall building performance and defines a uniform methodology of design to obtain the desired performance.

6.6.1. Overview of Performance Levels

FEMA 356 sets forth a menu of four rehabilitation objectives associated with four earthquake hazard levels. The rehabilitation objectives are

- Operational performance.
- Immediate occupancy performance.
- Life safety performance.
- Collapse prevention performance.

Each of these performance levels is associated with defined levels of damage to structural, architectural, mechanical, and electrical building components as well as tenant furnishings. The designer is referred to FEMA Tables C1.3–C1.7 for an overview of where each performance level falls within the overall spectrum of possible damage states. From these tables, the designer may infer, for example, a building designed for top-of-the-line performance using higher earthquake hazard levels is likely to come out scratch-free, delivering performance well above the code minimum for life safety level. On the other hand, much less is expected of a building rehabilitated to a collapse prevention performance level. It is deemed to have fulfilled its obligations if it remains standing during and after a large earthquake: Any other damage or loss is acceptable.

The four levels of earthquake levels hazard recognized in the development of design performance objectives are

- Frequent earthquakes, having a 50% chance of exceedence in 30 years (43-year mean return period).
- Occasional earthquakes, having a 50% chance of exceedence in 50 years (72-year mean return period).
- Rare earthquakes, having a 10% change of exceedence in 50 years (475-year mean return period). Also called basic safety earthquake (BSE-1) and design basis earthquake (DBE).
- Very rare earthquakes, having a 10% chance of exceedence in 100 years (950-year return period). Also called basic safety earthquake (BSE-2) and maximum considered earthquake (MCE).

In order to execute a performance-based design, a series of design parameters and acceptance criteria are given for each performance level for the various structural and nonstructural components. Design response parameters are defined at an element level in terms of element forces, interstory drifts, and plastic rotations. These can be derived from a structural analysis of building response to a particular design earthquake. Acceptance criteria are the limiting values for design parameters in order to attain a given performance level. For example, if interstory drift ratio is a design parameter used for a certain class of building, acceptance criteria would be the drift ratios defined for each performance level. Typical drift ratios normally considered in design are 0.020 for the near collapse level, 0.015 for the life safety level, 0.01 for the operational level, and 0.005 for the fully operational level. A wide variety of potential design parameters may need to be defined including deformation, strength, and energy-based parameters. The purpose of FEMA 356 is to provide a consensus-backed, professionally accepted, nationally applicable, seismic rehabilitation standard. It can be used as a tool by design professionals, a reference document by building regulatory officials, and a foundation for the future development and implementation of building code provisions and standards related specifically to existing buildings. The absence of such a standard has been the primary barrier to widespread seismic upgrading of buildings in the United States.

In new buildings, the structural system can be controlled to fit a set of preconditions or a configuration to satisfy the design objectives prescribed by building codes. The degree of nonlinear behavior can be designed to be consistent throughout the structural system, allowing a single seismic reduction factor, R, to be used for the entire building.

Experience in seismic design over the past 100 years has shown that buildings designed to resist ground shaking from an earthquake with a 10% chance of exceedence in 50 years, at a life safety level of performance, have been able to resist the strongest earthquake without collapse. This experience has given structural engineers enough confidence to design new structures in which ductile details are specified, properties of materials used in construction are controlled, and stringent requirements of testing and inspection are specified.

Assessing the seismic vulnerability of existing buildings is an entirely different problem. This is because, for existing buildings, structural details and the properties of materials must be confirmed or assumed from available information augmented by testing and inspection. Conservative assumptions consistent with the quality of the information available must be made prior to seismic evaluation. The engineer has no control over the structural system or its configuration. The existing building may not fit prescriptive details to permit code-type analysis. Nonlinear behavior of the components of the structural system will probably not be consistent. Thus, the properties of each component must be separately studied. Because of the inconsistent levels of reserve capacity in existing buildings and the differences between the 10% in a 50-year earthquake and the maximum considered earthquake (MCE) in various regions of the country, it is inappropriate that rehabilitated buildings be designed to resist a single level of earthquake shaking. Therefore, using an entirely different approach, FEMA 356 provides a basis of rehabilitation designs for a variety of structural performance levels, ranging from enhanced performance to collapse prevention. It emphasizes the idea that seismic rehabilitation should be directed to controlling deformation in order to minimize damage. Use of all existing seismic resistance is permitted in the evaluation. Acceptance criteria tailored to recognize the deformation capacity of all existing as well as enhanced or new components are provided.

The seismic loads used in the evaluation are based on a suite of USGS-developed acceleration maps including four key maps. Two of these are BSE-1 (basic safety earthquake-1) maps of acceleration response spectra having a 10% probability of exceedence in 50 years. The other two are BSE-2 (basic safety earthquake-2) maps of acceleration response spectra for the MCE—modified 2% probability of exceedence in 50-year maps: Both BSE-1 and BSE-2 maps are given for 0.2-second-period (short period) and 1-second-period buildings.

6.6.2. Permitted Design Methods

Two methods are permitted by FEMA 356, a simplified method and a systematic method. The simplified approach is for the rehabilitation design of small buildings of regular configuration, and is intended to fulfill limited objectives. Partial rehabilitation measures that seek to eliminate high-risk building deficiencies such as exterior falling hazards are included in the technique.

The systematic rehabilitation method discussed at length in this section is applicable to any building. It is a component- and element-based design. In this method, global seismic response of the building is sought with unreduced seismic loads (that is, with a global R-factor of unity). In the seismic evaluation, all components and seismic elements are considered with their individual deformation and force-resisting characteristics. It is a deformation-based design with the explicit rather than tacit acknowledgment that seismic elements and components behave in a nonlinear manner.

Any of the following analysis procedures may be used in the rehabilitation study and upgrade design:

- **Linear static procedure (LSP).** This procedure replaces the equivalent lateral force procedure included in most seismic design codes. It incorporates techniques for considering the nonlinear response of individual seismic elements. The distribution of forces is similar to equivalent lateral force procedures for new buildings.

- **Linear dynamic procedure (LDP).** In this method, the modeling and acceptance criteria are similar to those of LSP. However, calculations are carried out using modal spectra analysis or time history analysis using response spectra or time-history records that are not modified to account for inelastic response for distribution of forces.
- **Nonlinear static procedure (NSP).** This method is frequently referred to as a pushover analysis. It has been in use for some time without specific guidance from building codes and standards regarding modeling assumptions and acceptance criteria. This is now alleviated to some extent because FEMA 356 sets forth specific procedures.
- **Nonlinear dynamic procedure (NDP).** The modeling approaches and acceptance criteria for this method are similar to those of NSP. It differs from NSP in that response calculations are made using inelastic time history dynamic analysis to determine distribution of forces and corresponding internal forces and system displacements. Peer review by an independent engineer with experience in seismic design and nonlinear procedures is recommended because this method requires assumptions that are not included in FEMA 356.

6.6.3. Systematic Rehabilitation

The process of arriving at a systematic rehabilitation design includes the following steps:

1. Determination of seismic ground motions.
2. Determination of as-built conditions.
3. Classification of structural components into primary and secondary components.
4. Setting up of analytical models and determination of design forces.
5. Ultimate load combinations; combined gravity and seismic demand.
6. Component capacity calculations, Q_{CE} and Q_{CL}.
7. Capacity versus demand comparisons.
8. Development of seismic strengthening strategies.

First, the seismic hazard for the site is established by determining the probable ground shaking (spectral acceleration) from either seismic hazard maps or a site-specific investigation. Other site hazards such as liquefaction, lateral spreading, and land sliding are determined from site reconnaissance, existing documentation, or a subsurface investigation.

The desired performance level is then established. This requires close communication with the client, using damage descriptions for each performance level as a tool to get ideas across. The damage descriptions associated with each performance level can be used to inform and assist the client to make a decision of the preferred performance level.

Next, an analysis is performed after classifying building components as either primary or secondary. This distinction is required because the acceptance criteria are different for each type of component. The primary components are parts of the building's lateral-force-resisting system, whereas the secondary components are those not required for lateral-force resistance, although they may actually resist some lateral forces. The analysis is performed by considering general requirements such as $P\Delta$ effects, torsion, overturning, continuity, integrity of elements, and building separations. Cracked properties as given in Table 6.1 are used for concrete buildings.

New or modified components are evaluated using the same standards as existing components, and the designs are completed by comparing capacities with demands for

TABLE 6.1 Effective Stiffness Values

Component	Flexural rigidity[a]	Shear rigidity	Axial rigidity
Beams—nonprestressed	$0.5E_cI_g$	$0.4E_cA_w$	—
Beams—prestressed	E_cI_g	$0.4E_cA_w$	—
Columns with compression due to design gravity loads $\geq 0.5A_gf'_c$	$0.7E_cI_g$	$0.4E_cA_w$	E_cA_g
Columns with compression due to design gravity loads $\leq 0.3A_gf'_c$ or with tension	$0.5E_cI_g$	$0.4E_cA_w$	E_sA_s
Walls—uncracked (on inspection)	$0.8E_cI_g$	$0.4E_cA_w$	E_cA_g
Walls—cracked	$0.5E_cI_g$	$0.4E_cA_w$	E_cA_g
Flat Slabs—nonprestressed	[b]	$0.4E_cA_g$	—
Flat Slabs—prestressed	[b]	$0.4E_cA_g$	—

[a] It shall be permitted to take I_g for T-beams as twice the value of I_g of the web alone. Otherwise, I_g shall be based on the effective width as defined in Section 6.4.1.3. For columns with axial compression falling between the limits provided, linear interpolation shall be permitted. Alternatively, the more conservative effective stiffness shall be used.

[b] Slabs shall be modeled considering flexural, shear, and torsional‘tht stiffnesses.

(From Table 6.5 in FEMA 356.)

each component. The components and connections are redesigned where demand exceeds capacity and analysis is iterated to confirm the design. Nonstructural components are verified for the performance level and rehabilitation objective selected.

It should be noted that selection of a rehabilitation strategy follows confirmation of seismic deficiencies. From among many possible strategies, the strategy most likely to meet requirements is selected. Some possible strategies are modification of components, removal of irregularities and discontinuities, global strengthening and stiffening, mass reduction, seismic isolation and energy dissipation.

6.6.3.1. *Determination of Seismic Ground Motions*

Two characteristic earthquakes, referred to as BSE-1 and BSE-2, are of particular importance. These generally correspond to return periods of 474 and 2475 years, respectively, and are commonly referred to as earthquakes with a 10% chance of exceedence in 50 years and a 2% chance of exceedence in 50 years, respectively. At sites close to major faults, the probabilistic estimates of ground motion are capped by deterministic ones. The engineer has three choices for determining the acceleration response spectra corresponding to these earthquakes: 1) use spectral response acceleration contour maps developed by the USGS, available from the FEMA distribution center, and online; 2) use CD-ROM available from the USGS; or 3) engage a geotechnical engineer to develop site-specific response spectra based on the geologic, seismologic, and soil characteristics associated with the specific site. For some sites option three may be the only permitted method. However, to define as precise a seismic demand as possible, it is common practice to engage a geotechnical engineer to perform a site-specific study for developing response spectra corresponding to specific return periods. The geotechnical report also typically addresses other seismic hazards such as liquification, lateral spreading, or potential for land sliding at the site.

It should be noted that acceleration response spectra for earthquake hazard levels corresponding to probabilities of exceedence other than the BSE-1 and BSE-2 earthquakes can be determined by following procedures specified in FEMA 356.

6.6.3.2. Determination of As-Built Conditions

In this step the following tasks are performed:

- Field observation.
- Review of available documents, including plans, specifications geotechnical reports, shop drawings, test records, and maintenance histories.
- Review of information regarding material standards and construction practices for location and date of construction.
- Destructive and nondestructive testing of selected building components for determination of material properties and configuration of details.
- Interviews with people knowledgeable about the building (i.e., owners, tenants, maintenance personnel, architects, engineers, and builders).

As a measure of the knowledge gained from this investigation, engineers assign a numerical value to the knowledge coefficient k. ($k = 1.0$ if the available information is reliable; if not, $k = 0.75$.)

6.6.3.3. Classification of Structural Components into Primary and Secondary

Before setting up the analytical model, structural components are classified as either primary or secondary. Primary components are those that provide the structure's basic lateral resistance. Secondary components are those that do not, and as such are permitted to experience more damage and displace more than the primary components. Additionally, components are further classified as either deformation-controlled, if they are capable of sustaining the loads when strained inelastically, or force-controlled, if they are not capable of sustaining load when strained inelastically.

6.6.3.4. Setting Up Analytical Model and Determination of Design Forces

An analytical model of the building is set up to represent the structure's dynamic behavior. Although two-dimensional models may be adequate, current practice is to use three-dimensional models to account for torsion, plan and vertical irregularities, and non-uniform distribution of building mass. Only the primary components are modeled, with the stipulation that the secondary elements, if used in the model, cannot exceed 25% of the total structural stiffness. If they do, then some of the secondary components must be reclassified as primary components.

6.6.3.4.1. Calculation of Building Period T. The building period T is calculated by using either the modal analysis procedure, method 1, or empherical equations, method 2.

Method 1 is the preferred method. The fundamental period T is obtained by an eigenvalue analysis using the analytical model. This is the more commonly used method, particularly in seismic vulnerability studies.

In method 2, the period T is determined using the following equation:

$$T = C_t h_n^\beta \tag{6.1}$$

where

$\quad\quad C_t = 0.035$ for steel moment frames
$\quad\quad\quad = 0.018$ for concrete moment frames
$\quad\quad\quad = 0.030$ for eccentrically braced frames
$\quad\quad\quad = 0.060$ for wood buildings
$\quad\quad\quad = 0.020$ for all other framing systems

h_n = 0.035 height, in feet, above shear base to the building roof
β = 0.80 for steel moment frames
  = 0.90 for concrete moment frames
  = 0.75 for all other systems

However, there is a major difference worthy of note between building code procedures for new buildings and the FEMA 356 approach. Unlike the codes, there is no maximum limit on period calculated using method. 1. The intent of this omission is to encourage the use of more advanced analysis such as computer dynamic analysis. It is believed that sufficient controls on analysis and acceptance criteria are present within the FEMA standard to provide reasonably conservative results even though there is no upper limit for the period obtained by method 1.

6.6.3.4.2. Determination of Base Shear (Pseudolateral Load). The base shear, also referred to as pseudolateral load, for use in the design of new components and in the verification of existing components of the lateral-force-resisting system is given by

$$V = C_1 C_2 C_3 C_m S_a W \tag{6.2}$$

where

V = pseudo lateral load (the base shear)
C_1 = modification factor that accounts for the difference in the structure's elastic and inelastic displacement amplitude. Its value ranges from 1 to 1.5, depending upon the building's period T.
C_1 = 1.5 for $T < 0.10$ second.
C_1 = 1.0 for $T \geq T_s$ second.
T = building period
T_s = characteristic period of the response spectrum at which the constant acceleration region of the design response spectrum transitions to the constant velocity region
C_2 = modification factor that represents the effect of strength and stiffness degradation of the components on maximum displacement response. For linear procedures, $C_2 = 1.0$.
C_3 = modification factor that represents $P\Delta$ effects
C_m = effective mass factor to account for higher mode effects. $C_m = 1.0$ if building fundamental period T is greater than 1.0 sec
S_a = response spectrum acceleration at the fundamental period and damping ratio of the building.
W = effective seismic weight of the building, including the total dead load and applicable portions of other gravity loads listed below:

- In storage and warehouse occupancies, a minimum of 25% of floor live load.
- Where an allowance for partition load is included in the floor load design (the actual partition weight) or a minimum weight of 10 psf of floor area, whichever is greater.
- The total operating weight of permanent equipment.
- The effective snow load equal to 20% of the design snow load if design snow load exceeds 30 psf. If not, the effective snow load may be taken to be zero.

6.6.3.4.3. Vertical Distribution of Base Shear. The lateral force F_x, applied at any level x, is determined in accordance with Eqs. (6.3) and (6.4)

$$F_x = C_{vx} V \tag{6.3}$$

and $\quad C_{vx} = \dfrac{W_x h_x^k}{\displaystyle\sum_{i=1}^{n} W_i h_i^k}$ (6.4)

where

C_{vx} = vertical distribution factor
V = pseudolateral load (base shear)
w_i and w_x = the portion of the total gravity load of the building W located or assigned to level i or x
h_i and h_x = the height in feet from the base to level i or x
k = an exponent related to the building period as follows:
If the building period is 0.5 sec or less, $k = 1$.
If the buildings period is 2.5 sec or more, $k = 2$.
Linear interpretation is used for intermediate values of the period T.

6.6.3.4.4. Diaphragm Design Force F_{px}. Floor and roof diaphragms are designed to resist the combined effects of the inertial force F_{px} calculated in accordance with Eq. (6.5), and to resist the horizontal forces resulting from offsets in the vertical seismic elements above and below the diaphragm.

$$F_{px} = \sum_{i=x}^{n} F_i \dfrac{W_x}{\displaystyle\sum_{i=x}^{n} W_i}$$ (6.5)

where

F_{px} = total diaphragm inertial force at level x
F_i = lateral load at level i
w_i = portion of the effective seismic weight w
w_x = portion of the effective seismic weight w located at or assigned to floor level x

6.6.3.5. Ultimate Load Combinations: Combined Gravity and Seismic Demand

In this step the earthquake actions Q_E obtained in step 4 for the unreduced response spectra are combined with the gravity actions to determine the demand imposed on the component. When the effects of gravity and seismic loads are additive, an upper-bound value for gravity loads is estimated by using the following load combinations:

$$Q_G = 1.1(Q_D + Q_L + Q_s)$$ (6.6)

And when the effects of gravity and seismic loads are counteracting, a lower-bound value of the gravity load is estimated by using 90% of the dead load:

$$Q_G = 0.9Q_D$$ (6.7)

where

Q_D = dead load action
Q_L = effective live load action equal to 25% of the unreduced design live load but not less than the actual live load
Q_s = effective live load action equal to 20% of the design snow where the design snow load exceeds 30 pounds per square foot. No part of the load need be included if the design snow load is less than 30 psf.

Next, the gravity and seismic loads are combined using the following equations:
For deformation-controlled actions:

$$Q_{UD} = Q_G + G_E \qquad (6.8)$$

For force-controlled actions:

$$Q_{UF} = Q_G + \frac{Q_E}{C_1 C_2 C_3 J} \qquad (6.9)$$

where

Q_{UD} = deformation-controlled demand due to gravity loads and earthquake loads

Q_{UF} = force-controlled demand due to gravity loads in combination with earthquake loads

J = coefficient used to estimate the actual forces delivered to force-controlled components by other yielding components. The values of J are:

J = 2.0 in zones of high seismicity

= 1.5 in zones of moderate seismicity

= 1.0 in zones of losw seismicity

Alternatively, J may be taken as the smallest demand capacity ratio (DCR) for the components in the load path delivering force to the component being designed. The minimum value of J, the force-delivery reduction factor, is 1.0. See Sec. 6.6.3.4.2 for C_1, C_2, and C_3.

6.6.3.6. Component Capacity Calculations Q_{CE} and Q_{CL}

FEMA 356 specifies two different equations for evaluating component capacities depending upon whether the action of the component is deformation-controlled (Q_{CE}) or force-controlled (Q_{CL}). The subscript E in Q_{CE} stands for expected capacity, whereas L in Q_{CL} stands for lower-bound capacity. The subscript c in both Q_{CE} and Q_{CL} stands for capacity. FEMA uses the terminology *design actions* to define forces and moments in the components due to seismic and gravity effects.

The two types of actions—deformation-controlled actions and force-controlled actions—are defined to distinguish a ductile behavior from a brittle behavior.

6.6.3.6.1. Deformation-Controlled Actions. Deformation-controlled actions in simple terms refer to forces and moments in a component that has recognizable nonlinear deformation characteristics. Because of possible anticipated nonlinear response, the design forces and moments in the component are permitted to exceed their capacity. The acceptance criteria, Eq. (6.11), take this overload into account through the use of an *m*-factor, which in a conceptual sense is an indirect measure of the nonlinear deformation capacity of the component.

Some examples of deformation-controlled actions for steel and concrete components are as follows:

Steel Components

- Flexural moments at the ends of a frame–beams and columns.
- Columns panel zone shear.
- Link beams in eccentric braced frames (EBFs).
- Braces in compression (except EBF braces).
- Braces in tension (except EBF braces).
- Beams and columns in tension (except EBF beams and columns).
- Steel plate shear walls.
- Diaphragm components.

Concrete Components
- Beams controlled by flexure.
- Beams controlled by shear.
- Beams controlled by inadequate splicing along the span.
- Beams controlled by inadequate embedment into the beam–column joint.
- Columns controlled by flexure.

6.6.3.6.2. Force-Controlled Actions. Force-controlled actions differ from deformation-controlled actions in that they do not have a recognizable inelastic response. Therefore, demands for force-controlled actions must not exceed the calculated capacity (i.e., there are no *m*-factors in the acceptance criteria). It should be noted, however, that the calculated design force (demand) itself is reduced by the C_1, C_2, C_3, and J factors before demand is compared to capacity.

An ideal procedure for determining the magnitude of force-controlled actions is by identifying an inelastic limit state for the component and then, by statics, evaluation of the corresponding force-controlled action. For example, seismic shear in a frame–beam is determined from equilibrium considerations of a free-body diagram of the beam with a moment equal to the expected moment strength plus gravity moments.

However, it is acceptable to determine force-controlled actions from Eq. (6.14), where it is not possible to identify a well-defined limit state.

6.6.3.6.3. Capacity Q_{CE} of Steel Beam.

Given. A W27 × 194 frame–beam in a steel building constructed in 1992. The specified material is ASTM A-36, dual grade.

Required. Capacity Q_{CE} of the frame–beam.

Solution. First, determine Table 6.2 (FEMA Table 5.2) the lower-bound yield strength of steel manufactured in the year 1992 under ASTM A36 for structural size grouping 3 (see *AISC Manual of Steel Construction,* 9th ed., Table 2). This is equal to 52 ksi. Second, from Table 6.3 (FEMA Table 5.3), find the value of the factor that translates lower-bound steel properties to expected-strength properties. This is equal to 1.05. Next, the expected yield strength is obtained by multiplying the lower-bound value of 52 ksi and the translation factor of 1.05. Thus, for W27 × 194 frame–beam, $F_{ye} = 52 \times 1.05 = 54.6$ ksi.

The expected capacity Q_{CE} of the beam is given by:

$$Q_{CE} = M_{CE} = M_{PCE} = ZF_{ye}$$
$$= 628 \times 54.6 = 34288 \text{ k-in.}$$
$$= 2857.4 \text{ k-ft}$$

6.6.3.6.4. Capacity Q_{CE} of Concrete Beam.

Given. A reinforced concrete frame–beam in a building built in the year 1980. The beam has the following properties: b = 30", h = 48", d = 45", with 5 #11 top bars, ASTM A615, grade 60, at the negative zones.

$f'_c = 4$ ksi, $f_y = 60$ ksi

Required. The expected capacity Q_{CE} of the beam.

Solution. Reinforcement f_{ye}: From Table 6.4 (FEMA Table 6.2), the default lower-bound yield strength for ASTM A615, grade 60 reinforcement = 60 ksi. From Table 6.5 (FEMA Table 6.4), the value for the conversion factor = 1.25. Therefore, the expected strength of the reinforcement

$f_{yc} = 60 \times 1.25 = 75$ ksi

TABLE 6.2 Default Lower-Bound Material Strengths[a,b]

Date	Specification	Remarks	Tensile strength,[c] ksi	Yield strength,[c] ksi
1900	ASTM, A9	Rivet steel	50	30
	Buildings	Medium steel	60	35
1901–1908	ASTM, A9	Rivet steel	50	25
	Buildings	Medium steel	60	30
1909–1923	ASTM, A9	Structural steel	55	28
	Buildings	Rivet steel	46	23
1924–1931	ASTM, A7	Structural steel	55	30
		Rivet steel	46	25
	ASTM, A9	Structural steel	55	30
		Rivet steel	46	25
1932	ASTM, A140-32T issued	Plates, shapes, bars	60	33
	as a tentative revision to	Eyebar flats unannealed	67	36
	ASTM, A9 (buildings)			
1933	ASTM, A140-32T discontinued and ASTM, A9 (buildings) revised Oct. 30, 1933	Structural steel	55	30
	ASTM, A9 tentatively revised to ASTM, A9-33T (buildings)	Structural steel	60	33
	ASTM, A141-32T adopted as a standard	Rivet steel	52	28
1934 on	ASTM, A9	Structural steel	60	33
	ASTM, A141	Rivet steel	52	28
1961–1990	ASTM, A36/A36M-00	Structural steel		
	Group 1		62	44
	Group 2		59	41
	Group 3		60	39
	Group 4		62	37
	Group 5		70	41
1961 on	ASTM, A572, grade 50	Structural steel		
	Group 1		65	50
	Group 2		66	50
	Group 3		68	51
	Group 4		72	50
	Group 5		77	50
1990 on	A36/A36M-00 & dual grade	Structural steel		
	Group 1		66	49
	Group 2		67	50
	Group 3		70	52
	Group 4		70	49

[a] Lower-bound values for material prior to 1960 are based on minimum specified values. Lower-bound values for material after 1960 are mean minus one standard deviation values from statistical data.

[b] Properties based on ASTM and AISC structural steel specification streases.

[c] The indicated values are representative of material extracted from the flanges of wide flange shapes.

(From Table 5.2 in FEMA 356.)

TABLE 6.3 Factors to Translate Lower-Bound Steel Properties to Expected-Strength Steel Properties

Property	Year	Specification	Factor
Tensile strength	Prior to 1981		1.10
Yield strength	Prior to 1961		1.10
Tensile strength	1961–1990	ASTM A36/A36M-00	1.10
	1961–present	ASTM A572/A572M-89, Group 1	1.10
		ASTM A572/A572M-89, Group 2	1.10
		ASTM A572/A572M-89, Group 3	1.05
		ASTM A572/A572M-89, Group 4	1.05
		ASTM A572/A572M-89, Group 5	1.05
	1990–present	ASTM A36/A36M-00 & dual grade, Group 1	1.05
		ASTM A36/A36M-00 & dual grade, Group 2	1.05
		ASTM A36/A36M-00 & dual grade, Group 3	1.05
		ASTM A36/A36M-00 & dual grade, Group 4	1.05
Yield strength	1961–1990	ASTM A36/A36M-00	1.10
	1961–present	ASTM A572/A572M-89, Group 1	1.10
		ASTM A572/A572M-89, Group 2	1.10
		ASTM A572/A572M-89, Group 3	1.05
		ASTM A572/A572M-89, Group 4	1.10
		ASTM A572/A572M-89, Group 5	1.05
	1990–present	ASTM A36/A36M-00, rolled shapes	1.50
		ASTM A36/A36M-00, plates	1.10
		Dual grade, Group 1	1.05
		Dual grade, Group 2	1.10
		Dual grade, Group 3	1.05
		Dual grade, Group 4	1.05
Tensile strength	All	Not listed[a]	1.10
Yield strength	All	Not listed[a]	1.10

[a] For materials not conforming to one of the listed specifications.
(From Table 5.3 in FEMA 356.)

Structural concrete f'_{ce}**.** From Table 6.6 (FEMA Table 6.1), the default lower-bound compressive strength of structural concrete in beams built in 1980 varies from 3 to 5 ksi, with an average value of 4 ksi. From Table 6.5, the adjustment factor = 1.50. Therefore,

$$f'_{ce} = 1.50 \times 4 = 6 \text{ ksi}$$

The moment capacity M_{CE} is calculated using the following equation:

$$M_u = \phi \rho f_y bd^2 \left(1 - \frac{\rho f_y}{1.7 f'_c}\right) \tag{6.10}$$

except that $\phi = 1.0, f_y = f_{ye}, f'_c = f'_{ce},$ and $f_y = f_{ye}$. Hence, the expected flexural capacity M_{CE} of the beam is given by

$$Q_{CE} = M_{CE} = \rho f_{ye} bd^2 \left(1 - \frac{\rho f_{ye}}{1.7 f'_{ce}}\right) \qquad \rho = \frac{7.80}{30 \times 45} = 0.00578$$

$$Q_{CE} = 0.00578 \times 75 \times 30 \times 45^2 \left(1 - \frac{0.00578 \times 75}{1.7 \times 6}\right) = 25216 \text{ k-in.}$$

$$= 2101 \text{ k-ft}$$

TABLE 6.4 Default Lower-Bound Tensile and Yield Properties of Reinforcing Bars for Various ASTM Specifications and Periods[a]

ASTM Designation[e]	Steel type	Year range	Structural[b]	Intermediate[b]	Hard[b]			
		ASTM grade	33	40	50	60	70	75
		Minimum yield (psi)	33,000	40,000	50,000	60,000	70,000	75,000
		Minimum tensile (psi)	55,000	70,000	80,000	90,000	95,000	100,000
A15	Billet	1911–1966	x	x	x			
A16	Rail[c]	1913–1966			x			
A61	Rail[c]	1963–1966				x		
A160	Axle	1936–1964	x	x	x			
A160	Axle	1965–1966	x	x	x	x		
A408	Billet	1957–1966	x	x	x			
A431	Billet	1959–1966						x
A432	Billet	1959–1966				x		
A615	Billet	1968–1972		x		x		x
A615	Billet	1974–1986		x		x		
A615	Billet	1987–1997		x		x		x
A616[d]	Rail[c]	1968–1997			x	x		
A617	Axle	1968–1997		x		x		
A706	Low-alley	1974–1997					x	
A955	Stainless	1996–1997		x		x		x

[a] An entry of x indicates the grade was available in those years.
[b] The terms structural, intermediate, and hard became obsolete in 1968.
[c] Rail bars are marked with the letter R.
[d] Bars marked s (ASTM 616) have supplementary requirements for bend tests.
[e] ASTM steel is marked with the letter W.
(From Table 6.2 in FEMA 356.)

TABLE 6.5   Factors to Translate Lower-Bound Material Properties to Expected Strength Material Properties

Material property	Factor
Concrete compressive strength	1.50
Reinforcing steel tensile & yield strength	1.25
Connector steel yield strength	1.50

(From Table 6.4 in FEMA 356.)

6.6.3.7.   Capacity Versus Demand Comparisons

In this step, the component capacities are compared with the demand due to earthquake and gravity loads. If the capacity of a component exceeds the demand imposed on it by the seismic and gravity load combinations, the component is judged to satisfy the performance criteria. If not, a more refined technique such as a pushover analysis is performed before declaring the component deficient.

Two equations are given in FEMA 356 for verifying the acceptance criteria.

For deformation-controlled actions: $m\kappa Q_{CE} \geq Q_{UD}$. (6.11)

For force-controlled actions: $\kappa Q_{CL} \geq Q_{UF}$. (6.12)

where

m = modifier given in Tables 6.7–6.17 that takes into account the expected ductility of the component associated with the action being verified at the selected structural performance level

Q_{CE} = expected strength of component at the deformation level under consideration for deformation-controlled actions

κ = knowledge factor defined in Section 6.10

Q_{CL} = lower-bound strength of a component for force-controlled actions

Q_{UD} = deformation-controlled demand due to gravity and earthquake loads

Q_{UF} = force-controlled demand due to gravity and earthquake loads

Numerical values of m are given in FEMA 356 for steel, concrete, masonry, and wood components. Values are given separately for linear and nonlinear procedures and for primary and secondary components. An abbreviated version of the tables of m-values for primary components analyzed using linear procedures is given here for the following components:

TABLE 6.6   Default Lower-Bound Compressive Strength of Structural Concrete (psi)

Time frame	Footings	Beams	Slabs	Columns	Walls
1900–1919	1000–2500	2000–3000	1500–3000	1500–3000	1000–2500
1920–1949	1500–3000	2000–3000	2000–3000	2000–4000	2000–3000
1950–1969	2500–3000	3000–4000	3000–4000	3000–6000	2500–4000
1970–Present	3000–4000	3000–5000	3000–5000	3000–10000	3000–5000

(From Table 6.1 in FEMA 356.)

TABLE 6.7 Acceptance Criteria for Structural Steel Components: Beams—Flexure

	m-Factors[a]		
		Primary components	
	IO	LS	CP
Beams—flexure, deformation-controlled			
$\dfrac{b_f}{2t_f} \leq \dfrac{52}{\sqrt{F_{ye}}}$ and $\dfrac{h}{t_w} \leq \dfrac{418}{\sqrt{F_{ye}}}$	2	6	8
$\dfrac{b_f}{2t_f} \geq \dfrac{65}{\sqrt{F_{ye}}}$ or $\dfrac{h}{t_w} \geq \dfrac{640}{\sqrt{F_{ye}}}$	1.25	2	3

[a] For built-up members where the lacing plates do not meet the requirements of Section 5.6.2.4.2, divide m-factors by 2.0, but values need not be less than 1.0.
IO = immediate occupancy; LS = life safety; CP = collapse prevention.
(From Table 5.5 in FEMA 356.)

1. Steel Beams—Flexure, Deformation-Controlled   Table 6.7
2. Steel Columns—Flexure, and Column Panel Zones   Table 6.8
3. Steel Braces in Compression   Table 6.9
4. Steel EBF Link Beams, Braces, Beams and Columns in Tension, Steel Plate Shear Walls, and Diaphragm Components   Table 6.10
5. Reinforced Concrete Beams   Table 6.11
6. Reinforced Concrete Columns   Table 6.12
7. Reinforced Concrete Beam–Column Joints   Table 6.13
8. Flexure-Controlled Concrete Shear Walls, Columns, and Coupling Beams   Table 6.14
9. Shear-Controlled Concrete Walls and Coupling Beams   Table 6.15
10. Two-Way Slabs and Slab–Column Connections   Table 6.16
11. Reinforced Concrete Infilled Frames   Table 6.17

6.6.3.8. Development of Seismic Strengthening Strategies

If all of the components in the structure meet the basic acceptance criteria associated with their actions, no further analysis is necessary, and the building can be judged to meet the evaluation criteria. If not, typically a more refined study (including, perhaps, a pushover analysis) would be considered before deciding on a seismic rehabilitation program. The final evaluation should be based on a review of the qualitative and quantitative results. The evaluating engineer is urged to consider the issues carefully, to refrain from penalizing

TABLE 6.8 Acceptance Criteria for Structural Steel Components: Columns and Panel Zones

	m-Factors		
		Primary components	
	IO	LS	CP
Columns—Flexure[a,b]			
For $P/P_{CL} < 0.20$			
$\dfrac{b_f}{2t_f} \leq \dfrac{52}{\sqrt{F_{ye}}}$ and $\dfrac{h}{t_w} \leq \dfrac{300}{\sqrt{F_{ye}}}$	2	6	8
$\dfrac{b_f}{2t_f} \geq \dfrac{65}{\sqrt{F_{ye}}}$ or $\dfrac{h}{t_w} \geq \dfrac{460}{\sqrt{F_{ye}}}$	1.25	1.25	2
For $0.2 < P/P_{CL} < 0.50$			
$\dfrac{b_f}{2t_f} \leq \dfrac{52}{\sqrt{F_{ye}}}$ and $\dfrac{h}{t_w} \leq \dfrac{260}{\sqrt{F_{ye}}}$	1.25	c	d
$\dfrac{b_f}{2t_f} \geq \dfrac{65}{\sqrt{F_{ye}}}$ or $\dfrac{h}{t_w} \geq \dfrac{400}{\sqrt{F_{ye}}}$	1.25	1.25	1.5
Column panel zones—shear	1.5	8	11
Fully restrained moment connections[e]			
WUF[f]	1.0	4.3–0.083d	3.9–0.043d
Bottom haunch in WUF with slab	1.6	2.7	3.4
Reduced beam section[f]	2.2–0.008d	4.9–0.025d	6.2–0.032d

[a] Columns in moment or braced frames shall be permitted to be designed for the maximum force delivered by connecting members. For rectangular or square columns, replace $b_f/2t_f$ with b/t, replace 52 with 110, and replace 65 with 190.

[b] Columns with $P/P_{CL} > 0.5$ shall be considered force-controlled.

[c] $m = 9(1 - 1.7\ P/P_{CL})$.

[d] $m = 12(1 - 1.7\ P/P_{CL})$.

[e] Tabulated values shall be modified as indicated in Section 5.5.2.4.2, item 4, in FEMA 356.

[f] d is the beam depth; d_{bg} is the depth of the bolt group.

(From Table 5.5 in FEMA 356.)

TABLE 6.9 Acceptance Criteria for Structural Steel Components: Braces in Compression (Except EBF Braces)

| | | m-Factors | |
| | | Primary components | |
	IO	LS	CP
Braces in compression (except EBF braces)			
Double angles buckling in-plane	1.25	6	8
Double angles buckling out-of-plane	1.25	5	7
W or I shape	1.25	6	8
Double channels buckling in-plane	1.25	6	8
Double channels buckling out-of-plane	1.25	5	7
Concrete-filled tubes	1.25	5	7
Rectangular cold-formed tubes			
$\dfrac{d}{t} \leq \dfrac{90}{\sqrt{F_y}}$	1.25	5	7
$\dfrac{d}{t} \geq \dfrac{190}{\sqrt{F_y}}$	1.25	2	3
Circular hollow tubes			
$\dfrac{d}{t} \leq \dfrac{1500}{F_y}$	1.25	5	7
$\dfrac{d}{t} \geq \dfrac{6000}{F_y}$	1.25	2	3

IO = immediate occupancy; LS = life safety; CP = collapse prevention.
(From Table 5.5 in FEMA 356.)

the building due to fine technical points beyond those contained in the FEMA 356 evaluation methodology, and to visualize the building in its ultimate condition in an earthquake, being aware of the risks of brittle failure and buckling. Due consideration should be given to the mitigating influences of good workmanship, structural integrity, and the strengths and redundancies that are not explicitly considered to be part of the lateral-force-resisting system. Most important, engineering judgment based on sound seismic design principles should be exercised before pronouncing a building unsafe. The questions that review engineers should ask themselves before declaring a building noncompliant are many. Some of these are

1. What if the material properties are higher than assumed in the analysis?
2. What if we allow for a small amount of rocking and sliding at the base to absorb excess earthquake energy at little harm to structure?
3. What if we use gross properties for concrete components, particularly for T- and I-shaped beams?

TABLE 6.10 Acceptance Criteria for Structural Steel Components: Link Beams; Braces; Beams and Columns in Tension; Steel Plate Shear Walls; and Diaphragm Components

	m-Factors		
		Primary	
	IO	LS	CP
EBF link beam[a,b]			
$e \leq \dfrac{1.6 M_{CE}}{V_{CE}}$	1.5	9	13
$e \geq \dfrac{2.6 M_{CE}}{V_{CE}}$			
Braces in tension (except EBF braces)[c]	1.25	6	8
Beams, columns in tension (except EBF beams, columns)	1.25	3	5
Steel plate shear walls[d]	1.5	8	12
Diaphragm components			
Diaphragm shear yielding or panel or plate buckling	1.25	2	3
Diaphragm chords and collectors—full lateral support	1.25	6	8
Diaphragm chords and collectors—limited lateral support	1.25	2	3

[a] Values are for link beams with three or more web stiffeners. If no stiffeners, divide values by 2.0, but values need not be less than 1.25. Linear interpolation shall be used for one or two stiffeners.
[b] Assumes ductile detailing for flexural link, in accordance with AISC (1995) *LRFD Specifications*.
[c] For tension-only bracing, m-factors shall be divided by 2.0.
[d] Applicable if stiffeners, or concrete backing, is provided to prevent buckling.
(From Table 5.5 in FEMA 356.)

4. For a moment frame building, what if we reanalyze the frame using different size rigid joints in the frame model? Does inclusion of an elastic spring to represent the stiffness of the joint result in a more favorable demand/capacity ratio?

5. What if we use slightly higher values for the ductility factor m in verifying the acceptance criteria?

Although FEMA 356 has procedures to answer some of these questions the author recommends that a parametric study of the acceptance criteria be undertaken before declaring the building noncomplaint. This recommendation should not be constructed as sanctioning indiscriminate manipulation of the FEMA 356 procedure, but as a reminder for engineers to use that nonquantifiable, mysterious branch of engineering often called the art of design. It should be kept in mind that no matter how sophisticated an analysis is, it is hard to justify that its seismic behavior will be satisfactory if it has large vertical and horizontal discontinuities. Experience has taught time and again that unfavorable seismic characteristics arise in a poorly balanced structural system. The seismic retrofit should, then, focus on removing irregularities and discontinuities.

TABLE 6.11 Acceptance Criteria for Reinforced Concrete Beams

Conditions			m-Factors[a]		
			Performance level (primary components)		
			IO	LS	CP
i. Beams controlled by flexure[b]					
$\dfrac{\rho - \rho'}{\rho_{bal}}$	Trans. reinf.[c]	$\dfrac{V}{b_w d \sqrt{f_c'}}$			
≤0.0	C	≤3	3	6	7
≤0.0	C	≥6	2	3	4
≥0.5	C	≤3	2	3	4
≥0.5	C	≥6	2	2	3
≤0.0	NC	≤3	2	3	4
≤0.0	NC	≥6	1.25	2	3
≥0.5	NC	≤3	2	3	3
≥0.5	NC	≥6	1.25	2	2
ii. Beams controlled by shear[b]					
Stirrup spacing ≤ d/2			1.25	1.5	1.75
Stirrup spacing > d/2			1.25	1.5	1.75
iii. Beams controlled by inadequate development or splicing along the span[b]					
Stirrup spacing ≤ d/2			1.25	1.5	1.75
Stirrup spacing > d/2			1.25	1.5	1.75
iv. Beams controlled by inadequate embedment into beam-column joint[b]					
			2	2	3

[a] Linear interpolation between values listed in the table shall be permitted.
[b] When more than one of the conditions i, ii, iii, and iv occurs for a given component, use the minimum appropriate numerical value from the table.
[c] "C" and "NC" are abbreviations for conforming and nonconforming transverse reinforcement. A component is conforming if, within the flexural plastic hinge region, hoops are spaced at ≤ $d/3$, and if, for components of moderate and high ductility demand, the strength provided by the hoops (V_s) is at least three-fourths of the design shear. Otherwise, the component is considered nonconforming.
IO = immediate occupancy; LS = life safety; CP = collapse prevention.
(From Table 6.11 in FEMA 356.)

In the evaluation and upgrading of an existing structure, it is sometimes difficult to identify an existing lateral-force-resisting system. Innovative analytical procedures and reliance on existing materials and systems that are not generally considered for new construction are required to determine the load paths and capacities of the existing structures. When an existing structure is not adequate to resist the prescribed lateral forces, strengthening of the existing lateral-force-resisting system will be required.

The selection of an appropriate strengthening technique for the upgrading of an existing building that does not comply with the acceptance criteria will depend upon the type of structrual systems in the existing building and the nature of the deficiency. In some cases, the selection may be influenced by other than structural considerations. For example, a requirement that the building be kept operational during the structural modifications may dictate that the modification be restricted to the periphery of the building. On the other

TABLE 6.12 Acceptance Criteria for Reinforced Concrete Columns

			m-Factors[a]		
			Performance level (primary components)		
Conditions			IO	LS	CP
i. Columns controlled by flexure[b]					
$\dfrac{P}{A_g f'_c}$	Trans. reinf.[c]	$\dfrac{V}{b_w d \sqrt{f'_c}}$			
≤ 0.1	C	≤ 3	2	3	4
≤ 0.1	C	≥ 6	2	2.4	3.2
≥ 0.4	C	≤ 3	1.25	2	3
≥ 0.4	C	≥ 6	1.25	1.6	2.4
≤ 0.1	NC	≤ 3	2	2	3
≤ 0.1	NC	≥ 6	2	1.6	2.4
≥ 0.4	NC	≤ 3	1.25	1.5	2
≥ 0.4	NC	≥ 6	1.25	1.5	1.75
ii. Columns controlled by shear[b,d]					
Hoop spacing $\leq d/2$			—	—	—
or $\dfrac{P}{A_g f'_c} \leq 0.1$					
Other cases			—	—	—
iii. Columns controlled by inadequate development or splicing along the clear height[b,d]					
Hoop spacing $\leq d/2$			1.25	1.5	1.75
Hoop spacing $> d/2$			—	—	—
iv. Columns with axial loads exceeding $0.70P_o$[b,d]					
Conforming hoops over the entire length			1	1	2
All other cases			—	—	—

[a] Linear interpolation between values listed in the table shall be permitted.

[b] When more than one of the conditions i, ii, iii, and iv occurs for a given component, use the minimum appropriate numerical value from the table.

[c] "C" and "NC" are abbreviations for conforming and nonconforming transverse reinforcement. A component is conforming if, within the flexural plastic hinge region, hoops are spaced at $\leq d/3$, and if, for components of moderate and high ductility demand, the strength provided by the hoops (V_s) is at least three-fourths of the design shear. Otherwise, the component is considered nonconforming.

[d] To qualify, columns must have transverse reinforcement consisting of hoops. Otherwise, actions shall be treated as force-controlled.

(From Table 6.12 in FEMA 356.)

hand, it may be possible to temporarily relocate the occupants of a building that is to be upgraded. This, of course, provides more latitude in the selection of appropriate and cost-effective strengthening techniques. In many cases, seismic upgrading is accomplished concurrently with functional alterations, renovation, and/or energy retrofits. In these cases, the selected structural modification scheme should be the one that best suits the requirements of all the proposed alterations.

Determination of the seismic capacity of a structure includes consideration of all elements, structural and nonstructural, that contribute to the resistance of lateral forces.

TABLE 6.13 Numerical Acceptance Criteria for Linear Procedures: Reinforced Concrete Beam–Column Joints

				m-Factors[a]				
				Performance level (primary components)				
						Component type		
					Primary[b]		Secondary	
Conditions			IO	LS	CP	LS	CP
Interior joints[c,d]							
$\dfrac{P}{A_g f'_c}$	Trans. reinf.[e]	$\dfrac{V}{V_n}$					
≤0.1	C	≤1.2	—	—	—	3	4
≤0.1	C	≥1.5	—	—	—	2	3
≥0.4	C	≤1.2	—	—	—	3	4
≥0.4	C	≥1.5	—	—	—	2	3
≤0.1	NC	≤1.2	—	—	—	2	3
≤0.1	NC	≥1.5	—	—	—	2	3
≥0.4	NC	≤1.2	—	—	—	2	3
≥0.4	NC	≥1.5	—	—	—	2	3
Other joints[c,d]							
$\dfrac{P}{A_g f'_c}$	Trans. Reinf.[e]	$\dfrac{V}{V_n}$					
≤0.1	C	≤1.2	—	—	—	3	4
≤0.1	C	≥1.5	—	—	—	2	3
≥0.4	C	≤1.2	—	—	—	3	4
≥0.4	C	≥1.5	—	—	—	2	3
≤0.1	NC	≤1.2	—	—	—	2	3
≤0.1	NC	≥1.5	—	—	—	2	3
≥0.4	NC	≤1.2	—	—	—	1.5	2.0
≥0.4	NC	≥1.5	—	—	—	1.5	2.0

[a] Linear interpolation between values listed in the table shall be permitted.

[b] For linear procedures, all primary joints shall be force-controlled; *m*-factors shall not apply.

[c] P is the ratio of the design axial force on the column above the joint and A_g is the gross cross-sectional area of the joint.

[d] V is the design shear force and V_n is the shear strength for the joint. The design shear force and shear strength shall be calculated according to Section 6.5.2.3.

[e] "C" and "NC" are abbreviations for conforming and nonconforming transverse reinforcements. A joint is conforming if hoops are spaced at ≤ $h_c/3$ within the joint. Otherwise, the component is considered nonconforming.

(From Table 6.13 in FEMA 356.)

Physical properties are generally obtained from available data; otherwise, assumptions and/or tests must be made. The analysis must include the evaluation of the most rigid elements resisting the initial lateral forces, as well as the more flexible elements that resist the lateral distortions after the rigid elements yield or fail. Consideration must also be given to the interaction of various combinations of the structural framing systems and elements, which will contribute to the resistance of the lateral loads.

TABLE 6.14   Acceptance Criteria for Concrete Component Members Controlled by Flexure: Shear Walls, Columns, and Coupling Beams

			m-Factors		
			Performance-level (primary components)		
Conditions			IO	LS	CP
Shear walls and wall segments					
$\dfrac{(A_s - A'_s)f_y + P}{t_w l_w f'_c}$	$\dfrac{\text{Shear}^a}{t_w l_w \sqrt{f'_c}}$	Confined Boundary[b]			
≤0.1	≤3	Yes	2	4	6
≤0.1	≥6	Yes	2	3	4
≥0.25	≤3	Yes	1.5	3	4
≥0.25	≥6	Yes	1.25	2	2.5
≤0.1	≤3	No	2	2.5	4
≤0.1	≥6	No	1.5	2	2.5
≥0.25	≤3	No	1.25	1.5	2
≥0.25	≥6	No	1.25	1.5	1.75
Columns supporting discontinuous shear walls					
Transverse reinforcement[c]					
Conforming			1	1.5	2
Nonconforming			1	1	1
Shear wall coupling beams[d]					
Longitudinal reinforcement and transverse reinforcement[e]	$\dfrac{\text{Shear}^a}{t_w l_w \sqrt{f'_c}}$				
Conventional longitudinal reinforcement with conforming transverse reinforcement	≤3 ≥6		2 1.5	4 3	6 4
Conventional longitudinal reinforcement with nonconforming transverse reinforcement	≤3 ≥6		1.5 1.2	3.5 1.8	5 2.5
Diagonal reinforcement	n.a.		2	5	7

[a] Design shear shall be calculated using limit-state analysis procedures.

[b] Requirements for a confined boundary are the same as those given in *ACI 318*.

[c] Requirements for conforming transverse reinforcement in columns are: 1) hoops over the entire length of the column at a spacing ≤d/2; and 2) strength of hoops V_s ≥ required shear strength of column.

[d] For secondary coupling beams spanning <8'-0", with bottom reinforcement continuous into the supporting walls, secondary values shall be permitted to be doubled.

[e] Conventional longitudinal reinforcement consists of top and bottom steel parallel to the longitudinal axis of the coupling beam. Conforming transverse reinforcement consists of : 1) closed stirrups over the entire length of the coupling beam at a spacing ≤ d/3; and 2) strength of closed stirrups V_s ≥ 3/4 of required shear strength of the coupling beam.

(From Table 6.20 in FEMA 356.)

TABLE 6.15 Acceptance Criteria for Concrete Component Members Controlled by Shear:
Shear Walls and Coupling Beams

Conditions		*m*-Factors		
		Performance level (primary components)		
		IO	LS	CP
Shear walls and wall segments				
All shear walls and wall segments[a]		2	2	3
Shear wall coupling beams[b]				
Longitudinal reinforcement and transverse reinforcement[c]	$\dfrac{\text{Shear}}{t_w l_w \sqrt{f_c'}}$			
Conventional longitudinal reinforcement with conforming transverse reinforcement	≤ 3	1.5	3	4
	≥ 6	1.2	2	2.5
Conventional longitudinal reinforcement with nonconforming transverse reinforcement	≤ 3	1.5	2.5	3
	≥ 6	1.25	1.2	1.5

[a] For shear walls and wall segments where inelastic behavior is governed by shear, the axial load on the member must be $\leq 0.15 A_g f_c'$, the longitudinal reinforcement must be symmetrical, and the maximum shear demand must be $\leq 6\sqrt{f_c'}$; otherwise, the shear shall be considered to be a force-controlled action.

[b] For secondary coupling beams spanning <8'-0", with bottom reinforcement continuous into the supporting walls, secondary values shall be permitted to be doubled.

[c] Conventional longitudinal reinforcement consists of top and bottom steel parallel to the longitudinal axis of the coupling beam. Conforming transverse reinforcement consists of: 1) closed stirrups over the entire length of the coupling beam at a spacing $\leq d/3$; and 2) strength of closed stirrups $V_s \geq 3/4$ of required shear strength of the coupling beam.

(From Table 6.21 in FEMA 356.)

The results of the detailed structural analysis will identify the deficiencies with respect to the acceptance criteria of the various structural components and systems. These results should be carefully reviewed in the development of alternative upgrade concepts unless justification can be shown for a single solution. Each concept should be developed to the extent that will permit a reasonable cost estimate to be made. The extent of removal of existing construction should be considered, including the sizes and locations of new, replaced, or strengthened structural members. Typical structural connections with schematic details for upgrading nonstructural elements should be included in the study.

The following general considerations should be addressed in the development of the design concepts:

- Structural systems.
- Configuration.
- Horizontal diaphragms.
- Eccentricity.
- Deformation compatibility.
- Foundations.
- Basic isolation and passive energy dissipation.

TABLE 6.16 Acceptance Criteria for Two-Way Slabs and Slab–Column Connections

| | | *m*-Factors | | |
| | | Performance level (primary components) | | |
Conditions		IO	LS	CP
i. Slabs controlled by flexure, and slab–column connections[a]				
$\dfrac{V_g}{V_o}$ [b]	Continuity reinforcement[c]			
≤0.2	Yes	2	2	3
≥0.4	Yes	1	1	1
≤0.2	No	2	2	3
≥0.4	No	1	1	1
ii. Slabs controlled by inadequate development or splicing along the span[a]				
		—	—	—
iii. Slabs controlled by inadequate embedment into slab–column joint[a]				
		2	2	3

[a] When more than one of the conditions, i, ii, and iii occurs for a given component, use the minimum appropriate numerical value from the table.

[b] V_g = the gravity shear acting on the slab critical section as defined by *ACI 318*; V_o = the direct punching shear strength as defined by *ACI 318*.

[c] Under the heading "Continuity Reinforcement," use "Yes" where at least one of the main bottom bars in each direction is effectively continuous thrthe column cage. Where the slab is post-tensioned, use "Yes" where at least one of the post-tensioning tendons in each direction passes through the column cage. Otherwise, use "No."

(From Table 6.15 in FEMA 356.)

6.6.3.8.1. Structural Systems. The development of the structural upgrading concepts requires a complete understanding of the existing vertical and lateral-load-resisting systems of the existing building. The designer must be able to determine the consequences that the removal, addition, or modification of any structural or nonstrucutral element will have on the performance of the strengthened building.

An evaluation of the existing vertical load-carrying structural system should be made to determine the effects that the seismic upgrading may have on the performance of the building to resist gravity loads. Vertical load resisting elements such as columns and framing systems may also be affected by seismic upgrading. If these framing elements are not used for the lateral-force-resisting system, they must be analyzed for deformation compatibility. This analysis should include the effects of the lateral displacements due to extreme seismic motion on the vertical load-carrying capacity of the vertical structural elements.

6.6.3.8.2. Configuration. Severe problems may arise if the existing building is highly irregular in plan configuration or is composed of units with incompatible seismic response characteristic. An example is a flexible steel moment frame building connected to a relatively low rigid concrete shear wall building. If the resulting problem cannot be resolved by strengthening or upgrading the connection between two units, consideration should be given to separating them with a seismic joint. Each unit should have a complete system for resisting vertical as well as lateral loads. Structural members bridging the joint with sliding

TABLE 6.17 Acceptance Criteria for Reinforced Concrete Infilled Frames

| | *m*-Factors[a] | | |
| | Performance level (primary components) | | |
Conditions	IO	LS	CP
i. Columns modeled as compression chords[b]			
Columns confined along entire length[c]	1	3	4
All other cases	1	1	1
ii. Columns modeled as tension chords[b]			
Columns with well-confined splices, or no splices	3	4	5
All other cases	1	2	2

[a] Interpolation shall not be permitted.
[b] If load reversals will result in both conditions i and ii applying to a single column, both conditions shall be checked.
[c] A column may be considered to be confined along its entire ltength when the quantity of hoops along the entire story height including the joint is equal to three-quarters of that required by *ACI 318* for boundary elements of concrete shear walls. The maximum longitudinal spacing of sets of hoops shall not exceed either $h/3$ or $8d_b$.
IO = immediate occupancy, LS = life safety, CP = collapse prevention
(From Table 6.17 in FEMA 356.)

supports on the adjacent unit should be avoided. The criteria for new building separations apply to existing buildings. Seismic joints should provide for the three-dimensional uncoupled response of each of the separate units of a building, but need not extend through the foundations.

6.6.3.8.3. Horizontal Diaphragms. In most buildings, the horizontal framing systems (i.e., floors and roofs) will participate in the lateral-force-resisting system as diaphragms in addition to supporting the gravity loads. As part of the seismic upgrade, the floor and roof systems may require modifications (e.g., new topping or horizontal bracing), which will add to the dead load; thus, the capacity of the modified system must be evaluated for the new loading conditions. Every upgraded building should have either a rigid or a semirigid horizontal floor diaphragm. Roof diaphragms may be flexible or semiflexible.

6.6.3.8.4. Eccentricity. Provisions should be made for the increase in shear resulting from the horizontal torisonal moment due to an eccentricity between the center of mass and the center of rigidity. In the development of upgrading concepts, when the vertical shear-resisting elements must be strengthened, supplemented, or replaced with new elements, consideration should be given to location of new or strengthened elements so as to reduce eccentricity between the center of rigidity and the center of mass.

6.6.3.8.5. Deformation Compatibility. The compatibility of the deformation characteristics of existing elements and the new strengthening elements should be considered. When lateral forces are applied to a building, they will be resisted by the elements in proportion to their relative rigidities. If the structure is to be strengthened to resist seismic forces, the new structural elements must be more rigid than the existing elements if they are to take a major portion of the lateral forces and reduce the amount of force that is taken by the existing elements. Both the relative rigidities and strengths of all lateral-force-resisting elements must be considered.

Special consideration must be given in determination of relative rigidities of: 1) concrete components: cracked versus uncracked; 2) shear walls: participation of intersecting walls (e.g., effective flange widths) and the effects of openings; and 3) steel frames: participation of concrete floor slab and framing, and infill walls. Structural elements that are not part of the lateral-force-resisting system should be evaluated for the effects of the deformation that occurs in the lateral-force-resisting system. Brittle elements are particularly susceptible to damage if they are forced to conform to the deformations of the lateral-force-resisting system. In order to protect these elements from the possibility of being subjected to large distortions, provisions should be made to allow the structural system to distort without forcing distortion on the brittle elements. A good example is the isolation of a masonry wall from the slab soffit. When rigid walls are locked in between columns, a similar method of isolation may be required at each end of the wall.

6.6.3.8.6.   Foundations.   If the seismic upgrade adds weight or redistributes the gravity loads, the foundations must be analyzed for the additional gravity loads combined with the horizontal and overturning forces associated with the seismic lateral force. Existing foundation ties that do not provide for adequate load transfer must be strengthened or replaced, unless proper justification can be provided for waiving the deficiency.

6.6.3.8.7.   Base Isolation.   Design strategies that significantly modify the dynamic response of a structure at or near the ground level are generically termed base isolation. This is usually achieved by introduction of additional flexibility at the base of the structure. The objective is to force the entire superstructure to respond to vibratory ground motion as a rigid body with a new fundamental mode based on the stiffness of the isolation devices. This strategy is particularly effective for short buildings (i.e., buildings with a fundamental mode less than about 1 sec). For these buildings, it is feasible with the isolation devices to develop a new fundamental mode with a period of about 2 to 3 sec. For most sites (e.g., those with a predominant site period less than 1 sec), the new fundamental mode period will occur beyond the portion of the response spectrum that is subject to dynamic amplification, and the response of the structures will be greatly reduced.

A typical base isolation installation consists of large pads of natural or synthetic rubber layers bonded to steel plates in a sandwich assembly or sliding bearings with either a flat or a single curvature spherical sliding surface made of polytetra fluordethylene (PTFE) or PTFE-based composites in contact with polished stainless steel. The isolator assembly, as well as all connecting elements and building services, must be capable of resisting the design spectral displacement corresponding to the new fundamental mode (some installations have base isolation assemblies that can deflect elastically up to 24 in.). Certain base isolation assemblies may have a lead core or other device to increase damping and thus decrease the response at the isolator. Because of the uncertainties associated with ground motion predictions, seismic base isolators are designed with fall-safe provisions to arrest the motion of the building to development of instability due to excessive displacement of the isolator. Base isolation can be an effective strategy to reduce the seismic response of a building, provided careful consideration is given to the amplitude and frequency content of the expected ground motion, the design of the pipes and conduits providing services to accommodate the expected displacements, and provision of fail-safe mechanisms as described above. The ability of base isolation to reduce seismic response is even more attractive in application to existing buildings with inadequate seismic resistance.

However, in addition to the considerations just described, installation of base isolation in an existing building entails accurate determination of the magnitude and location of the vertical loads, a rigid diaphragm above the isolators to collect and distribute the lateral loads, and careful underpinning and jacking of the existing structure in order to effect a systemic transfer of the existing foundation loads to the base isolation device.

6.6.3.8.8.   Passive Energy Dissipation.   An effective means of providing substantial damping is through hysteretic energy dissipation. Some structures, for example, properly designed ductile steel and concrete frames, exhibit additional damping and reduced dynamic response as a result of the limited yielding of structural steel or concrete reinforcement.

In addition to the damping inherent in a ductile structure, passive energy-dissipating systems designed to increase structural damping have been in use for some time. This is an emerging technology that provides an alternate approach to conventional stiffening and strengthening schemes. The primary use of energy-dissipation devices is to reduce earthquake displacements in structure. These devices will also reduce the force in the structure, provided the structure is responding elastically, but would not be expected to reduce force in structures that are responding beyond yield.

Further discussion of base isolation and passive energy dissipation techniques is found in Chapter 8.

6.6.3.8.9.   Conclusion.   Before concluding this section, perhaps it is beneficial to reflect on some of the performance characteristics offered in FEMA 356, particularly those at the top-of-the line performance levels. It is the opinion of many engineers that building performance at the high end of the scale cannot be promised or achieved with 100% certainty. It promises to deliver performance that exceeds the code minimum, and top-of-the-line-peformance implies a near-perfect earthquake-proof building. Therefore, building owners and the public are likely to ask for it more frequently. Ask they should, but with the understanding that there is no such thing as earthquake-proof buildings, only earthquake-resistant buildings. It is therefore the structural engineers' responsibility to make this fact clear to the owners and to the public so that their expectations for building performance do not exceed those implied in the FEMA performance definition. Although major advances have been made in analytical capability and in the synthesizing of experimental and earthquake performance data, prediction of building performance, relative to future earthquakes is still a risky and dangerous business. Thus, seismic rehabilitation continues to challenge the very core of conventional thinking.

Seismic retrofit should be considered only if its entire cost is less than 70 to 80% of replacement cost. It should be noted that it is impossible to bring an existing structure into conformance with current code requirements. Therefore, the cost assigned to the retrofit should not exceed the expectations for its performance level.

It behooves the designer to consider more than one seismic retrofit strategy. One of these should be a conventional one. This ensures that designers are not carried away with a high-tech new approach when a more conventional retrofit strategy is more cost-effective.

Seismic retrofit design is invariably more expensive than new construction design. The extra design effort required for retrofit design should be communicated to the owner at the onset of the project. The cost of the retrofit design should be pegged to the complexity of the analysis required. Many designers do not assign sufficient design hours to projects that require NSP or NDP. The cost of developing and implementing material test recommendations should be considered at the start of the project. It is recommended that material test results be available to the designer before the design development phase is started. The design basis should already be stated and discussed with the owner and peer reviewers at the onset of the project.

It is emphasized that FEMA 356 is a "prestandard," not a code. A number of parameters used therein remain a matter of discussion and research. It must remain at the discretion of engineers to modify the parameters as they deem appropriate. Agreement on key issues must be reached with peer reviewers as the analysis progresses.

6.6.4. FEMA 356: Design Examples

6.6.4.1. *Steel Building with Moment Frames and Concentric Bracing*

Given.   An existing 5-story office building located in downtown Los Angeles, CA. The lateral system consists of a combination of moment frames and concentric braces. The building was built in 1990 and suffered damage during the 1994 Northridge earthquake. The damaged connections were repaired by using notch-tough welds for the beam bottom flanges. No seismic evaluations were made at the time of moment connection repairs.

In 2003, the building is being acquired by new owners who desire an assessment of the building's expected seismic performance before making the final deal. A structural engineer has been hired to evaluate the seismic vulnerability of the building, and, if required, to come-up with a seismic upgrade scheme. The selected design basis is that the building should be operational after a major earthquakes, i.e., the performance criterion is immediate occupancy (IO).

The engineer has selected the following components for a preliminary seismic evaluation:

1. Frame beams.
2. Frame columns.
3. Beam–column connection, including column panel zones.
4. Braces in concentric braced frames.
5. Diaphragm components.
6. Frame column-to-foundation connections.

However, for purposes of this example, compare the bending capacity of one sample frame-beam to the demand imposed by seismic and gravity loads. Assume the following:

Frame beam W30 × 116, F_y = 50 ksi,

Q_E = Action due to unreduced earthquake loads (i.e., R = 1), determined using a linear dynamic procedure (LDP). In our case, the beam action is bending; therefore, $M_E = Q_E$ = 3820 k-ft

Q_D = Dead-load action (i.e., dead load moment) $Q_D = M_D$ = 580 k-ft

Q_L = Effective live load action (equal of 25% of unreduced design live load, but not less than actual live load) $Q_L = M_L$ = 145 k-ft

Q_S = Effective snow load contribution = 0

Required.   Building seismic evaluation, and possibly an upgrade scheme that meets the requirements of FEMA 356 for enhanced performance objectives. To keep the numerical work simple, verify the acceptability of our W30 × 116 beam using the FEMA 356 procedures. Consider only the bending action.

Solution.

Step 1 is the determination of the characteristics of the ground motion likely to be experienced at the building site, because ground motions are the most common and significant cause of earthquake damage to buildings. Consequently, rehabilitation objectives are commonly established using earthquake ground shaking hazards, usually defined on a probabilistic basis. Performance characteristics that are functions of the severity of specified earthquakes are directly related to the extent of damage sustained by the building.

The owners of the building desire to determine the seismic vulnerability of the building and design a rehabilitation, if required, for an enhanced performance objective. Their intent is to have the building operational during and after the seismic events specified in FEMA 356 for the given rehabilitation objective. The engineer has, in nontechnical terms, described to the owners the broad range of expected building performance levels in terms of possible damage to both structural and nonstructural building components. Communication with the owner in lay terms is perhaps the most important step in a seismic rehabilitation study. The owners of our subject building are now well-informed about probable post-earthquake scenarios. Because the selected design is based on an operational performance level, they expect

- Overall damage to the building to be light.
- Structure to have no permanent lateral displacement.
- Structure's original strength and stiffness to remain substantially unchanged.
- Minor cracking in façade partitions and ceilings.
- Minor local yielding of structural elements at a few places, without fracture.
- Elevators and fire protection system remaining operable.
- In terms of nonstructural components, equipment and contents to be generally secure, but perhaps not operable due to mechanical failure or lack of utilities.

Prediction of building performance in a future earthquake is a dangerous and risky business. Consider for example the top-of-the-line performance as set forth in FEMA 356. It implies that a building designed for this performance level is likely to come out search-free after a high seismic event, giving the impression that the building is earthquake-proof. Structural engineers understand that buildings designed using the principle of ductile design are earthquake-resistant and not earthquake-proof. This important difference should be brought to the attention of building owners before they embark on a seismic upgrade.

Table 6.18 (FEMA 356 Table C1.1) displays in matrix format the characteristics of ground motion for three distinct earthquakes, represented by notations k + p + e. These are

TABLE 6.18 Rehabilitation Objectives[a,b]

	Target Building Performance Levels			
Earthquake Hazard Level	Operational Performance Level (1-A)	Immediate Occupancy Performance Level (1-B)	Life Safety Performance Level (3-C)	Collapse-Prevention Performance Level (5-E)
50%/50 year	a	b	c	d
20%/50 year	e	f	g	h
BSE-1 (≈10%/50 year)	i	j	k	l
BSE-2 (≈2%/50 year)	m	n	o	p

[a] Each cell in the matrix represents a discrete rehabilitation objective.
[b] The rehabilitation objectives in the matrix may be used to represent the three rehabilitation objectives defined in FEMA 356 Sections 1.4.1, 1.4.2, and 1.4.3 as follows:
    k + p = basic safety objective (BSO)
    k + p + any of a, c, i, b, f, j, or, n = enhanced objectives
    o, n, or m alone = enhanced objective
    k or p alone = limited objective
    c, e, d, h, l = limited objectives
(From Table C1.1 in FEMA 356.)

- BSE-1 earthquake with a 10% probability in 50 years (mean return period of 474 years, rounded to 500 years, used in most building codes).
- BSE-2 earthquake with a 2% probability in 50 years (mean recurrence interval of 2475 years, rounded to 2500 years).
- An earthquake with a 20% probability in 50 years (mean recurrence interval of 225 years).

For the example building, we assume that the project geotechnical engineer has, after conducting site-specific studies, developed acceleration response spectra for the earthquakes listed in the foregoing. Since the FEMA methodology is the same, we will verify the performance of the frame–beam for only the BSE-2 earthquake.

Step 2 is the determination of as-built conditions in order to arrive at a value for the reliability coefficient κ. The building is fairly new, built in 1995 after the 1994 Northridge earthquake. As-built information pertinent to its seismic performance, including construction documents and material test reports, is available, and a visual survey has indicated that there are no site-related concerns such as pounding from neighboring structures. Because of the abundance of as-built information, the engineer is able to gain a comprehensive knowledge and understanding of the behavior of structural components allowing a value of $\kappa = 1.0$ in the analyses. κ is a reliability coefficient used to reduce the component strength value for existing components. Because $\kappa = 1.0$, there is no need to reduce the computed strength values of the component when making demand capacity comparisons.

Step 3, the classification into primary and secondary components, is a straightforward task for the example building. Both frame–beams and columns are classified as primary because they are essential for providing the structure's basic lateral resistance. This step also includes the classification of the response of lateral-resisting components into either deformation-controlled or force-controlled actions. In our case it is obvious that the flexural action of the frame–beams is deformation-controlled. However, the classification of frame–columns subject to combined compression and bending is not so obvious. It could be either deformation-controlled or force-controlled, depending on the ratio of axial load in the column and its axial strength.

Step 4 entails setting up the analytical model; calculating the building period; and determining the base shear, its vertical distribution up the building height, and the forces in the floor and roof diaphragms. This task—an everyday occurrence in a design office— does not here require explanation except to point out that

1. If an LSP is used for seismic analysis, then the base shear V, also referred to as the pseudolateral load, is calculated using the unreduced spectral acceleration S_a without an upper limit on the building period, T.
2. If an LDP such as a modal superposition is used, then the analysis is carried out using a response spectrum that is not reduced to account for the anticipated inelastic response of the building (i.e., R or $R_W = 1.0$).

The purpose of an LSP or LDP is to determine the distribution of forces and deformations induced in a structure by the design ground motion. Although an LSP is permitted for simple buildings less than 100 ft in height, prevailing practice in most design offices is to use an LDP and the modal superposition method. Hence, we will assume that the forces and moments in the frame–beam have been evaluated by performing linear dynamic analyses for each of the three response spectra selected for the study.

Step 5 is where the seismic and gravity loads are combined to determine Q_{UD}, the demand imposed on elements due to seismic and gravity loads. Because the actions of both the frame–beams and frame–columns of the example building are considered deformation-controlled, we use the following equation to calculate demand:

$$Q_{UD} = Q_G + Q_E \tag{6.13}$$

If, on the other hand, the action of an element under consideration is force-controlled, the corresponding equation for verifying the acceptance criteria would have been

$$Q_{UF} = Q_G + \frac{Q_E}{C_1 C_2 C_3 J} \tag{6.14}$$

where

Q_{UF} = design action due to combinations of gravity and seismic loads
J = coefficient that estimates the maximum earthquake force that a component can sustain and deliver to other components

It is in the denominator of the second term of Eq. (6.14), related to earthquake force Q_E, that it is possible to recognize that in a nonlinear response, the actual force sustained by the component is likely to be less than earthquake force Q_E, determined by elastic analysis. The maximum value of J is 2. It is calculated by the relation

$J = 1$ in zones of low seismicity
$J = 1.5$ in zones of moderate seismicity
$J = 2$ in zones of high seismicity
C_1, C_2, and C_3 = modification coefficients explained in Section 6.6.3.4.2.

Step 6 is where the component capacities Q_{CE} or Q_{CL} are calculated, depending upon whether the action considered is deformation- or force-controlled. We will not dwell on this here, because it was explained in detail in the previous section.

Returning to the example, since the bending of the frame–beam is categorized as a deformation-controlled action, the demand Q_{UD} due to gravity and seismic is given by

$$Q_{UD} = Q_{UD} + Q_G + Q_E \tag{6.15}$$
$$Q_G = 1.1 \, (Q_D + Q_L + Q_s) \tag{6.16}$$
$$= 1.1 \, (580 + 145 + 0) = 797.5 \text{ k-ft (say, 800 k-ft)}$$
$$Q_E = 3820 \text{ k-ft}$$
$$Q_{UD} = 3820 + 800 = 4620 \text{ k-ft}$$

Step 7 is the final step, in which the acceptance criterion is verified for each component. Although we earmarked six distinct components for seismic assessment, to keep the presentation simple we will check the acceptance criterion for the frame–beam only.

In step 3 it was determined that the limit state for the example beam was flexure. Beam flexure (and for that matter, beam shear) with negligible axial loads is considered deformation-controlled action.

The design properties for the example frame–beam are as follows:

W30 × 116 $\quad$ F_y = 50 ksi, group 3, 1990 on.
$\qquad$ Default lower-bound yield strength = 52 ksi (Table 6.2)
$\qquad$ F_{ye} = expected steel strength = 1.05 × 52 = 54.6 ksi (Table 6.3)
$\qquad$ Expected capacity $Q_{CE} = ZF_{ye}$
$\qquad\qquad\qquad$ = 378 × 54.6
$\qquad\qquad\qquad$ = 20, 639 k-in
$\qquad\qquad\qquad$ = 1720 k-ft

A_s = 34.2 in.², d = 30.01 in., t_w = 0.565 in., b_f = 10.495 in., t_f = 0.85 in.

As previously discussed, the action of the beam is deformation-controlled, with $Q_{CE} >$ M_{PCE} due to lateral torsional buckling. Therefore, the m-factors given in the FEMA 356, Table 5.5 may be used directly, without calculating an equivalent value of m_e. Since we are verifying the acceptance criterion for IO performance, from Table 6.7, m is either 2 or 1.25, depending upon the ratio $b_f/2t_f$ and h/t_w.

In our case,

$$\frac{b_f}{2t_f} = \frac{10.495}{2 \times 0.35} = 6.17$$

$$\frac{52}{\sqrt{F_{ye}}} = \frac{52}{\sqrt{53.55}} = 7.10 > 6.17 \frac{b_f}{2t_f}$$

$$\frac{h}{t_w} = \frac{28.31}{0.565} = 50.1, \quad \frac{418}{\sqrt{F_{ye}}} = \frac{418}{\sqrt{53.55}} = 57.12 > 50.1$$

Therefore, the m-factors are

IO	LS	CP
2	6	8

The knowledge factor $\kappa = 1.0$ because the quality and extent of available information, as stated at the beginning of the design example, is comprehensive. Since we are verifying the component's acceptability for IO performance, $m = 2$, and

$$m\kappa Q_{CE} = 2 \times 1 \times 1720 = 3440 \text{ k-ft}$$
$$= 6 \times 1 \times 1416.7 = 8500 \text{ k-ft}$$

The demand Q_{UD} from step 6 is 4620 k-ft, which is greater than the expected capacity of 3440 k-ft, indicating noncompliance. A similar evaluation is made for other actions of the beam. The analysis is repeated for the other two earth quakes. The results are reviewed, keeping in mind that values of m are only an approximate indicator of seismic performance. Reevaluation of the building using nonlinear analysis procedures with reevaluated gravity and lateral loads is a prudent course of action before deciding on seismic rehabilitation.

The procedure for evaluating acceptance criteria is conceptually the same for other components of the moment frame, such as columns, panel zones, beam–column connections, and column-to-foundation connections.

Suppose the objective of the seismic, evaluation of our example building is basic life safety (LS), instead of immediate occupancy (IO). Does the procedure for seismic study differ from the preceding procedure for IO performance? What if the target performance is collapse prevention (CP) instead of IO?

The procedure is generally the same, irrespective of the selected target performance. The differences are in ground motion and in the values of the m-factors used in the acceptance criteria.

Consider, for example, the basic safety objective (BSO) as the target performance. A building that satisfies BSO will pose approximately the same earthquake risk for life safety traditionally considered acceptable in the United States. Therefore, a BSO study includes verification of the building's behavior for two distinctly different earthquakes:

1.   An earthquake with a 10% probability of occurrence in 50 years to satisfy LS
2.   An earthquake with a 2% probability of occurrence in 50 years to satisfy CP

These two earthquakes, defined as building safety earthquake (BSE)-1 and -2, are shown in Table 6-18.

The *m*-factors corresponding to BSO are typically larger than those for IO. For instance, *m* would equal 6, instead of 2, for the deformation-controlled steel beam studied in the illustrative example.

Collapse prevention refers to the building in the post-earthquake damage state that is on the verge of partial or total collapse but has not yet collapsed. Substantial damage to the structure has occurred, with considerable loss of stiffness and strength in the lateral-force-resisting system. However, all significant components of the gravity-load-resisting system continue to function.

If the building does not collapse, some engineers may wrongly consider that the life safety objective has been met. The life safety performance level includes a margin of safety against collapse for the lower-level earthquake. Significant risk of injury from falling structural debris may exist. It may not be practical to repair the structure and it may not be safe for reoccupancy, because aftershock activity may induce collapse.

To satisfy the limited safety objective of CP, the building must be evaluated for a single earthquake chosen from a range of specified earthquake hazard levels. Building safety earthquake-2, with a 2% probability of occurrence in 50 years, is one example. See Table 6.18 for other specified earthquakes.

The *m*-factors corresponding to CP are larger than those for LS. For instance, *m* would equal 8 instead of 6 for the beam investigated in the design example.

6.7. SUMMARY OF FEMA 356

The purpose is to predict, for a design earthquake, the force and deformation demands on the various components of the structure. The analysis allows for the evaluation of the acceptability of structural behavior (performance) through a series of demand versus capacity (D/C) checks.

FEMA 356 permits both linear and nonlinear analysis procedures, applicable for evaluation of existing construction, evaluation of rehabilitated construction, and design of new construction. It describes rehabilitation strategies, which include

- Global modifications such as
  - Increasing stiffness and strength by adding new elements.
  - Increasing damping using supplemental damping devices.
  - Isolating the structure from seismic ground motions by using seismic isolation.
  - Decreasing mass.
- Local modification of components consisting of
  - Local strengthening or weakening.
  - Jacketing.

The LSP is similar to the equivalent lateral procedure included in most building codes. However, the pseudolateral load $V = C_1 C_2 C_3 C_m S_a W$ incorporates techniques for considering the nonlinear response of individual elements and components, and is based on the unreduced spectral acceleration S_a.

The LDP may be used on either linear modal spectral analysis or linear time history analysis. In both cases, the results are modified with coefficients similar to those in the LSP. The acceptability criteria are the same as for LSP, including separating force- and displacement-controlled actions. The acceptance of performance is judged on a component action level. Each component action is defined as either deformation-controlled or force-controlled.

Permissible levels of inelastic displacement or strength demand are defined for each performance level.

In an NSP the analytical model consists of all elements having significant strength or stiffness. An analysis, commonly referred to as pushover analysis, is performed to develop the relationship between lateral forces and displacement at the roof or other convenient locations.

The elements that do not have significant lateral resistance can be designated secondary and removed from the model. Generally, a computer program with nonlinear analysis capability is used or a linear analysis with incremental loading is performed. Static lateral loads are applied incrementally and the element properties are adjusted for yielding or failure. The seismic global displacement demand is determined and deformation-controlled components are judged acceptable if their gravity plus earthquake deformation demand is less than or equal to the expected to permissible deformation capacity given in tabular form in the standard.

6.8. FIBER-REINFORCED POLYMER SYSTEMS FOR STRENGTHENING OF CONCRETE BUILDINGS

Composite materials made of fibers in a polymeric resin—also known as fiber-reinforced polymers (FRP)—have come into use as an alternative to traditional strengthening techniques such as steel plate bonding, section enlargement, and external post-tensioning. This technique has been used to strengthen many bridges and buildings around the world, and was first applied to concrete columns is Japan for providing additional confinement. The development of codes and standards for externally bonded FRP systems is ongoing in Europe, Japan, Canada, and the United States. Within the last 10 years, several documents related to the use of FRP materials in concrete structures have been published.

The FRP systems come in a variety forms including wet lay-up and precured systems. Wet lay-up systems consist of dry unidirectional or multidirectional fiber sheets impregnated with a saturating resin on-site. The saturating resin along with the compatible primer and putty is used to bond the FRP fabric to the concrete surface.

Prepregnation systems consist of uncured unidirectional or multidirectional fiber sheets or fabrics that are preimpregnated with a saturating resin in the manufacturing facility. They are bonded to the concrete surface with or without an additional resin application, depending upon specific system requirements.

Precured systems consist of a wide variety of manufactured composite shapes. The precured shapes are typically bonded to the concrete surface by an adhesive along with a primer and putty. There are three common types of precured systems:

- Unidirectional laminate sheets typically delivered to the site as thin ribbon strips coiled on a roll.
- Multidirectional grids, also typically delivered to the site coiled on a roll.
- Shell segments cut longitudinally so they can be opened and fitted around columns, beams, or other components of buildings.

6.8.1. Mechanical Properties and Behavior

Unlike steel reinforcement, FRP materials do not exhibit plastic behavior when loaded in tension. The stress-strain relationship is linearly elastic until failure, which is sudden and can be catastrophic.

The tensile property of the FRP material is governed by the type of fiber and its orientation and quantity. The tensile property of an FRP system should be characterized as a composite, based on not just the material properties of the individual fibers, but also on the efficiency of the fiber–resin system, the fabric design, and the method used to create the composite. The mechanical properties should be based on the testing of laminate samples with a known fiber content.

Externally bonded FRP systems should not be used as compression reinforcement. There has been very little testing to validate their use in resisting compressive forces. The failure mode for FRP laminates subjected to longitudinal compression can include transverse tensile, fiber micro-buckling, or shear failure.

The FRP materials subject to a constant load over time suddenly fail after a period referred to as endurance time, also referred to as creep–rupture. In general, carbon fibers are the least susceptible to creep–rupture, aramid fibers are moderately susceptible, and glass fibers are most susceptible.

Many FRP systems exhibit reduced mechanical properties after exposure to certain environmental factors, including temperature, humidity, and chemicals. The tensile properties reported by the manufacturers are based on tests conducted in a laboratory and do no reflect the effects of environmental exposure. Therefore, the properties should be adjusted to account for the anticipated service environment.

6.8.2. Design Philosophy

The design of FRP systems is based on traditional reinforced concrete design principles. The FRP strengthening systems are designed to resist tensile forces while maintaining strain compatibility with the concrete substrate. Unlike mild steel reinforcement, FRP systems should not be relied on to resist compressive forces. However, it is permissible for the FRP tension reinforcement to experience compression due to changes in moment patterns or moment reversals, as in members subjected to seismic forces, with the proviso that the compressive strength of the FRP system is neglected in calculating the member capacities.

In FRP design, certain limits are imposed to guard against collapse of the structure, should bond or some other type of failure occur due to vandalism, fire, or other causes. The designer is referred to ACI committee 440 recommendations for further details.

6.8.3. Flexural Design

An increase in the flexural strength of a concrete member can be achieved by bonding FRP reinforcement to its tension face with fibers oriented along the member's length. Although higher-strength increases are reported in test results, an increase of up to 40% of the original strength is considered reasonable in view of ductility and serviceability limits.

Flexural strengthening using FRP systems is not recommended for enhancing flexural capacity of members in the expected plastic regions. Cases in point are the plastic hinge regions of ductile moment frames resisting seismic loads. For such cases, the effect of cyclic load reversal on the FRP system should be investigated.

6.9. SEISMIC STRENGTHENING DETAILS

A thorough understanding of existing construction and seismic retrofit objectives acceptable to owners and to the building official is an important consideration before a seismic retrofit is undertaken. The importance of considering global and elemental deformations at expected levels of seismic forces, not at code or design levels, cannot be overstressed. This is because even with the use of amplification factors, the deformations are at best an approximation, particularly when applied to complex multistory and multidegree-of-freedom systems. It should be kept in mind that detailing in existing buildings often does not meet the requirements of new construction, and that the strength and stiffness of existing elements may not be comparable with new upgraded systems and elements. Thus, verification of elements for deformation compatibility becomes even more important. This criterion is secondary only to the requirement of providing a continuous load path that is sufficiently stiff and strong to resist realistic earthquake forces. Suggested rehabilitation measures listed by deficiencies are given in subsequent paragraphs.

1. Load path.
    Add elements to complete the load path. This may require adding new shear walls or frames to fill gaps in existing shear walls or frames that are not continued to the foundation. It also may require the addition of elements throughout the building to pick up loads from diaphragms that have no path into existing vertical elements.

2. Redundancy.
    Add new lateral-force-resisting elements in locations where the failure of a single element will cause an instability in the building. The added lateral-force-resisting elements should be of comparable stiffness as the elements they are supplementing.

3. Vertical irregularities.
    Provide new vertical lateral-force-resisting elements to eliminate vertical irregularity. For weak stories, soft stories, and vertical discontinuities, add new elements of the existing type.

4. Plan irregularities.
    Add lateral-force-resisting bracing elements that will support major diaphragm segments in a balanced manner. Verify whether it is possible to allow the irregularity to remain and instead strengthen those structural elements that are overstressed.

5. Adjacent buildings.
    Add braced frames or shear walls to one or both buildings to reduce the expected drifts to acceptable levels. With separate structures in a single building complex, it may be possible to tie them together structurally to force them to respond as a single unit. The relative stiffness of each and the resulting force interactions must be determined to ensure that additional deficiencies are not created. Pounding can also be eliminated by demolishing a portion of one building to increase the separation.

6. Lateral load path at pile caps.
    Typically, deficiencies in the load path at the pile caps are not a life safety concern. However, if it is determined that there is a strong possibility of a life safety hazard, piles and pile caps may be modified, supplemented, repaired, or, in the most severe condition, replaced in their entirety.

7. Deflection compatibility.

   Add vertical lateral-force-resisting elements to decrease the drift demand on the columns, or increase ductility of the columns. Jacketing the columns with steel or concrete is one way to increase their ductility.

8. Drift.

   The most direct mitigation approach is to add properly placed and distributed stiffening elements—new moment frames, braced frames, or shear walls—that can reduce the interstory drifts to acceptable levels. Alternatively, the addition of energy dissipation devices to the system may reduce the drift.

9. Noncompact members.

   Noncompact members can be made compact by adding steel plates. Lateral bracing can be added to reduce member unbraced lengths. Stiffening elements (e.g., braced frames, shear walls, or additional moment frames) can be added throughout the building to reduce the expected frame demands.

10. Strong column–weak beam.

    Steel plates can be added, or a steel column can be made composite by enclosing it with reinforced concrete, to increase the strength of the steel columns beyond that of the beams to eliminate this issue. Stiffening elements can be added to reduce the expected frame demands.

11. Connections.

    Add a stiffer lateral-force-resisting system to reduce the expected rotation demands. Connections can be modified by adding flange cover plates, vertical ribs, haunches, or brackets, or by removing beam flange material to initiate yielding away from the connection location (e.g., via a pattern of drilled holes or the cutting out of flange material). Partial penetration splices, which may become more vulnerable for conditions where the beam–column connections are modified to be more ductile, can be modified by adding plates and/or welds. Moment-resisting connection capacity can be increased by adding cover plates or haunches.

12. Frame and nonductile concerns.

    Add properly placed and distributed stiffening elements, such as shear walls, to supplement the moment frame system with a new lateral-force-resisting system. For eccentric joints, columns and beams may be jacketed to reduce the effective eccentricity. Jackets may be also be provided for shear critical columns.

    • Short captive columns.

      Columns may be jacketed with steel or concrete such that they can resist the expected forces and drifts. Alternatively, the expected story drifts can be reduced throughout the building by infilling openings or adding shear walls.

13. Cast-in-place concrete shear walls.

    • Shear stress.

      Add new shear walls and/or strengthen the existing walls to satisfy seismic demand criteria. New and strengthened walls must form a complete, balanced, and properly detailed lateral-force-resisting system for the building. Special care is needed to ensure that the connection of the new walls to the existing diaphragm is appropriate and of sufficient strength such that yielding will occur in the wall first. All shear walls must have sufficient shear and overturning resistance.

    • Overturning.

      Lengthening or adding shear walls can reduce overturning demand.

- Coupling beams.
  Strengthen the walls to eliminate the need to rely on the coupling beam. The beam should be jacketed only as a means of controlling debris. If possible, the existing opening should be infilled.
- Boundary component detailing.
  Splices may be improved by welding bars together after exposing them. The shear transfer mechanism can be improved by adding steel studs and jacketing the boundary components.

14. Steel Braced Frames
    - **System deficiency.** If the strength of the braced frames is inadequate, braced bays or shear wall panels can be added. The resulting lateral-force-resisting system must form a well-balanced system of braced frames that do not fail at their joints and are properly connected to the floor diaphragms, and whose failure mode is yielding of the braces rather than overturning.
    - **Stiffness of diagonals.** Diagonals with inadequate stiffness should be strengthened using supplemental steel plates, or replaced with a larger or different type of section. Global stiffness can be increased by the addition of braced bays or shear wall panels.
    - **Chevron or K-bracing.** Columns or horizontal girts can be added to support the tension brace when the compression brace buckles, or the bracing can be revised to another system throughout the building. The beam elements can be strengthened with cover plates to provide them with the capacity to fully develop the unbalanced forces created by tension brace yielding.
    - **Braced-frame connections**. Column splices or other braced-frame connections can be strengthened by adding plates and welds to ensure that they are strong enough to develop the connected members. Connection eccentricities that reduce member capacities can be eliminated, or the members can be strengthened to the required level by the addition of properly placed plates. Demand on the existing elements can be reduced by adding braced bays or shear wall panels.

6.9.1. Common Strategies for Seismic Strengthening

Techniques for strengthening or upgrading existing buildings will vary according to the nature and extent of the deficiencies, the configuration of the structural systems, and the structural materials used in construction. The following Figs. 6.11 through 6.25 show the seismic upgrading of typical structural members or systems. They provide guidelines for the application of designers' judgment and ingenuity in addressing specific situations. Many of the details shown are adapted from technical manual TM 5-809-10-12 published by the Departments of the U.S. Army, Navy, and Air Force.

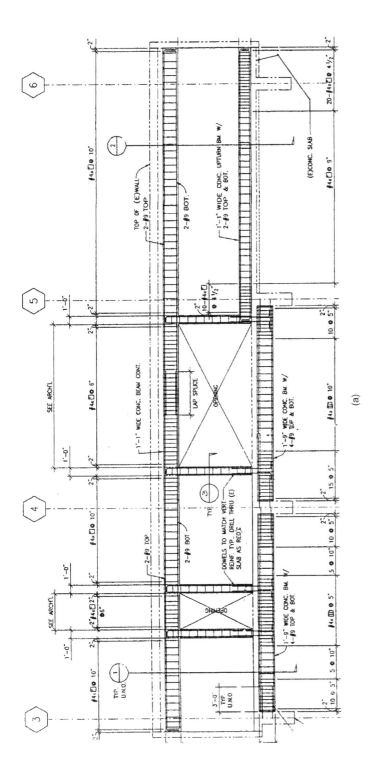

(a)

Figure 6.11. (N) openings in an (E) 3-story concrete shear wall building. The seismic upgrade consisted of providing concrete overlay to restore shear capacity of walls and adding boundary elements around (N) openings: (a) wall elevation; (b) concrete overlay with (N) beam below (E) slab; (c) concrete overlay with (N) beam above (E) slab; (d) plan detail at (N) boundary element.

Note: (E) = existing

(N) = new

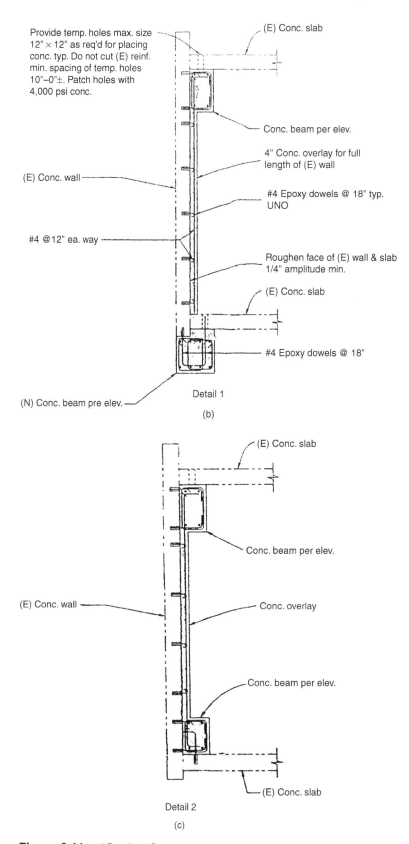

Provide temp. holes max. size 12" × 12" as req'd for placing conc. typ. Do not cut (E) reinf. min. spacing of temp. holes 10"–0"±. Patch holes with 4,000 psi conc.

(E) Conc. slab

Conc. beam per elev.

4" Conc. overlay for full length of (E) wall

(E) Conc. wall

#4 Epoxy dowels @ 18" typ. UNO

#4 @12" ea. way

Roughen face of (E) wall & slab 1/4" amplitude min.

(E) Conc. slab

#4 Epoxy dowels @ 18"

(N) Conc. beam pre elev.

Detail 1

(b)

(E) Conc. slab

Conc. beam per elev.

(E) Conc. wall

Conc. overlay

Conc. beam per elev.

(E) Conc. slab

Detail 2

(c)

Figure 6.11.   *(Continued)*

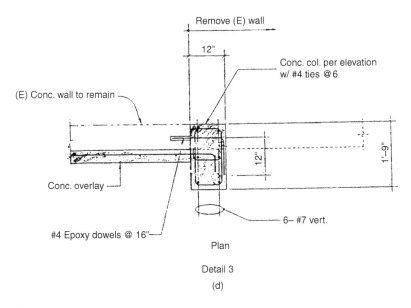

Detail 3

(d)

Figure 6.11. (*Continued*)

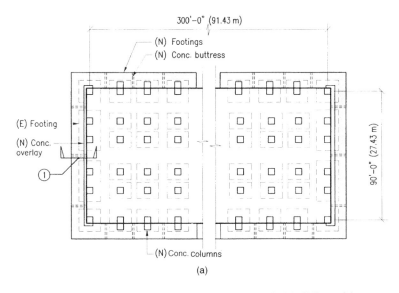

(a)

Figure 6.12. Seismic upgrade of a concrete hospital building with an external concrete moment frame. Modifications are restricted to the periphery of the building to keep the building operational with minimal interference to its functionality. (a) plan showing (N) foundations, (N) concrete overlay in the transverse direction, and (N) moment frames in the longitudinal direction.

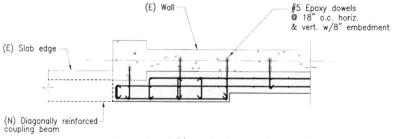

Figure 6.12b. Enlarged plan at (N) coupling beam and shear wall overlay.

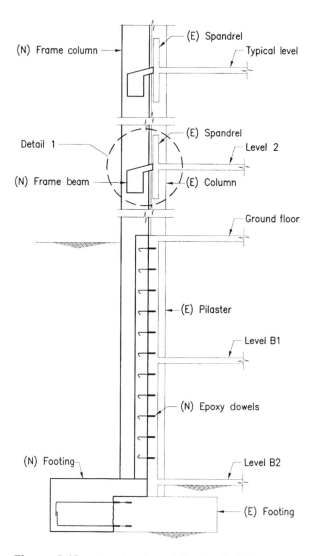

Figure 6.12c. Section through longitudinal frame.

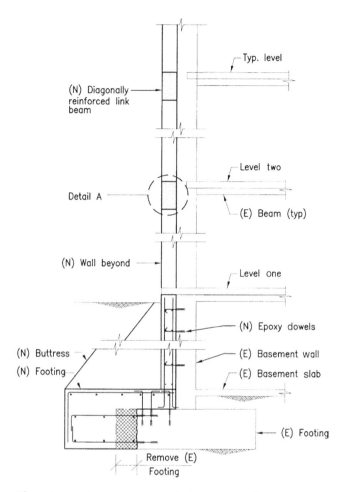

Figure 6.12d. Section through transverse wall.

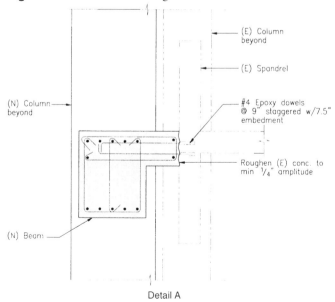

Detail A

Figure 6.12e. Connection between (N) and (E) frame.

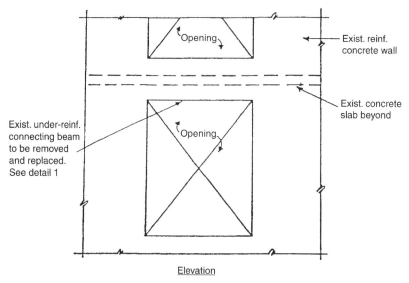

Exist. reinf.
concrete wall

Exist. concrete
slab beyond

Exist. under-reinf.
connecting beam
to be removed
and replaced.
See detail 1

Opening

Opening

<u>Elevation</u>

Existing connecting beam in concrete shear wall

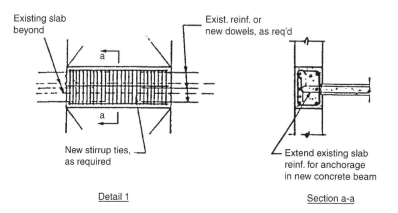

Existing slab
beyond

Exist. reinf. or
new dowels, as req'd

New stirrup ties,
as required

Extend existing slab
reinf. for anchorage
in new concrete beam

<u>Detail 1</u>   <u>Section a-a</u>

Figure 6.13.   Strengthening of existing connecting beams in reinforced concrete walls.

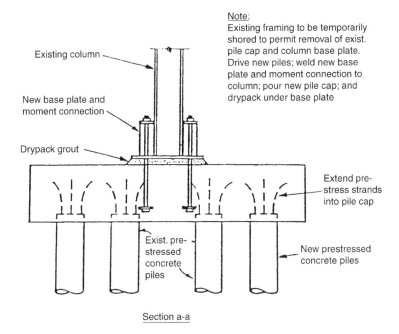

Note:
Existing framing to be temporarily
shored to permit removal of exist.
pile cap and column base plate.
Drive new piles; weld new base
plate and moment connection to
column; pour new pile cap; and
drypack under base plate

Existing column

New base plate and
moment connection

Drypack grout

Extend pre-
stress strands
into pile cap

Exist. pre-
stressed
concrete
piles

New prestressed
concrete piles

Section a-a

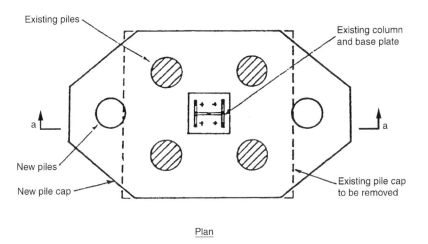

Existing piles

Existing column
and base plate

a

a

New piles

New pile cap

Existing pile cap
to be removed

Plan

Figure 6.14. Upgrading of an existing pile foundation. Add additional piles or piers, remove, replace, or enlarge existing pile caps.

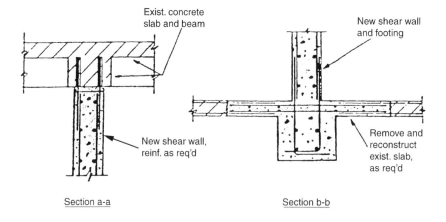

Section a-a

Section b-b

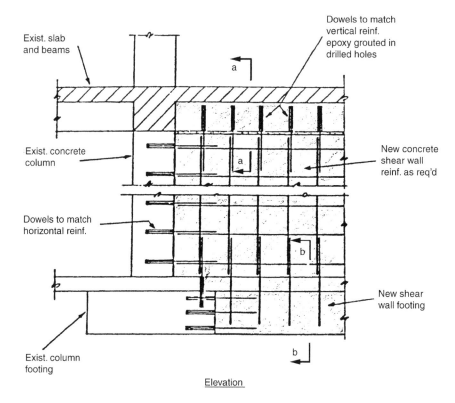

Elevation

Figure 6.15. Strengthening of an existing concrete frame building with a reinforced concrete shear wall.

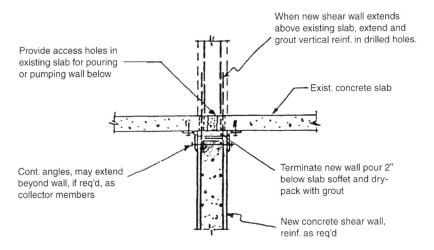

When new shear wall extends above existing slab, extend and grout vertical reinf. in drilled holes.

Provide access holes in existing slab for pouring or pumping wall below

Exist. concrete slab

Cont. angles, may extend beyond wall, if req'd, as collector members

Terminate new wall pour 2" below slab soffet and dry-pack with grout

New concrete shear wall, reinf. as req'd

Figure 6.15. (*Continued*) Section 9-9 (when new wall extends above existing slab).

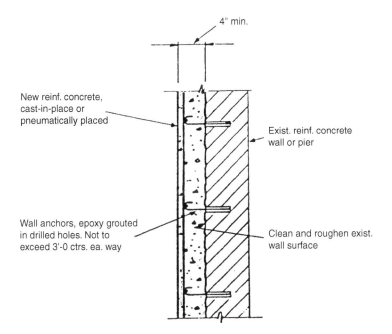

4" min.

New reinf. concrete, cast-in-place or pneumatically placed

Exist. reinf. concrete wall or pier

Wall anchors, epoxy grouted in drilled holes. Not to exceed 3'-0 ctrs. ea. way

Clean and roughen exist. wall surface

Figure 6.16. Strengthening of existing reinforced concrete walls or piers.

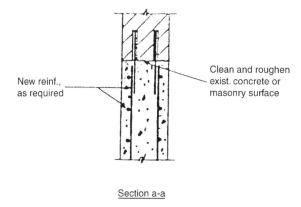

Section a-a

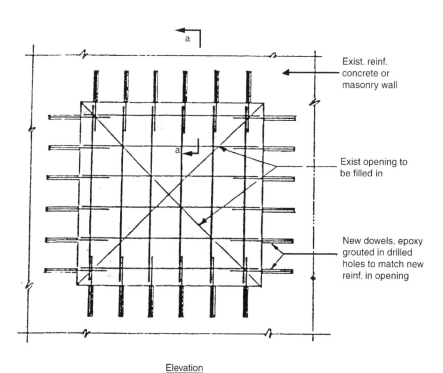

Elevation

Existing reinforced concrete
shear wall with opening to be filled in

Figure 6.17.   Strengthening of existing reinforced concrete walls by filling in of openings.

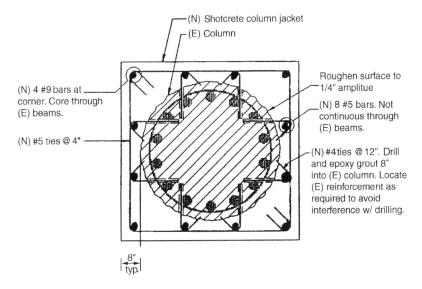

Figure 6.18. Jacketing of circular column.

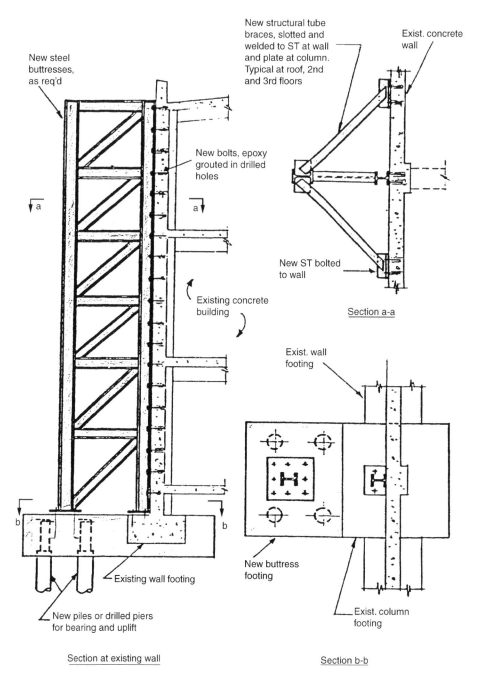

New structural tube braces, slotted and welded to ST at wall and plate at column. Typical at roof, 2nd and 3rd floors

Exist. concrete wall

New steel buttresses, as req'd

New bolts, epoxy grouted in drilled holes

Existing concrete building

New ST bolted to wall

Section a-a

Exist. wall footing

New buttress footing

Existing wall footing

New piles or drilled piers for bearing and uplift

Exist. column footing

Section at existing wall

Section b-b

Figure 6.19. Braced structural steel buttresses to strengthen an existing reinforced concrete building.

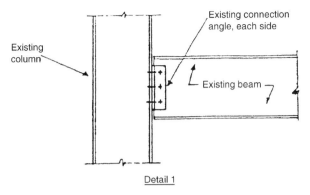

Existing connection angle, each side

Existing column

Existing beam

Detail 1

Existing simple beam connection

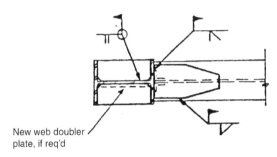

New web doubler plate, if req'd

Section a-a

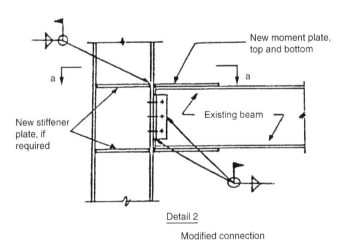

New moment plate, top and bottom

a

a

New stiffener plate, if required

Existing beam

Detail 2

Modified connection

Figure 6.20. Modification of an existing simple beam connection to a moment connection.

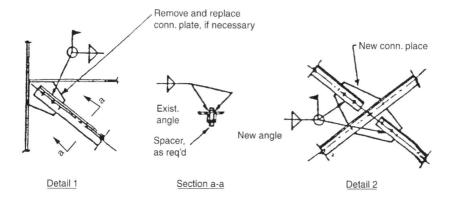

Detail 1   Section a-a   Detail 2

Upgrading of single angle bracing
to double angle bracing

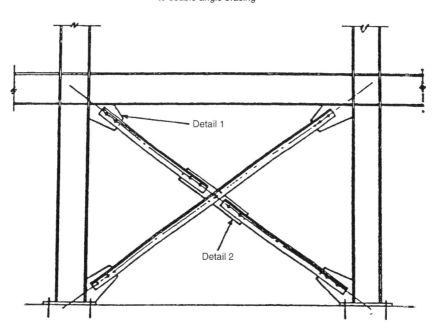

Elevation
Existing single angle bracing

Figure 6.21. Strengthening of existing bracing.

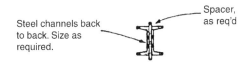

Steel channels back to back. Size as required.

Spacer, as req'd

Section a-a

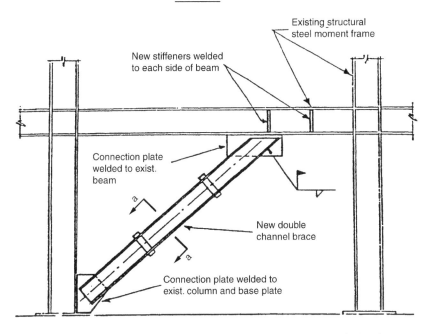

Existing structural steel moment frame

New stiffeners welded to each side of beam

Connection plate welded to exist. beam

New double channel brace

Connection plate welded to exist. column and base plate

Figure 6.22. Strengthening of an existing building with eccentric bracing.

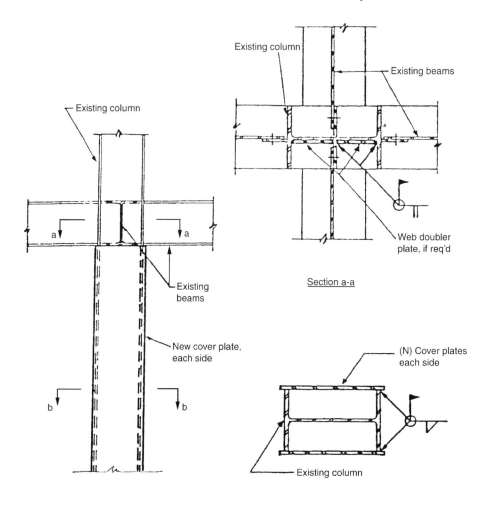

Figure 6.23. Strengthening of existing columns.

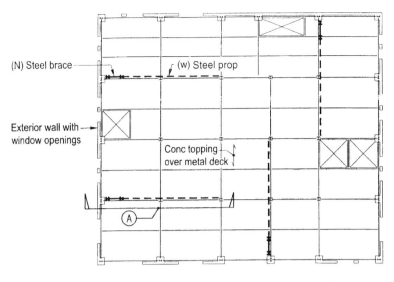

Building plan showing locations of (N) steel props

(a)

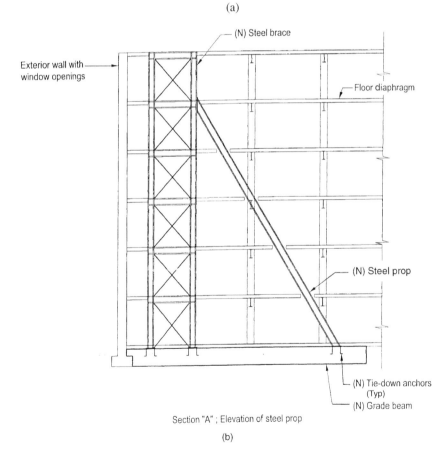

Section "A" ; Elevation of steel prop

(b)

Figure 6.24. (a) Building plan showing location of (N) steel props; (b) section A; elevation of (N) steel prop.

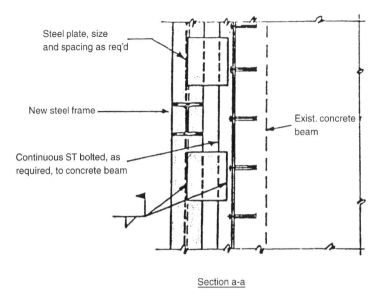

Steel plate, size
and spacing as req'd

New steel frame

Continuous ST bolted, as
required, to concrete beam

Exist. concrete
beam

Section a-a

Continuous
steel ST

Steel plate, spaced
as required for shear
transfer

Exist. concrete
frame building

New ductile moment-
resisting steel frame

Exist. column
footing beyond

New column footing

Section at existing wall

Figure 6.25. Upgrading an existing building with external frames.

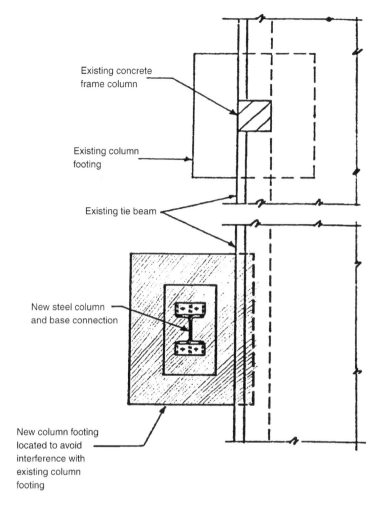

Existing concrete frame column

Existing column footing

Existing tie beam

New steel column and base connection

New column footing located to avoid interference with existing column footing

Section b-b

Figure 6.25. *(Continued)*

MTA RTD Gateway Plaza
Los Angeles, California

John A. Martin & Associates, Inc.
Structural Engineers
Los Angeles, CA

7
Gravity Systems

7.1. STRUCTURAL STEEL

There are basically three groups of structural steel available for use in bridges and buildings:

1. Carbon steel: American Society for Testing and Materials (ASTM) A36 and A529.
2. High-strength, low-alloy steels: ASTM-A440, A441, A572, and A588.
3. High-strength, treated, low-alloy steels: ASTM-A514.

In the A572 category, six grades of steel—40, 42, 45, 50, 60, and 65—are available for structural use. The grade numbers correspond to the minimum yield point in ksi, kilo pounds (kilos) per square inch of the specified steel. Carbon steel is available in 36 and 42 grades, whereas the A514 group consists of tempered steel with specified yield points ranging from 90 to 100 ksi (thousands of lbs per sq in.) (620.5 to 689.5 MPa).

The most commonly available type and grade of structural steel shapes in stock is grade 36, comprising approximately 75% of U.S. production of structural shapes. Until about a decade ago, it was a routine and economical choice because of availability for early delivery and maximum competition among bidders. Although low-alloy, high-strength steels are available with yeild points ranging from 40 to 65 ksi (276 to 448 MPa), the most common choice for high-strength steel is ASTM-A572, grade 50.

A relatively new type of steel is the ASTM A992. This has both minimum and maximum specified yield points, at 50 ksi and 65 ksi. It is routinely used in seismic designs requiring steel shapes with a specified upper limit on yield strength.

In the North American construction market, it is becoming more common for steel mills to sell dual certification ASTM A36, A572 steel meeting the requirements for both A36 and A572, grade 50. Since the cost of grade 50 is typically competitive with the cost of A36 steel, designers are able to take advantage of higher strength with little or no cost premium.

Steel buildings in the United States are designed per American Institute of Steel Construction (AISC) specifications, which were first published in 1923. The specifications are revised periodically to keep pace with new research findings and the availability of new materials. Steel construction for buildings is commonly referred to as steel skeleton framing, signifying that a majority of the members consist of linear structural elements such as beams and columns.

Skeleton framing is normally erected in 2-story increments, each increment being called a tier. Light-gauge steel decking, which serves as a permanent form and as positive reinforcement for concrete topping, is the most common method of slab construction.

The rules for the design of structural steel members subject to any one or a combination of stress conditions due to bending, shear, axial tension, axial compression, and web crippling are given in the AISC specifications. Members can be designed by the allowable stress method (ASD) or by the load resistance factor design (LRFD).

The functional needs of occupancy invariably dictate that floors be relatively flat. In a steel building this is most often achieved by horizontal subsystems consisting of beams, girders, spandrels, and trusses over which spans a light-gauge metal deck. Concrete topping over the metal deck completes the floor system.

7.1.1. Tension Members

Although any type of steel cross section may be used as a tension member, in practice the selection is usually influenced by the type of connections at the ends. The allowable stress in tension F_t is based on both the yield criteria over the gross section and the fracture criteria based on effective net area. Thus

$$F_t = 0.6F_y \qquad \text{(yield criteria, based on gross area)}$$

$$F_t = 0.5F_u \qquad \text{(fracture criteria, based on net area)}$$

For pin-connected members, the allowable tension is given by

$$F_t = 0.45F_y \qquad \text{(across the pin hole)}$$

$$F_t = 0.6F_y \qquad \text{(across the body of eye bar)}$$

In the above equations, F_y is the specified minimum yield stress of the type of steel being used, and F_u is the minimum tensile strength, both in the units of kips per square inch. A member selected for an axial tension T thus requires a gross area

$$A_g \geq \frac{T}{0.6F_y}$$

Alternatively, when applied to the effective net area,

$$A_e \geq \frac{T}{0.5F_u}$$

A_g is the gross area of the section and A_e is the effective net area, both as defined in the AISC. If end connections are welded and the area not otherwise reduced, $A_g = A_e$. For bolted members, determination of the reduced cross-section area A_n is required by searching for different possible chains of holes. When a chain of holes is straight and perpendicular to the member axis,

$$A_n = A_g - \phi N t$$

where

ϕ = effective diameter of the hold = $\phi_b + \frac{1}{8}$ in.: (ϕ_b is the bolt diameter)
N = number of bolts in the chain
t = thickness of material being connected

Although the slenderness ratio l/r for tension members is not critical, AISC recommends a nonmandatory limit of $l/r \leq 300$. Flexible members such as cables, round or square rods, or thin wide bars are excluded from this requirement.

7.1.1.1. Design Examples

7.1.1.1.1. Plates in Tension, Bolted Connections.

Given. A bolted joint connecting two 12"-wide × $^3/_4$"-thick plates with 1" ϕ bolts in standard holes. See Fig. 7.1a. The plate material is $F_y = 50$ ksi and $F_u = 65$ ksi. Assume block shear does not govern and that the bolts are satisfactory.

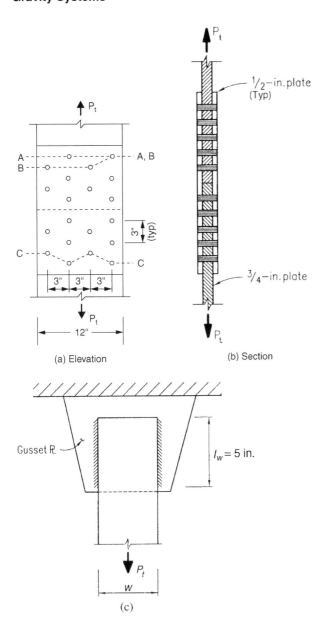

Figure 7.1. Bolted plate in tension: (a) elevation; (b) section; (c) welded plate in tension; design example.

Required. Calculate the tensile force that can be applied to the plates.
Solution.

Diameter of hole $d_h = 1 + \frac{1}{8} = 1.125$ in.

Gross area of plate $A_g = 12 \times 0.75 = 9$ in.^{2}

$0.85 A_g = A_{e(\max)} = 0.85 \times 9 = 7.65$ in.^{2}

For fracture line A-A, see Fig. 7.1a; the effective net area of the plate is

$$A_e = t(w - 2d_h) = 0.75(12 - 2 \times 1.125) = 7.31 \text{ in.}^2$$
$$< 0.85 A_g = 7.65 \text{ in.}^2 \qquad \text{OK}$$

The tensile capacity for this fracture condition along line A-A is

$$P_t = 0.5\, F_u\, A_e$$
$$= 0.5 \times 65 \times 7.31$$
$$= 237.6 \text{ kips.}$$

For the staggered fracture line B-B, the effective net area of the plate is

$$A_e = t\left(W - 3d_h + \frac{S^2}{4g} \right)$$

Noting that $S = 1.5$ in., and $g = 3$ in.,

$$A_e = 0.75\left(12 - 3 \times 1.125 + \frac{1.5^2}{4 \times 3} \right)$$
$$= 6.61 \text{ in.}^2 < 0.85A_g = 7.65 \text{ in.} \quad \text{OK}$$

Assuming equal distribution of tensile load among the given number of 10 bolts, each bolt resists 10% of the applied tensile force. Therefore, the fracture plane B-B is required to resist only 90% of the applied force. Expressed in another way, the equivalent net area along B-B is

$$A_{e(\text{equiv})} = \frac{A_e}{0.9} = \frac{6.61}{0.9} = 7.35 \text{ in.}^2$$

$$P_t = 0.5F_u - A_{e(\text{equiv})}$$
$$= 0.5 \times 65 \times 7.35 = 238.9 \text{ kips}$$

For the staggered fracture line c-c,

$$A_e = t\left(W - 4d_h + \frac{3S^2}{4g} \right)$$
$$= 0.75\left(12 - 4 \times 1.125 + \frac{3 \times 1.5^2}{4 \times 3} \right)$$
$$= 6.04 \text{ in.}^2 < A_{e(\text{max})} = 7.65 \text{ in.}^2 \quad \text{OK}$$

The tensile capacity is

$$P_t = 0.5\, F_h\, A_e$$
$$= 0.5 \times 65 \times 6.04$$
$$= 196.5 \text{ kips} \leftarrow \text{governs}$$

The tensile capacity based on yielding condition is

$$P_t = 0.6\, F_y A_g$$
$$= 0.6 \times 50 \times 9$$
$$= 270 \text{ kips}$$

Therefore, tension capacity of 196.5 kips, as calculated for fracture condition along c-c, controls the design.

7.1.1.1.2. Plate in Tension, Welded Connection.

Given. A $\tfrac{3}{4}$-in.-thick $\times$ 4-in.-wide plate welded to a gusset plate with two longitudinal welds (Fig 7.1c). The plate material is $F_y = 36$ ksi and $F_u = 58$ ksi. Assume that block shear does not control the design and that the welds are adequate.

Required. Calculate the tensile force that can be applied to the plate.

Solution. The design parameters are width of plate $W = 4$ in., thickness $t = 0.75$ in., and weld length $l = 5$ in.

$$\frac{l}{W} = \frac{5}{4} = 1.25$$

The reduction coefficient U that accounts for the effects of eccentricity and shear lag is

$U = 0.87$ (ASCE–B3)

The gross area of the plate is

$A_g = 0.75 \times 4 = 3$ in.^{2}

The effective net area is

$A_e = UA_g$
$\quad = 0.87 \times 3 = 2.61$ in.^{2} [ASCE Eq. (B3-2)]

The corresponding tensile capacity is

$P_t = 0.5 F_h A_e$
$\quad = 0.5 \times 58 \times 2.61$
$\quad = 75.69$ kips   (ASCE Sect. D1)

The tension capacity for yielding condition is

$P_t = 0.6 F_y A_g$
$\quad = 0.6 \times 36 \times 3$
$\quad = 64.80$ kips $\leftarrow$ governs

For other types of tension connections using rolled structural sections, such as W, M, and S, and angle shapes, the designer is referred to AISC section B3.

7.1.2. Members Subject to Bending

7.1.2.1. *Lateral Stability*

Consider a uniformly loaded continuous wide-flange beam as shown in Fig. 7.2. The beam segment between the points of contraflexure is subjected to positive bending with the top portion in compression throughout this region, acting in a manner similar to a column. Unless there are closely spaced restraints, i.e., $l_y \leq L_e$, the compression portion of the beam has a tendency to buckle laterally at some value of critical moment, M_{cr}, analogous to the critical load, P_{cr}, at which the column would buckle. The mode of lateral buckling is in torsion, partly due St. Venant's twisting, and partly due to warping torsion, the latter induced by the bending of beam flanges in opposite directions. Deep I-shaped open sections typically have large values for warping moment of inertia I_w. Consequently, the buckling mode for such beams is dominated by flange bending. On the other hand, for shallow I-beams, the St. Venant's torsion dominates the torsional response. These two considerations, the warping torsion and the St. Venant's torsion, lead to the two AISC equations, discussed presently for determining the allowable stress F_b for both compact and noncompact I-shaped sections with unbraced length $l_y > L_c$.

7.1.2.2. *Compact, Noncompact, and Slender Element Sections*

There are two main categories of beams, compact and noncompact. Compact beams by virtue for their special controls on their geometry are particularly stable. Compact section criteria are based on the yield strength of steel, the type of cross section, the ratios of

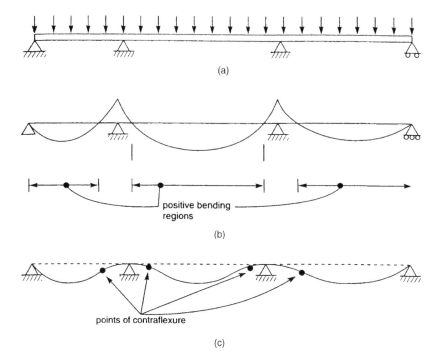

Figure 7.2. Lateral torsional buckling of beams: (a) continuous beam with uniformly distributed load; (b) bending moment diagram showing positive bending region; (c) points of contraflexure.

width to thickness of the elements of cross section. Members, whether they are rolled shapes or shapes made from plates, are compact if they fulfill the criteria. To meet all criteria, the member must meet certain stringent limits of b_f/t_f and h/t_w ratios, have unsupported length of compression flange less than L_c, and be bent about its major axis. Nonfulfillment of any of the compact criteria will degrade the member to noncompact. If the width: thickness ratios exceed the limiting width: thickness ratios of noncompact elements, then it is classified as a slender element section.

7.1.2.3. Allowable Bending Stresses

To obtain the allowable bending stress, the following criteria are used. For all I- and C-sections, the allowable major direction bending stress is computed based on the compactness criteria and the laterally unbraced compression flange length, l_y. If l_y is less than

$$\frac{76b_f}{\sqrt{F_y}} \quad \text{and} \quad \frac{20,000}{\left(\dfrac{d}{A_f}\right)F_y} \quad \text{(AISC F1-2)}$$

and the section is compact, the allowable major direction bending stress is taken as

$$F_{bx} = 0.66F_y \quad \text{(AISC F1-1)}$$

If l_y is less than the above limits and the section is noncompact,

$$F_{bx} = 0.60F_y \quad \text{(AISC F1-5)}$$

If the unbraced compression flange length l_y exceeds the above limits, then for both compact and noncompact sections, the following equations apply.

When

$$\sqrt{\frac{102 \times 10^3 C_b}{F_y}} \le \frac{l}{r_T} \le \sqrt{\frac{510 \times 10^3 C_b}{F_y}}$$

$$F_{bx} = \left[\frac{2}{3} - \frac{F_y\left(\dfrac{l}{r_T}\right)^2}{1530 \times 10^3 C_b}\right] F_y \le 0.60 F_y \qquad \text{(AISC F1-6)}$$

When

$$\frac{l}{r_T} \ge \sqrt{\frac{510 \times 10^3 C_b}{F_y}}, \qquad F_{bx} = \frac{170 \times 10^3 C_b}{\left(\dfrac{l}{r_T}\right)^2} \le 0.60 F_y \qquad \text{(AISC F1-7)}$$

For any value of l/r_T,

$$F_{bx} = \frac{12 \times 10^3 C_b}{l_y\left(\dfrac{d}{A_f}\right)} \le 0.6 F_y \qquad \text{(AISC F1-8)}$$

In the foregoing equations,

$$C_b = 1.75 + 1.05\left(\frac{M_1}{M_2}\right) + 0.3\left(\frac{M_1}{M_2}\right)^2 \le 2.3$$

where M_1 and M_2 are end moments of the unbraced segment and M_1 is less than M_2. The ratio M_1/M_2 is positive for double curvature bending and negative for single curvature bending. Also, if any moment within the segment is greater than M_2, C_b is taken as 1.0.

Figure 7.3 shows curves for F_{bx} for compact and noncompact sections. Note that there are two curves. The first (Fig. 7.3a) is based on the assumption that buckling of the unbraced compression flange is initiated by warping torsion. The second (Fig. 7.3c) shows F_{bx} controlled by St. Venant's torsion.

The minor direction allowable bending stress F_{by} is taken as

$$F_{by} = 0.60 F_y \qquad \text{(AISC F2-2)}$$

except in the case of compact I-sections where it is taken as

$$F_{by} = 0.75 F_y \qquad \text{(AISC F2-1)}$$

This is because a compact I-shape bent about its minor axis does not have to satisfy the criteria for unsupported length since the major axis stiffness provides a continuous lateral support. Fox box sections and rectangular tubes, the allowable bending stress in both the major and minor directions is taken as

$$F_b = 0.66 F_y \qquad \text{(AISC F3-1)}$$

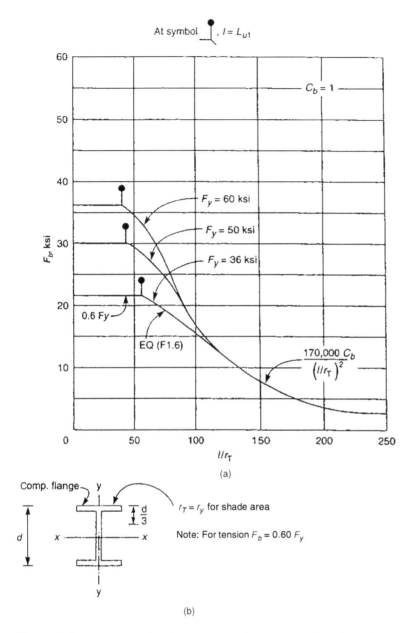

Figure 7.3. (a) Values of F_b for compact and noncompact I-sections: F_b versus l/r_t; F_b controlled by warping torsion; (b) beam section showing area for the calculation of r_T; (c) values of F_b for compact and noncompact I-sections: F_b controlled by St. Venant's torsion.

provided the section is compact and the unbraced length l_y is less than the greater of

$$\left(1950 + 1200 \frac{M_1}{M_2}\right)\frac{b}{F_y} \qquad \text{or} \qquad 1200\frac{b}{F_y}$$

where M_1 and M_2 have the same definitions as noted earlier in the formula for C_b.

If the unbraced compression flange length l_y exceeds the above limits or the section is noncompact,

$$F_b = 0.60F_y \qquad\qquad\qquad\qquad\qquad\qquad\qquad \text{(AISC F3-3)}$$

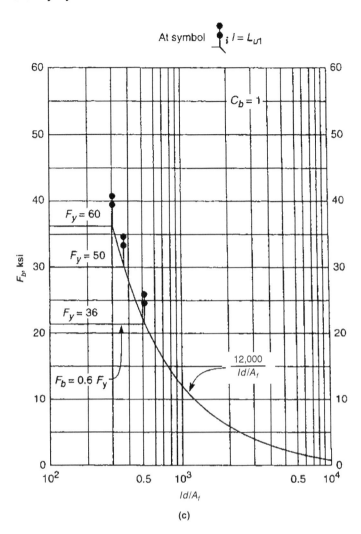

Figure 7.3. (*Continued*)

For pipe sections the allowable bending stress in all directions is taken as

$$F_b = 0.66F_y \qquad \text{(AISC F3-1)}$$

provided the section is compact, otherwise

$$F_b = 0.60F_y \qquad \text{(AISC F3-3)}$$

7.1.2.4. *Allowable Shear Stresses*

The allowable shear stress F_u is taken as $0.40F_y$ (AISC F4-1). For very slender webs, where $h/t_w > 380/\sqrt{F_y}$, a reduction in the allowable shear stress applies and must be seperately investigated (AISC F-4).

7.1.3. Members Subject to Compression

7.1.3.1. *Buckling of Columns*

In structural design, a column is considered slender if its cross-sectional dimensions are small compared to its length. The degree of slenderness is measured in terms of the ratio

l/r, where l is the unsupported length of the column and r is the radius of gyration. Whereas a stocky column fails by crushing or yielding, a slender column does so by buckling.

Before the AISC design equations are examined, a review of column behavior is useful for understanding the design parameters. Since the derivations of the column buckling formulas may be found in strength-of-material textbooks, the emphasis here is only on the column behavior as related to design.

Euler enunciated more than 200 years ago that a straight concentrically loaded pin-ended slender column fails by buckling at a critical load

$$P_c = \frac{\pi^2 EI}{l^2}$$

where E, I, and l are the familiar notations for Young's modulus, moment of inertia, and the unsupported length of the column. Dividing P_c by the cross-sectional area A of the column, the expression for the critical load may be written in terms of the critical average stress f_c on the gross section of the column

$$\frac{P_c}{A} = f_c = \frac{\pi^2 EI}{l^2 A}$$

Substituting $I = Ar^2$, where r is the radius of gyration, gives the critical stress equation

$$f_c = \frac{\pi^2 E}{(l/r)^2}$$

A plot of the critical stress versus the slenderness ratio, called a column curve, is shown in Fig. 7.4, illustrating the reduction in column strength as the slenderness increases. Stocky columns do not fail by buckling but do so by yielding or crushing of the material. There is a limiting slenderness ratio below which failure occurs by crushing, while for larger values, the mode of failure is by buckling.

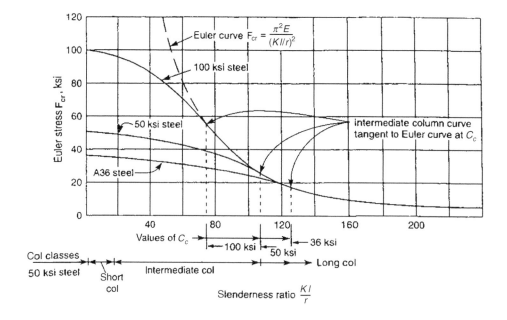

Figure 7.4.   Euler stress. P_{cr}/A versus Kl/r.

The expression for buckling load P_c is for an idealized column supported by frictionless supports, a condition that exists rarely in practice. Building columns are connected to beams which restrain column rotation, thereby inducing end moments. Aditionally, columns experience lateral deflections. Therefore, to determine the critical loads for practical cases, the idea of an effective length of column is used in design. The effective length is expressed as a product of actual length times a factor K, called the effective length factor. The critical load for practical cases is given by the relation

$$P_c = \frac{\pi^2 EI}{(Kl)^2}$$

7.1.3.2. Column Curves

To understand the performance of compression members, consider again the curves in Fig. 7.4, which show the failure stress versus the slenderness ratio Kl/r for three grades of steel. Three things are clear from the figure: the yield strength of steel is very significant for short columns, of decreasing significance through the intermediate range, and of no consequence in the performance of long columns. The most efficient use of the strength of steel is made by selecting columns in the intermediate range. To achieve large values of I and r for a given area A, a section that has the area distributed as far from its centroid as possible offers the best choice, other things being equal. The most efficient sections are those with $r_x/r_y = 1$. Of the available wide-flange sections, those with $b/d \approx 1$ are most efficient for columns.

Compression members are divided into two classes by their values of Kl/r, with the value of $C_c = \sqrt{2\pi^2 E/F_y}$ dividing the two classes. Short columns are defined by very low values of Kl/r. In this range, the Euler curve for critical load is approaching infinity. However, when the axial load becomes sufficient to cause yield stress, failure occurs by compression yielding although collapse is unlikely.

Failure for intermediate length columns is initiated by the tendency for buckling instability. The failure curve shows a smooth transition between the yield and the buckling conditions. The two curves become tangent at a value of $Kl/r = C_c$, somewhat arbitrarily chosen in the AISC specifications as $C_c = \sqrt{2\pi^2 E/F_y}$.

The allowable axial compressive stress value F_a for compact or noncompact sections is evaluated as follows:

when $\dfrac{Kl}{r} \leq C_c$

$$F_a = \frac{\left[1.0 - \dfrac{\left(\dfrac{Kl}{r}\right)^2}{2C_c^2} \right] F_y}{\dfrac{5}{3} + \dfrac{3\left(\dfrac{Kl}{r}\right)}{8C_c} - \dfrac{\left(\dfrac{Kl}{r}\right)^3}{8C_c^3}} \qquad \text{(AISC E2-1)}$$

where Kl/r is the larger of $K_x l_x/r_x$ and $K_y l_y/r_y$, and $C_c = \sqrt{(2\pi^2 E)/F_y}$. Otherwise, if $Kl/r > C_c$,

$$F_a = \frac{12\pi^2 E}{23(Kl/r)^2} \qquad \text{(AISC E2-2)}$$

The denominators in Eqs. (E2-1) and (E2-2) represents the safety factor. Note that: 1) for single angles, r_z is used in place of r_x and r_y; and 2) for members in compression, Kl/r must not be greater than 200 (AISC B7).

7.1.3.3. Stability of Frames: Effective Length Concept

Stability of frames is dealt with in AISC specification primarily by using the concept of effective length Kl. Frames are classified as braced and unbraced frames in their treatment of stability. As mentioned previously, the basic column formulas work only for pin-ended compression members with no lateral movement. Therefore, an effective length factor K is used to convert real cases to basic pin-ended cases. The term Kl represents the distance between points of theoretical zero moments.

There are two unsupported lengths l_x and l_y corresponding to instability in the major and minor directions of the column respectively. These are the lengths between the support points of the column in the corresponding directions.

Typically, in building design, all floor diaphragms are assumed to be lateral support points. Therefore, the unsupported length of a column is equal to the story height associated with the level. However, if a column is disconnected from any level, the unsupported length of the column is longer than the story height. In determining the values of l_x and l_y for the beam and column elements, the designer must recognize various aspects of the structure that have an effect on these lengths, such as member connectivity and diaphragm disconnections.

It should be noted that columns may have different unsupported lengths corresponding to the major and minor directions. For example, beams framing into columns in the column major and minor directions will give lateral support in both the directions. However, if a beam frames into only one direction of the column at a level where the column has been disconnected from the diaphragm, the beam gives lateral support only in that direction.

For beams, any column, brace, or wall support is generally assumed to be the location of the vertical support to the beam in the major direction as well as the lateral support to the beam in the minor direction. For brace elements, the unsupported length is generally assumed equal to the actual element length.

There are two K-factors, K_x and K_y, associated with each column. These values correspond to instability associated with the major and minor directions of the column, respectively. The calculation of the K-factor in a particular direction involves the evaluation of the stiffness ratios, G_{top} and G_{bot}, corresponding to the top and bottom support points of the column, in the direction under consideration:

$$G_{top} = \frac{\dfrac{E_{ca}I_{ca}}{L_{ca}} + \dfrac{E_{cb}I_{cb}}{L_{cb}}}{\displaystyle\sum_{n=1}^{n_b} \frac{E_{gn}I_{gn}}{L_{gn}} \cos^2 \theta_n}$$

where
E_{ca} = modulus of elasticity of column above top lateral support point
E_{cb} = modulus of elasticity of column below top lateral support point
I_{ca} = moment of inertia of column above top lateral support point
I_{cb} = moment of inertia of column below top lateral support point
L_{ca} = unsupported length of column in direction under consideration above top lateral support point
L_{cb} = unsupported length of column in direction under consideration below top lateral support point

E_{gn} = modulus of elasticity of beam, n, at top lateral support point

L_{gn} = major moment of inertia of beam, n, at top lateral support point

n_b = number of beams that connect to the column at lateral support level

θ = angle between the column direction under consideration and the beam, n

For the K-factor calculation, the unsupported lengths are generally based on full member lengths and do not consider any rigid end offsets.

The calculation for G_{bot} is similar, as it corresponds to the bottom lateral support point. The column K-factor for the corresponding direction is then calculated by solving the following relationship for α:

$$\frac{\alpha^2 G_{top} G_{bot} - 36}{6(G_{top} + G_{bot})} = \frac{\alpha}{\tan \alpha}$$

from which $K = \dfrac{\pi}{\alpha}$

This relationship is the mathematical formulation for K-factor evaluation assuming the sides way is uninhibited. The following are some important aspects associated with the column K-factor.

Cantilever beams and beams and columns having pin ends at the joint under consideration are excluded in the calculation of the stiffness EI/L summations because they do not contribute to the rotational stiffness of the joint. A column or beam that has a pin at the far end from the joint under consideration will contribute only 50% of the calculated EI/L value. If a pin release exists at a particular end of a column, the corresponding G-value is 10.0 in both directions. If there are no beams framing into a particular direction of a column, the associated G-value will be infinity. If rotational releases exist at both ends of a column, the corresponding K-factors are equal to unity.

Observe that the foregoing procedure for the calculation of K-factor can generate artificially high K-factors under certain circumstances. For example, in Fig. 7.5a, column line C2 has no beams framing in a direction parallel to the column minor direction. Similarly, column B3, shown in Figs. 7.5b, has no beams framing into the columns major direction. The G_{top} and G_{bot} values for these columns are infinity. Such columns are considered to be laterally supported by the floor diaphragms with column K-factor of unity. Now consider the conditions shown in Fig. 7.6a and b when the beams framing into a column are slightly inclined with the column major axis, as shown in Fig. 7.6b for column line C2. The small components of the beam stiffness in the column minor direction will generate small G_{top} and G_{bot} values for the column minor direction, resulting in a large minor direction K-factor. In general, such columns are laterally supported by the floor diaphragms in minor directions and should be assigned a K-factor of unity. For braced frames, the K-factors for the beam and brace elements are generally assumed to be unity.

7.1.3.4.   *Secondary Bending: PΔ Effects*

Frame columns in buildings are in effect "beam–columns," i.e., they are subject to simultaneous bending caused by lateral loads, and axial compression due to gravity loads. Consider the column shown in Fig. 7.7a subjected to simultaneous action of axial load and moments at the ends. At any point, the total moment M can be considered as a combination of the moment M_0 due to end moments plus the addition of the moment caused by P acting at an eccentricity y (Fig. 7.7b–d). Thus, $M = M_0 + Py$. Since the deflection is maximum at midheight, the secondary moment also reaches its maximum

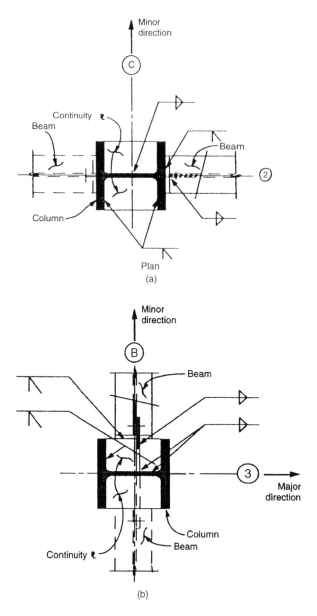

Figure 7.5. Beams framing into columns in one direction: (a) Beam framing into column flange; (b) beam framing into column web.

value at that height. A similar effect is caused when bending is produced by a lateral load as shown in Fig. 7.8. Since the deflection y and hence the magnitude of the secondary moment are functions of the end moments, a differential equation formulation is required for determining the stresses in beam–columns. Simple cases of beam–columns subjected to end moments and concentrated loads, uniformly distributed loads, etc., have been solved by differential equation techniques. In a practical structure, such a closed-form solution is extremely complicated if not impossible. Therefore, various design standards such as the ACI code and AISC specifications give provisions for approximate evaluation of the slenderness effect. The method in essence requires that the moments obtained by a so-called first-order analysis be magnified by a moment magnification factor.

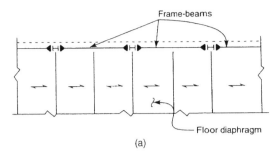

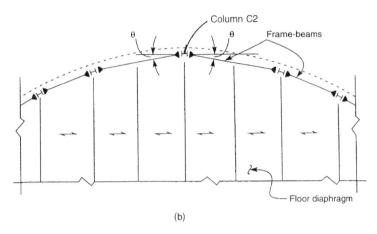

Figure 7.6. Beam framing conditions for evaluation of effective length factor, K: (a) beam framing into columns without skew; (b) skewed beamed framing into columns.

The direct addition of the maximum $P\Delta$ moment to the maximum primary moment is valid only when the beam–column is subjected to equal moments at the ends subjecting the column to bend in a single curvature. For all other cases, it represents an upper bound, giving a moment magnification factor much larger than that in a real structure. If the two end moments are unequal but of the same sign, producing single curvature, the primary

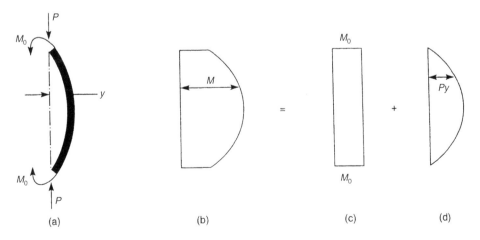

Figure 7.7. Moments in beam–columns: (a) column subjected to simultaneous axial load and bending moments; (b) combined moment diagram; (c) moment diagram due to equal end moments M_0; (d) moment due to $P\Delta$ effect.

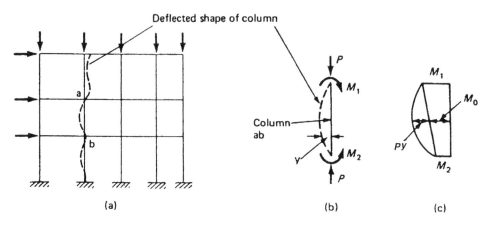

(a)   (b)   (c)

Figure 7.8. Behavior of building column: (a) building frame showing deflected shape of column; (b) column subjected to the simultaneous action of axial loads and moments; (c) moment diagram due to end moment and $P\Delta$ effect.

movement M_0 is certainly magnified but not to the same extent as when the moments are equal. If the end moments are of opposite sign, producing a reverse curvature in the column, the moment magnification effect will be very small. A moment magnification coefficient C_m is therefore used to take into account the relative magnitude and sense of the two end moments. It is given by the expression

$$C_m = 0.6 - 0.4 M_1/M_2$$

In this equation, M_1 and M_2 represent the smaller and larger end moments, respectively. The ratio M_1/M_2 is positive when the column bends in a reverse curvature and negative when the moments produce a single curvature. As can be expected, when $M_1 = M_2$, as in a column subjected to equal end moments, the value of C_m becomes equal to 1.0. The foregoing expression applies only to members braced against side sway. For columns which are part of the lateral resisting system, the maximum moment magnification occurs, i.e., $C_m = 1$, as illustrated in the following discussion.

Consider Fig. 7.9a, which shows the deflected shape of an unbraced portal frame subjected to the simultaneous action of gravity and lateral loads. Considering only the lateral

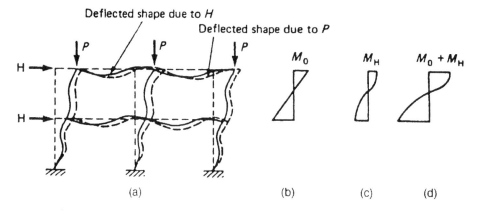

(a)   (b)   (c)   (d)

Figure 7.9. $P\Delta$ effect in laterally unbraced frames: (a) deflected shapes due to horizontal load H and vertical load P; (b) moment at column ends due to horizontal load H; (c) moment at column ends due to axial loads P; (d) combined moment diagram due to H and P. Maximum moment due to H and P occurs at the ends of columns resulting in $C_m = 1.0$.

loads, the deflection of the portal frame may be represented by solid lines as shown in Fig. 7.9a. The corresponding moments at the ends of a typical column are as shown in Fig. 7.9b.

When axial load is imposed on the deflected shape of the frame, additional sway occurs in the frame, as shown by dashed lines in Fig. 7.9a. This additional deflection imposes secondary moments in the column, as shown in Fig. 7.9c. It is seen that both the primary and secondary moments are of the same sign and have maximum values at the same locations, namely, at the two ends of the columns. They are, therefore, fully additive, as shown in Fig. 7.9d, meaning that the value of $C_m = 1$ for unbraced frames.

In American practice, for both steel and concrete buildings, the approach to the stability problem is to modify individual member design in a manner that approximately accounts for frame buckling effects. This is done by isolating a compression member together with its adjoining members at both ends and determining its critical load in terms of effective length factor K. The member is then analyzed as a beam–column by a simplified interaction equation which accounts for the moment magnification caused by the $P\Delta$ effect. Instead of frame analysis for the $P\Delta$ method, a member analysis is substituted.

We have seen earlier that using a total moment obtained by the direct addition of secondary and primary moments results in an overdesign if both these moments do not occur at the same location. The coefficient C_m in the interaction equation prevents over-design by reducing the design moment by taking into account the relative magnitude and sense of the moments occurring at the ends of columns.

Values of C_m less than 1.0 increase F_b, offsetting the effects of axial load when the shape of the elastic curve increases stability. When there is no joint translation and where the shape of the curve is not affected by transverse loading, reverse curvature bending may reduce C_m to as little as 0.4.

To prevent a dramatic increase in F_b, which can result in unsafe designs, an inter-action equation that does not contain the term C_m is also required to be satisfied.

The calculation of stress ratios in frame columns is essentially an exercise in the evaluation of stresses due to simultaneous axial and bending action.

7.1.3.5. *Interaction Equations*

Prior to 1963, structural engineers could have made peace with the entire design process of beam–columns by using the formula

$$\frac{f_a}{F_a} + \frac{f_b}{F_b} \leq 1.0$$

Since then, engineers have had to deal with many seemingly formidable factors that have been added onto the above interaction equation. For example, the allowable bending stress F_b now has a factor $(1 - f_a/F_e')$ to account for the reduction in the bending capacity because of axial loads. The more the axial load in the column, the greater the reduction of F_b. Reducing the allowable stress is mathematically equivalent to increasing the design moment for the $P\Delta$ effects. F_e' is the familiar Euler's stress divided by the same factor of safety, 23/12, that governs the allowable stress of long columns.

Consideration of only uniaxial bending reduces the AISC equations to the less intimidating format as follows:

$$\frac{f_a}{F_a} + \frac{C_m f_b}{(1 - f_a/F_e')F_b} \leq 1.0$$

$$\frac{f_a}{0.6F_y} + \frac{f_b}{F_b} \leq 1.0$$

where

f_a = axial stress in the column due to vertical loads
F_a = allowable axial stress
f_b = bending stress in the column
F_b = allowable bending stress
C_m = coefficient for modifying the actual bending moment to an equivalent moment diagram for purposes of evaluating secondary bending
F'_e = Euler's stress divided by safety factor, 23/12
F_y = yield stress of column steel

As mentioned previously, a stress ratio greater than 1.0 indicates overstress, requiring the redesign of the column.

For the general case of axial load plus biaxial bending, the interaction equations for calculating the stress ratios are as follows: If f_a is compressive and $f_a/F_a > 0.15$, the compressive stress ratio CR is given by the larger of CR_{1a} and CR_{1b}, where

$$CR_{1a} = \frac{f_a}{F_a} + \frac{C_{mx}f_{bx}}{\left(1 - \dfrac{f_a}{F'_{ex}}\right)F_{bx}} + \frac{C_{my}f_{by}}{\left(1 - \dfrac{f_a}{F'_{ey}}\right)F_{by}} \qquad \text{(AISC H1-1)}$$

and

$$CR_{1b} = \frac{f_a}{0.60F_y} + \frac{f_{bx}}{F_{bx}} + \frac{f_{by}}{F_{by}} \qquad \text{(AISC H1-2)}$$

If $f_a/F_a \leq 0.15$, $CR = CR_2$, where

$$CR_2 = \frac{f_a}{F_t} + \frac{f_{bx}}{F_{bx}} + \frac{f_{by}}{F_{by}} \qquad \text{(AISC H1-3)}$$

C_{mx} and C_{my} being coefficients that represent distribution of moment along member length. Although, as mentioned previously, their value could be as low as 0.4, in practice they are conservatively assumed equal to 1.0 in most cases.

If f_a is tensile or zero, the tensile stress ratio TR is given by the larger of TR_1 and TR_2, where

$$TR_1 = \frac{f_a}{F_t} + \frac{f_{bx}}{F_{bx}} + \frac{f_{by}}{F_{by}} \qquad \text{(AISC H2-1)}$$

and

$$TR_2 = \frac{f_{bx}}{F_{bx}} + \frac{f_{by}}{F_{by}}$$

In the calculation of the tensile ratio TR the allowable bending stresses F_{bx} and F_{by} have a minimum value of $0.60F_y$. For circular sections, a square root of sum of the squares (SRSS) combination is first made of the two bending components before adding the axial load component instead of the simple algebraic addition implied by the foregoing formulas for CR and TR.

7.1.3.6. *Direct Analysis of P∆ Effects*

It is important to realize that the moment magnification method using K-factors is an approximate method for evaluating the $P\Delta$ effects. With the availability of computer programs that can directly account for $P\Delta$ effects, it is not necessary to use the approximate moment magnification method for evaluating the effect of axial loads P acting through the lateral deflection Δ of the structure. Analysis of structures by using programs with $P\Delta$ capabilities is highly recommended for routine office use. With the results of the $P\Delta$ analysis in hand, the engineer need not worry about calculating the effective length factor K by using alignment charts or complicated equations. All columns whether they are gravity or frame columns can be designed by using the effective length factor K = 1. And, moreover, the $P\Delta$ method is applicable to all types of construction—steel, concrete, or composite—as a general procedure. Although the procedure itself is not codified by the ACI and AISC, it is highly endorsed in their commentaries.

The 1999 edition of the SEAOC Blue Book gives a drift ratio value of $0.02I/R$ as the threshold of lateral deformation beyond which the $P\Delta$ effects become significant. Consider, for example, a building with a seismic importance factor $I = 1.0$, and a typical floor-to-floor height of 13' 6" with special moment-resisting-frame as the lateral system. With an $R = 8.5$, the limiting drift ratio for this building is approximately equal to $^3/_8$". $P\Delta$ analysis is necessary in the lateral load analysis only when the drift ratio exceeds this value.

However, by using commercial programs it is easy to include $P\Delta$ effects in a single solution without having to use the iteration technique. Therefore, analysis of structures including $P\Delta$ effects is highly recommended for office practice. It should be noted that while $P\Delta$ effects are included in the programs, the effect of reduction of stiffness of columns due to axial loads, in general, is not accounted.

Observe that the columns of moment frames that are designed with $P\Delta$ effects included need not have their bending stresses amplified by the term $(1 - f_a/F_e)$ in AISC formula H1-1 or δ_s in ACI formula, since these factors were intended to account for $P\Delta$ effects.

It should be noted that $P\Delta$ effects are potentially much more significant in lower seismic zones than in higher seismic zones, because the relative stiffness of lateral-load-resisting systems in higher seismic zones is required to be greater than those in lower seismic zones.

7.2. CONCRETE SYSTEMS

A brief description of ACI 318-02 code revisions related to the design examples given in the following sections is summarized as follows:

- The release cycle for ACI 318 is typically three years. The most current edition is ACI 318-02, with the next edition planned for the year 2005. It is anticipated that revisions to the 2005 edition will be minor.
- The values for capacity reduction factors, ϕ, have been modified, as have been the load factors for calculating the required strengths. The load factors have been changed to achieve uniformity with other codes. Typically, there is a 10% reduction in the required strengths with a corresponding increase in ϕ values, such that there is parity between designs performed under ACI 318-99 and ACI 318-02 codes. However, the ϕ factor for tension-controlled sections such as for beams and slabs (typically designed for flexure) is not changed. Therefore, ACI 318-02 flexural reinforcement requirement is typically 10% less than that required by ACI 318-99. However, ductility requirements under the 20-02 code for these sections are somewhat more stringent. The maximum reinforcement

Table 7.1 Classification of Prestressed Flexural Members

Member condition	Class	Stress condition
Uncracked	U	$f_t \leq 7.5\sqrt{f_c'}$
Transition	T	$7.5 f_c' < f_c' \leq 12\sqrt{f_c'}$
Cracked	C	$f_t > 12\sqrt{f_c'}$

(From ACI 318-02, Table R18.3.3.)

permitted in a singly reinforced section is $0.714\rho_b$, compared to $0.75\rho_b$ of the 1999 Code.

- A significant revision has been made in the permissible design tension stress, f_t, for prestressed flexural members. Recall that ACI 318-99, like its predecessors, typically limits the extreme fiber stress in tension, f_t, in precompressed tensile zone to $6\sqrt{f_c'}$. This can be increased to $12\sqrt{f_c'}$ provided deflection analysis is based on transformed cracked sections and on bilinear moment-deflection relationships. Now, under provisions of the 2002 edition, prestressed members are classified into three classes with corresponding allowable tension stress as shown in Table 7.1. It should be noted that serviceability requirements get progressively stringent from class U to class C. (The designer is referred to Table R18.3.3 of the 2002 code for serviceability check requirements for each class.) Note that prestressed two-way slab systems must still be designed as class U. This restriction is to prevent the possibility of punching shear failure in two-way systems.

It should be noted that although the examples given here are based on ACI 318-99, they comply with the 2002 edition, because Appendix C of this edition continues to permit the use of load and strength reduction factors of ACI 318-99.

7.2.1. One-Way Slabs

One-way slabs are discussed here to illustrate the simplifications commonly made in a design office to analyze these systems.

Figure 7.10 shows a uniformly loaded floor slab with intermediate beams that divide the slab into a series of one-way slabs. If a typical 1-ft width of slab is cut out as a free body in the longitudinal direction, it is evident that the slab will bend with a positive curvature between the supporting beams, and a negative curvature at the supporting beams. The deflected shape is similar to that of a continuous beam spanning across transverse girders, which act as simple supports. The assumption of simple support neglects the torsional stiffness of the beams supporting the slab. If the distance between the beams is the same, and if the slabs carry approximately the same load, the torsional stiffness of the beams has little influence on the moments in the slab.

However, the slab twists the exterior beams, which are loaded from one side only. The resistance to the end rotation of the slab offered by the exterior beam is dependent on the torsional stiffness of the beam. If the beam is small and its torsional stiffness low, a pin support may be assumed at the exterior edge of slab. On the other hand, if the exterior beam is large with a high torsional rigidity, it will apply a significant restraining moment to the slab. The beam, in turn, will be subjected to a torsional moment that must be considered in design.

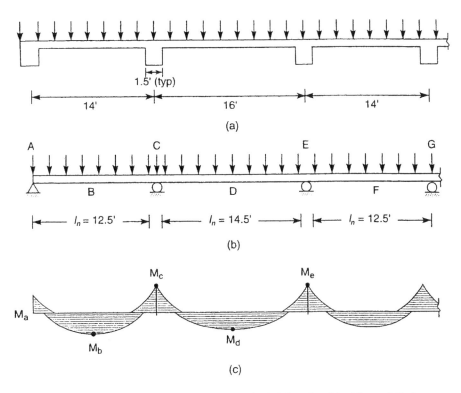

Figure 7.10. One-way slab example: (a) typical 1-ft strip; (b) slab modelled as a continuous beam; (c) design moments.

7.2.1.1. Analysis by ACI Coefficients

Analysis by this method is limited to structures in which: 1) the span lengths are approximately the same (with the maximum span difference between adjacent spans no more than 20%); 2) the loads are uniformly distributed; and 3) the live load does not exceed three times the dead load.

ACI values for positive and negative design moments are illustrated in Figs. 7.11 and 7.12. Observe that l_n equals the clear span for positive moment and shear, and the average of adjacent clear spans for negative moment.

Example. One-way mild steel reinforced slab.

Given. A one-way continuous slab as shown in Fig. 7.13.

$$f'_c = 4 \text{ ksi}, \qquad f_y = 60 \text{ ksi}$$

Ultimate load = 0.32 kip/ft

Required. Flexural reinforcement design for interior span between grids C and E.

Solution. Use Table 7.2 to determine the minimum slab thickness required to satisfy deflection limitations. Using l = center-to-center span = 16 ft,

$$h_{min} = \frac{l}{28} = \frac{12 \times 16}{28} = 6.86 \text{ in.} \quad \text{Use } 6\frac{1}{2} \text{ in. (165 mm)}$$

Analyze a 1-ft width of slab as a continuous beam using ACI coefficients to establish design moments for positive and negative steel (Fig. 7.13). Using a clear span $l_n = 12.5$ ft

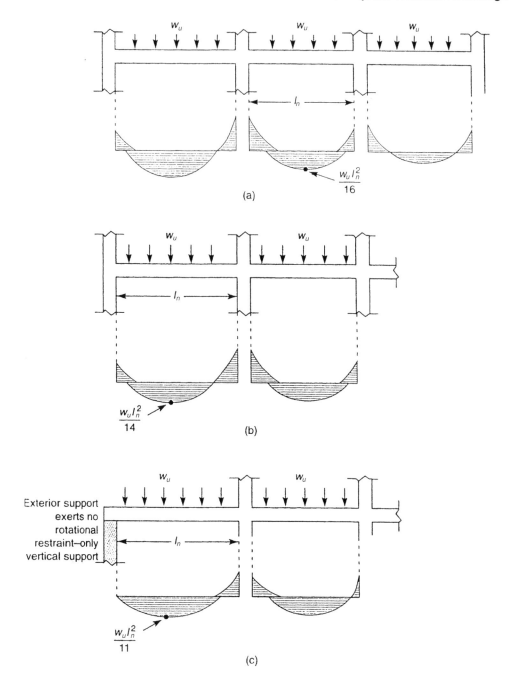

Figure 7.11. ACI positive moment coefficients: (a) interior span; (b) exterior span, discontinuous end integral with supports: (c) exterior span, discontinuous end unrestrained.

for the first bay,

$$M_a = \frac{w_u l_n^2}{24} = \frac{0.32 \times 12.5^2}{24} = 2.08 \text{ kip-ft}$$

$$M_b = \frac{w_u l_n^2}{11} = \frac{0.32 \times 12.5^2}{11} = 4.55 \text{ kip-ft}$$

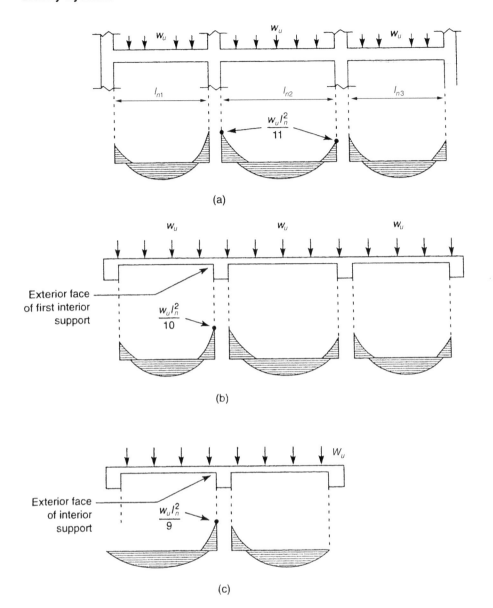

Figure 7.12. ACI coefficients for negative moments: (a) at interior supports; (b) at exterior face of first interior support, more than two spans; (c) at exterior face of first interior support, two spans.

At C, for negative moment, l_n is the average of adjacent clear spans: $l_n = (12.5 + 14.5)/2 = 13.5$ ft

$$M_c = \frac{w_u l_n^2}{10} = \frac{0.32 \times 13.5^2}{10} = 5.83 \text{ kip-ft}$$

$$M_d = \frac{w_u l_n^2}{16} = \frac{0.32 \times 14.5^2}{16} = 4.21 \text{ kip-ft}$$

$$M_e = \frac{w_u l_n^2}{11} = \frac{0.32 \times 14.5^2}{11} = 6.12 \text{ kip-ft}$$

Compute reinforcement A_s per foot width of slab at critical sections. For example, at the second interior support, top steel must carry $M_e = 6.12$ kip-ft. Note that ACI code

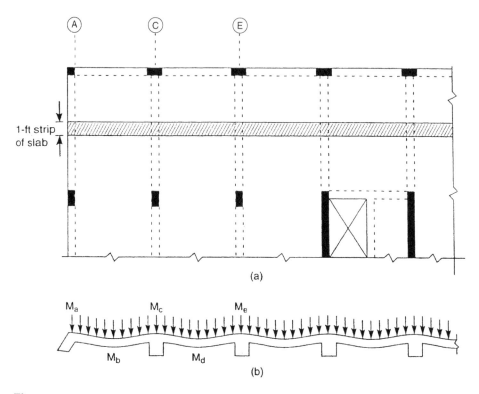

Figure 7.13. Design example, one-way slab: (a) partial floor plan; (b) section.

requires a minimum of $3/4$-in. cover for slab steel not exposed to weather or in contact with the ground.

We will use the trial method of determining the area of steel. In this method, the moment of the internal force couple is estimated. Next, the tension force T is evaluated by equating the applied moment to the internal force couple, i.e.,

$$M_u = \phi T \times \text{arm}$$

$$T = \frac{M_u}{\phi \times \text{arm}}$$

where $\phi = 0.9$ for flexure, and M_u = factored moment.

Table 7.2 Minimum Thickness of Beams or One-way Slabs Unless Deflections Are Computed[a]

	Minimum thickness h			
	Simply supported	One end continuous	Both ends continuous	Cantilever
Solid one-way slabs	$l/20$	$l/24$	$l/28$	$l/10$
Beams or ribbed one-way slabs	$l/16$	$l/18.5$	$l/21$	$l/8$

[a]Members not supporting or attached to partitions or other construction are likely to be damaged by large deflections. Span length l in inches. Values in the table apply to normal-weight concrete reinforced with steel of $f_y = 60,000$ lb/in.^{2} For lightweight concrete with a unit weight between 90 and 120 lb/ft^3, multiply the table values by $1.65-0.005w$, respectively, but by not less than 1.09; the unit weight w is in lb/ft^3. For reinforcement having a yield point other than 60,000 lb/in.^{2}, multiply the table values by $0.4 + f_y/100,000$ with f_y in lb/in.^{2} (From ACI 318-02 Table 9.5a.)

To start the procedure, the moment arm is estimated as $d - a/2$ by assuming a value of $a = 0.15d$, where d is the effective depth. The appropriate area of steel A_s is computed by dividing T by f_y.

To get a more accurate value of A_s, the components of the internal couple are equated to find a close estimate of the area A_c of the stress block. The compressive force C in the stress block is equated to the tension force T.

$$C = T$$
$$0.85 f_c' A_c = T$$
$$A_c = \frac{T}{0.85 f_c}$$

Once A_c has been evaluated, locate the position of C, which is the centroid of A_c, and recompute the arm between C and T. Using the improved value, find the second estimates of T and A_s. Regardless of the initial assumption for the arm, two cycles should be adequate for determining the required steel area.

For the example problem, the effective depth d for the slab is given by:

$$d = h - \left(0.75 + \frac{d_b}{2} \right) = 6.5 - (0.75 + 0.25) = 5.5 \text{ in.}$$
$$M_u = \phi T (d - a/2)$$

As a first trial, assume $a = 0.15d = 0.15 \times 5.5 = 0.83$ in.

$$6.12 \times 12 = 0.9 T \left(5.5 - \frac{0.83}{2} \right) = 4.58 T$$
$$T = 16.03 \text{ kips}$$
$$A_s = \frac{T}{f_y} = \frac{16.03}{60} = 0.27 \text{ in.}^2/\text{ft}$$

Repeat the procedure using an arm based on an improved value of a. Equate $T = C$:

$$16.03 = 0.85 f_c' A_c = 0.85 \times 4 \times a \times 12$$
$$a = 0.39 \text{ in.}$$
$$\text{Arm} = d - \frac{a}{2} = 5.5 - \frac{0.39}{2} = 5.31 \text{ in.}$$
$$T = \frac{M_u}{\phi \left(d - \frac{a}{2} \right)} = \frac{6.12 \times 12}{0.9 \times 5.31} = 15.37 \text{ kips}$$
$$A_s = \frac{15.37}{60} = 0.26 \text{ in.}^2$$

Check for temperature steel $= 0.0018 A_g$
$$= 0.0018 \times 6.5 \times 12 = 0.14 \text{ in.}^2/\text{ft} < 0.26 \text{ in.}^2/\text{ft}$$

Determine spacing of slab reinforcement to supply 0.26 in.^{2}/ft.

Using #4 rebars, $s = \dfrac{0.20}{0.26} \times 12$ in.   Say, 9 in.

Using #5 rebars, $s = \dfrac{0.31}{0.26} \times 12 = 14.31$ in.   Say, 14 in.

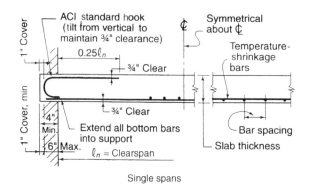

Single spans

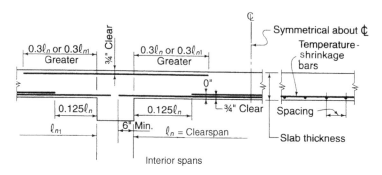

Interior spans

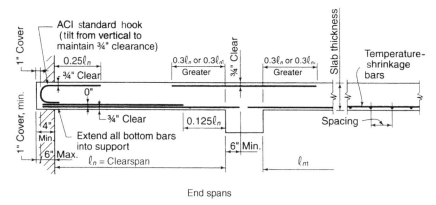

End spans

Figure 7.14. Recommended bar placement details, solid slabs. (Adapted from *CRSI Design Handbook* 2002.)

Use #4 @ 9 top at support *e*. Also by ACI code, the maximum spacing of flexural reinforcement should not exceed 18 in. or 3 times the slab thickness.

9 in. < 3(6.5 in.) = 19.5 in.   9-in. spacing is OK.

Recommeded bar placement details for solid slabs are shown in Fig. 7.14.

Note: The capacity reduction factor, ϕ, for determining tension reinforcement has remained unchanged at 0.9 in ACI 318-02. However, since there is a reduction in the foctored load combinations:

$$U = 1.2D + 1.6L \qquad \text{(ACI 318-02)}$$

versus

$$U = 1.4 + 1.7L \qquad \text{(ACI 318-99)}$$

we now get about 10% reduction in flexural reinforcement as compared to previous editions of ACI 318.

7.2.2. T-Beam Design

7.2.2.1. *Design for Flexure*

ACI 318-99.

$DL = 2.0$ k/ft (includes the self-weight of beam)

$LL = 1.3$ k/ft

$U = 1.4 \times 2.0 + 1.7 \times 1.3 = 5.0$ kip-ft   (ACI 318-99)

$U = 1.2 \times 2.0 + 1.6 \times 1.3 = 4.48$ kip-ft   (ACI 318-02)

See a design example of a simply supported T-beam in Fig. 7.15. The minimum depth of beam to control deflections from Table 7.2 is

$$h_{min} = \frac{l}{16} = \frac{30 \times 12}{16} = 22.5 \text{ in.} \quad \text{Use } 22.5 \text{ in.}$$

Try $b_w = 18$ in. The width must be adequate to carry shear and allow for proper spacing between reinforcing bars.

The effective width of the T beam b_{eff} is the smallest of

1. One-fourth the beam span:

$$\frac{30}{4} = 7.5 \text{ ft} = 90 \text{ in.} \quad \text{(controls)}$$

2. Eight times the slab thickness on each side of the stem plus the stem thickness:

$$8 \times 6.5 \times 2 + 18 = 122 \text{ in.}$$

3. Center-to-center spacing of the panel:

$$\frac{(16 + 14)}{2} \times 12 = 180 \text{ in.}$$

Select the flexural steel A_s for $M_u = 562.50$ kip-ft using the trial method.

The effective depth $d = h - 2.6 = 22.5 - 2.6 = 19.9$ in.

$$M_u = \phi T \left(d - \frac{a}{2} \right) \quad \text{Assume } a = 0.8 \text{ in.}$$

$$562.50 \times 12 = 0.9T \left(19.9 - \frac{0.8}{2} \right) = 17.557$$

$$T = 384.62$$

$$A_s = \frac{T}{f_y} = \frac{384.62}{60} = 6.41 \text{ in.}^2$$

Check value of a.

$$384.62 = T = C = ab_{eff}(0.85f_c)$$
$$= a(90)(0.85)4$$
$$a = 1.26$$

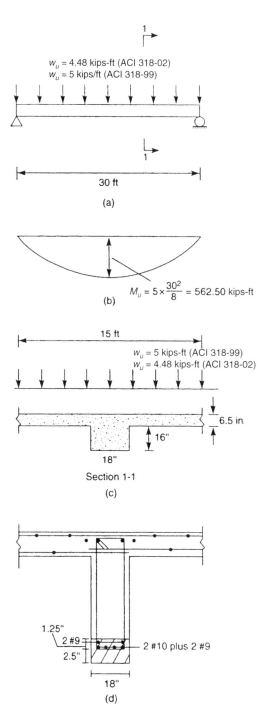

Figure 7.15.   Design example, simply supported T-beam.

Repeat the procedure using a moment arm based on the improved value of a.

$$M_u = 562.50 \times 12 = 0.9T\left(19.9 - \frac{1.26}{2}\right) = 17.34T$$

$$T = \frac{562.50 \times 12}{17.34} = 389.2 \text{ kips}$$

Check value of a.

$$389.2 = T = C = a \times 90 \times 0.85 \times 4$$

$$a = 1.27 \text{ in.}$$

$$A_s = \frac{T}{f_y} = \frac{389.2}{60} = 6.49 \text{ in.}^2$$

$$A_{s,min} = \frac{200 b_w d}{f_y} = \frac{200 \times 18 \times 19.9}{60,000} = 1.19 \text{ in.}^2 < 6.49 \text{ in.}^2$$

Since 6.49 in.^{2} controls, use two #10 and four #9 bars.

$$A_{s,provided} = 6.54 \text{ in.}^2$$

ACI 318-02. The ultimate load of 4.48 kip-ft by ACI 318-02 is about 10% less than that required by ACI 318-99. Therefore,

$$A_{s,required} = 6.49 \times \frac{4.48}{5.0} = 5.82 \text{ in.}^2$$

Use 6 #9, giving $A_s = 6 \times 1 = 6.0 \text{ in.}^2$

7.2.2.2. *Check Reinforcement Pattern for Crack Width*

ACI 318-95. ACI 318-95 limits crack widths to 0.013 in. (0.33 mm) and 0.016 in. (0.41 mm), respectively, for interior and exterior exposures. The corresponding value for the parameter Z given by the equation

$$z = f_s \sqrt[3]{d_c A}$$

is not to exceed 145 kip-in. (25.4 MN/m) and 175 kip-in. (30.6 MN/m), respectively. In this equation, f_s is the steel stress and may be taken as $0.6 f_y$ in kips per square inch, d_c is the distance from tension surface to the center of the row of reinforcing bars closest to outside surface, and A is the effective tension area of concrete divided by the number of reinforcing bars.

We now proceed with the example problem to verify whether the reinforcement pattern satisfies the ACI code requirements of crack control for exterior exposure.

Locate the center of gravity of steel by summing the moments of the bar areas about an axis through the base of beam steam.

$$A_{st}\bar{Y} = \sum A_n \bar{Y}_n$$

$$(6.54 \text{ in.}^2)\bar{Y} = (2 \text{ in.}^2)(2.5 \text{ in.}) + (2 \text{ in.}^2)(3.75 \text{ in.}) + (2.54 \text{ in.}^2)(2.63 \text{ in.})$$

$$\bar{Y} = 2.93 \text{ in.}$$

The effective tension area of concrete (Fig. 7.15d) is the product of the beam stem width and a web height equal to twice the distance between the steel centroid and the tension face. When the reinforcement consists of more than one bar size, as in the example, the number of bars is expressed by the size of the largest bar

$$\text{Number of bars} = \frac{\text{total area of steel}}{\text{area of largest bar}}$$

For the example, number of bars = 6.54/1.27 = 5.15. Therefore

$$A = \frac{18 \times 2.93 \times 2}{5.15} = 20.48$$

$$z = f_s \sqrt[3]{d_c A}$$

$$= 0.6 \times 60 \sqrt[3]{2.15 \times 20.48}$$

$$= 127 < 145 \text{ kip/in. Therefore, OK.}$$

ACI 318-02. The 1999 edition of ACI 318 includes a significant change in the method of verifying the distribution of flexural reinforcement for crack control. The new method limits the spacing s of flexural reinforcement closest to the tension face, instead of limiting the z factors. The 2002 edition was revised to calculate steel stresses at service load levels 10 to 20% higher than calculated in the 1999 edition. The value for default stress f_s remains at 0.6 f_y = 36 ksi for grade 60 reinforcement.

The spacing s of reinforcement closest to a tension surface must not exceed

$$s = (540/f_s) - 2.5 \, C_c \qquad\qquad\qquad \text{[ACI 318-02, Eq. (10.4)]}$$

and may not be greater than 12 $(36/f_s)$
where

$\qquad C_c$ = clear cover from the nearest surface in tension to the surface of the flexural
$\qquad\qquad$ tension reinforcement in inches.

For the example problem, using f_s = 36 ksi and C_c = 2.0 in., the minimum spacing is given by

$$s = (540/36) - 2.5 \times 2 = 10 \text{ in.} \leftarrow \text{controls}$$

$$s \leq 12 \times (36/36) = 12 \text{ in.}$$

For the example beam, the spacing provided is equal to

$$s = \frac{1}{3}\left[18 - 2\left(2.0 + 0.5 + \frac{1.128}{2}\right)\right] \cong 5 \text{ in.} < 10 \text{ in.} \quad \text{OK}$$

7.2.2.3. *Design for Shear*

ACI 318-99. The ACI procedure for shear design is an empirical method based on the assumption that a shear failure occurs on a vertical plane when shear force at that section due to factored service loads exceeds the concrete's fictitious vertical shear strength. The shear stress equation by strength of materials is given by

$$v = \frac{VQ}{Ib}$$

where
$\qquad v$ = shear stress at a cross section under consideration
$\qquad V$ = shear force on the member
$\qquad I$ = moment of inertia of the cross section about centroidal axis
$\qquad b$ = thickness of member at which v is computed
$\qquad Q$ = moment about centroidal axis of area between section at which v is computed
$\qquad\qquad$ and outside surface of member

This expression is not directly applicable to reinforced concrete beams. The ACI, therefore, uses a simple equation to calculate the average stress on the cross section,

$$v_c = \frac{V}{b_w d}$$

where

v_c = nominal shear stress
V = shear force
b_w = width of beam web
d = distance between centroid of tension steel and compression surface

To emphasize that v_c is not an actual stress but merely a measure of the shear stress intensity, it is termed a nominal shear stress.

For nonseismic design, ACI 318-02 assumes that concrete can carry some shear regardless of the magnitude of the external shearing force and that shear reinforcement must carry the remainder. Thus

$$V_u \leq \phi V_n = \phi(V_c + V_s)$$

where

V_u = factored or ultimate shear force
V_n = nominal shear strength provided by concrete and reinforcement
V_c = nominal shear strength provided by concrete
V_s = nominal reinforcement provided by shear reinforcement
ϕ = strength reduction factor = 0.85 for shear and torsion   (ACI 318-99)
$\quad\quad$ = 0.75   (ACI 318-02)

Shear design computations can be made in terms of shear force V or in terms of unit shear stress v. Stress is easier to compare with allowable values, and gives engineers a better frame of reference, thus reducing chances of error.

The shear strength equation in terms of shear stress is given by

$$v_u \leq \phi v_n = \phi(v_c + v_s)$$

A conservative value for v_c often used in design because of its simplicity, is $v_c = 2\sqrt{f'_c}$.
The nominal shear stress v_u can be calculated from

$$v_u = \frac{V_u}{\phi b_w d}$$

Including vertical stirrups,

$$v_u \leq v_c + \frac{A_v f_y}{b_w s}$$

where

A_v = area of vertical shear reinforcement
f_y = yield strength of shear reinforcement
s = spacing of shear reinforcement

Using ACI 318-99 for the example problem, we have

$$V_u = 5.0 \times 15 = 75 \text{ kips}$$

$$d = 19.9 \text{ in.,} \qquad b_w = 18 \text{ in.}$$

$$V_u \text{ at distance } d \text{ from the support} = 5.0\left(15 - \frac{19.4}{12}\right) = 70 \text{ kips}$$

$$v_u = \frac{70}{0.85 \times 18 \times 19.9} = 0.230 \text{ ksi}$$

$$v_c = 2\sqrt{f_c'} = 2\sqrt{4000} = 126 \text{ psi}$$

Shear stress to be carried by reinforcement:

$$v_s = v_u - v_c$$
$$= 0.230 - 0.126$$
$$= 0.110$$
$$s = \frac{A_v f_y}{v_s}$$

For two-legged #4 stirrups,

$$s = \frac{2 \times 0.2 \times 60}{0.104 \times 18} = 12.8 \text{ in.}$$

This should be checked for maximum spacing, as will be done presently.

We now calculate the stirrups using strength equation in terms of shear forces.

$$V_u = \phi(V_c + V_s)$$

For the example problem,

$$V_u = 70 \text{ kips}$$
$$V_c = 2\sqrt{f_c'}b_w d$$
$$= 2\sqrt{4000} \, 18 \times 19.9$$
$$= 45 \text{ kips}$$
$$\phi\frac{V_c}{2} = 0.85 \times \frac{45}{2} = 19.1 \text{ kips}$$

Since $V_u = 70$ kips exceeds $\phi V_c/2$, stirrups are required.

$$V_s = \frac{V_u}{\phi} - V_c$$
$$= \frac{70}{0.85} - 45 = 37.35 \text{ kips}$$

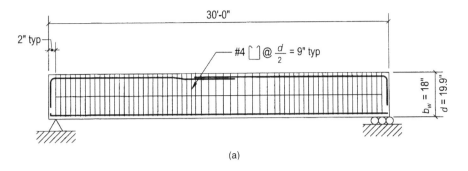

(a)

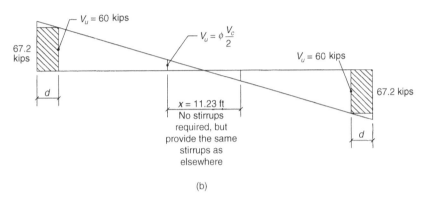

(b)

Figure 7.16. Schematic shear reinforcement.

Spacing for two-legged #4 stirrups,

$$s = \frac{A_v f_y d}{V_s}$$

$$= \frac{2 \times 0.2 \times 60 \times 19.9}{37.35^k}$$

$$= 12.8 \text{ in.}$$

Since V_s is less than $4\sqrt{f_c'} \, b_w d = 90$ kips,

$$s = \frac{d}{2} = \frac{19.9}{2} = 9.9, \text{ say, } 9 \text{ in.}$$

If $V_s \geq 4\sqrt{f_c'} \, b_w d$, the maximum spacing would have been $d/4$ but not to exceed 12 in.

$$A_{v,min} = \frac{50 b_w s}{f_y}$$

$$= \frac{50 \times 18 \times 9}{60,000}$$

$$= 0.135 \text{ in.}^2$$

$$A_{v,provided} = 0.4 \text{ in.}^2 > 0.135 \text{ in.}^2$$

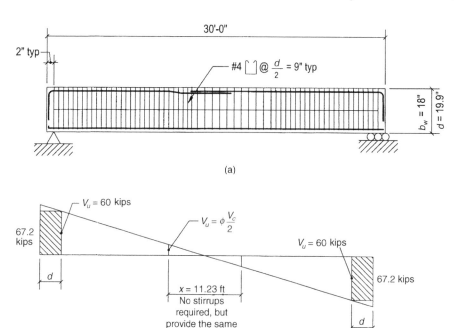

Figure 7.16. Schematic shear reinforcement.

Use #4 two-legged stirrups at 9 in. near the supports. A reduced spacing of stirrups equal to d may be used within the span where the calculated shear stress $v_u \leq v_c/2$. See Fig. 7.16 for placement of shear reinforcement.

ACI 318-02. V_u at a distance d from the support $= 4.48\left(15 - \frac{19.4}{12}\right) \cong 60$ kips

$$V_c = 2\sqrt{f_c'}\, b_w d$$

$$= 2\sqrt{4000} \times 18 \times 19.9$$

$$= 45 \text{ kips}$$

$$\frac{\phi V_c}{2} = 0.75 \times \frac{45}{2} = 16.9 \text{ kips}$$

Since $V_u = 60$ kips exceeds $\frac{\phi V_c}{2} = 16.9$ kips, stirrups are required.

$$V_s = \frac{V_u}{\phi} - V_c$$

$$= \frac{60}{0.75} - 45 = 35 \text{ kips}$$

Spacing of two-legged #4 stirrups,

$$s = \frac{A_v f_y d}{V_s}$$

$$= \frac{2 \times 0.2 \times 60 \times 19.9}{35} = 13.6 \text{ in.}$$

Since V_s is less than $4\sqrt{f_c'}\,b_w d = 90$ kips,

$$\text{max stirrup spacing } s = \frac{d}{2} = \frac{19.9}{2} = 9.95, \text{say, 9 in.}$$

$$A_{v,\,min} = 0.75\sqrt{f_c'}\,\frac{b_w s}{f_y} \geq \frac{50\,b_w s}{f_y}$$

$$= 0.75\sqrt{4000} \times \frac{18 \times 9}{60,000} = 0.128 \text{ in.}^2$$

$$A_{v,\,min} \geq \frac{50\,b_w \times 9}{60,000} = 0.135 \text{ in.}^2$$

$$A_{v,\,provided} = 0.4 \text{ in.}^2 > 0.135 \text{ in.}^2 \quad \text{OK}$$

Stirrups are not required if $V_u \leq \dfrac{\phi V_c}{2} = 0.75 \times \dfrac{45}{2} = 16.87$ kips. This occurs at a distance x from the supports, given by

$$67.2 - x \times 4.48 = 16.87$$

$$x = \frac{67.2 - 16.87}{4.48} = 11.23 \text{ ft}$$

Therefore, no shear reinforcement is required within the middle $(30 - 2 \times 11.23) = 7.54$ ft. However, it is a good practice to provide at least some shear reinforcement, even when not required by calculations.

For the example problem, we use #4 ⊔ at $(d/2) = (19.9/2) \cong 9$ in. for the entire span.

Observe that for perimeter beams, ACI 318-02 Section 7.13.2 requires the stirrups to have 135° hooks around continuous bars. As an alternate, one-piece closed stirrups ⬚ may be used.

7.2.2.4. Summary of Shear Design Provisions; ACI 318-02.

Using the most common loads—dead (D), live (L), wind (W), and earthquake (E)—the simplified load combinations are

$$\left.\begin{array}{l} U = 1.4D \\ U = 1.2D + 1.6L \end{array}\right\} \quad \text{Dead and live loads}$$

$$\left.\begin{array}{l} U = 1.2D + 1.6L + 0.8W \\ U = 1.2D + 1.0L + 1.6W \\ U = 0.9D + 1.6W \end{array}\right\} \quad \text{Dead, live, and wind loads}$$

$$\left.\begin{array}{l} U = 1.2D + 1.0L + 1.0E \\ U = 0.9D + 1.0E \end{array}\right\} \quad \text{Dead, live, and earthquake loads}$$

- Strength reduction factor for shear and torsion $\phi = 0.75$
- If $V_u - \phi V_c > \phi 8\sqrt{f_c'}\,b_w d$, increase f_c', b_w, or d, as required.
- If $V_u \leq \phi V_c/2$ no stirrups are required.

- If $\phi V_c \geq V_u > \phi V_c/2$, the required area of stirrups A_v is given by

$$A_v = 0.75\sqrt{f_c'}\,\frac{b_w s}{f_y} \geq \frac{50 b_w s}{f_y}$$

Stirrup spacing required $s = \dfrac{A_v f_y}{0.75\sqrt{f_c'}\,b_w} \leq \dfrac{A_v f_y}{50 b_w} \leq \dfrac{d}{2} \leq 24$ in.

- If $V_u > \phi V_c$, the required area of stirrups A_v is given by

$$A_v = \frac{(V_u - \phi V_c)s}{\phi\, f_y d}$$

Stirrup spacing required $s = \dfrac{\phi A_v f_y d}{V_u - \phi V_c}$

- Maximum spacing $s = \dfrac{d}{2} \leq 24$ in. for $(V_u - \phi V_c) \leq \phi\, 4\sqrt{f_c'}\,b_w d$

$$s = \frac{d}{4} \leq 12 \text{ in. for } (V_u - \phi V_c) > \phi 4\sqrt{f_c'}\,b_w d$$

7.2.3. Two-Way Slabs

Although two-way slabs may be designed by any method that satisfies the strength and serviceability requirements of the ACI code, most usually they are designed by the "equivalent-frame method" using computers. In this section, however, only the direct design method is discussed.

In this method the simple beam moment in each span of a two-way system is distributed as positive and negative moments at midspan and at supports. Since stiffness considerations, except at the exterior supports, are not required, computations are simple and can be carried out rapidly.

Three steps are required for the determination of positive and negative design moments.

1. Determine simple beam moment:

$$M_0 = \frac{w_u l_2 l_n^2}{8}$$

where

M_0 = simple beam moment
w_u = ultimate uniform load
l_2 = slab width between columns transverse to the span under consideration
l_n = clear span between face of columns or capitals

2. For interior spans, divide M_0 into M_c and M_s midspan and support moments as shown in Fig. 7.17; for exterior spans, use Fig. 7.18 to divide M_0 into moments M_1, M_2, and M_3.

3. Distribute M_c and M_s in the transverse direction across the width between column and middle strips by using Tables 7.3 and 7.4 which give percentage of moment in the column strips. The remainder is assigned to the middle strip.

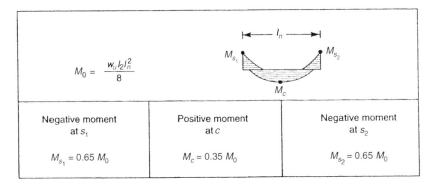

$$M_0 = \frac{w_u l_2 l_n^2}{8}$$

Negative moment at s_1	Positive moment at c	Negative moment at s_2
$M_{s_1} = 0.65\, M_0$	$M_c = 0.35\, M_0$	$M_{s_2} = 0.65\, M_0$

Figure 7.17. Assignment of moments at critical sections: interior span.

$$M_0 = \frac{w_u l_2 l_n^2}{8}$$

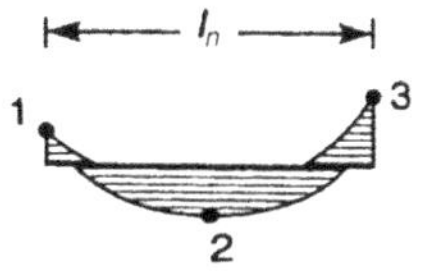

Edge restraint condition	Exterior negative moment at 1 M_1	Positive moment at 2 M_2	Interior negative moment at 3 M_3
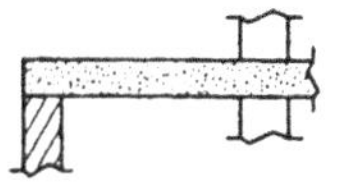 (a)	0	$0.63\, M_0$	$0.75\, M_0$
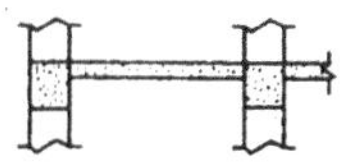 (b)	$0.16\, M_0$	$0.57\, M_0$	$0.70\, M_0$
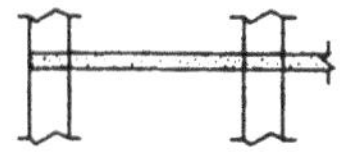 (c)	$0.26\, M_0$	$0.52\, M_0$	$0.70\, M_0$
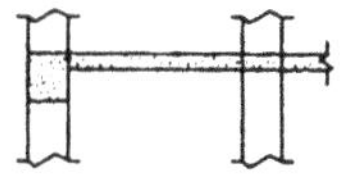 (d)	$0.30\, M_0$	$0.50\, M_0$	$0.70\, M_0$
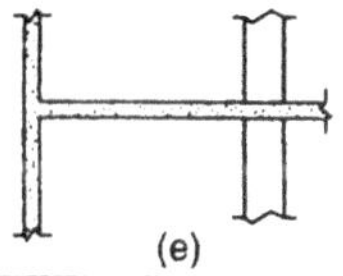 (e)	$0.65\, M_0$	$0.35\, M_0$	$0.65\, M_0$

Figure 7.18. Assignment of moments to critical sections—exterior span (ACI 318-02).

Table 7.3 Percentage of Positive Moment to
Column Strip, Interior Span

$\alpha_1 \dfrac{l_2}{l_1}$	l_2/l_1		
	0.5	1.0	2.0
0	60	60	60
≥ 1	90	75	45

Table 7.4 Percentage of Negative Moment to Column
Strip at an Interior Support

$\alpha_1 \dfrac{l_2}{l_1}$	l_2/l_1		
	0.5	1.0	2.0
0	75	75	75
≥ 1	90	75	45

Observe in Fig. 7.17 that, for an interior span, the positive moment M_c at midspan equals $0.35M_0$, and the negative moment M_s at each support equals $0.65M_0$, values that are approximately the same as for a uniformly loaded fixed-end beam. These values are based on the assumption that an interior joint undergoes no significant rotation, a condition that is assured by the ACI code restrictions that limit: 1) the difference between adjacent span lengths to one-third of the longer span; and 2) the maximum ratio of live load to dead load to 3.

The final step is to distribute the positive and negative moments in the transverse direction between column strip and middle strips. The distribution factors are tabulated (Tables 7.4 and 7.5) for three values (0.5, 1, 2) of panel dimensions l_2/l_1, and two values (0 and 1) of $\alpha_1(l_1/l_2)$. For intermediate values linear interpolation may be used. Table 7.4 is for interior spans while Table 7.5 is for exterior spans. For exterior spans, the distribution of moment is influenced by the torsional stiffness of the spandrel beam. Therefore an additional parameter β_t, the ratio of the torsional stiffness of the spandrel beam to flexural stiffness of the slab is given in Table 7.5.

For exterior spans, the distribution of total negative and positive moments between columns strips and middle strips is given in terms of the ratio l_2/l_1, the relative stiffness of the beam and slab, and the degree of torsional restraint provided by the edge beam. The parameter $\alpha = (E_{cb}I_b)/(E_{cs}I_s)$ is used to define the relative stiffness of the beam and slab spanning in either direction. The terms E_{cb} and E_{cs} are the moduli of elasticity of the beam

Table 7.5 Percentage of Negative Moment to Column Strip at an
Exterior Support

$\alpha_1 \dfrac{l_2}{l_1}$	β_t	l_2/l_1		
		0.5	1.0	2.0
0	0	100	100	100
0	≥ 2.5	75	75	75
≥ 1	0	100	100	100
≥ 1	≥ 2.5	90	75	45

and slab, respectively, and I_b and I_s are the moments of inertia, respectively. Subscripted parameters α_1 and α_2 are used to identify α for the directions of l_1 and l_2, respectively.

The parameter β_t in Table 7.5 defines the torsional restraint of edge beam. If there is no edge beam, i.e., $\beta = 0$, all of the exterior moment at 1 (Fig. 7.18) is apportioned to the column strip. For $\beta_t \geq 2.5$, i.e., for very stiff-edge beams, 75% of the moment at 1 is assigned to the column strip. For values of β between 0 and 2.5, linear interpolation is permitted. In most practical designs, distributing 100% of the moment at 1 to the column strip while using minimum slab reinforcement in the middle strip yields acceptable results.

7.2.3.1. Design Example

Given. A two-way slab system as shown in Fig. 7.19.

$$w_d = 150 \text{ psf}, \qquad w_l = 80 \text{ psf}$$

Determine the slab depth and design moments by the direct design method at all critical sections in the exterior and interior span along column line B.

Solution. From Tables 7.6a and 7.6b, for $f_y = 60$ ksi, and for slabs without drop panels, the minimum thickness of slab is determined to be $l_n/33$ for the interior panels. The same thickness is used for the exterior panels since the system has beams between the columns along the exterior edges.

For the example, the clear span in the long direction, $l_n = 24 - 2 = 22$ ft. The minimum thickness $h = (22 \times 12)/33 = 8$ in.

Interior Span

$$w_u = 1.4(0.15) + 1.7(0.08) = 0.346 \text{ ksf}$$

$$
\begin{aligned}
M_0 &= \frac{w_u l_2 l_n^2}{8} \\
&= \frac{0.346 \times 20 \times 22^2}{8} \\
&= 418.7 \text{ kip-ft}
\end{aligned}
$$

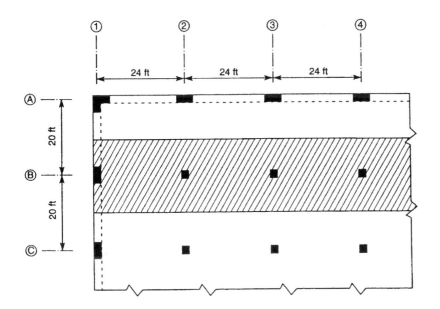

Figure 7.19. Design example, two-way slab.

Table 7.6a Minimum Thickness of Slabs Without Interior Beams

Yield strength, f_y, psi[b]	Without drop panels[a]			With drop panels[a]		
	Exterior panels		Interior panels	Exterior panels		Interior panels
	Without edge beams	With edge beams[c]		Without edge beams	With edge beams[c]	
40,000	$\dfrac{\ell_n}{33}$	$\dfrac{\ell_n}{36}$	$\dfrac{\ell_n}{36}$	$\dfrac{\ell_n}{36}$	$\dfrac{\ell_n}{40}$	$\dfrac{\ell_n}{40}$
60,000	$\dfrac{\ell_n}{30}$	$\dfrac{\ell_n}{33}$	$\dfrac{\ell_n}{33}$	$\dfrac{\ell_n}{33}$	$\dfrac{\ell_n}{36}$	$\dfrac{\ell_n}{36}$
75,000	$\dfrac{\ell_n}{28}$	$\dfrac{\ell_n}{31}$	$\dfrac{\ell_n}{31}$	$\dfrac{\ell_n}{31}$	$\dfrac{\ell_n}{34}$	$\dfrac{\ell_n}{34}$

[a] Drop panel is defined in 13.3.7.1 and 13.3.7.2.
[b] For values of reinforcement yield strength between the values given in the table, minimum thickness shall be determined by linear interpolation.
[c] Slabs with beams between columns along exterior edges. The value of α for the edge beam shall not be less than 0.8.
(From ACI 318-02, Table 9.5c.)

Divide M_0 between sections of positive and negative moments.
At midspan:

$$M_c = 0.35M_0$$
$$= 0.35 \times 418.7 = 146.5 \text{ kip-ft}$$

At supports:

$$M_s = 0.65M_0$$
$$= 0.65 \times 418.7 = 272.2 \text{ kip-ft}$$

For the distribution of the midspan moment M_c between column and middle strips, use Table 7.3. The value for α_1, the ratio of beam stiffness to slab stiffness for the example problem, is zero since there are no beams in the span direction under consideration. The

Table 7.6b Minimum Thickness[a] of Slabs Without Interior Beams

Yield strength, f_y, psi[b]	Without drop panels			With drop panels		
	Exterior panels		Interior panels	Exterior panels		Interior panels
	Without edge beams	With edge beams[c]		Without edge beams	With edge beams[c]	
40,000	$\dfrac{l_n}{33}$	$\dfrac{l_n}{36}$	$\dfrac{l_n}{36}$	$\dfrac{l_n}{36}$	$\dfrac{l_n}{40}$	$\dfrac{l_n}{40}$
60,000	$\dfrac{l_n}{30}$	$\dfrac{l_n}{33}$	$\dfrac{l_n}{33}$	$\dfrac{l_n}{33}$	$\dfrac{l_n}{36}$	$\dfrac{l_n}{36}$

[a] Minimum thickness for slabs without drop panels is 5 in. Minimum thickness for slabs with drop panels is 4 in.
[b] For values of reinforcement yield stress between 40,000 and 60,000 psi, minimum thickness shall be obtained by linear interpolation.
[c] Slabs with beams between columns along exterior edges. The value of α for the edge beam shall not be less than 0.8.
(From ACI 318-95, Table 9.5c)

ratio $l_2/l_1 = 20/24 = 0.833$. From Table 7.3, the column strip moment is 60% of the total moment.

Moment to column strip $= 0.60 \times 146.5 = 87.9$ kip-ft

Moment to middle strip $= 0.40 \times 146.5 = 58.6$ kip-ft

For the distribution of support moment M_s between column and middle strips, use Table 7.4. Since $\alpha_1 = 0$, and $l_2/l_1 = 0.833$, from Table 7.4, the column strip moment is 75% of the total moment.

Moment in column strip $= 0.75 \times 272.2 = 204$ kip-ft

Moment in middle strip $= 0.25 \times 272.2 = 68$ kip-ft

Exterior Span. The magnitude of the moments at critical sections in the exterior span is a function of both M_0, the simple beam moment, and α_{ec}, the ratio of stiffness of exterior equivalent column to the sum of the stiffness of the slab and beam framing into the exterior joint. Instead of computing α_{ec}, we use edge condition (d) given in Fig. 7.20 to evaluate the design moments at critical sections.

At the exterior column face:

$M_1 = 0.30 \times M_0 = 0.30 \times 418.7 = 125.6$ kip-ft

At midspan:

$M_2 = 0.50 \times M_0 = 0.50 \times 418.7 = 209.4$ kip-ft

At the interior column face:

$M_3 = 0.7 \times M_0 = 0.7 \times 418.7 = 293$ kip-ft

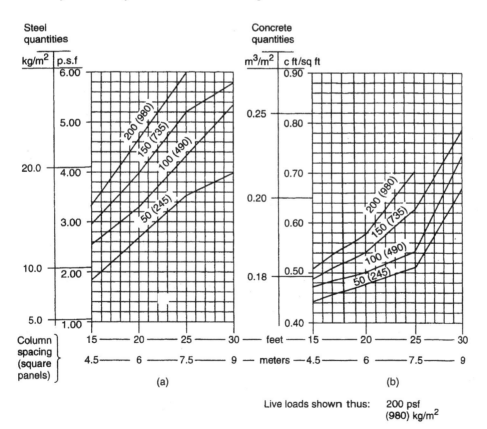

Figure 7.20. One-way solid slab, unit quantities: (a) reinforcement; (b) concrete.

At the exterior edge of the slab, the transverse distribution of the design moment to the column strip is given in Table 7.5. Instead of calculating the value of β_t, we conservatively assign 100% of the exterior moment to the column strip.

The moment to the column strip = 1×125.6 kip-ft. The middle strip is assumed to be controlled by the minimum steel requirements, an assumption which is satisfactory in almost all practical designs.

7.2.4. Unit Structural Quantities

Quantities used in cost estimates are physical items of construction to which unit costs are applied to arrive at a total construction cost. These are relatively easy to obtain once complete working drawings and specifications have been prepared. Prior to this point, however, the estimator or engineer must use "conceptual estimating" to determine approximate cost. Conceptual estimates require considerable judgment to correct so-called average unit costs to reflect complexity of construction operations, expected time required for construction, etc.

Typically, in the United States, units of structural quantities are dimensional, based on linear feet, square feet, or cubic feet. These result in unit quantities such as pounds per linear foot (plf), pounds per square foot (psf), etc.

Reinforcement and concrete unit quantities for various concrete floor-framing systems are shown in Figs 7.20–7.25. Live loads shown in the figures are working loads, and range

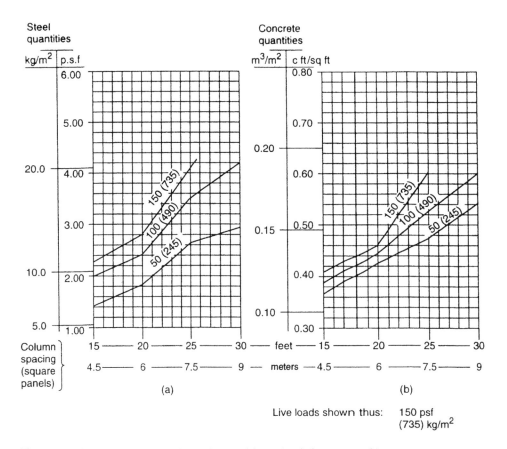

Figure 7.21. One-way pan joist, unit quantities: (a) reinforcement; (b) concrete.

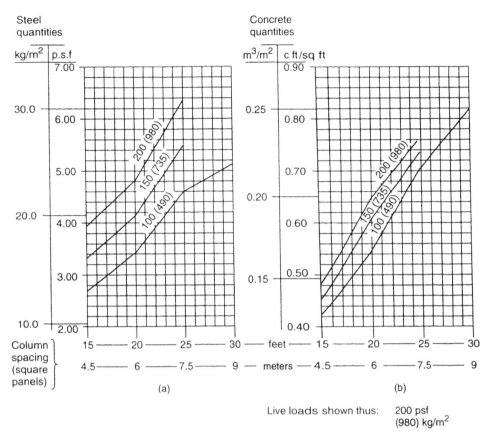

Steel quantities

kg/m² | p.s.f

Concrete quantities

m³/m² | c ft/sq ft

200 (980)
150 (735)
100 (490)

Column spacing (square panels)

15 —— 20 —— 25 —— 30 — feet — 15 —— 20 —— 25 —— 30

4.5—— 6 ——7.5—— 9 — meters — 4.5—— 6 ——7.5—— 9

(a)   (b)

Live loads shown thus: 200 psf
                          (980) kg/m²

Figure 7.22. Two-way slab, unit quantities: (a) reinforcement; (b) concrete.

from a typical office live load of 50 psf to a maximum of 200 psf appropriate for heavily loaded warehouse floors. The rebar quantities shown are for reinforcement required by design and do not include bars required for temperature and crack control, support bars, additional lengths required for laps, etc. The engineer should make allowances for these in the preliminary estimates by making appropriate notations on the preliminary drawings.

7.3. PRESTRESSED CONCRETE SYSTEMS

Prestressing boosts the span range of conventionally reinforced floor systems by about 30 to 40%. This is the primary reason for the increase in the use of prestressed concrete. Some of the other reasons are:

1. Prestressed concrete is generally crack-free and is therefore more durable.
2. Prestressing applies forces to members that oppose the service loads. Consequently, there is less net force to cause deflections.
3. Prestressed concrete is resilient. Cracks due to overloading completely close and deformations are recovered soon after removal of the overload.
4. Fatigue strength (though not a design consideration in building design) is considerably more than that of conventionally reinforced concrete because tendons are subjected to smaller variations in stress due to repeated loadings.

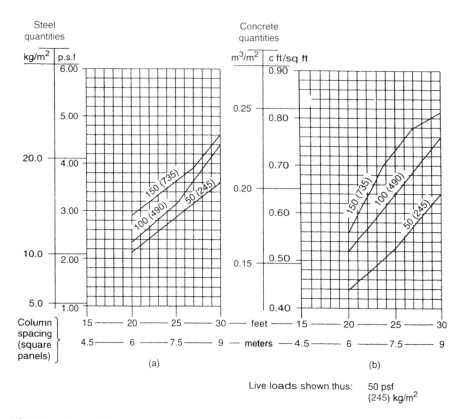

Figure 7.23. Waffle slab, unit quantities: (a) reinforcement; (b) concrete.

5. Prestressed concrete members are generally crack-free, and are therefore stiffer than conventional concrete members of the same dimensions.

6. The structural members are self-tested for materials and workmanship during stressing operations, thereby safeguarding against unexpected poor performance in service.

7. Prestress design is more controllable than mild steel design because a predetermined force is introduced in the system; the magnitude, location, and technique of introduction of such an additional force are left to the designer, who can tailor the design according to project requirements.

There are some disadvantages to the use of prestressed concrete, such as fire, the explosion resistance of unbonded systems, and difficulty in making penetrations due to the fear of cutting tendons.

A major motivation for the use of prestressed concrete comes from the reduced structural depth, which translates into lower floor-to-floor height and a reduction in the area of curtain wall and building volume, with a consequent reduction in heating and cooling loads.

In prestressed systems, the savings in mild steel reinforcement quantities resulting from prestress are just about offset by the higher unit cost of prestressing steel. The cost savings come from the reduction in the quantity of concrete combined with indirect nonstructural savings resulting from reduced floor-to-floor height. Although from an initial cost consideration prestressed concrete may be the least expensive, other costs associated with future tenant improvements, such as providing for large openings in floor slabs, must be considered before selecting the final scheme.

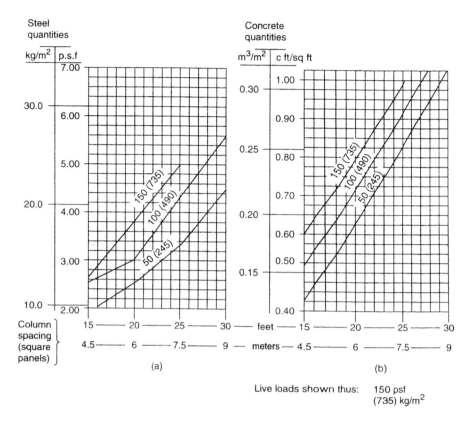

Figure 7.24.   Flat plate, unit quantities: (a) reinforcement; (b) concrete.

7.3.1. Prestressing Methods

Current methods of prestressing can be studied under two groups, pretensioning and post-tensioning. In pretensioning, the tendons are stretched and anchored against external bulkheads. Then concrete is placed around the tendons. After the concrete has hardened, the anchors are released, which imparts compression forces in the concrete as the tendon attempts to return to its original length.

In post-tensioning, the tendons are tensioned and anchored against the concrete after it has hardened. The tendons are stressed using hydraulic jacks after the concrete has reached a minimum of about 75% of the design strength. Tendon elongations are measured and compared against the calculated values; if satisfactory, the tendons projecting beyond the concrete are cut off. Form work is removed after post-tensioning. However, the floor is back-shored to support construction loads from the floors above.

Post-tensioning is accomplished using high-strength strands, wires, or bars as tendons. In North America, the use of strands by far leads the other two types. The strands are either bonded or unbonded depending upon the project requirements. In bonded construction, the tendons are installed in ducts that are filled with a mortar grout after stressing the tendons.

In building applications, unbonded construction is the preferred choice because it eliminates the need for grouting. Post-tensioned floor systems in buildings consist of slabs, joists, beams, and girders, with a large number of small tendons. Grouting each of the multitude of tendons is a time-consuming and expensive operation. Therefore, unbonded construction is more popular.

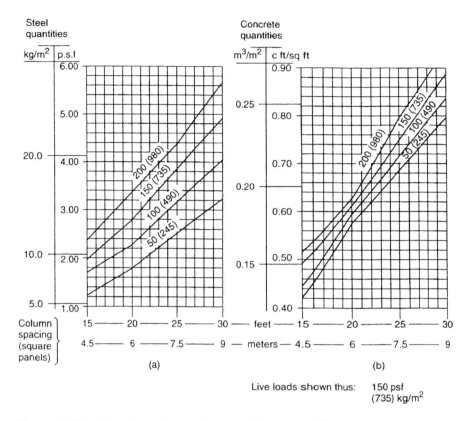

Figure 7.25.   Flat slab, unit quantities: (a) reinforcement; (b) concrete.

7.3.2.   Materials

7.3.2.1.   Post-Tensioning Steel

The basic requirement for post-tensioning steel is that the loss of tension in the steel due to shrinkage and creep of concrete and the effects of stress relaxation of the tendon should be a relatively small portion of the total prestress. In practice, the loss of prestress generally varies from a low of 15 ksi (103.4 MPa) to a high of 50 ksi (344.7 MPa). If mild steel having a yield of 60 ksi (413.7 MPa) were employed with an initial prestress of, say, 40 ksi, it is very likely that most of the prestress, if not the entire prestress, would be lost because of shrinkage and creep losses. To limit the prestress losses to a small percentage of, say, 20% of the applied prestress, the initial stress in the steel must be in excess of 200 ksi (1379 MPa). Therefore, high-strength steel is invariably used in prestressed concrete construction.

Although high-strength steel is generally produced using alloys such as carbon, manganese, and silicon, prestressing steel achieves its high-tensile strength by virtue of the process of cold-drawing, in which high-strength steel bars are drawn through a series of progressively smaller dyes. During this process, the crystallography of the steel is improved, because cold-drawing tends to realign the crystals.

High-strength steel in North America is available in three basic forms: 1) uncoated stress-relieved wires; 2) uncoated stress-relieved strands; and 3) uncoated high-strength steel bars. Stress-relieved wires and high-strength steel bars are not generally used for post-tensioning. High-strength strands are fabricated by helically twisting a group of six wires around a slightly larger center wire by a mechanical process called stranding. The resulting seven-wire strands are stress-relieved by a continuous heat treatment process to produce the required mechanical properties.

ASTM specification A416 specifies two grades of steel, 250 and 270 ksi (1724 and 1862 MPa), the higher strength being more common in the building industry. A modulus of elasticity of 27,500 ksi (189,610 MPa) is used for calculating the elongation of strands. To prevent the use of brittle steel, which would result in a failure pattern similar to that of an overreinforced beam, ASTM A-416 specifies a minimum elongation of 3.5% at rupture.

A special type of strand called low-relaxation strand is increasingly used because it has a very low loss due to relaxation, usually about 20 to 25% of that for stress-relieved strand. With this strand, less post-tensioning steel is required, but the cost is greater because of the special process used in its manufacture.

The corrosion of unbonded strand is possibile, but can be prevented by using galvanized strands. This is not, however, popular in North America because: 1) various anchorage devices in use for post-tensioned systems are not suitable for use with galvanized strand because of low coefficient of friction; 2) damage can result to the strand because the heavy bite of the anchoring system can ruin the galvanizing; and 3) galvanized strands are more expensive.

A little-understood and infrequent occurrence of great concern in engineering is the so-called stress corrosion that occurs in highly stressed strands. The reason for the phenomenon is little known, but chemicals such as chlorides, sulfides, and nitrates are known to start this type of corrosion under certain conditions. It is also known that high-strength steels exposed to hydrogen ions are susceptible to failure because of loss in ductility and tensile strength. This phenomenon is called hydrogen embrittlement and is best counteracted by confining the strands in an environment having a pH value greater than 8. Incidentally, the pH value of concrete is ±12.5. Therefore, it produces a good environment.

7.3.2.2. Concrete

Concrete with compressive strengths of 5000 to 6000 psi (34 to 41 MN) is commonly employed in the prestress industry. This relative high strength is desirable for the following reasons. First, high-strength concrete is required to resist the high stresses transferred to the concrete at post-tensioning anchors. Second, it is needed to develop rapid strength gain for productivity. Third, high-strength concrete has higher resistance in tension, shear, bond, and bearing, and is desirable for prestressed structures that are typically under higher stresses than those with ordinary reinforced concrete. Fourth, its higher modulus of elasticity and smaller creep result in smaller loss of prestress.

Post-tensioned concrete is considered a self-testing system because, if the concrete is not crushed under the application of prestress, it should withstand subsequent loadings in view of the strength gain that comes with age. In practice it is not the 28-day strength that dictates the mix design, but rather the strength of concrete at the transfer of prestress.

Although high–early-strength (type III) Portland cement is well-suited for post-tension work because of its ability to gain the required strength for stressing relatively early, it is not generally used because of higher cost. Invariably, type I cement conforming to ASTM C-150 is employed in buildings.

The use of admixtures and fly ash is considered good practice. However, use of calcium chlorides or other chlorides is prohibited because the chloride ion may result in stress corrosion of prestressing tendons. Fly ash reduces the rate of strength gain, and therefore increases the time until stresses can be transferred, leading to loss of productivity.

A slump of between 3 to 6 in. (76 to 127 mm) gives good results. The aggregate used in the normal production of concrete is usually satisfactory in prestressed concrete, including lightweight aggregates. However, care must be exercised in estimating volumetric changes so that a reasonable prestress loss can be calculated. Lightweight aggregates manufactured using expanded clay or shale have been used in post-tensioned buildings. Lightweight aggregates that are not crushed after burning maintain their coating and therefore absorb less water. Such aggregates have drying and shrinkage characteristics similar to the normal-weight aggregates, although the available test reports are somewhat conflicting. The size of aggregate, whether lightweight or normal weight, has a more profound effect on shrinkage. Larger aggregates offer more resistance to shrinkage and also require less water to achieve the same consistency, resulting in as much as 40% reduction in shrinkage when the aggregate size is increased from, say, $\frac{3}{4}$ to $1\frac{1}{2}$ in. (19 to 38 mm). It is generally agreed that both shrinkage and creep are more functions of cement paste than of the type of aggregate. Lightweight aggregate has been gaining acceptance in prestressed construction since about 1955 and has a good track record.

7.3.3. Design Considerations

The design involves the following steps:

1. Determination of the size of concrete member.
2. Establishment of the tendon profile.
3. Calculation of the prestressing force.
4. Verification of the section for ultimate bending and shear capacity.
5. Verification of the serviceability characteristics, primarily in terms of stresses and long-term deflections.

Deflections of prestressed members tend to be small because under service loads they are usually uncracked and are much stiffer than nonprestressed members of the same cross section. Also, the prestressing force induces deflections in an opposite direction to those produced by external loads. The final deflection, therefore, is a function of tendon profile and the magnitude of prestress. Appreciating this fact, the ACI code does not specify minimum depth requirements for prestressed members. However, as a rough guide, the suggested span-to-depth ratios given in Table 7.7 can be used to establish the depth of continuous flexural members. Another way of looking at the suggested span-to-depth ratios is to consider, in effect, that prestressing increases the span range by about 30 to 40% over and above the values normally used in nonprestressed concrete construction.

Table 7.7   Approximate Span Depth Ratios[a] for Post-Tensioned Systems

Floor system	Simple spans	Continuous spans	Cantilever spans
One-way solid slabs	48–48	42–50	14–16
Two-way flat slabs	36–45	40–48	13–15
Wide band beams	26–30	30–35	10–12
One-way joists	20–28	24–30	8–10
Beams	18–22	20–25	7–8
Girders	14–20	16–24	5–8

[a] The values are intended as a preliminary guide for the design of building floors subjected to a uniformly distributed superimposed live load of 50 to 100 psf (2394 to 4788 Pa). For the final design, it is necessary to investigate for possible effects of camber, deflections, vibrations, and damping. The designer should verify that adequate clearance exists for proper placement of post-tensioning anchors.

The tendon profile is established based on the type and distribution of load with due regard to clear cover required for fire resistance and corrosion protection. Clear spacing between tendons must be sufficient to permit easy placing of concrete. For maximum economy, the tendon should be located eccentric to the center of gravity of the concrete section to produce maximum counteracting effect to the external loads. For members subjected to uniformly distributed loads, a simple parabolic profile is ideal, but in continuous structures parabolic segments forming a smooth reversed curve at the support are more practical. The effect is to shift the point of contraflecture away from the supports. This reverse curvature modifies the load imposed by post-tensioning from those assumed using a parabolic profile between tendon high points.

The post-tension force in the tendon immediately after releasing the hydraulic jack is less than the jacking force because of: 1) slippage of anchors; 2) frictional losses along tendon profile; and 3) elastic shortening of concrete. The force is reduced further over a period of months or even years due to change in the length of concrete member resulting from shrinkage and creep of concrete and relaxation of the highly stressed steel. The effective prestress is the force in the tendon after all the losses have taken place. For routine designs, empirical expressions for estimating prestress losses yield sufficiently accurate results, but in cases with unusual member geometry, tendon profile, and construction methods it may be necessary to make refined calculations.

Prestressing may be considered as a method of balancing a certain portion of the applied loads. This method, first developed by T. Y. Lin, is applicable to statically indeterminate systems just as easily as to statically determinate structures. Also, the procedure gives a simple method of calculating deflections by considering only that portion of the applied load not balanced by the prestress. If the effective prestress completely balances the applied load, the post-tensioned member will undergo no deflection and will remain horizontal, irrespective of the modulus of rigidity or flexural creep of concrete.

A question that usually arises in prestress design is how much of the applied load is to be balanced. The answer, however, is not simple. Balancing all the dead load often results in too much prestressing, leading to uneconomical design. On the other hand, there are situations in which the live load is significantly heavier than the dead load, making it more economical to prestress not only for full dead loads but also for a significant portion of the live load. However, in the design of typical floor framing systems, the prestressing force is normally selected to balance about 70 to 90% of the dead load and, occasionally, a small portion of the live load. This leads to an ideal condition with the structure having little or no deflection under dead loads.

Limiting the maximum tensile and compressive stresses permitted in concrete does not in itself assure that the prestressed member has an adequate factor of safety against flexural failure. Therefore, its nominal bending strength is computed in a procedure similar to that of a reinforced concrete beam. Underreinforced beams are assumed to have reached the failure load when the concrete strain reaches a value of 0.003. Since the yield point of prestressing steel is not well defined empirical relations based on tests are used in evaluating the strain and hence the stress in tendons.

The shear reinforcement in post-tensioned members is designed in a manner almost identical to that of nonprestressed concrete members, with due consideration for the longitudinal stresses induced by the post-tensioned tendons. Another feature unique to the design of post-tensioned members is the high stresses in the vicinity of anchors. Prestressing force is transferred to concrete at the tendon anchorages. Large stresses are developed in the concrete at the anchorages, which requires provision of well-positioned reinforcement in the region of high stresses. At a cross section of a beam sufficiently far away (usually 2 to 3 times the larger cross-sectional dimensions of the beam) from the anchor zone, the axial and bending stresses in the beam due to an eccentric prestressing force are given by the usual *P/A* and *MC/I* relations. But in the vicinity of stress application, the stresses are distributed in a complex manner. Of importance are the transverse tensile forces generated at the end blocks for which reinforcement is to be provided. The tensile stress has a maximum value at 90° to the axis of the prestressing force. Its distribution depends on the location of bearing area and its relative proportion with respect to the areas of the end face.

Because of the indeterminate nature and intensity of the stresses, the design of reinforcement for the end block is primarily based on empirical expressions. It usually consists of closely spaced stirrups tied together with longitudinal bars.

7.3.3.1. *Rules of Thumb*

Certain rules of thumb such as span-to-depth ratios and the average value of post-tensioning stresses are useful in conceptual design. The span-to-depth for slabs usually works out between $L/40$ and $L/50$, whereas for joists it is between $L/25$ and $L/35$. Beams can be much shallower than joists, with a depth in the range of $L/20$ and $L/30$. Band beams, defined as those with a width-to-depth ratio in excess to 4, offer perhaps the least depth without using as much concrete as flat slab construction. Although a span-to-depth ratio approaching 35 is adequate for band beams from strength and serviceability considerations, clearance requirements for proper detailing of anchorages and for accessing stressing equipment may dictate a deeper section. As a rule of thumb, a minimum compression of 125 to 150 psi (862 and 1034 kPa) is a practical and economical range for slabs. For beams, the range is 250 to 300 psi (1724 to 2068 kPa). Compression stresses as high as 500 psi (3447 kPa) have been used in band–beam systems. Even higher stresses may be required for transfer girders.

7.3.3.2. *Example Buildings*

The first example shows a two-way post-tensioned flat plate system for a residential tower (Fig. 7.26). The tendons are $^1/_2$-diameter (12.7 mm) stands that are banded in the north–south direction. Uniformly distributed tendons run from left to right across the building width. Additional tendons are used in the end panels to resist increased moments due to lack of continuity at one end.

As a second example, Fig. 7.27 shows the framing plan for a post-tensioned band–beam–slab system. Shallow beams only 16 in. (0.40 m) deep span across two exterior bays of 40 ft (12.19 m) and an interior bay of 21 ft (6.38 in). Post-tensioned slabs 8 in. (203 mm) deep span between the band beams, typically spaced at 30 ft (9.14 m) on center. In the design of the slab, additional beam depth is considered as a haunch at each end. Primary tendons for the slab run across the building width, while the tendons that control the temperature and shrinkage are placed in the north–south direction between the band beams.

7.3.4. Cracking Problems in Post-Tensioned Floors

Cracking caused by restraint to shortening is one of the biggest problems associated with post-tensioned floor systems. The reason is that shortening of a floor state is a time-dependent complex phenomenon. Only subjective empirical solutions exist to predict the behavior.

Shrinkage of concrete is the biggest contributor to shortening in both prestressed and nonprestressed concrete. In prestressed concrete, out of the total shortening, only about 15% is due to elastic shortening and creep. Therefore the problem is not in the magnitude of shortening itself, but in the manner in which it occurs.

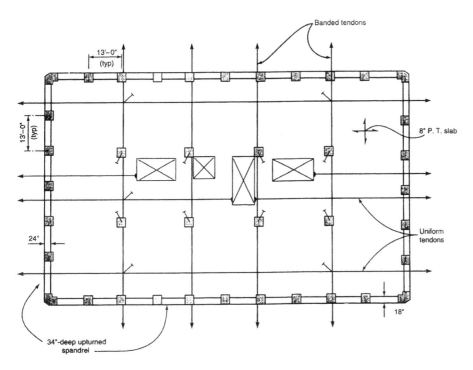

Figure 7.26 Two-way post-tensioned flat plate system for a residential tower, the Museum Tower, in Los Angeles, CA. (Structural engineers John A. Martin & Assoc., Inc., Los Angeles, CA.)

When a nonprestressed concrete slab tries to shorten, its movement is resisted internally by the bonded mild steel reinforcement. The reinforcement is put into compression while the concrete is in tension. As the concrete tension builds up, the slab cracks at fairly regular intervals allowing the ends of the slab to remain in the same position in which they were cast. In a manner of speaking, the concrete has shortened by about the same magnitude as a post-tensioned system, but not in overall dimensions. Instead of the total shortening occurring at the ends, the combined widths of many cracks which occur across the slab make up for the total shortening. The reinforcement distributes the shortening throughout the length of the slab in the form of numerous cracks. Thus reinforced

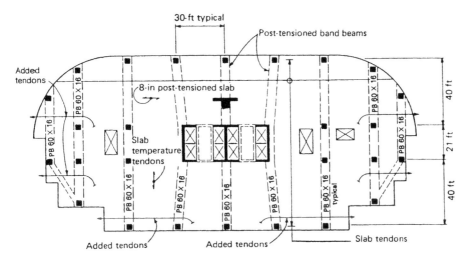

Figure 7.27. Typical floor plan of one-way post-tensioned slab system.

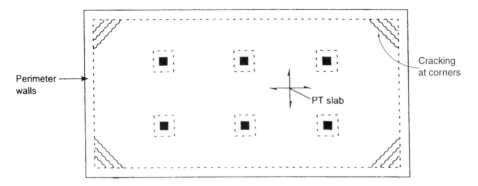

Figure 7.28.   Cracking in post-tensioned slab caused by restraint of perimeter walls.

concrete tends to take care of its own shortening problems internally by the formation of numerous small cracks, each small enough to be considered acceptable. Restraints provided by stiff vertical elements such as walls and columns tend to be of minor significance, since provision for total movement has been provided by the cracks in concrete.

This is not the case with post-tensioned systems in which shrinkage cracks, which would have formed otherwise, are closed by the post-tensioning force. Much less mild steel is present and consequently the restraint to the shortening provided is less. The slab tends to shorten at each end generating large restraining forces in the walls and columns particularly at the ends where the movement is greatest (Fig. 7.28). These restraining forces can produce severe cracking in the slab, walls, or columns at the slab extremities, causing problems to engineers and building owners alike. The most serious consequence is perhaps water leakage through the cracks.

The solution to the problem lies in eliminating the restraint by separating the slab from the restraining vertical elements. If a permanent separation is not feasible, cracking can be minimized by using temporary separations to allow enough of the shortening to occur prior to making the connection.

Cracking in a post-tensioned slab also tends to be proportional to initial pour size. Some general guidelines that have evolved over the years are as follows: 1) the maximum length between temporary pour strips (Fig. 7.29) is 150 ft (200 ft if restraint due to vertical elements is minimal); and 2) the maximum length of post-tensioned slab irrespective of the number of pour strips provided is 300 ft. The length of time for leaving the pour strips open is critical and can range anywhere from 30 to 60 days. A 30-day period is considered adequate for average restraint conditions with relatively centered, modest length walls, while a 60-day period is more the norm for severe shortening conditions with large pour sizes and stiff walls at the ends.

To minimize cracking caused by restraint to shortening, it is a good idea to provide a continuous mat of reinforcing steel in both directions of the slab. As a minimum, one-layer of #4 bars placed at mid-depth of slab, at 36 in. on centers both ways is recommended for typical conditions. For slab–pours in excess of 150 ft in length with relatively stiff walls at the ends, the minimum reinforcement should be increased to #4 bars at 24 in. on centers both ways.

7.3.5.   Concept of Secondary Moments

In a prestressed statically determinate beam, such as a single-span simply supported beam, the moment M_p due to prestress is given by the eccentricity e of prestress multiplied by

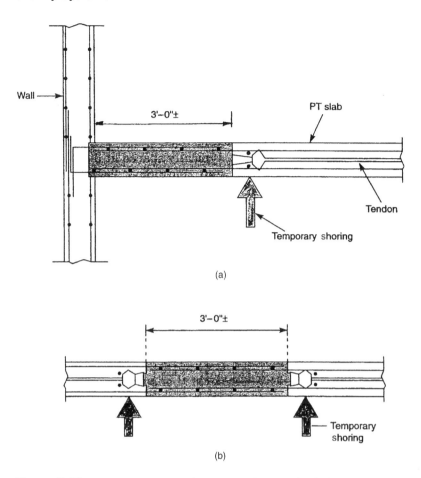

Figure 7.29. Temporary pour strip: (a) at perimeter of building; (b) at interior of slab.

the prestress P. In prestressed design, the moment $M_p = Pe$ is commonly referred to as the primary moment. In a simple beam or any other statically determinate beam, no support reactions can be induced by prestressing. No matter how much the beam is prestressed, only the internal stresses will be affected by the prestressing. The external reactions, being determined by statics, will depend on the dead and live loads, but are not affected by the prestress. Thus there are no secondary moments in a statically determinate beam. The total moment in the beam due to prestress is simply equal to the primary moment $M_0 = Pe$.

The magnitude and nature of secondary moments may be illustrated by considering a two-span, continuous, prismatic beam that is not restrained by its supports but remains in contact with them. The beam is prestressed with a straight tendon with force P and eccentricity e. See Fig. 7.30.

When the beam is prestressed, it bends and deflects. The bending of the beam can be such that the beam will tend to deflect itself away from B. Because the beam is restrained from deflection at B, a vertical reaction must be exerted to the beam to hold it there. The induced reaction produces secondary moments in the beam. These are called secondary because they are by-products of prestressing and do not exist in a statically determinate beam. The term secondary is misleading because the moments are secondary in nature, but not necessarily in magnitude.

One of the principal reasons for determining the magnitude of secondary moments is because they are required in the computations of ultimate flexural strength. An elastic

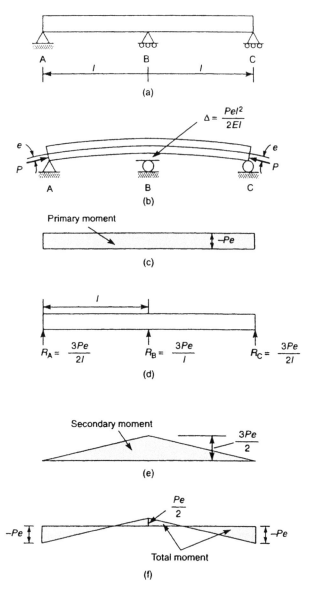

Figure 7.30. Concept of secondary moment: (a) two-span continuous beam; (b) vertical upward displacement due to PT; (c) primary moment; (d) reactions due to PT; (e) secondary moment; (f) final moments.

analysis of a prestressed beam offers no control over the failure mode or the factor of safety. To assure that prestressed members will be designed with an adequate factor of safety against failure, ACI 318-02, like its predecessors, requires that M_u, the moment due to factored service loads including secondary moments, not exceed ϕM_n, the flexural design strength of the member. The ultimate factored moment M_u is calculated by the following load combinations:

$$M_u = 1.2M_D + 1.6M_L + 1.0M_{sec} \qquad \text{(ACI 318-02)}$$

$$M_u = 1.4M_D + 1.7M_L + 1.0M_{sec} \qquad \text{(ACI 318-99)}$$

Since the factored load combination must include the effects due to secondary moments, its determination is necessary in prestress designs.

To further enhance our understanding of secondary moments, three numerical examples are given here:

1. A two-span continuous beam with a prestressd tendon at a constant eccentricity e.
2. The same beam as in the preceding example except the tendon is parabolic between the supports. There is no eccentricity of the tendon at the supports.
3. The same as in example 2, but the tendon has an eccentricity at the center support.

7.3.5.1. Design Examples.

Example 1.

Given. A two-span prestressed beam with a tendon placed at a constant eccentricity e from the C.G. of the beam. The prestress in the tendon is equal to P. (See Fig. 7.30.)

Required. Secondary moments in the beam due to prestress P.

Solution. The beam is statically indeterminate to the first degree because it is continuous at the center support B. It is rendered determinate by removing the support at B. Due to the moments $M_0 = Pe$ at the ends, the beam bends and deflects upward. The magnitude of vertical deflection δ_B due to moment M_0 is calculated using standard beam formulas such as the one that follows.

Beam Deflection Formula

Type of load	Slope as shown	Maximum deflection	Deflection equation
Simply supported beam		Bending moment applied at one end	
	$\theta_L = \frac{Ml}{6\,EI}$ $\theta_R = \frac{Ml}{3\,EI}$	$\delta = \frac{Ml^2}{9\sqrt{3}\,EI}$ at $x = l/\sqrt{3}$	$\delta = \frac{Mlx}{6\,EI}\left(1 - \frac{x^2}{l^2}\right)$

In our case, moment M is applied at both ends. Therefore

$$\delta = \frac{Mlx}{3EI}\left(1 - \frac{x^2}{l^2}\right)$$

$$\delta_{l/2} = \frac{Ml \times l}{3EI \times 2}\left(1 - \frac{l^2}{4l^2}\right)$$

$$= \frac{Ml^2}{8EI}$$

Also, for the example problem, $l = 2L$.

Therefore, deflection δ_B at support $B = \delta_L = \dfrac{M \times (2L)^2}{8EI}$

$$= \frac{ML^2}{2EI}$$

Since the beam is restrained from deflecting upward at B, a downward reaction $R_{B,sec}$ must be exerted to the beam to hold it there. The reaction $R_{B,sec}$ is given by

$$\delta_B = \frac{R_{B,sec}(2L)^3}{48\,EI}$$

$$R_{B,sec} = \frac{48EI}{(2L)^3}\,\delta_B$$

$$= \frac{48EI}{(2L)^3} \times \frac{M_0 L^2}{2EI}$$

$$= \frac{3M_0}{L}$$

The secondary moment induced due to the reaction $R_{B,sec}$ at the support B is given by

$$M_{B,sec} = R_{B,sec} \times \frac{2L}{4}$$

$$= \frac{3M_0}{L} \times \frac{2L}{4}$$

$$= \frac{3}{2}\,M_0$$

Observe that in this example, the secondary moment at B = 150% of the primary moment due to prestress. The secondary moment is thus secondary in nature, but not in magnitude.

Example 2A.

Given.   The two-span prestressed concrete beam shown in Fig. 7.31A(a) has a parabolic tendon in each span with zero eccentricity at the A and C ends, and at the center support B. Eccentricity of the tendon at the center of each span = 1.7 ft. The prestress force $P = 263.24$ kips.

Required.   Secondary reactions and moments.

Solution.   The approach here is similar to that typically used in commercially available computer programs. However, in the computer programs, statically indeterminate structures such as the example problem, are typically analyzed using a stiffness matrix approach. Here we take the easy street: We use beam formulas to analyze the two-span continuous beam. It should be noted that the analysis could be performed using other classical methods such as the moment distribution method or slop-deflection method.

First we determine the equivalent load due to prestress $P = 263.24$ kips acting at eccentricity $e = 1.7$ ft at the center of the two spans. The equivalent load consists of: 1) an upward uniformly distributed load W_p due to drape in the tendon; 2) a horizontal compression P equal to 263.24 kips at the ends; 3) downward loads at A, B, and C to equilibrate the upward load W_p; and 4) additional reactions at A, B, and C due to the restraining effect of support at B. The last set of loads need not be considered for this example, because the loads are implicitly included in the formulas for the statically indeterminate beam.

Of the equivalent loads shown in Fig. 7.31A(b), only the uniformly distributed load W_p corresponding to P acting at eccentricity e induces bending action in the beam. W_p is

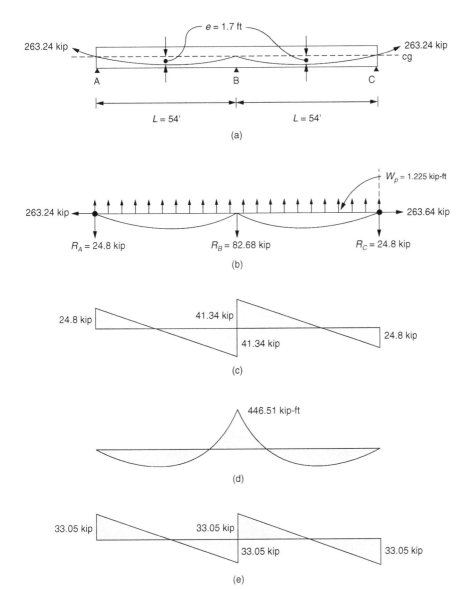

Figure 7.31A. Concept of secondary moments—example 2A: (a) two-span continuous prestressed beam; (b) equivalent loads due to prestress, consisting of upward UDL, horizontal compression due to prestress W_p, and downward loads at A, B, and C; (c) shear force diagram, statically indeterminate beam; (d) moment diagram, statically indeterminate beam; (e) primary shear force diagram; (f) primary moment diagram; (g) secondary shear forces; (h) secondary moments.

determined by the relation

$$Pe = \frac{W_p L^2}{8}$$

$$W_p = \frac{Pe \times 8}{L^2}$$

$$= \frac{263.24 \times 1.7 \times 8}{54^2}$$

$$= 1.227 \text{ kip-ft}$$

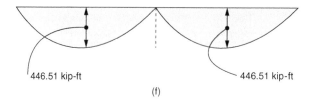

446.51 kip-ft   446.51 kip-ft

(f)

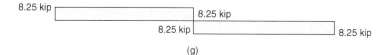

(g)

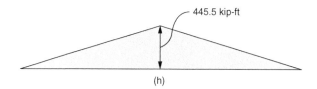

445.5 kip-ft

(h)

Figure 7.31A. *(Continued)*

Having determined the equivalent loads, we can proceed to determine the bending moments in our statically indeterminate beam, as for any continuous beam. As mentioned earlier, we use the formulas for continuous beams given in standard textbooks. One such formula follows.

Continuous Beam with Two Equal Spans and Uniform Load on Both Spans

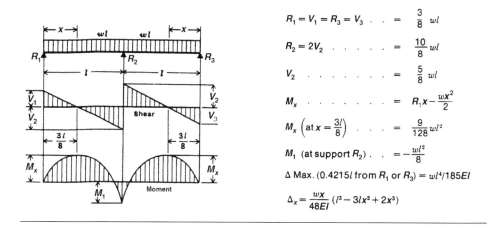

$$R_1 = V_1 = R_3 = V_3 \quad . \quad . \quad = \quad \frac{3}{8}wl$$

$$R_2 = 2V_2 \quad . \quad . \quad . \quad . \quad = \quad \frac{10}{8}wl$$

$$V_2 \quad . \quad . \quad . \quad . \quad . \quad . \quad = \quad \frac{5}{8}wl$$

$$M_x \quad . \quad . \quad . \quad . \quad . \quad . \quad = \quad R_1 x - \frac{wx^2}{2}$$

$$M_x\left(\text{at } x = \frac{3l}{8}\right) \quad . \quad . \quad . \quad = \quad \frac{9}{128}wl^2$$

$$M_1 \text{ (at support } R_2) . \quad . \quad = -\frac{wl^2}{8}$$

Δ Max. (0.4215*l* from R_1 or R_3) = $wl^4/185EI$

$$\Delta_x = \frac{wx}{48EI}(l^3 - 3lx^2 + 2x^3)$$

In our case, $w = W_p = 1.225$ kip-ft, $l = L = 54$ ft. Therefore,

$$V_1 = \frac{3}{8}W_pL$$

$$= \frac{3}{8} \times 1.225 \times 54$$

$$= 24.8 \text{ kips}$$

$$V_2 = \frac{5}{8} W_p L$$

$$= \frac{5}{8} \times 1.225 \times 54$$

$$= 41.34 \text{ kips}$$

$$M_1 = M_B = \frac{W_p L^2}{8} = \frac{1.225 \times 54^2}{8} = 446.50 \text{ kip-ft}$$

The shear force and bending moment diagrams are shown in Figs. 7.31A, parts (c) and (d).

Since the formulas account for the beam continuity, the resulting shear force and bending moments shown in Figs. 7.31A(c) and (d) include the effect of secondary moments. The resulting moment due to prestress, then, is the algebraic sum of the primary and secondary moments. Once the resulting moments are determined, the secondary moments can be calculated by the relation

$$M_{\text{bal}} = M_p + M_{\text{sec}}$$

where

M_{bal} = the resulting moment, also referred to as the total moment in the redundant beam due to equivalent loads

M_p = the primary moment that would exist if the beam were a statically determinate beam (M_p is given by the eccentricity of the prestress multiplied by the prestress.)

M_{sec} = the secondary moment due to redundant secondary reactions

With the known primary moment acting on the continuous beam, the secondary moment caused by induced reactions can be computed from the relation

$$M_{\text{sec}} = M_{\text{bal}} - M_p$$

A similar equation is used to calculate the shear forces.

The resulting secondary shear forces and bending moments are shown in Figs. 7.31A(g) and (h), while the primary shear forces and bending moments are shown in Figs. 7.31A(e) and (f).

Example 2B. Compatibility Method. To firm up our concept of secondary reactions and moments, perhaps it is instructive to redo the previous example using a compatibility approach. In this method the beam is rendered statically determinate by removing the redundant reaction at B. The net vertical deflection (which happens to be upward in our case) is calculated at B due to $W_p = 1.225$ kip-ft acting upward and a vertical downward load $= 1.225 \times 54 = 66.15$ kips acting downward at B. Observe that the reaction at B, along with those at A and C, equilibrates the vertical load of 1.225 kip-ft action on the tendon in its precise profile but does not necessarily guarantee compatibility at B.

Given. A two-span continuous beam analyzed previously, shown again for convenience in Fig. 7.31B.

Required. Secondary moments and shear forces using a compatibility approach.

Solution. The equivalent loads required to balance the effect of prestressed, draped tendons are shown in Fig. 7.31B(b). As before, $W_p = 1.225$ kip-ft. However, the reactions at A, B, and C do not include those due to secondary effects. The reactions are in equilibrium with load W_p, and do not necessarily assure continuity of the beam at support

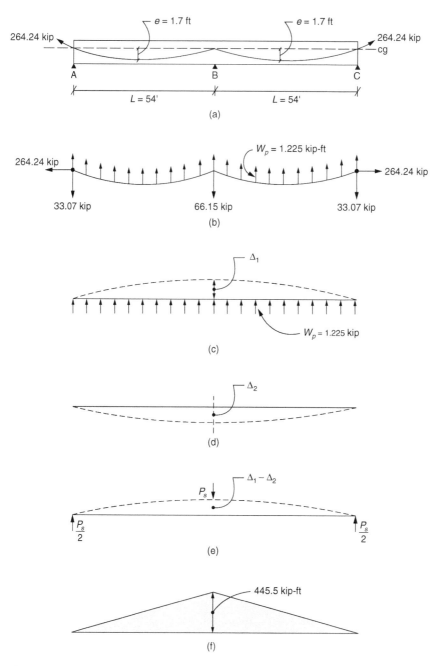

Figure 7.31B. Concept of secondary moments—example 2B, compatibility method: (a) two-span continuous beam; (b) equivalent loads; (c) upward deflection Δ_1 due to W_p (d) downward deflection Δ_2 due to a load of 66.15 kips at center span; (e) load P_s corresponding to $\Delta_1 - \Delta_2$; (f) secondary moments.

B. (If continuity were established, their magnitudes would have been the same as calculated in the previous example).

In determining the equivalent loads, we have not considered the effect of continuity at support B. Therefore, the beam has a tendency to move away from the support due to the upward-acting equivalent loads. Because the beam, by compatibility requirements, stays attached to support B, another set of reactions is needed to keep the beam in contact with support B. These are the secondary reactions, and the resulting moments are the secondary moments. Of the loads shown in Fig 7.31B(b), only the upward load $W_p =$ 1.225 kip-ft and the downward reaction $R_B = 66.15$ kips influence the vertical deflection at B. The upward deflection of the beam at B due to W_p is given by the standard formula

$$\Delta_{up} = \frac{5wl^4}{384EI} \qquad \text{(See Fig. 7.31B(c).)}$$

In our case, $w = W_p = 1.225$ kip-ft, $l = 2 \times 54 = 108$ ft. Therefore,

$$\Delta_{up} = \frac{5 \times 1.225 \times 108^4}{384EI}$$
$$= \frac{2,170,050}{EI} \qquad \uparrow \text{ upward}$$

The downward deflection at B due to reaction R_B is given by

$$\Delta_{down} = \frac{R_B L^3}{48EI}$$

In our case $R_B = 66.15$ kips, $L = 108$ ft.

$$\Delta_{down} = \frac{66.15 \times 108^3}{48EI}$$
$$= \frac{1,736,040}{EI} \qquad \downarrow \text{ downward} \qquad \text{(See Fig. 7.31B(d).)}$$

The net deflection at B

$$\Delta_B = \frac{2,170,50 - 1,736,040}{EI} = \frac{434,010}{EI} \qquad \uparrow \text{ upward} \qquad \text{(See Fig. 7.31B(e).)}$$

Because the beam is attached to support B, for compatibility the vertical deflection at B should be zero. This condition is satisfied by imposing a vertically downward secondary reaction $R_{B,sec}$ at B given by the relation

$$\frac{R_{B,sec} \times 108^3}{48EI} = \frac{434,010}{EI}$$
$$R_{B,sec} = 16.54 \text{ kips}$$

The resulting secondary reactions and moments shown in Fig. 7.31B(f) are exactly the same as calculated previously.

Example Problem #3.

Given. Same data as in problem #2. The only difference is that the tendon at the center support B has an eccentricity of 0.638′.

Required. Secondary moments using a compatibility-type analysis.

Solution. The equivalent loads balancing the effects of prestressed, draped tendons with eccentricities at the centers of spans and at the interior support B are shown in Fig. 7.31C.

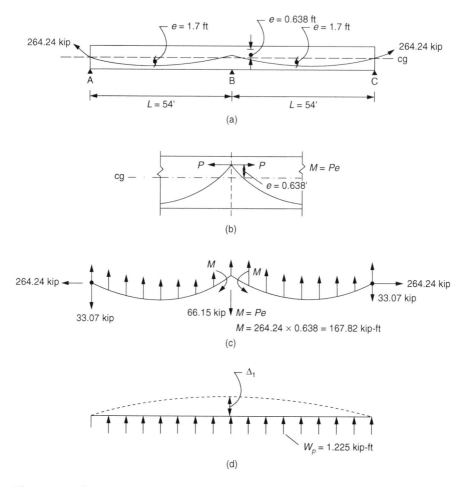

Figure 7.31C. Concept of secondary moments—example 3: (a) two-span continuous beam; (b) equivalent moment $M = Pe$ at center of span; (c) equivalent loads and moments; (d) upward deflection Δ_1 due to W_p; (e) downward deflection Δ_2 due to a load of 66.15 at center of span; (f) downward deflection Δ_3 due to moments $M = Pe$ at center of span; (g) load P_s corresponding to $\Delta_1 - \Delta_2 - \Delta_3$; (h) secondary moments.

Notice the two equal and opposite moments equal to the prestress of 263.24 kips times the eccentricity of 0.68 ft at the center support. See Fig. 7.31C(b) and (c). The solution follows the same procedure as used in the previous example, except that we include the effect of moments at B in deflection calculations.

As before, $W_p = 1.23$ kip-ft. Upward deflection at B due to W_p is given by

$$
\begin{aligned}
\Delta_{up} &= \frac{5}{384} \quad \frac{W_p(2L)^4}{EI} \\
&= \frac{5 \times 1.23 \times (2 \times 54)^4}{384 \times EI} \\
&= \frac{2{,}170{,}050}{EI} \qquad \uparrow \text{ upward} \qquad\qquad \text{(See Fig. 7.31C(d))}
\end{aligned}
$$

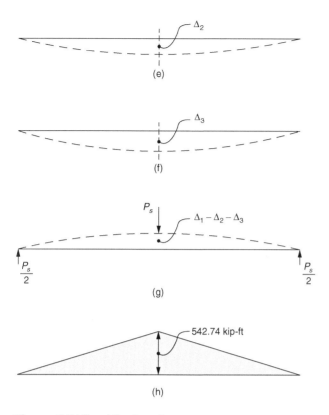

Figure 7.31C. (*Continued*)

The vertical reaction R_B at B to maintain vertical equilibrium is equal to $1.23 \times 54 = 66.42$ kips. The downward deflection at B due to this load is

$$
\begin{aligned}
\Delta_{\text{down}, R_B} &= \frac{R_B \times (54 \times 2)^3}{48EI} \\
&= \frac{66.42 \times (108)^3}{48EI} \\
&= \frac{1{,}743{,}126}{EI} \qquad \downarrow \text{ downward} \qquad \text{(See Fig. 7.31C(e).)}
\end{aligned}
$$

In addition to the upward and downward deflections at B, there is a third component to the vertical deflection due to the moment at B = $263.24 \times 0.638 = 167.95$ kip-ft.

For purposes of deflection calculations, moment M_B at B may be replaced by an equivalent point load equal to $\frac{2M_B}{L}$.

The downward deflection at B due to M_B, then, is

$$
\begin{aligned}
\Delta_{\text{down}, M_B} &= \frac{2M_B}{L} \times \frac{(2L)^3}{48EI} \\
&= \frac{M_B L^2}{3EI} \qquad \qquad \text{(See Fig. 7.31C(f).)}
\end{aligned}
$$

For the example, $M_B = 167.95$ kip-ft, $L = 54$ ft

$$\Delta_{\text{down},M_B} = \frac{167.95 \times 54^2}{3EI}$$

$$= \frac{163,150}{EI} \qquad \downarrow \text{ downward}$$

The net upward deflection due to W_p, R_B, and M_B is

$$\frac{1}{EI}(2,170,050 - 1,43,126 - 163,150) = \frac{263,774}{EI} \quad \uparrow \text{ upward} \qquad \text{(See Fig 7.31C(g).)}$$

The secondary reaction to establish vertical compatibility at B is given by

$$\frac{R_{B,\text{sec}} \times (2L)^3}{48EI} = \frac{263,774}{EI}$$

$$R_{B,\text{sec}} = \frac{48 \times 263,774}{(2 \times 54)^3} = 10.05 \text{ kips}$$

The secondary moments due to this redundant reaction are shown in Fig. 7.31C(h).

7.3.6. Step-by-Step Design Procedure

The aim of post-tension design is to determine the required prestressing force and hence the number, size, and profile of tendons for satisfactory behavior at service loads. The ultimate capacity must then be checked at critical sections to assure that prestressed members have an adequate factor of safety against failure.

The design method presented in this section uses the technique of load balancing in which the effect of prestressing is considered as an equivalent load. Take, for example, a prismatic simply supported beam with a tendon of parabolic profile, shown in Fig. 7.32. The tendon exerts a horizontal force equal to $P \cos \theta = P$ (for small values of θ) at the ends along with vertical components equal to $P \sin \theta$. The vertical component is neglected in design because it occurs directly over the supports. In addition to these loads, the parabolic tendon exerts a continuous upward force on the beam along its entire length. By neglecting friction between the tendon and concrete, we can assume that: 1) the upward pressure exerted is normal to the plane of contact; and 2) tension in tendon is constant. The upward pressure exerted by the tendon is equal to the tension in the tendon divided by the radius of curvature of the tendon profile. Due to the shallow nature of post-tensioned structures, the vertical component of the tendon force may be assumed constant. Considering one-half of the beam as a free body (Fig. 7.33b), the vertical load exerted by the tendon may be derived by summing moments about the left support. Thus the equivalent, load $W_p = 8Pe/L^2$. Equivalent loads and moments produced by other types of tendon profile are shown in Fig. 7.32b through d.

The step-by-step procedure is as follows:

1. Determine preliminary size of prestressed concrete members using the values given in Table 7.6 as a guide.
2. Determine section properties of the member: area A, moment of inertia I, and section moduli S_t and S_b.

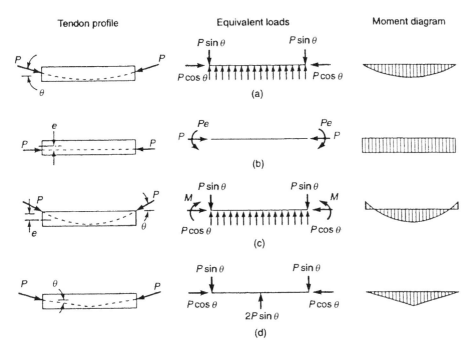

Figure 7.32. Equivalent loads and moments produced by prestressed tendons: (a) upward uniform load due to parabolic tendon; (b) constant moment due to straight tendon; (c) upward uniform load and end moments due to parabolic tendon not passing through the centroid at the ends; (d) vertical point-load due to sloped tendon.

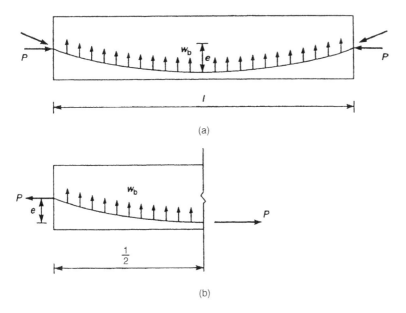

Figure 7.33. Load balancing concept: (a) beam with parabolic tendon; (b) free-body diagram.

3. Determine tendon profile with due regard to cover and location of mild steel reinforcement.
4. Determine effective span L_e by assuming $L_1 = 1/16$ to $1/19$ of the span length for slabs, and $L = 1/10$ to $1/12$ of the span length for beams. L_1 is the distance between the center line of support and the inflection point. The concept of effective length will be explained shortly.
5. Start with an assumed value for balanced load w_p equal to, say, 0.7 to 0.9 times the total dead load.
6. Determine the elastic moments for the total dead plus live loads (working loads). For continuous beams and slabs use a computer plane-frame analysis program, moment distribution method, or ACI coefficients, if applicable, in decreasing order of preference.
7. Reduce negative moments to the face of supports.
8. By proportioning the unbalanced load to the total load, determine the unbalanced moments at M_{ub} at critical sections such as at the supports and at the center of spans.
9. Calculate the bending stresses f_b and f_t at the bottom and top of the cross section due to M_{ub} at critical sections. Typically at supports the stresses f_t and f_b are in tension and compression, respectively. At center of spans the stresses are typically compression and tension at top and bottom, respectively.
10. Calculate the minimum required post-tension stress f_p by using the following equations.
   For negative zones of one-way slabs and beams:

$$f_p = f_t - 6\sqrt{f_c'}$$

   For positive moments in two-way slabs:

$$f_p = f_t - 2\sqrt{f_c'}$$

11. Find the post tension force P by the relation $P = f_p \times A$ where A is the area of the cross section of the beam.
12. Calculate the balanced load W_p due to P by the relation

$$w_p = \frac{8 \times Pe}{L_e^2}$$

   where

   e = drape of the tendon

   L_e = effective length of tendon between inflection points.

13. Compare the calculated value of W_p from step 12 with the value assumed in step 5. If they are about the same, the selection of post-tension force for the given loads and tendon profile is complete. If not, repeat steps 9–13 with a revised value of $W_p = 0.75W_{p1} + 0.25W_{p2}$. W_{p1} is the value of W_p assumed at the beginning of step 5, and W_{p2} is the derived value of W_p at the end of step 12. Convergence is fast requiring no more than three cycles in most cases.

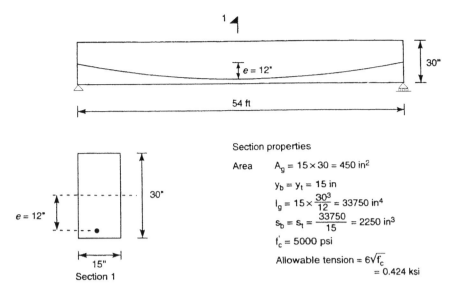

Figure 7.34. Preliminary design: simple span beam.

7.3.6.1. Simple Span Beam

The concept of preliminary design discussed in this section is illustrated in Fig. 7.34 where a parabolic profile with an eccentricity of 12 in. is selected to counteract part of the applied load consisting of a uniformly distributed dead load of 1.5 kip-ft and a live load of 0.5 kip-ft.

In practice, it is rarely necessary to provide a prestress force to fully balance the imposed loads. A value of prestress, often used for building system, is 75 to 95% of the dead load. For the illustrative problem, we begin with an assumed 80% of the dead load as the unbalanced load.

First Cycle   The load being balanced is equal to $0.80 \times 1.5 = 1.20$ kip-ft. The total service dead plus live load $= 1.5 + 0.5 = 2.0$ kip-ft, of which 1.20 kip-ft is assumed in the first cycle to be balanced by the prestressing force in the tendon. The remainder of the load equal to $2.0 - 1.20 = 0.80$ kip-ft acts vertically downward, producing a maximum unbalanced moment M_{ub} at center span given by

$$M_{ub} = 0.80 \times \frac{54^2}{8}$$
$$= 291.6 \text{ kip-ft}$$

The tension and compression in the section due to M_{ub} is given by

$$f_c = f_b = \frac{291.6 \times 12}{2250}$$
$$= 1.55 \text{ ksi}$$

The minimum prestress required to limit the tensile stress to $6\sqrt{f_c'} = 0.424$ is given by

$$f_p = 1.55 - 0.424 = 1.13 \text{ ksi}$$

Therefore, the required minimum prestressing force P = area of beam $\times$ 1.13 = 450 $\times$ 1.13 = 509 kips. The load balanced by this force is given by

$$W_p \times \frac{54^2}{8} = Pe = 509 \times 1$$

and so W_p = 1.396 kip-ft compared to the value of 1.20 used in the first cycle. Since these two values are not close to each other, we repeat the above calculations starting with a more precise value for W_p in the second cycle.

Second Cycle   We start with a new value of W_p by assuming a new value equal to 75% of the initial value + 25% of the derived value. The new value of

$$W_p = 0.75 \times 1.20 + 0.25 \times 1.396 = 1.25 \text{ kip-ft}$$

$$M_{ub} = (2 - 1.25) \times \frac{54^2}{8} = 273.3 \text{ kip-ft}$$

$$f_b = f_t = \frac{273.3 \times 12}{2250} = 1.458 \text{ ksi}$$

The minimum stress required to limit the tensile stress to $6\sqrt{f_c'} = 6\sqrt{5000} = 0.424$ ksi is given by

$$f_p = 1.458 - 0.424 = 1.03 \text{ ksi}$$

Minimum prestressing force $P = 1.03 \times 450 = 465$ kips. The balanced load corresponding to the prestress value of 465 is given by

$$W_p = \frac{8Pe}{L^2} = \frac{8 \times 465 \times 1}{54^2} = 1.27 \text{ kip-ft}$$

Therefore, W_p = 1.27 kip-ft, nearly equal to the value assumed in the second cycle. Thus the minimum prestress required to limit the tensile stress in concrete to $6\sqrt{f_c}$ is 465 kips.

To demonstrate how rapidly the method converges to the desired answer, we will rework the problem by assuming an initial value of W_p = 1.0 kip-ft in the first cycle.

First Cycle

$$W_p = 1.0 \text{ kip-ft}$$

$$M_{ub} = (2 - 1) \times \frac{54^2}{8} = 364.5 \text{ kip-ft}$$

$$f_b = f_t = \frac{364.5 \times 12}{2250} = 1.944 \text{ ksi}$$

$$f_p = 1.944 - 0.454 = 1.49 \text{ ksi}$$

$$P = 1.49 \times 450 = 670.5 \text{ kips}$$

$$W_p \times \frac{54^2}{8} = 670.5 \times 1$$

$$W_p = 1.84 \text{ kip-ft}$$

compared to 1.0 kip-ft used at the beginning of the first cycle.

Second Cycle

$$W_p = 0.75 \times 1 + 0.25 \times 1.84 = 1.21 \text{ kip-ft}$$

$$M_{ub} = (2 - 1.21) \times \frac{54^2}{8} = 288 \text{ kip-ft}$$

$$f_b = f_c = \frac{288 \times 12}{2250} = 1.536 \text{ ksi}$$

$$f_p = 1.536 - 0.454 = 1.082 \text{ ksi}$$

$$P = 1.082 \times 450 = 486.8 \text{ kips}$$

$$W_p = \frac{486.8 \times 1 \times 8}{54^2}$$
$$= 1.336 \text{ kip-ft}$$

compared to the value of 1.21 used at the beginning of second cycle.

Third Cycle

$$W_p = 0.75 \times 1.21 - 1.21 \times 0.25 \times 1.336 = 1.24 \text{ kip-ft}$$

$$M_{ub} = (2 - 1.24) \times \frac{54^2}{8} = 276.67 \text{ kip-ft}$$

$$f_b = f_c = \frac{276.47 \times 12}{2250} = 1.475 \text{ ksi}$$

$$f_p = 1.475 - 0.454 = 1.021 \text{ ksi}$$

$$P = 1.021 \times 450 = 459.3 \text{ kips}$$

$$W_p = 459.3 \times \frac{1 \times 8}{54^2} = 1.26 \text{ kip-ft}$$

compared to 1.24 assumed at the beginning of third cycle. The value of 1.26 kip-ft is considered close enough for design purposes.

7.3.6.2. *Continuous Spans*

The above example illustrates the salient features of load balancing. Generally, the prestressing force is selected to counteract or balance a portion of dead load, and under this loading condition the net stress in the tension fibers is limited to a value $= 6\sqrt{f_c'}$. If it is desired to design the member for zero stress at the bottom fiber at center span (or any other value less than the code allowed maximum value of $6\sqrt{f_c'}$), it is only necessary to adjust the amount of post-tensioning provided in the member.

    There are some qualifications to the foregoing procedure that should be kept in mind when applying the technique to continuous beams. Chief among them is the fact that it is not usually practical to install tendons with sharp break in curvature over supports, as shown in Fig. 7.35a. The stiffness of tendons requires a reverse curvature (Fig. 7.35b) in the tendon profile with a point of contraflexure some distance from the supports. Although this reverse curvature modifies the equivalent loads imposed by post-tensioning from those assumed for a pure parabolic profile between the supports, a simple revision to the effective length of tendon, as will be seen shortly, yields results sufficiently accurate for preliminary designs.

    Consider the tendon profiles shown in Figs. 7.36a,b for a typical exterior and an interior span. Observe three important features.

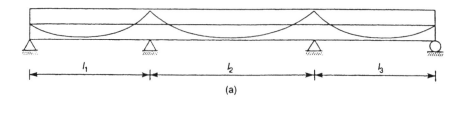

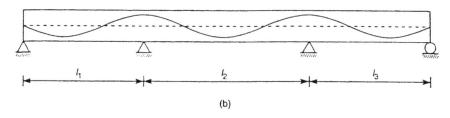

Figure 7.35. Tendon profile in continuous beams: (a) simple parabolic profile; (b) reverse curvature in tendon profile.

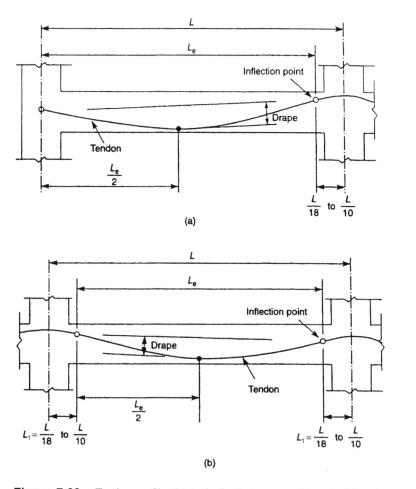

Figure 7.36. Tendon profile: (a) typical exterior span; (b) typical interior span.

1. The effective span L_e, the distance between the inflection points which is considerably shorter than the actual span.
2. The sag or drape of the tendon is numerically equal to average height of inflection points, less the height of the tendon midway between the inflection points.
3. The point midway between the inflection points is not necessarily the lowest point on the profile.

The upward equivalent uniform load produced by the tendon is given by

$$W_p = \frac{8Pe}{L_e^2}$$

where

W_p = equivalent upward uniform load due to prestress
P = prestress force
e = cable drape between inflection points
L_e = effective length between inflection points

Note that relatively high loads acting downward over the supports result from the sharply curved tendon profiles located within these regions (Fig. 7.37).

Since the large downward loads are confined to a small region, typically 1/10 to 1/8 of the span, their effect is secondary as compared to the upward loads. Slight differences occur in the negative moment regions between the applied load moments and the moment due to prestressing force. The differences are of minor significance and can be neglected in the design without losing meaningful accuracy.

As in simple spans the moments caused by the equivalent loads are subtracted from those due to applied loads, to obtain the net unbalanced moment that produces the flexural stresses. To the flexural stresses, the axial compressive stresses from the prestress are added to obtain the final stress distribution in the members. The maximum compressive and tensile stresses are compared to the allowable values. If the comparisons are favorable, an acceptable design has been found. If not, either the tendon profile or the force (and very rarely the cross-sectional shape of the structure) is revised to arrive at an acceptable solution.

In this method, since the moments due to equivalent loads are linearly related to the moments due to applied loads, the designer can bypass the usual requirement of determining the primary and secondary moments.

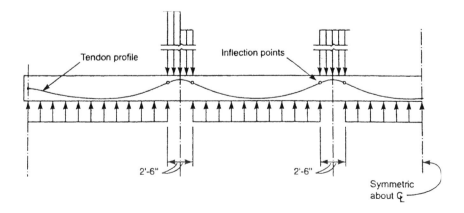

Figure 7.37. Equivalent loads due to prestress.

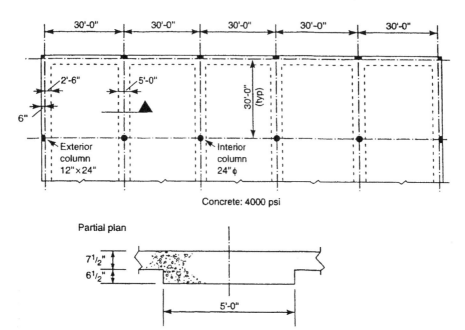

Figure 7.38.   Example 1: one-way post-tensioned slab.

7.3.6.2.1.   Example 1: One-Way Post-Tensioned Slab.

Given a 30'-0" column grid layout, design a one-way slab spanning between the beams shown in Fig. 7.38.

Slab and beam depths:

Clear span of slab = 30 – 5 = 25 ft

$$\text{Recommended slab depth} = \frac{\text{span}}{40} = \frac{25 \times 12}{40} = 7.5 \text{ in.}$$

Clear span for beams = 30 ft center-to-center span, less 2'-0" for column width = 30 – 2 = 28 ft

$$\text{Recommended beam depth} = \frac{\text{span}}{25} = \frac{28 \times 12}{25} = 13.44 \text{ in.} \quad \text{Use 14 in.}$$

Loading:

Dead load: 7.5" slab	= 94 psf
Mech. and lights	= 6 psf
Ceiling	= 6 psf
Partitions	= 20 psf
Total dead load	= 126 psf
Live load: Office load	= 100 psf

Code minimum is 50 psf
Use 100 psf per owner's request

Total $D + L$ = 226 psf

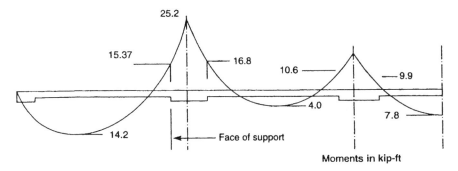

Figure 7.39. Example 1: one-way post-tensioned slab service load, $(D + L)$, moment diagram.

Slab design: Slab properties for 1'-0" wide strip:

$$I = \frac{bd^3}{12} = 12 \times \frac{7.5^3}{12} = 422 \text{ in.}^4$$

$$S_{top} = S_{bot} = \frac{422}{3.75} = 112.5 \text{ in.}^3$$

$$\text{Area} = 12 \text{ in.} \times 7.5 = 90 \text{ in.}^2$$

A 1-ft width of slab is analyzed as a continuous beam. The effect of column stiffness is ignored.

The moment diagram for a service load of 226 plf is shown in Fig. 7.39.

Moments at the face of supports have been used in the design instead of center line moments. Negative center line moments are reduced by a "$Va/3$" factor ($V =$ shear at that support, $a =$ total support width), and positive moments are reduced by $Va/6$ using average adjacent values for shear and support widths. A frame analysis may be course be used to obtain more accurate results.

The design of continuous strands will be based on the negative moment of 10.6 kip-ft. The additional prestressing required for the negative moment of 16.8 kip-ft will be provided by additional tendons in the end bays only.

Determination of Tendon Profile. Maximum tendon efficiency is obtained when the cable drape is as large as the structure will allow. Typically, the high points of the tendon over the supports and the low point within the span are dictated by concrete cover requirements and the placement of mild steel.

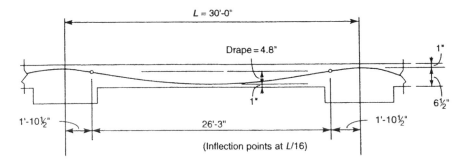

Figure 7.40. Example 1: one-way post-tensional slab tendon profile, interior bay.

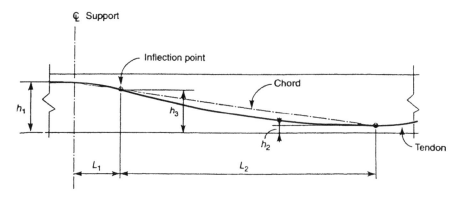

Figure 7.41. Dimensions for determining tendon drape.

The high and low points of tendon in the interior bay of the example problem are shown in Fig. 7.40. Next, the location of inflection points are determined. For slabs, the inflection points usually range within 1/16 to 1/19 of the span. The fraction of span length used is a matter of judgment, and is based on the type of structure. For this example, we choose 1/16 of span which works out to 1'-10 $\frac{1}{2}$".

An interesting property useful in determining the tendon profile shown in Fig. 7.41 is that, if a straight line (chord) is drawn connecting the tendon high point over the support and the low point midway between, it intersects the tendon at the inflection point. Thus, the height of the tendon can be found by proportion. From the height, the bottom cover is subtracted to find the drape.

Referring to Fig. 7.41,

$$\text{Slope of the chord line} = \frac{h_1 - h_2}{(L_1 + L_2)}$$

$$h_s = h_2 + L_2 \times (\text{Slope})$$

$$= h_2 + \frac{L_2(h_1 - h_2)}{(L_1 + L_2)}$$

$$\text{This simplifies to } h_3 = \frac{(h_1 L_2 + h_2 L_1)}{(L_1 + L_2)}$$

The drape h_d is obtained by subtracting h_2 from the foregoing equation. Note that notation e is also used in these examples to denote drape h_d.

In this case, the height of the inflection point is exact for symmetrical layout of the tendon about the center span. If the tendon is not symmetrical, the value is approximate but sufficiently accurate for preliminary design.

Returning to our example problem we have $h_1 = 6.5"$, $h_2 = 1"$, $L_1 = 1.875'$, and $L_2 = 13.125'$.

Height of tendon at the inflection point:

$$h_3 = \frac{(h_1 L_2 + h_2 L_1)}{(L_1 + L_2)}$$

$$= \frac{6.5 \times 13.125 + 1 \times 1.875}{(1.875 + 13.125)} = 5.813 \text{ in.}$$

Drape $h_d = e = 5.813 - 1 = 4.813"$ in.   Use 4.8"

Allowable stresses from ACI 318-99 are as follows:

f_t = tensile stress $6\sqrt{f_c'}$.

f_c = compressive stress = $0.45 f_c'$

For f_c' = 4000 psi concrete:

$f_t = 6\sqrt{4000} = 380$ psi

$f_c = 0.45 \times 4000 = 1800$ psi

Design of Through Strands. The design procedure is started by making an initial assumption of the equivalent load produced by the prestress. A first value of 65% of the total dead load is used.

First Cycle   Assume

$W_p = 0.65 W_d$

where

W_p = equivalent upward load due to post-tensioning, also denoted as W_{pt}
W_d = total dead load

Therefore, $W_p = 0.65 \times 126 = 82$ plf.
The balancing moment caused by the equivalent load is calculated from

$$M_{pt} = M_s \frac{W_{pt}}{W_s}$$

where

M_{pt} = balancing moment due to equivalent load
(also indicated by notation M_b)
M_s = moment due to service load, $D + L$
W_s = total applied load, $D + L$

In our example, M_s = 10.6 kip-ft for the interior span.

$$M_{pt} = 10.6 \times \frac{82}{226} = 3.85 \text{ kip-ft}$$

Next, M_{pt} is subtracted from M_s to give the unbalanced moment M_{ub}. The flexural stresses are then obtained by dividing M_{ub} by the section moduli of the structure's cross section at the point where M_s is determined. Thus

$$f_t = \frac{M_{ub}}{S_t}$$

$$f_b = \frac{M_{ub}}{S_b}$$

In our case, M_{ub} = 10.6 − 3.85 = 6.75 kip-ft. The flexural stress at the top of the section is found by

$$f_t = \frac{M_{ub}}{S_t} = \frac{6.75 \times 12}{112.5} = 0.72 \text{ ksi}$$

The minimum required compressive prestress is found by subtracting the maximum allowable tensile stress f_a given below, from the tensile stresses calculated above. The smallest required compressive stress is:

$$f_p = f_{ts} - f_a$$

where

f_{ts} = the computed tensile stress
$f_a = 6\sqrt{f_c'}$ for one-way slabs or beams for the negative zones
$f_a = 2\sqrt{f_c'}$ for positive moments in two-way slabs

In our case,

$$f_p = 0.720 - 0.380 = 0.34 \text{ ksi}$$

and

$$P = 0.34 \times 7.5 \times 12 = 30.60 \text{ kip-ft}$$

Use the following equation to find the equivalent load due to prestress:

$$W_p = \frac{8Pe}{L_e^2}$$

$$= 8 \times \frac{30.6 \times 4.81}{12 \times (26.25)^2} = 0.142 \text{ klf} = 142 \text{ plf}$$

This is more than 82 plf. N.G.

Since the derived value of W_p is not equal to the initial assumed value, the procedure is repeated until convergence is achieved. Convergence is rapid by using a new initial value for the subsequent cycle, equal to 75% of the previous initial value W_{p1} plus 25% of the derived value W_{p2}, for that cycle.

Second Cycle   Use the above criteria to find the new value of W_p for the second cycle.

$$W_p = 0.75W_{p1} + 0.25W_{p2} = 0.75 \times 82 + 0.25 \times 142 = 97 \text{ plf}$$

$$M_b = \frac{97}{226} \times 10.6 = 4.55 \text{ kip-ft}$$

$$M_{ub} = 10.6 - 4.55 = 6.05 \text{ kip-ft}$$

$$f_t = f_b = \frac{6.05 \times 12}{112.5} = 0.645 \text{ ksi}$$

$$f_p = 0.645 - 0.380 = 0.265 \text{ ksi}$$

$$P = 0.265 \times 90 = 23.89 \text{ kips}$$

$$W_p = \frac{8 \times 23.89 \times 4.81}{12 \times (26.25)^2} = 0.111 \text{ klf} = 111 \text{ plf}$$

This is more than 97 psf.   N.G.

Third Cycle

$$W_p = 0.75 \times 97 + 0.25 \times 111 = 100.5 \text{ plf}$$

$$M_b = \frac{100.5}{226} \times 10.6 = 4.71 \text{ kip-ft}$$

$$M_{ub} = 10.6 - 4.71 = 5.89 \text{ kip-ft}$$

$$f_t = f_b = \frac{5.89 \times 12}{112.5} = 0.629 \text{ ksi}$$

$$f_p = 0.629 - 0.380 = 0.248 \text{ ksi}$$

$$P = 0.248 \times 90 = 22.3 \text{ kips}$$

$$W_p = \frac{8 \times 22.3 \times 4.81}{12 \times (26.25)^2} = 0.104 \text{ klf} = 104 \text{ plf}$$

This is nearly equal to 100.5 plf. Therefore, satisfactory.
Check compressive stress at the section.

Bottom flexural stress = 0.629 ksi.
Direct axial stress due to prestress = 22.3/90 = 0.246 ksi
Total compressive stress = 0.629 + 0.246 = 0.876 ksi is less than $0.45f_c' = 1.8$ ksi.

Therefore, satisfactory.

End Bay Design. Design end bay prestressing using the same procedure for a negative moment of 15.37 kip ft.

Assume that at the left support, the tendon is anchored at the center of gravity of the slab with a reversed curvature. Assume further that the center of gravity of the tendon is at a distance 1.75 in. from the bottom of the slab. With these assumptions we have: $h_1 = 3.75''$, $h_2 = 1.75''$, $L_1 = 1.875'$, and $L_2 = 13.125'$.
The height of the tendon inflection point at left end:

$$h_3 = \frac{3.75 \times 13.125 + 1.75 \times 1.875}{15} = 3.25 \text{ in.}$$

The height of the right end:

$$h_3 = \frac{6.5 \times 13.125 + 1.75 \times 1.875}{15} = 5.906 \text{ in.}$$

$$\text{Average height of tendon} = \frac{3.25 + 5.906}{2} = 4.578'' \qquad \text{Use 4.6 in.}$$

Drape $h_d = e = 4.6 - 1.75 = 2.85$ in.

First Cycle We start with the first cycle, as for the interior span, by assuming $W_{pt} = 82$ plf.

$$M_{pt} = 15.37 \times \frac{82}{226} = 5.58 \text{ kip-ft}$$

$$M_{ub} = 15.37 - 5.58 = 9.79 \text{ kip-ft}$$

$$f_t = f_b = \frac{9.79 \times 12}{112.5} = 1.04 \text{ ksi}$$

$$f_p = 1.04 - 0.380 = 0.664 \text{ ksi}$$

$$P = 0.664 \times 90 = 59.7 \text{ kip-ft}$$

$$W_p = \frac{8 \times 59.7 \times 2.85}{12 \times (26.25)^2} = 0.165 \text{ klf} = 165 \text{ plf}$$

This is more than 82 plf.   N.G.

Second Cycle

$$W_p = 0.75 \times 82 + 0.25 \times 165 = 103 \text{ plf}$$

$$M_{pt} = 15.37 \times \frac{103}{226} = 7.0 \text{ kip-ft}$$

$$M_{ub} = 15.37 - 7.0 = 8.37 \text{ kip-ft}$$

$$f_t = f_b = \frac{8.37 \times 12}{112.5} = 0.893 \text{ ksi}$$

$$f_p = 0.893 - 0.380 = 0.513 \text{ ksi}$$

$$P = 0.513 \times 90 = 46.1 \text{ kips}$$

$$W_p = \frac{8 \times 46.1 \times 2.85}{12 \times (26.25)^2} = 0.127 \text{ klf} = 127 \text{ plf}$$

This is more than 103 plf.   N.G.

Third Cycle

$$W_p = 0.75 \times 103 + 0.25 \times 127 = 109 \text{ plf}$$

$$M_{pt} = 15.37 \times \frac{109}{226} = 7.41 \text{ kip-ft}$$

$$M_{ub} = 15.37 - 7.41 = 7.96 \text{ kip-ft}$$

$$f_t = f_b = \frac{7.96 \times 12}{112.5} = 0.849 \text{ ksi}$$

$$f_p = 0.849 - 0.380 = 0.469 \text{ ksi}$$

$$P = 0.469 \times 90 = 42.21 \text{ kips}$$

$$W_p = \frac{8 \times 42.21 \times 2.85}{12 \times (26.25)^2} = 0.116 \text{ klf} = 116 \text{ plf}$$

This is nearly equal to 109 plf used at the start of third cycle. Therefore, satisfactory.
Check compressive stress at the section:

$$f_b = 0.849 \text{ ksi}$$

$$\text{Axial stress due to prestress} = \frac{42.21}{90} = 0.469 \text{ ksi}$$

$$\text{Total compressive stress} = 0.849 + 0.469 = 1.381 \text{ ksi}$$

This is less than 1.8 ksi. Therefore, design is OK.

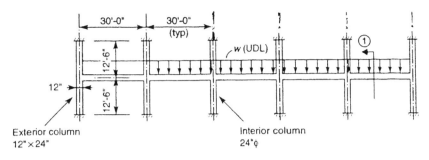

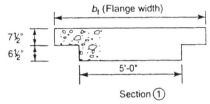

Section ①

Figure 7.42.   Example 2: post-tensioned beam, dimensions and loading.

Check the design against positive moment of 14.33 kip-ft:

$$W_p = 116 \text{ plf}$$

$$M_b = 14.33 \times \frac{116}{226} = 7.36 \text{ kip-ft}$$

$$M_{ub} = 14.33 - 7.36 = 6.97 \text{ kip-ft}$$

$$\text{Bottom flexural stress} = \frac{6.97 \times 12}{112.5} = 0.744 \text{ ksi (tension)}$$

$$\text{Axial compression due to prestress} = \frac{42.21}{12 \times 7.5} = 0.469 \text{ ksi}$$

$$\text{Tensile stress at bottom} = 0.744 - 0.469 = 0.275 \text{ ksi}$$

This is less than 0.380 ksi. Therefore, end bay design is OK.

7.3.6.2.2.   Example 2: Post-Tensioned Continuous Beam.   Refer to Fig. 7.42 for dimensions and loading. Determine flange width of beam using the criteria given in ACI 318-99.

The flange width b_f is the least of:

1. Span/4
2. Web width + 16 × (flange thickness)
3. Web width + $\frac{1}{2}$ clear distance to next web

Therefore

$$b_f = \frac{30}{4} = 7.5 \text{ ft (controls)}$$

$$= 5 + 16 \times \frac{7.5}{12} = 15 \text{ ft}$$

$$= 5 + \frac{25}{2} = 17.5 \text{ ft}$$

Section properties:

$I = 16,650$ in.^{4}
$Y = 7.69$ in.
$S_t = 2637$ in.^{8}
$S_b = 2166$ in.^{3}
$A = 1065$ in.^{2}

Loading:

$$\text{Dead load of } 7\tfrac{1}{2} \text{ in slab} = 94 \text{ psf}$$
$$\text{Mech. \& elec.} = 6 \text{ psf}$$
$$\text{Ceiling} = 6 \text{ psf}$$
$$\text{Partitions} = 20 \text{ psf}$$
$$\text{Additional dead load due to beam self wt} = \frac{615 \times 60 \times 150}{144 \times 30} = 13.5 = 14 \text{ psf}$$
$$\text{Total dead load} = 140 \text{ psf}$$
$$\text{Live load at owner's request} = 80 \text{ psf}$$
$$D + L = 220 \text{ psf}$$

Uniform load per ft of beam = $0.220 \times 30 = 6.6$ klf. The resulting service load moments are shown in Fig. 7.43. As before we design for the moments at the face of supports.

Interior Span. Calculate through tendons by using interior span moment of 427 kip-ft at the inside face of third column (Fig. 7.43).

Assume $h_1 = 11.5$ in., $h_2 = 2.5$ in., $L_1 = 2.5$ ft, and $L_2 = 12.5$ ft. Refer to Fig. 7.41 for notations.

The height of inflection point

$$h_3 = \frac{11.5 \times 12.5 + 2.5 \times 2.5}{15} = 10 \text{ in.}$$
$$h_d = e = 10 - 2.5 = 7.5 \text{ in.}$$

First Cycle   Assume

$$W_p = 3.5 \text{ klf}$$

$$M_p = \frac{3.5}{6.6} \times 427 = 226 \text{ kip-ft}$$

$$M_{ub} = 427 - 226 = 201 \text{ kip-ft}$$

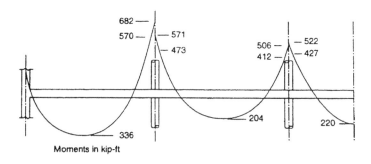

Figure 7.43.   Example 2: post-tensioned continuous beam; service load moments.

$$f_t = \frac{201 \times 12}{2637} = 0.915 \text{ ksi}$$

$$f_p = 0.915 - 0.380 = 0.535 \text{ ksi}$$

$$P = 0.535 \times 1065 = 570 \text{ kips}$$

$$W_p = \frac{8 \times 570 \times 7.5}{12 \times (27.5)^2} = 3.77 \text{ klf}$$

which is greater than 3.5 klf.   N.G.

Second Cycle   New value of

$$W_p = 0.75 \times 3.5 + 0.25 \times 3.77 = 3.57 \text{ klf}$$

$$M_p = \frac{3.57}{6.6} \times 427 = 231 \text{ kip-ft}$$

$$M_{ub} = 427 - 231 = 196 \text{ kip-ft}$$

$$f_t = \frac{196 \times 12}{2637} = 0.892 \text{ ksi}$$

$$f_p = 0.892 - 0.380 = 0.512 \text{ ksi}$$

$$P = 0.512 \times 1065 = 545 \text{ kips}$$

$$W_p = \frac{8 \times 545 \times 7.5}{12 \times (27.5)^2} = 3.60 \text{ klf}$$

which is nearly equal to 3.57 klf. Therefore, the design is satisfactory.
   Check design against positive moment of 220 kip-ft:

$$M_p = \frac{3.6 \times 220}{6.6} = 120 \text{ kip-ft}$$

$$M_{ub} = 220 - 120 = 100 \text{ kip-ft}$$

$$F_{bot} = \frac{100 \times 12}{2166} = 0.554 \text{ ksi (tension)}$$

$$\text{Axial comp. stress} = \frac{545}{1065} = 0.512 \text{ ksi (comp.)}$$

$$f_{total} = 0.554 - 0.512 = 0.042 \text{ ksi (tension)}$$

This is less than the allowable tensile stress of 0.380 ksi. Therefore, the design is satisfactory.
   End Span.   Determine end bay prestressing for a negative moment of 570 kip-ft at the face of first interior column (Fig. 7.43).

First Cycle   As before, assume

$$W_p = 3.5 \text{ klf}$$

$$M_p = \frac{3.5}{6.6} \times 570 = 302 \text{ kip-ft} \backslash$$

$$M_{ub} = 570 - 302 = 268 \text{ kip-ft}$$

$$f_t = \frac{268 \times 12}{2637} = 1.22 \text{ ksi}$$

$$f_p = 1.22 - 0.380 = 0.84 \text{ ksi}$$

$$P = 0.84 \times 1065 = 894 \text{ kips}$$

$$W_p = \frac{8 \times 894 \times 7.5}{12 \times (27.5)^2} = 5.912 \text{ klf}$$

which is greater than 3.5 klf.   N.G.

Second Cycle   New value of

$$W_p = 0.74 \times 3.5 + 0.25 \times 5.912 = 4.1 \text{ klf}$$

$$M_p = \frac{4.1}{6.6} \times 570 = 354 \text{ kip-ft}$$

$$M_{ub} = 570 - 354 = 216 \text{ kip-ft}$$

$$f_t = \frac{216 \times 12}{2637} = 0.983 \text{ ksi}$$

$$f_p = 0.983 - 0.380 = 0.603 \text{ ksi}$$

$$P = 0.603 \times 1065 = 642 \text{ kips}$$

$$W_p = \frac{8 \times 642 \times 7.5}{12 \times (27.5)^2} = 4.24 \text{ klf}$$

This is nearly equal to 4.1 klf. However, a more accurate value is calculated as follows:

$$W_p = 0.75 \times 4.1 + 0.25 \times 4.24 = 4.13 \text{ klf}$$

Check the design against positive moment of 336 kip-ft:

$$M_p = \frac{4.13}{6.6} \times 336 = 210 \text{ kip-ft}$$

$$M_{ub} = 336 - 210 = 126 \text{ kip-ft}$$

$$\text{Bottom flexural stress} = \frac{126 \times 12}{2166} = 0.698 \text{ ksi (tension)}$$

$$\text{Axial compressive stress due to post-tension} = \frac{642}{1065} = 0.603 \text{ ksi (comp.)}$$

$$f_{\text{total}} = 0.698 - 0.603 = 0.095 \text{ ksi}$$

This is less than the allowable tensile stress of 0.380 ksi. Therefore the design is OK.

    7.3.6.2.3.   Example 3: Flat Plate.   Figure 7.44 shows a schematic section of a two-way flat plate system. Design of post-tension slab for an office-type loading is required.
    Given.
    Specified compressive strength of concrete: $f'_c = 4000$ psi
    Modulus of elasticity of concrete: $E_c = 3834$ ksi

    Allowable tensile stress in precompressed tensile zone = $6\sqrt{f'_c} = 380$ psi

    Allowable fiber stress in compression = $0.45f'_c = 0.45 \times 4000 = 1800$ psi

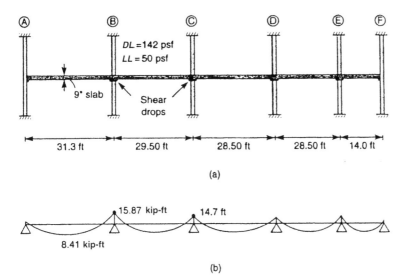

Figure 7.44. Example 3: flat plate: (a) span and loading; (b) elastic moments due to dead load plus live load.

Tendon cover: Interior spans	Top 0.75 in.
	Bot. 0.75 in.
Exterior spans	Top 0.75 in.
	Bot. 1.50 in.

Tendon diameter = 1/2 in.

Minimum area of bonded reinforcement:

    In negative moment areas at column supports:

$$A_s = 0.00075 \, A_{cf}$$

where

    A_{cf} = larger gross cross-sectional area of the slab beam strips of two orthogonal equivalent frames intersecting at a column of a two-way slab, in in.^{2}

    In positive moment areas where computed concrete stress in tension exceeds $2\sqrt{f_c'}$:

$$A_s = \frac{N_c}{0.5 f_y}$$

where

    N_c = tensile force in concrete due to unfactored dead load plus live load $(D + L)$, in lb

Rebar yield stress = 60 ksi. Max bar size = #5
Rebar cover 1.63 in. at top and bottom
Post-tension requirements:

    Minimum post-tensioned stress = 125 psi (See ACI 318-02, Sect. 18.12.4.)

    Minimum balanced load = 65% of total dead load

    Design. The flat plate is sized using the span:depth ratios given in Table 7.6. The maximum span is 31'-4" between grids A and B. Using a span:depth ratio of 40, the slab thickness is $\frac{31.33 \times 12}{40} = 9.4$ in., rounded to 9 in.

The flat plate has "shear drops" intended to increase only the shear strength and flexural support width. The shear heads are smaller than a regular drop panel as defined in the ACI code. Therefore shear heads cannot be included in calculating the bending resistance.

Loading: Dead load of 9" slab   112 psf
            Partitions   20 psf
            Ceiling and mechanical   10 psf
            Reduced live load   50 psf

Total service load = 112 + 20 + 10 + 50 = 192 psf

Ultimate load = $1.4 \times 142 + 1.7 \times 50 = 285$ psf

Slab properties (for a 1-ft-wide strip):

$$I = \frac{bh^3}{12} = 12 \times \frac{9^3}{12} = 729 \text{ in.}^4$$

$$S_{top} = S_{bot} = \frac{729}{4.5} = 162 \text{ in.}^3$$

$$\text{Area} = 12 \times 9 = 108 \text{ in.}^2$$

The moment diagram for a 1-ft-wide strip of slab subjected to a service load of 192 psf is shown in Fig. 7.45.

The design of continuous strands will be based on a negative moment of 14.7 kip-ft at the second interior span. The end bay prestressing will be based on a negative moment of 15.87 kip-ft.

Interior Span. Calculate the drape of tendon using the procedure given for the previous problem. See Fig. 7.41.

$$h_3 = \frac{h_1 L_2 + h_2 L_1}{L_1 + L_2}$$

$L_1 = 1.84$ ft $h_1 = 8$ in.

$L_2 = 12.90$ ft $h_2 = 1.25$ in.

$L_e = 12.9 \times 2 = 25.8$ ft

$$h_3 = \frac{8 \times 12.90 + 1.25 \times 1.84}{14.75} = 7.153 \text{ in.}$$

Tendon drape = $7.153 - 1.25 = 5.90$ in.

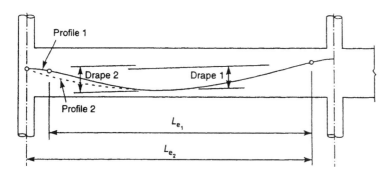

Figure 7.45.   End bay tendon profiles.

First Cycle

Minimum balanced load $= 0.65 \times$ (total DL)

$$= 0.65 \, (112 + 10 + 20) = 92 \text{ psf}$$

Moment due to balanced load $= \dfrac{92}{192} \times 14.7$

$$= 7.04 \text{ kip-ft}$$

This is subtracted from the total service load moment of 14.7 kip-ft to obtain the unbalanced moment M_{ub}.

$$M_{ub} = 14.7 - 7.04 = 7.66 \text{ kip-ft}$$

The flexural stresses at top and bottom are obtained by dividing M_{ub} by the section moduli of the structure's cross section.

$$f_t = \frac{7.66 \times 12}{162} = 0.567 \text{ ksi}$$

$$f_b = \frac{7.66 \times 12}{162} = 0.567 \text{ ksi}$$

The minimum required compressive prestress f_p is found by subtracting the maximum allowable tensile stress $f_a = 6\sqrt{f_c'}$ from the calculated tensile stress. Thus the smallest required compressive stress is:

$$f_p = f_t - f_a$$
$$= 0.567 - 380 = 0.187 \text{ ksi}$$

The prestress force is calculated by multiplying f_p by the cross-sectional area:

$$P = 0.187 \times 9 \times 12 = 20.20 \text{ kip-ft}$$

Determine the equivalent load due to prestress force P by the relation

$$W_p = \frac{8Pe}{L_e^2}$$

For the example problem,

$$P = 20.20 \text{ kip-ft}, \quad e = 5.90''$$
$$L_e = 2 \times 12.90 = 25.8 \text{ ft}$$

Therefore $W_p = \dfrac{8 \times 20.20 \times 5.90}{25.8^2 \times 12} = 0.120 \text{ klf} = 120 \text{ plf}$

Comparing this with the value of 93 plf assumed at the beginning of first cycle, we find the two values are not equal. Therefore, we assume a new value and repeat the procedure until convergence is obtained.

Second Cycle

$$W_p = 0.75 \times 92 + 0.25(120) = 99 \text{ plf}$$

$$M_b = \frac{99}{192} \times 14.7 = 7.58 \text{ kip-ft}$$

$$M_{ub} = 14.7 - 7.58 = 7.12 \text{ kip-ft}$$

$$f_t = f_b = \frac{7.12 \times 12}{162} = 0.527 \text{ ksi}$$

$$f_p = 0.527 - 0.380 = 0.147 \text{ ksi}$$

$$P = 0.147 \times 9 \times 12 = 15.92 \text{ kip-ft}$$

$$W_p = \frac{8 \times 15.92 \times 5.90}{25.8^2 \times 12} = 0.094 \text{ klf} = 94 \text{ plf}$$

This is less than 99 plf assumed at the beginning of second cycle. Therefore, we assume a new value and repeat the procedure.

Third Cycle

$$W_p = 0.75 \times 99 + 0.25(94) = 97.7 \text{ plf}$$

$$M_b = \frac{99.7 \times 14.7}{192} = 7.48 \text{ kip-ft}$$

$$M_{ub} = 14.7 - 7.48 = 7.22 \text{ kip-ft}$$

$$f_t = f_b = \frac{7.22 \times 12}{162} = 0.535 \text{ ksi}$$

$$f_p = 0.535 - 0.380 = 0.155 \text{ ksi}$$

$$P = 0.155 \times 9 \times 12 = 16.74 \text{ kip-ft}$$

$$W_p = \frac{8 \times 16.74 \times 5.90}{25.8^2 \times 12} = 0.99 \text{ klf} = 99 \text{ plf}$$

This is nearly equal to 97.7 plf assumed at the beginning of the third cycle. Therefore OK.
Check compressive stress at the support:

$$M_p = \frac{99 \times 14.7}{192} = 7.58 \text{ kip-ft}$$

$$M_{ub} = 14.7 - 7.58 = 7.12 \text{ kip-ft}$$

$$f_b = \frac{7.12 \times 12}{162} = 0.527 \text{ ksi} = 527 \text{ psi}$$

Axial compressive stress due to post-tension $= \dfrac{16.74 \times 1000}{9 \times 12} = 155 \text{ psi}$

Total compressive stress = 527 + 155 = 682 psi

This is less than the allowable compressive stress of 1800 psi. Therefore, the design is satisfactory.

End Bay Design. The placement of tendons within the end bay presents a few problems. The first problem is in determining the location of the tendon over the exterior support. Placing the tendon above the neutral axis of the member results in an increase in the total tendon drape, allowing the designer to use less prestress than would otherwise be required. Raising the tendon, however, introduces an extra moment that effectively cancels out some of the benefits from the increased drape. For this reason, the tendon is usually placed at a neutral axis at exterior supports.

The second problem is in making a choice in the tendon profile: whether to use a profile with a reverse curvature over each support (see Fig. 7.45, profile 1), or over the first interior support only (see Fig. 7.45, profile 2). A profile with the reversed curvature over the first interior support only gives a greater cable drape than the first profile, suggesting a larger equivalent load with the same amount of prestress. On the other hand, the effective length L_e between inflection points of profile 1 is less than that of profile 2 which suggests the opposite. To determine which profile is in fact more efficient, it is necessary to evaluate the amount of prestress for both profiles. More usually, a tendon profile with reverse curvature over both supports is 5 to 10% more efficient since the equivalent load produced is a function of the square of the effective length.

The last item addresses the extra end bay prestressing required in most situations. The exterior span in an equal span structure has the greatest moments due to support rotations. Because of this, extra prestressing is commonly added to end bays to allow efficient design of end spans. For design purposes, the extra end bay prestressing is considered to act within the end bay only. These tendons actually extend well into the adjacent span for anchorage, as shown in Fig. 7.46. Advantage can be taken of this condition by designing the through tendons using the largest moment found within the interior spans, including the moment at the interior face of the first support. The end bay prestress force is determined using the largest moment within the exterior span. The stress at the inside face of the first support is checked using the equivalent loads produced by the through tendons and the axial compression provided by both the through and added tendons. If the calculated stresses are less than the allowable values, the design is complete. If not, more stress is provided either by through tendons or added tendons or both.

The design of end bay using profiles 1 and 2 follows.

Profile 1: Reverse Curvature at the Right Support Only (Fig. 7.47b). Observe that the height of inflection point is exact if the tendon profile is symmetrical about the center of span. If it is not, as in span 1 of the example problem, sufficiently accurate value can be obtained by taking the average of the tendon inflection point at each end as follows.

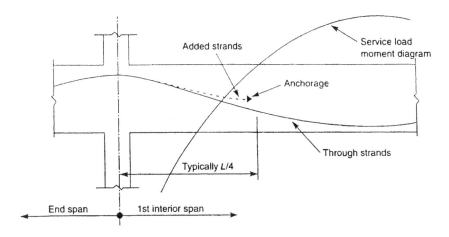

Figure 7.46. Anchorage of added tendons.

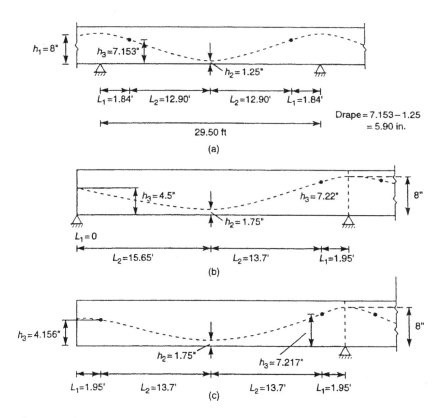

Figure 7.47. Example problem 3: flat plate, tendon profiles: (a) interior span; (b) exterior span, reverse curvature at right support; (c) exterior span, reverse curvature at both supports.

Left end:

$$h_3 = \frac{4.5 \times 15.65 + 1.75 \times 0}{0 + 15.67} = 4.5 \text{ in.}$$

Right end:

$$h_3 = \frac{8 \times 13.7 + 1.75 \times 1.95}{(1.95 + 13.70)} = 7.22 \text{ in.}$$

$$\text{Average } h_3 = \frac{4.5 + 7.22}{2} = 5.86 \text{ in.}$$

$$\text{Drape} = 5.86 - 1.75 = 4.11 \text{ in.}$$

First Cycle To show the quick convergence of the procedure, we start with a rather high value of

$$W_p = 0.75 \text{ DL} = 0.75 \times 142 = 106 \text{ plf}$$

$$M_b = \frac{106}{192} \times 15.87 = 8.76 \text{ kip-ft}$$

$$M_{ub} = 15.87 - 8.76 = 7.11 \text{ kip-ft}$$

$$f_t = f_b = \frac{7.11 \times 12}{162} = 0.527 \text{ ksi}$$

$$f_p = 0.527 - 0.380 = 0.147 \text{ ksi}$$

$$P = 0.147 \times 9 \times 12 = 15.87 \text{ kips}$$

$$W_p = \frac{8Pe}{L_e^2}$$

$$= \frac{8 \times 15.87 \times 4.11}{29.35^2 \times 12}$$

$$= 0.050 \text{ klf} = 50.0 \text{ plf}$$

This is less than 106 plf.   N.G.

Second Cycle

$$W_p = 0.75(106) + 0.25(50.0) = 92 \text{ plf}$$

$$M_b = \frac{92}{192} \times 15.87 = 7.60 \text{ kip-ft}$$

$$M_{ub} = 15.87 - 7.60 = 8.27 \text{ kip-ft}$$

$$f_t = f_b = \frac{8.27 \times 12}{162} = 0.612 \text{ ksi}$$

$$f_p = 0.612 - 0.380 = 0.233 \text{ ksi}$$

$$P = 0.233 \times 12 \times 9 = 25.16 \text{ kip-ft}$$

$$W_p = \frac{8 \times 25.16 \times 4.11}{29.35^2 \times 12}$$

$$= 0.080 \text{ klf} = 80 \text{ plf}$$

This is less than 91.5 psi used at the beginning of second cycle.   N.G.

Third Cycle

$$W_p = 0.75 \times 92 + 0.25 \times 80 = 89 \text{ plf}$$

$$M_b = \frac{89}{192} \times 15.87 = 7.356 \text{ kip-ft}$$

$$M_{ub} = 15.87 - 7.356 = 8.5 \text{ kip-ft}$$

$$f_t = f_b = \frac{8.5 \times 12}{162} = 0.631 \text{ ksi}$$

$$f_p = 0.631 - 0.380 = 0.251 \text{ ksi}$$

$$P = 0.251 \times 12 \times 9 = 27.10 \text{ kips}$$

$$W_p = \frac{8 \times 27.10 \times 4.11}{29.35^2 \times 12} = 0.086 \text{ klf} = 86 \text{ plf}$$

This is nearly equal to 89 plf used at the beginning of third cycle. Therefore OK.

Profile 2. Reverse Curvature Over Each Support (Fig. 7.47c).
Left end:

$$h_3 = \frac{4.5 \times 13.70 + 1.75 \times 1.95}{(13.70 + 1.95)} = 4.156 \text{ in.}$$

Right end:

$$h_3 = \frac{8 \times 13.70 + 1.75 \times 1.95}{(13.70 + 1.95)} = 7.221 \text{ in.}$$

$$\text{Average } h_3 = \frac{4.156 + 7.221}{2} = 5.689 \text{ in.}$$

$$e = h_d = 5.689 - 1.75 = 3.939 \text{ in.}$$

First Cycle   We start with an assumed balanced load of 0.65 DL = 92 plf.

$$\text{Balanced moment } M_b = 15.87 \times \frac{92}{192} = 7.60 \text{ klp-ft}$$

$$M_{ub} = 15.87 - 7.6 = 8.27 \text{ kip-ft}$$

$$f_t = f_b = \frac{8.27 \times 12}{162} = 0.613$$

$$f_p = 0.613 - 0.380 = 0.233 \text{ ksi}$$

$$P = 0.233 \times 9 \times 12 = 25.12 \text{ kips}$$

$$W_p = \frac{8 \times 25.12 \times 3.937}{(27.38)^2 \times 12} = 0.088 \text{ klf} = 88 \text{ plf}$$

This is less than 92 plf.   N.G.

Second Cycle

$$W_p = 0.75 \times 92 + 0.25 \times 88 = 91 \text{ plf}$$

$$M_b = 15.87 \times \frac{91}{192} = 7.52 \text{ kip-ft}$$

$$M_b = 15.87 \times \frac{91}{192} = 7.52 \text{ kip-ft}$$

$$M_{ub} = 15.87 - 7.52 = 8.348 \text{ kip-ft}$$

$$f_t = f_b = \frac{8.348}{162} \times 12 = 0.618 \text{ ksi}$$

$$f_p = 0.618 - 0.380 = 0.238 \text{ ksi}$$

$$P = 0.238 \times 9 \times 12 = 25.75 \text{ kips}$$

$$W_p = \frac{8 \times 25.75 \times 3.937}{(27.38)^2 \times 12} = 0.090 \text{ klf} = 90 \text{ plf}$$

This is nearly equal to the value at the beginning of second cycle. Therefore OK.
    Check the design against positive moment of 8.41 kip-ft:

$$W_p = 0.090 \text{ klf}$$

$$M_b = 8.41 \times \frac{0.090}{0.142} = 5.33 \text{ kip-ft}$$

$$M_{ub} = 8.41 - 5.33 = 3.08 \text{ kip-ft}$$

$$\text{Bottom flexural stress} = \frac{3.08 \times 12}{162} = 0.228 \text{ ksi (tension)}$$

$$\text{Axial compression due to post-tension } = \frac{25.75}{12 \times 9} = 0.238 \text{ ksi}$$

Total stress at bottom = 0.228 − 0.238 = −0.10 ksi (Compression)

This is less than allowable tension of 0.380 ksi. Therefore, design OK.

7.3.7. Strength Design for Flexure

In the design of prestress members it is not enough to limit the maximum values of tensile and compressive stresses within the permitted values at various loading stages. This is because although such a design may limit deflections, control cracking, and prevent crushing of concrete, an elastic analysis offers no control over the ultimate behavior or the factor-of-safety of a prestressed member. To ensure that prestressed members will be designed with an adequate factor-of-safety against failure, ACI 318-02, similar to its predecessors, requires that M_u, the moment due to factored service loads, not exceed ϕM_n, the flexural design strength of the member.

The nominal bending strength of a prestressed beam with bonded tendons is computed in nearly the same manner as that of a reinforced concrete beam. The only difference is in the method of stress calculation in the tendon at failure. This is because the stress–strain curves of high-yield-point steels used as tendons do not develop a horizontal yield range once the yield strength is reached. It continues upward at a reduced slope. Therefore, the final stress in the tendon at failure f_{ps} must be predicted by an empirical relationship.

The method of computing the bending strength of a prestressed beam given in the following section applies only to beams with bonded tendons. The analysis is performed using strain compatibility. Because by definition there is no strain compatibility between the tendon and concrete in an unbonded prestressed beam, this method cannot be used for prestressed beams with unbonded tendons; the empirical approach given in ACI 318-02, Section 18.7 is the recommended method.

The procedure for bonded tendons consists of assuming the location of the neutral axis, computing the strains in the prestressed and non-prestressed reinforcement, and establishing the compression stress block. Knowing the stress-strain relationship for the reinforcement, and assuming that the maximum strain in concrete is 0.003, the forces in the prestressed and nonprestressed reinforcement are determined and the sum of compression and tension forces are computed. If necessary, the neutral axis location is adjusted on a trial-and-error basis until the sum of the forces is zero. The moment of these forces is then computed to obtain the nominal strength of the section. To compute the stress in the prestressing strand, the idealized curve shown in Fig. 7.48 (adapted from *PCI Design Handbook*, 5th Edn.) is used.

The analysis presented here follows a slightly different procedure. Instead of assuming the location of the neutral axis, we assume a force in the prestressing strand, and compare it to the derived value. The analysis is continued until the desired convergence is reached.

7.3.7.1. *Examples*

7.3.7.1.1. **Example 1.**

Given. A rectangular prestressed concrete beam, as shown in Fig. 7.48a:

$f'_c = 5000$ psi

Mild steel reinforcement = 4 #5 bars at bottom, $f_y = 60$ ksi

Prestressed strands = 4 − 1/2 ϕ, $f_{ps} = 270$ ksi

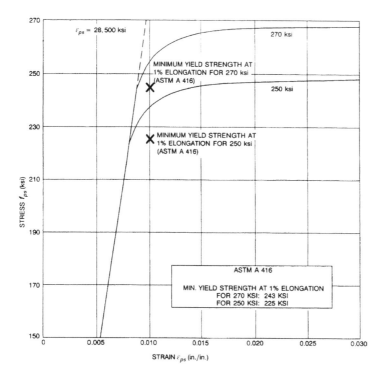

Figure 7.48. Typical stress-strain curve with seven-wire low-relaxation prestressing strand. These curves can be approximated by the following equations:

250 ksi

$$\varepsilon_{ps} \leq 0.0076 : f_{ps} = 28{,}500\,\varepsilon_{ps}\ (\text{ksi})$$

$$\varepsilon_{ps} > 0.0076 : f_{ps} = 250 - \frac{0.04}{\varepsilon_{ps} - 0.0064}\ (\text{ksi})$$

270 ksi

$$\varepsilon_{ps} \leq 0.0086 : f_{ps} = 28{,}500\,\varepsilon_{ps}\ (\text{ksi})$$

$$\varepsilon_{ps} > 0.0086 : f_{ps} = 270 - \frac{0.04}{\varepsilon_{ps} - 0.007}\ (\text{ksi})$$

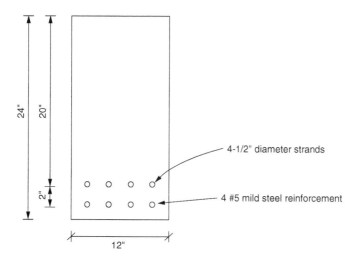

Figure 7.48a. Example 1: beam section.

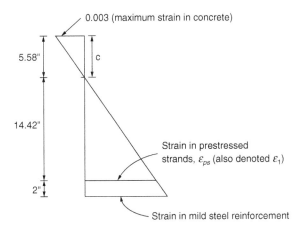

Figure 7.48b. Example 1: strain diagram, first trial.

Required. Ultimate flexural moment capacity of the beam

Solution. A trial-and-error procedure is used.

First Trial. For the first trial, assume the stress in the prestressed strands = 250 ksi, and the yield stress in the mild steel is 60 ksi.

The total tension T at the tension zone of the beam consists of T_1, the tension due to prestressed stands, plus T_2, the tension due to mild steel reinforcement.

Thus $T = T_1 + T_2$

T_1 = area of stands × assumed stress in prestressing steel
    = $4 \times 0.153 \times 250 = 153$ kips
T_2 = area of mild steel reinforcement × yield stress
    = $4 \times 0.31 \times 60 = 74.40$ kips
$T = 153 + 74.40 = 227.4$ kips

Draw a strain diagram for the beam at the nominal moment strength defined by a compressive strain of 0.003 at the extreme compression fiber. Using the strain diagram, find the compressive force $C = 0.85\ f'_c\ ab$. See Fig. 7.48b.

$C = T = 227.4$ kips

$$a = \frac{227.4}{0.85 f'_c b}$$

$$= \frac{227.4}{0.85 \times f'_c \times 12} = 4.46 \text{ in.}$$

$$c = \frac{a}{\beta_1} \qquad \qquad \beta_1 = 0.8 \text{ for } f'_c = 5 \text{ ksi}$$

$$= \frac{4.46}{0.8} = 5.58 \text{ in.}$$

Compute the strain in the prestressing steel and the corresponding stress.

$$\frac{0.003}{5.58} = \frac{\epsilon_1}{14.42}$$
$$\epsilon_1 \, 0.00775$$

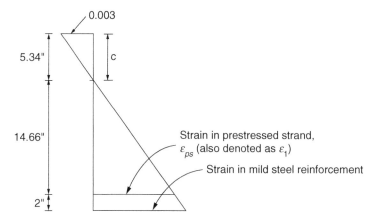

Figure 7.48c.   Example 1: strain diagram, second trial.

Since the strain = 0.00775, the corresponding stress is in the elastic region of the stress strain curve. See Fig, 7.48. The stress in the prestressed strand is given by:

$$f_{ps} = 28500 \times \epsilon_1$$
$$= 28500 \times 0.00775 = 221 \text{ ksi}$$
$$T_1 = 4 \times 0.153 \times 221 = 135.4 \text{ kips}$$
$$T_2 = 74.40 \text{ kips as before}$$
$$T = T_1 + T_2 = 135.4 + 74.40 = 209.8 \text{ kips}$$

Comparing this to $T = 227.4$ kips, by inspection we estimate that an improved value of $T = C$ = average of the two values.

$$= (227.4 + 209.8)/2 = 218.6 \text{ kips, say, 218 kips}$$

Use this value for the second trial. See Fig. 7.48c.
   Second Trial.

$$C = T = 218 \text{ kips}$$
$$a = \frac{218}{0.85 f_c' b} = \frac{218}{0.85 \times 5 \times 12} = 4.27 \text{ in.}$$
$$c = \frac{a}{\beta_1} = \frac{4.27}{0.8} = 5.34 \text{ in.}$$
$$\frac{0.003}{5.34} = \frac{\epsilon_1}{14.66}$$
$$\epsilon_1 = 0.0082 < 0.0086$$

Therefore

$$f_{ps} = 28500 \times 0.00824 \quad \text{(See Fig. 7.48)}$$
$$= 234.7 \text{ kips}$$
$$T_1 = 4 \times 0.153 \times 234.7 = 143.7$$
$$T = 143.7 + 74.4 = 218.1 \text{ kips}$$

This is practically the same as the value we assumed in the second trial. Therefore $T = 218$ kips may be used to compute the flexural strength of the beam.

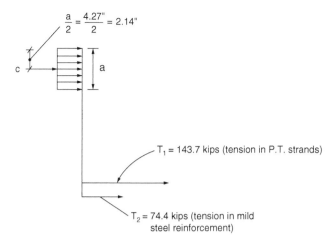

$$\frac{a}{2} = \frac{4.27"}{2} = 2.14"$$

$T_1 = 143.7$ kips (tension in P.T. strands)

$T_2 = 74.4$ kips (tension in mild steel reinforcement)

Figure 7.48d. Example 1: force diagram.

Flexural Strength. The nominal moment strength is obtained by summing the moments of T_1 and T_2 about the C.G. of compressive force C (see Fig. 7.48d).

$$M_n = 74.4 \times (20 - 2.14) + 143.7 (22 - 2.14)$$
$$= 4183 \text{ kip-in} = 348.6 \text{ kip-ft}$$

Usable capacity of the beam $= \phi M_n$
$$= 0.9 \times 348.6 \text{ kip-ft}$$
$$= 313.7 \text{ kip-ft}$$

7.3.7.1.2. Example 2.

Given. Same data as for example 1, except $3 - \frac{1}{2}" \phi$ strands are used instead of $4 - \frac{1}{2}" \phi$ strands. This example illustrates the calculation of stress in the strand in the nonelastic range of the stress strain curve shown in Fig. 7.48.

Required. Ultimate flexural capacity of the beam.

Solution. As before, we use a trial-and-error procedure.

First Trial. Assume stress in the strands = 240 ksi.

Total tension $T = T_1 + T_2$
$$= 3 \times 0.0153 \times 240 + 4 \times 0.31 \times 60$$
$$= 110.16 + 74.4 = 184.56 \text{ kips}$$

$$T = C = 0.85 \times 12 \times 5 \times a = 184.56 \text{ kips}$$

$$a = \frac{184.56}{0.85 \times 12 \times 5} = 3.62 \text{ in.}$$

$$c = \frac{3.62}{0.80} = 4.52 \text{ in.} \quad \text{(See Fig. 7.48e)}$$

$$\frac{0.003}{4.52} = \frac{\epsilon_{ps}}{15.48}$$

$$\epsilon_{ps} = 0.0102$$

$$f_{ps} = 270 - \frac{0.04}{0.0102 - 0.007} \quad \text{(See Fig. 7.48)}$$

$$= 270 - 12.22 = 257.8 \text{ ksi}$$

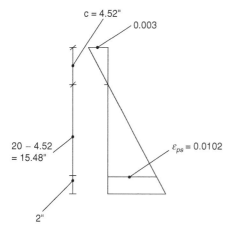

Figure 7.48e. Example 2: strain diagram, first trial.

$$T = T_1 + T_2$$
$$= 3 \times 0.0153 \times 257.8 + 74.4$$
$$= 192.7 \text{ kips compared to } 184.56 \text{ kips}$$

Use an average value $T = \frac{192.7+184.56}{2} = 188.6$, say, 189 kips for the second trial.

Second Trial.

$$C = T = 189 \text{ kips}$$
$$a = \frac{189}{0.85 \times 12 \times 5} = 3.71 \text{ in.}$$
$$c = \frac{3.71}{0.8} = 4.63 \text{ in.} \quad (\text{See Fig. 7.48f})$$
$$\frac{0.003}{4.63} = \frac{\epsilon_{ps}}{15.37}$$
$$\epsilon_{ps} = 0.00996$$
$$f_s = 270 - \frac{0.04}{0.00996 - 0.007}$$
$$= 270 - 13.5 = 256.5 \text{ kips}$$
$$T_1 = 0.459 \times 256.5 + 74.4$$
$$= 117.72 + 74.4 = 192 \text{ kips}$$

Compared to 189 kips used at the beginning of the second trial, this is considered sufficiently accurate for all practical purposes.

Calculate the nominal moment M_n by taking moments of T_1 and T_2 about the C.G. of compression block.

$$M_n = 117.72 \times 18.14 + 74.4 \times 20.14$$
$$= 2135.6 + 1498.42 = 3634 \text{ kip-in.}$$
$$= 302.84 \text{ kip-ft}$$

$$\phi M_n = 0.9 \times 302.84 = 272.6 \text{ kip-ft}$$

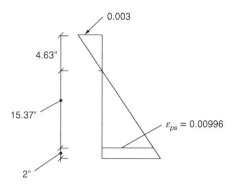

Figure 7.48f. Example 2: strain diagram, second trial.

7.3.7.1.3. Example 3: Prestressed T-beam.

Given. See Fig. 7.48g for the beam geometry. The area of prestressed strands = 2.4 in.^{2} $f_c' = 5$ ksi

Required. Ultimate flexural capacity of the T-beam.

First Trial. Assume $f_{ps} = 250$ ksi

$T = T_1$ $T_2 = 0$, since there is no mild steel reinforcement
  $= 2.4 \times 250 = 600$ kips

$C = T = 600$ kips

$$a = \frac{600}{0.85 \times 48 \times 5} = 2.94 \text{ in.}$$

$$c = \frac{a}{\beta_1} = \frac{2.94}{0.8} = 3.68 \text{ in.}$$

$$\frac{0.003}{3.68} = \frac{\epsilon_1}{20.32} \quad \text{(See Fig. 7.48h)}$$

$$\epsilon_1 = 0.01657$$

$$f_{ps} = 270 - \frac{0.04}{0.01657 - 0.007}$$

$$= 270 - 4.18 \approx 265.8 \text{ ksi}$$

$$T = 265.8 \times 2.4 = 638 \text{ kips} \quad \text{(compared to 600 kips)}$$

Use an average of the two, $\frac{638 + 600}{2} = 619$ kips, for the second trial.

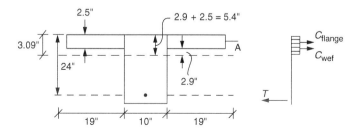

Figure 7.48g. Example problem 3: prestressed T-beam.

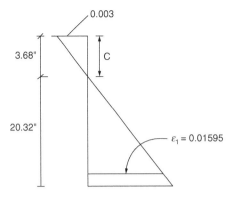

Figure 7.48h.   Example 3: strain diagram; first trial.

Second Trial.

$$T = 619 \text{ kips}$$

$$c = \frac{3.68}{600} \times 619 = 3.80 \text{ in.}$$

$$\frac{0.003}{3.80} = \frac{\epsilon_1}{20.20} \qquad \text{(See Fig. 7.48i)}$$

$$\epsilon_1 = 0.01595$$

$$f_{ps} = 270 - \frac{0.04}{0.01595 - 0.007} = 265.5 \text{ ksi} \qquad \text{(See Fig. 7.48)}$$

$$T = 265.5 \times 2.4 = 636 \text{ kips.}$$

Use $T = \dfrac{619 + 636}{2} = 627.5$, say, 630 kips for the third trial.

Third Trial.

$$T = 630 \text{ kips}$$

$$c = \frac{3.68}{600} \times 630 = 3.86 \text{ in.}$$

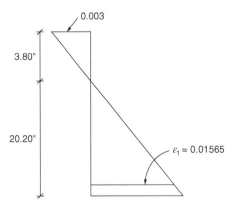

Figure 7.48i.   Example problem 3: strain diagram; second trial.

$$a = 0.8 \times 3.86$$
$$= 3.09 \text{ in.}$$

$$\frac{0.003}{3.86} = \frac{\epsilon_1}{20.14}$$

$$\epsilon_1 = 0.01565$$

$$f_{ps} = 270 - \frac{0.04}{0.01565 - 0.007} = 265.37 \text{ ksi} \qquad T = 265.37 \times 2.4 = 637 \text{ kips}$$

This value is nearly the same as the value of 630 kips used at the beginning of the third iteration. However, use average value equal to $(630 + 637)/2 = 633.5$ kips for calculating M_n.

Flexural Strength.

$$633.5 = A_c \, (0.85 \, f_c')$$

$$A_c = \frac{633.5}{0.85 \times 5} = 149 \text{ in.}^2$$

Since the area of flange $= 2.5 \times 48 = 120$ in.^{2} is less than 149 in.^{2}, the stress block extends into the web: $149 - 120 = \frac{29}{10} = 2.9$ in. into the web. (See Fig. 7.48g.) Compute M_n by separating the compression zone into two areas and summing moments of forces about the tendon force T.

$$M_n = C_{\text{flg}}\left(24 - \frac{2.5}{2}\right) + C_{\text{wlb}}\left(24 - \frac{5.4}{2}\right)$$

$$= 38 \times 2.5 \times 0.85 \times 5 \times 22.75 + 5.4 \times 10 \times 0.85 \times 5 \times 21.30$$

$$= 9185.3 + 4888 = 14073 \text{ kip-in.}$$

$$= 1172.8 \text{ kip-ft}$$

Usable flexural capacity of beam $= \phi \, M_n$
$$= 0.9 \times 1172.8$$
$$= 1055.5 \text{ kip-ft}$$

7.4. COMPOSITE GRAVITY SYSTEMS

Gravity systems in composite construction can be broadly classified into composite floor systems and composite columns. Composite floor systems can consist of simply supported prismatic or haunched structural steel beams, trusses, or stub girders linked via shear connectors to a concrete floor slab to form an effective T-beam flexural member. Formed metal deck supporting a concrete topping slab is an integral component in these floor systems used nearly exclusively in steel framed buildings in North America.

7.4.1. Composite Metal Deck

Metal deck is manufactured from steel sheets by a fully mechanized, high-speed cold-rolling process. Although it is possible to produce shapes up to $\frac{1}{2}$ in. (12.7 mm) and even

$^3/_4$ in. (19 mm) thick by cold forming, cold-formed steel construction is generally restricted to plates and sheets weighing from 0.5 psf (24 Pa) to a maximum of 9 psf (431 Pa).

Composite metal deck is manufactured with deformations specifically designed to produce composite action under flexure between the metal deck and concrete. Shear transfer between the two is achieved through lugs, corrugations, ridges, or embossments formed in the profile of the sheet to increase bond between the two materials. The steel deck cross-section is typically trapezoidal with relatively wide flutes suitable for through-deck welding of shear studs. Metal deck may also include closed cells to accommodate floor electrification system. Noncellular deck panels may be blended with cellular panels as part of the total floor system. Metal deck is commonly available in depths of $1^1/_2$, 2, and 3 in. (38, 51, and 76 mm) with rib spacings of 6, $7^1/_2$, 8, 9 and 12 in. (152, 190, 208, 228, and 305 mm).

A composite slab is usually designed as a simply supported reinforced concrete slab with the steel deck acting as positive reinforcement. Typical mesh used for control of temperature cracking does not provide enough negative reinforcement for typical beam spacing of 8 to 15 ft (2.44 to 4.57 m). Although the slab is designed as a simple span, it is a good practice to provide a nominal reinforcement of, say, #4 @ 18" c-c at the top to control excessive cracking of the slab. It is generally believed that cracking of the slab in the negative bending regions does not materially impair the composite beam strength.

The Steel Deck Institute (SDI), regarded as the industry standard by metal deck manufacturers, has published a manual which encompasses the design of composite decks, form decks, and roof decks. A brief description of the SDI specifications is given in the following section.

7.4.1.1.   The SDI Specifications

The SDI specification requires that steel used for fabrication of composite metal deck shall have a minimum yield point of 33 ksi (227.5 MPa). The specified yield point is the primary criterion for strength under static loading. The tensile strength is of secondary importance because fatigue strength and brittle fracture, which relate to tensile strength rather than yield point, are rarely of consequence; metal deck is rarely subjected to repetitive loads and the characteristic thinness invariably precludes the development of brittle fracture.

Considerable variations in thickness of metal deck may occur because of rolling tolerances. Therefore, SDI stipulates that the delivered thickness of bare steel without the finish such as phosphotising and galvanizing shall be not less than 95% of the specified thickness. The increase in the stiffness of deck due to galvanizing is not relied upon in the design of metal deck.

Opinions differ among engineers whether the metal deck used inside a building which is not directly exposed to weather needs galvanizing or not, and SDI does not mandate any particular type of finish. The appropriate finish is left to the discretion of the engineer with the recommendation that due consideration be given to the effects of environment to which the structure is subjected. However, SDI in its commentary recommends a galvanized coating conforming to ASTM A-525 G.60 requiring a minimum galvanizing of 0.75 ounce per square foot (2.24 Pa) of metal deck. Other salient features of SDI specifications are as follows:

1. Minimum compressive strength of concrete f'_c shall be 3.0 ksi (20.68 MPa). The compressive stress in concrete is limited to $0.4f'_c$ under the applied load for unshored construction and under the total dead and live loads for shored construction. The flexural or shear bond is to be based on ultimate strength

analysis with a minimum safety factor of 2. The minimum temperature and shrinkage reinforcement in a composite slab is a function of the area of concrete, as in ordinary reinforced concrete slab, but only the concrete area above the metal deck need be considered in calculating the area of concrete.

2. The use of admixtures containing chloride salts is prohibited because salts can corrode the steel deck.

3. When designing metal deck as form work, the bending section properties are to be calculated per AISI *Specification for the Design of Cold-Formed Steel Structural Members*.

4. Bending stress is limited to 0.6 times the yield strength of steel. An upper limit of 36.0 ksi (248.2 MPa) is imposed on the allowable stress. In addition to the weight of wet concrete and deck, allowance should be made for construction live loads of 20 psf (958 Pa) of uniform load or a 150-lb (667-N) concentrated load. This is to account for the weight of one person working on a 1-ft (305-mm) width of deck. It is a common practice to allow for a 200-lb (890-N) point load as an equivalent load. This is because the loading is considered temporary with a 33% increase in the stress, which is equivalent to reducing a 200-lb (890-N) load by 25%. Clear spans are to be used in the moment calculations.

5. For calculating deflections, it is not necessary to consider the construction loads since the deck, which is designed to remain elastic, will rebound after the removal of construction loads. The calculated deflection based on the weight of concrete is limited to the smaller value $L/180$ or $\frac{3}{4}$ in. (19 mm), in which L is the clear span of the deck. Deflections of composite slabs due to live loads of 50 to 80 psf (2394 to 3830 Pa) are seldom a design concern because the deflections are usually less than $L/360$, where L is the span of deck. Because the slab is assumed to have cracked at the supports, the deflections are best predicted by using the average of the cracked and uncracked moment of inertia of the transformed sections. Note that when slabs are cast level to compensate for the deflection of metal deck, a 10 to 15% of additional concrete is required.

6. A minimum bearing of $1\frac{1}{2}$ in. (38 mm) is required for proper deck seating on supports.

7. A maximum average spacing of 12 in. (305 mm) for arc-spot (puddle) welds is specified to obtain proper anchorage to supporting members. The maximum spacing between adjacent welds is limited to 18 in. (457 mm). Welding of decks with thickness less than 0.028 in. (0.71 mm) is not practical because of the likelihood of burning off the sheet. Therefore, SDI stipulates use of weld washers for floor decks less than 0.028 in. (0.71 mm) in thickness. Stud welding through the metal deck to the steel top flange can be used instead of puddle welds to satisfy the minimum spacing requirements. However, since it is possible to get uplift forces during wind storms, puddle welds should be used to prevent metal decks from being blown off buildings during construction.

8. Mechanical fasteners which satisfy the anchorage criteria can be used in lieu of puddle welds.

9. Side laps with proper fasteners are required between two longitudinal pieces of deck to: 1) prevent differential deflection; 2) provide sufficient diaphragm strength; and 3) sustain local construction loads without distortion or separation. Side laps may be fastened with seam weld or by button punching at 2 ft (0.61 m) on centers.

10. The edges of metal deck shall be connected to supports parallel to the deck with $^3/_4$-in. diameter arc-spot or fillet welds at a maximum spacing of 3 ft. throughout.

   To function as form work, decks supporting cantilevers should be proportioned to satisfy the following criteria: 1) dead load deflection should be limited to $L/90$ of overhang or $^3/_8$ in. (9.5 mm), whichever is smaller; 2) for decks with f_y = 33 ksi, steel stress should be limited to 26.7 ksi (184 MPa) for dead load plus 200 lb (890 N) concentrated load at the outer edge of overhang, or steel stresses limited to 20.0 ksi (138 MPa) for dead load plus 20 psf (958 Pa) of additional load, whichever is more severe; and 3) the deck should receive one seam weld at the cantilever end, and the spacing of welds throughout the cantilever span should not exceed 12 in. (0.30 m). Button punching can be used as an acceptable alternative to seam welding.

7.4.1.2. *Diaphragms*

In seismic design, the term diaphragm applies to a horizontal element that transfers earthquake-induced inertial forces to vertical elements of the lateral-force-resisting systems. To do so requires a collective action of diaphragm components including chords, collectors, and ties. In buildings, typically floors and roofs provide for the diaphragm action by connecting building masses to the primary vertical elements of the lateral-force-resisting system.

   A chord is a component of a diaphragm provided at each edge to develop the axial force due to bending. It may consist of either a continuous diaphragm chord, or of a combination of wall, frame, and chord elements. At reentrant corners, diaphragm chords are extended beyond the corners, a distance sufficient to develop the accumulated diaphragm boundary stresses into the diaphragm.

   For purposes of analysis, diaphragms are classified as either flexible or rigid depending upon their in-plane deformation relative to the average interstory drift of the vertical lateral-force-resisting elements of the story immediately below the diaphragm level. If the deformation of the diaphragm is twice the average interstory drift of the story below the diaphragm, then the diaphragm is considered flexible. If it is less, it is classified as rigid.

   A diaphragm collector may be defined as a horizontal element furnished to transfer accumulated diaphragm shear forces to the vertical lateral-force-resisting element. Its primary purpose is to deliver diaphragm forces that are in excess of the forces transferred directly to the vertical element.

   Metal deck diaphragms are composed of gauge-thickness steel sheets formed in a repeating pattern with ridges and valleys. These are attached to each other and to the structural steel supports by puddle welds or by mechanical means such as with screws or shot pins. Bare metal deck diaphragms are designed to resist seismic loads acting alone or in conjunction with supplementary horizontal diagonal bracing. Steel frame elements, to which bare metal deck diaphragms are attached at their boundaries, are considered as chord and collector elements.

   Capacities of steel deck diaphragms with and without concrete topping are given in metal deck manufacturer's literature or in the publications of the Steel Deck Institute (SDI).

   Metal deck diaphragms with structural concrete topping consist typically of a composite deck with indentations. The concrete topping is either a normal or lightweight structural concrete, with reinforcing consisting of wire mesh or reinforcing steel. Decking units are attached to each other and to structural steel supports using puddle welds typically at 1 or 2 ft. on center. For composite behavior of steel beams, shear studs are welded to beams and girders before the concrete is cast.

Diaphragm Design Summary: Buildings Assigned to SDC C and Above.
Typical steps in the seismic design of a diaphragm are as follows:

- Evaluate the diaphragm design force F_{px} at the floor and roof levels by the formula

$$F_{px} = \frac{\sum_{i=x}^{n} F_i}{\sum_{i=x}^{n} W_i} W_{px} \qquad \text{ASCE 7-02 Eq. (9.5.2.6.44)}$$

where
F_{px} = the diaphragm design force
F_i = the design force applied to level i
W_i = the weight tributary to level i
W_{Px} = the weight tributary to the diaphragm at level x

F_{px} need not exceed $0.4\,S_{DS}\,I w_{px}$ but shall not be less than $0.2\,S_{DS}\,I W_{px}$

- Observe that the force F_{px} computed from this equation is typically larger than the force F_x determined by

$$F_x = C_{vx}V \qquad \text{[ASCE 7-02 Eq. (9.5.5.4-1)]}$$

and

$$C_{vx} = \frac{W_x h_x^k}{\sum_{i=1}^{n} W_i h_i^k} \qquad \text{[ASCE 7-02 Eq. (9.5.5.4-2)]}$$

where
C_{vx} = vertical distribution factor
V = total design lateral force or shear at the base of the structure (kip or kN)
W_i and W_x = the position of the total gravity load of the structure (W) located or assigned to level i or x
h_i and h_x = the height (ft or m) from the base to level i or x
k = an exponent related to the structure period as follows:
For structures having a period of 0.5 sec or less, $k = 1$
For structures having a period of 2.5 sec or more, $k = 2$
For structures having a period between 0.5 and 2.5 seconds, k shall be 2 or shall be determined by linear interpolation between 1 and 2.

This is in recognition of the fact that higher-mode participation can result in larger forces at individual diaphragm levels than predicted by the preceding equation for F_{px}

- Perform a three-dimensional lateral load analysis of the building by applying F_{px} at the floor and roof levels. Include effects of torsion but ignore the effects if they reduce shear in the vertical elements of the lateral-load-resisting system.
- Determine the net shear in the vertical elements of the lateral-load-resisting system due to F_{px}. This is equal to the difference in shears resisted by the vertical elements immediately above and below the level of the diaphragm being designed. Conceptually the shear forces may be considered as reactions to the inertial forces of the diaphragm.
- Determine a set of equivalent loads at the diaphragm level that is in equilibrium with the shear forces by using both force and moment equilibrium conditions. The equivalent loads may be derived as a combination of primary action due to F_{px} and a secondary action due to torsional effects, as will be explained shortly in the numerical examples.
- Using the equivalent loads, determine shear and bending moment at critical sections of the diaphragm.
- Compute the shear per unit length to check the shear capacity of the diaphragm. Provide collectors, also referred to as drag beams, to carry the shear that is in excess of force transferred directly into the vertical elements. Use the special seismic load combinations with $\Omega_o Q_E$ for the design of collectors, if design is by strength design method. See AISC 7-02, Sect 9.5.2.6.3.1. Note that the published capacities of metal deck diaphragms are in working stress design (WSD) format.
- For reinforced concrete diaphragms, use the following equation to calculate the ultimate shear capacity of the diaphragm

$$V_u = \phi(V_c + V_s)$$
$$= \phi A_{cv}(2\sqrt{f_c'} + \rho_n f_y)$$

[ACI 318-02 Eq. (21.10)]

Note that the strength reduction factor for shear, ϕ, in diaphragm designs must not exceed the value used for the shear design of vertical elements of lateral-force-resisting systems.

- Check perimeter beams and their connections to columns for diaphragm chord forces.
- Extend chords into the diaphragm at reentrant corners, if any, to develop the forces calculated at the critical sections.

Example 1: Composite Diaphragm.

Given. A typical floor plan of a multistory steel building as shown in Fig. 7.49a1. The lateral-load-resisting system consists of a combination of rigid frames and braced frames. The floor and roof framing consist of a 3-in.-deep 20-gauge composite metal deck with a $3\frac{1}{4}$-in.-thick lightweight concrete topping.

By comparing the average interstory drift of the floor below the diaphragm to the deflection of the diaphragm itself, the engineer has determined that for analysis purposes, the composite floor and roof system may be considered as rigid (see AISC 7-02, Sect. 9.5.2.3.1).

The values of F_{px} at various floor levels and roof have been determined from the ASCE 7-02 Eq. (9.5.2.6.4.4).

Using these forces, a three-dimensional, lateral-load analysis of the building has been performed to determine the shears in various vertical elements of the lateral-load-resisting

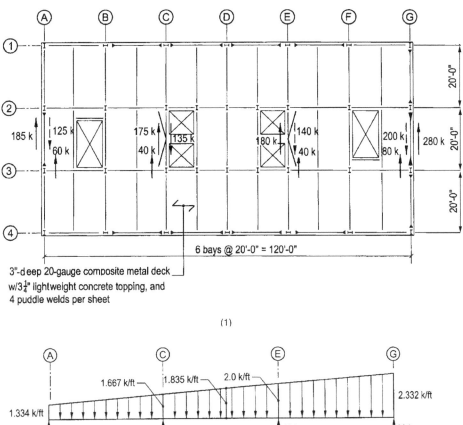

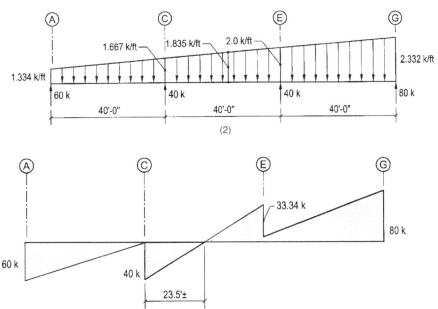

Figure 7.49a. Composite diaphragm design example: (1) floor plan; (2) equivalent lateral load; (3) shear force diagram; (4) bending moment diagram.

system (see Fig. 7.49a1). Solid arrows indicate the values of shear below the diaphragm while dashed arrows are for values above the diaphragm. All values are in WSD format calculated by dividing the ultimate values by a factor of 1.4. The loads are for seismic forces in the north–south direction.

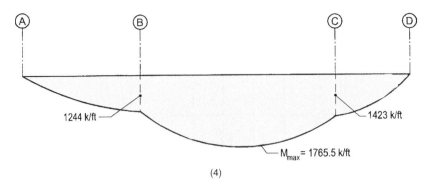

Figure 7.49a. (*Continued.*) Bending moment diagram.

It should be noted that a three-dimensional analysis that includes torsional effects always generates shear forces in a direction perpendicular to the applied load. However, these forces in the E–W direction do not affect our analysis performed for the N–S direction.

Required. Verification of shear capacity and design of the diaphragm chord and collector elements.

Solution. For analytical purposes, consider the diaphragm as a continuous beam spanning between grids A and G. Calculate the net shear forces in the frames by taking the difference in shears above and below diaphragm level. For example, the shear at grid A is equal to $185 - 125 = 60$ kips. Similarly, calculate the shears in other frames. The resulting shear forces, shown in Fig. 7.49a1, may be considered, for design purposes, as reactions to the diaphragm inertial loads.

Equivalent Loads. Because the shear in the frame on line G is larger than at A, by inception, the equivalent load distribution is trapezoidal as shown in Fig. 7.49a2. If w_1 and w_2 are the unit values of shear at A and G, by the force equilibrium:

$$w_1 \times 120 + \frac{1}{2}(w_2 - w_1) \times 120 = 220$$

or

$$w_1 + w_2 = 3.67 \tag{7.1}$$

Summing the moments about G due to w_1 and w_2 and the reactions at A, C, and E, we get

$$w_1 \times \frac{120}{2} \times \frac{2}{3} \times 120 + w_2 \times 120 \times \frac{120}{3}$$
$$= 60 \times 120 + 40 \times 80 + 40 \times 40 = 12000 \text{ kip-ft}$$

or

$$w_1 + 0.25w_2 = 2.50 \tag{7.2}$$

Solving Eqs. (7.1) and (7.2) simultaneously

$$w_1 = 1.334 \text{ kip-ft and } w_2 = 2.332 \text{ kip-ft}$$

Shear Design. The diaphragm shear forces and bending moments are shown in Figs. 7.51a3 and a4, and the information required for the design of collectors is summarized in Table 7.8. From this table the maximum shear per unit length of the diaphragm is at grid E and is equal to $80/60 = 1.33$ kips-ft. This compares with the allowable shear capacity of

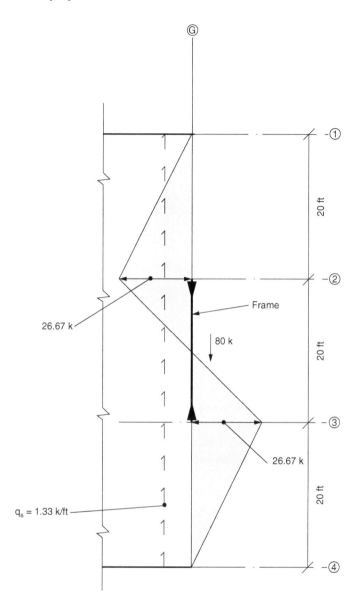

Figure 7.49b. Collector design forces.

1.74 kip-ft for the specified 3-inch-deep 20-gauge deck with a $3^1/_4$-inch lightweight concrete topping and puddle welds in every flute. See Table 7.9. Note that a one-third increase for seismic load is not permitted over and above the published values. Since the shear capacity of 1.74 kip-ft is greater than 1.67 kip-ft, the specified metal deck is OK for shear.

Drag-Strut (Collector) Design. Next we determine the drag-strut requirement by comparing the shear capacity of the metal deck with the calculated shear flow per ft length of frame given in row 6 of Table 7.8. The calculated values exceed the allowable capacity at grids A, B, and F. At these grids, drag struts must be designed to carry the shear in excess of the force transferred directly into the frames.

The two beams, each 20-ft long on each side of frames on grids A, B, and F, may be designed as drag struts. In addition to the gravity loads, these beams must be designed for the axial loads due to drag-strut action.

Table 7.8 Composite Deck Example: Collector Design

1	Grid	A	B		E		F
2	Diaphragm width, ft	60	60		60		60
3	Frame width, ft	20	20		20		20
4	Shear force, kips	60	0	40	33.4	6.64	80
5	Shear flow per ft of diaphragm	1.0	—	0.667	0.567	0.11	1.33
6	Shear flow per ft of frame	3.00	—	2.0	1.66	0.332	4.0
7	Drag requirement	Yes	No	Yes	No	No	Yes

Note: Shear capacity of diaphragm (3-in.-deep, 20-gauge metal deck, with a $3^1/_4$-inch lightweight concrete topping of 110 pcf density, and four puddle welds per sheet and a beam span 10 ft), from Table 7.9 = 1.74 kip-ft.

For illustration purposes, we will calculate the design axial force in the collector beams along grid G. The shear flow q_1 is

$$q_1 = \frac{80}{60} = 1.333 \text{ kips/ft}$$

As shown in Fig. 7.51b, the collector forces at grids 3 and 2 are

$F_3 = 1.33 \times 20 = 26.67$ kips
$F_2 = 1.33 \times 40 - 80 = -26.67$ kips

The maximum collector force = 26.67 kips, tension or compression. A drag beam may be designed for this force using the wide flange beams into the frame columns along grid G.

In lieu of designing the wide flange beams as collectors, supplemental reinforcement may be provided in the concrete topping to perform the same task of collecting and delivering the diaphragm shear to the frames. We will take this second approach using the strength design method to determine the required area of reinforcement. Observe that under Section 9.5.2.6.3.1 of ASCE 7-02, using the strength design method, collectors must be designed for special seismic load combination:

$$E = \Omega_o Q_E$$

where

Ω_o = over strength factor = 2.5 for a dual system consisting of moment and braced frames (ASCE 7-02, Table 9.5.2.2)
E = Effect of earthquake-induced forces due to special load combination
Q_E = Effect of horizontal earthquake-induced forces = 26.67 kips, tension (ASD level)
  = 1.4×26.67
  = 37.34 kips (ultimate)

Using the special seismic load combination, the ultimate design tension $T_u = 2.5 \times 37.34 = 93.35$ kips

$$A_s = \frac{93.35}{0.9 \times 60} = 1.73 \text{ in.}^2$$

Use 3 #7 bars continuous from grids 1 to 4 along line G, providing

$A_s = 3 \times 0.6 = 1.8$ in.^{2} > 1.73 in.^{2} OK

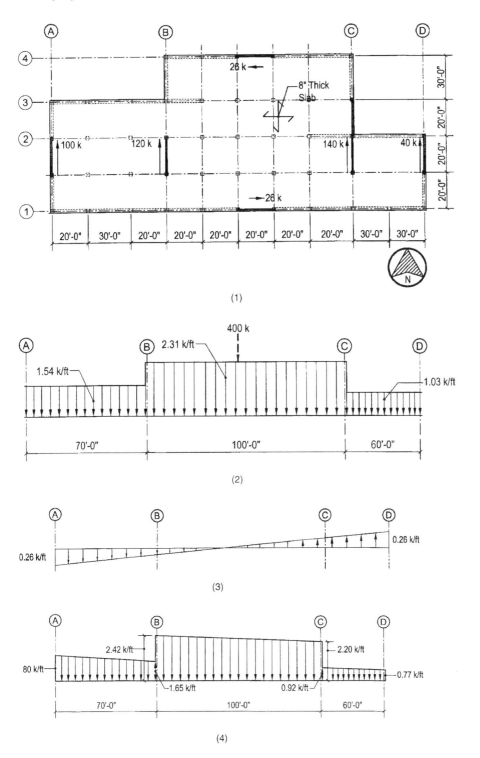

Figure 7.49c. Reinforced concrete diaphragm design example: (1) floor plan; (2) equivalent loads due to primary diaphragm action; (3) equivalent loads due to torsional effects; (4) final equivalent loads = (2) + (3); (5) shear diagram; (6) bending moment diagram.

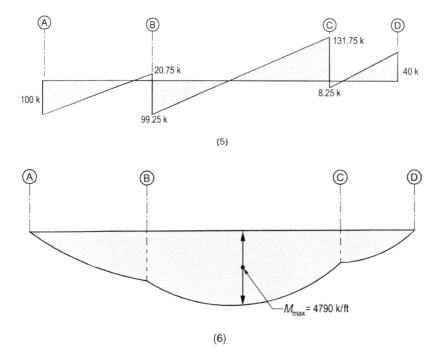

(5)

(6)

Figure 7.49c. (*Continued*)

Next, the collected load of 80 kips multiplied by the over strength factor Ω_o must be transferred from the reinforced slab into the frame on line G. This is done by welding headed studs. Using the special load combination, the ultimate load to be transferred is:

$$V_u = \Omega_o \times 80 \times 1.4$$
$$= 2.5 \times 80 \times 1.4 = 280 \text{ kips}$$

Assuming $^3/_4$"-diameter-headed studs, with a capacity governed by a concrete strength of 3000 psi,

$$\phi V_c = \phi \times 800 \times A_b \times \sqrt{f_c'}$$
$$= \frac{0.65 \times 800 \times 0.44 \times 0.75\sqrt{3000}}{1000}$$
$$= 9.4 \text{ kips/stud}$$

The required number of studs = 280/9.4 = 29.78. Use 30–$^3/_4$"-diameter studs at an approximate spacing of 7-in. over the entire 20-ft length of frame on line G.

Chord Design.   The maximum moment occurs at a distance of 63.5 ft from grid A and is equal to

$$M_{\max} = 1765.5 \text{ kip-ft}$$

See Fig. 7.49a3.

The corresponding axial force, compression, or tension in the edge beams B1 or B2 due to chord action is

$$C_u = T_u = \frac{1765.5}{60} = 29.4 \text{ kips}$$

TABLE 7.9 Diaphragm Shear Q and Flexibility Factor F for 3"-Deep Composite Metal Deck

4 Puddle welds		2½" Conc. fill: W = 110 pcf									3¼" Conc. fill: W = 110 pcf								
		Span																	
Gauge	Factor	8'-0"	9'-0"	10'-0"	11'-0"	12'-0"	13'-0"	14'-0"	15'-0"	16'-0"	8'-0"	9'-0"	10'-0"	11'-0"	12'-0"	13'-0"	14'-0"	15'-0"	16'-0"
16	Q	1820	1730	1660	1610	1560	1520	1490	1460	1440	2050	1970	1900	1850	1800	1760	1730	1700	1670
	F	.33	.34	.36	.37	.38	.39	.40	.41	.41	.29	.30	.31	.32	.33	.34	.34	.35	.36
18	Q	1700	1630	1570	1530	1490	1460	1440	1400	1390	1930	1890	1810	1770	1730	1700	1670	1650	1630
	F	.39	.41	.42	.43	.45	.45	.46	.47	.48	.34	.36	.37	.38	.38	.39	.40	.40	.41
20	Q	1590	1540	1500	1470	1440	1420	1400	1380	1360	1830	1780	1740	1700	1680	1650	1630	1620	1600
	F	.48	.50	.51	.52	.53	.54	.55	.56	.56	.42	.43	.44	.45	.46	.46	.47	.48	.48
22	Q	1550	1510	1470	1440	1420	1400	1390	1370	1360	1780	1740	1710	1680	1660	1640	1620	1600	1600
	F	.54	.56	.57	.58	.59	.60	.61	.61	.62	.47	.48	.49	.50	.51	.52	.52	.53	.53

The beams are designed for the combined axial force C_u or T_u and bending due to gravity loads. Additionally, their connections to the columns are verified for the combined gravity shear and axial tension. The reader is referred to standard structural steel design texts for design.

Example 2: Reinforced Concrete Diaphragm.

Given. A typical floor plan of a concrete building as shown in Fig. 7.49c1. The building's lateral-load-resisting system consists of special reinforced concrete shear walls in both directions. The floor framing is an 8-in.-thick, two-way flat slab system.

The wall forces above and below the given diaphragm have been determined by performing a three-dimensional analysis assuming a rigid diaphragm for the floors and roof. The differences between the two shears, which may be considered as reactions to the diaphragm inertial loads, are shown in Fig. 7.49k1. The shears in the E–W walls are due to torsional effects. These, however, will not affect the present diaphragm design for the N–S seismic forces.

In designing the shear walls for seismic loads, it was determined that their nominal shear strength was less than the shear corresponding to the development of the nominal flexural strength. Therefore, a strength reduction factor $\phi = 0.60$ was used for the shear design of the walls. And, because ACI 318-02 Section 9.3.4 (b) mandates the value of ϕ for shear design of diaphragms not to exceed the value used for the shear design of walls, we use $\phi = 0.60$.

Required. Diaphragm analysis including design of collectors and chords.

Solution.

Equivalent Loads. These may determined by considering the diaphragm inertial forces as a consequence of two actions. The first, the primary action, results from the inertial force F_{px} distributed along the length of the diaphragm in proportion to its mass. The second is due to the eccentricity of F_{px} with respect to the center of stiffness of the walls. This action results in a set of equal and opposite loads that establish moment equilibrium between the inertial forces and the reactions.

Because in most buildings, the mass per unit area of floor and roof system is constant over the entire area, F_{px} may be distributed along the length of the diaphragm in the same proportion as its width. For the example problem, this is shown in Fig. 7.49c2.

Total area of diaphragm $= 60 \times 70 + 90 \times 100 + 40 \times 60 = 15600$ ft^2
Inertial force per unit area $= \frac{400}{15600} = 0.0256$ kip-ft^2
Load per unit length of segment A $= 0.0256 \times 60 = 1.54$ kip-ft
 B $= 0.0256 \times 90 = 2.31$ kip-ft
 C $= 0.0256 \times 40 = 1.03$ kip-ft

If we were to draw a bending moment diagram corresponding to the primary equivalent loads shown in Fig.7.49c2, it would be seen that the moment diagram will not close. For example, the moment at D due to reactions at A, B, and C

$$M_D = 100 \times 230 + 120 \times 160 + 140 \times 60$$
$$= 50,600 \text{ kip-ft}$$

And the moment due to the equivalent inertial loads

$$M_D = 1.54 \times 70\left(\frac{70}{2} + 160\right) + 2.31 \times 100(50 + 60)$$

$$= 48267 \text{ kip-ft}$$

Thus there is a moment gap of 50,600 – 48,267 = 2333 kip-ft calculated at grid D. However, it should be noted that the gap is the same for the entire length of diaphragm and is essentially due to of torsion effects.

To close the gap, we modify the distribution of primary equivalent load. This is done by superimposing the equal and opposite inertia forces due to the secondary action on the primary action. Observe that the equivalent load corresponding to the secondary action is a self equilibrating system resulting only in an applied moment.

For the example problem, the moment gap of 2333 kip-ft is closed by imposing a triangular distribution of equal and opposite inertia forces as shown in Fig. 7.49c3. If w_T is the maximum value at the ends.

$$\frac{w_T}{2} \times \frac{L}{2} \times \frac{2L}{3} = 2333 \text{ kip-ft}$$

or

$$w_T = 0.26 \text{ kip-ft/ft}$$

The final equivalent load that determines the design shear force and bending moments is shown in Fig. 7.10.

Shear design. Shown in Figs. 7.49c4 and c5 are the shear force and bending moment diaphragms. The information for the design of collector elements is shown in Table 7.10.

In determining the shear capacity of diaphragms, the value for ϕ, the capacity reduction factor, should not exceed the value used in the design of vertical elements of the lateral-load-resisting systems. See ACI 318-02 Sect. 9.3.4 (b). For the example problem, ϕ for shear wall design is given as 0.6. Therefore, ϕ for diaphragm shear design should not exceed 0.6 Use $\phi = 0.6$.

The shear capacity per ft length of a concrete slab without shear reinforcement is given by

$$\phi V_c = \phi 2\sqrt{f_c'}bt$$

$$= 0.6 \times 2\sqrt{4000} \times 12 \times 8$$

$$= 7286 \text{ lbs}$$

$$= 7.3 \text{ kip-ft}$$

TABLE 7.10 Reinforced Concrete Slab: Collector Design

1	Grid	A	B		C		D
2	Diaphragm width, ft	60	60	90	90	40	40
3	Shear wall length, ft	20	20	20	40	40	20
4	Ultimate shear force, kips	100	20.75	99.25	131.75	8.25	40
5	Shear flow per ft of diap	1.6 7	0.3 5	1.10	1.46	0.20	1.0
6	Shear flow per ft of wall length	5.0	1.04	4.96	3.30	0.21	2.0
7	Drag requirement	Yes	No	Yes	No	No	No

The maximum ultimate shear flow equal to 1.67 kip-ft occurs in the diaphragm at line A. This is less than the capacity 7.3 kip-ft. Therefore, by calculations, no shear reinforcement is required. However, provide #4 @ 18 in. the N–S direction at middepth of slab for a width equal to 5 ft.

Collector (Drag–Strut) Design. For purposes of illustration, the drag strut along line A will be designed. The collected shear of 100 kips must be delivered from the slab to the shear wall on line A. This can be done by providing reinforcement perpendicular to the wall.

Using the special seismic load combination

$$v_u = \frac{\Omega_o V_u}{L_{wall}}$$

$$= \frac{2.5 \times 100}{20} = 12.5 \text{ kip-ft}$$

With one #4 reinforcement at top and bottom, and at 12-in. on centers,

$$\rho_v = \frac{0.2 \times 2 \times 12}{18 \times 8 \times 12} = 0.0028$$

$$\phi V_n = \phi(V_c + V_s)$$

$$= 0.6(2\sqrt{f_c'} + \rho_v f_y)bt$$

$$= \frac{0.6}{1000}(2\sqrt{4000} + 0.0028 \times 60{,}000) \times 12 \times 8$$

$$= 16.9 \text{ kip-ft} \quad > 12.5 \text{ kip-ftt} \quad \text{OK}$$

Chord Design. Chord design T_u at

$$\text{BA} = \frac{1244}{60} = 20.74 \text{ kips} \quad A_s = \frac{20.74}{0.9 \times 60} = 0.39 \text{ in.}^2, \text{ provide 2 \#6}$$

$$\text{BC} = \frac{1244}{90} = 13.82 \text{ kips} \quad A_s = \frac{13.82}{0.9 \times 60} = 0.26 \text{ in.}^2, \text{ provide 2 \#5}$$

$$\text{CB} = \frac{90}{90} = 15.92 \text{ kips} \quad A_s = \frac{15.92}{0.9 \times 60} = 0.30 \text{ in.}^2, \text{ provide 2 \#5}$$

$$\text{CD} = \frac{90}{40} = 35.82 \text{ kips} \quad A_s = \frac{35.82}{0.9 \times 60} = 0.67 \text{ in.}^2, \text{ provide 2 \#6}$$

Maximum moment occurs at center span.
The corresponding chord force

$$T_u = \frac{4790}{90} = 53.22 \text{ kips}$$

$$A_s = \frac{T_n}{\phi f_y} = \frac{53.22}{0.9 \times 60} = 0.98 \text{ in.}^2$$

Provide two additional #7 reinforcements giving $A_s = 2 \times 0.6 = 1.2$ in.^{2}, for chord action in the beams between grids B and C. At reentrant corners extend the beams one-bay into the slab system.

The maximum compressive force C_u occurs in span BC, and is equal to

$$C_u = T_u = 52.22 \text{ kips}$$

Assuming a 24 in. $\times$ 24 in. edge beam, the compressive stress due to chord action is equal to $52.22/24 \times 24 = 0.096$ ksi $= 96$ psi, which is less than $0.2 f_c = 0.2 \times 4000 = 800$ psi. Therefore, no additional compressive reinforcement is required in the edge beams.

7.4.2. Composite Beams

Two types of composite construction are recognized by the AISC specifications: 1) fully encased steel beams; and 2) steel beams with shear connectors. In fully encased steel beams, the natural bond between concrete and steel interface is considered sufficient to provide the resistance to horizontal shear provided that: 1) the concrete thickness is 2 in. (50.8 mm) or more on the beam sides and soffit, with the top of the beam at least $1^1/_2$ in. (38 mm) below the top and 2 in. (50.8 mm) above the bottom of the slab; and 2) the encasement is cast integrally with the slab and has adequate mesh or other reinforcing steel throughout the depth and across the soffit of the beam to prevent spalling of concrete.

Design of encased beams can be accomplished by two methods. In the first method for unshored construction, the stresses are computed by assuming that the steel beam alone resists all the dead load applied prior to hardening of concrete. The superimposed dead and live loads applied after hardening of concrete are assumed to be resisted by composite action. In addition to providing composite action, the concrete encasement is assumed to restrain the steel beam from both local and lateral torsional buckling. Therefore, an allowable stress of $0.66F_y$ instead of $0.60 F_y$ can be used when the analysis is based on the properties of transformed section. Thus, for positive bending moments we get

$$f_b = \frac{M_D}{S_s} + \frac{M_L}{S_{tr}} \leq 0.66F_y$$

where

f_b = computed stress in the bottom flange for positive bending moment
M_D = dead-load bending moment
M_L = superimposed dead- and live-load bending moment
S_s = section modulus of the steel section referred to its bottom flange
S_{tr} = section modulus of the transformed section referred to its bottom flange

The second method of design of encased beams is a recognition of a common engineering practice where it is desired to eliminate the calculation of composite section properties. This provision permits a higher stress of $0.76F_y$ in steel when the steel beam alone is designed to resist all loads. Thus

$$f_b = \frac{M_D + M_L}{S_a} \leq 0.76F_y$$

The second type of composite steel beam, namely, composite beams with shear connectors, is by far the more popular in the construction of buildings in North America. Typically, composite action is achieved by providing shear connectors between the steel top flange and the concrete topping. This method is more popular because encasing beams with concrete requires expensive form work.

Composite sections have greater stiffness than the summation of the individual stiffness of the slab and beam, and therefore can carry larger loads or similar loads with appreciably smaller deflection. Because of the greater stiffness, they are less prone to transient vibrations. Composite action results in an overall reduction of floor depth. Consequently, for high-rise buildings, because of reduced floor-to-floor height, the cumulative savings in curtain walls, electrical wiring, mechanical ductwork, interior walls, plumbing, etc., can be considerable.

Composite beams can be designed either for shored or unshored construction. For shored construction, the cost of shoring should be evaluated in relation to the savings achieved by the use of lighter beams. For unshored construction, steel is designed to support the wet weight of concrete and construction loads by itself. The steel section, therefore, is heavier than in shored construction.

In composite floor construction, the top flanges of the steel beams are attached to the concrete using shear connectors. Thus the concrete slab becomes part of the compression flange. As a result, the neutral axis of the section shifts upward, making the bottom flange of the beam more effective in tension.

Since the concrete already serves as part of the floor system, the only additional cost is that of the shear connectors. In addition to transmitting horizontal shear forces from the slab into the beam, the shear connectors prevent the tendency for the slab to rotate independently of the beam.

A typical shear connector is a short length of round steel bar welded to the steel beam at one end and having an anchorage provided in the form of a round head at the other end. The most common diameters are $1/_2$, $5/_8$, and $3/_4$ in. (12, 16, and 19 mm). The length is dependent on the depth of metal deck and should extend at least $1^1/_2$ in. (38 mm) above the top of the deck. The welding process typically reduces their length by about $3/_{16}$ in. (5 mm). The upset head thickness of the studs is usually $3/_8$ or $1/_2$ in. (9 to 12 mm), and the diameter $1/_2$ in. (12 mm) larger than the stud diameter. The studs are normally welded to the beam with an automatic welding gun, and when properly executed, the welds are stronger than the steel studs. Studs located closer to the beam support are more effective than studs located toward the beam centerline. The larger volume of concrete between the stud and the pushing side of the trough helps in the development of a larger failure cone in concrete, thus increasing its horizontal shear resistance.

The stud length has a definite effect on the shear resisted by it. As the length increases, so does the size of the shear cone, with a consequent increase in the shear value. The shear capacity of the stud also depends on the profile of the metal deck. To get a qualitative idea, consider the two types of metal decks shown in Figs 7.49f and 7.50. The deck in Fig. 7.49f has a narrow hump compared to the one in Fig. 7.50. When subjected to a load V, the concrete and the metal deck tend to behave as a portal frame. The concrete in the troughs can be thought of as columns with the concrete over the humps acting as beams (Fig. 7.50b). A narrow hump of the portal frame results in an equivalent beam of smaller span when compared to the one with a wider hump, meaning that a deck profile with the widest trough and narrowest hump will yield the highest connector strengths. However, other considerations such as volume of concrete, section modulus, and the stiffness of deck also influence the shear strength of the connector.

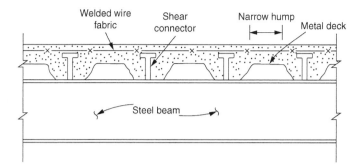

Figure 7.49f. Composite beam with narrow hump metal deck.

Metal decks for composite construction are available in the United States in three depths—$1\frac{1}{2}$ in. (38 mm), 2 in. (51 mm), and 3 in. (76 mm). The earlier types of metal deck did not have embossments, and the interlocking between concrete and metal deck was achieved by welding reinforcement transverse to the beam. Later developments of metal deck introduced embossments to engage the concrete and metal deck and dispensed

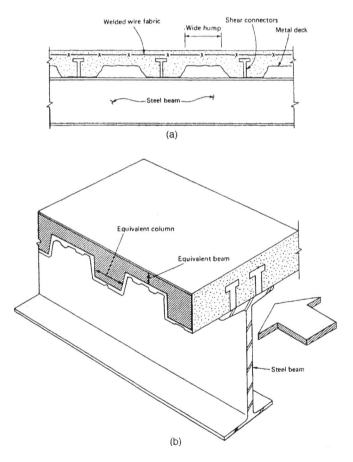

Figure 7.50. (a) Composite beam with wide hump metal deck; (b) simplified analytical model of composite metal deck subjected to horizontal shear.

with the transverse-welded reinforcement. Typical spans for composite metal deck are generally in the range of 8 to 15 ft (2.4 to 4.6 m).

In floor systems using $1^1/_2$-in. (38-mm)-deep decks, provision for electrical and telephone services is made by punching holes through the slab at various locations and passing the under-floor ducts through them. A deeper deck is required if the power distribution system is integrated as part of the structural slab; 2- or 3-in. (51- or 76-mm)-deep metal deck is sufficient. Tests have shown that there is very little loss of composite beam stiffness due to the ribbed configuration of metal deck in the depth range of $1^1/_2$ to 3 in. (38 to 76 mm). As long as the ratio of width to depth of the metal deck is at least 1.75, the entire capacity of the shear stud can be developed similar to that for beams with solid slabs. However, with deeper deck, a substantial decrease in shear strength of the stud occurs, which is attributed to a different type of failure mechanism. Instead of the failure of shear stud, the mode of failure is initiated by cracking of the concrete in the rib corners. Eventual failure takes place by separation of concrete from the metal deck. When more than one stud is used in a metal deck flute, a failure cone can develop over the shear stud group, resulting in lesser shear capacity per each stud. The shear stud strength is therefore closely related to the metal deck configuration and factors related to the surface area of the shear cone.

Often special considerations are required in composite design when openings interrupt slab continuity. For example, beams adjacent to elevator and stair openings may have full effective width for part of their length and perhaps half that value adjacent to the openings. Elevator sill details normally require a recess in the slab for door installations, rendering the slab ineffective for part of the beam length. A similar problem occurs in the case of trench header ducts, which require elimination of concrete, as opposed to the standard header duct, which is completely encased in concrete. When the trench is parallel to the composite beam, its effect can easily be incorporated into the design by suitably modifying the effective width of compression flange. The effect of the trench oriented perpendicular to the composite beam could range from negligible to severe depending upon its location. If the trench can be located in the region of minimum bending moment, such as near the supports in a simply supported beam, and if the required number of connectors could be placed between the trench and the point of maximum bending moment, its effect on the composite beam design is minimal. If, on the other hand, the trench must be placed in an area of high bending moment, its effect may be so severe as to require that the beam be designed as a noncomposite beam.

This slab thickness in composite construction is usually governed by fire-rating requirements rather than by the bending capacity of the slab. In certain parts of the United States it may be economical to use the minimum thickness required for strength and to use sprayed-on or some other method of fireproofing the deck to obtain the required ratings. Some major projects have used a $2^1/_2$-in. (63.5-mm)-thick concrete slab on 3-in. (76.2-mm)-deep metal deck spanning as much as 15 ft (4.57 m).

In continuous composite beams the negative moment regions can be designed such that: 1) the steel beam alone resists the negative moment; or 2) it acts compositely with mild steel reinforcement placed in the slab parallel to the beam. In the latter case, shear connectors must be provided through the negative moment region.

Careful attention should be paid to the deflection characteristics of composite construction because the slender not-yet-composite shape deflects as wet concrete is placed on it. There are three ways to alleviate the deflection problem.

1. Use relatively heavy steel beams to limit the dead-load deflection and place lens-shaped tapering slabs to obtain a nearly flat top. Although a reasonably flat surface results from this construction, the economic restraints of speculative

office buildings do not usually permit the luxury of the added cost of additional concrete and heavier steel beams.

2. Camber the steel beam to compensate for the deflection due to weight of steel beam and concrete. Place a constant thickness of slab by finishing the concrete to screeds set from the cambered steel. Continuous lateral bracing as provided by the metal deck is required to prevent the lateral torsion buckling of beam. If steel deck is not used, this system requires a substantial temporary bracing system to stabilize the beam during construction.

3. Camber and shore the steel beam. The beam is fabricated with a camber calculated to compensate for the deflection of the final cured composite section. Shores are placed to hold the steel at its curved position while the concrete is being poured. As in method 2, slab is finished to screeds set from cambered steel. Although methods 1 and 3 are occasionally used, the trend is to use method 2 because it is the least expensive.

7.4.2.1. *AISC Allowable Stress Design (ASD)*

Including provisions for solid slab, there are three categories of composite beams in the AISC specifications each with a differing effective concrete area.

7.4.2.1.1. Solid Slab. The total slab depth is effective in compression unless the neutral axis is above the top of the steel beam. (In typical floor systems with relatively thin slabs, the neutral axis of steel beams is invariably below the slab, rendering the total slab depth effective in compression).

7.4.2.1.2. Deck Perpendicular to Beam.

1. As illustrated in Fig. 7.51, concrete below the top of steel decking shall be neglected in computations of section properties and in calculating the number of shear studs, but the concrete below the top flange of deck may be included for calculating the effective width.

2. The maximum spacing of shear connectors shall not exceed 32 in. (813 mm) along the beam length.

3. The steel deck shall be anchored to the beam either by welding or by other means at a spacing not exceeding 16 in. (406 mm).

4. A reduction factor as given by the AISC formula I5-1

$$\left(\frac{0.85}{\sqrt{N_r}}\right)\left(\frac{w_r}{h_r}\right)\left(\frac{H_s}{h_r} - 1\right) \le 1.0$$

should be used for reducing the allowable horizontal shear capacity of stud connectors. In the above formula h_r is the nominal rib height in inches; H_s is length of stud connector after welding in inches. An upper limit of $(h_r + 3)$ is placed on the length of shear connectors used in computations even when longer studs are installed in metal decks. N_r is the number of studs in one rib. A maximum value of 3 can be used in computations although more than three studs may be installed. w_r is average width of concrete rib.

7.4.2.1.3. Deck Ribs Parallel to Beam.

1. The major difference between perpendicular and parallel orientation of deck ribs is that when the deck is parallel to the beam, the concrete below the top of the decking can be included in the calculations of section properties and must be included when calculating the number of shear studs, as illustrated in Fig. 7.52.

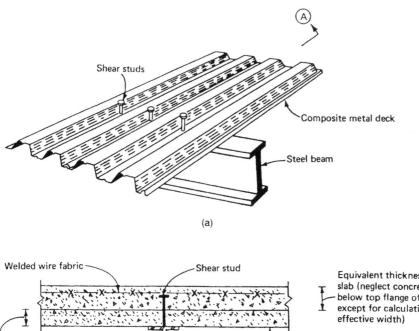

Figure 7.51. Composite beam with deck perpendicular to beam: (a) schematic view; (b) section A showing equivalent thickness of slab.

2. If steel deck ribs occur on supporting beam flanges, it is permissible to cut high-hat to form a concrete haunch.
3. When the nominal rib height is $1\frac{1}{2}$ in. (38.1 mm) or greater, the minimum average width of deck flute should not be less than 2 in. for the first stud in the transverse row plus four stud diameters for each additional stud. This gives minimum average widths of 2 in. (51 mm) for one stud, 2 in. plus $4d$ for two studs, 2 in. plus $8d$ for three studs, etc., where d is the diameter of the stud. Note that if a metal deck cannot accommodate this width requirement, the deck can be split over the girder to form a haunch.
4. A reduction factor as given by AISC formula I5-2

$$0.6\left(\frac{w_r}{h_r}\right)\left(\frac{H_s}{H_r} - 1.0\right) \le 1.0$$

shall be used for reducing the allowable horizontal shear capacity of stud connectors.

7.4.2.1.4.   ASCE Requirements for Formed Steel Deck Construction.

Certain specific ASCE requirements applicable to formed steel deck construction are shown schematically in Fig. 7.53. More general comments follow:

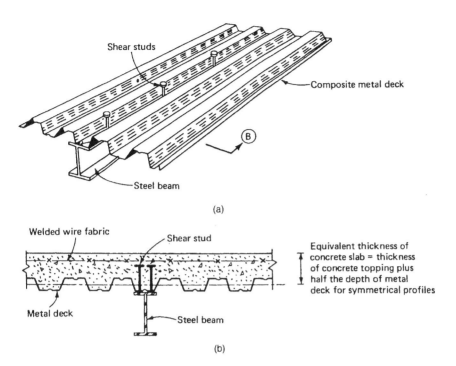

Figure 7.52. Composite beam with deck parallel to beam: (a) schematic view; (b) section B showing equivalent thickness of slab.

1. The deck rib height shall not exceed 3 in. (76.5 mm).
2. Rib average width shall not be less than 2 in. (51 mm). If the deck profile is such that the width at the top of the steel deck is less than 2 in. (51 mm), this minimum clear width shall be used in the calculation.
3. The section properties do not change a great deal from deck running perpendicular or parallel to the beam, but the change in the number of studs can be significant.
4. The reduction formula for stud length is based on rib geometry, number of studs per rib, and embedment length of the studs.
5. The equation for calculating the partial section modulus makes the choice of heavier, stiffer beams with fewer studs economically more attractive.
6. Higher shear values can be used in longer shear studs. Concrete cover over the top of the stud is not limited by the AISC specifications, but for practical reasons the author recommends a minimum of $\frac{1}{2}$ in. (12.7 mm).
7. Studs can be placed as close to the deck web as needed for installation and to maintain the necessary spacing.
8. Deck anchorages can be provided by the stud welds.
9. Maximum diameter of shear connectors is limited to $\frac{3}{4}$ in. (19 mm).
10. After installation, the studs should extend a minimum of $1\frac{1}{2}$ in. (38 mm) above the steel deck.
11. Total slab thickness including the ribs is used in determining the effective width without regard to the orientation of the deck with respect to the beam axis.
12. The slab thickness above the steel deck shall not be less than 2 in. (51 mm).

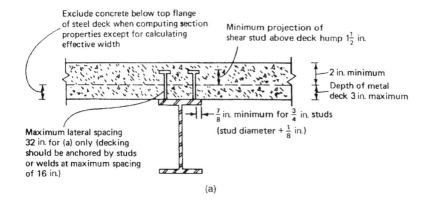

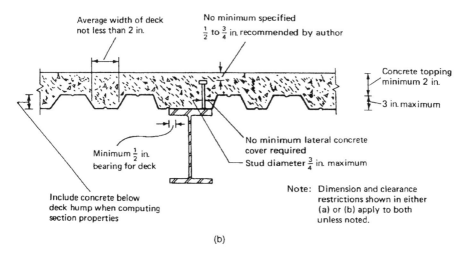

Figure 7.53. Composite beam, AISC requirements: (a) deck perpendicular to beam; (b) deck parallel to beam.

For design purposes, a composite floor system is assumed to consist of a series of T-beams, each made up of one steel beam and a portion of the concrete slab. The AISC limits on the width of slab that can be considered effective in the composite action are shown in Fig. 7.54. When the slab extends on one side of the beam only, as in spandrel beams and beams adjacent to floor openings, the effective width naturally is less than when the slab extends on both sides of the beam. For slabs extending on both sides of the beam, the maximum effective flange width b may not exceed: 1) one-fourth of the beam span L; or 2) one-half the clear distances to adjacent beams on both sides plus b_f, the width of steel beam flange. When the slab extends on only one side of the beam, the maximum effective width b may not exceed: 1) one-twelfth of the beam span L; or 2) one-half the clear distance to the adjacent beam plus b_f. Furthermore, the outboard effective width may not exceed the actual width of overhang, and the inboard effective width must not extend beyond the centerline between the edge beam and the adjacent interior span.

The design of composite beams is usually achieved by the transformed area method, in which the concrete effective area of the composite beam is transformed into an equivalent steel area. It is equally admissible to transform the steel area into an equivalent concrete area, but the calculations are somewhat simplified by the former method. The method

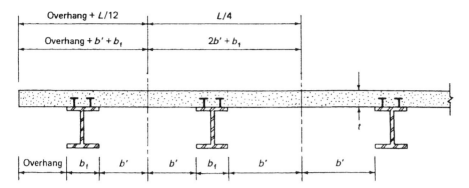

Figure 7.54. Effective width concept as defined in the AISC specifications.

assumes transverse compatibility at the concrete and steel interface. The unit stress in each material is equal to the strain times its modulus of elasticity. Because of strain compatibility, the stress in steel is n times the stress in concrete, where n is the modular ratio E_s/E_c. A unit area of steel is, therefore, mathematically equivalent to n times the concrete area. Thus, the effective area of concrete $A_c = bt$ can be replaced by an equivalent steel area A_c/n.

Concrete is neither linearly elastic nor ductile and its stress–strain curve exhibits a constantly changing slope with a sudden brittle failure. In spite of these characteristics, concrete is considered elastic within a stress–strain range of up to $0.50f'_c$, and the modulus of elasticity in pounds per square inch can be approximated by the relation $E_c = W_c^{1.5} \times 33\sqrt{f'_c}$, where W_c is the unit weight of concrete in pounds per cubic foot and f'_c is the compressive strength of concrete in pounds per square inch. The compressive strength f'_c of concrete normally used in floor construction is in the range of 3000 to 5000 psi (20.7 to 34.4 MPa) giving a value of E_c for normal weight concrete of $3.12 \times 10^6 < E_c < 4.03 \times 10^6$ psi ($21,512 < E_c < 27,787$ MPa), compared to E_s of steel at 29×10^6 psi (199,955 MPa). The value of $n = E_s/E_c$, therefore, lies between 9.3 and 7.2 and is usually approximated to the whole number in recognition of the error in the formula for E_c when compared to actual performance.

For strength calculations, the AISC specification uses the value of n for normal-weight concrete of the specified strength. However, for deflection computations, n depends not only on the specified strength but also on the unit weight of concrete, Therefore, in computing deflections, especially for beams subjected to heavy sustained loads, it is necessary to account for the effects of creep by using an appropriate value of n. This is even more important in shored construction when the dead load of the concrete is resisted by the composite action. Creep effect is accounted for in computing deflections by using a higher modular ratio, n. A factor of 2 for creep effects is typically adequate in building designs. Live loads are always resisted by the composite section. If they are of short duration, the deflections are computed using the short-term modular ratio.

The transformed steel section can be conveniently considered as the original steel beam with an added cover plate to the top flange of thickness t equal to slab thickness, and an equivalent width b/n. The composite properties of the transformed section are calculated by locating the neutral axis and the transformed moment of inertia I_{tr}. The maximum bending stress in the steel beam bottom flange is given by

$$f_{bs} = \frac{MY_{tr}}{I_{tr}}$$

where M is the total bending moment, Y_{tr} is the distance of the extreme bottom steel fibers from the neutral axis, and I_{tr} is the transformed moment of inertia. The maximum compressive stress in the concrete is given by

$$f_{bc} = \frac{MC_t}{nI_{tr}}$$

where C_t is the distance from the neutral axis to the extreme concrete fibers and n is the modular ratio. The value

$$S_{tr} = \frac{I_{tr}}{Y_{tr}}$$

is called the transformed section modulus of the beam referred to the bottom flange.

For construction without temporary shores, concrete compressive stress is based upon the load applied after it has reached 75% of the required strength. This compressive stress is limited to $0.45f'_c$, just as in the working stress design of reinforced concrete beams.

The total horizontal shear to be resisted between the point of maximum positive moment and point of zero moment is the smaller of the two values as determined by

$$V_h = \frac{0.85f'_c A_c}{2}$$

$$V_h = \frac{A_s F_y}{2}$$

where

f'_c = specified compressive strength of concrete
A_c = actual area of effective concrete flange
A_s = area of steel beam
F_y = specified yield stress of steel beam

Note that the formula $V_h = 0.85f'_c A_c/2$ assumes that there is no longitudinal reinforcing steel in the compression zone of composite beam. If the compressive zone is designed with mild steel reinforcement, the formula for horizontal shear is to be modified as follows:

$$V_h = \frac{0.85f'_c A_c}{2} + \frac{A'_s F_{yr}}{2}$$

where

A'_s = area of the longitudinal compressive steel
F_{yr} = yield stress of the reinforcing steel

AISC permits averaging of horizontal shear flow; that is, the total number of connectors between the point of maximum moment and point of zero moment must be sufficient to satisfy the total shear flow within that length. The shear connector formulas represent the horizontal shear at ultimate load divided by 2 to approximate conditions at working loads.

The number of shear connectors required for full composite action is determined by dividing the smaller value of V_h by the shear capacity of one connector. The number of connectors obtained represents the shear connectors required between the point of maximum positive moment and point of zero moment. For example, in a simply supported, uniformly loaded beam, this represents half the span; and in a simply supported beam with two equidistant concentrated loads, this represents the distance between the point load to the support point. The total number of connectors required for the entire span is thus double the number obtained earlier.

A composite beam subject to negative bending moment experiences tensile stresses in the concrete zone and loses much of its advantage. However, when reinforcement is placed parallel to the beam within the effective width of slab, and is anchored adequately to develop the tensile forces, the advantage of continuous construction is restored. The steel used in the tensile zone is included in computing the property of the composite section. Similarly, when the compressive stress in concrete subject to positive moment exceeds the allowable stress, it is permissible to use compressive steel in the effective width zone to reduce stresses.

Consider a continuous composite beam shown in Fig. 7.55. The total horizontal shear to be resisted by shear connectors between an interior support and each adjacent point of contraflexure (regions a, b, and c in Fig. 7.55c) is given as

$$V_h = \frac{A_{sr}F_{yr}}{2}$$

where

$\quad A_{sr}$ = area of reinforcing steel provided at the interior support within the effective flange width

$\quad F_{yr}$ = yield stress of the reinforcing steel

AISC permits uniform spacing of connectors between the points of maximum positive moment and the point of zero moment. Also, the connectors required in the region of negative bending can be uniformly distributed between the point of maximum moment and each point of zero moment. For concentrated loads, the numbers of shear connectors N_2 required between any concentrated load and the nearest point of zero moment is determined by the AISC formula

$$N_2 = \frac{N_1[(M\beta/M_{max}) - 1]}{\beta - 1}$$

where

$\quad M$ = moment at concentrated load point (less than the maximum moment)

$\quad N_1$ = number of connectors required between point of maximum moment and point of zero moment

$\quad \beta$ = ratio of transformed section modulus to steel section modulus

This relation is schematically shown in Fig. 7.56.

In the design of composite beams it is often unnecessary to develop the full composite action. A partial composite action with fewer studs is all that may be necessary to achieve the required strength and stiffness. AISC permits designs of less than 100% composite

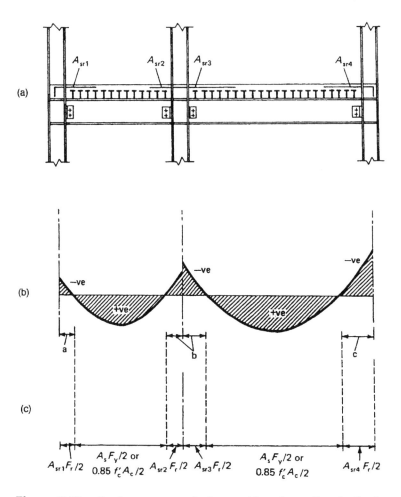

Figure 7.55. Continuous composite beam subjected to uniformly distributed load: (a) elevation; (b) moment diagram; (c) horizontal shear resisted by studs in the positive and negative moment regions.

action by introducing the concept of effective section modulus as determined by the relation

$$S_{\text{eff}} = S_s + \left(\frac{V_h'}{V_h} \right)^{1/2} (S_{tr} - S_s)$$

where S_s is the section modulus of the steel beam, and V_h' is the shear capacity provided by the shear connectors, obtained by multiplying the number of connectors used and the shear capacity of one connector. Transposing the above equation we get

$$V_h' = V_h \left(\frac{S_{\text{reqd}} - S_s}{S_{\text{avail}} - S_s} \right)^2$$

AISC stipulates that V_h' be not less than $0.25V_h$ to prevent excessive slip and loss of beam stiffness. This minimum requirement does not apply if shear studs are used for reasons other than increasing the flexural capacity, such as for increasing beam stiffness or for diaphragm connectivity.

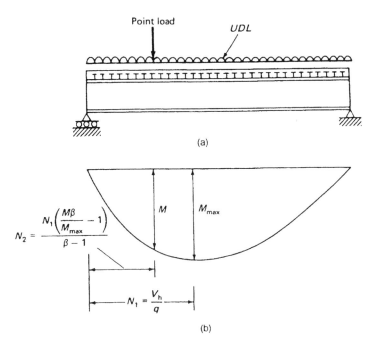

Figure 7.56. Shear connector requirements for composite beams subjected to concentrated loads: (a) schematic loading diagram; (b) shear connector requirements; (c) composite beam design example; (1) plan; (2) section.

The AISC specification gives three criteria for stud placement: 1) a minimum center-to-center spacing of six stud diameters between the studs in the longitudinal direction; 2) minimum spacing of four stud diameters in the transverse direction; and 3) maximum spacing in the longitudinal direction of 32 in. (813 mm). Note that if stud spacing exceeds 16 in. (406.4 mm), a plug weld between the studs is required to resist uplift forces.

If the required bending capacity is provided by the steel beam alone without relying on composite action, the maximum spacing requirement of 32 in. (813 mm) need not be met.

The recommended sequence for installing studs when the deck is perpendicular to the beam is as follows:

- Deck ribs at 6 in. (153 mm) on center. Start at beam ends and place a single stud at every fourth flute, working toward the center of beam. If studs remain, fill in empty ribs, again starting at beam ends and working toward the center without exceeding 30 in. (762 mm) for stud spacing.
- Deck ribs at 12 in. (305 mm) on center. Start at beam ends and place a single stud in every other flute working toward the center of beam. If studs remain, fill in empty ribs, again starting at beam ends and working toward the center of beam without exceeding 24 in. (610 mm) for stud spacing.
- If the number of studs is more than the number of ribs, place a double or triple row as needed, always starting from beam ends and working toward the beam center. In general, is studs cannot be uniformly spaced, the greatest number of studs should occur at the ends.

The recommended sequence for installing studs when the deck is parallel to the girder is as follows. Start at the girder ends by placing the first stud at approximately

12 in. (305 mm) from the centerline of support and work toward the center of girder with uniform spaces between the studs. If a double row of studs is required, it is a good practice to place them in a staggered pattern rather than side by side.

The allowable shear for stud connectors is influenced by several factors when used in metal deck construction. As in solid slabs, the strength and type of concrete, whether regular or lightweight, determines the allowable horizontal loads. The rib geometry of metal deck and the height of the stud above metal deck (when deck is parallel to the girder) are other factors influencing the allowable horizontal loads. For girders, the wider the rib opening and the greater the penetration of the stud above the deck, the more closely the allowable horizontal shear load will approach the published AISC value for studs in solid concrete slabs.

7.4.2.2. Design Example

Given.

(W18 × 40), 50 ksi steel beam, beam span = 40 ft (Fig. 7.57)

Beam sp10 ft

Tributary width for dead and live loads = 10 ft

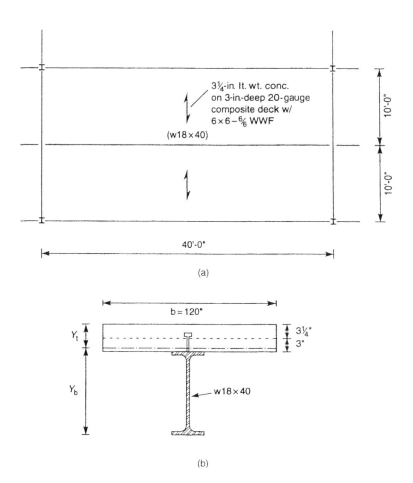

Figure 7.57.   Composite beam, design example: (a) plan; (b) section.

Composite floor construction: $3\frac{1}{4}$ in. of 115 pcf concrete slab over 3 in.
metal deck. Rib width = 6.0 in.

Compressive strength of concrete, $f'_c = 3.0$ ksi

Loading.
1. Dead load of slab with allowance for steel beam = 50 lb/ft²
2. Additional precomposite dead load due to extra concrete required for compensating beam deflection = 5 lb/ft² (ponding)
3. Additional composite dead loads: partitions = 20 lb/ft²
   ceiling plus miscellaneous = 10 lb/ft²
4. Live load = 50 lb/ft²

Required. Verification of (W18 × 40) for final design. *AISC Specifications* (AISCS), 9th Edition.

Solution. The ASD design is based on elastic analysis using the transformed section properties for composite beams. Because the compression flange is continuously braced, the allowable stress in the steel section is $0.66F_y$. Lateral-torsional buckling is not a concern for the completed structure, but it must be guarded against during construction.

The allowable stress in the concrete slab is $0.45f'_c$. In building design, typically the neutral axis is close to the top of the section. Therefore, stress in the steel is usually the controlling factor.

The section properties for unshored construction are computed by elastic theory. The bending stress in the steel beam is taken as the sum of: (1) the stress based on the assumption that steel section alone resists all loads applied prior to concrete reaching 75% of its specified strength; and (2) the stress based on the assumption that subsequent loads are resisted by the composite section.

For the example problem we have:
Uniform precomposite load = $(50 + 5) \times 10 = 550$ plf
Uniform postcomposite load = $(30 + 50) \times 10 = 800$ plf
Total   = 1350 plf

$$\text{Maximum moment} = \frac{(1350)(40)^2}{8 \times 1000} = 270 \text{ kip-ft}$$

$$\text{Required section modulus} = \frac{270 \times 12}{(0.66)(50)} = 98.18 \text{ in.}^3$$
for 50 ksi steel

Modulus of Elasticity of Concrete E_c. For stress check, AISCS permits the use of normal-weight concrete properties even for lightweight concrete topping. For deflection calculations, however, the use of actual properties is required.

$$E_c = 33W^{1.5}\sqrt{f'_c}$$

$$E_c \text{ for normal weight} = \frac{33 \times 145^{1.5}}{1000}\sqrt{3000} = 3156 \text{ ksi}$$

$$\text{Modular ratio} = \frac{29000}{3156} = 9.19$$

$$E_c \text{ for light weight} = \frac{33 \times 115^{1.5}}{1000}\sqrt{3000} = 2229 \text{ ksi}$$

$$\text{Modular ratio} = \frac{29000}{2229} = 13.06$$

Composite Beam Properties. The effective flange width b, of the composite section, is the smaller of

1. $L/4 = \dfrac{40}{4} \times 12 = 120$ in.
2. spacing of beams $= 10 \times 12 = 120$ in.

For 115-pcf concrete, $b/n = 120/13 = 9.22$ in. The composite beam properties of the transferred section are calculated by normal procedures.

The resulting values are

$Y_t = 5.46$ in., $Y_b = 18.69$ in.
$I = 2199$ in.^{4}

Section modulus for tension (at bottom), $S_{tr} = \dfrac{2199}{18.69} = 117.64$ in.^{3}

Section modulus for compression (at top), $S_t = \dfrac{2199}{5.46} = 402.75$ in.^{3}

For normal-weight concrete, $b/n = 120/9.19 = 13.06$ in.

$Y_t = 4.58$ in., $Y_b = 19.57$ in.
$I = 2199$ in.^{4}
$S_{tr} = 2351/19.57 = 120.1$ in.^{3}
$S_t = 2351/4.58 = 513.5$ in.^{3}

Stress Check. The allowable stress f_b in steel for unshored construction is verified for two conditions:

$$f_b = \frac{M_{D1+D2}}{S_s} + \frac{M_{D3+L}}{S_{tr}} \le 0.9F_y \quad \text{and}$$

$$f_b = \frac{M_{D1+D2} + M_{D3+L}}{S_{tr}} \le 0.66F_y$$

where

$\qquad S_s$ = section modulus of steel section
$\qquad S_{tr}$ = section modulus of composite section
$M_{D_1+D_2}$ = moment resisted by steel beam prior to composite action
M_{D_3+L} = moment resisted by composite section

In our case

$$f_b = \frac{110 \times 12}{68.38} + \frac{160 \times 12}{12 \times 120.1} = 35.12 \text{ ksi} \le 0.9F_y = 45 \text{ ksi} \qquad \text{OK}$$

$$f_b = \frac{(110 + 160) \times 12}{120.1} = 26.9 \text{ ksi} \le 0.6F_y = 33 \text{ ksi} \qquad \text{OK}$$

Allowable concrete compressive stress is determined for the composite section based on the load applied after the concrete has attained 75% of its required strength.

Moment due to postcomposite load $= \dfrac{800 \times 40^2}{8 \times 1000} = 160$ kip-ft

$$f_c' = \frac{160 \times 12}{9.19 \times 513.5} = 0.41 \text{ ksi} \quad < 0.45\sqrt{f_c'}$$

$$0.45\sqrt{3000} = 1.35 \text{ ksi} \qquad \text{OK}$$

Horizontal Shear V_h. It is the minimum of

1. $V_h = 0.85 \times f'_c \times \frac{A_c}{2} = 0.85 \times 3 \times 120 \times 3.25/2 = 497.25$ kips

2. $V_h = \frac{A_s F_y}{2} = 11.80 \times 50/2 = 295.00$ kips

$V_h = 295.00$ kips controls the design

Shear Connectors, Partial Composite Action. Usually economy is achieved by using fewer shear connectors than required for full composite action.

The required section modulus $= \dfrac{(110 + 160) \times 12}{0.66 \times 50} = 98.18$ in.^{3}

The modulus S_{tr} furnished $= 120.1$ in.^{3}

$$V'_h = V_h \left(\frac{S_{\text{eff}} - S_s}{S_{tr} - S_s} \right)^{1/2}$$

$$= 295 \left(\frac{98.18 - 68.38}{120.1 - 68.38} \right)^{1/2}$$

$$= 224 \text{ kips}$$

The percent of composition action is given by $\frac{V'_h}{V_h} \times 100$. In our case this is equal to $\frac{224}{295} \times 100 = 76\%$. This is greater than 25% stipulated in the AISCS. Therefore OK.

Using $\frac{3}{4}$-in. studs, the allowable shear in normal-weight concrete per connector is 11.5 kips. The reduction factor 115-pcf concrete from AISCS Table I4.2 is 0.86.

Number of shear studs required $= \dfrac{2 \times 224}{0.86 \times 11.5} = 46$

It is assumed that there are no reduction factors associated with deck geometry and stud layout.

Deflections. Effective moment of inertia, $I_{\text{eff}} = I_s + (I_{tr} - I_s) \times (\text{percent comp.})^{1/2}$

$$I_{\text{eff}} = 612 + (2199 - 612) \times \left(\frac{76}{100} \right)^{1/2} = 1995 \text{ in.}^4$$

For unshored construction,

Deflection under precomposite loads $= \dfrac{5WL^4 \times 12^3}{384EI_s}$

$$= \frac{5 \times 0.550 \times 40^4 \times 1728}{384 \times 29,000 \times 612} = 1.78 \text{ in.}$$

Camber beam for 75% of calculated unshored condition.
Therefore camber specified $= 1.78 \times 0.75 = 1.34$, say, 1.25 in.
Deflection under superimposed dead and live loads for

$$W = 850 \text{ lbs/ft}, \qquad I_{\text{eff}} = 1995 \text{ in.}^4$$

$$\Delta_{D+L} = 5W_{D+L} \times L^4 \times 12^3/384 \, EI_{\text{eff}}$$

$$= \frac{5 \times 0.85 \times 40^4 \times 1728}{384 \times 29,000 \times 1995} = 0.84 \text{ in.}$$

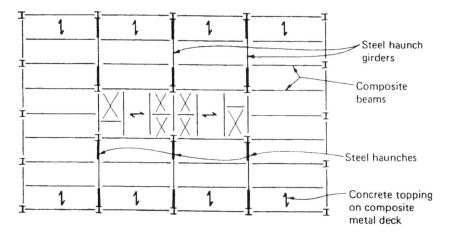

Figure 7.58.   Schematic floor plan showing haunch girders.

Compared to $L/360 = 40 \times 12/360 = 1.33$ in., the calculated deflection of 0.84 in. is small. Therefore the design is OK.

It should be noted that the lowest percentage of partial composite allowed by the AISC specifications is 25%. Some designers, however, will not allow partial composite action below 50%.

7.4.3.   Composite Haunch Girders

Composite haunch girders, although not often used as a floor framing system, merit mention because they help minimize the floor-to-floor height without requiring complicated fabrication. Fig. 7.58 shows a schematic floor plan in which composite haunch girders frame between exterior columns and interior core framing. The haunch girder typically consists of a shallow steel beam, 10- to 12-in. (254- to 305-mm) deep for spans in the 35- to 40-ft (10.6- to 12.19- m)-range. At each end of the beam a triangular haunch is formed by welding a diagonally cut wide-flange beam usually 24- to 27-in. (610- or 686-mm) deep (Fig. 7.59). The haunch is welded to the shallow beam and to the columns at each end of the girder. In this manner the last 8 or 9 ft (2.4 or 2.7 m) of the haunch girder at either end flares out toward the column with a depth varying from about 10 or 12 in. (254 or 305 mm) at the center to about 27 in. (686 mm) at the ends. This system uses less steel and provides greater flexibility for mechanical

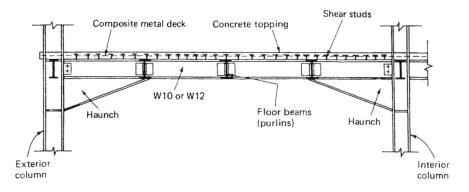

Figure 7.59.   Composite girder with tapered haunch.

ducts, which can be placed anywhere under the shallow central span. The reduction in floor-to-floor height further cuts costs of exterior cladding and of heating and cooling loads. This system, however, is not common because of higher fabricat\ion costs.

A variation of the same concept shown in Fig. 7.60 uses nontapered haunches at each end. The square haunch girder can be fabricated using a shallow-rolled section in the center and two deep-rolled sections, one at each end. Another method of fabricating the girder is to notch the bottom portion of the girder at midspan and reweld the flange to the web. The method requires more steel but comparatively less fabrication work.

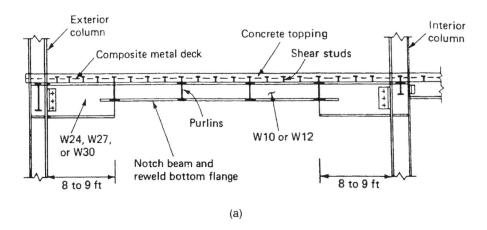

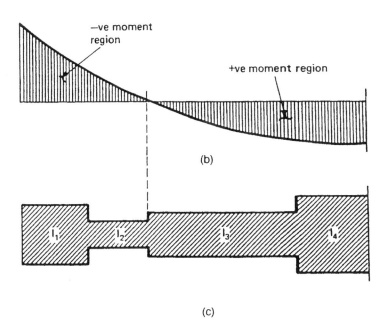

I_1 = Moment of inertia of unnotched steel section
I_2 = Moment of inertia of notched steel section
I_3 = Moment of inertia of composite notched section
I_4 = Moment of inertia of composite unnotched section

Figure 7.60. Composite girder with square haunch: (a) schematic elevation; (b) combined gravity and wind moment diagram; (c) schematic moment of inertia diagram.

In comparison to a shallow girder of constant depth, a haunch girder is significantly stiffer. Figure 7.60b shows the moment diagram in a haunch girder subjected to combined gravity and lateral loads. The corresponding stiffness properties including the effect of composite action in the positive moment regions is also shown in Fig. 7.60c.

7.4.4. Composite Trusses

Figure 7.61 shows a typical floor-framing plan with composite trusses. To keep the fabrication simple, the top and bottom chords consist of T-sections to which double-angled web members are welded directly without the use of gusset plates. The top chord is made to act compositely with the floor system by using welded shear studs. The space between the diagonals is used for the passage of mechanical and air-conditioning ducts. When the space between the diagonals is not sufficient, vertical members may be welded between the chords to form a vierendeel panel.

7.4.5. Composite Stub Girders

In building design, maximum flexibility is achieved if structural, mechanical, electrical, and plumbing trades have their own designated space in the ceiling. This is achieved in a conventional system by placing HVAC ducts, lights, and other fixtures under the beams. Where deep girders are used, penetrations are made in the girder webs to accommodate the ducts. In an office building the typical span between the core and the exterior is about 40 ft (12.2 m), requiring 18- to 21-in. (457 to 533-mm)-deep beams. Usual requirements of HVAC ducts, lights, sprinklers, and ceiling construction result in depths of 4 to 4.25 ft (1.21 to 1.3 m) between the ceiling and top of the floor slab. The depth can, however, be decreased at a substantial penalty either by providing penetrations in relatively deep beams or by using shallower, less economical beam depths.

The stub girder system shown in Fig. 7.62, invented by engineer Dr. Joseph Caloco, attempts to eliminate some of these shortcomings while at the same time reducing the floor steel weight. The key components of the system are short stubs welded intermittently to the top flange of a shallow steel beam. Sufficient space is left between stubs to accommodate mechanical ducts. Floor beams are supported on top of, rather than framed into, the shallow steel beam. Thus the floor beams are designed as continuous members which results in steel savings and reduced deflections. The stubs consist of short wide flange beams placed perpendicular to and between the floor beams. The floor system consists of concrete topping on steel decking connected to the top of stubs. The stub girders are spaced at 25 to 35 ft (7.62 to 10.7 m) on center, spanning between the core and the exterior of the building.

The behavior of a stub girder is akin to a vierendeel truss; the concrete slab serves as the compression chord, the full-length steel beam as the bottom tension chord, and the steel stubs as vertical web members. From an overall consideration, the structure allows installation of mechanical system within the structural envelope, thus reducing floor-to-floor height; the mechanical ducts run through and not under the floor.

7.4.5.1. Behavior and Analysis

The primary action of a stub girder is similar to that of a vierendeel truss; the bending moments are resisted by tension and compression forces in the bottom and top chords of the truss and the shearing stresses by the stub pieces. The bottom chord is a steel wide flange and the top chord is the concrete slab. The effective width of the concrete slab varies

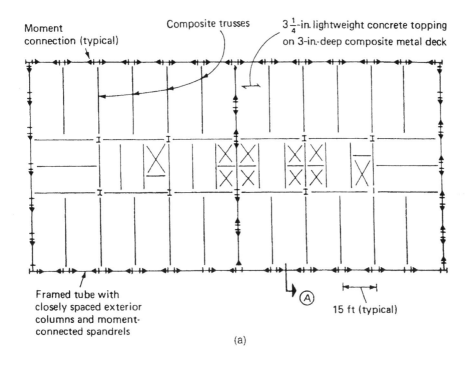

Moment connection (typical)

Composite trusses

$3\frac{1}{4}$-in. lightweight concrete topping on 3-in.-deep composite metal deck

Framed tube with closely spaced exterior columns and moment-connected spandrels

Ⓐ

15 ft (typical)

(a)

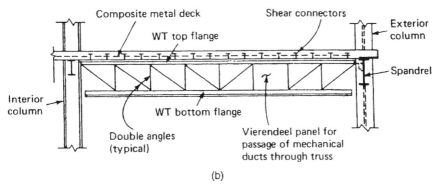

Composite metal deck

Shear connectors

Exterior column

WT top flange

Spandrel

Interior column

WT bottom flange

Double angles (typical)

Vierendeel panel for passage of mechanical ducts through truss

(b)

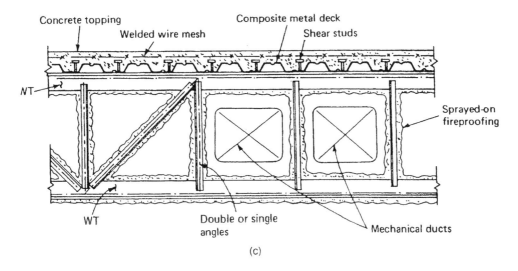

Concrete topping

Welded wire mesh

Composite metal deck

Shear studs

WT

Sprayed-on fireproofing

WT

Double or single angles

Mechanical ducts

(c)

Figure 7.61.   Composite truss: (a) framing plan; (b) elevation for truss, section A; (c) detail of truss.

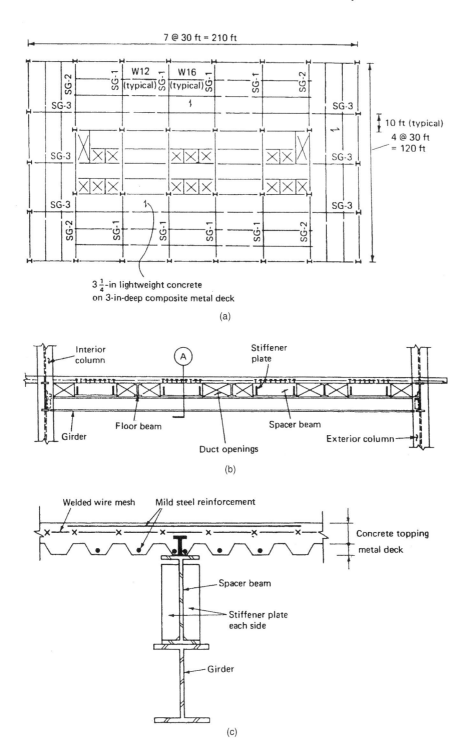

Figure 7.62.   Stub girder framing: (a) framing plan; (b) elevation of stub girder SG-1; (c) section A through stub girder; (d) photograph showing stub girder framing prior to placement of metal deck.

(d)

Figure 7.62d. Photograph showing stub girder framing prior to placement of metal deck.

from 6 to 7 ft (1.83 to 2.13 m), requiring additional reinforcement to supplement the compression capacity of the concrete. Stub peices are welded to the top flange of the steel beam and are connected to the metal deck and concrete topping through shear connectors.

Because the truss is a vierendeel truss as opposed to a diagonalized truss, bending of the top and bottom chords is significant. Therefore, it is necessary to consider the interaction between axial loads and bending stresses in the design.

Figure 7.62a shows a typical floor plan with stub girders SG1, SG2, etc. Consider stub girder SG1, spanning 40 ft (12.19 m) between the exterior and interior of the building (Fig. 7.62b). The deck consists of a 2-in. (51-mm)-deep 19-gauge composite metal deck with a $3\frac{1}{4}$-in. (82.5-mm) lightweight structural concrete topping. A welded wire fabric is used as crack control reinforcement in the concrete slab.

The first step in the analysis is to model the stub girder as an equivalent vierendeel truss. This is shown in Fig. 7.63a. A 14-in. (356-mm)-wide flange beam is assumed as the continuous bottom chord of the truss. The slab and the steel beam are modeled as equivalent top and bottom chords. Note the beam elements representing these members are at the neutral axes of the slab and beam, as shown in Fig. 7.63.

The stub pieces are modeled as a series of vertical beam elements between the top and the bottom chords of the truss with rigid panel zones at the top and bottom.

The various steps of modeling of stub girder are summarized as follows:

1. *Top chord of vierendeel truss.* As shown in Fig. 7.64, the top chord consists of an equivalent transformed area of the concrete topping, which is obtained by dividing the effective width of concrete slab by the modular ration $n = E_a/E_c$. The mild steel reinforcement in the concrete slab can be used in calculating the transformed properties. Although for strength calculations, the modulus of elasticity of normal-weight concrete is used even for lightweight concrete slabs in composite beam design, in stub girders the lower value of n for lightweight concrete is used both for deflection and strength calculations.

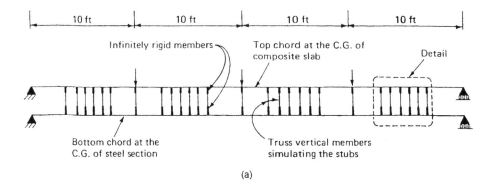

(a)

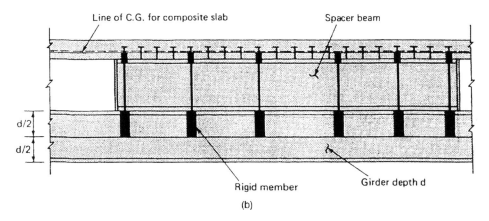

(b)

Figure 7.63.   (a) Elevation of vierendeel truss analytical model: (b) partial detail of analytical model.

The moment of inertia I_t of the top chord is obtained by multiplying the unit value of I of the composite slab, given in deck catalogs, by the effective width of the slab.

2. *Bottom chord.* The properties of the steel section are directly used for the bottom chord properties.

3. *Stub pieces.* The web area and moment of inertia of the stub in the plane of bending of the stub girder are calculated and apportioned to a finite number of vertical beam elements representing the stubs. The more elements employed to represent the stub pieces, the better will be the accuracy of the solution. As a minimum, the author recommends one vertical element for 1 ft (0.3 m) of

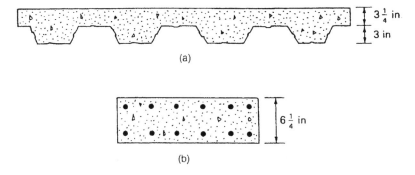

Figure 7.64.   Equivalent slab section.

stub width. The vertical segments between flanges of the stub and neutral axes of the top and bottom chords are treated as an infinitely rigid member. Stiffener plates used at the ends of stubs can be incorporated in calculating the moment of inertia of the stubs.

7.4.5.2. *Design Example*

A $3\frac{1}{4}$-in. (82.55 mm) lightweight, 3-ksi (20.7 MPa) concrete topping is to be used on 3-in.-(76.2-mm)-deep 18-gauge composite metal deck with nominal welded wire fabric reinforcement, with spans between steel purlins at 10-ft (3.05-m) centers. A W14 × 53 (356 mm × 773 N/m), 50-ksi (344.75-MPa) structural steel beam is used for the bottom continuous chord of the stub girder. Exterior and interior stub pieces consist of W16 × 26 (406 mm × 379 N/m), 36-ksi (248.3-MPa) steel beams with $6\frac{1}{2}$-and $3\frac{1}{2}$-ft (1.98 and 1.07-m) lengths, respectively. Model the stub girder as a vierendeel truss to design and check various elements under the AISC and ACI specifications.

As a first step, we compute the equivalent properties of the: 1) top chord; 2) bottom chord; and 3) stub pieces for setting up the computer model.

1. *Top chord*. The concrete slab extends on both sides of the stub girder. The effective width of concrete flange is determined by considering three values:

   $$b = 16t + b$$
   $$= L/4$$
   $$= \text{distance between stub girders}$$

   For the example, the least value of effective width for the top chord is

   $$16t + b = 16(1.5 + 3.25) + 5 = 81 \text{ in.} = 6.75 \text{ ft (2.06 m)}$$

   Area A of transformed section using $n = 14$

   $$A = (3.25 + 1.5) \times \frac{81}{4} = 27.48 \text{ in.}^2 (17,732 \text{ mm}^3)$$

   Equivalent moment of inertia I_e is the value of I for the particular metal deck and slab thickness given in the product catalog, multiplied by the effective width of compression chord. Assuming that the moment of inertia of $3\frac{1}{4}$-in. (82.55-mm) slab on 3-in. (76.2-mm) composite deck is 5.82 in.^{4}/ft, we get

   $$I = 5.82 \times 6.75 = 39.3 \text{ in.}^4 (0.164 \times 10^{-1} \text{ m}^4)$$

2. *Bottom chord*. The W14 × 53 steel wide-flange beam has the following section properties:

   $$A = 15.3 \text{ in.}^2 (9872 \text{ mm}^2), I = 541 \text{ in.}^4 (0.225 \times 10^{-3} \text{ m}^4)$$

   Shear area A_v = depth of web × web thickness

   $$= 13.92 \times 0.37 = 5.15 \text{ in.}^2 (3323 \text{ mm}^2)$$

3. (a) Exterior stub piece W16 × 26, $6\frac{1}{2}$-ft (1.98-m)-long

   $$\text{Area} = 78 \times 0.25 = 19.5 \text{ in.}^2 (12581 \text{ mm}^3)$$

   Moment of inertia in the plane of stub girder:

   $$I = \frac{bh^3}{12} = \frac{0.25 \times 78^3}{12} = 9886 \text{ in.}^4 (4.12 \times 10^{-3} \text{ m}^4)$$

We divide the area and moment of inertia of the stub piece into the six elements. These are used in the computer model to represent the stub piece. Therefore, the area of each vertical element is $19.5/6 = 3.25$ in.^{2} (2097 mm^3), and the moment of inertia I is $9886/6 = 1648$ in.^{4} (0.686×10^{-3} m^4).

(b) Interior stub piece W16 $\times$ 26, $3\frac{1}{2}$-ft (1.07-m)-long

$$A = 42 \times 0.25 = 10.5 \text{ in.}^2 \text{ (6775 mm}^2)$$

$$I = \frac{0.25 \times 42^3}{12} = 1543.5 \text{ in.}^4 \text{ (0.642} \times 10^{-3} \text{m}^4)$$

Since three vertical members are used to represent the interior stub piece, we divide the area and moment of inertia values by 3 to get the equivalent values for the computer model.

$$A = \frac{10.5}{3} = 3.5 \text{ in.}^2 \text{ (2258 mm}^2)$$

$$I = \frac{1548.5}{3} = 514.3 \text{ in.}^4 \text{ (0.214} \times 10^{-3} \text{ m}^4)$$

The next step is to set up the model of the equivalent vierendeel truss to obtain a computer solution for axial load, bending moment, and shear forces in all the members. The adequacy of each member under the action of combined forces is checked using the ACI and AISC procedures. A brief description of the procedure follows.

Bottom Chord. Assume that the maximum axial tension and bending moment obtained from the computer run are $T = 265$ kips (1178.8 kN) and $M = 90$ kip-ft (122.0 kN · m), respectively. Check W14 $\times$ 53 for combined tension and moment thus:

$$f_a = \frac{265}{15.3} = 17.2 \text{ ksi (118.6 MPa)}$$

$$f_b = \frac{90 \times 12}{77.8} = 13.75 \text{ ksi (94.8 MPa)}$$

$$F_a = 0.6F_y = 0.6 \times 50 = 30 \text{ ksi (206.8 MPa)}$$

$$\frac{d}{t} = \frac{13.92}{0.37} = 37.62 > \frac{257}{\sqrt{F_y}} = 36.3$$

$$\therefore F_b = 0.60F_y = 30 \text{ ksi (206.8 MPa)}$$

$$\frac{f_a}{F_a} + \frac{f_b}{F_b} = \frac{17.2}{30} + \frac{13.75}{30} = 1.03$$

This is very nearly equal to 1.0, and therefore is OK.

Top Chord. The top chord of the vierendeel truss is subjected to compression and bending moment and therefore is designed as a reinforced concrete column subjected to compressive forces and bending. In the opinion of the author, any rational method that does not violate the spirit of the ACI code can be used in the design. One procedure is to neglect the contribution of metal deck and design the slab section as an equivalent column. For purposes of calculation of moment magnification factor, the column can be conservatively assumed to have an effective length of 10 ft (3.04 m), which is equal to the distance between the purlins.

For the example problem, assume that the computer results for axial compression and bending moment at critical sections are 250 kips (1112.0 kN) and 10 kip-ft (13.56 kN · m), respectively.

We now proceed to design the equivalent section of the compression chord shown in Fig. 7.64 as a reinforced concrete column subject to axial compression and bending. First, calculate the slenderness ratio Kl_u/r by conservatively ignoring the restraint offered to the slab at the interface of stub pieces. The assumption that the equivalent column is hinged at the purlins gives a value of 1.0 for effective length factor K. The unsupported length l_u of the equivalent column can be considered equal to 10 ft (3.05 m), which is the distance between the purlins. The radius of gyration for the equivalent rectangular column is 0.3 times the overall dimension in the direction of bending, i.e., $0.3 \times 6.25 = 1.875$ in. (47.62 mm).

The slenderness ratio is expressed as

$$\frac{Kl_u}{r} = \frac{1 \times 10 \times 12}{1.875} = 64.0$$

Since this ratio is greater than 22, it is necessary to consider slenderness effects in the design of the column. The moment magnification procedure will be used to take the slenderness effects into account. A conservative approximation will be made by assuming that the value for the coefficient C_m (which relates the actual moment diagram to an equivalent moment diagram) is 1.0.

Since the axial load P and moment M obtained from the computer analysis are working stress values, these are converted to ultimate values by multiplying them with an average load factor of 1.5.

Therefore,

$$P_u = 1.5 \times 260 = 390 \text{ kips (1735 kN)}$$
$$M_u = 1.5 \times 10 = 15 \text{ kip ft (20.34 kN · m)}$$

The critical load P_c is given by the relation

$$P_C = \frac{\pi^2 E_C I}{(Kl_u)^2}$$

For the example problem, we have

$$E_C = w_C^{1.5} \times 33\sqrt{f_C'} = 110^{1.5} \times 33\sqrt{3000} = 2085 \text{ ksi (14378 MPa)}$$

$$I = \frac{40.5 \times 6.25^3}{12} = 824 \text{ in.}^4$$

Substituting, we get

$$P_C = \frac{3.14^2 \times 2085 \times 824}{(10 \times 12)^2} = 1176 \text{ kips (5230 kN)}$$

The moment magnification factor is given by

$$\delta_1 = \frac{C_m}{1 - (P_u/\phi P_C)}$$

$$= \frac{1}{1 - (390/0.7 + 1176)} = 1.90$$

Therefore, design $M_u = 15 \times 1.9 = 28.5$ kip-ft (38.65 kN · m).

The equivalent column is designed for $P_u = 390$ kips (1735 kN) and $M_u = 28.5$ kip-ft (38.65 kN · m). The required reinforcement is obtained using a procedure conforming to the ACI code. For the present example, longitudinal reinforcing bars of ten #5 are found to be adequate to carry the design axial load and bending moment.

Computation of Number of Shear Studs. The shear studs between the stub pieces and the concrete slab form the backbone of composite stub girders. Their design is similar to composite beam design for which the shear connector formulas represent the horizontal shear at ultimate load divided by 2 to approximate conditions at working loads. The total horizontal shear resisted by the connectors between the point of maximum moment and each end of the stub girder is the smaller of the values obtained from the following equations:

$$V_h = \frac{0.85 f_C' A_C}{2}$$

or

$$V_h = \frac{A_S F_y}{2}$$

For the example problem, a value of $V_h = 458$ kips (2037 kN) obtained from the first equation governs the design. Using a value of 9.5 kips (42.26 kN) as the allowable shear load, the number of shear studs $N = 458/9.5 = 48.2 \approx 50$, giving 32 and 18 shear connectors at the exterior and interior stubs.

Check Exterior Stub W16 × 26, $6^1/_2$ ft (1.98 m) Long. The design check is performed for shear and bending stresses per the AISC specifications. The summation of shear forces in the six elements used in the computer model to represent the exterior stub gives the design shear. The design moment for the stub is obtained by multiplying the accumulated shear by the stub height. Assume for the example problem that the accumulated shear = 210 kips (934 kN). The design moment then is $210 × 16/12$ or 280 kip-ft (380 kN · m). The shear stress is $210/78 × 25$ or 10.76 ksi (74.25 MPa). The allowable shear stress is $0.4 × F_v = 0.4 × 36 = 14.4$ ksi (99.3 MPa). Therefore, the stub is okay for shear.

To check the bending stresses, we calculate the moment of inertia and section modulus of the stub by including the contribution of the stiffener plates at the ends of stub. Without burdening the presentation with trivial calculations, let us assume that the section modulus of the stub piece and stiffener plate is equal to 300 in.[3] ($4.92 × 10^6$ mm³). The bending stress is expressed as

$$f_b = \frac{280 × 12}{300} = 11.2 \text{ ksi (77.3 MPa)}$$

This stress is checked against the allowable stresses per the AISC specifications.

A similar procedure is used to check the bending and shear stresses in the interior stub.

7.4.5.3. *Moment-Connected Stub Girder*

The stub girder system, due to its large overall depth of approximately 3 ft (0.92 m), has a very large moment of inertia and can be used as part of a lateral-force-resisting system. The model used for analysis is a vierendeel truss, where the concrete slab and the bottom steel beam are simulated as linear elements and each stub piece is divided into several elements. The gravity and lateral load shear forces and moments are introduced as additional load cases in the computer analysis, and the combined axial forces and moments in each section of the stub girder are obtained. all parts of the stub girder are checked for combined axial forces, shear, and moments as shown earlier. The controlling section for the slab is generally at the end of the first stub piece furthermost from the column. Particular care is required to transfer the moment at the column girder interfaces. If lateral moments are small, moment transfer can take place between the slab and the bottom steel beam. The slab needs to be attached to the column either by long deformed wire anchors or by welding reinforcing bars to the column. For relatively large moments, the solution for moment transfer is to extend the first stub piece to the column face. The top flange of the stub piece and the bottom flange of the W14 girder are welded to the column as in a typical moment connection. The design of the connection is, therefore, identical to welded beam-column moment connection. The girder should be checked along its full length for the critical combination of gravity and wind forces. Depending on the extent of stress reversals due to lateral load, bracing of the bottom chord may be necessary.

7.4.5.4.   *Strengthening of Stub Girder*

Strengthening of existing stub girders for tenant-imposed higher loads is more expensive than in conventional composite construction. A speculative type of investment building is usually designed for imposed loads of 50 psf (2.4 kN/m²) plus 20 psf (0.96 kN/m²) as partition allowance. For heavier loads, strengthening of local framing is required. The bottom girder, which is in tension and bending, is relatively easy to reinforce by welding additional plates or angles to the existing steel member. Reinforcing the top chord of the stub girder, which is in compression and bending, is somewhat tricky. The addition of structural steel angles using expansion anchors to the underside of metal deck and the welding of additional stub pieces to reduce the effective length of compression chord, which acts like a column, have been used with good results. From the point of view of ultimate load behavior, it is acceptable to strengthen the bottom chords to resist total load without the truss action. However, it is important to check the lateral bracing requirements for the top flange of the bottom chord.

7.4.6.   Composite Columns

The term composite column in the building industry is taken to represent a unique form of construction in which structural steel is made to interact compositely with concrete. The structural steel section can be a tubular section filled with structural concrete or it can be a steel wide-flange section used as a core surrounded by reinforced concrete.

Historically, composite columns evolved from the concrete encasement of structural steel shapes primarily intended as fire protection. Although the increase in strength and stiffness of the steel members due to concrete used as fireproofing was intuitively known, it was not until the 1940s that methods to actually incorporate the increases were developed. In fact, in earlier days, the design of the steel column was penalized by considering the weight of concrete as an additional dead load on the steel column. Later developments took into account the increased radius of gyration of the column because of the concrete encasement, and allowed for some reduction in the amount of structural steel. In some earlier high-rise designs, the concrete encasement was ignored for strength considerations, but the additional stiffness of concrete was included in calculating lateral defections.

After the development of sprayed-on contact fireproofing in the 1950s and 1960s, use of concrete for fireproofing of structural steel was no longer an economical proposition. The high form-work cost of concrete could not be justified for fireproofing.

Over the last 20 years, the use of encased structural steel columns has found application in buildings varying from as low as 10 stories to as high as 70-story or even taller buildings. These columns have been incorporated in an overall construction known as the composite system, which has successfully captured the essential advantages associated with steel and concrete construction: the speed of steel with the stiffness and moldability of concrete. Concrete columns with small steel-core columns used as erection columns were perhaps the earliest applications. Later much heavier columns were used, serving the dual purpose for both steel erection and load resistance. The heavier steel columns were used essentially to limit the size of composite vertical elements.

Another version consists of exterior concrete columns acting compositely with steel-plate or precast cladding. Yet another version popular in some countries uses laced columns fabricated from light structural shapes such as angles, T-sections, and channels. The concrete enclosure provides both fireproofing qualities and also provides additional stiffness to the light structural shapes, inhibiting their local buckling tendencies. Additional conventional reinforcement can be accommodated in the concrete encasement, as in conventionally reinforced concrete columns.

The ACI building code encompasses the design of all types of composite column under one unified method using the same general principles as for conventionally reinforced concrete columns.

The ACI procedure is based on an ultimate concrete strain of 0.3%. As in conventionally reinforced concrete design, the tensile stress in the concrete is ignored. Either a parabolic or an equivalent uniform concrete strain can be assumed in the compression zone.

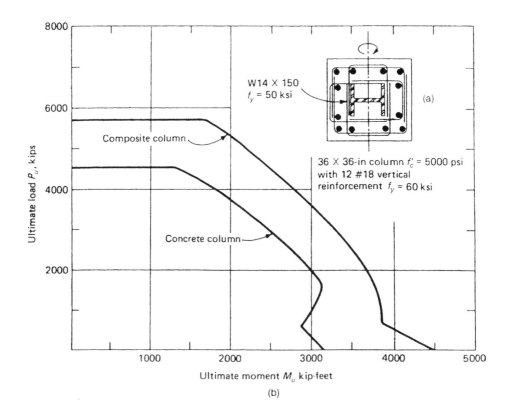

Figure 7.65.   Comparison of interaction diagrams: (a) column detail; (b) load moment interaction diagram.

The axial load assigned to the concrete portion of the composite column is required to be developed by direct bearing through studs, lugs, plates, or reinforcing bars welded to the structural steel plate prior to the casting of concrete. In other words, the code requires a positive method for the transfer of axial load between the steel core and the concrete encasement for strength calculations. For calculation of stiffness, however, merely wrapping the concrete around the steel core will suffice. Axial loads induced in the concrete section of the composite column due to column bending need not be transferred in direct bearing.

Tied composite columns are required by the ACI code to have more lateral ties than ordinary reinforced concrete columns. In fact, the ACI code stipulates twice as many ties, but this is based on somewhat questionable assumptions. First, it assumes that concrete that is laterally contained by ties is thin. Second, it assumes that concrete has a tendency to spall out from the smooth faces of the steel core. To prevent this separation, the lateral ties are specified to be vertically spaced no more than half the least dimension of the composite member. The ACI code does not permit the use of longitudinal bars in the evaluation of stiffness of columns on the premise that the longitudinal bars are rendered ineffective because of separation of concrete at high strains. They may, however, be included in the calculation of strength. Finally, the yield strength of the steel core is limited to 52 ksi (359 MPa) to correspond to the yielding strain of concrete of 0.0018.

A practical approach to the design of composite columns is to assume that the steel wide-flange section behaves as reinforcing steel. With this assumption, interaction diagrams can be generated for various combinations of concrete columns size, structural steel shape, and reinforcing steel. Figure 7.65 shows an interaction diagram generated for a 36 × 36 in. (915 × 915 mm) column with twelve #18 (57-mm diameter) reinforcing bars and a W14 × 150 (378 × 394 mm × 2188 N/m) structural steel shape. For comparison purposes, the interaction diagram for the same concrete column without the embedded structural steel shape is given. It can be seen that large increases in column capacity occur when structural steel shapes are included within the concrete envelope.

Paris Hotel, Casino & Eiffel Tower II
Las Vegas, Nevada

John A. Martin & Associates, Inc.
Structural Engineers
Los Angeles, CA

8
Special Topics

8.1. TALL BUILDINGS

Tall buildings have fascinated humans from the beginning of civilization as evidenced by the pyramids of Giza, Egypt; Mayan temples of Tikal, Guatemala; and Kutub Minar of Delhi, India. The motivation behind their construction was primarily for creating monumental rather than human habitats. By contrast, contemporary tall buildings are primarily a response to the demand by commercial activities, often developed for corporate organizations as prestige symbols in city centers.

The feasibility of tall buildings has always depended upon the available materials and the development of the vertical transportation necessary for moving people up and down the buildings. The ensuing growth that has occurred from time to time may be traced back to two major technical innovations that occurred in the middle to the end of the nineteenth century: the development of wrought iron and subsequently steel, and the incorporation of the elevator in high-rise buildings. The introduction of elevators made the upper floors as attractive to lease as the lower ones and, as a result, made the taller building financially successful.

During the last 120 years, three major types of structures have been employed in tall buildings. The first type was used in the cast iron buildings of the 1850s to 1910, in which the gravity load was carried mostly by the exterior walls. The second generation of tall buildings, which began with the 1883 Home Insurance Building, Chicago, and includes the 1913 Woolworth Building and the 1931 Empire State Building, are frame structures, in which a skeleton of welded or riveted steel columns and beams runs through, often encased in cinder concrete, and the exterior is a nonbearing curtain wall. Most high-rises erected since the 1960s use a third type of structure, in which the perimeter structure of these buildings resembles tubes consisting of either closely spaced columns or widely spaced megacolumns with braces. Inside the perimeter structure a core, made of steel, concrete, or a combination of the two, contains many of the services such as elevators, stairwells, mechanical equipment, and toilets.

The art of designing tall buildings in windy climates is to bestow them with enough strength to resist forces generated by windstorms and enough stiffness or energy dissipation so that people working on upper floors are not disturbed by the buildings periodic swaying.

In seismic regions of the world, including the most severe areas of California, the effects of earthquakes are relatively small for tall buildings. For example, using the provisions of ASCE 7-02, the calculated base shear for a 60-story steel moment frame building located in downtown Los Angeles, CA, would be about 4% of its mass, as compared to 9% for a five-story building. However, the taller building would move considerably more than its 5-story counterpart.

The intent in seismic design then is to limit building movements, not so much to reduce perception of motion but to maintain the building's stability and prevent danger to pedestrians due to breakage and falling down of nonstructural elements.

8.1.1. Structural Concepts

The adoration that skyscrapers command lies in their apparent freedom from gravity loads: they do no just stand tall; they do so effortlessly. The key idea in conceptualizing such a bewildering and yet efficient structural system is to think of the building as a beam cantilevering from the earth (Fig. 8.1). The laterally directed force generated due to either wind or seismic action tends both to snap it (shear) and to push it over (bending). Therefore, a building must have a system to resist shear as well as bending. In resisting shear forces, the building must not break by shearing off (Fig. 8.2a) and must not strain beyond the limit of elastic recovery (Fig. 8.2b). Similarly, in resisting bending, the building must not overturn from the combined forces of gravity and lateral loads; it must not break by premature failure of columns either by crushing or by excessive tensile forces; and its bending deflection should not exceed the limit of elastic recovery (Fig. 8.3). In addition, a building in seismically active regions must be able to resist realistic earthquake forces without losing its vertical load-carrying capacity.

In a structure's resistance to bending and shear, a tug-of-war ensues that sets the building in motion, thus creating a third engineering problem: motion perception or vibration. If the building sways too much, human comfort is sacrificed or, more importantly, nonstructural elements may break resulting in damage to building contents and causing danger to pedestrians.

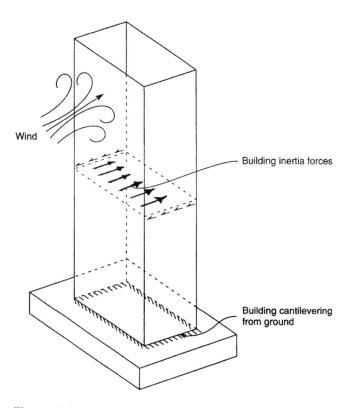

Figure 8.1.   Structural concept of a building subjected to lateral forces.

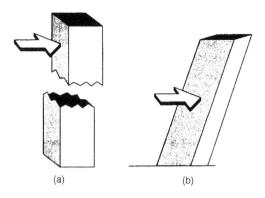

Figure 8.2. Shear resistance of building: (a) building must not break; (b) building must not deflect excessively due to shear.

A perfect structural form to resist effects of bending, shear, and excessive vibration is a system with vertically continuous elements ideally located at the farthest extremity from the geometric center of the building. A steel or concrete chimney is perhaps an ideal, if not an inspiring, engineering model for a rational super-tall structural form. The quest for the best solution lies in translating this form into a more practical skeletal structure.

Building structural design is governed by codes that specify the minimum loads that a building must have the strength to resist. However, in planning a new building, or in retrofitting an existing facility, an owner may request enhanced requirements in its design for events that are not anticipated in the building codes. Defense facilities, nuclear power plants, and overseas embassies are just a few examples where special strengthening features are requested by building owners in the design and engineering of their facilities. Therefore, designers must consider project-specific needs and owner expectations when determining building loads. The primary loads addressed in building codes are

- Gravity
- Wind
- Earthquake
- Snow and rain loads

Gravity load includes both the weight of the building and its content. The weight of the building is calculated based on material densities. The weight of the contents is not known specifically at the time of design and may vary depending upon the usage with time. Therefore, the codes

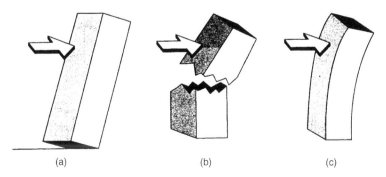

Figure 8.3. Bending resistance of building: (a) building must not overturn; (b) columns must not fail in tension or compression; (c) bending deflection must not be excessive.

specify minimum floor live loads on a per square foot basis. Wind load specified by codes is based on maps of design wind speed for different regions of the country. As wind speed increases, the wind pressure on the building increases proportionally with the square of the wind velocity. The wind speed, and therefore the pressure on the building, increases with height above ground and varies dynamically (turbulence) relative to the degree of shielding provided by other buildings and geographic features. Although not usually required by building codes, wind tunnel studies are frequently performed to more accurately determine wind loads on tall buildings where standard calculations may not be adequate. Building codes in the United States do not specify allowable lateral deflections caused by wind loading, but leave those to the engineer's judgment.

The earthquake hazard is also highly dependent on the geographic region. The effects of earthquake are relatively small for very tall buildings in all regions of the world, including the seismic area of California. The flexibility of a very tall building of, say, 80-plus stories generally allows the building to sway back and forth to the ground motions without developing forces nearly as large as those produced by design wind loads. Therefore, even in a severe seismic area, tall building design is generally controlled by wind loads. However, even then the detailing of the building components and connections should conform to seismic design requirements. This is because the actual seismic forces, when they occur, are likely to be significantly larger than code-prescribed forces; hence, the material limitations and seismic detailing in addition to strength requirements. In other words, for buildings in high-seismic zones, even when wind forces govern the design, the detailing and proportioning requirements for seismic resistance must also be satisfied. The requirements get progressively more stringent as the zone factor for seismic risk gets progressively higher.

8.1.2. Case Studies

Having noted that a building must have a system to resist both lateral bending and shear in addition to the ever-present gravity loads, let us take a trip around the world to explore how prominent engineers have exploited this concept. Although some of the case studies include run-of-the-mill designs that a large number of engineers solve on a day-to-day basis, others are once-in-a-lifetime high-profile projects, even daring in their engineering solutions. Many are examples of buildings constructed or proposed in seismically inactive regions, requiring careful examination of their ductile behavior and reserve strength capacity before they are applied in seismically active regions.

The main purpose of this section is to introduce the reader to various structural systems normally considered in the design of tall buildings. Presently it will be seen that design trend is toward using composite systems that include such components as megaframes, interior and exterior super-braced frames, spine structures, etc. The case studies highlight those aspects of conceptualization that are timeless constants of the design process and are as important for understanding structural design as is the latest computer software. The case histories are based on information contained in various technical publications and periodicals. Frequent use is made of personal information obtained from structural engineers-of-record.

8.1.2.1. *Empire State Building, New York*

We start our world tour in New York City to pay homage to the Empire State Building which was the tallest building in the world for more than 40 years, from the day of its completion in 1931 until 1972 when the Twin Towers of New York's World Trade Center

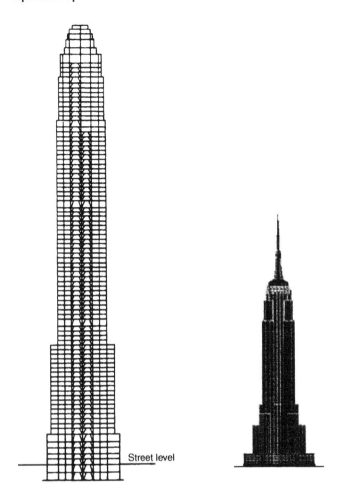

Street level

Figure 8.4. Empire State Building bracing system; riveted structural steel frame encased in cinder concrete.

exceeded its 1280-ft (381-m) height by almost 120 ft (37 m). The structural steel frame consisting of moment and braced frames with riveted joints, although encased in cinder concrete, was designed to carry 100% of the gravity and wind loads. The concrete encasement, although neglected in strength analysis, stiffened the frame considerably against wind loads. Measured frequencies of the building have estimated the actual stiffness at 4.8 times the stiffness of the bare frame. A schematic elevation of the structural framing is shown in Fig. 8.4.

8.1.2.2. Bank One Center, Indianapolis

This is a 52-story steel-framed office building that rises to a height of 623 ft (190 m) above the street level. In plan, the tower is typically 190 × 120 ft (58 × 37m) with set-backs at the 10th, 15th, 23rd, 45th, and 47th floors (Fig. 8.5).

The structural system for resisting lateral forces consists of two large vertical flange trusses in the north–south direction and two smaller core braces in the east–west direction acting as web trusses connecting the flange trusses. The flange trusses, while providing maximum lever arm for resisting the overturning moments, also serve to transfer gravity loads of the core to the exterior columns. The resulting equalization of axial stresses in the truss and the nontruss perimeter columns keeps the differential shortening between

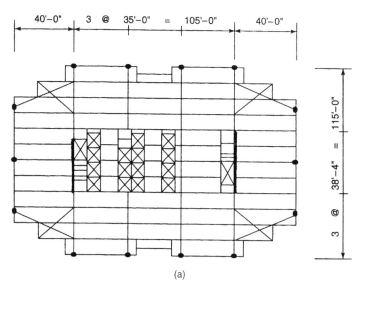

(a)

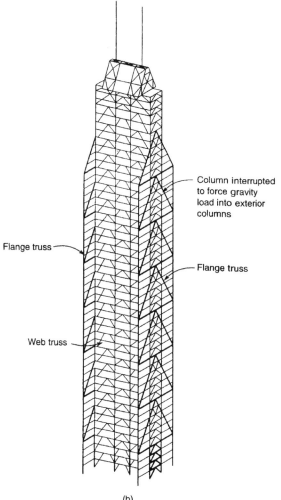

(b)

Figure 8.5. Bank One Center, Indianapolis: (a) plan; (b) lateral system.

the two to a minimum. To assure a direct load path for the transfer of gravity load from the core to the truss columns, a core column is removed below the level of braces at every 12th level, as shown in Fig. 8.5b. In addition, the step-back corners are cantilevered to maximize the tributary area of gravity load and thus compensate for the tensile force due to overturning moments. The structural design is by LeMessurier Consultants, Inc., Cambridge, MA.

8.1.2.3. MTA Headquarters, Los Angeles

This 28-story office building, shown in Fig. 8.6, has a four-level subterranean structure that will serve as a common base for the MTA Tower and two future office buildings. The basement extends beyond the footprint of the tower and consists of precast reinforced concrete columns and girders with a cast-in-place concrete slab.

(a)

Figure 8.6. MTA headquarters, Los Angeles: (a) building elevation; (b) typical floor framing plan. Architects: McLarand Vasquez & Partners, Inc.; Structural engineers: John A. Martin & Associates, Inc., Los Angeles.

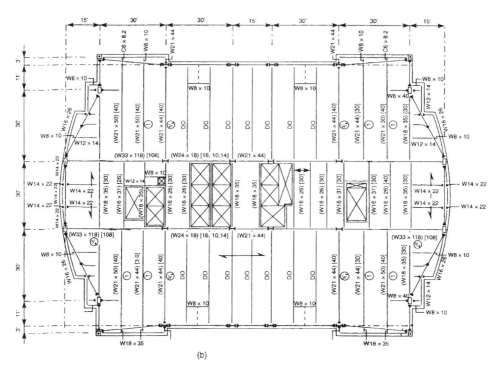

(b)

Figure 8.6.   *(Continued.)*

The plaza underneath the tower consists of a composite floor system with a $4\frac{1}{2}$-in. (114-mm) normal-weight concrete topping on a 3-in. (76-mm)-deep, 18-gauge composite metal deck. The metal deck spans between composite steel beams spaced typically at 7 ft 6 in. (2.29 m) outside the tower, which has a heavy landscape. The beam spacing is at 10 ft (3.04 m) on centers within the tower footprint.

The building is essentially rectangular in plan, 118×165 ft (36×50.3 m) with a slight radius on the short faces. The height is 400 ft (122 m), resulting in a fairly low height-to-width ratio of 3.39. Typical floor framing consists of 21-in. (0.54-m)-deep composite beams spanning 41 ft (12.5 m) from the core to the exterior. A 3-in. (76-mm)-deep metal deck with a $3\frac{1}{4}$-in. (83-mm)-thick lightweight concrete topping completes the floor system. See Fig. 8.6b.

The lateral system consists of a perimeter tube with widely spaced columns tied together with spandrel beams. The exterior columns on the broad faces vary from W30 × 526 at the plaza to W30 × 261 at the top. The spandrels vary from WTM36 × 286 at the plaza level to W36 × 170 at the top floors. The columns on the curvilinear faces are built-up, 34×16-in. (0.87×0.40-m) box columns while 24×24-in. (0.61×0.61-m) box columns are used at the corners. Plates varying in thickness from 4 in. (102 mm) at the bottom to 1 in. (25 mm) at the top are used for the built-up columns.

The architecture for the building is by McLarand Vasquez and Partners, Inc., while the structural engineering is by John A. Martin and Associates, Inc., both of Los Angeles, CA.

8.1.2.4.   *South Walker Tower, Chicago*

This tower, 946 ft (288.4 m) in height, has a changing geometry with the east face rising in a single plane from street level to 65th floor whereas the other three faces change shape. To the 14th level, the structure is basically a trapezoid in plan 135×225 ft (41.15×68.6 m)

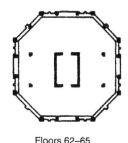

Floors 62–65

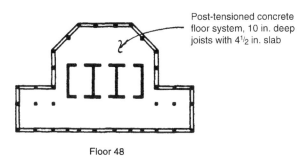

Post-tensioned concrete
floor system, 10 in. deep
joists with 4½ in. slab

Floor 48

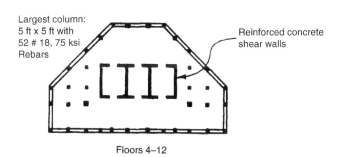

Largest column:
5 ft x 5 ft with
52 # 18, 75 ksi
Rebars

Reinforced concrete
shear walls

Floors 4–12

Figure 8.7.   South Walker Tower, Chicago; schematic plans.

overall. The building steps back at the 15th floor on three faces to provide ten corner offices on each floor. There are additional setbacks at the 47th floor. At the 51st floor, the sawtooth shape is dropped and the tower becomes an octagon in plan with 70-ft (21.4-m)-long sides. The slenderness ratio of the structures is 7.25:1. The schematic floor plans at various levels are shown in Fig. 8.7.

The core shear walls in the tower's lower floors carry much of the lateral loading with shear wall–frame interaction. There are four main shear walls—two I-shapes and two C-shapes—on a typical floor. These interact with the perimeter columns and perimeter spandrel beams through girders that span from core to the perimeter.

The girders have 39-in. (1.0-m)-deep haunches at the columns. Spandrels are 36 in. (0.92 m) deep. Core wall concrete design strength varies from 8000 psi (55.12 mPa) at the base to 4000 psi (27.6 mPa) at the upper levels.

There is a 40- to 48-ft (12.2- to 14.63-m) span between the core and the perimeter. The spacing between the perimeter columns is fairly short, about 14 ft (4.3 m), except at two corners where the spacing is 32 ft (9.76 m). Column loads range from 12,000 to 30,000 kips (53,376 to 133,440 kN). Concrete strengths range from 12,000 to 4000 psi (82.74 to 27.58 mPa).

(a)

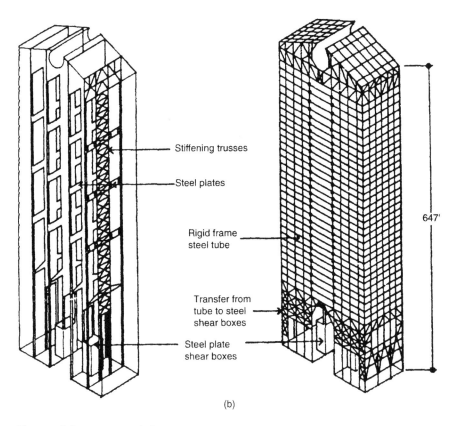

(b)

Figure 8.8. AT&T Building, New York: (a) building elevation; (b) lateral system. Perimeter steel tube interacts with interior braces. Steel plate outriggers interconnect the tube and braces.

The largest columns, which are 5×5 ft contain 52 #18, grade 75 rebars. The original design for the floor system had 16-in. (406.4-mm)-deep spans with 4-in. (101.6-mm)-thick slabs. This was changed to a post-tensioned system with a 10-in. (254 mm)-deep joist and a 4.5-in.-thick slab.

Structural design is by Brockette, Davis, Drake, Inc., Dallas, TX.

8.1.2.5. AT&T Building, New York City

The basic lateral-force-resisting structural system for the building shown in Fig. 8.8 consists of a rigid-frame steel tube at the building perimeter. Additional stiffness is added along the width of the building by means of four vertical steel trusses. At every eighth floor, two I-shaped steel plate walls, with holes cut for circulation, extend from the sides of the trusses to the exterior columns on the same column line. The steel walls act as outrigger trusses mobilizing the full width of the building in resisting lateral forces. The horizontal shear at the base of the building is transferred to two giant steel plate boxes (Fig. 8.8). Structural design is by Leslie Robertson and Associates, New York.

8.1.2.6. Miglin-Beitler Tower, Chicago

The proposed Miglin-Beitler Tower, designed by the New York Office of Thornton–Tomasetti Engineers, will rise to the height of 1486.5 ft (453 m) at the upper skyroom level, 1584.5 ft (483 m) at the top of the mechanical areas, and finally to 1999.9 ft (609.7 m) at the tip of the spire. An elevation and the schematic plan of the proposed building are shown in Fig. 8.9.

The structural system consists of five major components as shown in Fig 8.9c.

1. A 62 ft-6 in. $\times$ 62 ft-6 in. (19×19 m) concrete core with walls varying from a maximum thickness of 3 ft. (0.91 m) to a minimum thickness of 1 ft-6 in. (0.46 m).
2. A conventional structural steel composite floor system consisting of 18-in. (0.46-m) deep-rolled steel sections spaced 10 ft (3.05 m) on center with 3-in. (74-mm)-deep corrugated metal deck and a $3\frac{1}{2}$-in. (89-mm)-thick normal-weight concrete topping. The steel floor system is supported on light steel erection columns that allow the steel construction to proceed 8 to 10 floors ahead of concrete operation.
3. Concrete fin columns, each of which encases a pair of steel erection columns located at the face of the building. These fin columns, which extend 20 ft (6.10 m) beyond the 140×140-ft (42.7×42.7-m) footprint of the building, vary in dimension from $6\frac{1}{2} \times 33$ ft (2.0×10 m) at the base, $5\frac{1}{2} \times 15$ ft (1.68×4.6 m) at the middle, to $4\frac{1}{2} \times 13$ ft (1.38×4 m) near the top.
4. Concrete link beams that interconnect the four corners of the core to the eight fin columns at every floor. These beams tie the fin columns to the core, thus engaging the full structural width of the building to resist lateral loads. In addition to the link beams at each floor, there are three two-story-deep outrigger walls located at the 16th, 56th, and 91st stories. These outrigger walls further enhance the structural rigidity by linking the exterior fin columns to the concrete core.
5. Exterior vierendeel trusses comprising a horizontal spandrel and two columns at each of the 60-ft (18.3-m) faces on the four sides of the building. These vierendeels supplement the lateral force resistance and also improve the torsional resistance of the structural system. Additionally, these trusses transfer gravity loads to the exterior fin columns, thus minimizing uplift forces.

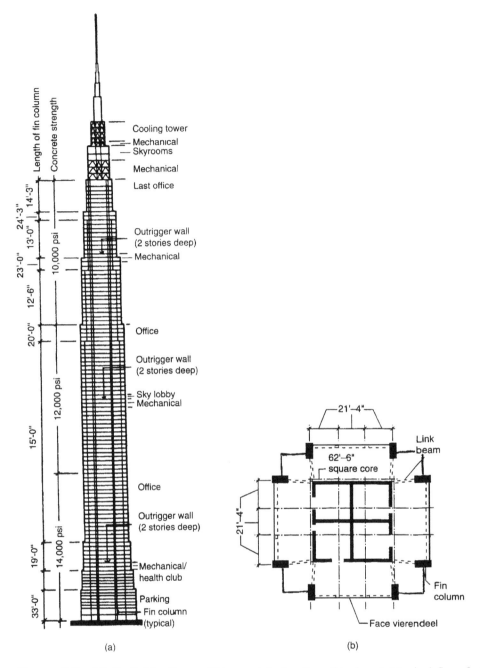

Figure 8.9. Miglin-Beitler Tower, Chicago: (a) elevation; (b) plan; (c) typical floor framing plan.

The proposed foundation system is rock caissons varying in diameter from 8 to 10 ft (2.44 to 3.0 m). The caisson will have a straight shaft steel casing and will be embedded into rock a minimum of 6 ft (1.88 m). The length of these caissons is 95 ft (29 m). A 4-ft (1.22-m)-thick concrete mat will tie the caissons and provide a means for resisting the shear forces at the base of the building. The bottom of the mat will be cast in a two-directional groove pattern to engage the soil in shear. Passive pressure on the

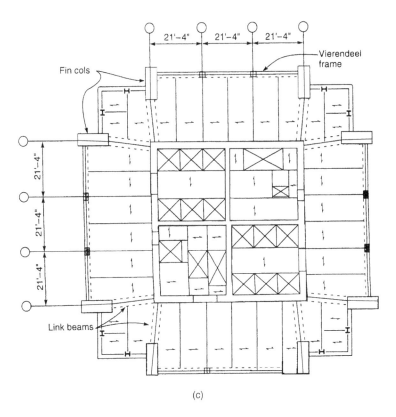

Figure 8.9. *(Continued.)*

edge of the mat and on the projected side surface of the caisson will provide additional resistance to shear at the base.

8.1.2.7. One Detroit Center

This is a 45-story office tower with a clear 45-ft-6-in. (13.87-m) span between the core and the exterior (Fig. 8.10). The structural system consists of eight composite concrete columns measuring 7 ft-6 in. × 4 ft-9 in. (2.28 × 1.45 m) at the base, placed 20 ft (6.1 m) away from the corners to provide column-free corner offices and also to optimize the free-span of the vierendeel frames. The composite columns are connected at each face by a system of perimeter columns and spandrels acting as vierendeel frames. The vierendeels are stacked four stories high and span between composite supercolumns to provide column-free entrances at the base of the tower. At each fourth level, the vierendeels are linked by hinges to transfer only horizontal shear between adjoining vierendeels and not gravity loads. The reason for this type of connection is to reduce: 1) the effect of creep and shrinkage of supercolumns on the members and connections of the vierendeel, and 2) gravity load transfer due to arch action of the vierendeel with associated horizontal thrusts. The four-story vierendeel achieves uniformity in the transfer of moment and shear between horizontal steel beams and composite supercolumns throughout the height of the tower.

A schematic representation of the structural system is shown in Fig. 8.10a–d. Figure 8.10d shows the connection details for the vierendeel frame. The structural design is by CBM Engineers, Inc., Houston, TX.

(a)

Figure 8.10. One Detroit Center: (a) building elevation; (b) typical floor framing plan; (c) free-spanning vierendeel elevations; (d) structural details for vierendeel frame, (1) partial elevation, (2) detail 1, (3) detail 2.

8.1.2.8.   Jin Mao Tower, Shanghai, China

This building consists of a 1381-ft (421-m) tower and an attached low-rise podium for a total gross building area of approximately 3 million sq ft (278,682 m^2). The building includes 50 stories of office space topped by 36 stories of hotel space with two additional floors for a restaurant and an observation deck. Parking for automobiles and bicycles is located below grade. The podium consists a retail spaces as well as an auditorium and exposition spaces.

The superstructure is a mixed use of structural steel and reinforced concrete with many major structural members composed of both steel and concrete. The primary components of the lateral system include a central reinforced concrete core linked to exterior composite megacolumns by outrigger trusses (Fig. 8.11). A central shear–wall core houses the primary building functions including elevators, mechanical fan rooms, and washrooms. The octagon-shaped core, nominally 90 ft (27.43 m) from centerline to centerline of perimeter flanges, is present from the foundation to level 87. Flanges of the core typically vary from 38 in. (84 cm) thick at the foundation to 18 in. (46 cm) at level 87 with concrete strengths varying from 7500 to 5000 psi (51.71 to 34.5 mPa). Four 18-in. (46-cm)-thick interconnecting core wall webs exist through the office floors. The central area of the core is open throughout the hotel floor, creating an atrium that leads into the spire with a total height of approximately 675 ft (206 m). The size of composite megacolumns varies from

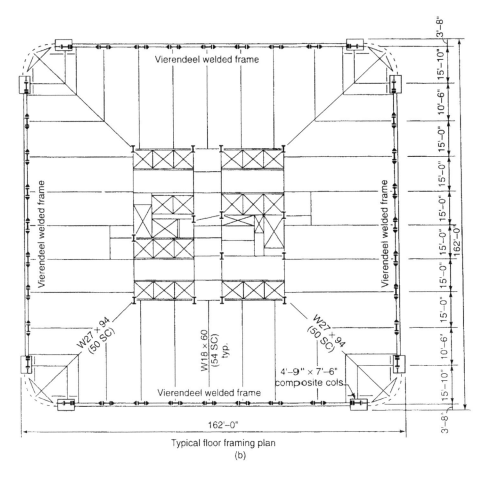

Figure 8.10. (*Continued.*)

5×16 ft (1.5×4.88 m) with a concrete strength of 7500 psi (51.71 mPa) at the foundation to 3×11 ft (0.91×3.53 m) with a concrete strength of 5000 psi (34.5 mPa) at level 87.

The shear–wall core is directly linked to the exterior composite megacolumns by structural steel outrigger trusses. The outrigger trusses resist lateral loads by maximizing the effective depth of the structure. Under bending, the building acts as a vertical cantilever with tension in the windward columns and compression in the leeward columns. Gravity load framing minimizes uplift in the exterior composite megacolumns. The octagon-shaped core provides exceptional torsional resistance, eliminating the need for any exterior belt or frame systems to interconnect exterior columns.

The outrigger trusses are located between levels 24 and 26, 51 and 53, and 85 and 87. The outrigger truss system between levels 85 and 87 is capped with a three-dimensional steel space that which provides for the transfer of lateral loads between the core and the exterior composite columns. It also supports gravity loads of heavy mechanical spaces located in the penthouse floors.

The structural elements for resisting gravity loads include eight structural steel built-up columns. Composite wide-flange beams and trusses are used to frame the floors. The floor-framing elements are typically 14 ft 6 in. (4.4 m) on center with a composite 3-in. (7.6-cm)-deep metal deck and a $3\frac{1}{4}$-in. (8.25-cm)-thick normal-weight concrete topping slab spanning between the steel members.

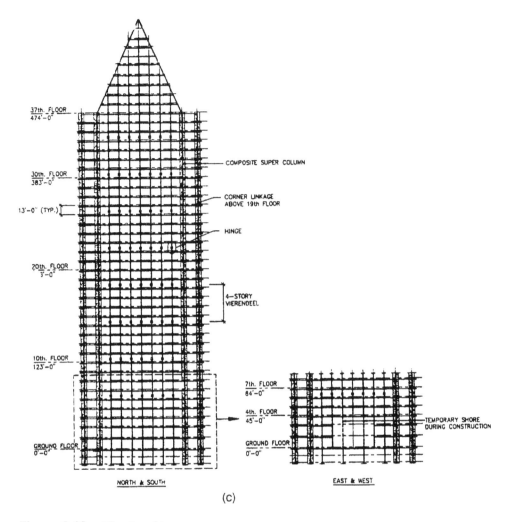

Figure 8.10. (*Continued.*)

The foundation system for the Tower consists of high-capacity piles capped with a reinforced concrete mat. High-water conditions required the use of a 3 ft-3 in. (1-m)-thick, 100-ft (30-m)-deep, continuous reinforced concrete slurry wall diaphragm along the 0.5-mile (805-m) perimeter of the site.

The high-capacity pile system consists of a 3-ft (0.91-m)-diameter structural steel open-pipe pile with a $\frac{7}{8}$-in. (2.22-cm)-thick wall typically spaced 9 ft (2.75 m) on center capped by a 13-ft (4-m)-deep reinforced concrete mat. Since soil conditions at the upper strata are so poor, the piles were driven into a deep, stiff sand layer located approximately 275 ft (84 m) below grade. The individual design-pile capacity is 1650 kips (7340 kN).

Strength design of the structure is based on a 100-year wind with a basic wind speed of 75 mph for a 10-min average time. The wind speed corresponds to a design wind pressure of approximately 14 psf (0.67 kN/m²) at the bottom of the building and 74 psf (3.55 kN/m²) at the top of the spire. Exterior wall-design pressures are in excess of 100 psf (4.8 kN/m²) at the top of the building.

Wind speeds can average 125 mph (56 m/s) at the top of the building over a 10-min time period during a typhoon event. The earthquake ground accelerations compare to 1994 UBC zone 2A. The overall building drift index for a 50-year return wind with a 2.5% structural damping is 1/1142. This increases to 1/887 for a future developed condition in

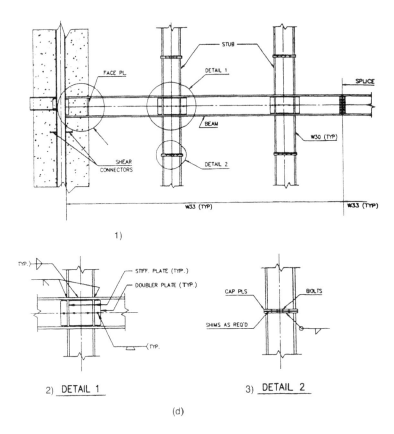

1)

2) DETAIL 1   3) DETAIL 2

(d)

Figure 8.10. (*Continued.*)

which two tall structures are proposed adjacent to the Jin Mao Building. The drift index based on specific Chinese code-defined winds, which were equivalent to a 3000-year wind, is 1/575.

The structural design for the tower is governed by its dynamic behavior under wind and not by its strength or its overall or interstory drift. The calculated fundamental translational periods are 5.7 sec. for each principal axis. The torsional period is 2.5 sec.

In a force-balance and aeroelastic wind-tunnel study, the accelerations at the top floors were evaluated using a value of 1.5% for structural damping. The accelerations measured in the wind tunnel were between 9 and 13 milli-g's for a 10-year return period, and between 3 and 5 milli-g's for a one-year return period—well within the generally accepted range of 20 to 25 milli-g's for a 10-year return. Only the passive characteristics of the structural system including its inherent mass, stiffness, and damping are required to control the dynamic behavior. Therefore, no mechanical damping systems are used.

Since the central core and composite megacolumns are interconnected by outrigger trusses at only three 2-story levels, the stresses in the trusses due to differential shortening of the core relative to the composite columns were of concern. Therefore, concrete stress levels in the core and megacolumns were controlled in an attempt to reduce relative movements. To further reduce the adverse effect of differential shortening, slotted connections were used in the trusses during the construction period of the building. Final bolting with hard connections was done after completion of construction to relieve the effect of differential shortening occurring during construction. The architecture and structural engineering of the building is by the Chicago office of Skidmore, Owings, and Merrill.

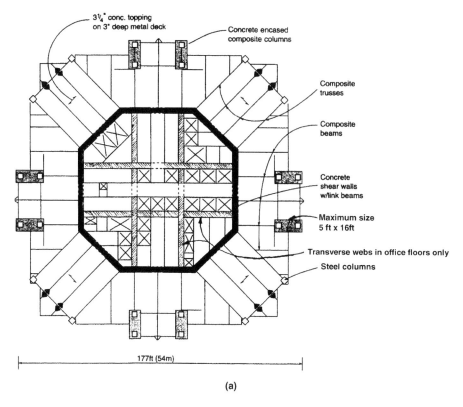

Figure 8.11.   Jin Mao Tower, Shanghai, China: (a) typical office floor framing plan.

8.1.2.9.   Petronas Towers, Malaysia

Two 1476-ft (450-m) towers, 33 ft (7 m) taller than Chicago's Sears Tower, and a sky bridge connecting the twin towers characterize the buildings in Kuala Lumpur, Malaysia (Fig. 8.12).

The towers have 88 numbered levels but are in fact equal to 95 stories when mezzanines and extra-tall floors are considered. In addition to 6,027,800 ft² (560,000 m²) of office space, the project includes 1,501,000 ft² (140,000 m²) of retail and entertainment space in a six-story structure linking the base of the towers, plus parking for 7000 vehicles in five below-ground levels.

The lateral system for the towers is of reinforced concrete consisting of a central core, perimeter columns, and ring beams using concrete strengths up to 11,600 psi (80 mPa). The foundation system consists of pile and friction barrette foundations with a foundation mat.

The typical floor system consists of wide-flange beams spanning from the core to the ring beams. A 2-in.-deep composite metal deck system with a $4^{1}/_{4}$-in. (110-mm) concrete topping completes the floor system.

Architecturally, the towers are cylinders 152 ft (46.2 in) in diameter formed by 16 columns. The facade between columns has pointed projections alternating with arcs, giving unobstructed views through glass and metal curtain walls on all sides. The floor plate geometry is composed of two rotated and superimposed squares overlaid with a ring of small circles. The towers have setbacks at levels 60, 72, 82, 85, and 88 and circular appendages at level 44. Concrete perimeter framing is used up to level 84. Above this level, steel columns and ring beams support the last few floors and a pointed pinnacle.

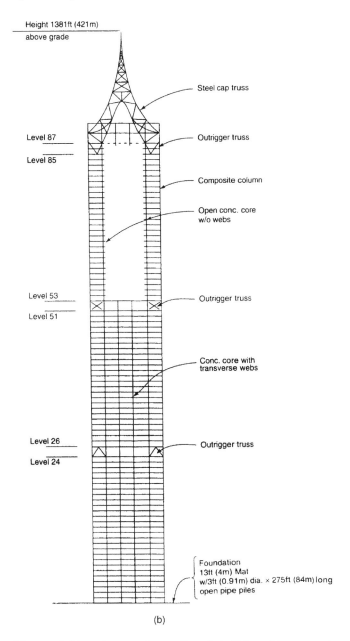

Height 1381ft (421m)
above grade

- Steel cap truss

Level 87

Level 85

- Outrigger truss

- Composite column

- Open conc. core
  w/o webs

Level 53

Level 51

- Outrigger truss

- Conc. core with
  transverse webs

Level 26

Level 24

- Outrigger truss

Foundation
13ft (4m) Mat
w/3ft (0.91m) dia. × 275ft (84m) long
open pipe piles

(b)

Figure 8.11. Jin Mao Tower, Shanghai, China: (b) structural system elevation.

The towers are slender with an aspect ratio of 8.64 (calculated to level 88). The design wind speed in Kuala Lumpur area is based on 65-mph (35.1-m/s) peak, 3-sec gusts at 33 ft (10 m) above grade for a 50-year return. In terms of the old U.S. standard of fastest mile wind, the corresponding wind speed is about 52 mph (28.1 m/s).

The mass and stiffness of concrete are taken advantage of in resisting lateral loads, whereas the advantages of speed of erection and long-span capability of structural steel are used in the floor framing system. The building density is about 18 lb/cu ft (290 kg/m³).

As is common for tall buildings of high aspect ratios, the towers were wind-tunnel tested to determine dynamic characteristics of the building in terms of occupant perception of wind movements and acceleration on the upper floors. The 10-year return

(c)

Figure 8.11.   Jin Mao Tower, Shanghai, China: (c) photograph. (Courtesy of architects Skidmore, Owings, and Merrill, LLP, Chicago; Design Partner Adrian D. Smith; and Gartner Photography.)

period acceleration is in the range of 20 mg, within the normally accepted criterion of 25 mg. The periods for the primary lateral modes are about 9 sec, while the torsional mode has a period of about 6 sec. The drift index for lateral displacement is of the order of 1/560.

Because the limestone bedrock lies 200 ft (60 m) to more than 330 ft (100 m) below dense salty sand formation, it was not feasible to extend the foundations to bedrock. A system of drilled friction piers was designed for the foundation, but barrettes (slurry-wall concrete segments) proposed as an alternative system by the contractor were installed. A 14.8-ft (4.5-m)-thick mat supports the 16 tower columns and 12 bustle columns. The floor corners of alternating right angles and arcs are cantilevered from the perimeter ring beams. Haunched ring beams varying from 46 in. (1.17 m) deep at columns to 31 in. (0.78 m) at midspan are used to allow for ductwork in office space outside of the ring beams. A similar approach with a midspan depth of 31 in. (0.78 m) is used in the bustles. The haunches are used primarily to increase the stiffness of the ring beams.

The central core for each tower houses elevators, exit stairs, and mechanical services, while the bustles have solid walls. The core and bustle walls carry about half the overturning moment at the foundation level.

Each core is 75-ft (23-m) square at the base, rising in four steps to 62×72 ft (18.8 $\times$ 22 m). Inner walls are a constant 14 in. (350 mm) thick while outer walls vary from 30 to 14 in. (750 to 350 mm). The concrete strength varies from 11,600 to 5800 psi (80 to 40 mPa).

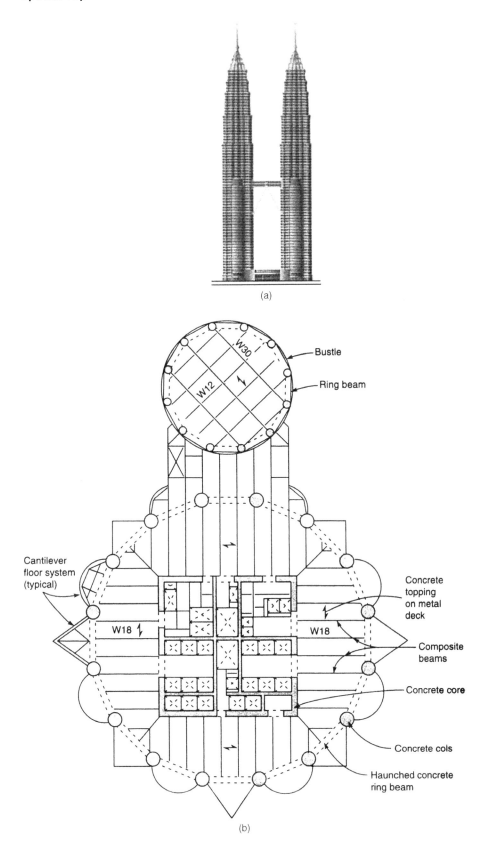

Figure 8.12. Petronas Twin Towers, Kuala Lumpur, Malaysia: (a) elevation; (b) structural system plan.

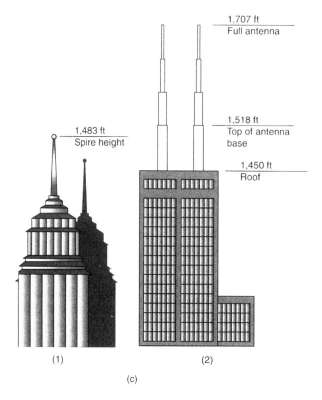

Figure 8.12.   Height comparison: (1) Petronas Towers; (2) Sears Tower, Chicago.

To increase the efficiency of the lateral system, the interior core and exterior frame are tied together by a two-story-deep outrigger truss at the mechanical equipment room (level 38). A vierendeel type of truss with three levels of relatively shallow beams connected by a midpoint column is used to give flexibility in planning of building occupancy.

The tower floors, Fig. 8.12b, typically consist of composite metal deck with concrete topping varying from 4½ in. (110 mm) in offices to 8 in. (200 mm) on mechanical floors, including a 2-in. (53-mm)-deep composite metal deck. Wide-flange beams frame the floors at spans up to 42 ft (12.8 m), and are W18 or shallower on most floors to provide room for ductwork, sprinklers, and lights.

Cantilevers for the points beyond the ring beams are 3.28-ft (1-m)-deep prefabricated steel trusses. For the arcs, the cantilevers are beams propped with kickers back to the columns. Trusses and beams are connected to tower columns by embedded high-strength bolts. The structural engineering is by Thornton–Tomasetti Engineers, and Ranhill Bersekutu Sdn. Bhd.

Although the Sears Tower's 110 stories dwarf the Malaysian twin skyscrapers' 88 floors (Fig. 8.12c), an engineering panel from the Council on Tall Buildings and Urban Habitat says that the Sears Tower is no longer the world's tallest building. This panel, which sets international building height standards, contends that the Petronas Towers' 242-foot-high ornamental spires are part of their height while the radio antennas of the Sears Tower are not. This is because traditionally the measurement from ground-floor entrance to the highest original structural point has been the criterion for assessing the height of skyscrapers for over a quarter of a century. Executives of the Chicago skyscrapers disagree, and say their building is actually 35 feet taller if the radio bases are considered as part of the height.

8.1.2.10. Central Plaza, Hong Kong

The building has 78 stories, with the highest office floor at 879 ft (268 m) above ground. Including the tower mast, the building is 1207.50 ft (368 m) tall (Fig. 8.13a). The building has a triangular floor plate with a sky lobby on the 46th floor.

The triangular design consisting of a typical floor area of 23,830 ft² (2214 m²), (Fig. 8.13b,c) was preferred over a more traditional square or rectangular plan because the triangular shape has very few dead corners and offers more views from the building interiors.

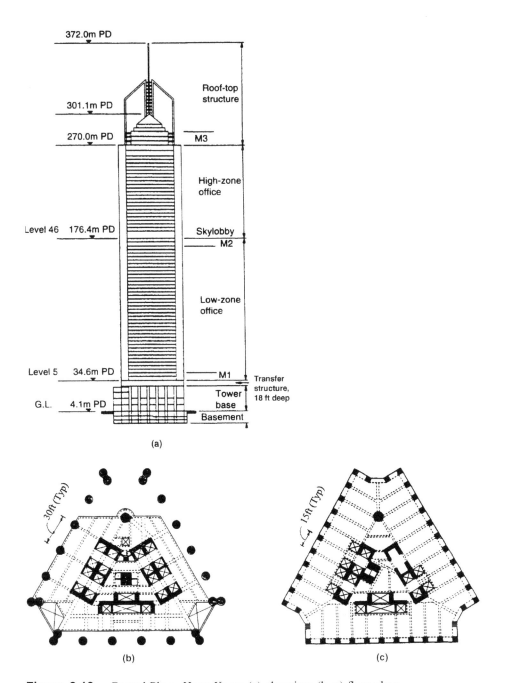

Figure 8.13.   Central Plaza, Hong Kong: (a) elevation; (b, c) floor plans.

The tower consists of three sections: 1) a 100-ft (30.5-m)-tall tower base forming the main entrance and public circulation spaces; 2) a 772.3-ft (235.4-m)-tall tower section containing 57 office floors, a sky lobby, and five mechanical floors; and 3) a top section consisting of six mechanical floors and a 334-ft (102-m)-tall tower mast.

The triangular building shape is not truly triangular because its three corners are chamfered to provide better internal office layout. The building facade is clad in insulated glass. The mast is constructed of structural steel tubes with diameters up to 6.1 ft (2 m).

The triangular core design provides a consistent structural and building services configuration. A column-free office space, with 30.84- to 44.3-ft (9.4- to 13.5-m) depth is provided between the core and the building perimeter.

To enhance the spatial quality of the tower at the base, the 15-ft (4.6-m) column grid of the tower is transformed to a 30-ft (9.2-m) column grid by eliminating every other column. An 18-ft (5.5-m)-deep transfer girder facilitates column termination.

The building site is typical of a recently reclaimed area in Hong Kong with sound bedrock lying between 82 and 132 ft (25 and 40 m) below ground level. This is overlaid by decomposed rock and marine deposits with the top 33 to 50 ft (10 to 15 m) consisting of a fill material. The allowable bearing pressure on sound rock is of the order of 480 ton/ft^2 (5.0 kN/m^2). The maximum water table is about 6.1 ft (2 m) below ground level.

Wind loading is the major lateral load criterion in Hong Kong, which is situated in an area subject to typhoon winds. The local wind design is based on a mean hourly wind speed of 100 mph (44.7 m/s), corresponding to a 3-sec gust of 158 mph (70.5 m/s). The resulting lateral design pressure is 86 psf (4.1 kN/m^2) at 656 ft (200 m) above ground level.

The basement consisting of a diaphragm slurry wall extends around the whole site perimeter and is constructed down to and grouted into rock. The diaphragm wall design allowed for the basement to be constructed by the "top down" method. This method typically has the following features:

1. Simultaneous construction of superstructure and basement, thus reducing the time required for construction.
2. Use of basement floor slabs for bracing of diaphragm walls, thereby reducing lateral tiebacks.
3. Construction of a watertight box within the site enabling installation of hand-dug caissons, traditional in some countries outside of North America.

The lateral system for the tower above the transfer girder consists of external facade frames acting as a tube. These consist of closely spaced 4.93-ft (1.5-m)-wide columns at 15-ft (4.6-m) centers and 3.6-ft (1.1-m)-deep spandrel beams. The floor-to-floor height is 11.82 ft (3.6 m). The core shear walls carry approximately 10% of the lateral load above the transfer level. The transfer girder located at the perimeter is 18 ft. (5.5 m) deep by 9.2 ft (2.8 m) wide. The increased column spacing, together with the elimination of spandrel beams in the tower base, results in the external frame no longer being able to carry the entire lateral load acting on the building. Therefore, the wind shears are transferred to the core through the diaphragm action of a 3.28-ft (1-m)-thick slab located at the transfer level. Structural engineering for the project is by Ove Arup and Partners.

8.1.2.11. *One-Ninety-One Peachtree, Atlanta*

This 50-story building (Fig. 8.14) uses the concept of composite partial tube, as shown in Fig. 8.14b. The partial tubes which extend uninterrupted from the foundation to the 50th floor consist of concrete columns encasing steel erection columns with cast-in-place

(a)

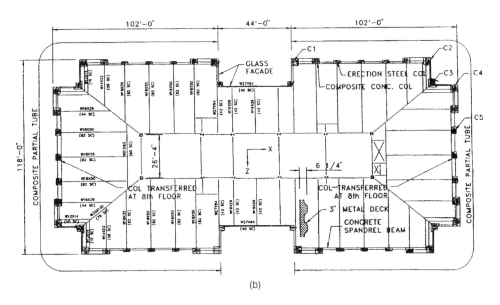

(b)

Figure 8.14. One-Ninety-One Peachtree, Atlanta: (a) building elevation; (b) typical floor framing plan.

concrete spandrels. The building interior is an all-steel structure with composite steel beams supported on steel columns (Fig. 8.14b).

Since the building did not achieve the lateral resistance until after the concrete had reached a substantial strength, a system of temporary bracing was provided in the core. The erected steel was allowed to proceed 12 floors above the completed composite frame with six floors of metal deck and six floors of concreted floors. The structural design is by CBM Engineers, Inc., Houston TX.

8.1.2.12.   Nations Bank Plaza, Atlanta

The 57-story office building has a square plan with the corners serrated to create the desired architectural appearance and to provide for more corner offices (Fig. 8.15). The typical floor plan (Fig. 8.15b) is 162×162-ft (49.87×49.87 m) with an interior core measuring 58 ft-8 in. $\times$ 66 ft-8 in. (17.89×20.32 m). A five-level basement provided below the tower is of reinforced concrete construction. The foundation consists of shallow drilled piers bearing on rock.

The gravity load is primarily supported by 12 composite supercolumns. Four of these are located at the corner of the core, and eight at the perimeter, as shown in Fig. 8.15b.

(a)

Figure 8.15.   Nations Bank Plaza, Atlanta, GA: (a) building elevation; (b) typical framing plan; (c) section.

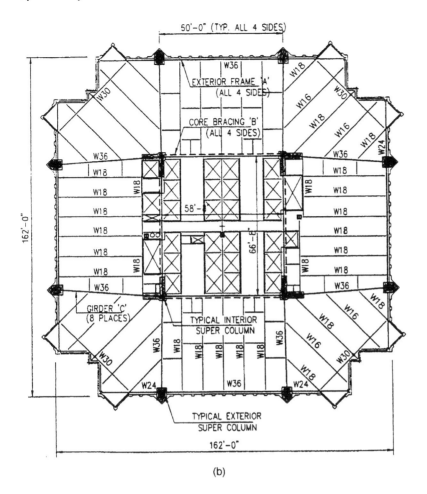

(b)

Figure 8.15. *(Continued.)*

The core columns are braced on all four sides with diagonal bracing as shown schematically in Fig. 8.15c. Since the braces are arranged to clear door openings in the core, their configuration is different on all four sides. Steel girders 36 in. (0.91 m) deep are moment-connected between the composite columns to transfer part of the overturning moment to the exterior columns. Because the girders are deeper than other gravity beams, openings have been provided in the girders to provide for the passage of mechanical ducts and pipes. A diagonal truss is used between levels 56 and 59 to tie the core columns to the perimeter supercolumns. These trusses transfer part of the overturning moment to the perimeter columns and also add considerable stiffness to the building. Above the 57th floor, the building tapers to form a 140-ft (42.68-m)-tall conehead which is used to house mechanical and telecommunication equipment. The structural design is by CBM Engineers, Inc., Houston, TX.

8.1.2.13. *First Interstate World Center, Los Angeles*

This 75-story granite-clad building (Fig. 8.16a) sports multiple step-backs. The structural system is a dual system consisting of an uninterrupted 73 ft 10 in. (22.5 m) square-braced spine (core) interacting with a perimeter ductile moment-resisting frame. The spine has a two-story-tall chevron bracing core, as shown in Fig. 8.16d.

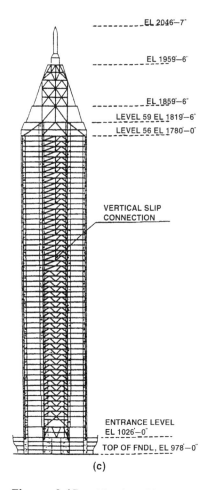

EL 2046'–7"

EL 1959'–6"

EL 1859'–6"
LEVEL 59 EL 1819'–6"
LEVEL 56 EL 1780'–0"

VERTICAL SLIP
CONNECTION

ENTRANCE LEVEL
EL 1026'–0"
TOP OF FNDL, EL 978'–0"

(c)

Figure 8.15. (*Continued.*)

The 55-ft (16.76-m) span for the floor beams coupled with the two-story-tall free-spanning core loads the corner core columns in such a way that the design is primarily governed by gravity design. To achieve overall economy and take advantage of the increase in allowable stresses permitted under combined gravity and lateral loads, the columns are widely spaced to collect gravity loads from large tributary areas (Fig. 8.16e). The column design is primarily for gravity loads with the additional loads due to seismic activity and wind resisted by the one-third increase in allowable stresses. As required by most seismic codes, the *strong column weak/beam* concept is maintained in the design of beam–column assemblies of the perimeter frame tube.

The sustained dead weight of the structure is 204,000 kips (927,272 kN), with the fundamental periods of vibration $T_x = 7.46$ sec, $T_y = 6.95$ sec, and $T_2 = 3.57$ sec. The interaction between the interior braced core and the perimeter ductile frame is typical of dual systems with the shear resistance of core increasing progressively from the top to the base of the building. Nearly 50% of the overturning moment is resisted by the core. The maximum calculated lateral deflection at the top under a 100-year wind is 23 in. (584 mm).

The structure is founded on shale rock with an allowable bearing capacity of 7.5 tons/ft² (720 kPa). The building core is supported on an 11.5-ft (3.5-m)-thick concrete mat while a perimeter ring footing supports the ductile frame. Typical floor framing consists

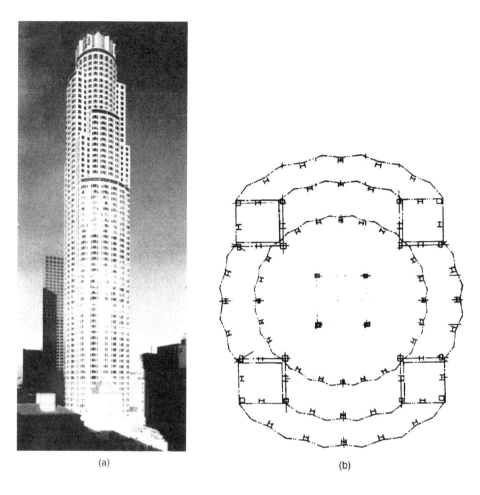

(a)   (b)

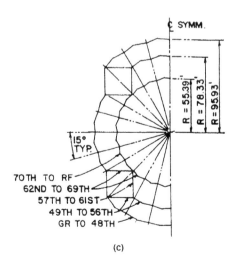

(c)

Figure 8.16. First Interstate World Center, Los Angeles: (a) elevation; (b) plan showing column transfers; (c) composite plan; (d) structural system; (e) framing plan.

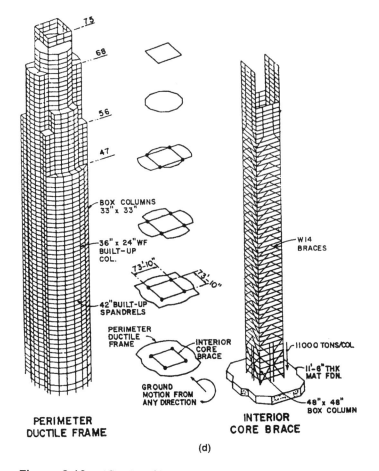

**PERIMETER
DUCTILE FRAME**   **INTERIOR
                                              CORE BRACE**

(d)

Figure 8.16. *(Continued.)*

of W24 wide-flange composite beams spaced at 13 ft centers, spanning a maximum of 55 ft (16.76 m) from the core to the perimeter. The structural design is by CBM Engineers, Inc., Houston, TX.

8.1.2.14. *Singapore Treasury Building*

This 52-story office tower, shown in Fig 8.17a, is unique in that every floor in the building is cantilevered from an inner cylindrical, 82-ft (25-m)-diameter core enclosing the elevator and service areas (Fig. 8.17b). Radial beams cantilever 38 ft (11.6 m) from the reinforced concrete core wall. Each cantilever girder is welded to a steel erection column embedded in the core wall. To reduce relative vertical deflections of adjacent floors, the steel beams are connected at their free ends by a 1 × 4-in. (25 × 100-mm) steel tie hidden in the curtain wall. A continuous perimeter ring–truss at each floor minimizes relative deflections of adjacent cantilevers on the same floor produced by uneven distribution of live load. Additionally the vertical ties and the ring beam provide a backup system for the cantilever beams.

Since there are no perimeter columns, all gravity and lateral loads are resisted solely by the concrete core. The thickness of core walls varies from 3.3 ft (1.0 m) at the top to 4 ft (1.2 m) at the sixteenth floor, and remains at 5.4 ft (1.65 m) below the sixteenth floor. The fundamental vibration period of this cylindrical tower is 5.6 seconds. Its foundation has six 8.0-m-diameter reinforced concrete caissons 35 m long, equally spaced on a 23.5-m-diameter circle, which transfer building loads to rock mainly via skin friction. Tops of

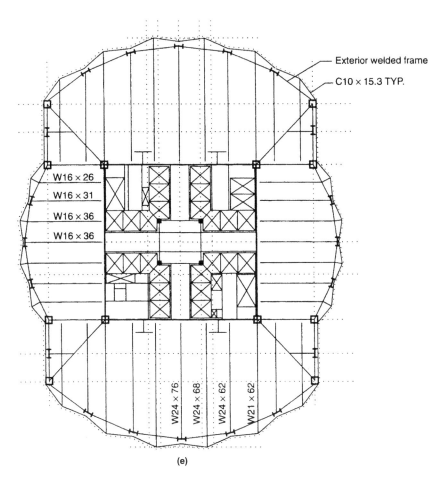

Exterior welded frame

C10 × 15.3 TYP.

W16 × 26
W16 × 31
W16 × 36
W16 × 36

W24 × 76
W24 × 68
W24 × 62
W21 × 62

(e)

Figure 8.16.   (*Continued.*)

caissons are connected by a 2.9-m-thick reinforced concrete mat. The structural engineering is by LeMessurier Consultants, Cambridge, MA, and Ove Arup Partners, Singapore.

8.1.2.15.   City Spire, New York

This 75-story office and residential tower, with a height-to-width ratio of 10:1, is one of the most slender buildings, concrete or steel, in the world today. The critical wind direction for this building is from the west, which produces maximum crosswind response. Wind studies indicated possible problems of vortex shedding as well as occupant perception of acceleration. This possibility was eliminated by adding mass and stiffness to the building.

The main structural system consists of shear walls in the spine connected to exterior jumbo columns with staggered rectangular concrete panels. The structure is subdivided into nine major structural subsystems with setbacks and column transfers as evident from the plans shown in Fig. 8.18. The structural design is by Robert Rosenwasser Associates, New York.

8.1.2.16.   21st Century Tower, China

Designed by the architectural firm of Murphy/Jahn, Inc., Chicago, the exterior form of this proposed building for Shanghai, China, is that of a rectangular tower; however, the building has a setback base and a series of nine-story-high wedge-shaped atria or "winter gardens,"

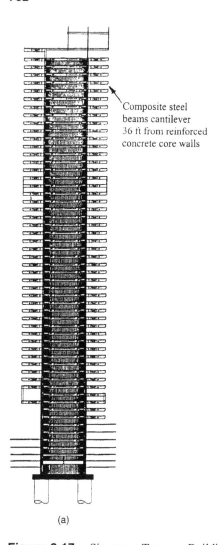

Composite steel
beams cantilever
36 ft from reinforced
concrete core walls

(a)

Figure 8.17.   Singapore Treasury Building: (a) schematic section; (b) typical floor framing plan.

which run for the full height of the tower. These features effectively remove one corner column over the full height of the building and an opposite column at both the top and the bottom of the tower. The structure therefore takes the form of a stack of nine-story-high chevrons. Other architectural features include a cable-suspended skylight canopy roof over a podium, exposed rod–truss curtain wall supports at the winter gardens, and exposed truss–stringer stairs. Structural elements, most notably the nine-story-high superbraces on each face of the tower, are boldly expressed in red, while blue and green solar glazing covers the office spaces. The winter gardens and the podium are enclosed in clear glass.

Although the building is expressed as a square, it is punctuated by a series of four nine-story-high wedge-shaped winter gardens cut into the northeast corner and two more at the southwest corner (Fig. 8.19a,b). The winter gardens have the effect of dividing the tower, both visually and structurally, into a stack of five modules outlined by the super-braces. At the lowest module, the northeast corner column is eliminated entirely. As a result, the tower has only a single axis of symmetry, which passes at 45 degrees through the corners; in addition, nine different floor plans are required within each module. Plan at a representative floor is shown in Fig. 8.19c.

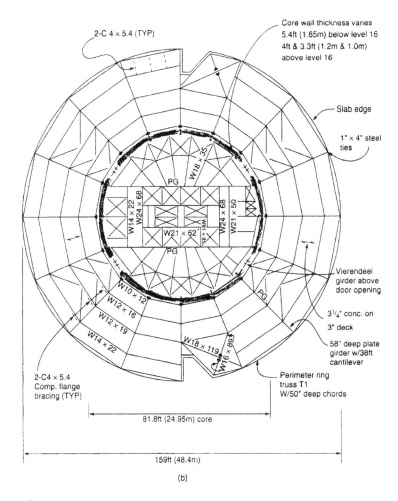

2-C 4 × 5.4 (TYP)

Core wall thickness varies
5.4ft (1.65m) below level 16
4ft & 3.3ft (1.2m & 1.0m)
above level 16

Slab edge

1" × 4" steel ties

W18 × 35

PG

W14 × 22
W24 × 68
W21 × 62
W24 × 68
W21 × 50
W18 × 18M

PG

Vierendeel girder above door opening

W10 × 12
W12 × 16
W12 × 19
W14 × 22
W18 × 119
W16 × 89

3¼" conc. on 3" deck

58" deep plate girder w/38ft cantilever

2-C4 × 5.4
Comp. flange bracing (TYP)

Perimeter ring truss T1
W/50" deep chords

81.8ft (24.95m) core

159ft (48.4m)

(b)

Figure 8.17. (*Continued.*)

The design of the tower is driven by the unique architectural treatment to the building envelope, by the arrangement of the winter gardens, and by the high wind speeds of up to 115 mph (185 km/hr) resulting in design pressures as high as 136 psf (6.5 kPa).

The structural solution for the tower consists of a superbrace system on the exterior skin supplemented by an eccentrically braced interior service core. The superbrace system extends for the full width of the tower to achieve maximum resistance to wind loading. Schematic interior core bracing is shown in Fig. 8.19e. Structural action in the primary columns and braces at the base due to lateral loads is shown in Fig.8.19d.

The braces generally consist of heavy W14 sections, field-spliced every three floors and connected at their ends to square steel box columns. The braces also act as inclined columns and carry vertical loads from the secondary columns above them. This arrangement maximizes the vertical load carried by the corner columns and minimizes uplift. Columns vary in size from 20 to 24 in. (520 to 610 mm) with plate thickness up to 5 in. (130 mm). The braces are arranged in five 9-story-high vees with a one-bay gap in the middle of each building face; stiffness in the gap is provided by a one-bay-wide rigid frame. Panel points for the superbraces occur at the ground, 5th, 15th, 24th, 33rd, 42nd, and roof levels. Horizontal members and diaphragms at these levels are stiffened to transfer horizontal brace forces. Service core bracing provides additional overall stiffness and gives lateral support to floors between the superbrace panel points. For architectural reasons,

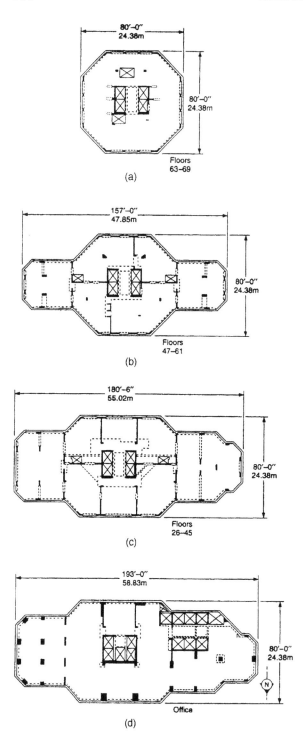

Figure 8.18.   City Spire, New York; floor plans.

braces at the center bay of each core-face are eccentric (Fig. 8.19e). Although most lateral loading is transferred at the ground level to the shear walls, core bracing is extended to the foundation of the substructure. The numerous corner cutouts of the tower structure effectively rotate the principal axes of the structure by 45 degrees, to pass through the corner columns. The lowest two modes of vibration of the tower are single-curvature

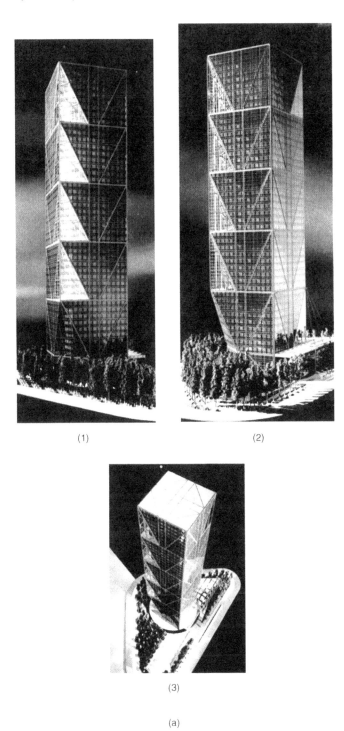

(1)　(2)

(3)

(a)

Figure 8.19. 21st Century Tower, China: (a) model photographs (1), (2), and (3); (b) bracing system; (c) framing plan, levels 19, 28, and 37; (d) structural action in primary columns and braces; (e) typical interior core bracing. Architects: Murphy/Jahn Inc., Chicago, structural engineers: John A. Martin & Associates Inc., Los Angeles, and Martin & Huang, International, Los Angeles.

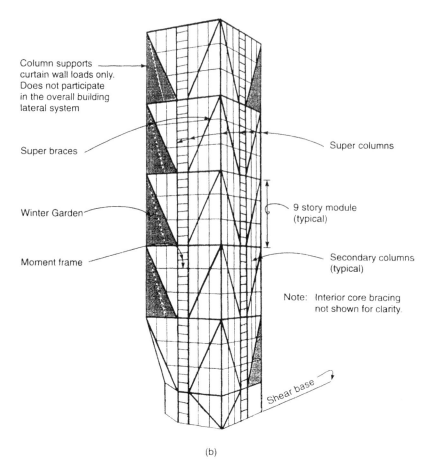

Column supports curtain wall loads only. Does not participate in the overall building lateral system

Super braces

Winter Garden

Moment frame

Super columns

9 story module (typical)

Secondary columns (typical)

Note: Interior core bracing not shown for clarity.

Shear base

(b)

Figure 8.19. (*Continued.*)

bending through these axes; the third mode is torsional. Periods of the first three modes are 4.98, 4.62, and 2.12 sec, respectively. The building's average steel weight is 29 psf (142 kg/in.²). The preliminary design is by John A. Martin & Associates and working drawings are by Martin & Huang, International, both of Los Angeles, CA.

8.1.2.17. Torre Mayor Office Building Mexico City

Constructed in one of the most seismically active regions in the world, this 738-ft (225-m)-tall building, shown in Fig. 8.20a and b, consists of 57 stories including a 13-story parking garage—4 stories below ground and 9 stories above. It rises from a 262.5 × 262.5-ft (80 × 80-m) footprint at the base, decreasing to 262.5 × 213.3 ft (80 × 65 m) in levels 4 through 10, and to 157.5 × 118 ft (48 × 36 m) in levels 11 through 53. The rectangular tower is juxtaposed with a curved facade. The tower is constructed on Mexico City's dry central lake bed or "bowl of Jello" consisting of high water tables and poor alluvial soils. The site is prone to very high seismic activity that can measure 8.2 on the Richter scale.

The design performance criterion established for the structure is that it remain operational immediately following a large seismic event. To fulfill this requirement, fluid viscous dampers have been installed in a structural system consisting of a perimeter moment frame interacting with an interior braced frame. A composite superbrace frame acting in conjunction with a steel tube is present at the perimeter. A braced core at the

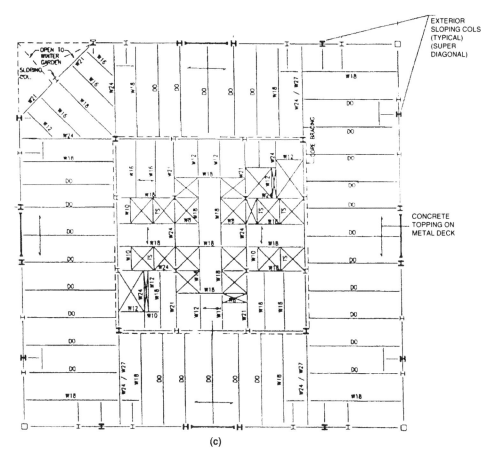

(c)

Figure 8.19.   (*Continued.*)

building interior completes the lateral resisting systems. The perimeter and core columns are encased in reinforced concrete up to the 30th floor. Viscous dampers, 74 of them in the core and 24 the perimeter framing, are installed to absorb and dissipate seismic energy.

Fluid viscous dampers have been installed in diamond-shaped superdiagonal bracing architecturally expressed on the building's perimeter moment frame (Fig. 8.20c). All four elevations of the building contain superdiagonals configured as diamonds rather than Xs. Broad south and north faces contain dampers, that resist seismic loads in the east–west direction. Each of these elevations has four steel diamonds with 137.8-ft (42-m) legs. The diamonds overlap each other at their peaks and valleys to form three smaller diamonds. Each small diamond has four 1200-kip-capacity dampers, one on each leg near the apex or valley. The building has a total of 98 dampers, 12 on each of the broad faces and 74 in the building core. The incorporation of dampers limits quake-induced damage to hung ceilings, fire sprinklers, partitions, mechanical systems, and cladding. Dampers of 600-kip capacity are used in the core in the north–south direction. Core dampers are located conventionally on diagonals of the vertical trusses that transverse the core, two in the end walls and two in between. Photographs of dampers are shown in Figs. 8.20d and e.

The columns in floors 1 through 10 are composite with structural steel columns encased in concrete, which limits the size of steel members. Damper clusters begin at the 11th floor. Rigid floor diaphragms that connect the perimeter frame to a 90.55 × 49.22-ft

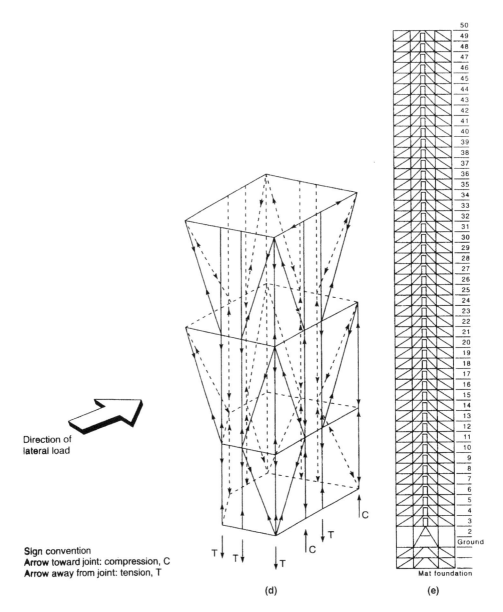

Direction of lateral load

Sign convention
Arrow toward joint: compression, C
Arrow away from joint: tension, T

(d)   (e)

Figure 8.19. *(Continued.)*

(27.6 × 15-m) structural steel core provide in-plane stiffness that ensures all structural elements respond simultaneously to a seismic event.

The building has a four-story basement that extends 49.2 ft (15 m) below ground. The foundation consists of 3.93-ft (1.2-m)-diameter caissons that extend 164 ft (50 m) to hard rock below the alluvial deposits. A reinforced concrete mat ranging in thickness from 3.28 to 8.2 ft (1 to 2.5 m) links the caissons. The structural engineering is by Cantor Seinuk Group, New York City, and Enrique Martinez Romero, Mexico City.

8.1.2.18. Fox Plaza, Los Angles

The structural system for resisting lateral loads for this 35-story building consists of special moment-resisting frames located at the building perimeter. The floor framing consists of W21 wide-flange composite beams spanning 40 ft (12.2 in.) between the core and the

(a)

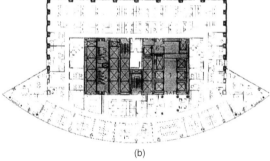

(b)

Figure 8.20. Torre Mayor Office Building, Mexico City: (a) building photograph; (b) plan; (c) schematic elevation showing viscous dampers on front elevation; (d) photograph showing bracing and dampers; (e) close-up view of dampers. (Photographs courtesy of Dr. Ahmad Rahimian, P.E., S.E., President, Cantor Seinuk Group, New York.)

perimeters. A 2-in. (51-mm)-deep 18-gauge composite metal deck with a $3\frac{1}{4}$-in. (83-mm) lightweight concrete topping is used for typical floor construction. A framing plan with sizes for typical members and a photograph of the building are shown in Fig. 8.21. Architects are Johnson, Fain, and Pereira, Inc.; the structural design is by John A. Martin & Associates, Los Angeles.

8.1.2.19. NCNB Tower, North Carolina

This building is an 870-ft (265.12-m)-tall, concrete office building with a 100-ft (30.5-m) crown of aluminum spires (Fig. 8.22). The building has a 12 ft-8 in. (3.87 m) floor-to-floor height and a 48-ft (14.63-m) column-free span from the perimeter to core.

The structural system for resisting lateral loads consists of a reinforced concrete perimeter tube with normal-weight concrete ranging in strength from 8000 psi (55.16 mPa)

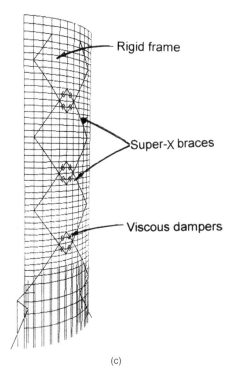

(c)

(d)

Figure 8.20. (*Continued.*)

near the building's base to 6000 psi (41.37 mPa) at the top. Typical column sizes range from 24 × 38 in. (0.61 × 0.97 m) at the base to 24 × 24 in. (0.61 × 0.61 m) at the top. The floor system (Fig. 8.22b) consists of a 4⅝-in. (118-mm)-thick lightweight concrete slab supported on 18-in. (458-mm)-deep post-tensioned beams spaced at 10 ft (3.05 m) on centers. Lightweight concrete was used to reduce the building weight and to achieve the required fire rating for the floor system.

The tower's columns are spaced 10 ft (3.05 m) on center and are connected by 40-in. (1.01-m)-deep spandrel beams. The building has a square plan at the base, but above the 13th floor it resembles a square set over a slightly larger cross, with the four major corners recessed and its four major faces bowed slightly outward. To maintain

(e)

Figure 8.20. (*Continued.*)

(a)

Figure 8.21. Fox Plaza, Los Angeles: (a) building photograph; (b) floor framing plan. Architects: Johnson, Fain & Perei; structural engineers: John A. Martin & Associates Inc., Los Angeles.

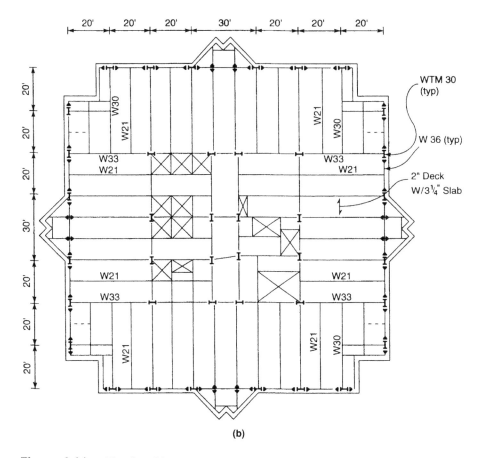

Figure 8.21.   (*Continued.*)

tube action between the 13th and 43rd floors, engineers used L-shaped vierendeel trusses to continue the tube around the corners. Instead of transfer girders at the building step-backs, the building's column-and-spandrel structure is used to create multilevel vier-endeel trusses on the building's main facades. These vierendeels transfer loads using another set of vierendeel trusses perpendicular to the facade at the edges of recessed corners. Differential shortening between the core and perimeter columns was a concern during design because the core columns will be under significantly higher stresses than the closely spaced perimeter columns, To compensate for this, the core columns were constructed slightly longer than the perimeter columns.

Both standard and lightweight concrete was used simultaneously. The normal-weight concrete was used for the perimeter columns, which ranged in size from 24×38 in. $(6.10 \times 965$ mm) at the bottom to 24×24 in. $(610 \times 610$ mm) at the top, as well as for the core columns, ranging from 2×18 ft $(0.61 \times 3.5$ m) at the base to 2×3 ft $(0.61 \times 0.92$ m) at the top.

Normal-weight concrete was also used for post-tensioned spandrels at the peri-meter of each floor, but 5000-psi (34.5-mPa) lightweight concrete was used for the $4\frac{5}{8}$-in. (118-mm)-thick floor slabs and the 18-in. (0.46-m)-deep post-tensioned beams. The two types of concrete were poured in quick succession and puddled to avoid a cold joint.

The foundation system for the Tower consists of high-capacity caissons under the perimeter columns and a reinforced concrete mat for the core columns. The high-capacity

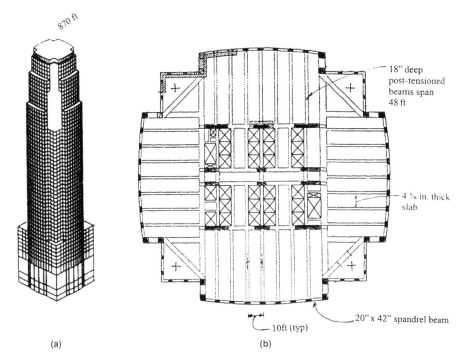

Figure 8.22. NCNB Tower, North Carolina: (a) schematic elevation; (b) typical floor framing plan.

caissons were designed for a total end-bearing pressure of 150 ksf (7182 kN/m²) and skin friction of 5 ksf (240 kN/m²). The high bearing pressure required that the caissons be advanced through the fractured and layered rock zones into high-quality bedrock. Full-length casing was provided to prevent intrusion of soil and ground water into the drilled hole and for the safety of inspectors.

The core columns are supported on a foundation mat bearing on partially weathered rock. The mat dimensions are 83 × 93 × 8 ft (25.3 × 28.35 × 2.44 m). The average total sustained bearing pressure under the mat is equal to 20 ksf (958 kN/m²). The structural design is by Walter P. Moore and Associates, Inc., Houston, TX.

8.1.2.20. *Museum Tower, Los Angeles*

This 22-story residential building, shown in Fig. 8.23, consists of a tubular ductile concrete frame with perimeter columns spaced at 13-ft (8.96-m) centers interconnected with upturned spandrel beams (Fig. 8.23b). The exterior frame is of exposed painted concrete.

The gravity system for the typical floor consists of an 8-in. (203-mm)-thick post-tensioned flat plate with banded and uniform tendons running in the short and long directions of the building, respectively, as shown in Fig. 8.23c.

Although the building is regular both in plan and elevation and is less than 240 ft (78 m) in height, because of transfers at the base (Fig. 8.23b), a dynamic analysis using site-specific spectrum was used in the seismic design. The dynamic base shear was scaled down to a value corresponding to the static base shear. To preserve the dynamic characteristics of the building, the spectral accelerations were scaled down without altering the story masses. The structural design is by John A. Martin & Associates, Inc., Los Angeles, CA.

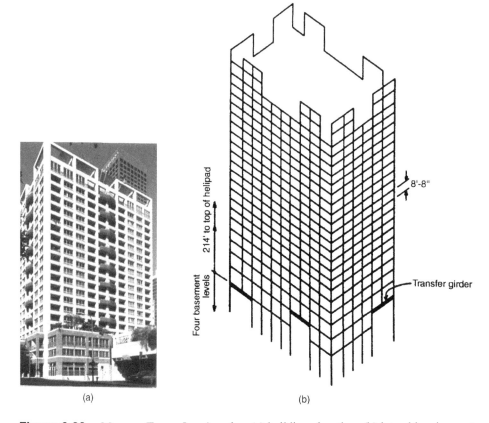

(a)   (b)

Figure 8.23. Museum Tower, Los Angeles: (a) building elevation; (b) lateral bracing system; (c) typical floor framing plan. Structural engineers: John A. Martin & Associates, Inc., Los Angeles.

8.1.2.21. *Figueroa at Wilshire, Los Angeles*

Floor framing plans at various step-backs and notches for the 53-story tower (Fig. 8.24a) are shown in Fig. 8.24d. The structural system, designed by CBM Engineers, Inc., Houston, TX, consists of eight steel supercolumns at the perimeter interconnected in a criss-cross manner to an interior-braced core with moment-connected beams acting as outriggers at each floor (Figs. 8.24b,c). The floor framing is structured such that the main columns participating in the lateral loading system are heavily loaded by gravity loads to compensate for the uplift forces due to overturning. The structural system consists of three major components:

1. Interior concentrically braced core.
2. Outrigger beams spanning approximately 40 ft from the core to the building perimeter. The beams perform three distinct functions. First, they support gravity loads. Second, they act as ductile moment-resisting beams between the core and exterior frame columns. Third, they enhance the overturning resistance of the building by engaging the perimeter columns to the core columns. To reduce the additional floor-to-floor height that might otherwise be required, these beams are notched at the center, and offset into the floor framing, as shown in Figs. 8.24e,f, to allow for mechanical duct work.
3. Exterior supercolumns loaded heavily by gravity loads to counteract the uplift effect of overturning moments.

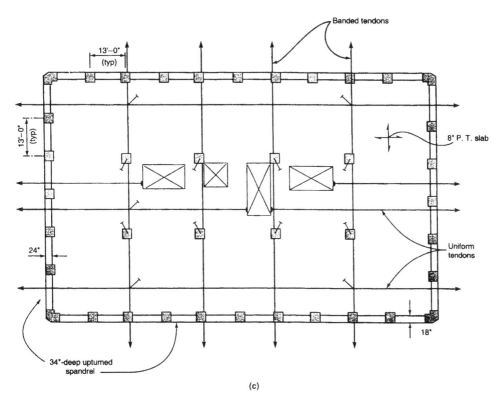

Figure 8.23. (*Continued.*)

8.1.2.22. California Plaza, Los Angeles

The project consists of a 52-story office tower rising above a base consisting of lobby and retail levels, and six levels of subterranean parking (Fig. 8.25a). A structural steel system consisting of a ductile moment-resisting frame at the perimeter resists the lateral loads. The parking areas outside the tower consist of a cast-in-place concrete system with waffle slab and concrete columns. Figure 8.25b shows a typical mid-rise floor plan for the tower with sizes for typical framing elements. Architects are Arthur Erickson, Inc., and structural design for the building is by John A. Martin & Associates, Inc., Los Angeles.

8.1.2.23. Citicorp Tower, Los Angeles

This 54-story tower rises to a height of 720 ft (219.50 m) above ground level and has a height-to-width ratio of 5.88:1 (Fig. 8.26a). It has two vertical setbacks of approximately 10-ft (3.05-m) depth at the 36th and 46th floors, as shown in the composite floor plan (Fig. 8.26b). As is common to most tall buildings in seismic zone 4, this building was designed for site-specific maximum probable and maximum credible response spectrums, which represent peak accelerations of 0.28 g and 0.35 g, respectively. The corresponding critical damping ratios are 5 and 7.5%. The structural system consists of a steel perimeter tube with WTM24 columns spaced at 10-ft (3.05-m) centers and 36-in. (0.91-m)-deep spandrels. The columns at the setback levels are carried by 48-in. (1.22-m)-deep transfer girders and by vierendeel action of the perimeter frame. Typical floor plans at the setback levels are shown in Figs. 8.26c,d.

The foundation for the tower consists of a 7-ft (2.14-m)-deep mat below a four-story basement. The structural design is by John A. Martin & Associates, Inc., Los Angeles.

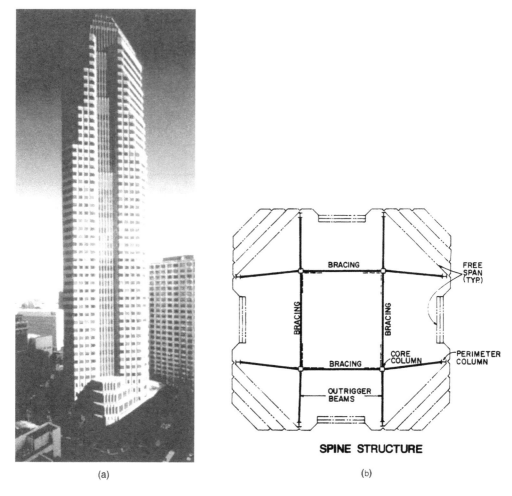

(a)   (b)

Figure 8.24. Figueroa at Wilshire, Los Angeles: (a) building elevation; (b) lateral system; (c) section; (d) framing plans; (e) design concepts; (f) reinforcement at beam notches.

8.1.2.24. Taipei Financial Center

The 101-story Taipei Financial Center in Taipei, Taiwan, at 1667 ft (508 m), is the world's record holder for tallness (Fig. 8.27a). The lowest 25 floors of the building taper gradually inward, forming a truncated pyramid. Above is a stack of eight 8-story-high modules with outward sloping wall, creating a "waist" at the 26th floor and setbacks at floors 34, 42, etc. (see Fig. 8.27b). The modules also have double-notched corners. A narrower tower segment and an architectural pinnacle top the eighth module. Because façade slopes and setbacks interrupt vertical continuity of columns, and doubly notched floor plans reduce the efficiency of an exterior moment frame around corners, a perimeter tubular-frame system was not used for this project.

    Instead, the lateral bracing consists of a dual system comprised of a braced core interconnected to a planar moment frame at each sloping face, through a system of outriggers and belt trusses. The braced core offers high-shear stiffness with chevron and diagonal braces of I-shaped sections in four planes in each direction. A mix of single-, double-, and triple-story outriggers is distributed every eight to ten floors along the building height (see Fig. 8.27c). Typically on each building face they engage two vertical supercolumns. Below

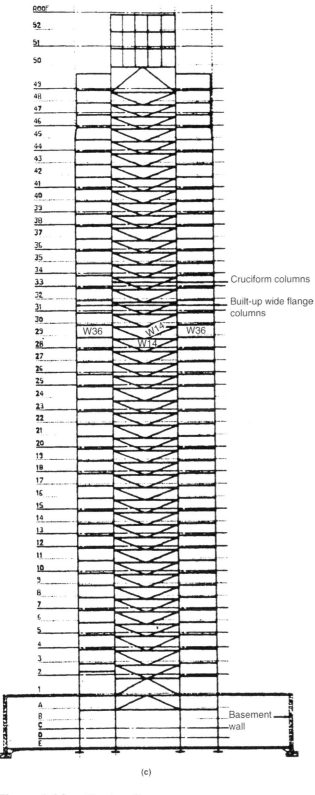

Cruciform columns

Built-up wide flange columns

W36 · W14 · W36

W14

Basement wall

(c)

Figure 8.24. (*Continued.*)

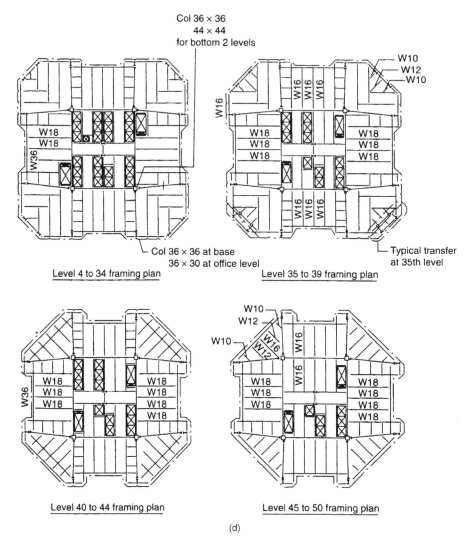

(d)

Figure 8.24. (*Continued.*)

the 26th floor, additional outriggers engage two more columns on each face with the belt trusses engaging corner columns as well. Steel box-core columns and perimeter supercolumns are filled with concrete to provide additional stiffness. The size of supercolumns' steel shell at the base is 8 × 10 ft (2.4 × 3.0 m).

The core is designed as a concentric braced frame (CBF). To clear architecturally required openings, some work points for adjacent braces are spread apart, creating eccentric links. But the system is not designed as an eccentric braced frame (EBF). The design is as for a CBF with the braces, not links, controlling the systems strength. Reduced beam section, RBS, also referred to as dogbone connection, is used at locations where the analyses showed plastic rotation demand in excess of 0.005 radian in a 950-year return-period earthquake. For added strength, beam-to-column connections within the braced core system are detailed as moment connections.

Wind engineering studies indicated that accelerations of the building's upper floors would be 30 to 40% higher than desired for this office building. Therefore, to improve the structure's ability to dissipate dynamic energy, a passive damping system consisting

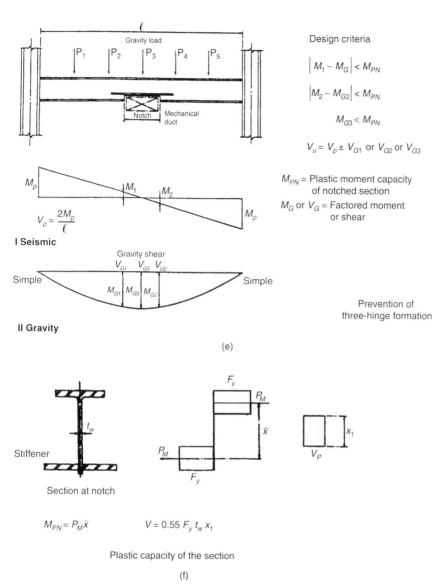

I Seismic

II Gravity

(e)

$M_{PN} = P_M \bar{x}$ $V = 0.55 F_y t_w x_1$

Plastic capacity of the section

(f)

Figure 8.24. (*Continued.*)

of a 730-ton tuned mass damper (TMD) has been installed near the top of the tower. The TMD consists of a massive steel sphere (Fig. 8.27d) suspended by flexible steel cables.

The design of the TMD is by Motioneering, Inc., a company in Ontario, Canada, that specializes in designing and supplying damping systems for dynamically sensitive structures. The building structural engineers are Evergreen Consulting Engineering, Taipei, and Thornton-Tomasetti Engineers, New York City.

8.1.2.25. *World Trade Center Towers, New York*

Of the seven buildings of the World Trade Center (WTC) complex of New York City, the WTC towers, known as WTC 1 and WTC 2, were the most visible and recognized tall buildings throughout the world (Fig. 8.28a). Each of the towers encompassed 110 stories above plaza level and seven levels below. WTC 1, the north tower, had a roof height of

Figure 8.25. Cal Plaza, Los Angeles: (a) building elevation; (b) mid-rise floor framing plan. Architects: Arthur Erickson, Inc.; structural engineers: John A. Martin & Associates Inc., Los Angeles.

1368 ft, while WTC 2 stood nearly as tall at 1362 ft. Each building had a square floor plate 207 ft 2 in. (63.14 m) long on each side, with chamfered corners measuring 6 ft 11 in. (2.10 m). A rectangular service core of approximately 137 by 87 ft (41.75 × 26.52 m) was present at the center of each building.

The buildings' signature architectural design feature was the vertical fenestration which featured a series of closely spaced built-up box columns, as shown in Fig. 8.28b. At typical floors, a total of 59 of these columns were present on each of the flat faces of the building, placed at 3 ft 4 in. (1.0 m) on centers. Adjacent perimeter columns were connected at each floor level typically by 52-in. (91.32-m)-deep spandrel plates (see Figs. 8.28d and). In alternate stories, an additional column was present at the center of each of the chamfered building corners. The resulting configuration of closely spaced columns interconnected with deep spandrel plates created a perforated perimeter tube (see Fig. 8.28c).

Twelve grades of steel, having yield strengths varying from 42 to 100 ksi (191 to 455 kN), were used to fabricate the perimeter columns and spandrel plates. In the upper stories of the buildings, plate thickness in the exterior wall was generally $\frac{1}{4}$ in. (6.35 mm), and at the base, column plates as thick as 4 in. (101.6 mm) were used.

The structural system was considered to constitute a tubular system, acting essentially as a cantilevered hollow tube with perforated walls. The side walls acting as stiff webs transfer shear between the windward and leeward walls, thus creating an efficient three-dimensional structure for resisting lateral loads. In the lower seven stories of the

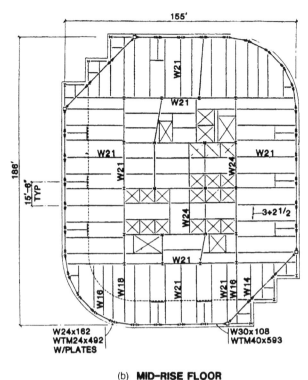

(b) **MID–RISE FLOOR**

Figure 8.25. (*Continued.*)

towers where there are fewer columns, vertical diagonal bracing in the building cores provided the lateral stiffness.

Floor construction typically consisted of 4 in. (101.6 mm) of concrete on $1\frac{1}{2}$-in. (38.1-mm)-deep, 22-gauge noncomposite metal deck. The slab thickness was 5 in. (127 mm) in the core area. Outside the central core, the floor deck was supported by a series of composite floor trusses, 29-in. (0.74-m)-deep, open-web joist-type trusses with ASTM A36 steel chord angles and steel rod diagonals. Composite behavior of the truss with the floor slab was achieved by extending the diagonal truss members above the top chord so that they would act much like shear stud (see Figs. 8.28g,h,i,j). Detailing of these trusses was similar to that typically used in open-web joist fabrication, but the floor system design was not typical of open-web-joist floor systems. It was considerably more redundant and was well-braced with transverse members. Trusses placed in pairs, at 6 ft-8-in. (2.03-m) spacing spanned approximately 60 feet (18.29 m) to the sides and 35 ft (10.67 m) at the ends of the central core. Metal deck spanned parallel to the main trusses and was supported by continuous transverse bridging trusses spaced at 13 ft 4 in. (4.06 m) from the transverse trusses.

In approximately 10,000 locations in each building, viscoelastic dampers extended between the lower chords of the trusses and gusset plates attached to exterior columns (see Fig. 8.28i). These dampers were provided to reduce occupant perception of wind-induced motion. Pairs of flat bars extended diagonally from the exterior wall to the top of chord of adjacent trusses. These diagonal flat bars, which were typically provided with shear studs, provided horizontal shear transfer between the floor slab and exterior wall, as well as out-of-plane bracing for perimeter columns not directly supporting floor trusses.

(a)

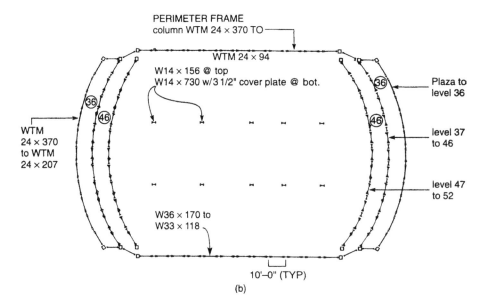

(b)

Figure 8.26. Citicorp Tower, Los Angles: (a) building photograph; (b) composite plan; (c) 36th floor framing plan; (d) 47th–52nd floor framing plan.

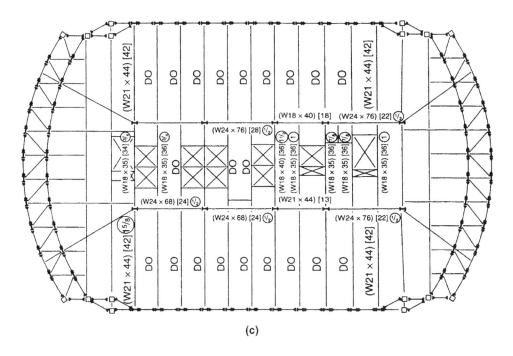

(c)

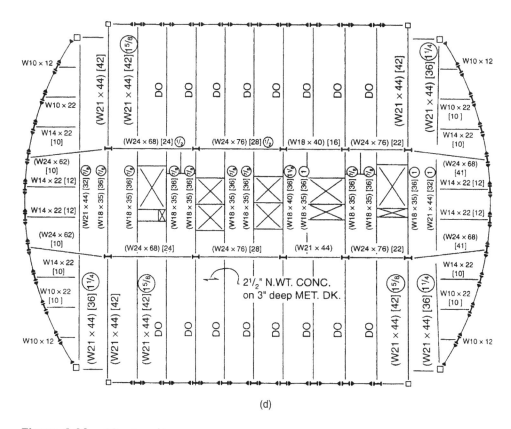

(d)

Figure 8.26. *(Continued.)*

Figure 8.27a.   Taipei 101 Financial Center. (Photograph courtesy of Mr. Hung Lee of John A. Martin & Associates, and Mr. David Lee.)

The core framing consisted of 5-in. (127-mm) concrete fill on metal deck supported by floor framing of rolled structural shapes, in turn supported by wide-flange shape and built-up box section columns. Some of these columns measured 14 by 36 in. (0.35 × 0.91 m). For the upper levels these box columns transitioned into wide-flange shapes. At the top a total of 10 outrigger trusses were present, six extending across the long direction of the core and four extending across the short direction (see Fig. 8.28f). In addition to providing support for a transmission tower (WTC 1 had a transmission tower; WTC 2 did not, but was designed to support such a tower), this outrigger system provided stiffening of the frame for wind resistance.

Prior to construction, the site was underlain by deep deposits of fill material, placed over a period of several hundred years to reclaim the shoreline. In order to construct the towers, perimeter walls for the subterranean structure were constructed using slurry wall and tieback technique. Tieback anchors were drilled diagonally down through the wall and grouted into position into the rock deep behind the walls. (For more information see Ref. 97).

Floors within the substructure were of reinforced concrete flat-slab construction, supported by structural steel columns. These floors also provided lateral support for the perimeter walls, holding back the earth and water pressures from the unexcavated side of the excavation. The tiebacks, which had been installed as temporary stabilizers, were decommissioned by cutting off their end anchorage hardware and repairing the pockets in the slurry wall where these anchors had existed.

In slurry wall construction, a trench is dug in the eventual location of the perimeter walls. A bentonite slurry is pumped into the trench as it is excavated, to keep the trench open against caving of the surrounding earth. Prefabricated reinforcing steel is lowered into the trench, and concrete is placed through a tremie to create a reinforced concrete

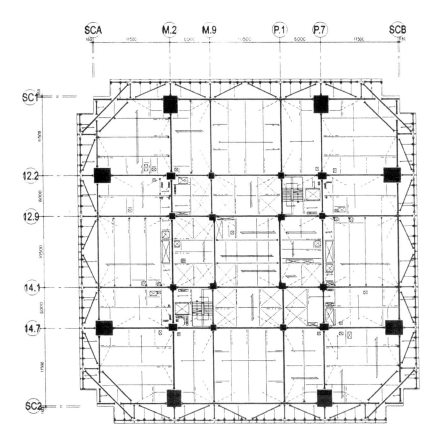

Figure 8.27b. Framing plan for level 50, Taipei Financial Center. The structure consists of a dual system of a braced core connecting to a perimeter sloping frame at each sloping face. The core diagonal and chevron braces are interconnected to vertical supercolumns via outrigger and belt trusses. The supercolumns at the base are 2.4 m × 3.0 m (approximately 8 ft × 10 ft).

wall around the site perimeter. After the concrete is cured, excavation of the substructure begins. As the excavation progresses below surrounding grade, tiebacks are drilled through the exposed concrete wall and through the surrounding soil into the rock below to provide stability for the excavation.

Tower foundations beneath the substructure consisted of massive spread footings, socketed into and bearing directly on the massive bedrock. Steel grillages, consisting of layers of orthogonally placed steel beams, were used to transfer the column loads in bearing to the reinforced concrete footings.

On September 11, 2001, two commercial airlines were hijacked, and one was flown into each of the towers. The structural damage sustained by each tower from the impact, combined with the ensuing fires, resulted in the total collapse of both buildings. The north tower was struck between floors 94 and 98, with the impact roughly centered on the north face. The south tower was hit between floors 78 and 84 toward the east side of the south face. Both planes banked steeply with estimated speeds of 470 mph and 590 mph at the time of impacting the north and south towers, respectively. The population on September 11, 2001, of the seven buildings of the WTC complex has been estimated at 58,000 people. Almost everyone in WTC 1 and 2 who was below the impact area was able to evacuate the buildings, due to the length of time between the impact and collapse of the individual towers.

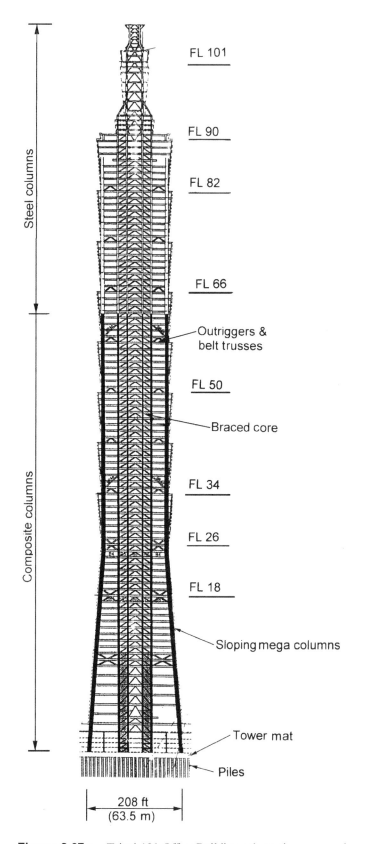

Figure 8.27c. Taipei 101 Office Building schematic cross section.

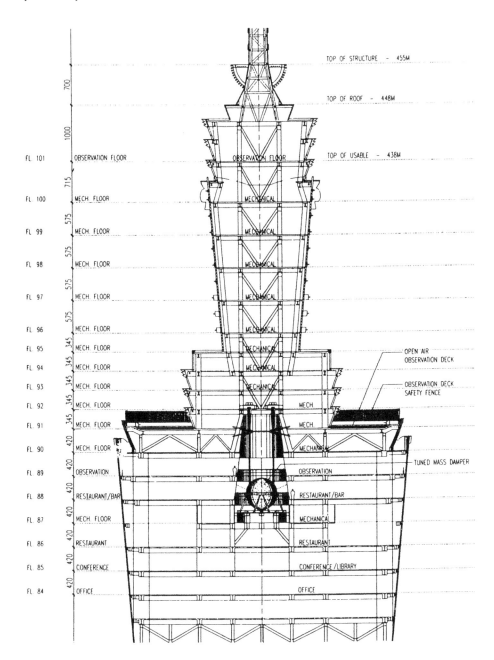

FL 101 — OBSERVATION FLOOR

FL 100 — MECH. FLOOR

FL 99 — MECH. FLOOR

FL 98 — MECH. FLOOR

FL 97 — MECH. FLOOR

FL 96 — MECH. FLOOR

FL 95 — MECH. FLOOR

FL 94 — MECH. FLOOR

FL 93 — MECH. FLOOR

FL 92 — MECH. FLOOR

FL 91 — MECH. FLOOR

FL 90 — MECH. FLOOR

FL 89 — OBSERVATION

FL 88 — RESTAURANT/BAR

FL 87 — MECH. FLOOR

FL 86 — RESTAURANT

FL 85 — CONFERENCE

FL 84 — OFFICE

TOP OF STRUCTURE – 455M

TOP OF ROOF – 448M

TOP OF USABLE – 438M

OBSERVATION FLOOR

MECHANICAL

MECHANICAL

MECHANICAL

MECHANICAL

MECHANICAL

MECHANICAL

MECHANICAL

MECHANICAL

MECH.

MECH.

MECHANICAL

OBSERVATION

RESTAURANT/BAR

MECHANICAL

RESTAURANT

CONFERENCE / LIBRARY

OFFICE

OPEN AIR
OBSERVATION DECK

OBSERVATION DECK
SAFETY FENCE

TUNED MASS DAMPER

Figure 8.27d. Tuned mass damper (TMD), Taipei Financial Center, currently the world's tallest building at 1667 ft (508 m). The 730-ton TMD, consisting of a steel sphere, is suspended by steel cables from level 92. In addition to the TMD for the tower itself, two additional TMDs have been installled for the 197-ft (60-m) spire. Structural engineering by Thornton-Tomasetti Engineers, New York, NY, and Evergreen Consulting Engineering, Inc., Taipei, Taiwan, Republic of China. TMD design by RWDI and Motion Engineering, Guelph, Ontario, Canada.

In each case, the aircraft impacts resulted in severe structural damage, including some localized partial collapse, but did not result in the initiation of global collapse. In fact, WTC 1 remained standing for a period of 1 hour 43 minutes following the initial impact, and WTC 2 for approximately 56 minutes. The fires heated the structural systems and, over a period of time, resulted in additional stressing of the damaged structure, as

(a.1)

(a.2)

Figure 8.28.   World Trade Center (WTC) Towers, New York: (a) photographs (1)–(6): photo (2) is a view looking from inside (photograph courtesy of Andrew Besirof, John A. Martin & Associates, Los Angeles, CA); (b) framing plan; (c) column axial loads due to wind force; (d) prefabricated column and spandrel assembly; (e) 1. Section A through spandrel, 2. Section B through perimeter column; (f) outrigger truss at tower roof, plan, and section; (g) floor framing system; (h) typical floor truss; (i) detail A, exterior wall end detail; (j) detail B, interior wall end detail; (k) height comparison of some contemporary tall buildings.

well as additional damage and strength loss to initiate a progressive sequence of failures that culminated in total collapse of both structures.

Design experts from ASCE and FEMA who investigated the WTC destruction have agreed that it would be futile to create a "terror code" to try to out-design terrorists. The WTC buildings were not required to protect their occupants during the disaster, but on 9/11 did so stunningly. Despite being subjected to stresses that never could have been anticipated, the structural design of the towers kept them standing long enough for more than 20,000 people to evacuate.

(a.3)

Figure 8.28.   (*Continued.*)

8.1.3.   Future of Tall Buildings

September 11, 2001, has not marked the end of the skyscraper era. Already there is talk of America reclaiming the crown with several of the recent proposals for the WTC site in Manhattan involving world-beating structures. But the race for tallness is happening not in America, but in the Far East (see Fig. 8.28k). This past 10 or 15 years (from the mid-1990s to, say, 2010) marks the tall building era of the Far East. Of the world's ten tallest buildings, eight are in Asia. Later this year Taipei will receive the crown from Kuala Lampur, only to pass it on to Shanghai, Hong Kong, Seoul, Tokyo, or New York.

What is the motivation behind the race? To be candid, the reasons are the same today as they were some 70 years ago: height now, as then, is an exhibition of technology and power. Nothing is more expressive than an upright symbol, particularly the one with high-tech items such as pressurized double-decker elevators, external damping devices to reduce sway caused by windstorms, and fiber optics incorporated into curtain walls that transform buildings into giant billboards. Tall buildings become instant icons, putting their cities on the map.

Given humanity's competitive nature, it is hard to believe that the Taipei 101 at 1667 ft (508 m) will wear the crown long. The quest for the title of world's tallest building is alive

(a.4)

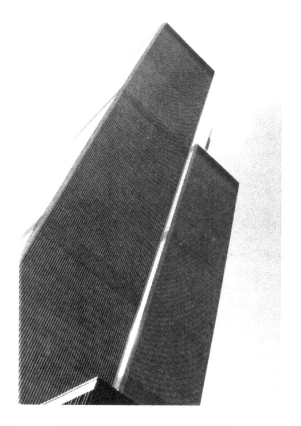

(a.5)

Figure 8.28.   (*Continued.*)

and well, as evidenced by an announcement in January 2003, by a group of multinational corporations, to add a 1772-ft (540-m) tower, by 2007, to a project already underway in Seoul. This begs the question, how tall can buildings go? Answer: No limit is in sight, at least from structural considerations. Humanity has an obsession with building super-tall structures, particularly when humans can live and work in them. While there are indeed lessons to be learned from the WTC catastrophe, the skyscraper will remain viable well into the foreseeable future.

(a.6)

Figure 8.28. *(Continued.)*

8.1.4. Unit Structural Quantities

Quantities are physical items of construction to which unit costs are applied to arrive at a total construction cost. These are relatively easy to obtain once complete working drawings and specifications have been prepared. Prior to this point, however, the estimator or engineer must use "conceptual estimating" to determine approximate quantities. Conceptual estimates require considerable judgment in addition to unit quantities in order to adjust so-called average unit costs to reflect complexity of construction operations, expected time required for construction, etc.

Typically, units of structural quantities are one-, two-, or three-dimensional, based on linear feet, square feet, or cubic feet. These result in unit quantities such as pounds per linear feet (plf), pounds per square foot (psf), etc.

8.1.4.1. Unit Weight of Structural Steel for Preliminary Estimate

The total quantity of structural steel divided by the gross area of the building has always been, and will always be, an item of great interest to building developers and designers alike. In U.S. practice, this unit quantity of steel is usually expressed in terms of pounds per square foot. In selecting a structural system, the usual practice is to look into several possible structural schemes that meet the basic architectural requirements. The deciding factor most often, then, is the unit quantity of material for the systems. Although the unit

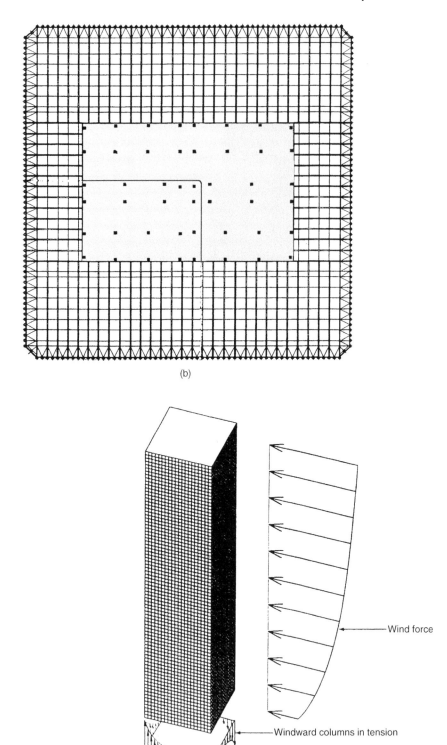

(b)

(c)

Figure 8.28.   (*Continued.*)

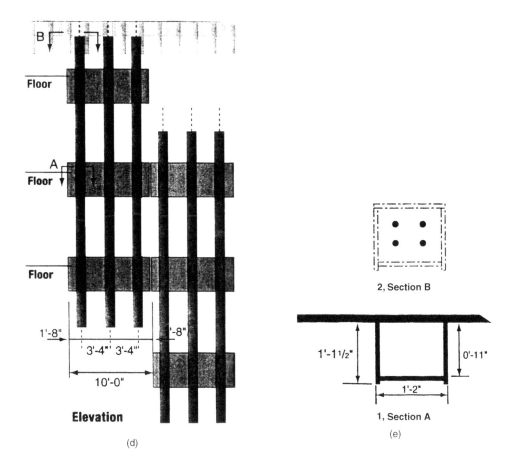

Figure 8.28. (*Continued.*)

quantity of steel in a steel building may not address other factors relevant to construction cost, such as unit price for steel for a particular scheme, whether or not the scheme has composite steel-concrete vertical systems, the cost of fireproofing, and the cost of borrowing money for the period of construction, it is the unit quantity that most often pushes a particular scheme to the forefront.

Prior to the advent of tubular and megaframe systems, most of the buildings were designed using braced or moment frames as lateral load-resisting systems. Consequently, their poundage is very heavy compared to present-day schemes.

Leaving aside the pre-1960 buildings, it is of interest for conceptual estimating purposes to assemble the unit structural steel quantities for buildings that have been built within the last three decades. Figure 8.29 shows the general trend in the increase of unit quantity of steel as building height is increased.

Four distinct regions A, B, C, and D are shown in the figure. Region A is for buildings up to 30 stories, B for buildings between 30 and 50 stories, C for buildings 50 to 70 stories, and D for high-rises in excess of 70 stories. Use of this figure is best explained with respect to the following examples.

Example 1. A proposed ten-story building in a wind-controlled, low-seismic risk area. The engineer is asked to come up with a unit quantity of structural steel for purposes of a conceptual cost estimate.

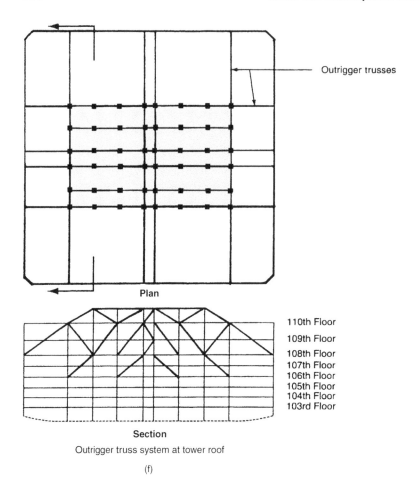

Plan

Section

Outrigger truss system at tower roof

(f)

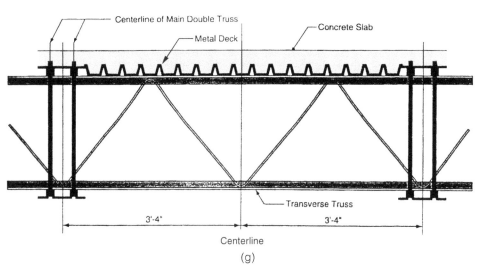

(g)

Figure 8.28. (*Continued.*)

For a ten-story building, it is seen from Fig. 8.29 that the lower and upper bounds for the unit quantity are 10.5 and 15 psf, respectively, which is a rather wide range. The engineer now has to make some judgment calls, depending on what is known about the building at this stage of the game. Some relevant questions are: 1) What are the typical

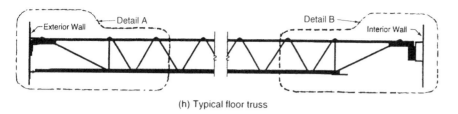

(h) Typical floor truss

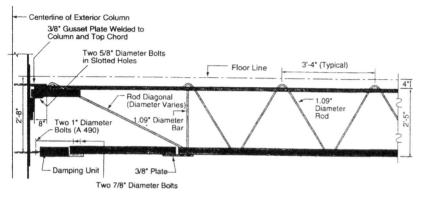

(i) Detail **A** - Exterior Wall End Detail

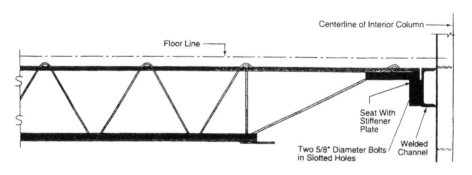

(j) Detail B - Interior Wall End Detail

Figure 8.28. (*Continued.*)

spans for the floor framing? 2) Are there any undue restrictions for beam and girder depths? 3) What is the likely lateral framing system? If none of these questions can be answered with any great certainty the engineer must make some judgment calls. For the current example, the designer decides that the subject building is an average building resulting in an average unit quantity equal to $(10.5 + 15)/2 = 12.75$, rounded to 13 psf.

Example 2. Fifty-story building. Using a similar procedure, the engineer determines that the lower- and upper-bound values are 21 and 32 psf, giving an average unit quantity of 26.5 psf.

Example 3. Twenty-two story building. Design controlled by seismic loads. The straight line designated as S in Fig. 8.29 represents an approximate unit quantity of steel for buildings located in high-risk seismic zones (zones 3 and 4). Observe that this line stops at 25 stories, implying that design of taller buildings, in general, is controlled by wind loads. The cut-off level of 25 stories with building periods in the range of 2–3 sec is considered the threshold where wind design requirements exceed those of seismic activity. For the example building, the unit quantity of steel is given by the straight line S as 22 psf.

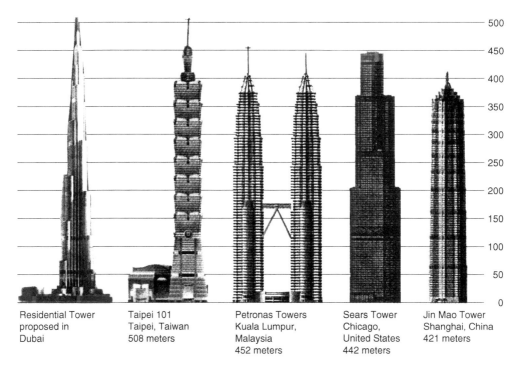

Figure 8.28k. Height comparison of some contemporary tall buildings. Dubai's proposed Residential Tower is reported to beat the height of Taipei's Financial Center tower.

Example 4. Now consider the same building in a seismically low-risk area such as Houston, TX. We go to the region designated as an average zone on the graph. From the graph the unit weight is seen to vary from a low of 13.5 psf to a maximum of 19.5 psf. As in the previous case we use an average unit quantity of 16.5 psf for the preliminary conceptual estimate.

Example 5. Seventy-story composite building. The line designated as CB1 is for approximate unit quantity of steel for buildings using composite steel columns as lateral-load-resisting frame columns. As a preliminary estimate, a unit quantity of 23 psf is appropriate for the building.

Example 6. Fifty-story composite building with composite columns and composite girders (buildings with composite columns and composite shear walls similar). From the graph noted as CB2, we get a unit quantity of 13.5 psf for this building. Note that design rules for composite construction in seismic zones 3 and 4 are not well-established in North America. Therefore, graphs CB1 and CB2 are valid for buildings in low seismic risk zones only.

8.2. DAMPING DEVICES FOR REDUCING MOTION PERCEPTION

Engineers have learned from building occupants and owners, and from wind tunnel studies, that designing a tall building to meet a given drift limit under code-specified equivalent static loads is not enough to make occupants comfortable during windstorms. However, they have only limited control over three intrinsic factors, namely, the height, the shape, and the mass, that influence the dynamic response of buildings. Additionally, the behavior of a tall building subjected to dynamic loads such as wind or seismic activity is difficult to predict

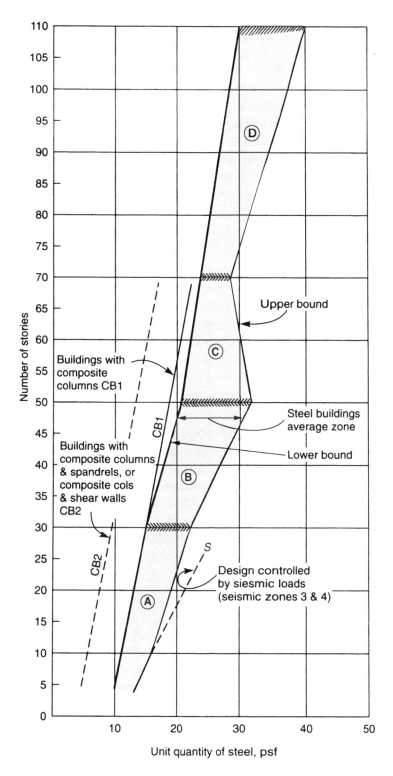

Figure 8.29. Structural steel unit quantities.

with any accuracy because of the uncertainty associated with the evaluation of a building's damping and stiffness, as well as the complicated nature of loading.

The present state of the art is such that an estimate of structural damping can be made with a plus or minus accuracy of only 30% until the building is constructed and the nonstructural elements are fully installed. It is well-known that wind-induced building response is inversely proportional to the square root of total damping, consisting of aerodynamic plus structural damping. So, if damping is quadrupled (increased by four times), a 50% response reduction is achieved, and if damping is doubled, the dynamic response is reduced by 29%. Because of the inherent damping of a building responding elastically to wind loads in the range of 0.5 to 1.5% of the critical response, it is impractical to increase the damping to, say, four times as much by use of modified structural materials.

Suppression of excessive vibrations can be dealt with limited success in a variety of ways. Additional stiffness can be provided to reduce the vibration period of a building to a less sensitive range. Changes in mass of a building can be effective in reducing excessive wind-induced excitation. Aerodynamic modifications to the building's shape, if agreeable to the building's owner and architect, can result in reduced vibrations caused by wind. However, these traditional methods can be implemented only up to a point beyond which the solutions may become unworkable because of other design constraints such as cost, space, or aesthetics. Therefore, to achieve reduction in response, a practical solution is to supplement the damping of the structure with a mechanical damping system external to the building's structure.

8.2.1. Passive Viscoelastic Dampers

Figure 8.30a shows schematics of a viscoelastic polymer damper. An early example of application of this type of damper is the World Trade Center Towers, conceived in the 1960s, constructed in the early seventies, and destroyed by terrorists on September 11, 2001. These buildings were designed with viscoelastic dampers distributed at approximately 10,000 locations in each building. The dampers extended between the lower chords of the floor joists and gusset plates mounted on the exterior columns beneath the stiffened seats (Fig. 8.28).

Viscoelastic dampers dissipate energy through deformation of polymers sandwiched between relatively stationary steel plates. Their energy dissipation depends on both relative shear deformation of the polymer and relative velocity within the device. The device is typically used to reduce occupants' perception of wind-induced motions. It does not require constant operational monitoring and is not dependent on electric power.

The Columbia Seafirst Center, a 76-story building in Seattle built in 1984, is another example of using this technology to reduce occupant perception of wind-induced building motion. The dampers used in this building consist of steel plates coated with a polymer compound. The plates are sandwiched between a system of relatively stationary plates. As the building sways under the action of wind loads, the steel plates which are attached to structural members are subjected alternately to compression and tension. In turn, the viscoelastic polymer subjected to shearing deformations absorbs and dissipates much of the strain energy into heat, thus reducing wind-induced motions.

8.2.2. Tuned Mass Damper

A typical application of a tuned mass damper (TMD) consists of a heavy mass installed near a building's top in such a way that it tends to remain still while the building moves beneath it. This strategy allows the mass at top to transmit its inertial force to the building

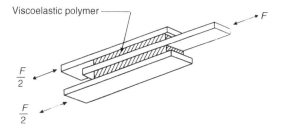

Figure 8.30a. Viscoelastic polymer damper. A building damping of about 4% can be attained using these dampers. Buildings equipped with viscoelastic dampers include the World Trade Center, New York, destroyed on Sept 11, 2001, and the Columbia Seafirst Center, Seattle.

in a direction opposite to the motions of the building itself, thereby reducing the building's oscillations.

The mass itself need weigh only a small fraction—0.25 to 0.70%—of the building's total weight, which corresponds to about 1 to 2% of first modal mass. "Tuned" simply means the mass can be adjusted to move in a fundamental period equal to the building's natural period so that it will be more effective in counteracting the building oscillations. In addition to the initial tuning when it is first installed, the TMD may be fine-tuned as the building period changes with time. The period may increase as the building occupancy changes, as nonstructural partitions are added, or as elements contributing nonstructural stiffness "loosen-up" after initial wind storms.

Thus a TMD may be considered as a small damped mass of single-degree-of-system riding "piggy-back" atop a building. Although its mass is a small fraction of the building's mass, its vibration characteristics are adjusted to mimic those of the building's. For example, if a tall building sways, say, 24 in. to the right at a fundamental frequency of 0.16 Hz, the TMD is designed to move to the left at the same frequency.

The idea of using the inertia of a floating mass to tame the sway of a tall building is not entirely new. In fact, the invention of the TMD as an energy-dissipative vibration absorber is credited to Frahm, who developed the concept in 1909. The theory was later described by Den Hertog in his classic textbook in 1956, and since then has been applied in automotive and aircraft engines to reduce vibrations. Since the wind force–time relationship is not harmonic (sinusoidal), the basic ideas developed by Den Hartog have been modified in building applications to account for the random nature of wind.

When activated during windstorms, the TMD becomes free-floating by rising on a nearly frictionless film of oil. To dissipate energy, the TMD must be allowed to move with respect to the building. In the earlier TMDs installed in tall buildings, spring-like devices connecting the mass to the building pull the building back to center, as the building sways away from its equilibrium position. The mass is also connected to the building with a damping device, in the form of a hydraulic actuator, which is controlled to provide a predetermined percentage of critical damping. This limits the lateral displacements of the mass relative to the building.

The TMD's advantages become academic in a power failure. It needs electricity to work and if that's lost in a heavy windstorm, when the TMD would most be needed, it wouldn't work. So it is advisable to have the TMD wired to an emergency power system.

During a major wind storm, the mass will move in relation to the building some 2 to 5 ft. The system is controlled to activate when a predetermined building lateral acceleration occurs. This motion is registered on an accelerometer and, if the allowable limit is reached, the mass is activated automatically.

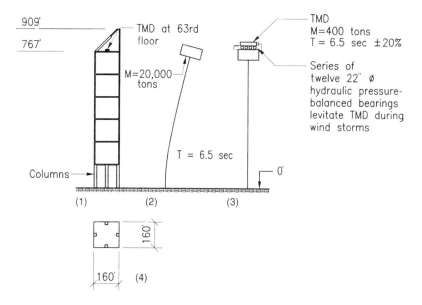

Figure 8.30b. Tuned mass damper for Citicorp Tower, New York: (1) building elevation; (2) plan; (3) first-mode response; (4) TMD atop the building.

8.2.2.1. Citicorp Tower, New York

The Citicorp Tower (shown in schematic view in Fig. 8.30b) consists of a unique structural system of perimeter-braced tubes elevated on four 112-ft-high columns and a central core. It rises approximately 914 ft above grade. The tower is square in cross section with plan dimensions of approximately 157 by 157 ft. The top 140-ft portion of the tower slopes downward from north to south.

The TMD designed for the building consists of a concrete block 29 × 29 × 9 ft that weighs 410 tons (820 kips). It is attached to the buiding with two nitrogen-charged pneumatic spring devices and two hydraulic actuators that are controlled to provide damping to the TMD and linearize the "springs." One set counters north–south building dynamic motion and the other set counters east–west motion. The spring stiffness, and thereby the TMD frequency, is adjusted (tuned) by changing the pneumatic pressure. It also has an antiyaw device to prevent twisting of the block, and snubbers to prevent excessive motion of the block.

The TMD is capable of a 45-in. operating stroke in each orthogonal direction. The operating period is adjustable independently in each axis. The mass block is supported with twelve 22-in.-diameter pressure-balanced bearings connected to a hydraulic pump.

The block positioned at the building's 63rd floor (780 ft high) represents approximately 2% of first-period modal mass of the building. The motions of the block are controlled by pneumatic devices and servohydraulics resulting in a system that has the characteristics of a spring-mass-damper system, as shown schematically in Fig. 8.30c.

To dissipate energy, the TMD is allowed to move with respect to the building. It is continuously on standby, and is designed to start up automatically whenever the accelerations exceeds a predetermined value. The TMD kicks in whenever the accelerations for two successive cycles of building motion exceed 3 milli-g (1 milli-g = 1/1000 of acceleration due to gravity. Therefore, 3 milli-g corresponds to an acceleration of approximately 1.16 in./sec²).

The system continues to operate as long as building motions continue and stops only a half-hour after the last pair of building cycles for which maximum acceleration is

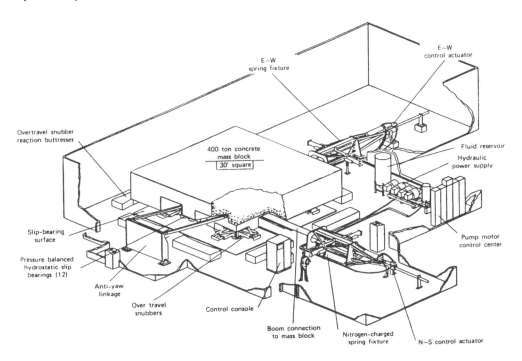

Figure 8.30c. Schematic view of a TMD operating on top of the Citicorp Center. The TMD consists of a 400-ton concrete block bearing on a thin film of oil. The structural stiffness of the TMD is aided by pneumatic springs tuned to the frequency of the building. The TMD damping system is aided by shock absorbers.

greater than 0.75 milli-g. The TMD provides the building with an effective structural damping of about 4% of critical. This is a significant increase above the inherent damping estimated to be just under 1% of critical. Since wind-induced accelerations of a building are approximately proportional to the inverse of the square root of the damping, when in operation the TMD reduces the building sway oscillations by over 40%.

The Citicorp TMD is installed on the 63rd floor. At this elevation, the building may be represented by a single-degree-of-freedom system with a modal mass of 40,000 kips resonating biaxially at a 6.8-sec period with a critical damping factor of 1%. The TMD is designed with a moving mass of 820 kips, biaxially resonant with a period of 6.7 seconds plus or minus 20%, and an adjustable damping of 8 to 14% of critical. Observe that the moving mass represents approximately 2% of the first-period modal mass, which typically corresponds to about 0.6 to 0.7% of the total mass.

8.2.2.2. John Hancock Tower, Boston, MA

The TMD for the John Hancock Mutual Life Insurance Co.'s glass-clad landmark in Boston is somewhat different from that for Citicorp Tower. It was added as an afterthought to prevent occupant discomfort. Second, Hancock Tower is rectangular in plan and consists of moment frames unlike Citicorp's diagonally braced frame (Fig. 8.30d). Because of the building's shape, location, and vibration properties, its dynamic wind response is mainly in the east–west direction and in torsion about its vertical axis. There is a TMD near each end of an upper floor. They are tuned to a vibration period of approximately 7.5 sec. The total east–west moving mass represents about 1.4% of the building first-mode generalized mass, while in the twist direction the moving masses represents about 2.1% of the building's

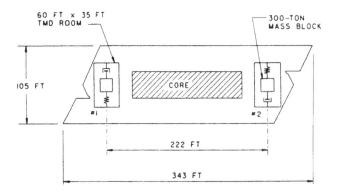

Figure 8.30d. Dual TMD system: John Hancock Tower Boston, MA. Two 60,000-pound masses at each end of the building reduce expected motion by 50%. Effective damping is increased from about 1% to 4%.

generalized torsional inertia. The dampers, then, move only in an east-west direction and work together to resist sway motions in the short direction, or in opposition to stabilize torsional rotations of the building. They are located 220 ft apart, and when moving in opposition act in effect as a 220-ft lever arm to resist twisting. Hancock's dampers each have a 300-ton mass consisting of lead blocks contained in a steel coffer box. They also activate at 3 milli-g of acceleration. In operation the masses may move up to six feet with an operating cycle of about 7.5 sec. Each mass block is supported on sixteen 22-in.-diameter pressure-balanced bearings connected to a hydraulic pump.

The TMDs in both of these towers are used only to assure occupants' comfort. Their beneficial effects in reducing wind-induced dynamic forces are not relied upon for structural integrity under extreme wind loads.

Both the John Hancock Tower and Citicorp Tower TMDs are called passive-powered because, although the reduction in the buildings sway response comes from the inertial force of the dampers, initially power is required to activate the masses. The sliding masses installed in these towers cannot move until their oil bearings are pressurized to levitate the masses.

8.2.2.3. *Design Considerations for the TMD*

There are a number of practical considerations in the design of the TMD. One of these is the need to limit the motions of the TMD mass under very high wind loading such as will occur in the design storm or under ultimate load conditions. One way of doing this is to use a nonlinear hydraulic damper in the TMD. By employing such a damper, the motions of the TMD mass can be greatly reduced under very high wind loading conditions or under strong seismic excitation. A further safeguard against excessive TMD motion is to install hydraulic buffers around the mass. When the mass comes into contact with the buffers, high velocities are quickly reduced.

Both the Citicorp and John Hancock TMD systems have sensors and feedback and electronic control systems, but these were designed to make the TMD operate like a passive tuned mass damper. Tuned mass dampers can in principle be readily converted to be an active system by incorporating sensors and feedback systems that can drive the TMD mass to produce more effective damping than is possible in a purely passive mode. As a result, a larger effective damping can be obtained from a given mass. This approach has been used in several commercially available ready-to-install systems. The TMD is thus made more efficient, a benefit to be weighed against the increased cost, complexity, and maintenance requirements that are entailed with an active system.

8.2.3. Sloshing Water Damper

A simple sloshing type of damper consists of a tuned rectangular tank filled to a certain level with water. The tuning of the system consists of matching the tank's natural period of wave oscillation to the building's period by appropriate geometric design of the tank. If obstacles such as screens and baffles are placed in the tank, dissipation of the waves takes place when water sloshes across these obstructions resulting in a behavior similar to that of a TMD, and the result is again that the tank behaves as a TMD. However, analysis indicates that a sloshing water tank does not make as efficient use of the water mass as a tuned liquid column damper.

8.2.4. Tuned Liquid Column Damper

A tuned liquid column damper (TLCD) is in many ways similar to a TMD that uses a heavy concrete block or steel as the tuned mass. The difference is that the mass is now water or some other liquid. The damper is essentially a tank in the shape of a U. It has two vertical columns connected by a horizontal passage and filled up to a certain level with water or other liquid. Within the horizontal passage, screens or a partially closed sluice gate are installed to obstruct flow of water, thus dissipating energy due to motion of water. The TLCD is mounted near the top of a building, and when the building moves, the inertia of the water causes the water to oscillate into and out of the columns, travelling in the passage between them. The columns of water have their own natural period of oscillation which is determined purely by the geometry of the tank. If this natural period is close to that of the building's period then the water motions become substantial. Thus the building's kinetic energy is transferred to the water. However, as the water moves past the screens or partially open sluice gate in the horizontal portion of the tank, the drag of these obstacles to the flow dissipates the energy of the motion. The end result is added damping to reduce building ocillations.

8.2.4.1. *Wall Center, Vancouver, British Columbia*

Shown in Fig. 8.30e is the plan for the mechanical penthouse of a building called Wall Center, a 48-story residential tower in Vancouver, British Columbia. From wind-tunnel tests, predicted 10-year accelerations were in the range of 40-milli-g, depending on the structural systems considered in the preliminary design. To minimize occupants' perception of motion due to wind excitations, a limit of 15 milli-g was chosen as the design criterion for a 10-year acceleration. A damper using water serves a dual purpose by also providing a large supply of water high up in the tower for fire suppression. Initially, a sloshing water damper was considered but the TLCD was found preferable due to its greater efficiency in using the available water mass. The design turned out to be a remarkably economical solution considering the saved cost of having to install a high-capacity water pump and emergency generator in the base of the building as initially required by fire officials. The total mass required was on the order of 600 tons which corresponds to a large volume of water. However, sufficient space was available. Also a helpful factor was that the motions of the tower were primarily in one direction only. Therefore only motions in one direction needed to be damped, which simplified the design. Figure 8.30f illustrates the TLCD design consisting of two identical U-shaped concrete tanks. Since the building was concrete, it was relatively easy to incorporate the tanks into the design and to construct them as a simple addition to the main structure. The structural design is by Glotman Simpson Engineers, Vancouver, British Columbia, Canada. The

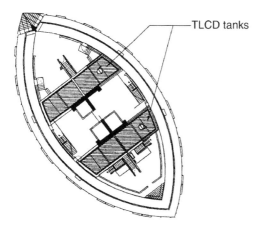

Figure 8.30e. Mechanical penthouse cross section of the Wall Center, a 48-story building in Vancouver, British Columbia. Two specially shaped tanks containing 50,000 gallons of water provide the mass for the building's TLCD. Structural engineering by Glotman Simpson Consulting Engineers, Vancouver, British Columbia, Canada; TLCD by Rowan, Williams, Davies, & Irwin, Inc., and Motion Engineering, Inc., Guelph, Ontario, Canada.

design of the TCLD is by Rowan, Williams, Davis, and Irwin, Inc., Guelph, Ontario, Canada.

8.2.4.2. Highcliff Apartment Building, Hong Kong

Another example of a tall building that uses TLMD to control accelerations and provide enhanced structural performance during typhoon conditions, is the 73-story Highcliff apartment building in Hong Kong, one of the windiest places on earth. The building soars to a height of 705 ft (215 m) with an astonishing slenderness ratio of 20:1. A unique

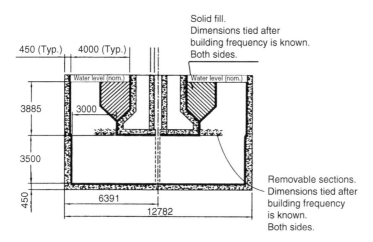

Figure 8.30f. TLCD for Wall Center, Vancouver, British Columbia. The motions of the tower were primarily in one direction only. Therefore only one direction needed to be damped. Two TLCDs extend nearly the full width of the tower. Within each tank is a long horizontal chamber at the bottom and two columns of water at each end. The dampers work by allowing the water to move back and forth along the bottom chamber of the tank and up into the columns of water.

Figure 8.30g.   Highcliff apartment building, Hong Kong.

structural system that incorporates all vertical elements as part of the lateral system, in combination with a series of tuned liquid mass dampers, ensures the safety and comfort of the buildings occupants.

Photographs of the building are shown in Fig. 8.30g. The structural engineering is by the Seattle firm of Magnusson Klemencie Associates.

8.2.5.   Simple Pendulum Damper

The principle feature of the system shown in Fig. 8.30h is a mass block slung from cables with adjustable lengths. The mass typically represents approximately 1.5 to 2% of the building's generalized mass in the first mode of vibration. The mass is connected to hydraulic dampers that dissipate energy while reducing the swinging motions of the pendulum.

The adjustable frame is used as a tuning device to tailor the natural period of vibration of the pendulum. The frame can be moved up and down and clamped on the cables to allow the natural period of the pendulum to be adjusted. The mass is connected to an antiyaw device to prevent rotations about a vertical axis. Below the mass there is a bumper ring connected to hydraulic buffers to prevent travel beyond the hydraulic cylinder's stroke length.

(1)

(2)

Figure 8.30h.   (1) Simple pendulum damper; (2) Hydraulic dampers attached to mass block. (Photograph courtesy of Dr. Peter Irwin of Rowan, Williams, Davis, & Irwin, Inc., Guelph, Ontario, Canada.)

8.2.5.1.   Taipei Financial Center

An example of a tuned mass pendulum damper (TMPD) architecturally expressed as a building feature is shown in Fig. 8.30i. At a height of 1667 feet (508 m), consisting of 101 stories, the building, called Taipei Financial Center, is poised to steal the crown from the twin Malaysian Petronas Towers as the tallest building in the world. A special space has

Figure 8.30i.   Spherical damper, Taipei Financial Center, Taiwan. A 20-ft (6-m)-diameter steel ball assembled on site in layers of 5-in. (12-cm)-thick steel plate is suspended from level 92 by four sets of cables. Eight hydraulic pistons, each 6.5 ft (2 m) long, attached to the ball, dissipate dynamic energy as heat.

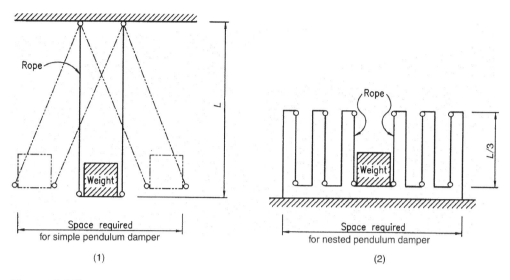

Figure 8.30j. (1) Simple pendulum damper; (2) nested pendulum damper.

been allocated for the TMPD near the top of the building and people will be able to walk around it and view it from a variety of angles. The TMPD, consisting of a 730-ton steel ball, will be brightly colored, and special lighting effects are planned. The architecture of the building is by C.Y. Lee and Partners, Taiwan; structural engineering is by Evergreen Consulting Engineering, Inc., Taipei, Taiwan, and Thornton-Tomasetti Engineers, New York; and the design of the TMPD is by Motioneering, Inc., a company in Ontario, Canada, that specializes in designing and supplying damping systems for dynamically sensitive structures.

8.2.6. Nested Pendulum Damper

In situations where the height available in a building is insufficient to allow installation of a simple pendulum system, a nested TMD may be designed as illustrated in Fig. 8.30j. The design shown is for a North American residential tower. The total vertical space occupied by the damper, which has a natural period of about 6 sec and a mass of 600 tons, is less than 25 ft (7.62 m), as compared to 30 ft (9.14 m) required for a simple pendulum. The design of the damper is by Rowan, Williams, Davis, and Irwin, Inc., Guelph, Ontario, Canada.

A nested pendulum damper is installed at the top of the 70-story, 971-ft-tall Landmark Tower, Yokohama, Japan. The damper requires only a one-story-high space, and is semi-actively controlled. Wind-induced lateral accelerations are expected to be reduced at least 60%. The damper design is by Mitsubishi Heavy Industries, Ltd., Tokyo, Japan.

8.3. PANEL ZONE EFFECTS

Structural engineers involved in the design of high-rise structures are confronted with many uncertainties when calculating lateral drifts. For example, they must decide the magnitude of appropriate wind loads and the limit of allowable lateral deflections and accelerations. Even assuming that these are well defined, another question that often comes up in modeling of building frames is whether or not one should consider the panel zones at the beam–column intersections as rigid.

The panel zone can be defined as that portion of the frame whose boundaries are within the rigid connection of two or more members with webs lying in a common plane.

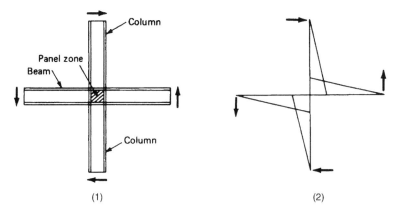

Figure 8.31a. Typical frame element: (1) free-body diagram; (2) bending moments due to shear in beam and columns.

It is the entire assemblage of the joint at the intersection of moment-connected beams and columns. It could consist of just two orthogonal members as at the intersection of a roof girder and an exterior column, or it may consist of several members coming together as at an interior joint, or any other valid combination. In all of these cases, the panel zone can be looked upon as a link for transferring loads from horizontal members to vertical members, and vice versa. For example, consider the free-body diagram of a frame element consisting of an assemblage of two identical beams and columns with points of zero moment at the ends (Fig. 8.31a). These zero-moment ends are, in fact, representative of points of inflection in the members.

Consider the frame element subjected to lateral loads. It is easy to see that, because of these loads, the columns are subjected to horizontal shear forces and corresponding bending moments, as shown in Fig. 8.31a(2). Equilibrium considerations result in vertical shear forces in the beams at the inflection points and corresponding bending moments in the beams. The panel zone thus acts as a device for transferring the moments and forces between columns and beams. In providing for this mechanism, the panel zone itself is subjected to large shear stresses.

The presence of high shear forces in a panel zone is best explained with reference to the connection shown in Fig. 8.31b(1). The bending moment in the beam can be considered as being carried as tensile forces in the top flange and compressive forces in the bottom flange, and the shear stresses can be assumed as being carried by the column web. In the panel zone, the tensile force in the top flange is carried into the web by horizontal shear forces and, by a similar action, is converted back into a tensile force in the outer flange of the column. The distribution of the actual state of stress in the panel zone is highly indeterminate, but a reasonable approximation can be obtained by assuming that the tensile stresses are reduced linearly from a maximum at the edge of the corner B or D to zero at the external corner. If members AB and CD are assumed as stiffeners, a distinct load path can be visualized for the compressive and tensile forces in the beam flange. Consideration of equilibrium of forces within the panel zone results in shear stress and a corresponding shear deformation as shown in Fig. 8.31b(4). It is this deformation that is of considerable interest in the calculation of drift of multistory buildings.

Before proceeding with an explanation of the behavior of panel zones and their influence on building drift, it is instructive to discuss some of the assumptions commonly made in the analysis of building frames. Prior to the availability of commercial analysis

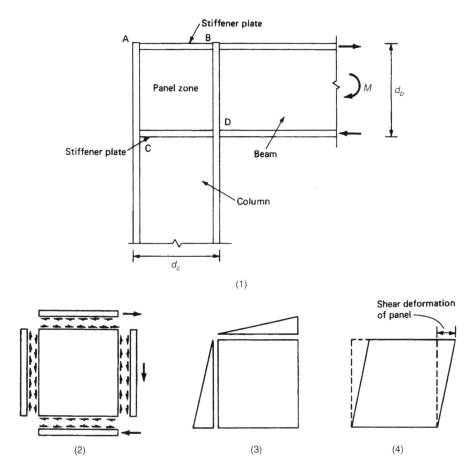

Figure 8.31b. Panel zone behavior: (1) corner panel; (2) schematic representation of shear forces in panel zone; (3) linear distribution of tensile stresses; (4) shear deformation of panel zone.

programs with built-in capability of treating panel zones as rigid joints, it was common practice to ignore their effects; the frame was usually modeled using actual properties along the centerlines of beams and columns.

If the size and number of joints in a frame were relatively large, an effort was made to include the effect of joint rigidity by artificially increasing the moments of inertia of beams and columns; the actual properties were usually multiplied by a square of the ratio of centerline dimensions to clear-span dimensions.

It is now relatively easy to model the panel zone as a rigid element because of the availability of a large number of computer programs which include this feature. Flexibility of panel zones can also be considered in some of these programs, although somewhat awkwardly, by artificially decreasing the size of panel zones.

Computations of beam, column, and panel zone contributions to frame drift can be carried out by hand calculation using a virtual work method. For this purpose consider again the typical frame element subjected to horizontal shear forces P_c and vertical shear forces P_b at the inflection points (Fig. 8.31c(1)).

The notations used in the development of the method are as follows:

d_b = depth of panel zone
d_c = width of panel zone

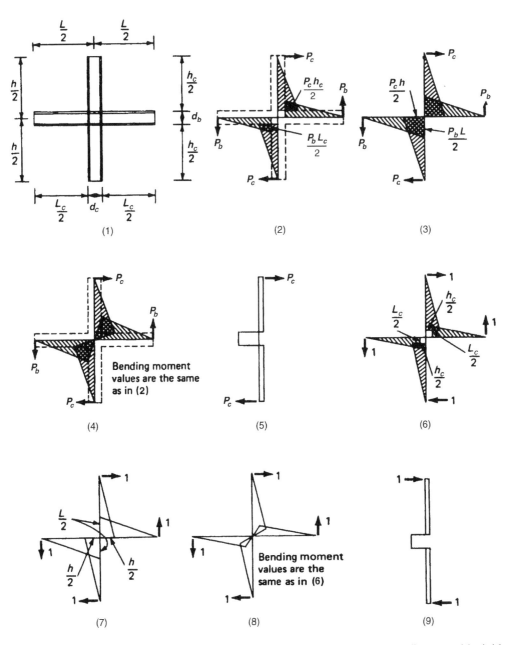

Figure 8.31c. Typical frame segment: (1) geometry; (2) bending moment diagram with rigid panel zone; (3) bending moment diagram without panel zone; (4) bending moment diagram with flexible panel zone; (5) shear force diagram; (6–9) unit load diagrams.

h_c = clear height of column
L_c = clear span of beam
L = center-to-center span of beam
h = center-to-center height of column
I_c = moment of inertia of column
I_b = moment of inertia of beam
E = modulus of elasticity

G = shear modules

Δ_b = frame drift due to beam bending

Δ_c = frame drift due to column bending

Δ_p = frame drift due to panel zone shear deformation

The bending moment diagrams for the typical frame element can be obtained under three different assumptions.

1. The first assumption corresponds to ignoring the rigidity of panel zone; the bending moment diagrams for the external and unit loads can be assumed as shown in Figs. 8.31c(3) and (7). The bending moments increase linearly from the point of contraflexure to the centerline of the joint. By integrating the moment diagrams shown in Figs. 8.31c(3) and (7), the column- and beam-bending contributions to the frame drift are given by:

$$\Delta_c = \frac{P_c h^3}{12 E I_c}$$

$$\Delta_b = \frac{P_b L^3}{12 E I_b}$$

2. In the second case, which corresponds to assuming that the panel zone is completely rigid, we get bending moment diagrams for external and unit loads as shown in Figs. 8.31c(2) and (6). The bending moments increase linearly from the points of contraflexure but stop at the face of beams and columns. Integration of moment diagrams gives the expressions for Δ_c and Δ_b as follows:

$$\Delta_c = \frac{P_c h_c^3}{12 E I_c}$$

$$\Delta_b = \frac{P_b L_c^3}{12 E I_b}$$

3. The third assumption, which attempts to account for the flexibility of panel zones, results in bending moment and shear force diagrams for external and unit loads as shown in Figs. 8.31c(4), (8), (5), and (9). Integration of bending moment and shear force diagrams leads to the following expressions in Δ_c, Δ_b, and Δ_p:

$$\Delta_c = \frac{1}{12 E I_c} \left(P_c h_c^3 + P_c h_c^2 d_b \right)$$

$$\Delta_b = \frac{1}{12 E I_b} \left(P_b L_c^3 + P_b L_c^2 d_c \right)$$

$$\Delta_P = \frac{P_c h_c^2}{d_b t_w d_c}$$

The effect of panel zone continuity plates may be determined by performing a finite element analysis of a typical frame unit, as shown in Fig. 8.31d. A series of finite element analyses can be performed to relate the effect of panel zone to basic section properties of beam and columns of the typical unit. Halvorson (Ref. 64) indicates that for a typical

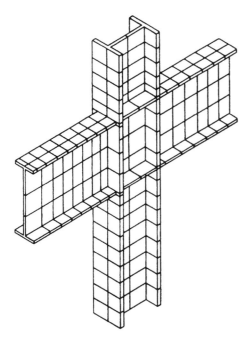

Figure 8.31d.   Finite element idealization of typical frame unit.

13.08-ft-high by 15-ft-long (4- by 4.57-m) unit consisting of $W\,36 \times 300$ columns and W 36×230 beams, the frame stiffness is approximately 8, 15, or 22% stiffer than for a stick element model depending upon whether no, AISC minimum-, or full-continuity plates are provided, respectively.

Using the virtual work expressions given above, or by performing a finite element analysis, it is relatively easy to compute the contribution of panel zone deformation to frame drift. The author recommends that before undertaking the analysis of large tubelike frames, representative frame elements, say, at one-fourth, one-half, and three-fourths the height of the building, be analyzed to get a feel for the contribution of panel zone deformation to frame drift. Armed with the results, it is relatively easy to modify the properties of beam and columns such that the overall behavior of the frame is properly represented in the model.

8.4.   DIFFERENTIAL SHORTENING OF COLUMNS

Columns in tall buildings experience large axial displacements because they are relatively long and accumulate gravity loads from a large number of floors. A 60-story interior column of a steel building, for example, may shorten as much as 2 to 3 in. (50 to 76 mm) at the top, while a concrete column of similar height may experience an additional 2 to 3 in. (50 to 76 mm) of shortening due to creep and shrinkage of concrete. If such shortening is not given due consideration, problems may develop in the performance of building cladding systems. Proper awareness of this problem is necessary on the part of structural engineer, architect, and curtain wall supplier to avoid unwelcome arguments, lost time, and money.

The maximum shortening of a column occurs at the roof level, reducing to zero at the base. In a concrete frame it may take several years for the shortening to occur because of the long-term effect of creep, although a major part of it occurs within the first few months of construction. Very little can be done to minimize shortening, but the design team should be aware of the magnitude of frame shortening so that soft joints are properly detailed between the building frame and cladding to prevent axial load from being transferred into the building facade. Before fabrication of cladding, the in-place elevations of structural frame should be measured and used in the fabrication of cladding. The design should provide for sufficient space between the cladding panels to allow for the movement of the structure. Insufficient space may result in bowed cladding components or, in extreme cases, the cladding panels may even pop out of the building.

A similar problem occurs when mechanical and plumbing lines are attached rigidly to the structure. Frame shortening may force the pipes to act as structural columns resulting in their distress. A general remedy is to make sure that nonstructural elements are not brought in to bear the vertical loads by separating them from the structural elements.

The axial loads in all columns of a building are seldom the same, giving rise to the problem of so-called differential shortening. The problem is more acute in a composite structure because steel columns that are later encased in concrete are typically slender, and are therefore subject to large axial loads during construction. Determining the magnitude of axial shortening in a composite system is complicated because many of the variables that contribute to the shortening cannot be predicted with sufficient accuracy. Consider, for example, the lower part of the composite column that is continually undergoing creep. The steel column during construction is partly enclosed in concrete at the lower floors, with the bare steel column projecting beyond the concreted levels by as many as 8 or 10 floors. Another factor that is difficult to predict is the gravity load redistribution due to frame action of columns and, if the building is founded on compressible material, foundation settlement is another factor that influences the relative changes in the elevations of the columns. The magnitude of load imbalance continually changes, making an accurate assessment of column shortening rather challenging. For concrete buildings, the method of construction more or less takes care of the immediate column shortening and, to a limited extent, the creep effects. This is because as each floor is leveled at the time of its construction, the column shortening that has occurred prior to the construction of that floor is compensated. Also, the creep and shrinkage effects tend to be small because dead load accumulates incrementally over a 12- to 15-month construction period.

Creep is difficult to quantify because it is time-dependent. Initially the rate of creep is significant; it diminishes as time progresses until it eventually reaches zero. Because of sustained loads, the stress in concrete gradually gets transferred to the reinforcement with a simultaneous decrease in concrete stress.

Columns with different percentages of reinforcement and different volume-to-surface ratios creep and shrink differently. An increase in the percentage of reinforcement and volume-to-surface ratio reduces the strain due to creep and shrinkage under similar stresses. Differential shortening of columns induces moments in frame beams, resulting in gravity load transfer to adjacent columns. A column that has shortened less receives more load, thus compensating for the initial imbalance.

Differential rather than the *absolute* shortening of column is more significant. Relative displacement between columns occurs because of the difference between the P/A ratios of columns. P is the axial load on and A is the area of the column under consideration. If all columns in a building have the same area and are sized for gravity load requirement only, there will be no relative vertical movement between the columns. All columns will undergo the same displacement because the P/A ratio is nearly constant for all columns. In a building, this condition is seldom present. This is because typically in building design, not all columns are designed for the same combination of loads. For example, the design of frame column is governed by the combined gravity and lateral loads while nonframe columns are designed for gravity loads only. This results in a large difference in the P/A ratios between the two sets of columns. Differential column shortening between perimeter and interior columns can produce floors that slope excessively. Since architectural partition walls, doors, and ceilings are normally built plumb and level, respectively, problems will result. Also, see Section 8.5, Floor-Leveling Problems.

Consider, for example, a steel tubular system with closely spaced exterior columns and widely spaced interior columns. High-strength steel up to 65 ksi is used for the interior column design, and because of large tributary areas and the desire to minimize column sizes, the resulting P/A ratios are large. The exterior columns, on the other hand, usually have a small P/A ratio for two reasons. First, their tributary areas are small because of their close spacing of usually 5 to 12 ft (2.44 to 3.66 m). Second, the columns are sized to limit lateral displacements, resulting in areas much in excess of those required for strength consideration alone. Because of this imbalance in the gravity stress level, these two groups of columns undergo different axial shortenings; the interior columns shorten much more than the exterior columns.

A reversed condition occurs in buildings with interior-braced core columns and widely spaced exterior columns; the exterior columns experience more axial shortening than the interior columns. The behavior of columns in buildings with other types of structural systems, such as interacting core and exterior frames, tends to be somewhere in between these two limiting cases.

In all these cases, it is relatively easy to evaluate the shortening of columns. The procedure requires a step-by-step manipulation of the basic PL/AE equation.

Having obtained the axial shortening values of all columns in a building, the next step is to assign column length correction Δ_c for each column. The objective is to attain as level a floor as practical. Δ_c is thus the difference between the specified theoretical height of a given column and its actual height after it has shortened. The magnitude of correction in a typically tall building of 30 to 60 stories is rather small, perhaps $\frac{1}{8}$ in. (3.17 mm) per floor, at the most. Therefore, instead of specifying this small correction at each level, in practice it is usual to lump the corrections of a few floors to stipulate the required correction. For example, in lieu of $\frac{1}{8}$-in. correction at every level one would specify 1 in. (25.4 mm) at every eighth floor.

Let us consider a typical column of a tall building with variations in story heights, gravity loads, and areas up the height, as shown in Fig. 8.32a. The axial shortening of the column at level n, denoted as Δ_n, is given by the following equation:

$$\Delta_n = \frac{1}{E} \sum_{k=1}^{n} \frac{L_k}{A_k} \sum_{i=k}^{NS} P_i$$

where

Δ_n = axial shortening at level n
P_i = axial load increment
L_k = column height at story k
A_k = column area at story k
NS = number of stories

To illustrate this rather trivial procedure, consider a column that is N stories high, with a constant cross-sectional area A, subjected to a constant load P at each floor. See Fig. 8.32b. The above simplifications are not valid in a practical column but keep the explanation simple.

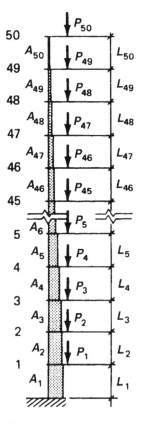

Figure 8.32a. Axial shortening computations for a practical column.

It is evident that the axial shortening Δ_1 at level 1 is equal to the total load at that level multiplied by $\frac{L}{AE}$.

Thus,

$$\Delta_1 = \frac{L}{AE}(NS)P$$

Similarly,

$$\Delta_2 = \Delta_1 + \frac{L}{AE}(NS - 1)P$$

$$\Delta_3 = \Delta_2 + \frac{L}{AE}(NS - 2)P$$

$$- - - - - - - - - - - -$$

$$- - - - - - - - - - - -$$

$$\Delta_n = \Delta_{n-1} + \frac{L}{AE}(NS + 1 - n)P$$

$$\Delta_{ns} = \Delta_{ns-1} + \frac{L}{AE}P$$

In a similar manner, expressions for $\Delta_1, \Delta_2, \Delta_3, \ldots \Delta_n \ldots \Delta_{ns}$ for a practical column shown in Fig. 8.32a with a cross-sectional area decreasing up the height in a stepwise manner,

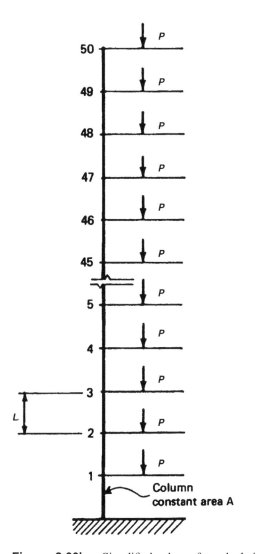

Figure 8.32b. Simplified column for calculation column shortening.

and subjected to different axial loads, may be derived as follows:

$$\Delta_1 = \frac{L_1}{A_1 E} \sum_{i=1}^{ns} P_i$$

$$\Delta_2 = \Delta_1 + \frac{L_2}{A_2 E} \sum_{i=2}^{NS} P_i$$

$$\Delta_3 = \Delta_2 + \frac{L_3}{A_3 E} \sum_{i=3}^{NS} P_i$$

$$- - - - - - - - - -$$
$$- - - - - - - - - -$$

$$\Delta_n = \Delta_{n-1} + \frac{L_n}{A_n E} \sum_{i=n}^{NS} P_i$$

$$- - - - - - - - - -$$
$$- - - - - - - - - -$$

$$\Delta_{NS} = \Delta_{NS-1} + \frac{L_{ns}}{A_{ns} E} \sum_{i=ns}^{ns} P_i$$

Table 8.1 shows in a tabular form the computations performed using the above procedure. The assumed column sections, lengths, and variation of axial loads are given in the table. The last column shows the lumped corrections at levels 2, 10, 20, 30, and 40, and the roof. Basically, these corrections represent the lengths to be added to the theoretical lengths of columns to achieve equal heights after the columns have shortened due to gravity loads. For example, $\Delta_c = 1\frac{1}{4}$ in. (31.75 mm) at the tenth level means that the actual fabricated length of column from its base to the tenth level should be made $1\frac{1}{4}$ in. longer than the theoretical length. This overlength could be achieved by increasing the length of column in each tier by $\frac{1}{4}$ in. (6.35 mm) (ten stories equal five tiers; therefore, $\frac{1}{4}$ in. times 5 gives $1\frac{1}{4}$ in.). However, the fabricator may elect to increase the column length in each story by $\frac{1}{8}$ in. (3.2 mm) instead of $\frac{1}{4}$ in. per tier. This and other similar options are, of course, permissible because the end result of achieving a desired Δ_c at the tenth floor will be the same.

The value of $\Delta_c = 2$ in. (50.8 mm) at the 20th floor means the overlength of columns between levels 1 and 20 should be 2 in. However, an overlength of $1\frac{1}{4}$ in. (31.75 mm) up to the tenth level has already been achieved by specifying $\Delta_c = 1\frac{1}{4}$ in. at the tenth level. Therefore, the increment between the 10th and 20th levels should be 2 in. less $1\frac{1}{4}$ in. $= \frac{3}{4}$ in. (19.0 mm).

8.4.1. Simplified Method

In a steel building, typically the cross-sectional area of a column increases in two-story increments from a minimum at top to a maximum at the base as shown in Fig. 8.32c. The incremental steps in column areas are due to the finite choice of column shapes. Similarly, the axial load on a column increases at each floor in a stepwise manner up the building height. In tall buildings the significance of these incremental steps diminishes rapidly, allowing us to make the following assumptions which can be used to derive a simplified formulation for axial shortening of columns. The first relates to the axial load variation due to gravity loads that may be assumed to increase linearly from top to the bottom. The second is similar to the first but applies to variation of column areas. However, a linear variation using the actual column area at the bottom appears to underestimate the actual shortening of columns. A slight modification in which the column area at the bottom is taken as 0.9 times the actual area $(0.9 \times A_b)$ appears to work well in predicting axial shortenings. (See Figs. 8.32d and 8.32e.)

8.4.1.1. Derivation of Closed-Form Solution

The notations used in the derivation of a closed-form solution (Fig. 8.32f) are as follows:

L = height of the building (note previously in the longhand method, notation L was used to denote story height)

Δ_z = axial shortening at a height x (also denoted as z) above foundation level

A_t = column area at top

A_b = modified column area at bottom equal to $0.9 \times$ actual area of column at bottom
    $= 0.9 \times A_B$

A_x = area of column at height x (also denoted as z) above foundation level

α = rate of change of area of column

P_t = axial load at top

P_b = axial load at bottom

P_x = axial load at height x above foundation

β = rate of change of axial load

TABLE 8.1   Axial Shortening Computations for Practical Column

Level	$\sum P_i$ Accumulated load, kips	Column section	L_k Story height, in.	Δ_n Column shortening, in.	Column length correction each level, in.	Lumped column length correction, in.	Column shortening, in.
50	53	W14 × 43	156	5.14	0.023	0.73	5.11
49	106	43	210	5.12	0.061		5.08
48	159	53	168	5.05	0.051		5.02
47	212	53	156	5.00	0.073		4.95
46	265	68	156	4.93	0.071		4.89
45	318	68	156	4.86	0.086		4.82
44	371	84	156	4.77	0.081		4.75
43	424	84	156	4.69	0.092		4.67
42	477	95	156	4.60	0.092		4.59
41	530	95	156	4.51	0.102		4.50
40	583	111	156	4.41	0.09	1.02	4.42
39	636	111	156	4.32	0.105		4.33
38	689	127	156	4.21	0.09		4.24
37	742	127	156	4.12	0.107		4.15
36	795	142	156	4.01	0.103		4.04
35	848	142	156	3.91	0.109		3.96
34	901	167	156	3.80	0.09		3.86
33	954	167	156	3.71	0.105		3.76
32	1007	176	156	3.62	0.105		3.66
31	1060	176	156	3.50	0.110		3.56
30	1113	202	156	3.39	0.101	1.07	3.46
29	1166	202	156	3.29	0.106		3.36
28	1219	211	156	3.19	0.106		3.26
27	1272	211	156	3.08	0.110		3.16
26	1325	228	156	2.97	0.106		3.12
25	1378	228	156	2.86	0.111		2.95
24	1431	246	156	2.75	0.107		2.85
23	1484	246	156	2.65	0.111		2.74
22	1537	264	156	2.53	0.107		2.64
21	1590	264	156	2.43	0.110		2.53
20	1643	287	156	2.32	0.104	1.06	2.42
19	1696	287	156	2.20	0.108		2.32
18	1749	314	156	2.10	0.101		2.21
17	1802	314	156	2.00	0.105		2.10
16	1855	314	156	1.90	0.108		1.99
15	1908	314	156	1.79	0.111		1.88
14	1961	342	156	1.68	0.104		1.78
13	2014	342	156	1.58	0.107		1.67
12	2067	370	156	1.47	0.101		1.56
11	2120	370	156	1.37	0.104		1.45
10	2173	370	156	1.26	0.107	0.53	1.34
9	2226	370	156	1.16	0.109		1.23
8	2279	398	156	1.05	0.104		1.12
7	2332	398	156	0.94	0.107		1.01
6	2385	398	156	0.84	0.11		0.89
5	2488	398	210	0.73	0.11		0.78
4	2491	426	168	0.62	0.11		0.67
3	2544	426	156	0.51	0.11		0.56
2	2597	500	156	0.40	0.09		0.45
Mezzanine	2650	500	240	0.31	0.15		0.17
1	2770	W14 × 500	240	0.16	0.16		0.17

NS = Number of stories = 50.

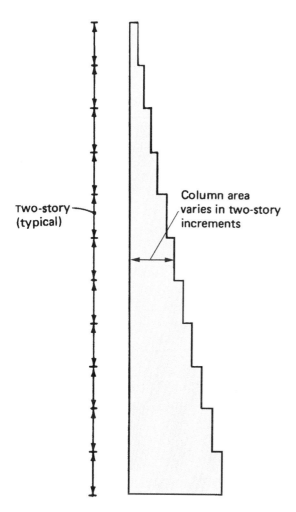

Figure 8.32c. Variation of cross-sectional area of a high-rise column.

ε_x = axial strain at height x

E = modulus of elasticity

The area of column at height z is given by

$$A_x = A_t \frac{x}{L} + A_b\left(1 - \frac{x}{L}\right)$$

$$= A_b - (A_b - A_t)\frac{x}{L}$$

$$= A_b - \alpha x$$

where

$$\alpha = \frac{A_b - A_t}{L}$$

The axial load at height z above foundation is given by

$$P_x = P_t \frac{x}{L} + P_b\left(1 - \frac{x}{L}\right)$$

$$= P_b - \beta x$$

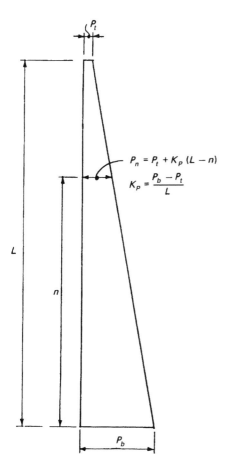

Figure 8.32d. Idealized gravity load distribution on a column.

where

$$\beta = \frac{P_t - P_t}{L}$$

The axial strain

$$\varepsilon_x = \frac{Px}{A_x E}$$

Using vertical work:

$$P_z^1 \Delta_z = \int_0^z P_x^1 \varepsilon_x dx$$

with

$$P_z^1 = 1 = P_x^1,$$

$$\Delta_z = \frac{1}{E} \int_0^z \frac{P_b - \beta x}{A_b - \alpha x} dx$$

$$= \frac{P_b}{E} \int_0^z \frac{dx}{A_b - \alpha x} - \frac{\beta}{E} \int_0^z \frac{x dx}{A_b - \alpha x}$$

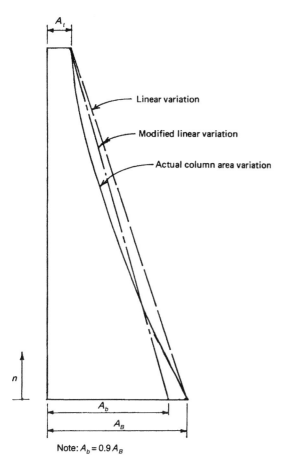

Figure 8.32e. Idealized column cross-sectional areas.

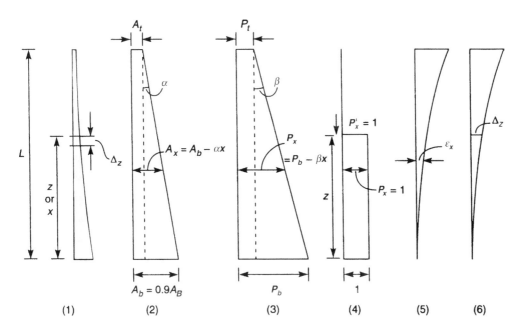

Figure 8.32f. Axial shortening of columns; closed-form solution: (1) axial shortening Δ_z; (2) column area; (3) column axial load; (4) unit load at height z; (5) axial strain; (6) axial displacement.

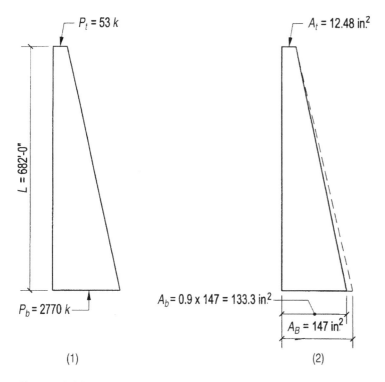

Figure 8.32g. Example 1, column-shortening calculations: (1) axial load variation; (2) actual and assumed variation in column cross-sectional areas.

Evaluating the integrals we get the following final expression for Δ_z.

$$\Delta_z = \frac{P_b}{E}\left[-\frac{1}{\alpha}\ln\left(1-\frac{\alpha z}{A_b}\right)\right] - \frac{\beta}{E}\left[-\frac{1}{\alpha^2}\left\{az + A_b\ln\left(1-\frac{\alpha z}{A_b}\right)\right\}\right]$$

Example 1.

Given. (See Fig. 8.32g)
Height of building: $L = 682$ ft $= 8184$ in. (207.8 m)
Modulus of elasticity: $E = 29,000$ ksi (200×10^3 MPa)
Axial load at top: $P_t = 53$ kips (237.5 kN)
Area of column at top: $A_t = 12.48$ in.^{2} (8052 mm^2)
Axial load at base: $P_b = 2770$ kips (12.32×10^3 kN)
Actual column area at base: $A_B = 147$ in^2. (94.84×10^3 mm^2)
Reduced column area at base: $A_b = 0.9 \times 147 = 133.3$ in^2. (86.0×10^3 mm^2)
Required. Axial shortening of column at top.
Solution. Since column shortening is calculated at top, $z = L$.

$$\alpha = \frac{A_b - A_t}{L} = \frac{133 - 12.48}{8184} = 0.01476 \text{ in.}^2/\text{in.}$$

$$\beta = \frac{P_b - P_t}{L} = \frac{2770 - 53}{8184} = 0.332 \text{ kip/in.}$$

$$\ln\left(1 - \frac{\alpha L}{A_b}\right) = \ln\left(1 - \frac{0.01476 \times 8184}{133.3}\right)$$
$$= \ln(0.09362)$$
$$= -2.36847$$

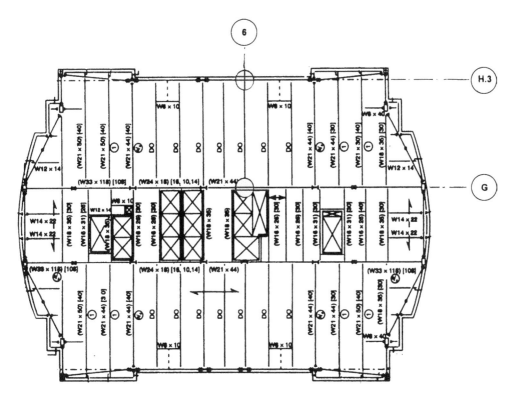

Figure 8.32h. Example 2, differential shortening of columns H.3–6 and and G.6; schematic framing plan.

$$\Delta_L \text{ at top} = \frac{2770}{29000}\left\{-\frac{1}{0.01476}\times(-2.36847)\right\} - \frac{0.332}{29000}$$
$$\times\{-4590.15(0.01476\times 8184 + 133.3\times -2.36847)\}$$
$$= 15.327 - 10.2$$
$$= 5.127 \text{ in.}$$

Similarly, the axial shortening is calculated at various heights by substituting appropriate values for z. The results given in column 8 of Table 8.1 agree closely with those from the longhand method. The appropriateness of the closed-form solution is obvious.

Example 2.

Given. A steel building 403 ft tall with 31 framed levels including the roof. Tributary areas for gravity load calculations for the exterior column H.3 = 472 ft^2 per floor, and for the interior column G.6 is 810 ft^2. See Fig. 8.32h for a schematic framing plan, and Table 8.2 for an abbreviated column schedule.

Typical loads for estimating axial shortening columns are as follows:

Interior Col G.6

3 ¼ lt. wt. on 3" deck	= 50 psf
Partitions	= 10 psf
Allowance for floor finishes, ceiling, mech., etc.	= 10 psf
Structural frame	= 10 psf
Live load	= 15 psf

$$95 \text{ psf} \times \frac{810}{1000} = 76.9 \text{ kips/floor}$$

use 77 kips/floor

TABLE 8.2 Example 2: Column Schedule

Level	Interior column	Area (in.²)	Exterior column	Area (in.²)
R	W14 × 68	20	W30 × 211	62
28	W14 × 90	26.5	W30 × 235	69
26	W14 × 120	35.3	W30 × 261	76.7
24	W14 × 132	38.8	W30 × 292	85.7
22	W14 × 145	42.7	W30 × 326	95.7
20	W14 × 193	56.7	W30 × 326	95.7
18	W14 × 233	68.5	W30 × 357	104
16	W14 × 257	75.6	W30 × 357	104
14	W14 × 283	83.3	W30 × 391	114
12	W14 × 311	91.4	W30 × 391	114
10	W14 × 342	101	W30 × 433	127
8	W14 × 370	109	W30 × 433	127
6	W14 × 398	117	W30 × 477	140
4	W14 × 426	125	W30 × 477	140
2	W14 × 455	134	W30 × 526	154
G	W14 × 500	147	W30 × 581	170

Exterior Col H.3–6

3 ¼ lt. wt. on 3″ deck	= 50 psf
Partitions	= 10 psf
Allowance for floor finishes, ceiling, mech., etc.	= 10 psf
Structural frame	= 15 psf
Live load	= 15 psf
Exterior cladding	= $\underline{5}$ psf

$$105 \times \frac{472}{1000} = 49.56 \text{ kips/floor}$$

use 50 kips/floor

Required. Compute axial shortening of columns H.3-6 and G-6 at the roof due to gravity loads using the closed-form equation given earlier. Provide column length corrections at levels 8, 16, 24, and the roof.

Solution. The variation of axial loads and cross-sectional areas for the two columns are shown in Figs. 8.32i and 8.32j. Observe that A_b is the actual area of the column at the foundation level multiplied by a factor of 0.9.

Thus A_b for the interior column = 0.9 × 147 = 132.3 in.²

A_b for the exterior column = 0.9 × 170 = 153. in.²

The loads P_b for the exterior and interior columns at foundation level are:

P_b = 31 × 50 = 1550 kips (exterior column), and
P_b = 31 × 77 = 2387 kips (interior column)

Load P_t = 50 kips for the exterior, and 77 kips for the interior column.

Column Length-Shortening Computations for Column G.6 (Interior Column)

L = 4836 in.
E = 29000 ksi
P_t = 77 kips

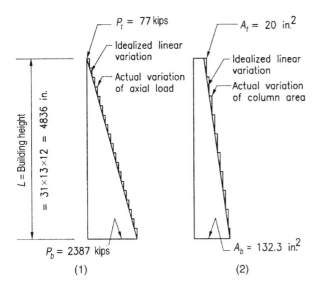

Figure 8.32i. Example 2, interior column G-6: (1) axial load variation; (2) variation of cross-sectional areas.

$P_b = 2387$ kips
$A_t = 20$ in.^{2}
$A_b = 0.9 \times 147 = 132.3$ in.^{2}

$$\alpha = \frac{A_b - A_t}{L} = \frac{132.3 - 20}{4836} = 0.02322 \text{ in.}^2/\text{in.}$$

$$\beta = \frac{P_b - P_t}{L} = \frac{2387 - 77}{4836} = 0.4777 \text{ kip/in.}$$

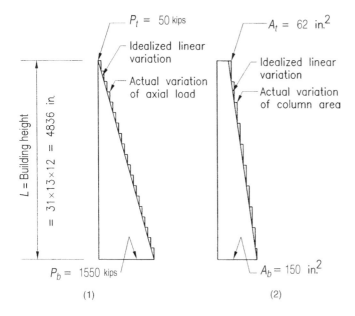

Figure 8.32j. Example 2, exterior column H.3–6: (1) axial load variation; (2) variation of cross-sectional areas.

$$l_n\left(1 - \frac{\alpha L}{A_b}\right) = l_n\left(1 - \frac{0.02322}{132.3} \times 4836\right)$$

$$= l_n(0.151233)$$

$$= -1.8889$$

$$\Delta_L = \frac{2387}{29000}\left[-\frac{1}{0.02322} \times -1.8889\right]$$

$$-\frac{0.4777}{29000}[-1854.7\{0.02322 \times 4836 + 132.3\,(-1.8889)\}]$$

$$= 6.6958 - 4.1869$$

$$= 2.50 \text{ in.}$$

Column Length-Shortening Calculations for Column H.3–6 (Exterior Column)

$L = 31 \times 13 \times 12 = 4836$ in.
$E = 29000$ ksi
$P_t = 50$ kips
$P_b = 1550$ kips
$A_t = 62$ in.^{2}
$A_b = 0.9 \times 170 = 153$ in.^{2}

$$\alpha = \frac{A_b - A_t}{L} = \frac{153 - 62}{4836} = 0.018817 \text{ in.}^2/\text{in.}$$

$$\beta = \frac{P_b - P_b}{L} = \frac{1550 - 50}{4836} = 0.3102 \text{ kip/in.}$$

$$l_n = \left(1 - \frac{\alpha L}{A_b}\right) = l_n\left(1 - \frac{0.018817 \times 4836}{153}\right)$$

$$= l_n(0.4052)$$

$$= -0.90328$$

$$\Delta_L = \frac{1550}{29000}\left[-\frac{1}{0.018817} \times -0.90328\right]$$

$$-\frac{0.3102}{29000}[-2824.2\{0.018817 \times 4836 + 153(-0.90328)\}]$$

$$= 2.56 - 1.42 = 1.134 \text{ in.}$$

The required column length corrections are shown in Table 8.3.

8.4.2. Column Shortening Verification During Construction

This concept is best explained with reference to Fig. 8.32k that shows a framing plan for a hypothetical building, say, some 48 stories tall. Identified therein are two columns: C_1, an interior column with a large tributary area; and C_2, an exterior column of framed tube with a relatively small tributary area. Under gravity loads, C_1 would shorten more than C_2 because: 1) C_1, designed only for gravity loads, has a P/A ratio that is relatively high; and 2) C_2, designed as a frame column, has its P/A ratio significantly less than that for C, because it is lightly loaded under gravity loads.

TABLE 8.3 Column Length Correction

Level	Interior column		Exterior column	
	Column shortening	Correction to scheduled length	Column shortening	Correction to scheduled length
Roof	2.50 in.	4@ ³⁄₁₆ in. = 0.75 in.	1.13 in.	2@ ⅛ in. = 0.25 in.
24	1.80 in.	4@ ⅛ in. = 0.50 in.	0.75 in.	2@ ⅛ in. = 0.25 in.
16	1.25 in.	4@ ³⁄₁₆ in. = 0.75 in.	0.50 in.	2@ ³⁄₁₆ in. = 0.375 in.
8	0.625 in.	4@ ⅛ in. = 0.50 in.	0.25 in.	2@ ⅛ in. = 0.25 in.

Assume that you as the engineer for the project have specified column length corrections to C_1 at levels 8, 16, 24, 32, 40, and 48 with correction of 2 in. specified at level 24. Let us say that when steel erection is at that level, i.e., at level 24, the contractor takes an elevation survey of the columns, reports the top of column C_1 is 1 in. higher than the top of C_2, and requests the engineer to confirm if this is acceptable in view of the fact that additional shortening of the column is yet to occur.

Further calculations are needed to verify that this 1-in. overlength of C_1 will take place after the application deadloads at levels 24 through roof. This concept of verifying the overlength of columns during construction is shown in Fig. 8.32l. Note that Δ_{Rn} shown therein corresponds to the 1 in. discussed here for the hypothetical building. For a column with a gradually decreasing area up the height, subjected to floor loads $P_1, P_2 \ldots P_{Roof}$, Δ_{Rn} at level n can be shown to be:

$$\Delta_{Rn} = \frac{1}{E} \sum_{i=n}^{NS} P_i \sum_{i=1}^{n} \frac{L_i}{A_i}$$

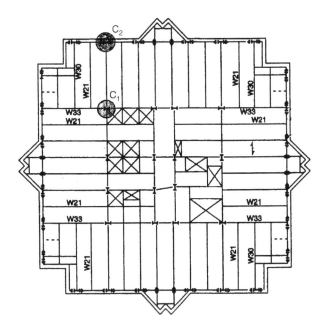

Figure 8.32k. Framing plan. Column C_1, designed for gravity loads only, shortens more than C_2, designed for both gravity and lateral loads. Compensating for relative elevation difference between these two categories of columns is of imortance in tall buidings.

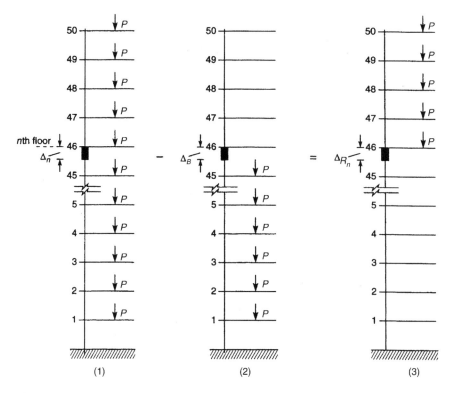

Figure 8.32l. Physical interpretation of column overlength: (1) column shortening due to loads at all floors; (2) column shortening due to loads below the nth level; (3) column residual overlength $\Delta_{R_n} = \Delta_n - \Delta_B$.

8.5. FLOOR-LEVELING PROBLEMS

Although not a safety issue, achieving level floors has become more arduous in modern buildings because stronger materials combined with more refined designs have resulted in lighter floor construction that is more prone to deflections than in the earlier heavier buildings. Engineers face considerable challenge in providing level floors because many of the variable factors encountered in practice are difficult to define and defy exact numerical solutions.

In steel office buildings, composite floor beams for typical 30–42-ft spans are often specified and delivered with a predetermined camber, while in concrete buildings the camber is built into the form work. Usually the specified camber for composite beams ranges from a minimum of $\frac{1}{2}$ in. (12.7 mm) to a maximum of 2.5 in. (63.5 mm). Cambers smaller than $\frac{1}{2}$ in. (12.7 mm) are difficult to achieve, while cambers substantially greater than 2.5 in. (63.5 mm) will result in other constructability problems. Cambers are specified anticipating that dead loads imposed on the floors will overcome the camber, resulting in a level floor. This is not always the case because: 1) Steel beam rolling and construction tolerances combined with long-term effect of creep of concrete affect the final result up or down in both steel and concrete construction; 2) Usually, camber is calculated as if a beam were pin-connected or completely fixed, depending upon the type of connections specified at the beam ends (Actual conditions vary. For instance, even with simple shear tab connections, composite steel beams experience partial fixity. Depending upon the degree of fixity, the final result could vary again up or down); 3) Columns in buildings shorten elastically due to gravity loads, and the magnitude of shortening between an interior and exterior column or between any two adjacent exterior columns is likely to be different, compounding floor-leveling problems further.

Because of these variable factors, combined with the fact that none of these is mathematically determinable, it is almost an accident if the floor turns out to be perfectly level. The problem comes to light at the time of interior finishing of the space when ceiling and partitions are being installed. One sure method of obtaining a level floor is to float the floor to remove the bumps and fill the low spots. Cement-based self-leveling underlayments are used for this purpose. In a floor built to commercially acceptable tolerance, the average fill over the entire floor area should not exceed $\frac{1}{2}$ in. (12.7 mm), which translates into an additional dead load of 6 psf. Depending upon the type of construction, this additional load may represent an increase of 3 to 6% of the total allowable stress design (ASD) or service loads. It is recommended that an allowance be made for this additional load in the design, irrespective of whether or not the floor is floated with the underlayment.

The most commonly specified tolerance for finished floor slab surfaces is $\frac{1}{8}$ in. (3.7 mm) in 10 ft (3.048 m), which is considered too stringent for most uses. The reasons for unlevelness are many, including form-work sagging, deflection of members due to dead and live loads, finishing irregularities, or tolerances allowed in setting of steel beams or form work. As a result the as-built surface of the floor always exhibits bumps and dips.

In recognition of this problem, the American Concrete Institute has revised its "Standard Tolerances for Concrete Construction and Materials" (ACI 117). The standard includes floor finish tolerances based on two measuring methods: the F-number system and the straight-edge method. F numbers describe floor flatness. The larger the F number, the flatter the floor. An F-60 floor is roughly twice as flat as an F-30 floor.

8.6. FLOOR VIBRATIONS

8.6.1. General Discussion

Building floors are subjected to a variety of vibrational loads that come from building occupancy. Although almost all loads except dead loads are nonstatic, internal sources of vibration that might be a cause of concern in an office or a residential building are the oscillating machinery, passage of vehicles, and various types of impact loads such as those caused by dancing, athletic activities, and even pedestrian traffic. The trend in the design of floor framing systems of high rises is for long spans using structural systems of minimum weight. To this end, high-strength steel with lightweight concrete topping is routinely employed. With the use of lightweight concrete, most building codes allow for a reduction in the thickness of slab required for fire rating. This results in a further reduction in the mass and stiffness of the structural system, thereby increasing the period of the structure, which at times may approach the period of the source causing the vibration. Resonance may occur, causing large forces and amplitudes of vibration.

The performance of floor systems can be greatly improved by adding nonstructural elements such as partitions and ceilings, which contribute greatly to the damping of vibrations. Nonstructural elements may also add to mass and stiffness to produce the desired degree of solidity. Although the essential requirement in establishing the adequacy of a floor system is its strength, large deflections and strongly perceptible vibrations can be objectionable for several reasons: 1) Excessive deflections and vibrations may give the user the negative impression that the building is not solid. In retail areas, for example, the china may rattle every time someone goes by, or mirrors in dressing rooms of clothing stores may shake, giving the customer the somewhat nebulous but real feeling that the structure is not solid. In extreme cases, vibration may cause damage to the structure as a result of loosening of connections, brittle fracture of welds, etc. It is therefore important that the structure be able to absorb impact forces and vibrations but not respond with humanly perceptible shaking or bouncing. Monolithic concrete buildings are more solid in this respect as compared to light-framed buildings with steel or precast concrete; 2) Excessive deflection may result in curvature or misalignments perceptible to the eye; 3) Large deflections may result in fracture of more recently installed architectural elements such as plaster or masonry; and 4) Large deflections may result in the transfer of load to nonstructural elements such as curtain wall frames.

It is difficult to establish general criteria related to perception of vibrations. Feeling of bounciness varies from person to person, and what is objectionable to some may be barely noticeable to others. Among the criteria employed in the design of floor systems are limitations on the span-to-depth ratio and flexibility, which normally lead to deeper sections than would be required from strength considerations alone. It is somewhat dubious that these limitations assure occupants' comfort.

Recognizing that there is no single scale by which the limit of tolerable deflection can be defined, the AISC specification does not specify any limit on the span-to-depth ratios for floor framing members. However, as a guide, the commentary on the specification recommends that the depth of fully stressed beams and girders in floors should not be less than $(F_y/800)$ times the span. If beams of lesser depth are used, it is recommended that the allowable bending stresses be decreased in the same ratio as the depth. Where human comfort is the criterion for limiting motion, the commentary recommends that the depth of steel beams supporting large open floor areas free of partitions and other sources of damping should not be less than one-twentieth of the span, to minimize perception of transient vibration due to pedestrian traffic.

Thus there is no clear-cut requirement on the flexibility to limit the perception of vibration by occupants. Flexibility limits are given, however, from other considerations such as fracture of architectural elements like plaster ceilings. The rule-of-thumb limitations are 1/150 to 1/180 of the span for visibly perceptible curvature and 1/240 to 1/360 of the span for curvature likely to result in fracture of applied ceiling finishes.

In the design of floor systems, fatigue damage due to transient vibrations is not a consideration because it is tacitly assumed that the number of cycles to which the floor system is subjected is well within the fatigue limitations. However, damage due to fatigue can be a cause of concern in floors subjected to aerobic exercise activities.

Human response is directly related to the characteristics of the vertical motion of the floor system. Users perceive floor vibrations more strongly when standing or sitting on the floor than when walking across it. Human response to vibration seems to be a factor for consideration in design only when a significant proportion of the users will be standing, walking slowly, or seated.

Most of the experiments done on human response to vibrations are related to the physical safety and performance abilities of physically conditioned young subjects in a vibrating environment such as the research supported by NASA and various defense agencies. Very little information is available on the comfort of humans subjected to unexpected vibrations during the course of their normal duties such as slowly walking across a floor or sitting at a desk. Comfort is a subjective human response and defies scientific quantification. Different people report the same vibrations to be perceptible, unpleasant, or even intolerable. A measure for human response to steady sinusoidal vibration (taken from Ref. 60) is shown in Fig. 8.33a. Although there is no simple physical characteristic of vibration that completely defines the human response, there is enough evidence to suggest that acceleration associated in the frequency range of 1 to 10 Hz is

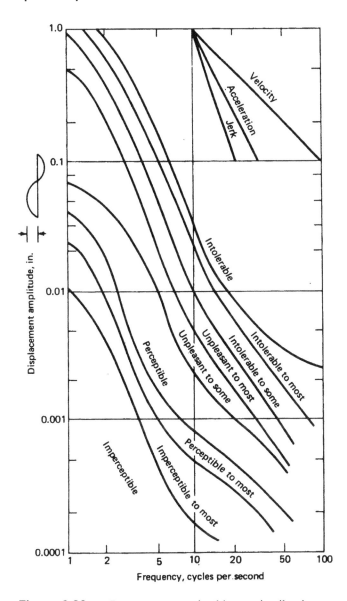

Figure 8.33a.   Response to sustained harmonic vibration.

the preferable criterion. This is the range for normally encountered natural frequencies of floor beams. Investigations have shown that human susceptibility to building floor vibrations is influenced by the rate at which the vibrations decay; people tend to be less sensitive to vibrations that decay rapidly. In fact, experiments have shown that people do not react to vibrations that persist for fewer than five cycles.

8.6.2.   Response Calculations

The procedures for evaluating floor vibration response given in the following sections are based on methods available prior to the publication of AISC/CISC Design Guide No. 11, *Floor vibrations Due to Human Activity*, by T. M. Murray, D. E. Allen, and E. E. Ungar (1997). The designer is referred to this publication for the most current recommendations.

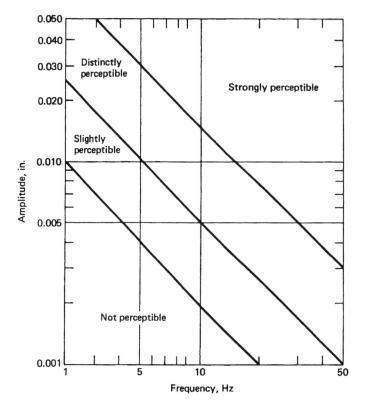

Figure 8.33b.   Reiher-Meister vibration criteria.

Human response ratings to a steady state of vibrations as originally documented by two researchers, Reiher and Meister (Ref. 60), have been found to be too severe for the design of building floors subject to transient vibrations caused by human activity. Lenzen (Ref. 61) has modified the Reiher and Meister rating scale by multiplying the amplitude scale by 10 to account for the nonsteady state of vibrations. The modified curves that account implicitly for damping are shown in Fig. 8.33b. In this figure the natural frequency f is plotted on the horizontal scale and the amplitude A_0 is plotted on the vertical scale.

The natural frequency f for a simply supported beam is given by the relation

$$f = 1.57 \sqrt{\frac{EI_b g}{W_d l^4}}$$

where
   f = frequency in cycles per sec
   E = the modulus of elasticity of the system in ksi
   I_b = transformed moment of inertia of the beam assuming full interaction with slab system in in.^{4}
   g = acceleration due to gravity, 386.4 in./s
   W_d = dead load tributary to beam in kips/in.
   l = effective span of beam, in in.

The design amplitude A_0 is obtained by modifying the initial amplitude of vibration of a simply supported beam subjected to the impact load of a 190-lb person executing a heel drop .The initial amplitude for the most common value of E of 29,000 ksi is given by the relation:

$$A_{0t} = (\text{DLF})_{\text{max}} \times \frac{l^3}{80 EI_b}$$

where $(\text{DLF})_{\text{max}}$ is the maximum dynamic factor which can be obtained from a graph given in Ref. 62.

Since a floor system usually consists of a number of parallel beams, Ref. 62 suggests that the design amplitude be obtained by dividing the initial amplitude by a factor N_{eff} to account for the action of multiple beams. Methods of estimating N_{aff} are given in Ref. 62.

The design procedure can thus be summarized as follows:

1. Compute the transformed moment of inertia of the beam under investigation. Use full composite action regardless of method of construction and assume an effective width equal to the sum of half the distances to adjacent beams. For composite beams on metal deck, use an effective slab depth that is equal in weight to the actual slab including concrete in valleys of decking and the weight of decking itself.
2. Compute the frequency from the relation $f = 1.57\sqrt{EI_b g/W_d l^4}$.
3. Compute the heel drop amplitude of a single beam by using the relation $A_{0t} = (\text{DLF})_{\text{max}} \times l^3/80EI_b$.
4. Estimate the effective number of beams, N_{eff} (Ref. 63), and compute the design amplitude by the relation $A_0 = A_{0t}/N_{\text{eff}}$.
5. Plot on the modified Reiher-Meister scale (Fig. 8.33b) the computed frequency f and the amplitude A_0.
6. Redesign if necessary.

Another response rating based on experimental data has been developed by Wiss and Parmelee (Ref. 47). In their method the response rating R is given as a function of frequency, peak amplitude, and damping. Based upon the computed value of R, the expected human response is classified into one of the five following categories:

1. Imperceptible $R<1.5$
2. Barely perceptible $1.5<R<2.5$
3. Distinctly perceptible $2.5<R<3.5$
4. Strongly perceptible $3.5<R<4.5$
5. Severe $R>4.5$

The response factor R is given by

$$R = 5.08(FA_0/D^{0.217})^{0.265}$$

where

R = response rating
F = frequency, in cycles per sec
A_0 = Displacement in in.
D = Damping ratio expressed as a ratio of actual damping to critical damping

The damping coefficient D, among other things, depends on the inherent characteristics of the floor, such as ceiling, duct work, flooring, furniture, and partitions. It should be noted that D cannot be determined theoretically but can only be estimated in relation to existing floors and their contents. For a rough estimate, the Canadian Standards Association suggests the following values:

Bare floors	$D = 0.03$
Finished floor with ceiling, mechanical ducts, flooring, and furniture	$D = 0.06$
Finished floor with partitions	$D = 1.13$

(The designer is reffered to Table 4.1 of AISC Design Guide No. 11 for revised values.)

Floor structures subjected to rhythmic activities such as dancing, aerobics, and other jumping exercises have been a source of annoyance to owners and engineers alike. Unlike vibration problems encountered in office occupancies, the vibrations due to rhythmic activities are continuous. These vibrations can be greatly amplified when periodic forces are synchronized with the floor frequency, a condition called resonance. Unlike transient vibrations, continuous vibrations may not decay. The National Building Code of Canada (NBC) in its commentary recommends that floor frequencies less than 5 Hz should be avoided for light residential floors, schools, auditoriums, gymnasiums, and other similar occupancies. It recommends a frequency of 10 Hz or more for very repetitive activities because of the possibility of getting resonance when the rhythmic beat is on every second cycle of vibration.

In a paper titled "Vibration Criteria for Assembly Occupancies," Allen, Rainer, and Pernica present a procedure for designing floor structures subjected to rhythmic activities. Briefly, the procedure is as follows:

1. Determine the density of occupancy based on type of activity. For example, if the floor area is 30 by 60 ft (9.15 by 18.3 m) and has an aerobic class of 50 people of average weight of 120 lb, the equivalent density of occupancy works out to be

$$\frac{50 \times 120}{30 \times 60} = 3.33 \text{ psf} \quad (159.6 \text{ Pa})$$

2. Choose an appropriate forcing frequency f and a dynamic load factor α. For aerobic exercises, the value of f suggested in the paper is between 1.5 and 3 Hz, while the value for α is given as 1.5.
3. Choose an acceptable limiting acceleration ratio, α_0/g, at the center of the floor. The suggested value for physical exercise activity is 0.05.
4. Determine the lowest acceptable fundamental frequency f_0 of the floor system by the relation:

$$f_0 \geq f \sqrt{1 + \frac{1.3}{\alpha_0/g} \frac{\alpha W_P}{W_t}}$$

where
w_p = weight per unit area of participants
w_t = total weight per unit area of structure, participants, furniture, etc.

5. Determine the natural frequency f_0 of the floor structure. In addition to the weight of the floor structure itself, weights of participants and furniture, if any, are to be included in the computation of f_0.
6. The frequency f_0 should be greater than or equal to the frequency obtained in step 4. If not, the options are to stiffen the floor system, install passive

tuned mass dampers between floor beams, relocate the activity, or persuade the owner to accept a higher limiting acceleration by pointing out that no serious safety-related problems are known to have occurred for floors with frequencies higher than 6 Hz.

Increasing the frequency of the floor system by increasing the stiffness is usually cost-prohibitive. The most prudent course is to make building owners aware of vibration-related problems during the early design phase.

8.7.   SEISMIC ISOLATION

Seismic isolation is a viable design strategy that has been used for seismic rehabilitation of existing buildings and in the design of a number of new buildings. In general, this system will be applicable to the rehabilitation and design of buildings whose owners desire superior earthquake performance and can afford the special costs associated with the design, fabrication, and installation of seismic isolators. The concepts are relatively new and sophisticated, and require more extensive design and detailed analysis than do most conventional schemes. In California, peer review of these new concepts is required for all designs that use seismic isolation.

Conceptually, isolation reduces response of the superstructure by "decoupling" the building from seismic ground motions. Typical isolation systems reduce seismic forces transmitted to the superstructure by lengthening the period of the building and adding some amount of damping. Added damping is an inherent property of most isolators, but may also be provided by supplemental energy dissipation devices installed across the isolation interface. Under favorable conditions, the isolation system reduces drift in the superstructure by a factor of at least two—and sometimes by as much as factor of five—from that which would occur if the building were not isolated. Accelerations are also reduced in the structure, although the amount of reduction depends on the force-deflection characteristics of the isolators and may not be as significant as the reduction of drift.

Reduction of drift in the superstructure protects structural components and elements as well as nonstructural components sensitive to drift-induced damage. Reduction of acceleration protects nonstructural components that are sensitive to acceleration-induced damage.

To understand the design principles for base-isolated buildings, consider Fig. 8.34a, which shows four distinct response curves A, B, C, and D. Let us examine the design of a building, say, some five stories tall, with a fixed-base fundamental period of 0.6 sec. Curve A, the lowest, shows lateral design forces resulting from loads prescribed in building codes such as IBC 2003 and ASCE 7-02. Curve B, the second lowest, represents the probable strength of the structure. This strength is generally greater than the design strength because of several factors. Chief among them are: 1) actual material strengths are almost always higher than those assumed in design; 2) use of load factors typically overestimates the actual loads imposed on the structure; 3) some conservatism is used in sizing of structural members; 4) designs are often based on drift limits; and 5) members are designed to have at least some ductility. It is estimated that the probable strength of a structure designed to code-level forces is about 1.5 to 2.0 times larger than the design strength.

Curve D at top shows the forces our fixed-base building would experience if it were to remain elastic for the entire duration of a design earthquake. However, in earthquake-resistant design, it is assumed that the lateral-force-resisting system will make

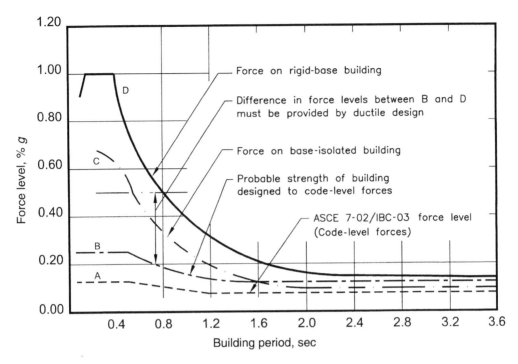

Figure 8.34a.   Design concept for base-isolated buildings: Top curve D shows the forces in the structure if it were to remain elastic during an earthquake. The conventional design approach is to build ductility into the structure to absorb the difference in forces between B and D. By providing seismic isolation, the maximum force experienced by the building, curve C, is reduced to its probable strength, curve B.

excursions well into the nonlinear inelastic capacities of the structural materials. Therefore, typical buildings are designed to resist only a fraction of the full linear elastic demands of major earthquakes. Heavy reliance is placed on special prescribed details that are presumed to provide ductility for the extreme nonlinear inelastic demands. The difference between the linear elastic demand, Curve D, and the probable capacity of the building, Curve B, conceptually represent the magnitude of energy dissipation expected of the structure.

Let us compare this to the energy dissipation required of the building, if it is seismically isolated. The elastic forces experienced by a seismically isolated building are significantly reduced for two reasons. First, the flexibility of the base isolators shifts the period of the building toward the low end of the spectrum. For instance, our example building with a fixed-base period of 0.6 sec would probably now have a period in the neighborhood of, say, 2 to 2.5 sec. The drop in the elastic design force, as seen in the graph, is considerable.

The second factor contributing to the reduction in force level is the additional damping provided by the dampers. Depending on the type of base isolater and supplemental viscous damper (if any) chosen for the building, the damping may increase from a generally assumed value of 5% of critical to as much as 20% or more. Together, these two factors help to reduce the ductility demand expected of the structure during a large seismic event. In fact, it is quite likely that our base-isolated structure may never be pushed beyond its elastic limit. In other words, in the 2.0- to 2.5-sec-period range, the probable strength of the building is very nearly the same as the maximum unreduced elastic demand. Therefore,

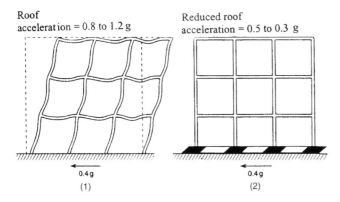

Roof
acceleration = 0.8 to 1.2 g

Reduced roof
acceleration = 0.5 to 0.3 g

0.4g
(1)

0.4g
(2)

Figure 8.34b. Comparison of response of a fixed-base and a base-isolated building: (1) fixed-base; (2) base-isolated. Base isolation typically reduces roof acceleration of low-rise buildings by about 60 to 80%.

the building need not take excursions into nonlinear inelastic range, and can remain elastic for the entire duration of a design earthquake.

In simple terms, seismic isolation involves placing a building on isolators that have great flexibility in the horizontal plane (Fig. 8.34b). The system consists of:

- A flexible mounting to increase the building period which, in turn, reduces seismic forces in the structure above.
- A damper or energy dissipater to reduce relative deflections between a building and the ground it rests upon.
- A mounting that is sufficiently rigid to control the building lateral deflection during minor earthquakes and wind storms.

To decrease base shear, flexibility can be introduced into the building by many devices, including elastomeric bearings, rollers, sliding plates, cable suspension, sleeved piles, and rocking foundations. However, decrease in base shear due to lengthening of a building's period comes at a price; the flexibility at the base gives rise to large relative displacements across the flexible mount. Hence, the necessity of providing additional damping at the base isolation level.

While a flexible mounting is required to isolate a building from seismic loads, its flexibility under frequently occurring wind and minor earth tremors is undesirable. Therefore, the device at the base must be stiff enough at these loads, such that the building's response is as if it were on a fixed base.

Generally one isolator per column is used. However, more than one isolator may be required in certain buildings. For isolation of shear walls, one or more isolators are used at each end, and if the wall is long, isolators may be placed along its entire length, the spacing depending upon the spanning ability of the wall between the isolators.

8.7.1. Salient Features

1. Access for inspection and replacement of bearings should be provided at bearing locations.
2. Stub-walls or columns to function as backup systems should be provided to support the building in the event of isolator failure.

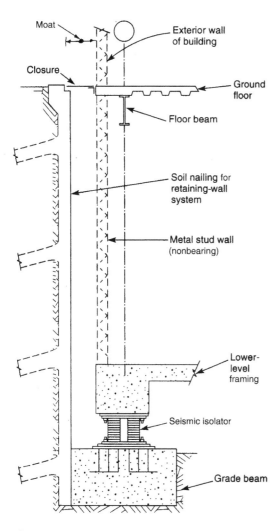

Figure 8.34c. Moat around base-isolated building.

3. A diaphragm capable of delivering lateral loads uniformly to each bearing is preferable. If the shear distribution is unequal, the bearings should be arranged such that larger bearings are under stiffer elements.
4. A moat to allow free movement for the maximum predicted horizontal displacement must be provided around the building (Figs. 8.34c and 8.34d).
5. The isolator must be free to deform horizontally in shear and must be capable of transferring maximum seismic forces between the superstructure and the foundation.
6. The isolators should be tested to ensure that they have lateral stiffness properties that are both predictable and repeatable. The tests should show that over a wide range of shear strains, the effective horizontal stiffness and area of the hysteresis loop are in agreement with values used in the design.

When earthquakes occur, the elastomeric bearings used for base isolation are subjected to large horizontal displacements, as much as 15 in. or greater in a 10-story steel-framed building. They must therefore be designed to carry the vertical loads safely at these displacements.

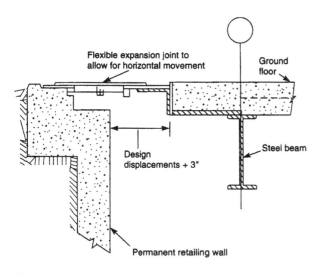

Figure 8.34d.   Moat detail at ground level.

Isolation systems should be considered for achieving the immediate occupancy structural performance level and operational nonstructural performance level. Conversely, isolation will likely not be an appropriate design strategy for achieving the collapse prevention structural performance level. In general, isolation systems provide significant protection to the building structure, nonstructural components, and contents, but at a cost that precludes practical application when the budget and design objectives are modest.

8.7.2.   Mechanical Properties of Seismic Isolation Systems

A seismic isolation system is the collection of all individual seismic isolators and may be composed entirely of one type of seismic isolator, a combination of different types of seismic isolators, or a combination of seismic isolators acting in parallel with energy dissipation devices (i.e., a hybrid system).

The most popular devices for seismic isolation in the United States may be classified as either elastomeric or sliding. Examples of elastomeric isolators include high-damping rubber bearings (HDR), low-damping rubber bearings (RB), or low-damping rubber bearings with a lead core (LRB). Sliding isolators include flat assemblies or those with a curved surface, such as the friction-pendulum system (FPS).

8.7.2.1.   Elastomeric Isolators

Elastomeric bearings represent a common means for introducing flexibility into structure. They consist of thin layers of natural rubber that are vulcanized and bonded to steel plates (see Fig. 8.34e). Natural rubber exhibits a complex mechanical behavior that can be described simply as a combination of viscoelastic and hysteretic behavior. Low-damping natural rubber bearings exhibit essentially linearly elastic and linearly viscous behavior at large shear strains. The effective damping is typically less than or equal to 0.07 for shear strains in the range of 0 to 2.0.

Lead-rubber bearings are generally constructed of low-damping natural rubber with a preformed central hole into which a lead core is press-fitted (see Figs. 8.34f and 8.34g). Under lateral deformation, the lead core deforms in almost pure shear, yields at low levels of stress (approximately 1160–1450 psi (8 to 10 Mpa) in shear at normal temperature),

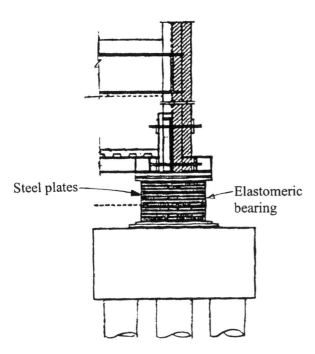

Figure 8.34e. High-damping rubber bearing, made by bonding sheets of rubber to thin steel plates. The steel plate increases vertical compressive stiffness of the unit while maintaining the desired low lateral stiffness.

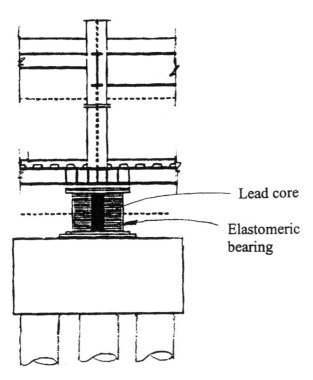

Figure 8.34f. Installation of a lead-rubber bearing under interior columns. A lead-rubber bearing (LRB) consists of one or more lead plugs inserted into holes preformed in low-damping rubber bearings. The lead core provides for energy dissipation by deforming plastically at a stress of 1500 psi (10 Mpa).

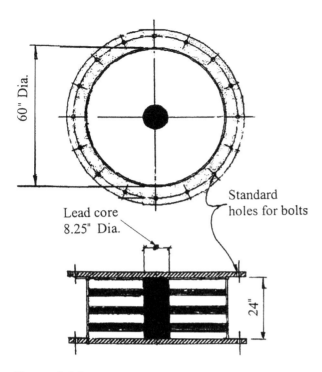

60" Dia.

Lead core
8.25" Dia.

Standard
holes for bolts

24"

Figure 8.34g. Lead-rubber bearing (LRB) for an interior column of a five-story steel framed building; approximate dimensions.

and produces hysteretic behavior that is stable over many cycles. Unlike mild steel, lead recrystallizes at normal temperature (about 20°C), so that repeated yielding does not cause fatigue failure. Lead-rubber bearings generally exhibit characteristic strength that ensures rigidity under service loads.

High-damping rubber bearings are made of specially compounded rubber that exhibits effective damping between 0.10 and 0.20 of critical. The increase in effective damping of high-damping rubber is achieved by the addition of chemical compounds that may also affect other mechanical properties of rubber.

Scragging is the process of subjecting an elastomeric bearing to one or more cycles of large amplitude displacement. The scragging process modifies the molecular structure of the elastomer and results in more stable hysteresis at strain levels lower than that to which elastomer was scragged. Although it is usually assumed that the scragged properties of an elastomer remain unchanged with time, recent studies suggest that partial recovery of unscragged properties is likely. The extent of this recovery is dependent on the elastomer compound.

8.7.2.2. Sliding Isolators

Sliding isolaters with either a flat or a single-curvature spherical sliding surface are typically made of PTFE or PTFE-based composites in contact with polished stainless steel. The shape of the sliding surface allows large contact areas that, depending on the materials used, are loaded to average bearing pressures in the range of 1015 to 10150 psi (7 to 70 Mpa).

Sliding isolaters tend to limit the transmission of force to an isolated structure to a predetermined level. While this is desirable, the lack of significant restoring force can result in significant variations in the peak displacement response, and can result in permanent offset displacements. To avoid these undesirable features, sliding isolators are typically used in combination with a restoring force mechanism.

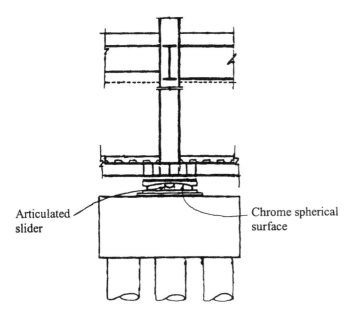

Figure 8.34h. Sliding bearing; friction-pendulum system (FPS). An FPS consists of an articulated slider that glides on a polished spherical concave chrome surface. Whereas in elastomeric base-isolated buildings $P\Delta$ effects are equally distributed between superstructure and foundation, in sliding base-isolated buildings the entire $P\Delta$ effect can be accommodated in either the superstructure or the foundation, depending on whether the spherical surface is attached to the foundation or the superstructure.

Combined elastomeric-sliding isolation systems have been used in buildings in the United States. Japanese engineers have also used elastomeric bearings in combination with mild steel elements designed to yield in strong earthquakes and enhance the energy dissipation capability of the isolation systems.

Details of a spherical sliding system commonly referred to as a friction-pendulum system (FPS) are shown in Figs. 8.34h, 8.34i, 8.34j(1), 8.34j(2), and 8.34j(3). Figure 8.34k shows a schematic of base-isolation devices acting in conjuction with viscoelastic dampers.

8.7.3. Seismically Isolated Structures: ASCE 7-02 Design Provisions

The procedures and limitations for the design of seismically isolated structures is determined considering zoning, site characteristics, vertical acceleration, cracked section properties of concrete and masonry members, seismic use group, configuration, structural system, and height. Both the lateral force-resisting system and the isolation system must be designed to resist the deformations and stresses produced by the effects of ground motions. The stability of the vertical load-carrying elements of the isolation system must be verified by analysis and tested for lateral seismic displacement equal to the total maximum displacement. All portions of the structure, including the structure above the isolation system, must be assigned a seismic use group based on ASCE 7-02 provisions with an occupancy importance factor taken as 1.0 regardless of its seismic use group categorization. Each structure must be designated as being regular or irregular on the basis of the structural configuration above the isolation system.

Three procedures are permissible: static analysis, response spectrum analysis, and time–history analysis. The static analysis procedure is generally used to start the design

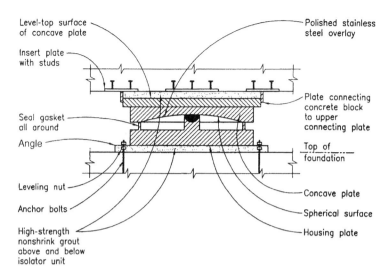

Level-top surface of concave plate

Insert plate with studs

Seal gasket all around

Angle

Leveling nut

Anchor bolts

High-strength nonshrink grout above and below isolator unit

Polished stainless steel overlay

Plate connecting concrete block to upper connecting plate

Top of foundation

Concave plate

Spherical surface

Housing plate

Figure 8.34i. Friction pendulum system details. The $P\Delta$ effect in this arrangement, where $\Delta =$ earthquake-induced displcement, is accounted for in the design of the superstructure. If the spherical plate is attached to the foundation, the $P\Delta$ effect is accounted for in the design of the foundation.

process and to calculate benchmark values for key design parameters (displacement and base shear) evaluated using either response spectrum or time–history analysis procedures.

The static analysis procedure is straightforward. However, the procedure cannot be used when the spectral demands cannot be adequately characterized using the assumed spectral shape. Typically this occurs for:

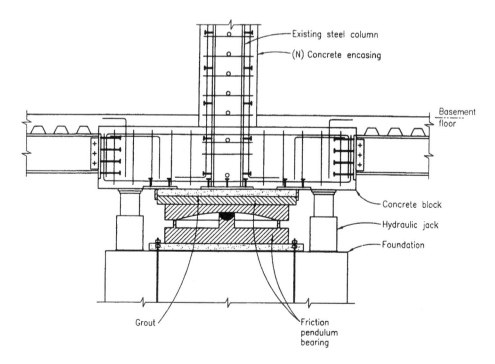

Existing steel column

(N) Concrete encasing

Basement floor

Concrete block

Hydraulic jack

Foundation

Grout

Friction pendulum bearing

Figure 8.34j(1). Installation details of FPS bearing under existing interior columns.

Figure 8.34j(2).   FPS bearing. (Photograph courtesy of Anoop Mokha, Ph.D., S.E., Vice President, Earthquake Protection Systems, Vallejo, CA.)

Figure 8.34j(3).   FPS bearing in new construction.

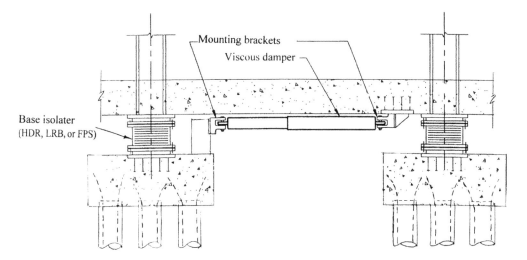

Figure 8.34k. Base isolator operating in concert with a viscous damper.

1. Isolated buildings located in the near field.
2. Isolated buildings on soft soil sites.
3. Long-period isolated buildings (beyond the constant velocity domain).

Further, the static procedure cannot be used for nonregular superstructures or for highly nonlinear isolation systems.

Response spectrum analysis is permitted for the design of all isolated buildings except for those buildings located on very soft soil sites (for which site-specific spectra should be established), buildings supported by highly nonlinear isolation systems for which the assumptions implicit in the definitions of effective stiffness and damping break down, or buildings located in the very near field of major active faults where response spectrum analysis may not capture pulse effects adequately.

Time–history analysis is the default analysis procedure: it must be used when the restrictions set forth on static and response spectrum analysis cannot be satisfied, and may be used for the analysis of any isolated building. Arguably the most detailed of the analysis procedures, the results of time–history analysis must be carefully reviewed to avoid any gross design errors.

8.7.3.1. *Equivalent Lateral Force Procedure*

This procedure is permitted when the following restrictions are met:

1. The structure is located at a site with S_1 less than or equal to 0.60 g.
2. The structure is located on a class A, B, C, or D site.
3. The structure above the isolation interfaces is less than or equal to four stories or 65 ft (19.8 in.) in height.
4. The effective period of the isolated structure at maximum displacement T_m is less than or equal to 3.0 sec.
5. The effective period of the isolated structure at the design displacement T_p is greater than three times the elastic, fixed-base period of the structure above the isolation system.
6. The structure above the isolation system is of regular configuration.

7. The isolation system meets all the following criteria:
    a. The effective stiffness of the isolation system at the design displacement is greater than one-third of the effective stiffness at 20% of the design displacement.
    b. The isolation system is capable of producing a restoring force such that the lateral force at the total design displacement D_T is at least $0.025w$ greater than the lateral force at 50% of the total design displacement.
    c. The isolation system has force-deflection properties that are independent of the rate of loading.
    d. The isolation system has force-deflection properties that are independent of vertical load and bilateral load.
    e. The isolation system does not limit maximum considered earthquake displacement to less than S_{M1}/S_{D1} times the total design displacement.

8.7.3.1.1 Lateral Displacements. There are as many as six definable displacements in base isolation terminology. Three of these are defined in Fig. 8.34l, while the others, related to certain prescribed formulas, are explained in the text.
Design Displacement. The isolation system must be designed and constructed to withstand design lateral earthquake displacements D_D, calculated to occur in the direction of each of the main horizontal axes of the structure in accordance with the following equation:

$$D_D = \frac{gS_{D1}T_D}{4\pi^2 B_D}$$

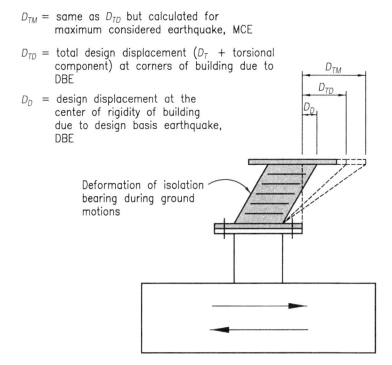

D_{TM} = same as D_{TD} but calculated for maximum considered earthquake, MCE

D_{TD} = total design displacement $(D_T$ + torsional component) at corners of building due to DBE

D_D = design displacement at the center of rigidity of building due to design basis earthquake, DBE

Deformation of isolation bearing during ground motions

Figure 8.34l. Isolator displacement terminology. Note: MCE = earthquake corresponding to 2% probability of exceedence in a 50-year period (2500-year return period); DBE = earthquake corresponding to 10% probability of exceedence in a 50-year period (475-year return period).

where

D_D = design displacement of the isolation system

g = acceleration of gravity

S_{D1} = design 5% damped spectral acceleration at 1-sec period

T_D = effective period of seismically isolated structure in seconds at the design displacement in the direction under consideration

B_D = numerical coefficient related to the effective damping of the isolation system at the design displacement, D_D as set forth in Table 9.13.3.3.1

Effective Period at Design Displacement. The effective period of the isolated structure at design displacement T_D shall be determined using the deformational characteristics of the isolation system in accordance with the following equation:

$$T_D = 2\pi \sqrt{\frac{W}{k_{Dmin}g}}$$

where

T_D = effective period of the isolated structure at design displacement D_D

W = total seismic dead load weight of the structure above the isolation interface

K_{Dmin} = minimum effective stiffness in kips/in. (kN/mm) of the isolation system at the design displacement in the horizontal direction under consideration

g = acceleration due to gravity

Maximum Lateral Displacement. The maximum displacement of the isolation system D_M in the most critical direction of horizontal response shall be calculated in accordance with the formula:

$$D_M = \frac{gS_{M1}T_M}{4\pi^2 B_M}$$

where

D_M = maximum displacement of the isolation system

g = acceleration of gravity

S_{M1} = maximum considered 5% damped spectral acceleration at 1-sec period

T_M = effective period of seismic-isolated structure at the maximum displacement in the direction under consideration

B_M = numerical coefficient related to the effective damping of the isolation system at the maximum displacement D_M.

Effective Period at Maximum Displacement. The effective period of the isolated structure, T_M, at maximum displacement D_M shall be determined using the deformational characteristics of the isolation system in accordance with the equation:

$$T_M = 2\pi \sqrt{\frac{W}{k_{Mmin}g}}$$

where

T_M = the effective period of the isolated structure at maximum displacement D_M

W = total seismic dead load weight of the structure above the isolation interface

K_{Mmin} = minimum effective stiffness of the isolation system at the maximum displacement in the horizontal direction under consideration

g = acceleration of gravity

Total Lateral Displacement. The total design displacement D_{TD} and the total maximum displacement D_{TM} of elements of the isolation system shall include additional displacement due to actual and accidental torsion calculated from the spatial distribution of the lateral stiffness of the isolation system and the most disadvantageous location of mass eccentricity. The total design displacement D_{TD} and the total maximum displacement D_{TM} of elements of an isolation system with uniform spatial distribution of lateral stiffness shall not be taken as less than that prescribed by the following equations:

$$D_{TD} = D_D\left[1 + y\,\frac{12e}{b^2 + d^2}\right]$$

$$D_{TM} = D_M\left[1 + y\,\frac{12e}{b^2 + d^2}\right]$$

where

D_D = design displacement, in in. (mm), at the center of rigidity of the isolation system in the direction under consideration

D_M = maximum displacement, in in. (mm), at the center of rigidity of the isolation system in the direction under consideration

y = the distance, in ft (mm), between the centers of rigidity of the isolation system and the element of interest measured perpendicular to the direction of seismic loading under consideration.

e = the actual eccentricity, in ft (mm), measured in plan between the center of mass of the structure above the isolation interface and the center of rigidity of the isolation system, plus accidental eccentricity, in ft (mm), taken as 5% of the longest plan dimension of the structure perpendicular to the direction of force under consideration

b = the shortest plan dimension of the structure, in ft (mm), measured perpendicular to d

d = the longest plan dimension of the structure, in ft (mm)

8.7.3.1.2   Minimum Lateral Forces.
Isolation System and Structural Elements at or Below Isolation System.   The isolation system, the foundation, and all structural elements below the isolation system shall be designed and constructed to withstand a minimum lateral seismic force V_b using all of the appropriate provisions for a nonisolated structure according to

$$V_b = k_{D\max}D_D$$

where

V_b = the minimum lateral seismic design force or shear on elements of the isolation system or elements below the isolation system

$K_{D\max}$ = maximum effective stiffness of the isolation system at the design displacement in maximum effective stiffness of the isolation system at the design displacement in the horizontal direction under consideration the horizontal direction under consideration

D_D = design displacement at the center of rigidity of the isolation system in the direction under consideration.

V_b shall not be taken as less than the maximum force in the isolation system at any displacement up to and including the design displacement.

Structural Elements Above Isolation System. The structure above the isolation system shall be designed and constructed to withstand a minimum shear force V_s using all of the appropriate provisions for a nonisolated structure according to

$$V_s = \frac{k_{D\max}D_D}{R_I}$$

where

$K_{D\max}$ = maximum effective stiffness of the isolation system at the design displacement in the horizontal direction under consideration.

D_D = design displacement at the center of rigidity of the isolation system in the direction under consideration.

R_I = numerical coefficient related to the type of lateral force-resisting system above the isolation system.

The R_I factor shall be based on the type of lateral force-resisting system used for the structure above the isolation system and shall be three-eighths of the R value a of nonisolated structure with an upper-bound value not exceeding 2.0 and a lower-bound value not less than 1.0.

Limits on V_s. The value of V_s shall not be taken as less than the following:

- The lateral seismic force of a fixed-base structure of the same weight W, and a period equal to the isolated period T_D.
- The base shear corresponding to the factored design wind load.
- The lateral seismic force required to fully activate the isolation system (e.g., the yield level of a softening system, the ultimate capacity of a sacrificial wind-restraint system, or the breakaway friction level of a sliding system) factored by 1.5.

Vertical Distribution of V_s. The total force shall be distributed over the height of the structure above the isolation interface in accordance with the following equation:

$$F_x = \frac{V_s w_x h_x}{\sum_{i=1}^{n} w_i h_i}$$

where

F_x = lateral force at level x

V_s = total lateral seismic design force or shear on elements above the isolation system

w_x = portion of W that is located at or assigned to Level i, n, or x, respectively

h_x = height above the base Level i, n, or x, respectively

w_i = portion of W that is located at or assigned to Level i, n, or x, respectively

h_i = height above the base Level i, n, or x, respectively

At each level designated as x, the force F_x shall be applied over the area of the structure in accordance with the mass distribution at the level. Stresses in each structural element shall be calculated as the effect of force F_x applied at the appropriate levels above the base.

8.7.3.1.3 Drift Limits. The story drift Δ is computed as the difference of deflections at the top and bottom of the story under consideration. It should be noted that Δ is computed using ASCE 7 ultimate earthquake loads, even though the design of the building, as for a steel building, may be in ASD, allowable stress design.

The maximum interstory drift Δ permitted by ASCE 7-02 is a function of the method of analysis. If Δ is calculated by response spectrum analysis, the maximum drift permitted = $0.015 h_{sx}$. If it is calculated by time–history analysis based on the force deflection characteristics of nonlinear elements of the lateral-force resisting systems, then the maximum Δ permitted = $0.020\ h_{sx}$. The term h_{sx} denotes the story height below level x.

The deflection δ_x at level x at the center of mass is determined by the equation:

$$\delta_x = \frac{c_d \delta_{xe}}{I}$$

where

C_d = the deflection amplification factor. C_d need not be greater than 2.0.
δ_{xe} = deflection determined by an elastic analysis, as denoted by the suffix e.
I = occupancy importance factor ranging from 1.0 to 1.5.

For example, $I = 1.5$ if the building is placed in seismic use group III.

It should be noted that for structures in SDC C, D, E, or F having torsional or extreme torsional irregularities, the story drift Δ should be computed at the building corners, and not at the center of the mass. The calculation of interstory drift should include vertical deformation of the isolation systems and $P\Delta$ effects where required.

8.7.3.2. Dynamic Analysis

Both the response spectrum and the time–history analyses are permitted under dynamic analysis procedure.

8.7.3.2.1. Response Spectrum Analysis. This analysis is permitted subject to the following stipulations:

1. The structure is located on a class A, B, C, or D site
2. The isolation system meets the criteria of Item 7 of the equivalent lateral force procedure. (Section 8.7.3.1)

Response spectrum analysis should be performed using a modal damping value for the fundamental mode in the direction of interest not greater than the effective damping of the isolation system or 30% of critical, whichever is less. Damping values for higher modes should be selected consistent with those appropriate for response spectrum analysis of the structure above the isolation system on a fixed base.

Response spectrum analysis used to determine the total design displacement and the total maximum displacement should include simultaneous excitation of the model by 100% of the most critical direction of ground motion and 30% of the ground motion on the orthogonal axis. The maximum displacement of the isolation system must be calculated as the vectorial sum of the two orthogonal displacements.

The isolated building should be represented by a three-dimensional linear elastic structural model. The isolators should be represented by linear springs with stiffness K_{eff}. The calculation of K_{eff} may require multiple iterations, as K_{eff} will be a function of the target displacement.

8.7.3.2.2. Time–History Analysis. Time–history analysis is permitted for the design of any seismically isolated structure. It is mandated for the design of all seismically isolated structures located in a class E or F site, and for isolation systems not meeting the criteria of Item 7 of Section 8.7.3.1.

Time–history analysis must be performed with at least three matched pairs of horizontal time–history components. Parameters of interest should be calculated for each time–history analysis pair. The parameters of interest should include member forces, connection forces, interstory drift, isolator displacements, and overturning forces. Each matched pair of horizontal ground motion records should be simultaneously applied to the mathematical model, considering the most disadvantageous location of mass eccentricity, to calculate the maximum displacements in the isolation system. Where orthogonal forces are applied simultaneously, such as in time–history analysis, the required 5% displacement of the center of mass should be applied for only one of the orthogonal forces at a time. An analysis where the 5% displacement is applied for both orthogonal directions concurrently would result in a double application of the torsional effect of accidental eccentricity and would not be consistent with the original intent for the use of accidental eccentricity.

Common design practice is to: 1) impose the orthogonal components along each of the principal axes of the building separately; and 2) repeat step 1 after changing the signs of the ground motion components (separately), for a total of eight analyses per matched pair per mass eccentricity.

To reduce the computational effort, preliminary analysis may be undertaken to identify: 1) the most advantageous location of the mass eccentricity; 2) the critical matched pair of ground motion records; and 3) the critical orientation of the matched pair identified in foregoing Item 2.

Much of the analysis and design work may be completed with this substantially reduced set of parameters. Once the analysis and design effort is near completion, the final design(s) may be analyzed using the unreduced set of parameters, if deemed necessary. Site-specific ground-motion spectra of the design earthquake and the maximum considered earthquake are required for design and analysis of all seismically isolated structures if any one of the following conditions apply:

1. The structure is located on a class F site.
2. The structure is located at a site with S_1 greater than 0.60 g.

Site-specific spectra must be prepared for the design of long-period base-isolation systems, base-isolated buildings on a soft soils, and base-isolated buildings located either near an active fault or in seismic zones. This requirement stems from the uncertainties associated with the spectral shapes set forth in ASCE 7-02 design provisions.

Although the development of a site-specific spectrum is encouraged, the ordinates of the spectrum are not permitted to be less than 80% of the ordinates of the standard design spectrum to guard against the use of inappropriately generated site-specific spectra. This limit is imposed on the ordinates of both site-specific DBE and MCE.

8.7.3.2.3. Mathematical Model. Several modeling procedures have been developed for the analysis and design of seismic-isolated buildings. These procedures may not adequately capture the secondary forces that develop as a function of the horizontal displacement (often large) of the isolators. One key example is the moment, equal to the product of the load on the isolator and the isolator displacement, commonly known as the P-delta ($P\Delta$) effect, that must be resisted by the isolator, the connections of the isolators to the structural framing above and/or below the isolator, and the structural framing above and/or below the isolator.

Three-dimensional elastic models of isolated buildings are commonly used for response spectrum analysis. The model of the superstructure should include all significant structural members in the building frame and accurately account for their stiffness and mass for both static and response spectrum analysis. The isolators are modeled as linear springs with stiffness equal to the effective stiffness, requiring an a priori estimate of the likely displacement in the isolators.

Another common procedure used for the design of isolated buildings using time–history analysis assumes nonlinear isolators and a linear elastic superstructure. The procedure is appropriate for buildings in which the superstructure is assumed to undergo none-to-minimal inelastic response. The recent development of analysis software packages that include three-dimensional elastic modeling of the superstructure and three-dimensional nonlinear modeling of the isolators has made the analysis of such structures simpler and more efficient.

Another procedure involves the development of a complete three-dimensional non-linear model of the building. This level of effort is computationally intensive and likely rarely justified. This procedure should probably be used only for isolated buildings in which the superstructure is likely to experience substantial inelastic response, an assumption at odds with the stated performance goals for seismic-isolated buildings.

8.7.3.3. *Design and Construction Review*

Design review of both the analysis and the design of the isolation system, and the isolator testing program, is mandated by ASCE 7-02 for three key reasons:

- The consequences of isolator failure could be catastrophic.
- Isolator design, fabrication, testing methods, and technology are evolving rapidly, perhaps utilizing technologies unfamiliar to many design professionals.
- Isolation system analysis and design often involve use of complex procedures, e.g., nonlinear time–history analysis, which can be highly sensitive to assumptions and idealizations made during the analysis and design process.

Design review aims to minimize the possibility of inappropriate assumptions and procedures in the analysis and the design process. The review should be performed by: 1) a team independent of the design team and the project contractors; and 2) a review team composed of individuals with special expertise in one or more aspects of the design and implementation of seismic isolation systems. The review teams should be formed prior to the development of ground motion criteria and isolator design options. Further, the review team should be given complete access to all pertinent information such that the review team can work closely with all consultants and regulatory agencies involved in the project.

8.7.3.4. *Required Tests for Isolation Systems*

For each cycle of testing, the force-deflection behavior of the prototype test specimen must be recorded so that the data can be used to determine whether the isolation system complies with both these requirements and the specifications prepared by the engineer of record. The engineer of record and the independent review team should review all raw data from the prototype tests.

The total number of testing cycles of substantial response will likely be greater for soft sites and systems with small damping values. If the mechanical characteristics of the isolation system are dependent on the rate of loading, additional dynamic tests must be performed to characterize this dependence. Rate-dependence behavior will be exhibited by most sliding isolation systems (velocity-dependent) and selected elastomeric isolation systems (strain rate-dependent). Reduced-scale models of isolators can be used to capture rate effects on stiffness and damping values, provided that the reduced-scale isolators are fabricated using the same processes and quality control procedures as the full-size isolators. Dimensional relationships between full- and reduced-scale units should be established and verified prior to finalizing the testing program.

The implementation of a quality control program is key to the production of isolators of uniform quality with consistent mechanical properties. This quality control program should be implemented for both prototype and production isolators. If the production (quality control) testing results are to be based in any part on the results of the prototype tests, the production testing program should be completed on each of the prototype isolators prior to starting the prototype tests. Qualified and independent inspection/ monitoring of the testing and manufacture process is an important element of an adequate quality control.

The criteria used to judge whether the properties of an isolation system are dependent on the rate of loading are specified: Namely, the isolators are to be considered rate-dependent if the test data demonstrate that the effective stiffness of the isolator changes by more than plus or minus 10% when the cycling rate is varied from the effective frequency at the design displacement to any frequency within the range of 0.1 to 2.0 times the effective frequency. If the effective stiffness and damping of any of the isolators in the isolation system are dependent on the magnitude of the imposed orthogonal displacement, additional testing is required to quantify this dependence. Reduced-scale isolators may be used to substantiate this dependence, provided the reduced-scale isolators are fabricated using the same processes and quality control procedures as the full-size isolators. Dimensional relationships between full- and reduced-scale units should be established and verified prior to finalizing the testing program. The properties of the isolators may be considered to be independent of bilateral displacement if the effective stiffness at 100% bilateral displacement does not differ from the effective stiffness at 0% bilateral displacement by more than plus or minus 10%.

The static vertical load test is used to verify isolator stability at the total maximum displacement under maximum and minimum vertical loads. The maximum vertical load is calculated using $1.2\,DL + 1.0\,LL$ and the maximum downward seismic overturning load from the MCE. The minimum vertical load is calculated using 0.8 DL and the maximum upward seismic overturning load from the MCE. This is a static stability test; no cycling is required. Prototype tests are not required if the isolator unit is of similar dimensional characteristics, of the same type and material, and constructed using the same processes as a prototype isolator unit that has been previously tested using the specified sequence of tests. The independence engineering team should determine whether the results of previously tested units are suitable, sufficient, and acceptable.

8.7.3.5. *Illustrative Example: Static Procedure*

Up to this point we have discussed the basic principles of seismic isolation and the design provisions of ASCE 7-02. It should be clear from the discussions that a dynamic analysis is mandatory for almost all buildings because buildings that meet the requirements of regularity are indeed rare, even in high seismic zones. However, the design principles are best understood by working through a static example. We will do so here using design provisions given in ASCE 7-02. As mentioned previously, the design provisions also apply to IBC-03, since ASCE 7-02 has been adopted by IBC. Ample interpretation of ASCE 7-02 provisions is repeated to present the solution in a stand-alone format.

Given. A new four-story hospital building to be located in the outskirts of Los Angeles, CA. The owners of the facility have desired a building of superior earthquake performance and are willing to incur the special costs associated with the design, fabrication, and installation of seismic isolaters. A target building performance level of immediate occupancy or better is sought.

The structure is expected to outperform a comparable fixed-base building in moderate and large earthquakes. The intent is to limit damage to the structure and its contents by using seismic isolation that, in effect, permits an elastic response of the structure, while limiting the floor accelerations to low levels even in a large earthquake event.

Building Characteristics

- A single basement, four-story, regular configuration steel building. The building has no vertical or plan irregularities.
- Seismic bracing consists of steel eccentric-braced frame with nonmoment-resisting connections away from links.
- Response modification coefficient $R = 7$ (ASCE 7-02, Table 9.5.2.2).
- Building is located in the outskirts of Los Angeles, CA.
- From seismic hazard maps $S_s = 1.5$ g and $S_1 = 0.60$ g for the building site.
- Importance factor $I = 1.0$. Observe that importance factor I for a seismic-isolated building is taken as 1.0, regardless of the occupancy category, since there is no design ductility demand on the structure.
- Building period calculated as a fixed-base building = 0.9 sec.
- Building plan dimensions are 120×120 ft.
- Calculated distance between the center of mass and the center of rigidity is 5 ft at each floor and at the roof.
- The project geotechnical engineer has established the building site as site class D.
- Building weight for seismic design = 7200 kips.
- The project structural engineer has established that, to achieve immediate occupancy performance goals, the isolation system should provide effective isolated periods of $T_D = 2.5$ and $T_M = 3.0$ sec, and a damping of 20% of the critical. A margin of $\pm 15\%$ variation in stiffness of isolators from the mean values is considered acceptable.

Required. A preliminary design using the provisions of ASCE 7-02 for base-isolation of the building. For purposes of illustration, a friction pendulum system, FPS, is selected as the base-isolation system. It should be noted that, in practice, building ownership, particularly if it is a public entity, requires that the design accommodate alternative systems to secure competitive bids. However, for illustration purposes we will consider only the FPS, it being understood that other isolation systems such as high-damping rubber and lead-rubber isolators are equally viable alternatives.

As part of preliminary design determine

- Minimum design displacements D_D and D_M under DBE and MCE. Also total displacements D_{TD} and D_{TM} which include effects of torsion.
- Base shear V_b for designing the structure below the isolation surface.
- Base shear V_s for designing the structure above the isolation surface.
- Maximum dimension of the isolators.

Solution. The restrictions placed on the use of the static lateral response procedures effectively require dynamic analysis for most isolated structures. Therefore one might ask, "Why perform, in this day and age of computers, a static analysis of a building with a sophisticated system such as base isolation?" The answer is quite simple: to establish a minimum level of design forces and displacement. Lower-bound limits on design displacements and design forces are specified in ASCE 7-02 as a percentage of the values prescribed by the static procedure. These lower-bound limits on key design parameters ensure consistency in the design of isolated structures and serve as a safety net against gross undersign.

As mentioned previously, seismic isolation, also referred to as base isolation, is a design concept based on the premise that a structure can be substantially "decoupled" from potentially damaging earthquake ground motions. By decoupling the structure from ground shaking, isolation reduces the level of response in the structure from a level that would otherwise occur in a conventional fixed-base building. Typically, decoupling is accomplished using an isolation system that makes the effective period of the isolated structure several times greater than the period of the structure above the isolation system.

In our case, the four-story example building with a fixed-base period of 0.9 sec and a standard damping of 5% would have experienced a first-mode acceleration of 0.48 g (see Fig. 8.34a). By decoupling the building from the ground, the period of the building is expected to increase to 2.7 sec. Additionally, the base isolation is counted upon to increase the damping from a standard 5% to about 20% of the critical. Together, these two factors reduce the first mode acceleration to 0.12g, as shown in Fig. 8.34a.

The underlying philosophy behind isolated structures may be characterized as a combination of primary performance objective for fixed-base buildings, which is the provision of life safety in a major earthquake, and the additional performance objective of damage protection, an attribute provided by isolated structures. The design criteria are then a combination of life safety and damage protection goals summarized as follows:

- Two levels of earthquake, the design basis earthquake DBE and the maximum considered earthquake MCE, are typically considered in the design of isolated structures. The DBE is the same level of ground shaking as that recommended for design of fixed-base structures. The MCE is a higher level of earthquake ground motion defined as the maximum level of ground shaking that may be expected at the building site within the known geological framework.
- The isolators must be capable of sustaining loads and displacements corresponding to the MCE without failure.
- The structure above the isolation system must remain "essentially elastic" for the DBE.

From the criteria given above, it is seen that the performance objectives and design requirements for fixed-base and isolated buildings vary significantly. The performance objective for fixed-base construction is life safety in a DBE; the intent is to prevent substantial loss of life rather than control damage. For isolated buildings, the performance objectives are

1. Minimal to no damage in the design earthquake (thus providing *life safety*).
2. A stable isolation system in the maximum capable earthquake.

The performance of an isolated building in a design basis earthquake will likely be much better (less interstory drift, smaller floor accelerations) than its fixed-base counterpart. Further, isolated buildings can be designed to provide continued function following a design earthquake: a level of performance that is very difficult to achieve with conventional fixed-base construction.

Fixed-base buildings are generally designed using large response modification factors to reduce elastic spectral demands to a design level, a strategy predicated on significant inelastic deformation of the framing system and damage to nonstructural building element. Such buildings are checked for response in the design earthquake only; there is no design check for the MCE. In contrast, isolated buildings are designed using a dual level approach, namely, the framing system is designed to remain essentially elastic (no damage) in the design earthquake, and the isolators are designed (and tested) to remain stable in the MCE.

The subject building is a steel-braced frame building. Using the post-earthquake scenario given in FEMA 356 as a guide, our building is expected to have

- No permanent drift. Structure substantially retains original strength and stiffness.
- Negligible damage to nonstructural components.
- Minor hairline cracking in concrete frames. No crushing of concrete.
- Minor local yielding at a few places in steel frames. No fracture.
- Minor yielding or buckling of braces.
- Connections between deck units and framing intact. Minor distortions.
- Cladding connections may yield. No failure.
- Some cracked panes in glazing. None broken.
- Negligible damage in stairs and fire escapes.
- Elevators operate.
- Fire alarm systems and electrical equipment functional.
- Computer units undamaged and operable.

Before proceding with the illustrative example, certain design requirements touched upon briefly in the preceding sections will be explained in greater detail. The purpose is to delve into the design intent behind these provisions.

8.7.3.5.1. Effective Stiffness of Isolators.   Typically, isolation systems are non-linear, meaning that their effective stiffness is displacement- and/or velocity-dependent, as shown by an idealized force-deflection relationship in Fig. 834m.

The effective stiffness k_{eff} of a seismic isolator is calculated using the forces in the isolator at the maximum and minimum displacements as given in the following equation:

$$k_{eff} = \frac{|F^+| + |F^-|}{|\Delta^+| + |\Delta^-|}$$

where $F+$ and F^- are the positive and negative forces at $\Delta+$ and Δ^-, respectively.

For isolators whose properties are independent of velocity, the forces in the isolator at the maximum and minimum displacements will generally be maximum and minimum forces, respectively. For isolators whose properties exhibit velocity-dependence, the forces in the isolator at the maximum and minimum displacements will generally be less than the maximum and minimum forces, respectively. However, it is usually assumed that maximum and minimum forces in an isolator are attained at maximum and minimum displacements, respectively. For most types of isolator, this assumption is reasonable.

The deformational characteristics of an isolation system determine: 1) the design displacements; and 2) the maximum forces transmitted to the isolated structure. Deformational characteristics are represented by the effective (secant) stiffness of the isolation system. Recognizing that force–displacement hysteresis of an isolation system may change over the course of an earthquake, the maximum effective stiffness is used to calculate the maximum force transmitted by the isolators, and the minimum effective stiffness is used to calculate the fundamental period of the isolated building. The reason for using minimum is to arrive at a conservative estimate of the design displacement. The limiting values are generally established in the design phase and are required to be confirmed by testing.

Effective stiffness of the isolation system is determined from the force-displacement (hysteresis) loops based on the results of cyclic testing of a selected sample of isolator. The values of maximum effective stiffness and minimum effective stiffness can be calculated, as shown in Fig. 8.34m, for both design and maximum displacement levels.

8.7.3.5.2. Effective Damping. The effective damping β_{eff} is used to quantify the energy dissipation furnished by the isolation system. The maximum effective stiffness of the isolation system is used to provide a lower-bound, i.e., conservative, estimate of the effective damping.

For the purpose of design, energy dissipation is characterized as an equivalent viscous damping. The following equation defines the equivalent viscous damping β_{eff} for

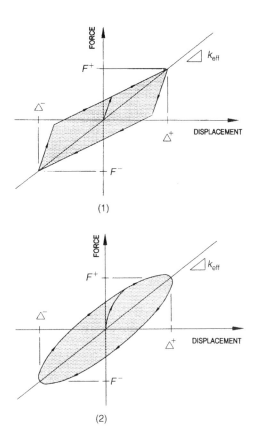

Figure 8.34m. Idealized force–displaement relationships for base-isolation systems: (1) hysteretic system; (2) viscous system. In seismic isolation, hysteretic behavior is a term that describes intrinsic damping due to inelastic deformation of base isolators. Energy is dissipated through work done by the inelastic actions in the isolators. Viscoelastic or viscous behavior, on the other hand, typifies damping action of external devices that use viscous liquids to absorb energy. No inelastic deformations are involved: Energy is dissipated as heat. Effective stiffness k_{eff} of an isolator is calculated from test data by measuring forces F^+ and F^-, and corresponding deformations Δ^+ and Δ^-. The area enclosed by the force–displacement loop is used to calculate effective damping β_{eff}.

a single isolator:

$$\beta_{\text{eff}} = \frac{2}{\pi} \frac{E_{\text{loop}}}{k_{\text{eff}}(|\Delta^+| + |\Delta^-|)^2}$$

where

β_{eff} = effective damping of the isolation system and isolator unit

E_{loop} = area enclosed by the force-displacement loop of a single isolator in a complete cycle of loading to maximum positive and maximum negative displacement, $\Delta+$ and Δ^-

$\Delta+$ = maximum positive displacement of isolator during prototype testing

Δ^- = maximum negative displacement of isolator during prototype testing

8.7.3.5.3. Total Design Displacement. The design of isolated structures must consider additional displacements due to actual and accidental eccentricity, similar to those prescribed for fixed-base structures. Equations given below provide a simple means to combine translational and torsional displacement in terms of the gross plan dimensions of the building (i.e., dimensions b and d), the distance from the center of the building to the point of interest (i.e., dimension y), and the actual plus the accidental eccentricity, as follows:

$$D_{TD} = D_D\left[1 + y\,\frac{12e}{b^2 + d^2}\right]$$

$$T_{TM} = D_M\left[1 + y\,\frac{12e}{b^2 + d^2}\right]$$

where e = the sum of the actual and accidental eccentricities.

Notice that the design displacement D_D at the center of the building has been modified to account for additional displacement at the corners or edges of the building due to torsion. It is assumed that the stiffness of the isolation system is distributed in plan proportional to the distribution of the supported weight of the building.

Smaller values of D_{TD} can be used for design if the isolation system is configured to resist torsion (e.g., if stiffer isolator units are positioned near the edges and corners of the building). However, the minimum value of D_{TD} is set equal to 1.1 D_D for all types of isolation systems. The total displacement D_{TM} is calculated in a manner similar to the calculation of D_{TD}. The eccentricity e used for calculating torsional displacements is the actual eccentricity of the isolation system plus an allowance of 5% of the width of building to account for accidental torsion. The parameter y is the distance between the center of rigidity of the isolation system and farthest corners of the building.

It should be noted that the stiffness values $K_{D\text{min}}$ and $K_{m\text{min}}$ are not known to the designer during the preliminary design stage, but are derived from the known or expected values of periods of the building. Since the expected periods may not turn out to be equal to the final values, the derived stiffness values are also preliminary. After completing a satisfactory preliminary design, typically prototype isolaters are tested to obtain values of $K_{D\text{min}}$, $K_{D\text{max}}$, $K_{M\text{min}}$, and $K_{M\text{max}}$.

8.7.3.5.4.   Minimum Design Lateral Forces.

1.   Isolation System and Structural Elements at or Below Isolation Interface.
The design actions for elements at or below the isolation interface are based on the maximum forces delivered by the isolation system during the design basis earthquake. The building's foundation, the isolation system, and all structural elements at or below the isolation interface are required to be designed and constructed to withstand a minimum lateral force.

$$V_b = K_{D\text{max}}\, D_D$$

The maximum force V_b is the product of the maximum stiffness of the isolation system at the design displacement $K_{D\max}$ and the design displacement D_D. The design force V_b represents strength level forces.

The previous equation for V_b is for use in regions of high seismicity, such as UBC zone 4, wherein the difference between the total design displacement and total maximum displacement is relatively small; that is, if a supporting element was designed for design basis earthquake forces at the strength level, it is probable that such a supporting element could resist the forces associated with the MCE without failure.

There are significant differences in values of M_M between regions of high and low seismicity: Values of M_M may be less than 1.25 in regions of high seismicity, but may exceed 2.5 in regions of low seismicity. As such, in a region of low seismicity, a supporting element designed for DBE-induced forces may be unable to sustain forces associated with the MCE without significant distress or failure. Therefore, in these regions it may be prudent to consider MCE-level forces to check the design of the isolation system and the structural elements at or below the isolation interface.

Isolation interface is the boundary between the upper portion of the building, which is isolated, and the lower portion, which is rigidly attached to the foundation or ground. The isolation interface can be assumed to pass through the midheight of elastomeric bearings or the sliding surface of sliding bearings. Observe that the isolation interface need not be a horizontal plane, but could change elevation if the isolators are positioned at different elevations throughout the building.

The isolation system includes the isolator units, connections of isolator units to the structural system, and all structural elements required for isolator stability. Isolator units include bearings that support the building's weight and provide lateral flexibility. Typically, isolation system bearings provide damping and wind restraint as an integral part of the bearing. Isolator systems may also include supplemental damping devices. For example, an FPS of basic isolation may include viscous dampers.

Structural elements that are required for structural stability include all structural elements necessary to resist design forces at the connection of the structure to isolator units. For example, a column segment and a beam immediately above an isolator constitute elements of the isolation system because they are necessary to resist forces due to the lateral earthquake displacement of the isolators.

2. Structural Elements above Isolation System. The design of the framing above the isolation system is based on the maximum force delivered by the isolation system divided by a response reduction factor, R_I. The values assigned to R_I reflect system overstrength only and no expected ductility demand. By using these values for R_I, a significant measure of damage control is afforded in the design earthquake, since the structure remains essentially elastic.

The minimum base shear for the design of the structure above the isolation is given by

$$V_s = \frac{K_{D\max}D_D}{R_I}$$

Three limits are imposed for the calculation of V_s.

- V_s shall not be less than the base shear required for a fixed-base structure of the same weight w and a period equal to the isolated period.
- V_s shall not be less than the total shear corresponding to the design wind load. (In wind design, engineers seldom use the term base shear to define the total shear due to wind. However, base shear and total shear are one and the same.)
- V_s shall not be less than 150% of the lateral seismic force required to fully activate the system.

Thus there are three lower-bound limits set on the minimum seismic shear to be used for the design of the framing above the isolation system. The first limit requires design base shear to be at least that of a fixed-base building of comparable period. The second limit ensures that the elements above the isolation system remain elastic during a design windstorm. The third limit is designed to prevent the elements above the isolation system from deforming inelastically before the isolation system is activated.

3. **Vertical Distribution of V_s.** The vertical distribution of the seismic base shear is similar to that used for fixed-base buildings, namely, a distribution that approximates the first-mode shape of the fixed-base building. This distribution conservatively approximates the inertia force distributions measured from time–history analyses.

Continuation of Illustrative Problem. The effective periods T_D and T_M of the isolated building are

$$T_D = 2\pi \sqrt{\frac{W}{K_{Dmin}g}}$$

$$T_D^2 = \frac{4\pi^2 W}{K_{Dmin}g} \qquad T_D = 2.5 \text{ sec (given)}$$

$$W = 7200 \text{ kips (given)}$$
$$g = 386.4 \text{ in./sec}^2$$

$$
\begin{aligned}
K_{Dmin} &= \frac{4\pi^2 W}{T_D^2 g} \\
&= \frac{4 \times \pi^2 \times 7200}{2.5^2 \times 386.4} \\
&= 117.7 \text{ kips/in., say, } 118 \text{ kips/in.}
\end{aligned}
$$

Similarly,

$$K_{Mmin} = \frac{4 \times \pi^2 \times 7200}{3^2 \times 386.4} \qquad T_M = 3 \text{ sec (given)}$$
$$= 81.7 \text{ kips/in., say, } 82 \text{ kips/in.}$$

As stated in the problem, a plus or minus 15% variation in stiffness from the mean values is permitted. Therefore, use a factor of 0.85 to determine K_{Dmax} and K_{Mmax}.

$$K_{Dmax} = \frac{1.15 \times 118}{0.85} = 159.5 \text{ kips/in., say, } 160 \text{ kips/in.}$$

$$K_{Mmax} = \frac{1.15 \times 82}{0.85} = 111 \text{ kips/in.}$$

From ASCE 7-02, Table 9.13.3.31, for a 20% effective damping, i.e., B_D or $B_M = 20\%$, the value of damping coefficient $B = 1.5$.

Observe that the same damping coefficient is applied to both DBE and MCE events. The value of F_V as a function of site class and mapped 1-sec period MCE spectral acceleration is given in Table 9.4.1.2.4b of ASCE 7-02. From this table, for site class D and $S_1 = 0.60 > 0.50$, we get $F_V = 1.5$. The spectral response acceleration S_{M1} at a period of 1 sec, adjusted for site class D, is equal to

$$
\begin{aligned}
S_{M1} &= F_V S_1 \\
&= 1.5 \times 0.6 \\
&= 0.9 \text{ g}
\end{aligned}
$$

The design spectral response acceleration S_{D1} is given by

$$S_{D1} = \frac{2}{3} S_{M1}$$
$$= \frac{2}{3} \times 0.9$$
$$= 0.6 \text{ g}$$

Similarly,

$$S_S = 1.5$$
$$S_{MS} = F_a S_S$$
$$= 1 \times 1.5$$
$$= 1.5$$
$$S_{DS} = \frac{2}{3} \times 1.5$$
$$= 1.0 \text{ g}$$

The minimum design displacements are obtained as follows:

$$D_D = \left(\frac{386.4}{4\pi^2}\right) \frac{0.60 \times 2.5}{1.5}$$
$$= 9.79 \text{ in.}$$
$$D_M = \left(\frac{386.4}{4\pi^2}\right) \frac{0.9 \times 3}{1.5}$$
$$= 17.62 \text{ in.}$$

The eccentricity for calculating torsional effects is equal to the actual eccentricity plus 5% of the building width. Thus,

$$e = 60 + 0.05 \times 120 \times 12$$
$$= 132 \text{ in.}$$

The displacements including the torsional effects are

$$D_{TD} = D_D\left(1 + y\frac{12e}{b^2 + d^2}\right)$$
$$= 9.79\left\{1 + \frac{120}{2} \times \frac{132}{(120^2 + 120^2)}\right\}$$
$$= 12.48 \text{ in.}$$
$$D_{TM} = 17.62 \times 1.275$$
$$= 22.47 \text{ in.}$$
$$V_b = K_{D\max}D_D$$
$$= 160 \times 9.79$$
$$= 1566.4 \text{ kips}$$

Given a seismic weight of $W = 7200$ kips, the seismic base shear coefficient for the design of isolation system and structural elements below it corresponds to

$$\frac{1566.4}{7200} = 0.218 \text{ or } 21.8\% \text{ of gravity}$$

Before calculating the base shear for the super structure, we need to calculate R_I. However, R_I need not be greater than 2.0.

$$R_I = \frac{3}{8} R = \frac{3}{8} \times 7 = 2.657 > 2.0$$

Therefore, $R_I = 2.0$

$$V_S = \frac{K_{D\,max} D_D}{R_I}$$

$$= \frac{V_D}{2}$$

$$= \frac{1566.4}{2}$$

$$= 783.2 \text{ kips   or   } \frac{783.2}{7200} \times 100 = 10.9\% \text{ of gravity}$$

Using the equivalent lateral procedure of ASCE 7-02, we now calculate the base shear required for a fixed-base structure of weight $W = 7200$ kips, and a period $T = T_D = 2.5$ sec, equal to the period of the isolated building.

1. $V_{min} = \dfrac{S_{DS}}{\left(\dfrac{R}{I}\right)} W$

$$= \frac{1}{\left(\dfrac{7}{1}\right)} W = 0.143\, W = 14.3\% \text{ g}$$

2. $V_{max} = \dfrac{S_{D1}}{T\left(\dfrac{R}{I}\right)} W$

$$= \frac{0.6}{2.5 \times 7} W = 0.034\, W = 3.4\% \text{ g}$$

3. $V_{min} = 0.044\, S_{DS} I W$

$$= 0.044 \times 1 \times 1 \times W = 0.044\, W = 4.4\% \text{ g}$$

4. $V_{min} = \dfrac{0.5 S_1}{\left(\dfrac{R}{I}\right)}$

$$= \frac{0.5 \times 0.6}{\left(\dfrac{7}{1}\right)} = 0.043\, W = 4.3\% \text{ g}$$

The base shear $V_{max} = 4.4\%$ g, obtained from the third of the four equations above, yields the design base shear for the fixed-based building. However, a base shear equal to 10.9% of gravity, obtained from the calculations for a base-isolated building, controls the design of the subject building. Using this base shear, the structural elements above the isolation system are designed by applying the appropriate provisions of a nonisolated structure.

Preliminary Design of Friction Pendulum System. Recall that the period T of a pendulum is inversely proportional to the square root of its length, and does not depend on the mass $m = w/g$. Similarly, the period of an FPS depends only on the square root of its radius R of the dish and not the supported mass of the building above. To increase the period of a pendulum we increase the length; to increase the apparent period of the building we increase the radius of the dish.

If the weight of the building above is W, and the radius of FPS dish is R, then the horizontal stiffness of the isolator is given by

$$K_h = \frac{W}{R}$$

The period of the isolated system is a function of its radius R only, and is given by

$$T = 2\pi \sqrt{\frac{R}{g}}$$

For our building, the effective isolated period $T_D = 2.5$ sec, as given in the statement of the problem.

Therefore,

$$2.5 = 2\pi \sqrt{\frac{R}{386.4}} \quad \text{giving}$$

$$R = 61.23 \text{ in.}$$

The effective stiffness of a FPS is given by

$$K_{eff} = \frac{W}{R} + \frac{\mu W}{D}$$

where the new term μ = friction coefficient.

The friction coefficient μ for an FPS may be assumed to be independent of velocity for pressures of 20 ksi or more. The damping β provided by the system is given by

$$\beta = \frac{2}{\pi} \frac{\mu}{\mu + \dfrac{D}{R}}$$

Assuming $\mu = 0.06$ and a design displacement of 10 in., the effective damping is calculated from

$$\beta_{eff} = \frac{2 \times 0.06}{\pi \left(0.06 + \dfrac{10}{61.23} \right)}$$

$$= 0.10 \quad \text{or} \quad 10\% \text{ of the critical}$$

The selected value of $D = 10$ in. satisfies the minimum code displacement of $D_D = 9.79$ in., calculated earlier for $T = 2.5$ sec, $\beta = 20\%$, and $B = 1.5$.

The effective stiffness is calculated from

$$K_{\text{eff}} = \frac{W}{R} + \frac{\mu W}{D}$$

$$= \frac{7200}{61.23} + \frac{0.06 \times 7200}{10} = 161 \text{ kips/in.}$$

This is almost exactly the same as $K_{D\text{max}}$ of 160 kips/in. derived earlier. Therefore, no further iterations are necessary.

With regard to the example problem, the following observations are appropriate for preliminary design purposes.

- An FPS of approximately 5-ft radius is required underneath each column.
- The required stiffness of each FPS is approximately equal to 100 kips/in.
- A moat equal to about 23 in. around the building is required to accommodate the calculated displacement $D_{\text{TM}} = 22.47$ in.
- The torsional contribution to the displacement is equal to D_{TM} minus D_m. In our case this is equal to $(22.47 - 17.62) = 4.85$ in. A possible solution to reducing the torsion contribution is to use a stiffer FPS at the building perimeter.
- As mentioned previously, other competing isolation systems are generally evaluated to achieve competitive bids. Usually a performance type of specifications for a base isolation system accompanies structural drawings to encourage competitive bids.

8.8. PASSIVE ENERGY DISSIPATION SYSTEMS

Passive energy dissipation is an emerging technology that enhances the performance of a building by adding damping (and in some cases, stiffness) to the building. The primary use of energy dissipation devices is to reduce earthquake displacement of the structure. Energy dissipation devices will also reduce force in the structure, provided the structure is responding elastically, but would not be expected to reduce force in structure that is responding beyond yield.

For most applications, energy dissipation provides an alternative approach to conventional stiffening and strengthening schemes, and would be expected to achieve comparable performance levels. In general, these devices are expected to be good candidates for projects that have a target building performance level of life safety or perhaps immediate occupancy, but would be expected to have only limited applicability to projects with a target building performance level of collapse prevention. Other objectives may also influence the decision to use energy dissipation devices, since these devices can also be useful for control of building response to small earthquakes and wind loads.

A wide variety of passive energy dissipation devices is available, including fluid viscous dampers, viscoelastic materials, and hysteretic devices. Ideally, energy dissipation devices dampen earthquake excitation of the structure that would otherwise cause higher levels of response and cause damage to components of the building. Under favorable conditions, energy dissipation devices reduce drift of the structure by a factor of about two to three (if no stiffness is added) and by larger factors if the devices also add stiffness to the structure.

Unlike base isolation, passive energy dissipation does not intercept earthquake energy entering the structure. It allows earthquake energy into the building. However, the energy is directed toward energy dissipation devices located within the lateral resisting elements. Earthquake energy is transformed into heat by these devices and dissipated into the structure.

A fluid viscous damper attached to diagonals of a braced frame, shown in Fig. 8.35a, is one such energy dissipation device. It dissipates energy by forcing a fluid through an orifice, similar to the shock absorbers of an automobile (Fig. 8.35b). The fluid used is usually of high viscosity, such as a silicone. The unique feature of these devices is that

(1)

(2)

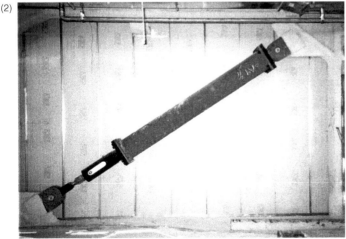

Figure 8.35a. Fluid viscous dampers attached to diagonals: (1) Diagonals with dampers; (2) Close-up of a diagonal; (3) Close-up of a damper. (Photos courtesy of Bob Schneider, Taylor Devices, Inc., New York.)

(3)

Figure 8.35a. (*Continued.*)

their damping characteristics, and hence the amount of energy dissipated, can be made proportional to the velocity. The response of a fluid viscous damper is considered to be out-of-phase with those due to seismic activity. This is because the damping force provided by the device varies inversely with the dynamic lateral displacements of a building. To understand the concept, consider a building shaking laterally back and forth during a seismic event. The stress in a lateral-load-resisting element such as a frame–column is at its maximum when the building deflection is also at maximum. This is also the point at which the building reverses direction to move back in the opposite direction. The damping force of a fluid viscous damper will drop to zero at this point of maximum deflection. This is because the damper stroking velocity goes to zero as the building reverses direction. As the building moves back in the opposite direction, a maximum damper force occurs at the maximum velocity which happens when the building goes through its normal upright position. This is also the point when the stresses in the lateral-load-resisting elements are at a minimum. Therefore, the damping provided by the device

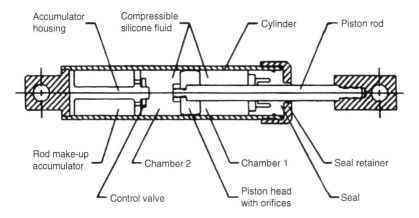

Figure 8.35b. Viscous fluid damper, consisting of a piston in a damping housing filled with a compound of silicone or similar type of oil. The piston contains small orifices through which the fluid passes from one side of the piston to the other. The damper thus dissipates energy through the movement of the piston in a highly viscous fluid.

Figure 8.35c. Fluid viscous damper installed in an existing building.

varies from a maximum to a minimum as the building moves from an at-rest position to its maximum lateral deflection position. This out-of-phase response is considered a desirable feature in seismic designs.

A photograph of a fluid viscous damper installed in an existing building as part of seismic upgrade is shown in Fig. 8.35c.

8.9. BUCKLING-RESTRAINED BRACED FRAME

Unbonded Brace System

Buckling-restrained brace frames (BRBFs) have a high degree of ductility (energy absorbing capability) and good lateral stiffness, and are relatively simple to repair after a major earthquake. Unbonded braced frames, which may be considered a special class of BRBFs, consist of a steel core installed within an outer shell with mortar infill between the plate and the shell. An unbonding agent is applied to the core plate to prevent it from transmitting axial load to the buckling-restraining mechanism (see Fig. 8.36a). The unbonded brace element, typically a diagonal member, consists of a restrained yielding segment, nonyielding restrained steel segments, and nonyielding unrestrained segments (see Fig. 8.36b.). The yielding segment commonly referred to as the core typically consists of a steel plate. The nonyielding segments are typically of cruciform shape. The entire assembly is generally procured as a preassembled unit manufactured to meet the performance objectives specified by the design engineer.

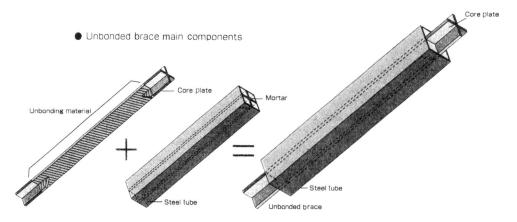

Figure 8.36a.   Main components of unbonded brace.

Because the braces are able to yield without buckling in compression, well-defined, stable, fairly symmetric hysteretic loops are generated when the braces are subjected to reversed cyclic loading, resulting in excellent energy dissipating characteristics.

Use of a buckling-restrained braced frame as a seismic-lateral-resisting system is relatively new in the United States. However, it is similar to a special concentrically braced frame in that it also has a triangulated vertical framework of members that resist lateral loads through axial tension and compression. The main difference is that the buckling-restrained braces achieve significantly higher ductility and energy dissipation characteristics by effectively eliminating buckling and the poor hysteretic performance associated with it. Because of this, the tension and compression behaviors of the brace are very similar.

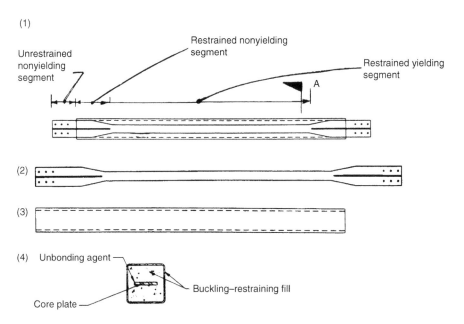

Figure 8.36b.   Components of unbonded brace. (1) buckling restrained brace; (2) core; (3) sleeve; (4) section. The yielding of core plates in compression without buckling results in a stable hysteretic loop with excellent energy-dissipating characteristics.

In buckling-restrained braces, a fairly long segment can yield in compression as well as in tension. The yielding segment is part of an axial-force-resisting steel core. Its effective slenderness is extremely low due to the lateral restraint provided by a surrounding casing of steel infilled with mortar. For buckling to be precluded, this casing must be kept free from axial forces. Several methods of confining the axial force to the steel core are in use in the United States. Most of these are developed around proprietary specifications, and some are patented.

Since BRBFs are a recent development, they are not yet addressed by building codes such as IBC-03 or AISC seismic provisions (AISC 341-02). Therefore, to facilitate wider use of this system, the Structural Engineers Association of Northern California (SEAONC) has developed, in conjunction with the SEAOC seismology committee and AISC TC9, a set of design provisions. Their work has resulted in a document, *Recommended Provisions for Buckling-Restrained Braced Frames* (SEAONC 2001). The provisions are currently under review for inclusion in future building codes and seismic provisions.

The design approach for this system typically follows the same force-based approach that is used for other types of braced frames. The method for computing the base shear is similar to that for the special concentrically braced frame, with differences in the values of certain seismic coefficients because BRBFs are more flexible than SCBFs. Consequently, buildings with BRBFs with longer periods may warrant use of a larger value of the coefficient C_T in the calculation of base shear. Similarly, use of larger values of the response reduction coefficient R may be appropriate in the design of BRBFs. SEAONC 2001 proposes a value of 8.0, while a value of 7.0 is being considered for inclusion in the 2003 NEHRP provisions.

As buckling-restrained braces are typically a specification item, the required brace strengths are generally specified by the design engineer. Customarily the manufacturer designs the braces to comply with the given requirements using the material and grade specified for the element. Since the material grade has a significant effect on the brace stiffness, the lower the yield stress, the greater the required area of steel, resulting in a stiffer brace. Decreasing the yield length concentrates the inelastic strain, reducing the cumulative energy dissipation capacity.

Because the tension and compression strengths of buckling-restrained braces are similar, a chevron configuration (see Fig. 8.36c) does not penalize the design of the beam connected to the chevron braces.

With certain simplifications, gusset plates at the connections are designed similar to those for SCBF. However, BRBF gussets are not required to accommodate buckling of the brace; hinge zones are therefore not required, nor are the gussets required to have flexural strength in excess of those of the brace. Small eccentricities may be

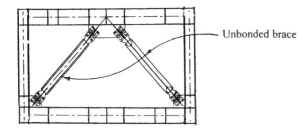

— Unbonded brace

Figure 8.36c. Buckling-restrained brace frame, BRBF, with chevron braces. Since the braces yield both in tension and compression, a chevron configuration does not penalize design of the beam connected to the braces.

Figure 8.36d. Unbonded brace frame elevation.

Figure 8.36e. Unbonded brace frame connection (photographs courtesy of Edwin Shlemon, S.E., Associate Partner, ARUP Partners, Los Angeles; CA).

permissible in the connection design if the resulting brace rotations are still within the tested limits.

The maximum connection force is calculated using the brace strength and over-strength factors β and ω determined from testing. The factor β represents the overstrength in compression (buckling-restrained braces tend to be somewhat stronger in compression than in tension), and ω represents strain-hardening within the expected deformation range. The factor R_y, representing expected yield strength as compared to nominal yield strength, is assumed not to be applicable in sizing of the braces because the final cross-sectional size of a buckling-restrained brace is typically determined considering the material yield strength as measured from coupon tests. The brace yield strength can thus be calculated without guesswork from the required strength and resistance factor.

Column design forces are determined using the special seismic load combinations specified in the codes. Although this can be done in a manner similar to that for the method presented for SCBF, BRBFs tend to have much lower overstrength. Therefore, an explicit consideration of brace capacity can usually result in lower column design forces, resulting in savings in the columns and foundations.

Figures 8.36d and 8.36e show an unbonded braced frame elevation and connection.

MGM City Center Vdara Tower,

Las Vegas, Nevada

At 572 feet tall above the basement and 1.5 million square feet, the Vdara Tower is the second phase of MGM City Center. It contains a condominium hotel tower, retail areas, spa facilities, restaurants and parking. The lateral load resisting system for the tower incorporates cast-in-place reinforced concrete shear walls. A two-way, post-tensioned concrete flat plate is used to frame the floors.

(Owner: MGM Mirage Design Group; Architect: Rafael Vinõly Architects – Leo A. Daly Company; Structural Engineer: DeSimone Consulting Engineers)

Selected References

1. Farzad Naeim, ed., *The Seismic Design Handbook.* New York: Van Nostrand Reinhold, 1989.
2. *Uniform Building Code*, International Code Council, 1991, 1994, and 1997.
3. *American National Standards Institute (ANSI) A58.1*, Washington, DC, 1982.
4. "Structuring Tall Buildings," *Progressive Architecture,* vol. 61 no. 12, December 1980.
5. J. E. Cermak, "Applications of Fluid Mechanics to Wind Engineering," *Journal of Fluids Engineering*, vol. 97, p. 9, March 1975.
6. "Wind Effects on High Rise Buildings," Symposium Proceedings, March 1970, Northwestern University, Illinois.
7. T. Tschanz, "Measurement of Total Dynamic Loads Using Elastic Models With High Natural Frequencies," in Timothy J. Reinhold ed., *Wind Tunnel Modeling for Civil Engineering Applications*. Cambridge, UK: Cambridge University Press, 1982.
8. "Tall Buildings," Conference Proceedings, 1984, Singapore.
9. Robert A. Coleman, *Structural System Designs.* Englewood Cliffs, NJ: Prentice-Hall, 1983.
10. *Post-Tension Manual*, 4th ed. Post-Tensioning Institute, phoenix, AZ, 1986.
11. T. Y. Lin and S. D. Stotesbury, *Structural Concepts and Systems for Architects and Engineers.* New York: John Wiley, 1981.
12. James R. Libby, *Modern Pre-Stressed Concrete.* New York: Van Nostrand Reinhold, 1984.
13. T. Y. Lin, *Design of Prestressed Concrete Structure,* 3rd ed. New York: John wiley, 1981.
14. George Winter and Arthur H. Nilson, *Design of Concrete Structures.* New York: McGraw-Hill, 1979.
15. Frederick S. Merritt, *Building Design and Construction Handbook*, 4th ed. New York: McGraw-Hill, 1982.
16. Edwin H. Gaylord, Jr., and Charles N. Gaylord, *Design of Steel Structures*, 2nd ed. New York: McGraw-Hill, 1972.
17. N. M. Newmark and E. Rosenblueth, *Fundamentals of Earthquake Engineering.* Englewood Cliffs, NJ: Prentice-Hall, 1971.
18. Alex Coull and Bryan Stafford Smith, *Tall Buildings with Particular Reference to Shear Wall Structures.* New York: Pergamon Press, 1967.
19. Monoru Wakabayashi, *Design of Earthquake Resistant Buildings,* New York: McGraw-Hill, 1986.
20. Bungale S. Taranath, *Structural Analysis and Design of Tall Buildings.* New York: McGraw-Hill, 1988.
21. Kenneth Leet, *Reinforced Concrete Design.* New York: McGraw-Hill, 1991.
22. John M. Biggs, *Introduction to Structural Dynamics.* New York: McGraw-Hill, 1964.
23. Edwin H. Gaylord, Jr., and Charles N. Gaylord, *Structural Engineering Handbook*, 3rd ed. New York: McGraw-Hill, 1990.
24. Sol. E. Cooper and Andrew C. Chen, *Designing Steel Structures—Methods and Cases.* Englewood Cliffs, NJ: Prentice-Hall, 1985.
25. John Karlberg, *Preliminary Design for Post-Tensioned Structures, Structural Engineering Practice*, vol. 2. New York: Marcel Dekker, Inc., 1983.
26. Wolfgang Schueller, *High-Rise Building Structures.* New York: John Wiley and Sons, 1977.
27. Witold Zbirohowski-Koscia, *Thin-Walled Beams.* London: Crosby Lockwood, 1967.

28. Milo S. Ketchum ed., *Structural Engineering Practice.* Vol. 1 & 2, 1982; vol. 4, 1982–83; vol. 1, 2, & 3, 1983; vol. 4, 1983–84. New York: Marcel Dekker.

29. *National Building Code of Canada* and NBCC *Supplement for the 1977 Code.* National Research Council Canada, Ottawa, 1995.

30. N. W. Murray, *Introduction to the Theory of Thin-Walled Structures.* New York: Oxford University Press, 1984.

31. Timothy J. Reinhold (ed.), *Wind Tunnel Modeling for Civil Engineering Applications.* Cambridge, UK: Cambridge University Press, 1982.

32. M. B. Kanchi, *Matrix Methods of Structural Analysis.* New York: Wiley Eastern Limited, 1994.

33. Mario Paz, *Structural Dynamics.* New York: Van Nostrand Reinhold, 1985.

34. J. R. Choudhury, "Analysis of Plain and Spatial Systems of Interconnected Shear Walls," Ph.D. Thesis, University of Southampton, Southampton, UK, 1968.

35. V. Z. Vlasov, *Thin-Walled Elastic Beams.* Washington, DC: National Science Foundation, 1961.

36. Bungale S. Taranath, "Torsional Behavior of Open-Section Shear Wall Structures," Ph.D. Thesis, University of Southampton, Southampton, UK.

37. Aslam Qadeer, "Interaction of Floor Slabs and Shear Walls," Ph.D. Thesis, University of Southampton, Southampton, UK, 1968.

38. Bungale S. Taranath, "A New Look at Composite High-Rise Construction," Our World in Concrete and Structures, Singapore, 1983.

39. Bungale S. Taranath *et al.,* "A Practical Computer Method of Analysis for Complex Shear Wall Structure," *8th ASCE Conference in Electronic Computation,* 1983.

40. Bungale S. Taranath, "Composite Design of First City Tower," *The Structural Engineer,* vol. 60, no. 9, pp. 271–281, 1982.

41. Bungale S. Taranath, "Differential Shortening of Columns in High-Rise Buildings," *Journal of Torsteel,* 1981.

42. Bungale S. Taranath, "Analysis of Interconnected Open Section Wall Structures," *ASCE Journal,* 1986.

43. Bungale S. Taranath, "The Effect of Warping on Interconnected Shear Wall Flat Plate Structures," *Proceedings of the Institution of Civil Engineers,* 1976.

44. Bungale S. Taranath, "Torsion Analysis of Braced Multi-Storey Buildings," *The Structural Engineer,* vol. 53, no. 8, pp. 345–347, 1975.

45. Bungale S. Taranath, "Optimum Belt Truss Locations for High-Rise Buildings," *AISC. Engineering Journal,* vol. 11, no. 1, pp. 18–21,

46. Kenneth H. Lenzen, "Vibrations of Steel Joist–Concrete Slab Floors," *AISC Engineering Journal* vol. 3, no. 3, pp. 133–136, July 1966.

47. J. F. Wiss and R. H. Parmalee, "Human Perception of Transient Vibrations," *Journal of Structural Division ASCE,* vol. 100, pp. 773–783 April 1974.

48. Thomas M. Murray, "Design to Prevent Floor Vibrations," *Engineering Journal American Institute of Steel Construction,* vol. 12, no. 3, p. 82, 1975.

49. R. Halvorson and N. Isyumov, "Comparison of Predicted and Measured Dynamic Behavior of Allied Bank Plaza," in N. Isyumov and T. Tschanz, eds. *Building Motion in Wind.* New York: ASCE, 1982.

50. Bryan Stafford Smith and Alex Coull *Tall Building Structures, Analysis and Design.* New York: John Wiley & Sons, Inc., 1991.

51. EMA-222 and 223. *NEHRP Recommended Provisions for the Development of Seismic Regulations for New Buildings, Part 1, Provisions, Part 2, Commentary.* Federal Emergency Management Agency, Washington, DC. 1991.

52. FEMA-140, *Guide to Application of the NEHRP Recommended Provisions in Earthquake-resistant Building Design.* Federal Emergency Management Agency, Washington, DC, 1990.

53. FEMA-178, *NEHRP Handbook for the Seismic Evaluation of Existing Buildings.* Federal Emergency Management Agency, Washington, DC, 1992.

54. FEMA-172, *NEHRP Handbook for Seismic Rehabilitation of Existing Buildings.* Federal Emergency Management Agency, Washington, DC, 1992.

55. Ronald O. Hamburger, Anthony B. Court and Jeffrey R. Soulages Vision 2000: A Framework for Performance-Based Engineering of buildings, in Proceedings of 6th SEAOC Convention, pp. 127–146, 1995.

56. Alexander Coull and J. R. Choudhury, "Analysis of Coupled Shear Walls," *ACI Journal,* vol. 64. no. 9, pp. 587–593, September 1967.

57. ETABS Version 6, Steeler, *Stress Check of Steel Frames;* Conker, *Design of Concrete Frames by Computers and Structures, Inc.,* January 1995.

58. M. Russell Nester and Allen R. Porush, "A Rational System for Earthquake Risk Management," In *SEAOC Conference Proceedings,* 1991.

59. *Seismic Design Guidelines for Upgrading Existing Buildings,* PB 89-220453. Departments of the Army, the Navy, and the Air Force, Washington, DC, 1988.

60. *Building Code Requirements for Structural Concrete* (ACI 318-95) *and Commentary* (ACI 318R-95). American Concrete Institute, Farmington Hills, MI, 1996.

61. *Manual of Steel Construction: Allowable Stress Design,* 9th ed. American Institute of Steel Construction. Chicago, IL, 1989.

62. *Recommended Lateral Force Requirements and Commentary* (Blue Book), 7th ed. Seismology Committee, Structural Engineers Association of California, Sacramento, CA, 1999.

63. Bungale S. Taranath, *Steel, Concrete, and Composite Design of Tall Buildings.* New York: McGraw-Hill, 1997.

64. *Notes on ACI 318-02 Building Code Requirements for Structural Concrete,* 8th ed. Portland Cement Association, Skokie, IL, 2002.

65. *2003 International Building Code.* ICC Publications.

66. *2003 International Building Code Commentary,* vol. 2. ICC Publications.

67. *2000 IBC Structural/Seismic Design Manual,* vols. 1, 2, and 3. Structural Engineers Association of California. ICC Publications.

68. S. K. Ghosh, "Seismic Design Using Structural Dynamics," in *2000 International Building Code.* ICC Publications.

69. Narendra Taly, *Loads and Load Paths in Buildings: Principles of Structural Design.* ICC Publications.

70. S. K. Ghosh, "Seismic and Wind Design of Concrete Buildings," in *2000 International Building Code.* ICC Publications.

71. Alan Williams, *Seismic and Wind Forces: Structural Design Examples.* ICC Publications.

72. Wai-Fah Chen and Charles Scawthorn, eds., *Earthquake Engineering Handbook.* ICC Publications.

73. Yousef Bozorgnia and Vitelmo V. Bertero, eds., *Earthquake Engineering: From Engineering Seismology to Performance-Based Engineering.* ICC Publications, 2004.

74. Farzad Maeim, ed., *The Seismic Design Handbook,* 2nd ed. Boston: Kluwer Academic Publishers, 2001.

75. *CRSI Design Handbook*, 9th ed. Concrete Reinforcing Steel Institute, Schaumberg, IL, 2002.

76. FEMA-178, *NEHRP Handbook for the Seismic Evaluation of Existing Buildings,* developed by the Building Seismic Safety Council for the Federal Emergency Management Agency, Washington, DC, 1992.

77. FEMA-267, *Interim Guidelines, Inspection, Evaluation, Repair, Upgrade and Design of Welded Moment-Resisting Steel Structures,* prepared by the SAC Joint Venture for the Federal Emergency Management Agency, Washington, DC, 1995.

78. FEMA-267A, *Interim Guidelines Advisory No. 1,* prepared by the SAC Joint Venture for the Federal Emergency Management Agency, Washington, DC, 1996.

79. FEMA-267B, *Interim Guidelines Advisory No. 2,* prepared by the SAC Joint Venture for the Federal Emergency Management Agency, Washington, DC, 1999.

80. FEMA-273, *NEHRP Guidelines for the Seismic Rehabilitation of Buildings,* prepared by the Applied Technology Council for the Building Seismic Safety Council, published by the Federal Emergency Management Agency, Washington, DC, 1997.

81. FEMA-274, *NEHRP Commentary on the Guidelines for the Seismic Rehabilitation of Buildings,* prepared by the Applied Technology Council for the Building Seismic Safety Council, published by the Federal Emergency Management Agency, Washington, DC, 1997.

82. FEMA-302, *NEHRP Recommended Provisions for Seismic Regulations for New Buildings and Other Structures, Part 1—Provisions,* prepared by the Building Seismic Safety Council for the Federal Emergency Management Agency, Washington, DC, 1997.

83. FEMA-303, *NEHRP Recommended Provisions for Seismic Regulations for New Buildings and Other Structures, Part 2—Commentary,* prepared by the Building Seismic Safety Council for the Federal Emergency Management Agency, Washington, DC, 1997.

84. FEMA-310, *Handbook for the Seismic Evaluation of Buildings—A Prestandard,* prepared by the American Society of Civil Engineers for the Federal Emergency Management Agency, Washington, DC, 1998.

85. FEMA-350, *Recommended Seismic Design Criteria for New Steel Moment-Frame Buildings,* prepared by the SAC Joint Venture for the Federal Emergency Management Agency, Washington, DC, 2000.

86. FEMA-351, *Recommended Seismic Evaluation and Upgrade Criteria for Existing Welded Steel Moment-Frame Buildings,* prepared by the SAC Joint Venture for the Federal Emergency Management Agency, Washington, DC, 2000.

87. FEMA-352, *Recommended Postearthquake Evaluation and Repair Criteria for Welded Steel Moment-Frame Buildings* prepared by the SAC Joint Venture for the Federal Emergency Management Agency, Washington, DC, 2000.

88. FEMA-353, *Recommended Specifications and Quality Assurance Guidelines for Steel Moment-Frame Construction for Seismic Applications,* prepared by the SAC Joint Venture for the Federal Emergency Management Agency, Washington, DC, 2000.

89. FEMA-354, *A Policy Guide to Steel Moment-Frame Construction,* prepared by the SAC Joint Venture for the Federal Emergency Management Agency, Washington, DC, 2000.

90. FEMA-355A, *State-of-the-Art Report on Base Metals and Fracture,* prepared by the SAC Joint Venture for the Federal Emergency Management Agency, Washington, DC.

91. FEMA-355B, *State-of-the-Art Report on Welding and Inspection,* prepared by the SAC Joint Venture for the Federal Emergency Management Agency, Washington, DC, 2000.

92. FEMA-355C, *State-of-the-Art Report on Systems Performance of Steel Moment-Frames Subject to Earthquake Ground Shaking,* prepared by the SAC Joint Venture for the Federal Emergency Management Agency, Washington, DC, 2000.

93. FEMA-355D, *State-of-the-Art Report on Connection Performance,* prepared by the SAC Joint Venture for the Federal Emergency Management Agency, Washington, DC, 2000.

94. FEMA-355E, *State-of-the-Art Report on Past Performance of Steel Moment-Frame Buildings in Earthquakes,* prepared by the SAC Joint Venture for the Federal Emergency Management Agency, Washington, DC, 2000.

95. FEMA-355F, *State of the Art Report on Performance Prediction and Evaluation of Steel Moment-Frame Buildings,* prepared by the SAC Joint Venture for the Federal Emergency Management Agency, Washington, DC, 2000.

96. FEMA-356, *Prestandard and Commentary for the Seismic Rehabilitation of Buildings,* prepared by the SAC Joint Venture for the Federal Emergency Management Agency, Washington, DC, 2000.

97. FEMA-403, World Trade Center Building Performance Study: Data Collection, Preliminary Observations, and Recommendations. Federal Insurance Mitigation Association, Washington, DC, and the Federal Emergency Management Agency, Region II, New York, 2000.

98. Ghosh, S. K., "Design of Reinforeced Concrete Buildings under the 1997 UBC," *Building Standards,* International Conference of Building Officials, Whittier, CA, May–June 1998, pp. 20–24.

99. Ghosh, S. K., "Needed Adjustments in 1997 UBC," Proceedings, 1998 Convention of the Structural Engineers Association of California, Reno/Sparks, Nevada, October 1985, pp. 9.1–9.15.

100. Ghosh, S. K., "Major Changes in Concrete-Related Provisions—1997 UBC and Beyond," *Earthquake Spectra,* Earthquake Engineering Research Institute, Oakland, CA, 1999.

Appendix A

Conversion Factors: U.S. Customary to SI Units

	Multiply		by		to Obtain
Length	inches	×	25.4	=	millimeters
	feet	×	0.3048	=	meters
	yards	×	0.9144	=	meters
	miles (statute)	×	1.609	=	kilometers
Area	square inches	×	645.2	=	square millimeters
	square feet	×	0.0929	=	square meters
	square yards	×	0.8361	=	square meters
Volume	cubic inches	×	16,387	=	cubic millimeters
	cubic feet	×	0.028,32	=	cubic meters
	cubic yards	×	0.7646	=	cubic meters
	gallons (U.S. liquid)	×	0.003,785	=	cubic meters
Force	pounds	×	4.448	=	newtons
	kips	×	4448	=	newtons
Force per unit length	pounds per foot	×	14.594	=	newtons per meter
	kips per foot	×	14,594	=	newtons per meter
Load per unit volume	pounds per cubic foot	×	0.157,14	=	kilonewtons per cubic meter
Bending moment or torque	inch-pounds	×	0.1130	=	newton meters
	foot-pounds	×	1.356	=	newton meters
	inch-kps	×	113.0	=	newton meters
	foot-kips	×	1356	=	newton meters
	inch-kips	×	0.1130	=	kilonewton meters
	foot-kips	×	1.356	=	kilonewton meters
Stress, pressure, loading (force) per unit area	pounds per sq inch	×	6895	=	pascals
	pounds per sq inch	×	6.895	=	kilopascals
	pounds per sq inch	×	0.006,895	=	megapascals
	kips per sq inch	×	6.895	=	megapascals
	pounds per sq foot	×	47.88	=	pascals
	pounds per sq foot	×	0.047,88	=	kilopascals
	kips per sq foot	×	47.88	=	kilopascals
	kips per sq foot	×	0.047,88	=	megapascals
Mass	pounds	×	0.454	=	kilograms
Mass per unit volume (density)	pounds per cubic foot	×	16.02	=	kilograms per cubic meter
	pounds per cubic yard	×	0.5933	=	kilograms per cubic meter
Moment of inertia	inches	×	416.231	=	millimeters
Mass per unit length	pounds per foot	×	1.488	=	kilograms per meter

Index

T - #0452 - 071024 - C10 - 254/178/41 - PB - 9780367393496 - Gloss Lamination